OV

# Springer Texts in Statistics

*Series Editors*
G. Casella
S. Fienberg
I. Olkin

For further volumes:
http://www.springer.com/series/417

Brani Vidakovic

# Statistics for Bioengineering Sciences

## With MATLAB and WinBUGS Support

 Springer

Brani Vidakovic
Department of Biomedical Engineering
Georgia Institute of Technology
2101 Whitaker Building
313 Ferst Drive
Atlanta, Georgia 30332-0535
USA
brani@bmc.gatech.edu

*Series Editors:*

George Casella
Department of Statistics
University of Florida
Gainesville, FL 32611-8545
USA

Stephen Fienberg
Department of Statistics
Carnegie Mellon University
Pittsburgh, PA 15213-3890
USA

Ingram Olkin
Department of Statistics
Stanford University
Stanford, CA 94305
USA

ISSN 1431-875X
ISBN 978-1-4614-0393-7        e-ISBN 978-1-4614-0394-4
DOI 10.1007/978-1-4614-0394-4
Springer New York Dordrecht Heidelberg London

Library of Congress Control Number: 2011931859

Springer is part of Springer Science+Business Media (www.springer.com)

# Preface

This text is a result of many semesters of teaching introductory statistical courses to engineering students at Duke University and the Georgia Institute of Technology. Through its scope and depth of coverage, the text addresses the needs of the vibrant and rapidly growing engineering fields, bioengineering and biomedical engineering, while implementing software that engineers are familiar with.

There are many good introductory statistics books for engineers on the market, as well as many good introductory biostatistics books. This text is an attempt to put the two together as a single textbook heavily oriented to computation and hands-on approaches. For example, the aspects of disease and device testing, sensitivity, specificity and ROC curves, epidemiological risk theory, survival analysis, and logistic and Poisson regressions are not typical topics for an introductory engineering statistics text. On the other hand, the books in biostatistics are not particularly challenging for the level of computational sophistication that engineering students possess.

The approach enforced in this text avoids the use of mainstream statistical packages in which the procedures are often black-boxed. Rather, the students are expected to code the procedures on their own. The results may not be as flashy as they would be if the specialized packages were used, but the student will go through the process and understand each step of the program. The computational support for this text is the MATLAB© programming environment since this software is predominant in the engineering communities. For instance, Georgia Tech has developed a practical introductory course in computing for engineers (CS1371 – Computing for Engineers) that relies on MATLAB. Over 1,000 students take this class per semester as it is a requirement for all engineering students and a prerequisite for many upper-level courses.

In addition to the synergy of engineering and biostatistical approaches, the novelty of this book is in the substantial coverage of Bayesian approaches to statistical inference.

I avoided taking sides on the traditional (classical, frequentist) vs. Bayesian approach; it was my goal to expose students to both approaches. It is undeniable that classical statistics is overwhelmingly used in conducting and reporting inference among practitioners, and that Bayesian statistics is gaining in popularity, acceptance, and usage (FDA, Guidance for the Use of Bayesian Statistics in Medical Device Clinical Trials, 5 February 2010). Many examples in this text are solved using both the traditional and Bayesian methods, and the results are compared and commented upon.

This diversification is made possible by advances in Bayesian computation and the availability of the free software WinBUGS that provides painless computational support for Bayesian solutions. WinBUGS and MATLAB communicate well due to the free interface software MATBUGS. The book also relies on stat toolbox within MATLAB.

The World Wide Web (WWW) facilitates the text. All custom-made MATLAB and WinBUGS programs (compatible with MATLAB 7.12 (2011a) and WinBUGS 1.4.3 or OpenBUGS 3.2.1) as well as data sets used in this book are available on the Web:

```
http://springer.bme.gatech.edu/
```

To keep the text as lean as possible, solutions and hints to the majority of exercises can be found on the book's Web site. The computer scripts and examples are an integral part of the text, and all MATLAB codes and outputs are shown in blue typewriter font while all WinBUGS programs are given in red-brown typewriter font. The comments in MATLAB and WinBUGS codes are presented in green typewriter font.

The three icons 🖼, 🛪 , and 🛉 are used to point to data sets, MATLAB codes, and WinBUGS codes, respectively.

The difficulty of the material in the text necessarily varies. More difficult sections that may be omitted in the basic coverage are denoted by a star, *. However, it is my experience that advanced undergraduate bioengineering students affiliated with school research labs need and use the "starred" material, such as functional ANOVA, variance stabilizing transforms, and nested experimental designs, to name just a few. Tricky or difficult places are marked with Donald Knut's "bend" ⌇.

Each chapter starts with a box titled WHAT IS COVERED IN THIS CHAPTER and ends with chapter exercises, a box called MATLAB AND WINBUGS FILES AND DATA SETS USED IN THIS CHAPTER, and chapter references.

The examples are numbered and the end of each example is marked with ✐.

I am aware that this work is not perfect and that many improvements could be made with respect to both exposition and coverage. Thus, I would welcome any criticism and pointers from readers as to how this book could be improved.

**Acknowledgments.** I am indebted to many students and colleagues who commented on various drafts of the book. In particular I am grateful to colleagues from the Department of Biomedical Engineering at the Georgia Institute of Technology and Emory University and their undergraduate and graduate advisees/researchers who contributed with real-life examples and exercises from their research labs.

Colleagues Tom Bylander of the University of Texas at San Antonio, John H. McDonald of the University of Delaware, and Roger W. Johnson of the South Dakota School of Mines & Technology kindly gave permission to use their data and examples. I also acknowledge Mathworks' statistical gurus Peter Perkins and Tom Lane for many useful conversations over the last several years. Several MATLAB codes used in this book come from the MATLAB Central File Exchange forum. In particular, I am grateful to Antonio Truillo-Ortiz and his team (Universidad Autonoma de Baja California) and to Giuseppe Cardillo (Merigen Research) for their excellent contributions.

The book benefited from the input of many diligent students when it was used either as a supplemental reading or later as a draft textbook for a semester-long course at Georgia Tech: BMED2400 Introduction to Bioengineering Statistics. A complete list of students who provided useful comments would be quite long, but the most diligent ones were Erin Hamilton, Kiersten Petersen, David Dreyfus, Jessica Kanter, Radu Reit, Amoreth Gozo, Nader Aboujamous, and Allison Chan.

Springer's team kindly helped along the way. I am grateful to Marc Strauss and Kathryn Schell for their encouragement and support and to Glenn Corey for his knowledgeable copyediting.

Finally, it hardly needs stating that the book would have been considerably less fun to write without the unconditional support of my family.

BRANI VIDAKOVIC
*School of Biomedical Engineering*
*Georgia Institute of Technology*
brani@bme.gatech.edu

# Contents

# Chapter 1
# Introduction

*Many people were at first surprised at my using the new words "Statistics" and "Statistical," as it was supposed that some term in our own language might have expressed the same meaning. But in the course of a very extensive tour through the northern parts of Europe, which I happened to take in 1786, I found that in Germany they were engaged in a species of political inquiry to which they had given the name of "Statistics".... I resolved on adopting it, and I hope that it is now completely naturalised and incorporated with our language.*

— Sinclair, 1791; Vol XX

<br>

### WHAT IS COVERED IN THIS CHAPTER

• What is the subject of statistics?
• Population, sample, data
• Appetizer examples

The problems confronting health professionals today often involve fundamental aspects of device and system analysis, and their design and application, and as such are of extreme importance to engineers and scientists.

Because many aspects of engineering and scientific practice involve non-deterministic outcomes, understanding and knowledge of statistics is important to any engineer and scientist. Statistics is a *guide to the unknown*. It is a science that deals with designing experimental protocols, collecting, summarizing, and presenting data, and, most importantly, making inferences and aiding decisions in the presence of variability and uncertainty. For example, R. A. Fisher's 1943 elucidation of the human blood-group system Rhesus in terms of the three linked loci $C$, $D$, and $E$, as described in Fisher (1947) or Edwards (2007), is a brilliant example of building a coherent structure of new knowledge guided by a statistical analysis of available experimental data.

The uncertainty that statistical science addresses derives mainly from two sources: (1) from observing only a part of an existing, fixed, but large population or (2) from having a process that results in nondeterministic outcomes. At least a part of the process needs to be either a *black box* or inherently stochastic, so the outcomes cannot be predicted with certainty.

A *population* is a statistical universe. It is defined as a collection of existing attributes of some natural phenomenon or a collection of potential attributes when a process is involved. In the case of a process, the underlying population is called hypothetical, for obvious reasons. Thus, populations can be either finite or infinite. A subset of a population selected by some relevant criteria is called a subpopulation.

Often we think about a population as an assembly of people, animals, items, events, times, etc., in which the attribute of interest is measurable. For example, the population of all US citizens older than 21 is an example of a population for which many attributes can be assessed. Attributes might be *a history of heart disease, weight, political affiliation, level of blood sugar*, etc.

A sample is an observed part of a population. Selection of a sample is a rich methodology in itself, but, unless otherwise specified, it is assumed that the sample is selected at random. The randomness ensures that the sample is representative of its population.

The sampling process depends on the nature of the problem and the population. For example, a sample may be obtained via a retrospective study (usually existing historical outcomes over some period of time), an observational study (an observer monitors the process or population in real time), a sample survey, or a designed study (an observer makes deliberate changes in controllable variables to induce a cause/effect relationship), to name just a few.

*Example 1.1.* **Ohm's Law Measurements.** A student constructed a simple electric circuit in which the resistance $R$ and voltage $E$ were controllable. The output of interest is current $I$, and according to Ohm's law it is

$$I = \frac{E}{R}.$$

This is a mechanistic, theoretical model. In a finite number of measurements under an identical $R,E$ setting, the measured current varies. The population here is hypothetical – an infinite collection of all potentially obtainable measurements of its attribute, current $I$. The observed sample is finite. In the presence of sample variability one establishes an empirical (statistical) model for currents from the population as either

$$I = \frac{E}{R} + \epsilon \quad \text{or} \quad I = \epsilon\frac{E}{R}.$$

On the basis of a sample one may first select the model and then proceed with the inference about the nature of the discrepancy, $\epsilon$.

*Example 1.2.* **Cell Counts.** In a quantitative engineering physiology laboratory, a team of four students was asked to make a LabVIEW$^{\copyright}$ program to automatically count MC3T3-E1 cells in a hemocytometer (Fig. 1.1). This automatic count was to be compared with the manual count collected through an inverted bright field microscope. The manual count is considered the gold standard.

The experiment consisted of placing 10 $\mu$L of cell solutions at two levels of cell confluency: 20% and 70%. There were $n_1 = 12$ pairs of measurements (automatic and manual counts) at 20% and $n_2 = 10$ pairs at 70%, as in the table below.

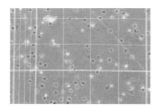

**Fig. 1.1** Cells on a hemocytometer plate.

|  | 20% confluency |  |  |  |  |  |  |  |  |  |  |  |
|---|---|---|---|---|---|---|---|---|---|---|---|---|
| Automated | 34 | 44 | 40 | 62 | 53 | 51 | 30 | 33 | 38 | 51 | 26 | 48 |
| Manual | 30 | 43 | 34 | 53 | 49 | 39 | 37 | 42 | 30 | 50 | 35 | 54 |
|  | 70% confluency |  |  |  |  |  |  |  |  |  |  |  |
| Automated | 72 | 82 | 100 | 94 | 83 | 94 | 73 | 87 | 107 | 102 |  |  |
| Manual | 76 | 51 | 92 | 77 | 74 | 81 | 72 | 87 | 100 | 104 |  |  |

The students wish to answer the following questions:

(a) Are the automated and manual counts significantly different for a fixed confluency level? What are the confidence intervals for the population differences if normality of the measurements is assumed?

(b) If the difference between automated and manual counts constitutes an error, are the errors comparable for the two confluency levels?

We will revisit this example later in the book (Exercise 10.17) and see that for the 20% confluency level there is no significant difference between the automated and manual counts, while for the 70% level the difference is significant. We will also see that the errors for the two confluency levels significantly differ. The statistical design for comparison of errors is called a difference of differences (DoD) and is quite common in biomedical data analysis.

*Example 1.3.* **Rana Pipiens.** Students in a quantitative engineering physiology laboratory were asked to expose the gastrocnemius muscle of the northern leopard frog (*Rana pipiens*, Fig. 1.2), and stimulate the sciatic nerve to observe contractions in the skeletal muscle. Students were interested in modeling the length–tension relationship. The force used was the active force, calculated by subtracting the measured passive force (no stimulation) from the total force (with stimulation).

**Fig. 1.2** *Rana pipiens.*

The active force represents the dependent variable. The length of the muscle begins at 35 mm and stretches in increments of 0.5 mm, until a maximum length of 42.5 mm is achieved. The velocity at which the muscle was stretched was held constant at 0.5 mm/sec.

| Reading | Change in Length (in %) | Passive force | Total force |
|---------|-------------------------|---------------|-------------|
| 1  | 1.4  | 0.012 | 0.366 |
| 2  | 2.9  | 0.031 | 0.498 |
| 3  | 4.3  | 0.040 | 0.560 |
| 4  | 5.7  | 0.050 | 0.653 |
| 5  | 7.1  | 0.061 | 0.656 |
| 6  | 8.6  | 0.072 | 0.740 |
| 7  | 10.0 | 0.085 | 0.865 |
| 8  | 11.4 | 0.100 | 0.898 |
| 9  | 12.9 | 0.128 | 0.959 |
| 10 | 14.3 | 0.164 | 0.994 |
| 11 | 15.7 | 0.223 | 0.955 |
| 12 | 17.1 | 0.315 | 1.019 |
| 13 | 18.6 | 0.411 | 0.895 |
| 14 | 20.0 | 0.569 | 0.900 |
| 15 | 21.4 | 0.751 | 0.905 |

The correlation between the active force and the percent change in length from 35 mm is –0.0941. Why is this correlation so low?

The following model is found using linear regression (least squares):

$$\hat{F} = 0.0618 + 0.2084 \cdot \delta - 0.0163 \cdot \delta^2 + 0.0003 \cdot \delta^3$$
$$- 0.1732 \cdot \sin\left(\frac{\delta}{3}\right) + 0.1242 \cdot \cos\left(\frac{\delta}{3}\right),$$

where $\hat{F}$ is the fitted active force and $\delta$ is the percent change. This model is nonlinear in variables but linear in coefficients, and standard linear regression methodology is applicable (Chap. 16). The model achieves a coefficient of determination of $R^2 = 87.16$.

A plot of the original data with superimposed model fit is shown in Fig. 1.3a. Figure 1.3b shows the residuals $F - \hat{F}$ plotted against $\delta$.

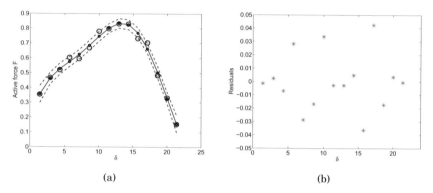

(a)                      (b)

**Fig. 1.3** (a) Regression fit for active force. Observations are shown as *yellow* circles, while the smaller *blue* circles represent the model fits. *Dotted (blue) lines* are 95% model confidence bounds. (b) Model residuals plotted against the percent change in length $\delta$.

Suppose the students are interested in estimating the active force for a change of 12%. The model prediction for $\delta = 12$ is 0.8183, with a 95% confidence interval of $[0.7867, 0.8498]$.

*Example 1.4.* **The 1954 Polio Vaccine Trial.** One of the largest and most publicized public health experiments was performed in 1954 when the benefits of the Salk vaccine for preventing paralytic poliomyelitis was assessed. To ensure that there was no bias in conducting and reporting, the trial was blind to doctors and patients. In boxes of 50 vials, 25 had active vaccines and 25 were placebo. Only the numerical code known to researchers distinguished the well-mixed vials in the box. The clinical trial involved a large number of first-, second-, and third-graders in the USA.

The results were convincing. While the numbers of children assigned to active vaccine and placebo were approximately equal, the incidence of polio in the active group was almost four times lower than that in the placebo group.

|  | Inoculated with vaccine | Inoculated with placebo |
|---|---|---|
| Total number of children inoculated | 200,745 | 201,229 |
| Number of cases of paralytic polio | 33 | 115 |

On the basis of this trial, health officials recommended that every child be vaccinated. Since the time of this clinical trial, the vaccine has improved;

Salk's vaccine was replaced by the superior Sabin preparation and polio is now virtually unknown in the USA. A complete account of this clinical trial can be found in Francis et al.'s (1955) article or Paul Meier's essay in a popular book by Tanur et al. (1972).

The numbers are convincing, but was it possible that an ineffective vaccine produced such a result by chance?

In this example there are two hypothetical populations. The first consists of all first-, second-, and third-graders in the USA who would be inoculated with the active vaccine. The second population consists of US children of the same age who would receive the placebo. The attribute of interest is the presence/absence of paralytic polio. There are two samples from the two populations. If the selection of geographic regions for schools was random, the randomization of the vials in the boxes ensured that the samples were random.
✐

The ultimate summary for quantifying a population attribute is a statistical model. The statistical model term is used in a broad sense here, but a component quantifying inherent uncertainty is always present. For example, random variables, discussed in Chap. 5, can be interpreted as basic statistical models when they model realizations of the attributes in a sample. The model is often indexed by one, several, or sometimes even an infinite number of unknown parameters. An inference about the model translates to an inference about its parameters.

Data are the specific values pertaining to a population attribute recorded from a sample. Often, the terms sample and data are used interchangeably. The term *data* is used as both singular and plural. The singular mode relates to a set, a collection of observations, while the plural is used when referring to the observations. A single observation is called a *datum*.

The following table summarizes the fundamental statistical notions that we discussed.

| | |
|---:|:---|
| attribute | Quantitative or qualitative property, feature(s) of interest |
| population | Statistical universe; an existing or hypothetical totality of attributes |
| sample | A subset of a population |
| data | Recorded values/realizations of an attribute in a sample |
| statistical model | Mathematical description of a population attribute that incorporates incomplete information, variability, and the nondeterministic nature of the population |
| population parameter | A component (possibly multivariate) in a statistical model; the models are typically specified up to a parameter that is left unknown |

The term *statistics* has a plural form but is used in the singular when it relates to methodology. To avoid confusion, we note that *statistics* has another meaning and use. Any sample summary will be called a *statistic*. For example, a sample mean is a statistic, and sample mean and sample range are statistics. In this context, statistics is used in the plural.

## CHAPTER REFERENCES

Edwards, A. W. F. (2007). R. A. Fisher's 1943 unravelling of the Rhesus blood-group system. *Genetics*, **175**, 471–476.

Fisher, R. A. (1947). The Rhesus factor: A study in scientific method. *Amer. Sci.*, **35**, 95–102.

Francis, T. Jr., Korns, R., Voight, R., Boisen, M., Hemphill, F., Napier, J., and Tolchinsky, E. (1955). An evaluation of the 1954 poliomyelitis vaccine trials: Summary report.*American Journal of Public Health*, **45**, 5, 1–63.

Sinclair, Sir John. (1791). The Statistical Account of Scotland. Drawn up from the communications of the Ministers of the different parishes. Volume first. Edinburgh: printed and sold by William Creech, Nha. V27.

Tanur, J. M., Mosteller, F., Kruskal, W. H., Link, R. F., Pieters, R. S. and Rising, G. R., eds. (1989). *Statistics: A Guide to the Unknown,* Third Edition. Wadsworth, Inc., Belmont, CA.

# Chapter 2
# The Sample and Its Properties

*When you're dealing with data, you have to look past the numbers.*

– Nathan Yau

**WHAT IS COVERED IN THIS CHAPTER**

• MATLAB Session with Basic Univariate Statistics
• Numerical Characteristics of a Sample
• Multivariate Numerical and Graphical Sample Summaries
• Time Series
• Typology of Data

## 2.1 Introduction

The famous American statistician John Tukey once said, "Exploratory data analysis can never be the whole story, but nothing else can serve as the foundation stone – as the first step." The term *exploratory data analysis* is self-defining. Its simplest branch, *descriptive statistics*, is the methodology behind approaching and summarizing experimental data. No formal statistical training is needed for its use. Basic data manipulations such as calculating averages of experimental responses, translating data to pie charts or histograms, or assessing the variability and inspection for unusual measurements are all examples of descriptive statistics. Rather than focusing on the population using information from a sample, which is a staple of statistics, descriptive statistics is concerned with the description, summary, and presentation of the sample itself. For example, numerical summaries of a sample could be measures of location (mean, median, percentiles, mode, extrema), measures of variability (sample standard deviation/variance, robust versions of the variance, range of data, interquartile range, etc.), higher-order statistics ($k$th moments, $k$th central moments, skewness, kurtosis), and functions of descriptors (coefficient of variation). Graphical summaries of samples involve various visual presentations such as box-and-whisker plots, pie charts, histograms, empirical cumulative distribution functions, etc. Many basic data descriptors are used in everyday data manipulation.

Ultimately, exploratory data analysis and descriptive statistics contribute to the principal goal of statistics – inference about population descriptors – by guiding how the statistical models should be set.

It is important to note that descriptive statistics and exploratory data analysis have recently regained importance due to ever increasing sizes of data sets. Some complex data structures require several terrabytes of memory just to be stored. Thus, preprocessing, summarizing, and dimension-reduction steps are needed to prepare such data for inferential tasks such as classification, estimation, and testing. Consequently, the inference is placed on data summaries (descriptors, features) rather than the raw data themselves.

Many data managing software programs have elaborate numerical and graphical capabilities. MATLAB provides an excellent environment for data manipulation and presentation with superb handling of data structures and graphics. In this chapter we intertwine some basic descriptive statistics with MATLAB programming using data obtained from real-life research laboratories. Most of the statistics are already built-in; for some we will make a custom code in the form of m-functions or m-scripts.

This chapter establishes two goals: (i) to help you gently relearn and refresh your MATLAB programming skills through annotated sessions while, at the same time, (ii) introducing some basic statistical measures, many of which should already be familiar to you. Many of the statistical summaries will be revisited later in the book in the context of inference. You are encouraged to continuously consult MATLAB's online help pages for support since many programming details and command options are omitted in this text.

## 2.2 A MATLAB Session on Univariate Descriptive Statistics

In this section we will analyze data derived from an experiment, step by step with a brief explanation of the MATLAB commands used. The whole session can be found in a single annotated file ◀ carea.m available at the book's Web page.

The data can be found in the file ⬛ cellarea.dat, which features measurements from the lab of Todd McDevitt at Georgia Tech: http://www.bme.gatech.edu/groups/mcdevitt/.

This experiment on cell growth involved several time durations and two motion conditions. Here is a brief description:

Embryonic stem cells (ESCs) have the ability to differentiate into all somatic cell types, making ESCs useful for studying developmental biology, in vitro drug screening, and as a cell source for regenerative medicine and cell-based therapies. A common method to induce differentiation of ESCs is through the formation of multicellular spheroids termed embryoid bodies (EBs). ESCs spontaneously aggregate into EBs when cultured on a nonadherent substrate; however, under static conditions, this aggregation is uncontrolled and EBs form in various sizes and shapes, which may lead to variability in cell differentiation patterns. When rotary motion is applied during EB formation, the resulting population of EBs appears more uniform in size and shape.

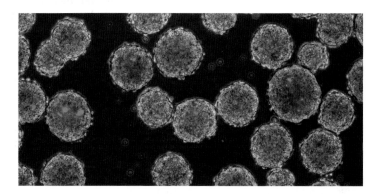

**Fig. 2.1** Fluorescence microscopy image of cells overlaid with phase image to display incorporation of microspheres (*red stain*) in embryoid bodies (*gray clusters*) (courtesy of Todd McDevitt).

After 2, 4, and 7 days of culture, images of EBs were acquired using phase-contrast microscopy. Image analysis software was used to determine the area of each EB imaged (Fig. 2.1). At least 100 EBs were analyzed from three separate plates for both static and rotary cultures at the three time points studied.

Here we focus only on the measurements of visible surface areas of cells (in $\mu m^2$) after growth time of 2 days, $t = 2$, under the static condition. The data are recorded as an ASCII file ⬛ cellarea.dat. Importing the data set into MATLAB is done using the command

```
load('cellarea.dat');
```

given that the data set is on the MATLAB path. If this is not the case, use addpath('foldername') to add to the search path foldername in which the file resides. A glimpse at the data is provided by histogram command, hist:

```
hist(cellarea, 100)
```

After inspecting the histogram (Fig. 2.2) we find that there is one quite unusual observation, inconsistent with the remaining experimental measurements.

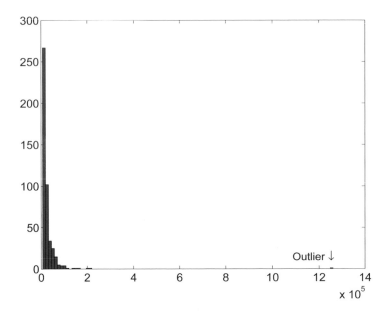

**Fig. 2.2** Histogram of the raw data. Notice the unusual measurement beyond the point $12 \times 10^5$.

We assume that the unusual observation is an outlier and omit it from the data set:

```
car = cellarea(cellarea ~= max(cellarea));
```

(Some formal diagnostic tests for outliers will be discussed later in the text.)

Next, the data are rescaled to more moderate values, so that the area is expressed in thousands of $\mu m^2$ and the measurements have a convenient order of magnitude.

```
car = car/1000;
n = length(car); %n is sample size
%n=462
```

Thus, we obtain a sample of size $n = 462$ to further explore by descriptive statistics. The histogram we have plotted has already given us a sense of the distribution within the sample, and we have an idea of the shape, location, spread, symmetry, etc. of observations.

Next, we find numerical characteristics of the sample and first discuss its location measures, which, as the name indicates, evaluate the relative location of the sample.

## 2.3 Location Measures

**Means.** The three averages – arithmetic, geometric, and harmonic – are known as Pythagorean means.

The arithmetic mean (mean),

$$\overline{X} = \frac{X_1 + \cdots + X_n}{n} = \frac{1}{n}\sum_{i=1}^{n} X_i,$$

is a fundamental summary statistic. The geometric mean (geomean) is

$$\sqrt[n]{X_1 \times X_2 \times \cdots \times X_n} = \left(\prod_{i=1}^{n} X_i\right)^{1/n},$$

and the harmonic mean (harmmean) is

$$\frac{n}{1/X_1 + 1/X_2 + \cdots 1/X_n} = \frac{n}{\sum_{i=1}^{n} 1/X_i}.$$

For the data set $\{1,2,3\}$ the mean is 2, the geometric mean is $\sqrt[3]{6} = 1.8171$, and the harmonic mean is $3/(1/1 + 1/2 + 1/3) = 1.6364$. In standard statistical practice geometric and harmonic means are not used as often as arithmetic means. To illustrate the contexts in which they should be used, consider several simple examples.

*Example 2.1.* You visit the bank to deposit a long-term monetary investment in hopes that it will accumulate interest over a 3-year span. Suppose that the investment earns 10% the first year, 50% the second year, and 30% the third year. What is its average rate of return? In this instance it is not the arithmetic mean, because in the first year the investment was multiplied by 1.10, in the second year it was multiplied by 1.50, and in the third year it was multiplied by 1.30. The correct measure is the geometric mean of these three numbers, which is about 1.29, or 29% of the annual interest. If, for example, the ratios are averaged (i.e., ratio = new method/old method) over many experiments, the geometric mean should be used. This is evident by considering an example. If one experiment yields a ratio of 10 and the next yields a ratio of 0.1, an

arithmetic mean would misleadingly report that the average ratio was near 5. Taking a geometric mean will report a more meaningful average ratio of 1.
✎

*Example 2.2.* Consider now two scenarios in which the harmonic mean should be used.

(i) If for half the distance of a trip one travels at 40 miles per hour and for the other half of the distance one travels at 60 miles per hour, then the average speed of the trip is given by the harmonic mean of 40 and 60, which is 48; that is, the total amount of time for the trip is the same as if one traveled the entire trip at 48 miles per hour. Note, however, that if one had traveled for half the time at one speed and the other half at another, the arithmetic mean, in this case 50 miles per hour, would provide the correct interpretation of average.

(ii) In financial calculations, the harmonic mean is used to express the average cost of shares purchased over a period of time. For example, an investor purchases $1000 worth of stock every month for 3 months. If the three spot prices at execution time are $8, $9, and $10, then the average price the investor paid is $8.926 per share. However, if the investor purchased 1000 shares per month, then the arithmetic mean should be used.
✎

**Order Statistic.** If the sample $X_1, \ldots, X_n$ is ordered as $X_{(1)} \le X_{(2)} \le \cdots \le X_{(n)}$ so that $X_{(1)}$ is the minimum and $X_{(n)}$ is the maximum, then $X_{(1)}, X_{(2)}, \ldots X_{(n)}$ is called the *order statistic*. For example, if $X_1 = 2$, $X_2 = -1$, $X_3 = 10$, $X_4 = 0$, and $X_5 = 4$, then the order statistic is $X_{(1)} = -1$, $X_{(2)} = 0$, $X_{(3)} = 2$, $X_{(4)} = 4$, and $X_{(5)} = 10$.

**Median.** The median[1] is the middle of the sample sorted in numerical order. In terms of order statistic, the median is defined as

$$
Me = \begin{cases} X_{((n+1)/2)}, & \text{if } n \text{ is odd,} \\ (X_{(n/2)} + X_{(n/2+1)})/2, & \text{if } n \text{ is even.} \end{cases}
$$

If the sample size is odd, then there is a single observation in the middle of the ordered sample at the position $(n+1)/2$, while for the even sample sizes, the ordered sample has two elements in the middle at positions $n/2$ and $n/2+1$ and the median is their average. The median is an estimator of location robust to extremes and outliers. For instance, in both data sets, $\{-1, 0, 4, 7, 20\}$ and $\{-1, 0, 4, 7, 200\}$, the median is 4. The means are 6 and 42, respectively.

**Mode.** The most frequent (fashionable[2]) observation in the sample (if such exists) is the mode of the sample. If the sample is composite, the observation $x_i$ corresponding to the largest frequency $f_i$ is the mode. Composite samples consist of realizations $x_i$ and their frequencies $f_i$, as in $\begin{pmatrix} x_1 & x_2 & \cdots & x_k \\ f_1 & f_2 & \cdots & f_k \end{pmatrix}$.

---

[1] Latin: *medianus* = middle

[2] *Mode* (fr) = fashion

Mode may not be unique. If there are two modes, the sample is bimodal, three modes make it trimodal, etc.

**Trimmed Mean.** As mentioned earlier, the mean is a location measure sensitive to extreme observations and possible outliers. To make this measure more robust, one may trim $\alpha \cdot 100\%$ of the data symmetrically from both sides of the ordered sample (trim $\alpha/2 \cdot 100\%$ smallest and $\alpha/2 \cdot 100\%$ largest observations, Fig. 2.3b).

If your sample, for instance, is $\{1,2, 3,4, 5,6,7, 8,9,100\}$, then a 20% trimmed mean is a mean of $\{2,3, 4,5, 6,7, 8,9\}$.

Here is the command in MATLAB that determines the discussed locations for the cell data.

```
location = [geomean(car) harmmean(car) mean(car) ...
    median(car) mode(car) trimmean(car,20)]
%location = 18.8485 15.4211  24.8701 17 10 20.0892
```

By applying $\alpha 100\%$ trimming, we end up with a sample of reduced size $[(1-\alpha)100\%]$. Sometimes the sample size is important to preserve.

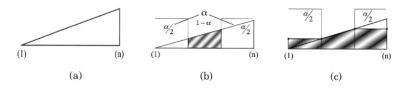

**Fig. 2.3** (a) Schematic graph of an ordered sample; (b) Part of the sample from which $\alpha$-trimmed mean is calculated; (c) Modified sample for the winsorized mean.

**Winsorized mean.** A robust location measure that preserves sample size is the winsorized mean. Similarly to a trimmed mean, a winsorized mean identifies outlying observations, but instead of trimming them the observations are replaced by either the minimum or maximum of the trimmed sample, depending on if the trimming is done from below or above (Fig. 2.3c).

The winsorized mean is not a built-in MATLAB function. However, it can be calculated easily by the following code:

```
alpha=20;
sa = sort(car);
sa(1:floor( n*alpha/200 )) = sa(floor( n*alpha/200 ) + 1);
sa(end-floor( n*alpha/200 ):end) = ...
        sa(end-floor( n*alpha/200 ) - 1);
winsmean = mean(sa) % winsmean = 21.9632
```

Figure 2.3 shows schematic graphs of of a sample

## 2.4 Variability Measures

Location measures are intuitive but give a minimal glimpse at the nature of a sample. An important set of sample descriptors are variability measures, or measures of spread. There are many measures of variability in a sample. Gauss (1816) already used several of them on a set of 48 astronomical measurements concerning relative positions of Jupiter and its satellite Pallas.

**Sample Variance and Sample Standard Deviation.** The variance of a sample, or sample variance, is defined as

$$s^2 = \frac{1}{n-1} \sum_{i=1}^{n} (X_i - \overline{X})^2.$$

Note that we use $\frac{1}{n-1}$ instead of the "expected" $\frac{1}{n}$. The reasons for this will be discussed later. An alternative expression for $s^2$ that is more suitable for calculation (by hand) is

$$s^2 = \frac{1}{n-1} \left( \sum_{i=1}^{n} (X_i^2) - n(\overline{X})^2 \right),$$

see Exercises 2.6 and 2.7.

In MATLAB, the sample variance of a data vector x is var(x) or var(x,0) Flag 0 in the argument list indicates that the ratio $1/(n-1)$ is used to calculate the sample variance. If the flag is 1, then var(x,1) stands for

$$s_*^2 = \frac{1}{n} \sum_{i=1}^{n} (X_i - \overline{X})^2,$$

which is sometimes used instead of $s^2$. We will see later that both estimators have good properties: $s^2$ is an unbiased estimator of the population variance while $s_*^2$ is the maximum likelihood estimator. The square root of the sample variance is the *sample standard deviation*:

$$s = \sqrt{\frac{1}{n-1} \sum_{i=1}^{n} (X_i - \overline{X})^2}.$$

In MATLAB the standard deviation can be calculated by `std(x)=std(x,0)` or `std(x,1)`, depending on whether the sum of squares is divided by $n-1$ or by $n$.

```
%Variability Measures
var(car)    % standard sample variance, also var(car,0)
            %ans = 588.9592
var(car,1) % sample variance with sum of squares
            % divided by n
   %ans =   587.6844
std(car)    % sample standard deviation, sum of squares
            % divided by (n-1), also std(car,0)
   %ans = 24.2685
std(car,1) % sample standard deviation, sum of squares
            % divided by n
   %ans = 24.2422
sqrt(var(car))   %should be equal to std(car)
   %ans = 24.2685
sqrt(var(car,1)) %should be equal to std(car,1)
   %ans = 24.2422
```

**Remark.** When a new observation is obtained, one can update the sample variance without having to recalculate it. If $\bar{x}_n$ and $s_n^2$ are the sample mean and variance based on $x_1, x_2, \ldots, x_n$ and a new observation $x_{n+1}$ is obtained, then

$$s_{n+1}^2 = \frac{(n-1)s_n^2 + (x_{n+1} - \bar{x}_n)(x_{n+1} - \bar{x}_{n+1})}{n},$$

where $\bar{x}_{n+1} = (n\bar{x}_n + x_{n+1})/(n+1)$.

**MAD-Type Estimators.** Another group of estimators of variability involves absolute values of deviations from the center of a sample and are known as MAD estimators. These estimators are less sensitive to extreme observations and outliers compared to the sample standard deviation. They belong to the class of so-called robust estimators. The acronym MAD stands for either *mean absolute difference from the mean* or, more commonly, *median absolute difference from the median*. According to statistics historians (David, 1998), both MADs were already used by Gauss at the beginning of the nineteenth century.

MATLAB uses `mad(car)` or `mad(a,0)` for the first and `mad(car,1)` for the second definition:

$$\mathrm{MAD}_0 = \frac{1}{n}\sum_{i=1}^{n}|X_i - \overline{X}|, \quad \mathrm{MAD}_1 = \mathrm{median}\{|X_i - \mathrm{median}\{X_i\}|\}.$$

A typical convention is to multiply the $MAD_1$ estimator mad(car,1) by 1.4826, to make it comparable to the sample standard deviation.

```
mad(car) % mean absolute deviation from the mean;
         % MAD is usually referring to
         % median absolute deviation from the median
  %ans = 15.3328
realmad = 1.4826 * median( abs(car - median(car)))
         %real mad in MATLAB is 1.4826 * mad(car,1)
  %realmad = 10.3781
```

**Sample Range and IQR.** Two simple measures of variability, or rather the spread of a sample, are the range $R$ and interquartile range (IQR), in MATLAB range and iqr. They are defined by the order statistic of the sample. The range is the maximum minus the minimum of the sample, $R = X_{(n)} - X_{(1)}$, while IQR is defined by sample quantiles.

```
range(car)  %Range, span of data, Max - Min
  %ans = 212
iqr(car)       %inter-quartile range, Q3-Q1
  %ans = 19
```

If the sample is bell-shape distributed, a robust estimator of variance is $\hat{\sigma}^2 = (IQR/1.349)^2$, and this summary was known to Quetelet in the first part of the nineteenth century. It is a simple estimator, not affected by outliers (it ignores 25% of observations in each tail), but its variability is large.

**Sample Quantiles/Percentiles.** Sample quantiles (in units between 0 and 1) or sample percentiles (in units between 0 and 100) are very important summaries that reveal both the location and the spread of a sample. For example, we may be interested in a point $x_p$ that partitions the ordered sample into two parts, one with $p \cdot 100\%$ of observations smaller than $x_p$ and another with $(1-p)100\%$ observations greater than $x_p$. In MATLAB, we use the commands quantile or prctile, depending on how we express the proportion of the sample. For example, for the 5, 10, 25, 50, 75, 90, and 95 percentiles we have

```
%5%, 10%, 25%, 50%, 75%, 90%, 95% percentiles are:
prctile(car, 100*[0.05, 0.10, 0.25, 0.50, 0.75, 0.90, 0.95] )
%ans =   7     8    11    17    30    51    67
```

The same results can be obtained using the command

```
qts = quantile(car,[0.05 0.1 0.25 0.5 0.75 0.9 0.95])
%qts =   7     8    11    17    30    51    67
```

In our dataset, 5% of observations are less than 7, and 90% of observations are less than 51.

Some percentiles/quantiles are special, such as the median of the sample, which is the 50th percentile. Quartiles divide an ordered sample into four parts; the 25th percentile is known as the first quartile, $Q_1$, and the 75th percentile is known as the third quartile, $Q_3$. The median is $Q_2$, of course.[3]

---

[3] The range is equipartitioned by a single median, two terciles, three quartiles, four quintiles, five sextiles, six septiles, seven octiles, eight naniles, or nine deciles.

In MATLAB, `Q1=prctile(car,25); Q3=prctile(car,75)`. Now we can define the IQR as $Q_3 - Q_1$:

```
prctile(car, 75)- prctile(car, 25) %should be equal to iqr(car).
%ans = 19
```

The five-number summary for univariate data is defined as ($Min$, $Q_1$, $Me$, $Q_3$, $Max$).

$z$-**Scores.** For a sample $x_1, x_2, \ldots, x_n$ the $z$-score is the standardized sample $z_1, z_2, \ldots, z_n$, where $z_i = (x_i - \overline{x})/s$. In the standardized sample, the mean is 0 and the sample variance (and standard deviation) is 1. The basic reason why standardization may be needed is to assess extreme values, or compare samples taken at different scales. Some other reasons will be discussed in subsequent chapters.

```
zcar = zscore(car);
mean(zcar)
%ans = -5.8155e-017
var(zcar)
%ans = 1
```

**Moments of Higher Order.** The term *sample moments* is drawn from mechanics. If the observations are interpreted as unit masses at positions $X_1, \ldots, X_n$, then the sample mean is the first moment in the mechanical sense – it represents the balance point for the system of all points. The moments of higher order have their corresponding mechanical interpretation. The formula for the $k$th moment is

$$m_k = \frac{1}{n}(X_1^k + \cdots + X_n^k) = \frac{1}{n}\sum_{i=1}^{n} X_i^k.$$

The moments $m_k$ are sometimes called *raw* sample moments. The *power $k$ mean* is $(m_k)^{1/k}$, that is,

$$\left(\frac{1}{n}\sum_{i=1}^{n} X_i^k\right)^{1/k}.$$

For example, the sample mean is the first moment and power 1 mean, $m_1 = \overline{X}$. The *central* moments of order $k$ are defined as

$$\mu_k = \frac{1}{n} \sum_{i=1}^{n} (X_i - m_1)^k.$$

Notice that $\mu_1 = 0$ and that $\mu_2$ is the sample variance (calculated by var(.,1) with the sum of squares divided by $n$). MATLAB has a built-in function moment for calculating the central moments.

```
%Moments of Higher Orders
    %kth (row) moment: mean(car.^k)
mean(car.^3) %third
    %ans = 1.1161e+005
    %kth central moment mean((car-mean(car)).^k)
mean( (car-mean(car)).^3 ) %ans=5.2383e+004
    %is the same as
moment(car,3) %ans=5.2383e+004
```

**Skewness and Kurtosis.** There are many uses of higher moments in describing a sample. Two important sample measures involving higher-order moments are *skewness* and *kurtosis*.

Skewness is defined as

$$\gamma_n = \mu_3/\mu_2^{3/2} = \mu_3/s_*^3$$

and measures the degree of asymmetry in a sample distribution. Positively skewed distributions have longer right tails and their sample mean is larger than the median. Negatively skewed sample distributions have longer left tails and their mean is smaller than the median.

Kurtosis is defined as

$$\kappa_n = \mu_4/\mu_2^2 = \mu_4/s_*^4.$$

It represents the measure of "peakedness" or flatness of a sample distribution. In fact, there is no consensus on the exact definition of kurtosis since flat but fat-tailed distributions would also have high kurtosis. Distributions that have a kurtosis of <3 are called *platykurtic* and those with a kurtosis of >3 are called *leptokurtic*.

```
%sample skewness mean(car.^3)/std(car,1)^3
mean( (car-mean(car)).^3 )/std(car,1)^3 %ans = 3.6769
skewness(car) %ans = 3.6769
%sample kurtosis
mean( (car-mean(car)).^4 )/std(car,1)^4 % ans = 22.8297
kurtosis(car)% ans = 22.8297
```

A robust version of the skewness measure was proposed by Bowley (1920) as

$$\gamma_n{}^* = \frac{(Q_3 - Me) - (Me - Q_1)}{Q_3 - Q_1},$$

and ranges between −1 and 1. Moors (1988) proposed a robust measure of kurtosis based on sample octiles:

$$\kappa_n{}^* = \frac{(O_7 - O_5) + (O_3 - O_1)}{O_6 - O_2},$$

where $O_i$ is the $i/8 \times 100$ percentile ($i$th octile) of the sample for $i = 1, 2, \ldots, 7$. If the sample is large, one can take $O_i$ as $X_{(\lfloor i/8 \times n \rfloor)}$. The constant 1.766 is sometimes added to $\kappa_n^*$ as a calibration adjustment so that it is comparable with the traditional measure of kurtosis for samples from Gaussian populations.

```
%robust skewness
(prctile(car, 75)+prctile(car, 25) - ...
2 * median(car))/(prctile(car, 75) - prctile(car, 25))
%0.3684
```

```
%robust kurtosis
(prctile(car,7/8*100)-prctile(car,5/8*100)+prctile(car,3/8*100)- ...
  prctile(car,1/8*100))/(prctile(car,6/8*100)-prctile(car,2/8*100))
%1.4211
```

**Coefficient of Variation.** The coefficient of variation, CV, is the ratio

$$CV = \frac{s}{\overline{X}}.$$

The CV expresses the variability of a sample in the units of its mean. In other words, a CV equal to 2 would mean that the variability is equal to $2\,\overline{X}$. The assumption is that the mean is positive. The CV is used when comparing the variability of data reported on different scales. For example, instructors A and B teach different sections of the same class, but design their own final exams individually. To compare the effectiveness of their respective exam designs at creating a maximum variance in exam scores (a tacit goal of exam designs), they calculate the CVs. It is important to note that the CVs would not be related to the exam grading scale, to the relative performance of the students, or to the difficulty of the exam.

```
%sample CV [coefficient of variation]
std(car)/mean(car)
%ans = 0.9758
```

The reciprocal of CV, $\overline{X}/s$, is sometimes called the signal-to-noise ratio, and it is often used in engineering quality control.

**Grouped Data.** When a data set is large and many observations are repetitive, data are often recorded as grouped or composite. For example, the data set

$$
\begin{array}{l}
4\ 5\ 6\ 3\ 4\ 3\ 6\ 4\ 5\ 4\ 3 \\
7\ 3\ 5\ 2\ 5\ 6\ 4\ 2\ 4\ 3\ 4 \\
7\ 7\ 4\ 2\ 2\ 5\ 4\ 2\ 5\ 3\ 8
\end{array}
$$

is called a simple sample, or raw sample, as it lists explicitly all observations. It can be presented in a more compact form, as grouped data:

$$
\begin{array}{c|cccccccc}
X_i & 2 & 3 & 4 & 5 & 6 & 7 & 8 \\
\hline
f_i & 5 & 6 & 9 & 6 & 3 & 3 & 1
\end{array}
$$

where $X_i$ are distinctive values in the data set with frequencies $f_i$, and the number of groups is $k = 7$. Notice that $X_i = 5$ appears six times in the simple sample, so its frequency is $f_i = 6$.

The function ◢ [xi fi]=simple2comp(a) provides frequencies fi for a list xi of distinctive values in a.

```
a=[  4  5  6  3  4  3  6  4  5  4  3 ...
     7  3  5  2  5  6  4  2  4  3  4 ...
     7  7  4  2  2  5  4  2  5  3  8];
[xi fi] = simple2comp( a )
% xi =
%      2    3    4    5    6    7    8
% fi =
%      5    6    9    6    3    3    1
```

Here, $n = \sum_i f_i = 33$.

When a sample is composite, the sample mean and variance are

$$
\overline{X} = \frac{\sum_{i=1}^{k} f_i X_i}{n}, \quad s^2 = \frac{\sum_{i=1}^{k} f_i (X_i - \overline{X})^2}{n-1}
$$

for $n = \sum_i f_i$. By defining the $m$th raw and central sample moments as

$$
\overline{X^m} = \frac{\sum_{i=1}^{k} f_i X_i^m}{n} \quad \text{and} \quad \mu_m = \frac{\sum_{i=1}^{k} f_i (X_i - \overline{X})^m}{n-1},
$$

one can express skewness, kurtosis, CV, and other sample statistics that are functions of moments.

**Diversity Indices for Categorical Data.**  If the data are categorical and numerical characteristics such as moments and percentiles cannot be defined, but the frequencies $f_i$ of classes/categories are given, one can define Shannon's diversity index:

$$H = \frac{n \log n - \sum_{i=1}^{k} f_i \log f_i}{n}. \tag{2.1}$$

If some frequency is 0, then $0 \log 0 = 0$. The maximum of $H$ is $\log k$; it is achieved when all $f_i$ are equal. The normalized diversity index, $E_H = H/\log k$, is called Shannon's homogeneity (equitability) index of the sample.

Neither $H$ nor $E_H$ depends on the sample size.

*Example 2.3.* **Homogeneity of Blood Types.**    Suppose the samples from Brazilian, Indian, Norwegian, and US populations are taken and the frequencies of blood types (ABO/Rh) are obtained.

| Population | O+ | A+ | B+ | AB+ | O– | A– | B– | AB– | total |
|---|---|---|---|---|---|---|---|---|---|
| Brazil | 115 | 108 | 25 | 6 | 28 | 25 | 6 | 1 | 314 |
| India | 220 | 134 | 183 | 39 | 12 | 6 | 6 | 12 | 612 |
| Norway | 83 | 104 | 16 | 8 | 14 | 18 | 2 | 1 | 246 |
| US | 99 | 94 | 21 | 8 | 18 | 18 | 5 | 2 | 265 |

Which county's sample is most homogeneous with respect to blood type attribute?

```
br = [115    108    25     6     28    25     6     1];
in = [220    134    183    39    12     6     6    12];
no = [ 83    104    16     8     14    18     2     1];
us = [ 99     94    21     8     18    18     5     2];

Eh = @(f) (sum(f)*log(sum(f)) - ...
      sum( f.*log(f)))/(sum(f)*log(length(f)))

Eh(br)   % 0.7324
Eh(in)   % 0.7125
Eh(no)   % 0.6904
Eh(us)   % 0.7306
```

Among the four samples, the sample from Brazil is the most homogeneous with respect to the blood types of its population as it maximizes the statistic $E_H$. See also Exercise 2.13 for an alternative definition of diversity/homogeneity indices.

## 2.5 Displaying Data

In addition to their numerical descriptors, samples are often presented in a graphical manner. In this section, we discuss some basic graphical summaries.

**Box-and-Whiskers Plot.** The top and bottom of the "box" are the 25th and 75th percentile of the data, respectively, with the distances between them representing the IQR. The line inside the box represents the sample median. If the median is not centered in the box, it indicates sample skewness. Whiskers extend from the lower and upper sides of the box to the data's most extreme values within 1.5 times the IQR. Potential outliers are displayed with red "+" beyond the endpoints of the whiskers.

The MATLAB command `boxplot(X)` produces a box-and-whisker plot for $X$. If $X$ is a matrix, the boxes are calculated and plotted for each column. Figure 2.4a is produced by

```
%Some Graphical Summaries of the Sample
figure;
boxplot(car)
```

**Histogram.** As illustrated previously in this chapter, the histogram is a rough approximation of the population distribution based on a sample. It plots frequencies (or relative frequencies for normalized histograms) for interval-grouped data. Graphically, the histogram is a barplot over contiguous intervals or bins spanning the range of data (Fig. 2.4b). In MATLAB, the typical command for a histogram is `[fre,xout] = hist(data,nbins)`, where `nbins` is the number of bins and the outputs `fre` and `xout` are the frequency counts and the bin locations, respectively. Given the output, one can use `bar(xout,n)` to plot the histogram. When the output is not requested, MATLAB produces the plot by default.

```
figure;
hist(car, 80)
```

The histogram is only an approximation of the distribution of measurements in the population from which the sample is obtained.

There are numerous rules on how to automatically determine the number of bins or, equivalently, bin sizes, none of them superior to the others on all possible data sets. A commonly used proposal is Sturges' rule (Sturges, 1926), where the number of bins $k$ is suggested to be

$$k = 1 + \log_2 n,$$

where $n$ is the size of the sample. Sturges' rule was derived for bell-shaped distributions of data and may oversmooth data that are skewed, multimodal, or have some other features. Other suggestions specify the bin size as $h = 2 \cdot \text{IQR}/n^{1/3}$ (Diaconis–Freedman rule) or, alternatively, $h = (7s)/(2n^{1/3})$ (Scott's rule; $s$ is the sample standard deviation). By dividing the range of the data by $h$, one finds the number of bins.

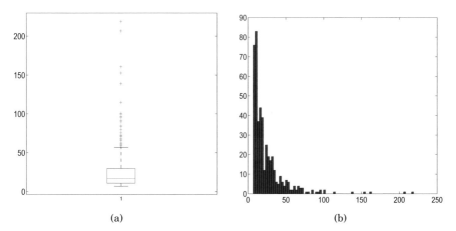

**Fig. 2.4** (a) Box plot and (b) histogram of cell data `car`.

For example, for cell-area data `car`, Sturges' rule suggests 10 bins, Scott's 19 bins, and the Diaconis–Freedman rule 43 bins. The default `nbins` in MATLAB is 10 for any sample size.

The histogram is a crude estimator of a probability density that will be discussed in detail later on (Chap. 5). A more esthetic estimator of the population distribution is given by the *kernel smoother density* estimate, or `ksdensity`. We will not go into the details of kernel smoothing at this point in the text; however, note that the spread of a kernel function (such as a Gaussian kernel) regulates the degree of smoothing and in some sense is equivalent to the choice of bin size in histograms.

Command `[f,xi,u]=ksdensity(x)` computes a density estimate based on data `x`. Output `f` is the vector of density values evaluated at the points in `xi`. The estimate is based on a normal kernel function, using a window parameter `width` that depends on the number of points in `x`. The default width `u` is returned as an output and can be used to tune the smoothness of the estimate, as is done in the example below. The density is evaluated at 100 equally spaced points that cover the range of the data in `x`.

```
figure;
[f,x,u] = ksdensity(car);
plot(x,f)
hold on
[f,x] = ksdensity(car,'width',u/3);
plot(x,f,'r');
[f,x] = ksdensity(car,'width',u*3);
plot(x,f,'g');
legend('default width','default/3','3 * default')
hold off
```

**Empirical Cumulative Distribution Function.** The empirical cumulative distribution function (ECDF) $F_n(x)$ for a sample $X_1, \ldots, X_n$ is defined
as

$$F_n(x) = \frac{1}{n} \sum_{i=1}^{n} \mathbf{1}(X_i \leq x) \tag{2.2}$$

and represents the proportion of sample values smaller than $x$. Here $\mathbf{1}(X_i \leq x)$
is either 0 or 1. It is equal to 1 if $\{X_i \leq x\}$ is true, 0 otherwise.

The function ◢ empiricalcdf(x,sample) will calculate the ECDF based on
the observations in sample at a value x.

```
xx = min(car)-1:0.01:max(car)+1;
yy = empiricalcdf(xx, car);
plot(xx, yy, 'k-','linewidth',2)
xlabel('x');   ylabel('F_n(x)')
```

In MATLAB, [f xf]=ecdf(x) is used to calculate the proportion f of the
sample x that is smaller than xf. Figure 2.5b shows the ECDF for the cell area
data, car.

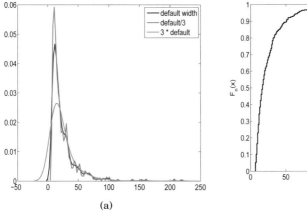
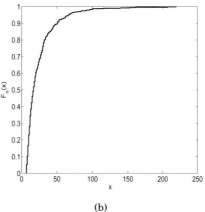

(a)                                                        (b)

**Fig. 2.5** (a) Smoothed histogram (density estimator) for different widths of smoothing kernel; (b) Empirical CDF.

**Q–Q Plots.** Q–Q plots, short for quantile–quantile plots, compare the distribution of a sample with some standard theoretical distribution, such as
normal, or with a distribution of another sample. This is done by plotting the
sample quantiles of one distribution against the corresponding quantiles of the
other. If the plot is close to linear, then the distributions are close (up to a scale

and shift). If the plot is close to the 45° line, then the compared distributions are approximately equal. In MATLAB the command qqplot(X,Y) produces an empirical Q–Q plot of the quantiles of the data set $X$ vs. the quantiles of the data set $Y$. If the data set $Y$ is omitted, then qqplot(X) plots the quantiles of $X$ against standard normal quantiles and essentially checks the normality of the sample.

Figure 2.6 gives us the Q–Q plot of the cell area data set against the normal distribution. Note the deviation from linearity suggesting that the distribution is skewed. A line joining the first and third sample quartiles is superimposed in the plot. This line is extrapolated out to the ends of the sample to help visually assess the linearity of the Q–Q display. Q–Q plots will be discussed in more detail in Chap. 13.

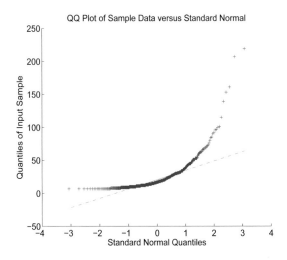

**Fig. 2.6** Quantiles of data plotted against corresponding normal quantiles, via qqplot.

**Pie Charts.** If we are interested in visualizing proportions or frequencies, the pie chart is appropriate. A pie chart (pie in MATLAB) is a graphical display in the form of a circle where the proportions are assigned segments.

Suppose that in the cell area data set we are interested in comparing proportions of cells with areas in three regions: smaller than or equal to 15, between 15 and 30, and larger than 30. We would like to emphasize the proportion of cells with areas between 15 and 30. The following MATLAB code plots the pie charts (Fig. 2.7).

```
n1 = sum( car <= 15 ); %n1=213
n2 = sum( (car > 15 ) & (car <= 30) ); %n2=139
n3 = sum( car > 30 ); %n3=110
```

```
% n=n1+n2+n3 = 462
% proportions n1/n, n2/n, and n3/n are
%              0.4610,0.3009 and 0.2381
explode = [0 1 0]
pie([n1, n2, n3], explode)
pie3([n1, n2, n3], explode)
```

Note that option `explode=[0 1 0]` separates the second segment from the circle. The command `pie3` plots a 3-D version of a pie chart (Fig. 2.7b).

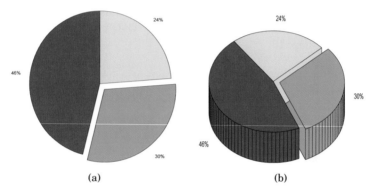

(a)                                                    (b)

**Fig. 2.7** Pie charts for frequencies 213, 139, and 110 of cell areas smaller than or equal to 15, between 15 and 30, and larger than 30. The proportion of cells with the area between 15 and 30 is emphasized.

## 2.6 Multidimensional Samples: Fisher's Iris Data and Body Fat Data

In the cell area example, the sample was univariate, that is, each measurement was a scalar. If a measurement is a vector of data, then descriptive statistics and graphical methods increase in importance, but they are much more complex than in the univariate case. The methods for understanding multivariate data range from the simple rearrangements of tables in which raw data are tabulated, to quite sophisticated computer-intensive methods in which exploration of the data is reminiscent of futuristic movies from space explorations.

Multivariate data from an experiment are first recorded in the form of tables, by either a researcher or a computer. In some cases, such tables may appear uninformative simply because of their format of presentation. By simple rules such tables can be rearranged in more useful formats. There are several guidelines for successful presentation of multivariate data in the form of tables. (i) Numbers should be maximally simplified by rounding as long as it does not affect the analysis. For example, the vector (2.1314757, 4.9956301, 6.1912772) could probably be simplified to (2.14, 5, 6.19); (ii) Organize the numbers to compare columns rather than rows; and (iii) The user's cognitive load should be minimized by spacing and table lay-out so that the eye does not travel long in making comparisons.

**Fisher's Iris Data.** An example of multivariate data is provided by the celebrated Fisher's iris data. Plants of the family *Iridaceae* grow on every continent except Antarctica. With a wealth of species, identification is not simple. Even iris experts sometimes disagree about how some flowers should be classified. Fisher's (Anderson, 1935; Fisher, 1936) data set contains measurements on three North American species of iris: *Iris setosa canadensis, Iris versicolor*, and *Iris virginica* (Fig. 2.8a-c). The 4-dimensional measurements on each of the species consist of sepal and petal length and width.

(a)                    (b)                    (c)

**Fig. 2.8** (a) Iris setosa, C. Hensler, The Rock Garden, (b) Iris virginica, and (c) Iris versicolor, (b) and (c) are photos by D. Kramb, SIGNA.

The data set `fisheriris` is part of the MATLAB distribution and contains two files: `meas` and `species`. The `meas` file, shown in Fig. 2.9a, is a 150 × 4 matrix and contains 150 entries, 50 for each species. Each row in the matrix `meas` contains four elements: sepal length, sepal width, petal length, and petal width. Note that the convention in MATLAB is to store variables as columns and observations as rows.

The data set `species` contains names of species for the 150 measurements. The following MATLAB commands plot the data and compare sepal lengths among the three species.

```
load fisheriris
s1 = meas(1:50, 1);    %setosa,     sepal length
s2 = meas(51:100, 1);  %versicolor, sepal length
s3 = meas(101:150, 1); %virginica,  sepal length
s = [s1 s2 s3];
figure;
imagesc(meas)
```

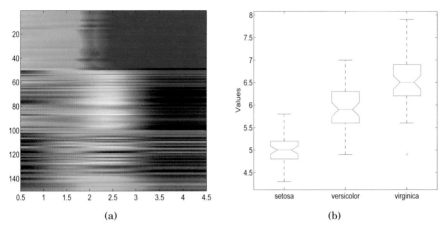

**Fig. 2.9** (a) Matrix meas in fisheriris, (b) Box plots of Sepal Length (the first column in matrix meas) versus species.

```
figure;
boxplot(s,'notch','on',...
        'labels',{'setosa','versicolor','virginica'})
```

**Correlation in Paired Samples.** We will briefly describe how to find the correlation between two aligned vectors, leaving detailed coverage of correlation theory to Chap. 15.

Sample correlation coefficient $r$ measures the strength and direction of the linear relationship between two paired samples $X = (X_1, X_2, \dots, X_n)$ and $Y = (Y_1, Y_2, \dots, Y_n)$. Note that the order of components is important and the samples cannot be independently permuted if the correlation is of interest. Thus the two samples can be thought of as a single bivariate sample $(X_i, Y_i)$, $i = 1, \dots, n$.

The correlation coefficient between samples $X = (X_1, X_2, \dots, X_n)$ and $Y = (Y_1, Y_2, \dots, Y_n)$ is

$$r = \frac{\sum_{i=1}^{n}(X_i - \overline{X})(Y_i - \overline{Y})}{\sqrt{\sum_{i=1}^{n}(X_i - \overline{X})^2 \cdot \sum_{i=1}^{n}(Y_i - \overline{Y})^2}}.$$

The summary $\mathbb{C}\mathrm{ov}(X,Y) = \frac{1}{n-1} \sum_{i=1}^{n}(X_i - \overline{X})(Y_i - \overline{Y}) = \frac{1}{n-1}\left(\sum_{i=1}^{n} X_i Y_i - n\overline{XY}\right)$ is called the sample covariance. The correlation coefficient can be expressed as a ratio:

$$r = \frac{\mathbb{C}\mathrm{ov}(X,Y)}{s_X \, s_Y},$$

where $s_X$ and $s_Y$ are sample standard deviations of samples $X$ and $Y$.

Covariances and correlations are basic exploratory summaries for paired samples and multivariate data. Typically they are assessed in data screening before building a statistical model and conducting an inference. The correlation ranges between $-1$ and 1, which are the two ideal cases of decreasing and increasing linear trends. Zero correlation does not, in general, imply independence but signifies the lack of any linear relationship between samples.

To illustrate the above principles, we find covariance and correlation between sepal and petal lengths in Fisher's iris data. These two variables correspond to the first and third columns in the data matrix. The conclusion is that these two lengths exhibit a high degree of linear dependence as evident in Fig. 2.10. The covariance of 1.2743 by itself is not a good indicator of this relationship since it is scale (magnitude) dependent. However, the correlation coefficient is not influenced by a linear transformation of the data and in this case shows a strong positive relationship between the variables.

```
load fisheriris
X = meas(:, 1);     %sepal length
Y = meas(:, 3);     %petal length
cv = cov(X, Y); cv(1,2)   %1.2743
r = corr(X, Y)             %0.8718
```

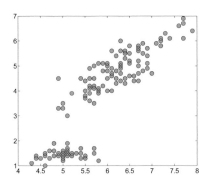

**Fig. 2.10** Correlation between petal and sepal lengths (columns 1 and 3) in iris data set. Note the strong linear dependence with a positive trend. This is reflected by a covariance of 1.2743 and a correlation coefficient of 0.8718.

In the next section we will describe an interesting multivariate data set and, using MATLAB, find some numerical and graphical summaries.

*Example 2.4.* **Body Fat Data.** We also discuss a multivariate data set analyzed in Johnson (1996) that was submitted to  http://www.amstat.

`org/publications/jse/datasets/fat.txt` and featured in Penrose et al. (1985). This data set can be found on the book's Web page as well, as  `fat.dat`.

**Fig. 2.11** Water test to determine body density. It is based on underwater weighing (Archimedes' principle) and is regarded as the gold standard for body composition assessment.

Percentage of body fat, age, weight, height, and ten body circumference measurements (e.g., abdomen) were recorded for 252 men. Percent of body fat is estimated through an underwater weighing technique (Fig. 2.11).

The data set has 252 observations and 19 variables. Brozek and Siri indices (Brozek et al., 1963; Siri, 1961) and fat-free weight are obtained by the underwater weighing while other anthropometric variables are obtained using scales and a measuring tape. These anthropometric variables are less intrusive but also less reliable in assessing the body fat index.

| –       |          | Variable description                                                             |
|---------|----------|----------------------------------------------------------------------------------|
| 3–5     | casen    | Case number                                                                      |
| 10–13   | broz     | Percent body fat using Brozek's equation: 457/density – 414.2                     |
| 18–21   | siri     | Percent body fat using Siri's equation: 495/density – 450                         |
| 24–29   | densi    | Density (gm/cm$^3$)                                                               |
| 36–37   | age      | Age (years)                                                                       |
| 40–45   | weight   | Weight (lb.)                                                                      |
| 49–53   | height   | Height (in.)                                                                      |
| 58–61   | adiposi  | Adiposity index = weight/(height$^2$) (kg/m$^2$)                                  |
| 65–69   | ffwei    | Fat-free weight = $(1 - \text{fraction of body fat}) \times$ weight, using Brozek's formula (lb.) |
| 74–77   | neck     | Neck circumference (cm)                                                           |
| 81–85   | chest    | Chest circumference (cm)                                                          |
| 89–93   | abdomen  | Abdomen circumference (cm)                                                        |
| 97–101  | hip      | Hip circumference (cm)                                                            |
| 106–109 | thigh    | Thigh circumference (cm)                                                          |
| 114–117 | knee     | Knee circumference (cm)                                                           |
| 122–125 | ankle    | Ankle circumference (cm)                                                          |
| 130–133 | biceps   | Extended biceps circumference (cm)                                               |
| 138–141 | forearm  | Forearm circumference (cm)                                                        |
| 146–149 | wrist    | Wrist circumference (cm) "distal to the styloid processes"                        |

**Remark**: There are a few false recordings. The body densities for cases 48, 76, and 96, for instance, each seem to have one digit in error as seen from the two body fat percentage values. Also note the presence of a man (case 42) over 200 lb. in weight who is less than 3 ft. tall (the height should presumably be 69.5 in., not 29.5 in.)! The percent body fat estimates are truncated to zero when negative (case 182).

```
load('\your path\fat.dat')
casen = fat(:,1);
broz = fat(:,2);
siri = fat(:,3);
densi = fat(:,4);
age = fat(:,5);
weight = fat(:,6);
height = fat(:,7);
adiposi = fat(:,8);
ffwei = fat(:,9);
neck = fat(:,10);
chest = fat(:,11);
abdomen = fat(:,12);
hip = fat(:,13);
thigh = fat(:,14);
knee = fat(:,15);
ankle = fat(:,16);
biceps = fat(:,17);
forearm = fat(:,18);
wrist = fat(:,19);
```

We will further analyze this data set in this chapter, as well as in Chap. 16, in the context of multivariate regression.

## 2.7 Multivariate Samples and Their Summaries*

Multivariate samples are organized as a data matrix, where the rows are observations and the columns are variables or components. One such data matrix of size $n \times p$ is shown in Fig. 2.12.

The measurement $x_{ij}$ denotes the $j$th component of the $i$th observation. There are $n$ row vectors $x_1', x_2', \ldots, x_n'$ and $p$ columns $x_{(1)}, x_{(2)}, \ldots, x_{(n)}$, so that

$$X = \begin{bmatrix} x_1' \\ x_2' \\ \vdots \\ x_n' \end{bmatrix} = \begin{bmatrix} x_{(1)}, x_{(2)}, \ldots, x_{(n)} \end{bmatrix}.$$

Note that $x_i = (x_{i1}, x_{i2}, \ldots, x_{ip})'$ is a $p$-vector denoting the $i$th observation, while $x_{(j)} = (x_{1j}, x_{2j}, \ldots, x_{nj})'$ is an $n$-vector denoting values of the $j$th variable/component.

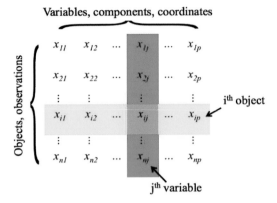

**Fig. 2.12** Data matrix $X$. In the multivariate sample the rows are observations and the columns are variables.

The mean of data matrix $X$ is a vector $\overline{x}$, which is a $p$-vector of column means

$$\overline{x} = \begin{bmatrix} \frac{1}{n}\sum_{i=1}^{n} x_{i1} \\ \frac{1}{n}\sum_{i=1}^{n} x_{i2} \\ \vdots \\ \frac{1}{n}\sum_{i=1}^{n} x_{ip} \end{bmatrix} = \begin{bmatrix} \overline{x}_1 \\ \overline{x}_2 \\ \vdots \\ \overline{x}_p \end{bmatrix}.$$

By denoting a vector of ones of size $n \times 1$ as $\mathbf{1}$, the mean can be written as $\overline{x} = \frac{1}{n} X' \cdot \mathbf{1}$, where $X'$ is the transpose of $X$.

Note that $\overline{x}$ is a column vector, while MATLAB's command mean(X) will produce a row vector. It is instructive to take a simple data matrix and inspect step by step how MATLAB calculates the multivariate summaries. For instance,

```
X = [1 2 3; 4 5 6];
[n p]=size(X)  %[2 3]: two 3-dimensional  observations
meanX = mean(X)'      %or mean(X,1), along dimension 1
     %transpose of meanX needed to be a column vector
meanX = 1/n * X' * ones(n,1)
```

> For any two variables (columns) in $X$, $x_{(i)}$ and $x_{(j)}$, one can find the sample covariance:
>
> $$s_{ij} = \frac{1}{n-1}\left(\sum_{k=1}^{n} x_{ki}x_{kj} - n\overline{x}_i\overline{x}_j\right).$$
>
> All $s_{ij}$s form a $p \times p$ matrix, called a *sample covariance matrix* and denoted by $S$.

A simple representation for $S$ uses matrix notation:

$$S = \frac{1}{n-1}\left(X'X - \frac{1}{n}X'JX\right).$$

Here $J = 11'$ is a standard notation for a matrix consisting of ones. If one defines a *centering matrix* $H$ as $H = I - \frac{1}{n}J$, then $S = \frac{1}{n-1}X'HX$. Here $I$ is the identity matrix.

```
X = [1 2 3; 4 5 6];
[n p]=size(X);
J = ones(n,1)*ones(1,n);
H = eye(n) - 1/n * J;
S = 1/(n-1) * X' * H * X
S = cov(X) %built-in command
```

An alternative definition of the covariance matrix, $S^* = \frac{1}{n}X'HX$, is coded in MATLAB as `cov(X,1)`. Note also that the diagonal of $S$ contains sample variances of variables since $s_{ii} = \frac{1}{n-1}\left(\sum_{k=1}^{n} x_{ki}^2 - n\bar{x}_i^2\right) = s_i^2$.

Matrix $S$ describes scattering in data matrix $X$. Sometimes it is convenient to have scalars as measures of scatter, and for that purpose two summaries of $S$ are typically used: (i) the determinant of $S$, $|S|$, as a generalized variance and (ii) the trace of $S$, $\mathrm{tr}S$, as the total variation.

The sample correlation coefficient between the $i$th and $j$th variables is

$$r_{ij} = \frac{s_{ij}}{s_i\, s_j},$$

where $s_i = \sqrt{s_i^2} = \sqrt{s_{ii}}$ is the sample standard deviation. Matrix $R$ with elements $r_{ij}$ is called a sample correlation matrix. If $R = I$, the variables are uncorrelated. If $D = diag(s_i)$ is a diagonal matrix with $(s_1, s_2, \ldots, s_p)$ on its diagonal, then

$$S = DRD, \quad R = D^{-1}RD^{-1}.$$

Next we show how to standardize multivariate data. Data matrix $Y$ is a standardized version of $X$ if its rows $y_i'$ are standardized rows of $X$,

$$Y = \begin{bmatrix} y_1' \\ y_2' \\ \vdots \\ y_n' \end{bmatrix}, \quad \text{where } y_i = D^{-1}(x_i - \bar{x}), \ i = 1, \ldots, n.$$

$Y$ has a covariance matrix equal to the correlation matrix. This is a multivariate version of the z-score For the two-column vectors from $Y$, $y_{(i)}$ and $y_{(j)}$, the correlation $r_{ij}$ can be interpreted geometrically as the cosine of angle $\varphi_{ij}$ between the vectors. This shows that correlation is a measure of similarity because close vectors (with a small angle between them) will be strongly positively correlated, while the vectors orthogonal in the geometric sense will be uncorrelated. This is why uncorrelated vectors are sometimes called orthogonal.

Another useful transformation of multivariate data is the Mahalanobis transformation. When data are transformed by the Mahalanobis transformation, the variables become decorrelated. For this reason, such transformed data are sometimes called "sphericized."

$$
\mathbf{Z} = \begin{bmatrix} \mathbf{z_1}' \\ \mathbf{z_2}' \\ \vdots \\ \mathbf{z_n}' \end{bmatrix}, \quad \text{where } \mathbf{z}_i = \mathbf{S}^{-1/2}(\mathbf{x}_i - \overline{\mathbf{x}}),\ i = 1,\ldots,n.
$$

The Mahalanobis transform decorrelates the components, so $\mathrm{Cov}(\mathbf{Z})$ is an identity matrix. The Mahalanobis transformation is useful in defining the distances between multivariate observations. For further discussion on the multivariate aspects of statistics we direct the student to the excellent book by Morrison (1976).

*Example 2.5.* ⚛ The Fisher iris data set was a data matrix of size $150 \times 4$, while the size of the body fat data was $252 \times 19$. To illustrate some of the multivariate summaries just discussed we construct a new, 5 dimensional data matrix from the body fat data set. The selected columns are broz, densi, weight, adiposi, and biceps. All 252 rows are retained.

```
X = [broz densi weight adiposi biceps];
varNames = {'broz'; 'densi'; 'weight'; 'adiposi'; 'biceps'};

varNames =
    'broz'    'densi'    'weight'    'adiposi'    'biceps'

Xbar = mean(X)

Xbar = 18.9385 1.0556 178.9244 25.4369 32.2734

S = cov(X)

S =
    60.0758    -0.1458    139.6715    20.5847    11.5455
    -0.1458     0.0004     -0.3323    -0.0496    -0.0280
   139.6715    -0.3323    863.7227    95.1374    71.0711
    20.5847    -0.0496     95.1374    13.3087     8.2266
    11.5455    -0.0280     71.0711     8.2266     9.1281

R = corr(X)
```

```
R =
    1.0000    -0.9881     0.6132     0.7280     0.4930
   -0.9881     1.0000    -0.5941    -0.7147    -0.4871
    0.6132    -0.5941     1.0000     0.8874     0.8004
    0.7280    -0.7147     0.8874     1.0000     0.7464
    0.4930    -0.4871     0.8004     0.7464     1.0000

% By ''hand''
[n p]=size(X);
H = eye(n) - 1/n * ones(n,1)*ones(1,n);
S = 1/(n-1) * X' * H * X;
stds = sqrt(diag(S));
D = diag(stds);
R = inv(D) * S * inv(D);
%S and R here coincide with S and R
%calculated by built-in functions cov and cor.

Xc= X - repmat(mean(X),n,1);  %center X
%subtract component means
%from variables in each observation.

%standardization
Y =  Xc * inv(D);  %for Y, S=R

%Mahalanobis transformation
M = sqrtm(inv(S))  %sqrtm is a square root of matrix

%M =
%     0.1739     0.8423    -0.0151    -0.0788     0.0046
%     0.8423   345.2191    -0.0114     0.0329     0.0527
%    -0.0151    -0.0114     0.0452    -0.0557    -0.0385
%    -0.0788     0.0329    -0.0557     0.6881    -0.0480
%     0.0046     0.0527    -0.0385    -0.0480     0.5550

Z = Xc * M;  %Z has uncorrelated components
cov(Z)       %should be identity matrix
```

Figure 2.13 shows data plots for a subset of five variables and the two transformations, standardizing and Mahalanobis. Panel (a) shows components broz, densi, weight, adiposi, and biceps over all 252 measurements. Note that the scales are different and that weight has much larger magnitudes than the other variables.

Panel (b) shows the standardized data. All column vectors are centered and divided by their respective standard deviations. Note that the data plot here shows the correlation across the variables. The variable density is negatively correlated with the other variables.

Panel (c) shows the decorrelated data. Decorrelation is done by centering and multiplying by the Mahalanobis matrix, which is the matrix square root of the inverse of the covariance matrix. The correlations visible in panel (b) disappeared.

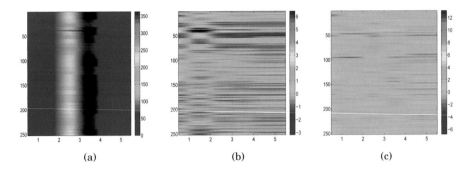

(a)                          (b)                          (c)

**Fig. 2.13** Data plots for (a) 252 five-dimensional observations from Body Fat data where the variables are broz, densi, weight, adiposi, and biceps. (b) $Y$ is standardized $X$, and (c) $Z$ is a decorrelated $X$.

## 2.8 Visualizing Multivariate Data

The need for graphical representation is much greater for multivariate data than for univariate data, especially if the number of dimensions exceeds three.

For a data given in matrix form (observations in rows, components in columns), we have already seen a quite an illuminating graphical representation, which we called a data matrix.

One can extend the histogram to bivariate data in a straightforward manner. An example of a 2-D histogram obtained by m-file hist2d is given in Fig. 2.14a. The histogram (in the form of an image) shows the sepal and petal lengths from the fisheriris data set. A scatterplot of the 2-D measurements is superimposed.

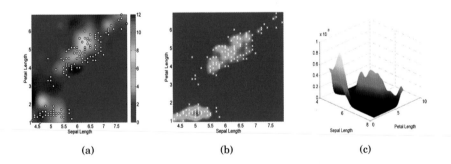

(a)                          (b)                          (c)

**Fig. 2.14** (a) Two-dimensional histogram of Fisher's iris sepal ($X$) and petal ($Y$) lengths. The plot is obtained by hist2d.m; (b) Scattercloud plot – smoothed histogram with superimposed scatterplot, obtained by scattercloud.m; (c) Kernel-smoothed and normalized histogram obtained by smoothhist2d.m.

Figures 2.14b-c show the smoothed histograms. The histogram in panel (c) is normalized so that the area below the surface is 1. The smoothed histograms are plotted by ◀ scattercloud.m and ◀ smoothhist2d.m (S. Simon and E. Ronchi, MATLAB Central).

If the dimension of the data is three or more, one can gain additional insight by plotting pairwise scatterplots. This is achieved by the MATLAB command gplotmatrix(X,Y,group), which creates a matrix arrangement of scatterplots. Each subplot in the graphical output contains a scatterplot of one column from data set $X$ against a column from data set $Y$.

In the case of a single data set (as in body fat and Fisher iris examples), $Y$ is omitted or set at Y=[ ], and the scatterplots contrast the columns of $X$. The plots can be grouped by the grouping variable group. This variable can be a categorical variable, vector, string array, or cell array of strings.

The variable group must have the same number of rows as $X$. Points with the same value of group appear on the scatterplot with the same marker and color. Other arguments in gplotmatrix(x,y,group,clr,sym,siz) specify the color, marker type, and size for each group. An example of the gplotmatrix command is given in the code below. The output is shown in Fig. 2.15a.

```
X = [broz densi weight adiposi biceps];
varNames = {'broz'; 'densi'; 'weight'; 'adiposi'; 'biceps'};
agegr = age > 55;
gplotmatrix(X,[],agegr,['b','r'],['x','o'],[],'false');
text([.08 .24 .43 .66 .83],  repmat(-.1,1,5), varNames, ...
    'FontSize',8);
text(repmat(-.12,1,5), [.86 .62 .41 .25 .02], varNames, ...
    'FontSize',8, 'Rotation',90);
```

**Parallel Coordinates Plots.** In a *parallel coordinates plot*, the components of the data are plotted on uniformly spaced vertical lines called component axes. A $p$-dimensional data vector is represented as a broken line connecting a set of points, one on each component axis. Data represented as lines create readily perceived structures. A command for parallel coordinates plot parallelcoords is given below with the output shown in Fig. 2.15b.

```
parallelcoords(X, 'group', age>55, ...
                'standardize','on', 'labels',varNames)
set(gcf,'color','white');
```

Figure 2.16a shows parallel cords for the groups age > 55 and age <= 55 with 0.25 and 0.75 quantiles.

```
parallelcoords(X, 'group', age>55, ...
    'standardize','on', 'labels',varNames,'quantile',0.25)
set(gcf,'color','white');
```

**Andrews' Plots.** An *Andrews plot* (Andrews, 1972) is a graphical representation that utilizes Fourier series to visualize multivariate data. With an

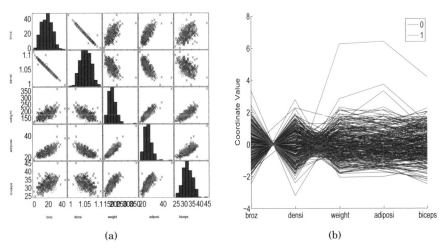

|                                                      |                                                       |
| :--------------------------------------------------: | :---------------------------------------------------: |
|                         (a)                          |                          (b)                          |

**Fig. 2.15** (a) gplotmatrix for broz, densi, weight, adiposi, and biceps; (b) parallelcoords plot for $X$, by age>55.

observation $(X_1, \ldots, X_p)$ one associates the function

$$F(t) = X_1/\sqrt{2} + X_2 \sin(2\pi t) + X_3 \cos(2\pi t) + X_4 \sin(2 \cdot 2\pi t) + X_5 \cos(2 \cdot 2\pi t) + \ldots,$$

where $t$ ranges from $-1$ to $1$. One Andrews' curve is generated for each multi-variate datum – a row of the data set. Andrews' curves preserve the distances between observations. Observations close in the Euclidian distance sense are represented by close Andrews' curves. Hence, it is easy to determine which observations (i.e., rows when multivariate data are represented as a matrix) are most alike by using these curves. Due to the definition, this representation is not robust with respect to the permutation of coordinates. The first few variables tend to dominate, so it is a good idea when using Andrews' plots to put the most important variables first. Some analysts recommend running a principal components analysis first and then generating Andrews' curves for principal components. The principal components of multivariate data are linear combinations of components that account for most of the variability in the data. Principal components will not be discussed in this text as they are beyond the scope of this course.

An example of Andrews' plots is given in the code below with the output in Fig. 2.16b.

```
andrewsplot(X, 'group', age>55, 'standardize','on')
set(gcf,'color','white');
```

**Star Plots.** The star plot is one of the earliest multivariate visualization objects. Its rudiments can be found in the literature from the early nineteenth

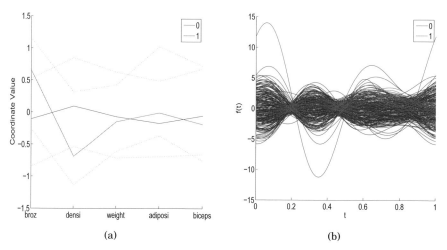

(a)                             (b)

**Fig. 2.16** (a) $X$ by age>55 with quantiles; (b) andrewsplot for $X$ by age>55.

century. Similar plots (rose diagrams) are used in Florence Nightingale's Notes on Matters Affecting the Health, Efficiency and Hospital Administration of the British Army in 1858 (Nightingale, 1858).

The star glyph consists of a number of spokes (rays) emanating from the center of the star plot and connected at the ends. The number of spikes in the star plot is equal to the number of variables (components) in the corresponding multivariate datum. The length of each spoke is proportional to the magnitude of the component it represents. The angle between two neighboring spokes is $2\pi/p$, where $p$ is the number of components. The star glyph connects the ends of the spokes.

An example of the use of star plots is given in the code below with the output in Fig. 2.17a.

```
ind = find(age>67);
strind = num2str(ind);
h = glyphplot(X(ind,:), 'glyph','star', 'varLabels',...
      varNames,'obslabels', strind);
set(h(:,3),'FontSize',8); set(gcf,'color','white');
```

**Chernoff Faces.** People grow up continuously studying faces. Minute and barely measurable differences are easily detected and linked to a vast catalog stored in memory. The human mind subconsciously operates as a super computer, filtering out insignificant phenomena and focusing on the potentially important. Such mundane characters as :), :(, :0, and >:p are readily linked in our minds to joy, dissatisfaction, shock, or affection.

Face representation is an interesting approach to taking a first look at multivariate data and is effective in revealing complex relations that are not visible in simple displays that use the magnitudes of components. It can be used to aid in cluster analysis and discrimination analysis and to detect substantial changes in time series.

Each variable in a multivariate datum is connected to a feature of a face. The variable-feature links in MATLAB are as follows: variable 1 – size of face; variable 2 – forehead/jaw relative arc length; variable 3 – shape of forehead; variable 4 – shape of jaw; variable 5 – width between eyes; variable 6 – vertical position of eyes; variables 7–13 – features connected with location, separation, angle, shape, and width of eyes and eyebrows; and so on. An example of the use of Chernoff faces is given in the code below with the output in Fig. 2.17b.

```
ind = find(height > 74.5);
strind = num2str(ind);
h = glyphplot(X(ind,:), 'glyph','face', 'varLabels',...
varNames,'obslabels', strind);
set(h(:,3),'FontSize',10);   set(gcf,'color','white');
```

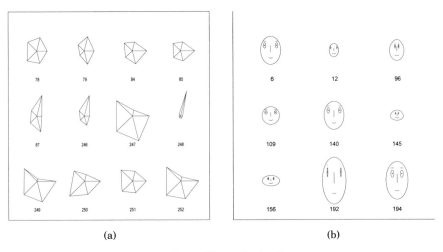

(a)                                                              (b)

**Fig. 2.17** (a) Star plots for $X$; (b) Chernoff faces plot for $X$.

## 2.9 Observations as Time Series

Observations that have a time index, that is, if they are taken at equally spaced instances in time, are called *time series*. EKG and EEG signals, high-frequency bioresponses, sound signals, economic indices, and astronomic and geophysical measurements are all examples of time series. The following example illustrates a time series.

*Example 2.6.* **Blowflies Time Series.** The data set 📇 blowflies.dat consists of the total number of blowflies (*Lucilia cuprina*) in a population under controlled laboratory conditions. The data represent counts for every other day. The developmental delay (from egg to adult) is between 14 and 15 days for insects under the conditions employed. Nicholson (1954) made 361 bi-daily recordings over a 2-year period (722 days), see Fig. 2.18a.
✒

In addition to analyzing basic location, spread, and graphical summaries, we are also interested in evaluating the degree of autocorrelation in time series. Autocorrelation measures the level of correlation of the time series with a time-shifted version of itself. For example, autocorrelation at lag 2 would be a correlation between $X_1, X_2, X_3, \ldots, X_{n-3}, X_{n-2}$ and $X_3, X_4, \ldots, X_{n-1}, X_n$. When the shift (lag) is 0, the autocorrelation is just a correlation. The concept of autocorrelation is introduced next, and then the autocorrelation is calculated for the blowflies data.

Let $X_1, X_2, \ldots, X_n$ be a sample where the order of observations is important. The indices $1, 2, \ldots, n$ may correspond to measurements taken at time points $t, t + \Delta t, t + 2\Delta t, \ldots, t + (n-1)\Delta t$, for some start time $t$ and time increments $\Delta t$. The autocovariance at lag $0 \le k \le n-1$ is defined as

$$\hat{\gamma}(k) = \frac{1}{n} \sum_{i=1}^{n-k} (X_{i+k} - \overline{X})(X_i - \overline{X}).$$

Note that the sum is normalized by a factor $\frac{1}{n}$ and not by $\frac{1}{n-k}$, as one may expect.

The autocorrelation is defined as normalized autocovariance,

$$\hat{\rho}(k) = \frac{\hat{\gamma}(k)}{\hat{\gamma}(0)}.$$

Autocorrelation is a measure of self-affinity of the time series with its own shifts and is an important summary statistic. MATLAB has the built-in functions autocov and autocorr. The following two functions are simplified versions illustrating how the autocovariances and autocorrelations are calculated.

```
function acv = acov(ts, maxlag)
%acov.m: computes the sample autocovariance function
%           ts    = 1-D time series
%           maxlag = maximum lag ( < length(ts))
%usage: z = autocov (a,maxlag);
n = length(ts);
ts = ts(:) - mean(ts); %note overall mean
suma = zeros(n,maxlag+1);
suma(:,1) = ts.^2;
for h = 2:maxlag+1
    suma(1:(n-h+1), h) = ts(h:n);
```

```
    suma(:,h) = suma(:,h) .* ts;
end
acv = sum(suma)/n; %note the division by n
                   %and not by expected (n-h)


function [acrr] = acorr(ts , maxlag)
  acr =  acov(ts, maxlag);
  acrr = acr ./ acr(1);
```

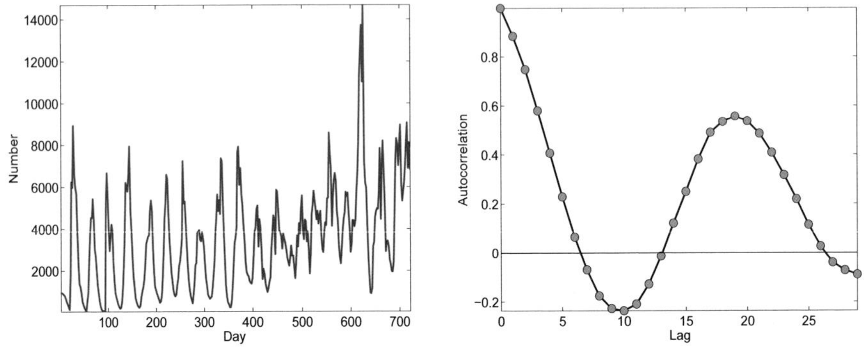

(a)                                                      (b)

**Fig. 2.18** (a) Bi-daily measures of size of the blowfly population over a 722-day period, (b) The autocorrelation function of the time series. Note the peak at lag 19 corresponding to the periodicity of 38 days.

Figure 2.18a shows the time series illustrating the size of the population of blowflies over 722 days. Note the periodicity in the time series. In the autocorrelation plot (Fig. 2.18b) the peak at lag 19 corresponding to a time shift of 38 days. This indicates a periodicity with an approximate length of 38 days in the dynamic of this population. A more precise assessment of the periodicity and related inference can be done in the frequency domain of a time series, but this theory is beyond the scope of this course. Good follow-up references are Brillinger (2001), Brockwell and Davis (2009), and Shumway and Stoffer (2005). Also see Exercise 2.12.

## 2.10 About Data Types

The cell data elaborated in this chapter are *numerical*. When measurements are involved, the observations are typically *numerical*. Other types of data encountered in statistical analysis are categorical. Stevens (1946), who was influenced by his background in psychology, classified data as nominal, ordinal, interval, and ratio. This typology is loosely accepted in other scientific circles. However, there are vibrant and ongoing discussions and disagreements, e.g., Veleman and Wilkinson (1993). *Nominal data*, such as race, gender, political affiliation, names, etc., cannot be ordered. For example, the counties in northern Georgia, Cherokee, Clayton, Cobb, DeCalb, Douglas, Fulton, and Gwinnett, cannot be ordered except that there is a nonessential alphabetical order of their names. Of course, numerical attributes of these counties, such as size, area, revenue, etc., can be ordered.

*Ordinal data* could be ordered and sometimes assigned numbers, although the numbers would not convey their relative standing. For example, data on the Likert scale have five levels of agreement: (1) Strongly Disagree, (2) Disagree, (3) Neutral, (4) Agree, and (5) Strongly Agree; the numbers 1 to 5 are assigned to the degree of agreement and have no quantitative meaning. The difference between Agree and Neutral is not equal to the difference between Disagree and Strongly Disagree. Other examples are the attributes "Low" and "High" or student grades A, B, C, D, and F. It is an error to treat ordinal data as numerical. Unfortunately this is a common mistake (e.g., GPA). Sometimes T-shirt-size attributes, such as "small," "medium," "large," and "x-large," may falsely enter the model as if they were measurements 1, 2, 3, and 4.

Nominal and ordinal data are examples of *categorical* data since the values fall into categories.

*Interval data* refers to numerical data for which the differences can be well interpreted. However, for this type of data, the origin is not defined in a natural way so the ratios would not make sense. Temperature is a good example. We cannot say that a day in July with a temperature of 100°F is twice as hot as a day in November with a temperature of 50°F. Test scores are another example of interval data as a student who scores 100 on a midterm may not be twice as good as a student who scores 50.

*Ratio data* are at the highest level; these are usually standard numerical values for which ratios make sense and the origin is absolute. Length, weight, and age are all examples of ratio data.

Interval and ratio data are examples of *numerical* data.

MATLAB provides a way to keep such heterogeneous data in a single structure array with a syntax resembling C language.

Structures are arrays comprised of structure elements and are accessed by named fields. The fields (data containers) can contain any type of data. Storage in the structure is allocated dynamically. The general syntax for a structure format in MATLAB is structurename(recordnumber).fieldname=data

For example,

```
patient.name = 'John Doe';
patient.agegroup = 3;
patient.billing = 127.00;
patient.test = [79 75 73; 180 178 177.5; 220 210 205];
patient
%To expand the structure array, add subscripts.
patient(2).name = 'Ann Lane';
```

```
patient(2).agegroup =  2;
patient(2).billing = 208.50;
patient(2).test = [68 70 68; 118 118 119; 172 170 169];
patient
```

## 2.11 Exercises

2.1. **Auditory Cortex Spikes.** This data set comes from experiments in the
lab of Dr. Robert Liu of Emory University[4] and concerns single-unit elec-
trophysiology in the auditory cortex of nonanesthetized female mice. The
motivating question is the exploration of auditory neural differences between female parents vs. female virgins and their relationship to cortical
response.

Researchers in Liu's lab developed a restrained awake setup to collect single neuron activity from both female parent and female naïve mice. Multiple trials are performed on the neurons from one mother and one naïve
animal.

The recordings are made from a region in the auditory cortex of the mouse
with a single tungsten electrode. A sound stimulus is presented at a time
of 200 ms during each sweep (time shown is 0–611 and 200 is the point at
which a stimulus is presented). Each sweep is 611 ms long and the duration of the stimulus tone is 10 to 70 ms. The firing times for mother and
naïve mice are provided in the data set 🖳 spikes.dat, in columns 2 and 3.
Column 1 is the numbering from 1 to 611.

(a) Using MATLAB's diff command, find the inter-firing times. Plot a histogram for both sets of interfiring times. Use biplot.m to plot the histograms
back to back.

(b) For inter-firing times in the mother's response find descriptive statistics
similar to those in the cell area example.

2.2. **On Average.** It is an anecdotal truth that an average Australian has less
than two legs! Indeed, there are some Australians that have lost their leg(s);
thus the number of legs is less than twice the number of people. In this
exercise, we compare several sample averages.

A small company reports the following salaries: 4 employees at 20K, 3 employees at 30K, the vice-president at 200K, and the president at 400K. Calculate the arithmetic mean, geometric mean, median, harmonic mean, and
mode. If the company is now hiring, would an advertising strategy in which
the mean salary is quoted be fair? If not, suggest an alternative.

2.3. **Contraharmonic Mean and $f$-Mean.** The contraharmonic mean for
$X_1, X_2, \ldots, X_n$ is defined as

---
[4] http://www.biology.emory.edu/research/Liu/index.html

$$C(X_1,\ldots,X_n) = \frac{\sum_{i=1}^{n} X_i^2}{\sum_{i=1}^{n} X_i}.$$

(a) Show that $C(X_1,X_2)$ is twice the sample mean minus the harmonic mean of $X_1,X_2$.

(b) Show that $C(x,x,x,\ldots,x) = x$.

The generalized $f$-mean of $X_1,\ldots,X_n$ is defined as

$$X_f = f^{-1}\left(\frac{1}{n}\sum_{i=1}^{n} f(X_i)\right),$$

where $f$ is suitably chosen such that $f(X_i)$ and $f^{-1}$ are well defined.

(c) Show that $f(x) = x, \frac{1}{x}, x^k, \log x$ gives the mean, harmonic mean, power $k$ mean, and geometric mean.

2.4. **Mushrooms.** The unhappy outcome of uninformed mushroom picking is poisoning. In many cases, such poisoning is due to ignorance or a superficial approach to identification. The most dangerous fungi are Death Cap (*Amanita phalloides*) and two species akin to it, *A. verna* and Destroying Angel (*A. virosa*). These three toadstools cause the majority of fatal poisoning.

One of the keys to mushroom identification is the spore deposit. Spores of *Amanita phalloides* are colorless, nearly spherical, and smooth. Measurements in microns of 28 spores are given below:

| | | | | | | |
|---|---|---|---|---|---|---|
| 9.2 | 8.8 | 9.1 | 10.1 | 8.5 | 8.4 | 9.3 |
| 8.7 | 9.7 | 9.9 | 8.4 | 8.6 | 8.0 | 9.5 |
| 8.8 | 8.1 | 8.3 | 9.0 | 8.2 | 8.6 | 9.0 |
| 8.7 | 9.1 | 9.2 | 7.9 | 8.6 | 9.0 | 9.1 |

(a) Find the *five-number summary* $(Min,Q_1,Me,Q_3,Max)$ for the spore measurement data.

(b) Find the mean and the mode.

(c) Find and plot the histogram of $z$-scores, $z_i = (X_i - \overline{X})/s$.

2.5. **Manipulations with Sums.** Prove the following algebraic identities involving sums, useful in demonstrating properties of some sample summaries.

| | |
|---|---|
| (a) $\sum_{i=1}^{n}(x_i - \overline{x}) = 0$ | (b) If $y_1 = x_1 + a, y_2 = x_2 + a, \ldots, y_n = x_n + a$, then $\sum_{i=1}^{n}(y_i - \overline{y})^2 = \sum_{i=1}^{n}(x_i - \overline{x})^2$ |
| (c) If $y_1 = c \cdot x_1, y_2 = c \cdot x_2, \ldots, y_n = c \cdot x_n$, then $\sum_{i=1}^{n}(y_i - \overline{y})^2 = c^2 \sum_{i=1}^{n}(x_i - \overline{x})^2$ | (d) If $y_1 = c \cdot x_1 + a, y_2 = c \cdot x_2 + a, \ldots, y_n = c \cdot x_n + a$, then $\sum_{i=1}^{n}(y_i - \overline{y})^2 = c^2 \sum_{i=1}^{n}(x_i - \overline{x})^2$. |
| (e) $\sum_{i=1}^{n}(x_i - \overline{x})^2 = \sum_{i=1}^{n} x_i^2 - n()^2$ | (f) $\sum_{i=1}^{n}(x_i - \overline{x})(y_i - \overline{y}) = \sum_{i=1}^{n} x_i y_i - n(\overline{x})(\overline{y})$ |
| (g) $\sum_{i=1}^{n}(x_i - a)^2 = \sum_{i=1}^{n}(x_i - \overline{x})^2 + n(\overline{x} - a)^2$ | (h) For any constant $a$, $\sum_{i=1}^{n}(x_i - \overline{x})^2 \le \sum_{i=1}^{n}(x_i - a)^2$ |

2.6. **Emergency Calculation.** Graduate student Rosa Juliusdottir reported the results of an experiment to her advisor who wanted to include them in his grant proposal. Before leaving to Reykjavik for a short vacation, she left the following data in her advisor's mailbox: sample size $n = 12$, sample mean $\overline{X} = 15$, and sample variance $s^2 = 34$.

The advisor noted with horror that the last measurement $X_{12}$ was wrongly recorded. It should have been 16 instead of 4. It would be easy to fix $\overline{X}$ and $s^2$, but the advisor did not have the previous 11 measurements nor the statistics training necessary to make the correction. Rosa was in Iceland, and the grant proposal was due the next day. The advisor was desperate, but luckily you came along.

2.7. **Sample Mean and Standard Deviation After a Change.** It is known that $\overline{y} = 11.6$, $s_y = 4.4045$, and $n = 15$. The observation $y_{12} = 7$ is removed and observation $y_{13}$ was misreported; it was not 10, but 20. Find $\overline{y}_{new}$ and $s_{y(new)}$ after the changes.

2.8. **Surveys on Different Scales.** We are interested in determining whether UK voters (whose parties have somewhat more distinct policy positions than those in the USA) have a wider variation in their evaluations of the parties than voters in the USA. The problem is that the British election survey takes evaluations scored 0–10, while the US National Election Survey gets evaluations scored 0–100. Here are two surveys.

$$
\begin{vmatrix}
\text{UK} & 6 & 7 & 5 & 10 & 3 & 9 & 9 & 6 & 8 & 2 & 7 & 5 \\
\text{US} & 67 & 65 & 95 & 86 & 44 & 100 & 85 & 92 & 91 & 65 & &
\end{vmatrix}
$$

Using CV compare the amount of variation without worrying about the different scales.

2.9. **Merging Two Samples.** Suppose $\overline{X}$ and $s_X^2$ are the mean and variance of the sample $X_1, \ldots, X_m$ and $\overline{Y}$ and $s_Y^2$ of the sample $Y_1, \ldots, Y_n$. If the two samples are merged into a single sample, show that its mean and variance are

$$\frac{m\overline{X} + n\overline{Y}}{m+n} \quad \text{and} \quad \frac{1}{m+n-1}\left[(m-1)s_X^2 + (n-1)s_Y^2 + \frac{mn}{m+n}(\overline{X}-\overline{Y})^2\right].$$

2.10. **Fitting the Histogram.** The following is a demonstration of MATLAB's built-in function histfit on a simulated data set.

```
dat = normrnd(4, 1,[1 500]) + normrnd(2, 3,[1 500]);
figure; histfit(dat(:));
```

The function histfit plots the histogram of data and overlays it with the best fitting Gaussian curve. As an exercise, take Brozek index broz from the data set ![] fat.dat (second column) and apply the histfit command. Comment on how the Gaussian curve fits the histogram.

2.11. **QT Syndrome.** The QT interval is a time interval between the start of the Q wave and the end of the T wave in a heart's electrical cycle (Fig. 2.19). It measures the time required for depolarization and repolarization to occur. In long QT syndrome, the duration of repolarization is longer than normal, which results in an extended QT interval. An interval above 440 ms is considered prolonged. Although the mechanical function of the heart could be normal, the electrical defects predispose affected subjects to arrhythmia, which may lead to sudden loss of consciousness (syncope) and, in some cases, to a sudden cardiac death.

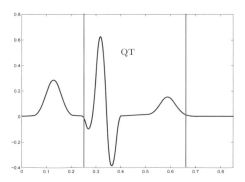

**Fig. 2.19** Schematic plot of ECG, with QT time between the red bars.

The data set ![] QT.dat|mat was compiled by Christov et al. (2006) and is described in http://www.biomedical-engineering-online.com/content/5/1/31. It provides 548 QT times taken from 293 subjects. The subjects include healthy controls (about 20%) and patients with various diagnoses, such as myocardial infarction, cardiomyopathy/heart failure, bundle branch block, dysrhythmia, myocardial hypertrophy, etc. The Q-onsets and T-wave ends are evaluated by five independent experts, and medians of their estimates are used in calculations of the QT for a subject.
Plot the histogram of this data set and argue that the data are reasonably "bell-shaped." Find the location and spread measures of the sample. What proportion of this sample has prolonged QT?

2.12. **Blowfly Count Time Series.** For the data in Example 2.6 it was postulated that a major transition in the dynamics of blowfly population size appeared to have occurred around day 400. This was attributed to biological evolution, and the whole series cannot be considered as representative of the same system. Divide the time series into two data segments with indices 1–200 and 201–361. Calculate and compare the autocorrelation functions for the two segments.

2.13. **Simpson's Diversity Index.** An alternative diversity measure to Shannon's in (2.1) is the Simpson diversity index defined as

$$D = \frac{n^2}{\sum_{i=1}^{k} f_i^2}.$$

It achieves its maximum $k$ when all frequencies are equal; thus Simpson's homogeneity (equitability) index is defined as $E_D = D/k$.

Repeat the calculations from Example 2.3 with Simpson's diversity and homogeneity indices in place of Shannon's. Is the Brazilian sample still the most homogeneous, as it was according to Shannon's $E_H$ index?

2.14. **Speed of Light.** Light travels very fast. It takes about 8 min to reach Earth from the Sun and over 4 years to reach Earth from the closest star outside the solar system. Radio and radar waves also travel at the speed of light, and an accurate value of that speed is important to communicate with astronauts and orbiting satellites. Because of the nature of light, it is very hard to measure its speed. The first reasonably accurate measurements of the speed of light were made by A. Michelson and S. Newcomb. The table below contains 66 transformed measurements made by Newcomb between July and September 1882. Entry 28, for instance, corresponds to the actual measurement of 0.000024828 s. This was the amount of time needed for light to travel approx. 4.65 miles.

| 28 | 22 | 36 | 26 | 28 | 28 | 26 | 24 | 32 | 30 | 27 |
|----|----|----|----|----|----|----|----|----|----|----|
| 24 | 33 | 21 | 36 | 32 | 31 | 25 | 24 | 25 | 28 | 36 |
| 27 | 32 | 34 | 30 | 25 | 26 | 26 | 25 | −44 | 23 | 21 |
| 30 | 33 | 29 | 27 | 29 | 28 | 22 | 26 | 27 | 16 | 31 |
| 29 | 36 | 32 | 28 | 40 | 19 | 37 | 23 | 32 | 29 | −2 |
| 24 | 25 | 27 | 24 | 16 | 29 | 20 | 28 | 27 | 39 | 23 |

You can download 📥 light.data|mat and read it in MATLAB.

If we agree that outlier measurements are outside the interval $[Q_1 - 2.5 \ IQR, Q_3 + 2.5 \ IQR]$, what observations qualify as outliers? Make the data "clean" by excluding outlier(s). For the cleaned data find the mean, 20% trimmed mean, real MAD, std, and variance.

Plot the histogram and kernel density estimator for an appropriately selected bandwidth.

2.15. **Limestone Formations in Jamaica.** This data set contains 18 observations of nummulited specimens from the Eocene yellow limestone formation in northwestern Jamaica (📥 limestone.dat). The use of faces to represent points in $k$-dimensional space graphically was originally illustrated on this data set (Chernoff, 1973). Represent this data set graphically using Chernoff faces.

| ID | $Z_1$ | $Z_2$ | $Z_3$ | $Z_4$ | $Z_5$ | $Z_6$ | ID | $Z_1$ | $Z_2$ | $Z_3$ | $Z_4$ | $Z_5$ | $Z_6$ |
|---|---|---|---|---|---|---|---|---|---|---|---|---|---|
| 1 | 160 | 51 | 10 | 28 | 70 | 450 | 45 | 195 | 32 | 9 | 19 | 110 | 1010 |
| 2 | 155 | 52 | 8 | 27 | 85 | 400 | 46 | 220 | 33 | 10 | 24 | 95 | 1205 |
| 3 | 141 | 49 | 11 | 25 | 72 | 380 | 81 | 55 | 50 | 10 | 27 | 128 | 205 |
| 4 | 130 | 50 | 10 | 26 | 75 | 560 | 82 | 70 | 53 | 7 | 28 | 118 | 204 |
| 6 | 135 | 50 | 12 | 27 | 88 | 570 | 83 | 85 | 49 | 11 | 19 | 117 | 206 |
| 41 | 85 | 55 | 13 | 33 | 81 | 355 | 84 | 115 | 50 | 10 | 21 | 112 | 198 |
| 42 | 200 | 34 | 10 | 24 | 98 | 1210 | 85 | 110 | 57 | 9 | 26 | 125 | 230 |
| 43 | 260 | 31 | 8 | 21 | 110 | 1220 | 86 | 95 | 48 | 8 | 27 | 114 | 228 |
| 44 | 195 | 30 | 9 | 20 | 105 | 1130 | 87 | 95 | 49 | 8 | 29 | 118 | 240 |

2.16. **Duchenne Muscular Dystrophy.** *Duchenne muscular dystrophy* (DMD), or Meryon's disease, is a genetically transmitted disease, passed from a mother to her children (Fig. 2.20). Affected female offspring usually suffer no apparent symptoms and may unknowingly carry the disease. Male offspring with the disease die at a young age. Not all cases of the disease come from an affected mother. A fraction, perhaps one third, of the cases arise spontaneously, to be genetically transmitted by an affected female. This is the most widely held view at present. The incidence of DMD is about 1 in 10,000 male births. The population risk (prevalence) that a woman is a DMD carrier is about 3 in 10,000.

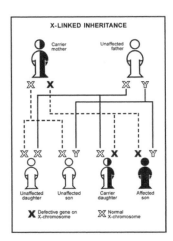

**Fig. 2.20** Each son of a carrier has a 50% chance of having DMD and each daughter has a 50% chance of being a carrier.

From the text page download data set ▥ dmd.dat|mat|xls. This data set is modified data from Percy et al. (1981) (entries containing missing values excluded). It consists of 194 observations corresponding to blood samples collected in a project to develop a screening program for female relatives of boys with DMD. The program was implemented in Canada and its goal was to inform a woman of her chances of being a carrier based on serum markers as well as her family pedigree. Another question of interest was whether age should be taken into account. Enzyme levels were measured in known carriers (67 samples) and in a group of noncarriers (127 samples). The first two serum markers, creatine kinase and hemopexin (ck,h), are inexpensive to obtain, while the last two, pyruvate kinase and lactate dehydroginase (pk,ld), are expensive.

The variables (columns) in the data set are

| Column | Variable | Description |
|--------|----------|-------------|
| 1 | age | Age of a woman in the study |
| 2 | ck | Creatine kinase level |
| 3 | h | Hemopexin |
| 4 | pk | Pyruvate kinase |
| 5 | ld | Lactate dehydroginase |
| 6 | carrier | Indicator if a woman is a DMD carrier |

(a) Find the mean, median, standard deviation, and *real MAD* of pyruvate kinase level, pk, for all cases (carrier=1).

(b) Find the mean, median, standard deviation, and *real MAD* of pyruvate kinase level, pk, for all controls (carrier=0).

(c) Find the correlation between variables pk and carrier.

(d) Use MATLAB's gplotmatrix to visualize pairwise dependencies between the six variables.

(e) Plot the histogram with 30 bins and smoothed normalized histogram (density estimator) for pk. Use ksdensity.

2.17. **Ashton's Dental Data.** The evolutionary status of fossils (Australopithecinae, Proconsul, etc.) stimulated considerable discussion in the 1950s. Particular attention has been paid to the teeth of the fossils, comparing their overall dimensions with those of human beings and of the extant great apes. As "controls" measurements have been taken on the teeth of three types of modern man (British, West African native, Australian aboriginal) and of the three living great apes (gorilla, orangutan, and chimpanzee).

The data in the table below are taken from Ashton et al. (1957), p. 565, who used 2-D projections to compare the measurements. Andrews (1972) also used an excerpt of these data to illustrate his methodology. The values in the table are not the original measurements but the first eight *canonical variables* produced from the data in order to maximize the sum of distances between different pairs of populations.

| A. West African | −8.09 | 0.49 | 0.18 | 0.75 | −0.06 | −0.04 | 0.04 | 0.03 |
| B. British | −9.37 | −0.68 | −0.44 | −0.37 | 0.37 | 0.02 | −0.01 | 0.05 |
| C. Au. aboriginal | −8.87 | 1.44 | 0.36 | −0.34 | −0.29 | −0.02 | −0.01 | −0.05 |
| D. Gorilla: male | 6.28 | 2.89 | 0.43 | −0.03 | 0.10 | −0.14 | 0.07 | 0.08 |
| E.      Female | 4.82 | 1.52 | 0.71 | −0.06 | 0.25 | 0.15 | −0.07 | −0.10 |
| F. Orangutan: Male | 5.11 | 1.61 | −0.72 | 0.04 | −0.17 | 0.13 | 0.03 | 0.05 |
| G.      Female | 3.60 | 0.28 | −1.05 | 0.01 | −0.03 | −0.11 | −0.11 | −0.08 |
| H. Chimpanzee: male | 3.46 | −3.37 | 0.33 | −0.32 | −0.19 | −0.04 | 0.09 | 0.09 |
| I.      Female | 3.05 | −4.21 | 0.17 | 0.28 | 0.04 | 0.02 | −0.06 | −0.06 |
| J. *Pithecanthropus* | −6.73 | 3.63 | 1.14 | 2.11 | −1.90 | 0.24 | 1.23 | −0.55 |
| K.      *pekinensis* | −5.90 | 3.95 | 0.89 | 1.58 | −1.56 | 1.10 | 1.53 | 0.58 |
| L. *Paranthropus robustus* | −7.56 | 6.34 | 1.66 | 0.10 | −2.23 | −1.01 | 0.68 | −0.23 |
| M. *Paranthropus crassidens* | −7.79 | 4.33 | 1.42 | 0.01 | −1.80 | −0.25 | 0.04 | −0.87 |
| N. *Meganthropus paleojavanicus* | −8.23 | 5.03 | 1.13 | −0.02 | −1.41 | −0.13 | −0.28 | −0.13 |
| O. *Proconsul africanus* | 1.86 | −4.28 | −2.14 | −1.73 | 2.06 | 1.80 | 2.61 | 2.48 |

Andrews (1972) plotted curves over the range $-\pi < t < \pi$ and concluded that the graphs clearly distinguished humans, the gorillas and orangutans, the chimpanzees, and the fossils. Andrews noted that the curve for a fossil (*Proconsul africanus*) corresponds to a plot inconsistent with that of all other fossils as well as humans and apes.

Graphically present this data using (a) star plots, (b) Andrews plots, and (c) Chernoff faces.

2.18. **Andrews Plots of Iris Data.** Fisher iris data are 4-D, and Andrews plots can be used to explore clustering of the three species (*Setosa, Versicolor*, and *Virginica*). Discuss the output from the code below.

```
load fisheriris
andrewsplot(meas,'group',species);
```

What species clearly separate? What species are more difficult to separate?

2.19. **Cork Boring Data.** Cork is the bark of the cork oak (*Quercus suber L*), a noble tree with very special characteristics that grows in the Mediterranean. This natural tissue has unique qualities: light weight, elasticity, insulation and impermeability, fire retardancy, resistance to abrasion, etc. The data measuring cork boring of trees given in Rao (1948) consist of the weights (in centigrams) of cork boring in four directions (north, east, south, and west) for 28 trees. Data given in Table 2.1 can also be found in cork.dat|mat.

(a) Graphically display the data as a data plot, pairwise scatterplots, Andrews plot, and Chernoff faces.

(b) Find the mean $\bar{x}$ and covariance matrix $S$ for this data set. Find the trace and determinant of $S$.

(c) Find the Mahalanobis transformation for these data. Check that the covariance matrix for the transformed data is identity.

2.20. **Balance.** When a human experiences a balance disturbance, muscles throughout the body are activated in a coordinated fashion to maintain an

**Table 2.1** Rao's data. Weights of cork boring in four directions (north, east, south, west) for 28 trees.

| Tree | N | E | S | W | Tree | N | E | S | W |
|------|----|----|----|----|------|----|----|-----|----|
| 1 | 72 | 66 | 76 | 77 | 15 | 91 | 79 | 100 | 75 |
| 2 | 60 | 53 | 66 | 63 | 16 | 56 | 68 | 47 | 50 |
| 3 | 56 | 57 | 64 | 58 | 17 | 79 | 65 | 70 | 61 |
| 4 | 41 | 29 | 36 | 38 | 18 | 81 | 80 | 68 | 58 |
| 5 | 32 | 32 | 35 | 36 | 19 | 78 | 55 | 67 | 60 |
| 6 | 30 | 35 | 34 | 26 | 20 | 46 | 38 | 37 | 38 |
| 7 | 39 | 39 | 31 | 27 | 21 | 39 | 35 | 34 | 37 |
| 8 | 42 | 43 | 31 | 25 | 22 | 32 | 30 | 30 | 32 |
| 9 | 37 | 40 | 31 | 25 | 23 | 60 | 50 | 67 | 54 |
| 10 | 33 | 29 | 27 | 36 | 24 | 35 | 37 | 48 | 39 |
| 11 | 32 | 30 | 34 | 28 | 25 | 39 | 36 | 39 | 31 |
| 12 | 63 | 45 | 74 | 63 | 26 | 50 | 34 | 37 | 40 |
| 13 | 54 | 46 | 60 | 52 | 27 | 43 | 37 | 39 | 50 |
| 14 | 47 | 51 | 52 | 43 | 28 | 48 | 54 | 57 | 43 |

upright stance. Researchers at Lena Ting Laboratory for Neuroengineering at Georgia Tech are interested in uncovering the sensorimotor mechanisms responsible for coordinating this automatic postural response (APR). Their approach was to perturb the balance of a human subject standing upon a customized perturbation platform that translates in the horizontal plane. Platform motion characteristics spanned a range of peak velocities (5 cm/s steps between 25 and 40 cm/s) and accelerations (0.1 g steps between 0.2 and 0.4 g). Five replicates of each perturbation type were collected during the experimental sessions. Surface electromyogram (EMG) signals, which indicate the level of muscle activation, were collected at 1080 Hz from 11 muscles in the legs and trunk.

The data in ⌨ balance2.mat are processed EMG responses to backward-directed perturbations in the medial gastrocnemius muscle (an ankle plantar flexor located on the calf) for all experimental conditions. There is 1 s of data, beginning at platform motion onset. There are 5 replicates of length 1024 each collected at 12 experimental conditions (4 velocities crossed with 3 accelerations), so the data set is 3-D 1024 × 5 × 12.

For example, data(:,1,4) is an array of 1024 observations corresponding to first replicate, under the fourth experimental condition (30 cm/s, 0.2 g). Consider a fixed acceleration of 0.2 g and only the first replicate. Form 1024 4-D observations (velocities 25, 30, 35, and 40 as variables) as a data matrix. For the first 16 observations find multivariate graphical summaries using MATLAB's gplotmatrix, parallelcoords, andrewsplot, and glyphplot.

2.21. **Cats.** Cats are often used in studies about locomotion and injury recovery. In one such study, a bundle of nerves in a cat's legs were cut and then surgically repaired. This mimics the surgical correction of injury in people. The recovery process of these cats was then monitored. It was monitored quantitatively by walking a cat across a plank that has force plates, as well as by monitoring various markers inside the leg. These markers provided data for measures such as joint lengths and joint moments. A variety of data was collected from three different cats: Natasha, Riga, and Korina. Natasha (cat = 1) has 47 data entries, Riga (cat = 2) has 39 entries, and Korina (cat = 3) has 35 entries.

The measurements taken are the number of steps for each trial, the length of the stance phase (in milliseconds), the hip height (in meters), and the velocity (in meters/second). The researchers observe these variables for different reasons. They want uniformity both within and between samples (to prevent confounding variables) for steps and velocity. The hip height helps monitor the recovery process. A detailed description can be found in Farrell et al. (2009).

The data set, courtesy of Dr. Boris Prilutsky, School of Applied Physiology at, Georgia Tech, is given as the MATLAB structure file ☒ cats.mat. Form a data matrix

`X = [cat.nsteps cat.stancedur cat.hipheight cat.velocity cat.cat];`

and find its mean and correlation matrix. Form matrix $Z$ by standardizing the columns of $X$ (use zscore). Plot the image of the standardized data matrix.

2.22. **BUPA Liver Data.** The BUPA liver disorders database (courtesy of Richard Forsyth, BUPA Medical Research Ltd.) consists of 345 records of male individuals. Each record has 7 attributes,

| Attribute | Name | Meaning |
|---|---|---|
| 1 | mcv | Mean corpuscular volume |
| 2 | alkphos | Alkaline phosphotase |
| 3 | sgpt | Alamine aminotransferase |
| 4 | sgot | Aspartate aminotransferase |
| 5 | gammagt | Gamma-glutamyl transpeptidase |
| 6 | drinks | Number of half-pint equivalents of alcoholic beverages drunk per day |
| 7 | selector | Field to split the database |

The first five variables are all blood tests that are thought to be sensitive to liver disorders that might arise from excessive alcohol consumption.

The variable selector was used to partition the data into two sets, very likely into a training and validation part.

Using gplotmatrix explore the relationship among variables 1 through 6 (exclude the selector).

2.23. **Cell Circularity Data.** In the lab of Dr. Todd McDevitt at Georgia Tech, researchers wanted to elucidate differences between the "static" and "rotary" culture of embrionic bodies (EBs) that were formed under both conditions with equal starting cell densities. After 2, 4, and 7 days of culture, images of EBs were acquired using phase-contrast microscopy. Image analysis software was used to determine the circularity (defined as

$4\pi(Area/Perimeter^2)$) of each EB imaged. A total of $n = 325$ EBs were analyzed from three separate plates for both static and rotary cultures at the
three time points studied. The circularity measures were used to examine
differences in the shape of EBs formed under the two conditions as well as
differences in their variability.

The data set  circ.dat|mat consists of six columns corresponding to six
treatments (2d, rotary), (4d, rotary), (7d, rotary), (2d, static), (4d, static),
and (7d, static). Note that this is not an example of multivariate data since
the columns are freely permutable, but rather six univariate data sets.

(a) For rotation and static 2d measurements, plot back-to-back histograms
( bihist.m) as well as boxplots.

(b) For static 7d measurements graph by pie chart (pie) the proportion of
EBs with circularity smaller than 0.75.

---

**MATLAB FILES AND DATA SETS USED IN THIS CHAPTER**

http://springer.bme.gatech.edu/Ch2.Descriptive/

            acorr.m, acov.m, ashton.m, balances.m, bat.m, bihist.m,
biomed.m, blowfliesTS.m, BUPAliver.m, carea.m, cats.m, cats1.m,
circular.m, corkrao.m, crouxrouss.m, crouxrouss2.m, diversity.m,
ecg.m, empiricalcdf.m, fisher1.m, grubbs.m, hist2d.m, histn.m,
lightrev.m, limestone.m, mahalanobis.m, meanvarchange.m,
multifat.m, multifatstat.m, mushrooms.m, myquantile.m, mytrimmean.m,
piecharts.m, scattercloud.m, simple2comp.m, smoothhist2D.m, spikes.m,
surveysUKUS.m

            ashton.dat, balance2.mat, bat.dat, blowflies.dat|mat,
BUPA.dat|mat|xlsx, cats.mat, cellarea.dat|mat, circ.dat|mat,
coburn.mat, cork.dat|mat, diabetes.xls, dmd.dat|mat|xls, fat.dat,
light.dat, limestone.dat, QT.dat|mat, raman.dat|mat, spikes.dat

---

# CHAPTER REFERENCES

Anderson, E. (1935). The Irises of the Gaspe Peninsula. *Bull. Am. Iris Soc.*, **59**, 2–5.

Andrews, F. D. (1972). Plots of high dimensional data. *Biometrics*, **28**, 125–136.

Bowley, A. L. (1920). *Elements of Statistics*. Scribner, New York.

Brillinger, D. R. (2001). *Time Series: Data Analysis and Theory*. Classics Appl. Math. **36**, SIAM, pp 540.

Brockwell, P. J. and Davis, R. A. (2009). *Introduction to Time Series and Forecasting*. Springer, New York.

Brozek, J., Grande, F., Anderson, J., and Keys, A. (1963). Densitometric analysis of body composition: revision of some quantitative assumptions. *Ann. New York Acad. Sci.*, **110**, 113–140.

Chernoff, H. (1973). The use of faces to represent points in $k$-dimensional space graphically. *J. Am. Stat. Assoc.*, **68**, 361–366.

Christov, I., Dotsinsky, I. , Simova, I. , Prokopova, R., Trendafilova, E., and Naydenov, S. (2006). Dataset of manually measured QT intervals in the electrocardiogram. *BioMed. Eng. OnLine*, **5**, 31 doi:10.1186/1475-925X-5-31. The electronic version of this article can be found online at:
http://www.biomedical-engineering-online.com/content/5/1/31

David, H. A. (1998). Early sample measures of variability. *Stat. Sci.*, **13**, 4, 368–377.

Farrell B., Bulgakova M., Hodson-Tole E.F., Shah S., Gregor R.J., Prilutsky B.I. (2009). Short-term locomotor adaptations to denervation of lateral gastrocnemius and soleus muscles in the cat. In: Proceedings of the Society for Neuroscience meeting, 17–21 October 2009, Chicago.

Fisher, R.A. (1936). The use of multiple measurements in taxonomic problems. Ann. Eugen. 7, Pt. II, 179–188.

Gauss, C. F. (1816). Bestimmung der Genauigkeit der Beobachtungen. *Zeitschrift Astron.*, **1**, 185–197.

Johnson, R. W. (1996). Fitting percentage of body fat to simple body measurements. *J. Stat. Educ.*, **4**, 1.
http://www.amstat.org/publications/jse/v4n1/datasets.johnson.html

Kaufman, L. and Rock, I. (1962). The moon illusion. *Science*, **136**, 953–961.

Moors, J. J. A. (1988). A Quantile Alternative for Kurtosis. *Statistician*, **37**, 25–32.

Morrison, D. F. (1976). *Multivariate Statistical Methods*, 2nd edn. McGraw-Hill, New York

Nicholson, A. J. (1954). An Outline of the Dynamics of Animal Populations. *Aust. J. Zool.* , **2**, 1, 9–65.

Nightingale, F. (1858). Notes on matters affecting the health, efficiency, and hospital administration of the British army. Founded chiefly on the experience of the late war. Presented by request to the Secretary of State for War. Privately printed for Miss Nightingale, Harrison and Sons.

Penrose, K., Nelson, A., and Fisher, A. (1985). Generalized body composition prediction equation for men using simple measurement techniques (abstract). *Med. Sc. Sports Exerc.*, **17**, 2, 189.

Percy, M. E., Andrews, D. F., Thompson,M. W., and Opitz J. M. (1981). Duchenne muscular dystrophy carrier detection using logistic discrimination: Serum creatine kinase and hemopexin in combination. *Am. J. Med. Genet.*, **8**, 4, 397–409.

Rao, C. R. (1948). Tests of significance in multivariate analysis. *Biometrika*, **35**, 58–79.

Shumway, R. H. and Stoffer, D. S. (2005). *Time Series Analysis and Its Applications*. Springer Texts in Statistics, Springer, New York.

Siri, W. E. (1961). Body composition from fluid spaces and density: Analysis of methods. In *Techniques for Measuring Body Composition*, Eds. J. Brozek and A. Henzchel. National Academy of Sciences, Washington, 224–244.

Stevens, S. S. (1946). On the theory of scales of measurement. *Science*, **103**, 2684, 677–680. PMID 17750512.

Sturges, H. (1926). The choice of a class-interval. *J. Am. Stat. Assoc.*, **21**, 65–66.

Velleman, P. F. and Wilkinson, L. (1993). Nominal, ordinal, interval, and ratio typologies are misleading. *Am. Stat.*, **47**, 1, 65–72.

# Chapter 3
# Probability, Conditional Probability, and Bayes' Rule

*Misunderstanding of probability may be the greatest of all impediments to scientific literacy.*

– Stephen Jay Gould

**WHAT IS COVERED IN THIS CHAPTER**

• Events, Sample Spaces, and Classical Definition of Probability
• Probability of Unions and Intersections
• Independence of Events and Conditional Probability
• Total Probability and Bayes' Rule

## 3.1 Introduction

If statistics can be defined as the science that studies uncertainty, then probability is the branch of mathematics that quantifies it. One's intuition of chance and probability develops at a very early age (Piaget and Inhelder, 1976). However, the formal, precise definition of probability is elusive. There are several competing definitions for the probability of an event, but the most practical one uses its relative frequency in a potentially infinite series of experiments.

Probability is a part of all introductory statistics programs for a good reason: It is the theoretical foundation of statistics. The basic statistical concepts, random sample, sampling distributions, statistic, etc., require familiarity with probability to be understood, explained, and applied.

Probability is critical for the development of statistical concepts. Despite this fact, it will not be a focal point of this course for reasons of time. There is a dangerous temptation to dwell on urns, black and white balls, and combinatorics for so long that more important statistical concepts such as regression or ANOVA fall into a zeitnot (a term used in chess to describe the pressure felt from having little remaining time).

Many students taking a university-level introductory statistics course have already been exposed to probability and statistics previously in their education. With this in mind, we will use this chapter as a survey of probability using a range of examples. The more important concepts of independence, conditioning, and Bayes' rule will be covered in more detail and repeatedly used later in various contexts. Ross (2009) is recommended for a review and comprehensive coverage.

## 3.2 Events and Probability

If an experiment has the potential to be repeated an infinite number of times, then the probability of an outcome can be defined through its relative frequency of appearing. For instance, if we rolled a die a number of times, we could construct a table showing how many times each face came up. These individual frequencies ($n_i$) can be transformed into proportions or relative frequencies by dividing them by the total number of tosses $n : f_i = n_i/n$. If we were to see the outcome ⚅ in 53 out of 300 tosses, then that face's proportion, or relative frequency, would be $f_6 = 53/300 = 0.1767$. As more tosses are made, we would "expect" the proportion of ⚅ to stabilize around $\frac{1}{6}$. The "experiments" in the next example are often quoted in the literature on elementary probability.

*Example 3.1.* **Famous Coin Tosses.** Buffon tossed a coin 4,040 times. Heads appeared 2,048 times. K. Pearson tossed a coin 12,000 times and 24,000 times. The heads appeared 6,019 times and 12,012, respectively. For these three tosses the relative frequencies of heads are $2048/4040 \approx 0.5049$, $6019/12000 \approx 0.5016$, and $12012/24000 \approx 0.5005$.

What if the experiments cannot be repeated? For example, what is the probability that "Squiki" the guinea pig survives its first treatment by a particular drug? Or in the "experiment" of taking a statistics course this semester, what is the probability of getting an *A*? In such cases we can define probability *subjectively* as a measure of strength of belief. Here is another example.

*Example 3.2.* **Tutubalin's Problem.** In a desk drawer in the house of Mr. Jay Parrino of Kansas City there is a coin, a 1913 Liberty Head nickel (Fig. 3.1). What is the probability that the coin is heads up? This is an example where

**Fig. 3.1** A gem-proof-condition 1913 Liberty Head nickel, one of only five known, and the finest of the five. Collector Jay Parrino of Kansas City bought the elusive nickel for a record $1,485,000, the first and only time an American coin has sold for over $1 million.

equal levels of uncertainty for the two sides lead to the subjective answer of 1/2.

The *symmetry* of the experiment led to the classical definition of probability. An ideal die is symmetric. All sides are "equiprobable." When rolling a fair die, the probability of outcome ⚄ is a ratio of the number of *favorable* outcomes (in our example only one outcome is favorable) to the number of all possible outcomes, 1/6.[1]

Among several possible ways to define probability, three are outlined below.

> **Frequentist.** An event's *probability* is the proportion of times that we would expect the event to occur if the experiment were repeated a large number of times.
>
> **Subjectivist**. A subjective *probability* is an individual's degree of belief in the occurrence of an event.
>
> **Classical.** An event's *probability* is the ratio of the number of favorable outcomes to possible outcomes in a (symmetric) experiment.

A formal definition of probability is axiomatic (Kolmogorov, 1933) and is a special case of measure theory in mathematics.

The events that are assigned probabilities can be considered as sets of outcomes. The following table uses a rolling die experiment to introduce the set notation among events.

---

[1] This definition is criticized by philosophers because of the fallacy called a vicious circle in definition (*circulus vitiosus in definiendo*). One defines the notion of probability in terms of equi*probable* outcomes.

| Term | Description | Example |
|------|-------------|---------|
| Experiment | A phenomenon, action, or procedure where the outcomes are uncertain | A single roll of a balanced six-sided die |
| Sample space | Set of all possible outcomes in an experiment | $\mathscr{S} = \{\boxed{\cdot},\boxed{\cdot\cdot},\boxed{\cdot\cdot\cdot},\boxed{::},\boxed{:\cdot:},\boxed{:::}\}$ |
| Event | A collection of outcomes; a subset of $\mathscr{S}$ | $A = \{\boxed{\cdot\cdot\cdot}\}$ (3 dots show), $B = \{\boxed{\cdot\cdot\cdot},\boxed{::},\boxed{:\cdot:},\boxed{:::}\}$ (at least three dots show), $C = \{\boxed{\cdot},\boxed{\cdot\cdot}\}$ |
| Probability | A number between 0 and 1 assigned to an event. | $P(A) = \frac{1}{6}, \quad P(B) = \frac{4}{6} = \frac{2}{3}, P(C) = \frac{2}{6} = \frac{1}{3}$ |

To understand the probabilities from the above table, consider a simple MATLAB code that will simulate rolling a fair die. A random number from $(0,1)$ is generated and multiplied by 6. This becomes a random number between 0 and 6. When this number is rounded up to the closest integer, the outcomes $\boxed{\cdot},\boxed{\cdot\cdot},\ldots,\boxed{:::}$ are simulated. They are all equally likely. For example, the outcome $\boxed{::}$ comes from the original number, which is in the range (3,4), and this interval is one-sixth part of (0,6). Formal justification of this fact requires the concept of uniform distribution, which will be covered in Chap. 5.

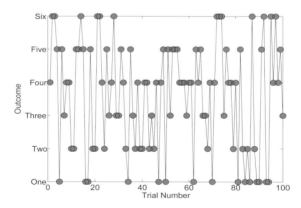

**Fig. 3.2** MATLAB simulation of rolling a fair die. The first 100 outcomes $\{4,6,6,5,1,\ldots,4,5,3\}$ are shown.

The MATLAB code ⚓ rollongdie1.m generates 50,000 outcomes and checks the proportion of those equal to 3, probA, those outcomes greater than or equal to 3, probB, and those smaller than 3, probC. The relative frequencies of these outcomes tend to their theoretical probabilities of 1/6, 2/3, and 1/3. Figure 3.2

shows the outcomes of the first 100 simulations described in the MATLAB code below.

```
% rollingdie1.m
outcomes = []; %keep outcomes here
M=50000 %# of rolls
for i= 1:M
    outcomes = [outcomes ceil( 6*rand )];
    % ceil(6*rand) rounds up (takes ceiling) of random
    % number from (0,6), thus the outcomes 1,2,3,4,5,and 6
    % are equally likely
end
probA = sum((outcomes == 3))/M
    % probA = 0.1692
probB = sum((outcomes >= 3))/M
    % probB = 0.6693
probC = sum((outcomes < 3))/M
    % probC = 0.3307
```

Events in an experiment are sets containing the elementary outcomes, that is, distinctive outcomes of the experiment. Among all events in an experiment, two are special: a sure event and an impossible event. A *sure event* occurs *every time* an experiment is repeated and has a probability of 1. It consists of all outcomes and is equal to the sample space of the experiment, $\mathscr{S}$. An *impossible event never* occurs when an experiment is performed and is usually denoted as $\emptyset$. It contains no elementary outcomes and its probability is 0.

For any event $A$, the probability that $A$ will occur is a number between 0 and 1, inclusive:

$$0 \le \mathbb{P}(A) \le 1.$$

Also,

$$\mathbb{P}(\emptyset) = 0, \text{ and } \mathbb{P}(\mathscr{S}) = 1.$$

The *intersection* $A \cap B$ of two events $A$ and $B$ occurs if both events $A$ *and* $B$ occur. The key word in the definition of the intersection is *and*. The intersection of two events $A \cap B$ is often written as a product $AB$. We will use both the $\cap$ and product notations.

The product of the events translates into the product of their probabilities only if the events are independent. We will see later that relationship $\mathbb{P}(AB) = \mathbb{P}(A)\mathbb{P}(B)$ is the definition of the independence of events $A$ and $B$.

Events are said to be *mutually exclusive* if they have no common elementary outcomes. In other words, it is impossible for both events to occur in a single trial of the experiment. For mutually exclusive events, $\mathbb{P}(A \cdot B) = \mathbb{P}(\emptyset) = 0$.

In the die-toss example, events $A = \{\boxed{\cdot\cdot}\}$ and $B = \{\boxed{\cdot\cdot},\boxed{\cdot},\boxed{\cdot\cdot},\boxed{\vdots}\}$ are not mutually exclusive, since the elementary outcome $\{\boxed{\cdot\cdot}\}$ belongs to both of them. On the other hand, the events $A = \{\boxed{\cdot\cdot}\}$ and $C = \{\boxed{\cdot},\boxed{\cdot}\}$ are mutually exclusive.

The *union* $A \cup B$ of two events $A$ and $B$ occurs if at least one of the events $A$ or $B$ occurs. The key word in the definition of the union is *or*.

For mutually exclusive events, the probability that at least one of them occurs is

$$\mathbb{P}(A \cup C) = \mathbb{P}(A) + \mathbb{P}(C).$$

For example, if the probability of event $A = \{\boxed{\cdot\cdot}\}$ is 1/6, and the probability of the event $C = \{\boxed{\cdot},\boxed{\cdot}\}$ is 1/3, then the probability of A or C is

$$\mathbb{P}(A \cup C) = \mathbb{P}(A) + \mathbb{P}(C) = 1/6 + 1/3 = 1/2.$$

The *additivity* property is valid for any number of mutually exclusive events $A_1, A_2, A_3, \ldots$:

$$\mathbb{P}(A_1 \cup A_2 \cup A_3 \cup \ldots) = \mathbb{P}(A_1) + \mathbb{P}(A_2) + \mathbb{P}(A_3) + \ldots.$$

What is $\mathbb{P}(A \cup B)$ if events $A$ and $B$ are not mutually exclusive?

For any two events $A$ and $B$, the probability that either $A$ or $B$ will occur is given by the *inclusion-exclusion* rule:

$$\mathbb{P}(A \cup B) = \mathbb{P}(A) + \mathbb{P}(B) - \mathbb{P}(A \cdot B). \tag{3.1}$$

If events $A$ and $B$ are exclusive, then $\mathbb{P}(A \cdot B) = 0$, and we get the familiar result $\mathbb{P}(A \cup B) = \mathbb{P}(A) + \mathbb{P}(B)$.

The inclusion-exclusion rule can be generalized to unions of an arbitrary number of events. For example, for three events $A, B,$ and $C$, the rule is

$$\mathbb{P}(A \cup B \cup C) = \mathbb{P}(A) + \mathbb{P}(B) + \mathbb{P}(C) - \mathbb{P}(A \cdot B) - \mathbb{P}(A \cdot C) - \mathbb{P}(B \cdot C) + \mathbb{P}(A \cdot B \cdot C).$$
$$\tag{3.2}$$

For every event defined on a space of elementary outcomes, $\mathscr{S}$, we can define a counterpart event called its *complement*. The complement $A^c$ of an event $A$ consists of all outcomes that are in $\mathscr{S}$ but are not in $A$. The key word in the definition of a complement is *not*. In our example, $A^c$ consists of the outcomes $\{\boxed{\cdot},\boxed{\cdot},\boxed{\cdot\cdot},\boxed{\cdot\cdot},\boxed{\vdots}\}$.

Events $A$ and $A^c$ are mutually exclusive by definition. Consequently,

$$\mathbb{P}(A \cup A^c) = \mathbb{P}(A) + \mathbb{P}(A^c).$$

Since we also know from its definition that $A^c$ includes all outcomes in the sample space, $\mathscr{S}$, that are not in $A$, so that $\mathscr{S} = A \cup A^c$, it follows that

$$\mathbb{P}(A) + \mathbb{P}(A^c) = \mathbb{P}(\mathscr{S}) = 1.$$

For any pair of complementary events $A$ and $A^c$,
$$\mathbb{P}(A) + \mathbb{P}(A^c) = 1, \quad \mathbb{P}(A) = 1 - \mathbb{P}(A^c), \text{ and } \quad \mathbb{P}(A^c) = 1 - \mathbb{P}(A).$$

These equations simplify the solutions of some probability problems. If $\mathbb{P}(A^c)$ is easier to calculate than $\mathbb{P}(A)$, then the equations above let us obtain $\mathbb{P}(A)$ indirectly.

Having defined the complement, we can prove (3.1). The argument is easy if event $B$ is written as a union of two exclusive events, $B = (B \cap A^c) \cup (A \cap B)$. From this and the additivity property,

$$\mathbb{P}(B \cap A^c) = \mathbb{P}(B) - \mathbb{P}(A \cap B).$$

Since $A \cup B$ is equal to a union of exclusive events, $A \cup B = A \cup (B \cap A^c)$, by the additivity property of probability we obtain

$$\mathbb{P}(A \cup B) = \mathbb{P}(A) + \mathbb{P}(B \cap A^c) = \mathbb{P}(A) + \mathbb{P}(B) - \mathbb{P}(A \cap B).$$

This and some other probability properties are summarized in the table below.

| Property | Notation |
|---|---|
| If event S will *always* occur, its probability is 1. | $\mathbb{P}(\mathscr{S}) = 1$ |
| If event $\emptyset$ will *never* occur, its probability is 0. | $\mathbb{P}(\emptyset) = 0$ |
| Probabilities are always between 0 and 1, inclusive. | $0 \le \mathbb{P}(A) \le 1$ |
| If $A,B,C,\dots$ are all mutually exclusive then $\mathbb{P}(A \cup B \cup C \dots)$ can be found by addition. | $\mathbb{P}(A \cup B \cup C\dots) = \mathbb{P}(A) + \mathbb{P}(B) + \mathbb{P}(C) + \dots$ |
| The general *addition rule* for probabilities | $\mathbb{P}(A \cup B) = \mathbb{P}(A) + \mathbb{P}(B) - \mathbb{P}(A \cdot B)$ |
| Since $A$ and $A^c$ are mutually exclusive and between them include all outcomes from $\mathscr{S}$, $\mathbb{P}(A \cup A^c)$ is 1. | $\mathbb{P}(A \cup A^c) = \mathbb{P}(A) + \mathbb{P}(A^c) = \mathbb{P}(\mathscr{S}) = 1$, and $\mathbb{P}(A^c) = 1 - \mathbb{P}(A)$ |

Of particular importance in assessing the probability of composite events are De Morgan's laws which are simple algebraic relationships between events. The laws are named after Augustus De Morgan, British mathematician and logician (Fig. 3.3).

For any set of $n$ events $A_1, A_2, \ldots, A_n$,

$$(A_1 \cup A_2 \cup \cdots \cup A_n)^c = A_1^c \cap A_2^c \cap \cdots \cap A_n^c,$$
$$(A_1 \cap A_2 \cap \cdots \cap A_n)^c = A_1^c \cup A_2^c \cup \cdots \cup A_n^c.$$

De Morgan's laws can be readily demonstrated using Venn diagrams, discussed in Sect. 3.4.

**Fig. 3.3** Augustus De Morgan (1806–1871).

The following example shows how to apply De Morgan's laws.

*Example 3.3.* **Nanotubules and Cancer Cells.** One technique of killing cancer cells involves inserting microscopic synthetic rods called carbon nanotubules into the cell. When the rods are exposed to near-infrared light from a laser, they heat up, killing the cell, while cells without rods are left unscathed (Wong et al., 2005). Suppose that five nanotubules are inserted in a single cancer cell. Independently of each other they become exposed to near-infrared light with probabilities 0.2, 0.4, 0.3, 0.6, and 0.5. What is the probability that the cell will be killed?

Let $B$ be an event where a cell is killed and $A_i$ an event where the $i$th nanotubule kills the cell. The cell is killed if $A_1 \cup A_2 \cup \cdots \cup A_5$ happens. In other words, the cell is killed if nanotubule 1 kills the cell, or nanotubule 2 kills the cell, etc. We consider the event where the cell is not killed and apply De Morgan's laws. De Morgan's laws state that $A_1^c \cup A_2^c \cup \cdots \cup A_n^c = (A_1 \cap A_2 \cap \cdots \cap A_n)^c$,

$$\mathbb{P}(B) = 1 - \mathbb{P}(B^c) = 1 - \mathbb{P}((A_1 \cup A_2 \cup \cdots \cup A_5)^c) = 1 - \mathbb{P}(A_1^c \cap A_2^c \cap \cdots \cap A_5^c)$$
$$= 1 - (1 - 0.2)(1 - 0.4)(1 - 0.3)(1 - 0.6)(1 - 0.5) = 0.9328.$$

Thus, the cancer cell will be killed with a probability of 0.9328.

*Example 3.4.* As an example of the algebra of events and basic rules of probability, we derive the Bonferroni inequality. It will be revisited later in the text when calculating the significance level in simultaneous testing of multiple hypotheses (p. 342).

The Bonferroni inequality states that for arbitrary events $A_1, A_2, \ldots, A_n$,

$$\mathbb{P}(A_1 \cap A_2 \cap \cdots \cap A_n) \geq \mathbb{P}(A_1) + \mathbb{P}(A_2) + \cdots + \mathbb{P}(A_n) - n + 1. \tag{3.3}$$

Start with $n$ events $A_i$, $i = 1, \ldots, n$ and the event $A_1^c \cup A_2^c \cup \cdots \cup A_n^c$. The probability of any union of events is always smaller than the sum of probabilities of individual events:

$$\mathbb{P}(A_1^c \cup A_2^c \cup \cdots \cup A_n^c) \leq \mathbb{P}(A_1^c) + \mathbb{P}(A_2^c) + \cdots + \mathbb{P}(A_n^c).$$

De Morgan's laws state that $A_1^c \cup A_2^c \cup \cdots \cup A_n^c = (A_1 \cap A_2 \cap \cdots \cap A_n)^c$ and

$$1 - \mathbb{P}((A_1 \cap A_2 \cap \cdots \cap A_n)^c) \leq (1 - \mathbb{P}(A_1)) + (1 - \mathbb{P}(A_2)) + \cdots + (1 - \mathbb{P}(A_n)),$$

leading to the inequality in (3.3).

**Circuits.** The application of basic probability rules involving unions, intersections, and complements of events can be quite useful. An example is the application in the reliability of a complex system consisting of many components that work independently. If a complex system can be expressed as a configuration of simple elements that are linked in a "serial" or "parallel" fashion, the reliability of such a system can be calculated by knowing the reliabilities of its constituents.

Let a system $S$ consist of $n$ constituent elements $E_1, E_2, \ldots, E_n$ that can be interconnected in either a serial or a parallel fashion (Fig. 3.4). Suppose that elements $E_i$ work in time interval $T$ with probability $p_i$ and fail with probability $q_i = 1 - p_i$, $i = 1, \ldots, n$. The following table gives the probabilities of working for elements in $S$.

| Connection | Notation | Works with prob | Fails with prob |
|---|---|---|---|
| Serial | $E_1 \cap E_2 \cap \cdots \cap E_n$ | $p_1 p_2 \ldots p_n$ | $1 - p_1 p_2 \ldots p_n$ |
| Parallel | $E_1 \cup E_2 \cup \cdots \cup E_n$ | $1 - q_1 q_2 \ldots q_n$ | $q_1 q_2 \ldots q_n$ |

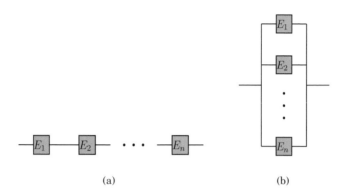

**Fig. 3.4** (a) Serial connection modeled as $E_1 \cap E_2 \cap \cdots \cap E_n$. (b) Parallel connection modeled as $E_1 \cup E_2 \cup \cdots \cup E_n$.

If the system has both serial and parallel connections, then the probability of the system working can be found by the subsequent application of the probabilities for the union and intersection of events.

Here is an example.

*Example 3.5.* **Circuit.** A complex system $S$ is defined via

$$S = E_1 \cap [(E_2 \cap E_3) \cup (E_4 \cap (E_5 \cup E_6))] \cap E_7,$$

where the unreliable components $E_i$, $i = 1, \ldots, 7$ work and fail independently. The system is depicted in Fig. 3.5. The components are operational in some

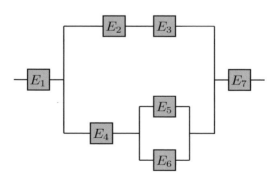

**Fig. 3.5** Circuit $E_1 \cap [(E_2 \cap E_3) \cup (E_4 \cap (E_5 \cup E_6))] \cap E_7$.

fixed time interval $[0, T]$ with probabilities given in the following table.

| Component | $E_1$ $E_2$ $E_3$ $E_4$ $E_5$ $E_6$ $E_7$ |
|---|---|
| Probability of functioning well | 0.9 0.5 0.3 0.1 0.4 0.5 0.8 |

We will find the probability that system $S$ will work in $[0,T]$ first analytically and then find an approximation by simulating the circuit in MATLAB and WinBUGS.

To find the probability that system $S$ works/fails, it is useful to create a table with probabilities $p_i = \mathbb{P}(\text{component } E_i \text{ works})$ and their complements $q_i = 1 - p_i$, $i = 1,\ldots,7$:

| Component | $E_1$ $E_2$ $E_3$ $E_4$ $E_5$ $E_6$ $E_7$ |
|---|---|
| $p_i$s | 0.9 0.5 0.3 0.1 0.4 0.5 0.8 |
| $q_i$s | 0.1 0.5 0.7 0.9 0.6 0.5 0.2 |

and then calculate step by step the probabilities of subsystems that ultimately add up to the final system. For example, we calculate the probability of working/failing for $S_1 = E_2 \cap E_3$, then $S_2 = E_5 \cup E_6$, then $S_3 = E_4 \cap S_2$, then $S_4 = S_1 \cup S_3$, and finally $S = E_1 \cap S_4 \cap E_7$.

| Component | Probability of working | Probability of failing |
|---|---|---|
| $S_1 = E_2 \cap E_3$ | $p_{s1} = 0.5 \cdot 0.3 = 0.15$ | $q_{s1} = 1 - 0.15 = 0.85$ |
| $S_2 = E_5 \cup E_6$ | $p_{s2} = 1 - 0.3 = 0.7$ | $q_{s2} = 0.6 \cdot 0.5 = 0.3$ |
| $S_3 = E_4 \cap S_2$ | $p_{s3} = 0.1 \cdot 0.7 = 0.07$ | $q_{s3} = 1 - 0.07 = 0.93$ |
| $S_4 = S_1 \cup S_3$ | $p_{s4} = 1 - 0.7905 = 0.2095$ | $q_{s4} = 0.85 \cdot 0.93 = 0.7905$ |
| $S = E_1 \cap S_4 \cap E_7$ | $p_S = 0.9 \cdot 0.2095 \cdot 0.8 = \mathbf{0.15084}$ | $q_S = 1 - 0.15084 = 0.84916$ |

Thus the probability that the system will work in the time interval $[0,T]$ is 0.15084.

The MATLAB code that approximates this probability uses a random number generator to simulate the case where the simple elements "work" and binary operations to simulate intersections and unions. For example, the fact that $e_1$ is functioning well (working) with a probability of 0.9 is modeled by e1 = rand < 0.9. Note that the left-hand side of the equation e1 = rand < 0.9 is a logical expression that takes values TRUE (numerical value 1) and FALSE (numerical value 0). Given that the event {rand < 0.9} is true 90% of the time, the value e1 represents the status of component $E_1$. This will be 0 with a probability of 0.1 and 1 with a probability of 0.9. The unions and intersections of $e_1, e_2, \ldots, e_n$ are modeled as $(e_1 + e_2 + \cdots + e_n > 0)$ and $e_1 * e_2 * \cdots * e_n$, respectively. Equivalently, they can be modeled as $\max\{e1, e2, \ldots, en\}$ and $\min\{e1, e2, \ldots, en\}$. Indeed, the former is 1 if at least one $e_i$ is 1, and the latter is 1 if all $e_i$s are 1, thus coding the union and the intersection.

To assess the probability that the system is operational, subsystems are formed and gradually enlarged, identical to the method used to find the analytic solution ( ◀ circuit.m).

```
% circuit.m
M=1000000;
s = 0;
for i = 1:M
e1 = rand < 0.9; e2 = rand < 0.5; e3 = rand < 0.3;
e4 = rand < 0.1; e5 = rand < 0.4; e6 = rand < 0.5;
e7 = rand < 0.8;
% ===============
s1 = min(e2,e3);  %   or s1 = e2*e3;
s2 = max(e5,e6);  %   or s2= e5+e6>0;
s3 = min(e4,s2);  %   or s3 = e4*s2;
s4 = max(s1,s3);  %   or s4 = s1+s3 > 0;
st = min([e1;s4;e7]); %   or st=e1*s4*e7;
s = s + st;
end
works = s/M
fails = 1 - works

% works = 0.150944
% fails = 0.849056
```

Next we repeat this simulation in WinBUGS. There are many differences between MATLAB and WinBUGS that go beyond the differences in the syntax. In MATLAB we had an explicit loop to generate $10^6$ runs; in WinBUGS this is done via the Model>Update tool and is not a part of the code. Also, the $e_i$s in MATLAB are 0 and 1; in WinBUGS they are 1 and 2 since the outcomes are realizations of a categorical discrete random variable dcat, and this variable is coded by nonnegative integers: 1, 2, 3, . . . . For this reason we adjusted the probability of a system working as ps <- s - 1.

```
# circuit1.odc
model
for (i in 1:7)
e[i] ~ dcat(p[i,])

s1 <- min(e[2],e[3])
s2 <- max(e[5],e[6])
s3 <- min(e[4],s2)
s4 <- max(s1,s3)
s <- min( min(e[1],s4) , e[7] )
ps <- s-1

DATA IN:
list(
 p = structure(.Data =
 c(0.1,0.9,       0.5,0.5,
   0.7,0.3,       0.9,0.1,
   0.6,0.4,       0.5,0.5,
   0.2,0.8) , .Dim = c(7,2) ) )
```

```
INITS NONE,  just 'gen inits'
```

The result of the simulations is close to the theoretical value.

| | mean | sd | MC error | val2.5pc | median | val97.5pc | start | sample |
|---|---|---|---|---|---|---|---|---|
| ps | **0.1508** | 0.3578 | 3.528E–4 | 0.0 | 0.0 | 1.0 | 10001 | 1000000 |

This is the first WinBUGS program in the text, and the reader is advised to consult Chap. 19, which discusses how communication with the WinBUGS program is structured and carried out. This comment has the mark "dangerous bend" since many students initially find the BUGS interface and programming intimidating.

## 3.3 Odds

Odds are alternative measures for the likelihood of events. If an event $A$ has a probability $\mathbb{P}(A)$, then the odds of $A$ are defined as

$$\mathbb{O}dds(A) = \frac{\mathbb{P}(A)}{\mathbb{P}(A^c)}, \qquad \mathbb{P}(A) = \frac{\mathbb{O}dds(A)}{\mathbb{O}dds(A)+1}.$$

From the classical definition of probability $\mathbb{P}(A) = \frac{\text{\# of favorable for A}}{\text{\# in the sample space}} = n_A/n$, the odds of $A$ are defined as $\mathbb{O}dds(A) = n_A/(n-n_A)$. For instance, the odds of event $A = \{⚄\}$ are $1/(6-1)$, one in five.

In economic decision theory, epidemiology, game theory, and some other areas, odds and odds ratios are preferred measures of quantifying and comparing events.

*Example 3.6.* The odds that the circuit $S$ in Example 3.5 is working are 17.76%, since $\mathbb{P}(S) = 0.15084$ and $\mathbb{O}dds(S) = 0.15084/(1-0.15084) = 0.17763$.

## 3.4 Venn Diagrams*

Venn diagrams help in graphically presenting the algebra of events and in determining the probability of composite events involving unions, intersections, and complements. The diagrams are named after John Venn (Fig. 3.6), the English logician who introduced the diagrams in his 1880 paper (Venn, 1880).

Venn diagrams connect sets and events in a graphical way – the events are represented as circles (squares, rectangles) and the notions of unions, intersections, complements, exclusiveness, implication, etc. among the events translate directly to the corresponding relations among the geometric areas.

**Fig. 3.6** John Venn (1834–1923), English logician.

Exclusive events are represented by nonoverlapping circles, while the notion of causality among the events translates to the subset relation. The geometric areas representing the events are plotted in a large rectangle representing the sample space (sure event).

Panels (a) and (b) in Fig. 3.7 show the union and intersection of events $A$ and $B$, while panel (c) shows the complement of event $A$.

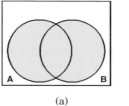 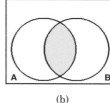 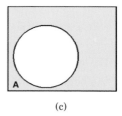

(a)                    (b)                    (c)

**Fig. 3.7** (a) Union and (b) intersection of events $A$ and $B$ and (c) complement of event $A$.

It is possible to define more exotic operations with events. For example, the difference between events $A$ and $B$, denoted as $A \backslash B$, is shown in Fig. 3.8a. It is obvious from the diagram that $A \backslash B = A \cap B^c$. The symmetric difference (or exclusive union) of events $A$ and $B$, denoted as $A \Delta B$, is an event in which either $A$ or $B$ happens, but not both (Fig. 3.8b). From the Venn diagram it is easy to see that $A \Delta B = (A \cap B^c) \cup (B \cap A^c) = (A \backslash B) \cup (B \backslash A)$.

Sometimes, the evidence for more complex algebraic relations between events can be established by Venn diagrams. Usually a Venn diagram of the left-hand side in a relation is compared with the Venn diagram of the right-hand side, and if the resulting sets coincide, we have a "proof." Proofs of this kind can be formalized with the help of mathematical logic and tautologies.

For example, one of De Morgan's laws for three events, $(A \cup B \cup C)^c = A^c \cap B^c \cap C^c$, can be demonstrated by Venn diagrams. Panel (a) in Fig. 3.9 shows $A \cup B \cup C$, while panel (b) shows $A^c \cap B^c \cap C^c$. It is obvious that the sets in the two panels are complementary and De Morgan's law is "demonstrated."

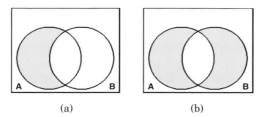

**Fig. 3.8** Difference $A \setminus B$ and symmetric difference $A \Delta B$.

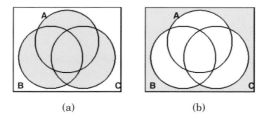

**Fig. 3.9** De Morgan's Law: $(A \cup B \cup C)^c = A^c \cap B^c \cap C^c$.

Likewise, if we want to demonstrate the distributive law $A \cup (B \cap C) = (A \cup B) \cap (A \cup C)$, the Venn diagram argument is shown in Fig. 3.10a-c. The set $A \cup (B \cap C)$ is shown in panel (a). Panels (b) and (c) show sets $A \cup B$ and $A \cup C$, respectively. Their intersection coincides with the set in panel (a).

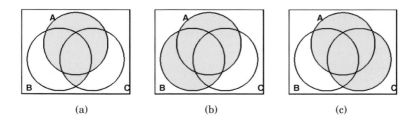

**Fig. 3.10** Distributive law among events, $A \cup (B \cap C) = (A \cup B) \cap (A \cup C)$.

In addition to algebraic relations among events, Venn diagrams can help in finding the probability of the complex algebraic composition of events. The probability can be informally connected with the area of a set in a Venn diagram, and this connection is extremely useful. For example, for the result $\mathbb{P}(A \cup B \cup C) = \mathbb{P}(A) + \mathbb{P}(B) + \mathbb{P}(C) - \mathbb{P}(AB) - \mathbb{P}(AC) - \mathbb{P}(BC) + \mathbb{P}(ABC)$, the formal proof is quite involved. An informal "proof" based on areas in a Venn diagram is simple and intuitive. The argument is as follows. If the probability is thought of as an area, then the area of $A \cup B \cup C$ can be obtained by adding the areas of $A$, $B$, and $C$, respectively. However, by adding the three areas there is an excess in the total area, and the regions counted multiple times should be subtracted. Thus areas of $A \cap B$, $A \cap C$, and $B \cap C$ are subtracted from the sum $\mathbb{P}(A) + \mathbb{P}(B) + \mathbb{P}(C)$. In this subtraction, the area of $A \cap B \cap C$ is subtracted three times and should be "patched back." Alternatively, one can think about *painting* the set $A \cup B \cup C$ with a *single layer of paint,* and the total *amount of paint* used is the probability. Of course, the amount of paint needed to paint the universal event $\mathscr{S}$ is 1. Although very informal, such a discursion can be quite useful.

## 3.5 Counting Principles*

Many experiments can be modeled by a sample space with a finite number of equally likely outcomes. We discussed the experiment of rolling a die, in which the sample space had six equally likely outcomes. In finding the probability of an event defined on this sample space we divided the number of outcomes favorable to $A$ by 6. For example, the event $A = \{$ ⚀,⚁,⚂ $\}$ (the number is even) has a probability of 3/6=1/2. But what if 10 dice are simultaneously rolled and we were interested in the probability that the sum of numbers will be equal to 55? The problem here is to count how many of $6^{10} = 60,466,176$ possible equally likely outcomes produce the sum of 55, and a simple inspection of the sample space applicable for one or two dice is not feasible. In situations like this, combinatorial and counting principles help. We will briefly illustrate the most important principles and introduce mathematical notions (factorial, $n$-choose-$k$, etc.) needed later in the course. A comprehensive coverage and a wealth of examples can be found in Ross (2009).

We start with definitions and basic properties of factorials and $n$-choose-$k$ operations.

Factorial $n!$ is defined as the product

$$n! = n(n-1)(n-2)\ldots2\cdot1 = \prod_{i=1}^{n} i.$$

For example, $5! = 5\cdot4\cdot3\cdot2\cdot1 = 120$. By definition $0! = 1$.

An $n$-choose-$k$ operation (or binomial coefficient) is defined as follows:

$$\binom{n}{k} = \frac{n(n-1)\ldots(n-k+1)}{k!} = \frac{n!}{(n-k)!k!}.$$

As the name indicates, $n$-choose-$k$ is the number of possible subsets of size $k$ from a set of $n$ elements. For example, the number of different committees of size 3 formed from a group of 8 students is $\binom{8}{3} = \frac{8\times7\times6}{3\times2\times1} = 56$. In MATLAB the command for $\binom{n}{k}$ is nchoosek(n,k). For example, nchoosek(8,3) results in 56.

The following properties follow directly from the definition of $\binom{n}{k}$:

$$\binom{n}{k} = \binom{n}{n-k},$$

$$\binom{n}{0} = 1 \quad \text{and} \quad \binom{n}{1} = n,$$

$$\binom{n}{k} + \binom{n}{k+1} = \binom{n+1}{k+1}.$$

**Fundamental Counting Principle** . If an experiment consists of $k$ actions, and the $i$th action can be performed in $n_i$ different ways, then the whole experiment can be performed in $n_1 \times n_2 \times \cdots \times n_k$ different ways. This is called the *multiplication counting rule* or *fundamental counting principle*.

*Example 3.7.* Out of 15 items, 4 are defective. The items are inspected one by one. What is the probability that the ninth item was the last defective one?

Consider the arrangement of 11 conforming and 4 defective items. The number of all possible arrangements is $\binom{15}{4} = \binom{15}{11} = 1316$, as one chooses 4 places out of 15 to place defective items or, equivalently, 11 places out of 15 to place conforming items.

The number of favorable outcomes can be found by the multiplication rule. Favorable outcomes are defined as follows: among the first-selected eight items three are defective, the ninth position is occupied by a defective item, and none of the remaining six items is defective:

$$\binom{8}{3} \times 1 \times 1 = 56.$$

Note that the number of ways in which a defective item falls at the ninth position, and the number of ways where six fair items occupy positions 10 to 15, are 1 each. Thus, the required probability is $56/1365 = 0.041$.

There is also an *addition counting rule* that mimics the additive property of probability: If $k$ events are exclusive and have $n_1, n_2, \ldots, n_k$ outcomes, then their union has $n_1 + n_2 + \cdots + n_k$ outcomes. If the events are not exclusive, this rule is known as the *inclusion-exclusion principle*. For instance, if two events are arbitrary, the inclusion-exclusion rule count for outcomes in their union is $n_1 + n_2 - n_{12}$, where $n_{12}$ is the number of common outcomes. For three events the inclusion-exclusion rule is $n_1 + n_2 + n_3 - n_{12} - n_{13} - n_{23} + n_{123}$; cf. (3.2).

If the population has $N$ subjects and a sample of size $n$ is needed, then Table 3.1 summarizes the number of possible samples, given the sampling policy and importance of ordering.

We first introduce the necessary notation. When the order is important, the samples are called *variations* or *permutations*. One can think about variations as words in an alphabet, since for words the order of letters is important. By the fundamental counting principle, the number of variations with repetitions of $N$ elements of length $n$ is $\overline{V}_N^n = N^n$ since each of $n$ places can be selected in $N$ ways. The number of variations without repetition of $N$ elements of length $n$ is $V_N^n = N \times (N-1) \times \cdots \times (N-n+1) = N^{(n)}$, $n \leq N$. Note that $V_N^N = N!$ is the number of permutations of $N$ distinct elements.

In combinations, the order in the sample is not important. If there is no repetition of elements, then $C_N^n = \binom{N}{n}$. If the repetition is possible, then $\overline{C}_N^n = \binom{N+n-1}{n}$.

**Table 3.1** Number of variations/combinations when the selection of $n$ from $N$ elements is done with/without the repetition.

|  | Order important (variations or permutations) | Order not important (combinations) |
|---|---|---|
| Sampling w/ repetition | $\overline{V}_N^n = N^n$ | $\overline{C}_N^n = \binom{N+n-1}{n}$ |
| Sampling w/o repetition | $V_N^n = N(N-1)\ldots(N-n+1)$ | $C_N^n = \binom{N}{n}$, $n \leq N$ |

The number of permutations of $N$ distinctive elements is $N!$, but if among $N$ elements there are only $k$ different elements, $n_1$ of type 1, $n_2$ of type 2, $\ldots$, $n_k$ of type $k$, $(n_1 + n_2 + \cdots + n_k = N)$, then the number of different permutations is

$$\binom{N}{n_1, n_2, \ldots n_k} = \frac{N!}{n_1! n_2! \cdots n_k!}. \tag{3.4}$$

The number in (3.4) is also called the multinomial coefficient.

*Example 3.8.* **Probability by Counting.** What is the probability that in a six-digit licence plate of a randomly selected car
  (a) All digits will be different?
  (b) Exactly two digits will be equal?
  (c) At least three digits will be different?
  (d) There will be exactly two pairs of equal digits?
  We assume that any digit from 0 to 9 can be at any of the six positions in the six-digit plate number.

This example is solved by using the classical definition of probability. For each event in (a)–(d), the number of favorable outcomes will be divided by the number of possible outcomes. The number of all possible outcomes is common, $\overline{V}_{10}^6 = 10^6$.

  (a) To find the number of favorable outcomes for the event where all digits are different, consider forming a six-digit number position by position. For the first position there are ten digits available, for the second nine (the digit used in the first position is eliminated as a choice for the second position), for the third eight, etc., for the last five. By the fundamental counting principle, the number of all favorable outcomes is the product $10 \times 9 \times 8 \times 7 \times 6 \times 5 = 10^{(6)} = V_{10}^6$, and the probability is

$$\frac{10^{(6)}}{10^6} = 0.1512.$$

(b) Out of ten digits choose one and place it on any two positions out of six available. This can be done in $10 \times \binom{6}{2} = 150$ ways. The remaining four positions could be chosen in $9 \times 8 \times 7 \times 6$ ways. Thus, the number of favorable outcomes is $150 \times 9 \times 8 \times 7 \times 6$, and the required probability is 0.4536.

(c) The opposite event for "at least three digits are different" is "all digits are the same" or "exactly two digits are the same," and we will find its probability first. The number of cases where all digits are the same is ten, while the number of cases where there are exactly two different digits is $\binom{10}{2} \times (2^6 - 2)$. Digits $a$ and $b$ can be selected in $\binom{10}{2}$ ways. Given the fixed selection, there are $2^6 - 2$ words in alphabet $\{a, b\}$ of length 6 where the words $aaaaaa$ and $bbbbbb$ are excluded. The probability of "at least three different digits" is

$$1 - \frac{10 + \binom{10}{2}(2^6 - 2)}{10^6} = 1 - 0.0028 = 0.9972.$$

(d) From ten digits first select two for the two pairs, and then an additional two digits for the remaining two places. There are four different digits: $a, b$ for the two pairs and $c, d$ for the remaining two places. The selection can be done in $\binom{10}{2} \times \binom{8}{2}$ ways. Once selected, the digits can be arranged in $\binom{6}{2,2,1,1} = \frac{6!}{2!2!1!1!}$ ways, by using permutations with repetitions as in (3.4).

Thus, the probability is

$$\frac{\binom{10}{2} \times \binom{8}{2} \times \frac{6!}{2!\, 2!\, 1!\, 1!}}{10^6} = 0.2268.$$

The following two important equations for probability calculation are a direct consequence of the combinatorial properties discussed in this section. They will be used later in the text when discussing the binomial and hypergeometric distributions and their generalizations.

**Multinomial and Multihypergeometric Trials.** Suppose that an experiment can result in $m$ possible outcomes, $A_1, A_2, \ldots, A_m$, that have probabilities $\mathbb{P}(A_1) = p_1, \ldots, \mathbb{P}(A_m) = p_m$, $p_1 + \cdots + p_m = 1$. If the experiment is independently repeated $n$ times, then the probability that event

$A_1$ will appear exactly $n_1$ times, $A_2$ exactly $n_2$ times, ..., $A_m$ exactly $n_m$ times $(n_1 + \cdots + n_m = n)$ is

$$\binom{n}{n_1, n_2, \ldots n_m} p_1^{n_1} p_2^{n_2} \cdots p_m^{n_m}.$$

If a finite population of size $m$ has $k_1$ subjects of type 1, ..., $k_p$ subjects of type $p$, $(k_1 + \cdots + k_p = m)$, and $n$ subjects are sampled at random, then the probability that $x_1$ will be of type 1, ..., $x_p$ will be of type $p$ $(x_1 + \cdots + x_p = n)$, is

$$\frac{\binom{k_1}{x_1}\binom{k_2}{x_2}\cdots\binom{k_l}{x_l}}{\binom{m}{n}}.$$

## 3.6 Conditional Probability and Independence

Important contemporary applications of probability in bioengineering, medical diagnostics, system biology, bioinformatics, etc. concern modeling and prediction of causal relationships in complex systems. The methodologies include influence diagrams, Bayesian networks, Granger causality, and related methods for which the notions of conditional probability, causality, and independence are fundamental. In this section we discuss conditional probabilities and independence.

A *conditional probability* is the probability of one event if we have information that another event, typically from the same sample space, has occurred. In the die-toss example, the probability of event $A = \{\boxdot\}$ is $\mathbb{P}(A) = \frac{1}{6}$. But what if we knew that event $B = \{\boxdot, \boxdot, \boxdot, \boxdot\}$ occurred? There are only four possible outcomes, only one of which is favorable for $A$. Thus, the probability of $A$ given $B$ is $\frac{1}{4}$. The conditional probability of $A$ given $B$ is denoted as $\mathbb{P}(A|B)$.

In general, the conditional probability of an event $A$ given that $B$ has occurred is equal to the probability of their intersection $\mathbb{P}(AB)$ divided by the probability of the event that we are conditioning upon, $\mathbb{P}(B)$. Of course, event $B$ has to have a positive probability, $\mathbb{P}(B) > 0$, because conditioning upon an event of zero probability is equivalent to the indeterminacy 0/0, as $0 \le \mathbb{P}(AB) \le \mathbb{P}(B)$:

$$\mathbb{P}(A|B) = \frac{\mathbb{P}(A \cdot B)}{\mathbb{P}(B)}, \text{ for } \mathbb{P}(B) > 0.$$

Figure 3.11 gives a graphical description of the conditional probability $\mathbb{P}(A|B)$. Once event $A$ is conditioned by $B$, $B$ "becomes the sample space" and $B$'s Venn diagram expands by a factor of $\frac{1}{\mathbb{P}(B)}$. The intersection $AB$ in the expanded $B$ becomes event $A|B$.

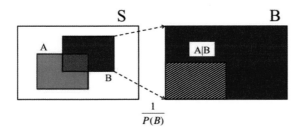

**Fig. 3.11** Graphical illustration of conditional probability.

An event $A$ is *independent* of $B$ if the conditional probability of $A$ given $B$ is the same as the probability of $A$ alone.

Events $A$ and $B$ are independent if

$$\mathbb{P}(A|B) = \mathbb{P}(A). \qquad (3.5)$$

In the die-toss example, $\mathbb{P}(A) = \frac{1}{6}$ and $\mathbb{P}(A|B) = \frac{1}{4}$. Therefore, events $A$ and $B$ are not independent.

We saw that the probability of the union $A \cup B$ was $\mathbb{P}(A \cup B) = \mathbb{P}(A) + \mathbb{P}(B) - \mathbb{P}(AB)$. Now we are ready to introduce the general rule for the probability of an intersection.

The probability that events $A$ and $B$ will both occur is obtained by applying the *multiplication rule*:

$$\mathbb{P}(A \cdot B) = \mathbb{P}(A)\mathbb{P}(B|A) = \mathbb{P}(B)\mathbb{P}(A|B), \qquad (3.6)$$

where $\mathbb{P}(A|B)$ and $\mathbb{P}(B|A)$ are conditional probabilities of $A$ given $B$ and of $B$ given $A$, respectively.

Only for independent events, equation (3.6) simplifies to

$$\mathbb{P}(A \cdot B) = \mathbb{P}(A)\mathbb{P}(B). \qquad (3.7)$$

Relationship (3.7) is also used to define independence, but (3.5) and (3.7) are equivalent.

With the repeated application of the multiplication rule, one can easily show

$$\mathbb{P}(A_1A_2\ldots A_n) = \mathbb{P}(A_1|A_2\ldots A_n)\,\mathbb{P}(A_2|A_3\ldots A_n)\,\ldots\mathbb{P}(A_{n-1}|A_n)\,P(A_n),$$

which is sometimes referred to as the *chain rule*. Here is one example of the use of the chain rule.

*Example 3.9.* **3+3 Dose Escalation Scheme.** In a dose finding stage of clinical trials (Phase I) patients are given the drug at some dose, and if there is no dose limiting toxicity (DLT), the dose is escalated. A version of the popular 3+3 method is implemented as follows. At a particular dose level three patients are randomly selected and given the drug. If there is no DLT, the dose is escalated to the next higher one. If there are two or more DLTs, the escalation process is stopped. If there is exactly one DLT among the three patients, three new patients are selected at random and given the drug at the same dose. If there are no DLTs among these three new patients, the dose is escalated. If there is at least one DLT, the escalation process is stopped.

Assume that 30 patients are available for the trial at some fixed dose. If among them 4 will exhibit DLT at that dose, what is the probability that in the described step of the 3+3 procedure the dose will be escalated?

We assume that patients are selected and given the drug one by one. Denote by $A_i$ the event that the $i$th patient will exhibit no DLT.

Then the dose will be escalated if the event

$$B = A_1A_2A_3 \,\cup\, A_1^cA_2A_3A_4A_5A_6 \,\cup\, A_1A_2^cA_3A_4A_5A_6 \,\cup\, A_1A_2A_3^cA_4A_5A_6$$

happens. Here, for example, the event $A_1A_2^cA_3A_4A_5A_6$ means that among the first three subjects the second experienced DLT, and that in the second group of three there was no DLT. Since the events $A_1A_2A_3$, $A_1^cA_2A_3A_4A_5A_6$, $A_1A_2^cA_3A_4A_5A_6$, and $A_1A_2A_3^cA_4A_5A_6$ are exclusive, the probability of their union is the sum of probabilities:

$$\mathbb{P}(B) = \mathbb{P}(A_1A_2A_3) + \mathbb{P}(A_1^cA_2A_3A_4A_5A_6)$$
$$+ \mathbb{P}(A_1A_2^cA_3A_4A_5A_6) + \mathbb{P}(A_1A_2A_3^cA_4A_5A_6).$$

For each of the probabilities the chain rule is needed. For example,

$$\mathbb{P}(A_1^cA_2A_3A_4A_5A_6) = 0.0739,$$

since

$$\mathbb{P}(A_1^c) \cdot \mathbb{P}(A_2|A_1^c) \cdot \mathbb{P}(A_3|A_1^c A_2) \cdot \mathbb{P}(A_4|A_1^c A_2 A_3)$$
$$\cdot \mathbb{P}(A_5|A_1^c A_2 A_3 A_4) \cdot \mathbb{P}(A_6|A_1^c A_2 A_3 A_4 A_5)$$
$$= \frac{4}{30} \cdot \frac{26}{29} \cdot \frac{25}{28} \cdot \frac{24}{27} \cdot \frac{23}{26} \cdot \frac{22}{25}.$$

Thus,

$$\mathbb{P}(B) = \frac{26}{30} \cdot \frac{25}{29} \cdot \frac{24}{28} + \frac{4}{30} \cdot \frac{26}{29} \cdot \frac{25}{28} \cdot \frac{24}{27} \cdot \frac{23}{26} \cdot \frac{22}{25}$$
$$+ \frac{26}{30} \cdot \frac{4}{29} \cdot \frac{25}{28} \cdot \frac{24}{27} \cdot \frac{23}{26} \cdot \frac{22}{25} + \frac{26}{30} \cdot \frac{25}{29} \cdot \frac{4}{28} \cdot \frac{24}{27} \cdot \frac{23}{26} \cdot \frac{22}{25}$$
$$= 0.8620.$$

The dose will be escalated with probability 0.8620.

If counting rules are applied, the solution can be expressed as

$$\frac{\binom{26}{3}\binom{4}{0}}{\binom{30}{3}} + \frac{\binom{26}{2}\binom{4}{1}}{\binom{30}{3}} \times \frac{\binom{24}{3}\binom{3}{0}}{\binom{27}{3}}.$$

The conditional odds of $A$ given that $B$ occurred is

$$\mathbb{O}dds(A|B) = \frac{\mathbb{P}(A|B)}{\mathbb{P}(A^c|B)} = \frac{\mathbb{P}(AB)}{\mathbb{P}(A^c B)}.$$

If events $A$ and $B$ are independent, then $\mathbb{O}dds(A|B) = \mathbb{O}dds(A)$.

For two events $A$ and $B$ the notions of exclusiveness $AB = \emptyset$ and independence $\mathbb{P}(AB) = \mathbb{P}(A)\mathbb{P}(B)$ are often considered equivalent by some students. Their argument can be summarized as follows. If two events do not share outcomes and their intersection is empty, then they must be independent. The contrary is true. If the events are exclusive and none are impossible, then they *must* be dependent. This can be demonstrated with the simple example of a coin-flipping experiment.

If $A$ denotes tails up and $B$ denotes heads up, then $A$ and $B$ are exclusive but dependent. If we have information that $A$ happened, then we also have complete information that $B$ did not happen.

If the sample spaces are different and the events are well separated in either time or space, their independence is intuitive. However, if the events share the sample space, it could be difficult to discern whether or not they are independent without resorting to the definition. The following example shows this.

*Example 3.10.* Let an experiment consist of drawing a card at random from a standard deck of 52 playing cards. Define events $A$ and $B$ as "the card is a ♠"

and "the card is a queen." Are the events $A$ and $B$ independent? By definition, $P(A \cdot B) = P(Q\spadesuit) = \frac{1}{52}$. This is the product of $P(\spadesuit) = \frac{13}{52}$ and $P(Q) = \frac{4}{52}$, and events $A$ and $B$ in question are independent. In this situation, intuition provides no help. Now, pretend that the $2\heartsuit$ is drawn and excluded from the deck prior to the experiment. Events $A$ and $B$ become dependent since

$$\mathbb{P}(A) \cdot \mathbb{P}(B) = \frac{13}{51} \cdot \frac{4}{51} \neq \frac{1}{51} = \mathbb{P}(A \cdot B).$$

The multiplication rule tells us how to find the probability for a composite event $(A \cdot B)$. The probability of $(A \cdot B)$ is used in the general *addition rule* for finding the probability of $(A \cup B)$.

| Rule | Notation |
|---|---|
| **Definitions** <br> The *conditional probability* of $A$ given $B$ is the probability of event $A$ if event $B$ occurred. | $\mathbb{P}(A\|B)$ |
| $A$ is *independent* of $B$ if the conditional probability of $A$ given $B$ is the same as the unconditional probability of $A$. | $\mathbb{P}(A\|B) = \mathbb{P}(A)$ |
| **Multiplication rule** <br> The general *multiplication rule* for probabilities. | $P(A \cdot B) = \mathbb{P}(A)\mathbb{P}(B\|A) = \mathbb{P}(B)\mathbb{P}(A\|B)$ |
| For *independent events* only, the multiplication rule is simplified. | $\mathbb{P}(A \cdot B) = \mathbb{P}(A)\mathbb{P}(B)$ |

### 3.6.1 Pairwise and Global Independence

If three events $A, B$, and $C$ are such that any pair of them is exclusive, i.e., $AB = \emptyset$, $AC = \emptyset$, or $BC = \emptyset$, then the events are mutually exclusive, $ABC = \emptyset$. However, an analogous result does not hold for independence. Even if the events are *pairwise* independent for all three pairs $A, B$; $A, C$; and $B, C$, i.e., $\mathbb{P}(AB) = \mathbb{P}(A)\mathbb{P}(B)$, $\mathbb{P}(AC) = \mathbb{P}(A)\mathbb{P}(C)$, and $\mathbb{P}(BC) = \mathbb{P}(B)\mathbb{P}(C)$, they may not be independent in their totality. That is, it could happen that $\mathbb{P}(ABC) \neq \mathbb{P}(A)\mathbb{P}(B)\mathbb{P}(C)$.

Here is one example of such a triple.

*Example 3.11.* The four sides of a tetrahedron (regular three-sided pyramid with four sides consisting of isosceles triangles) are denoted by 2, 3, 5, and 30, respectively. If the tetrahedron is "rolled," the number on the bottom side is the outcome of interest. The three events are defined as follows: $A$ – the number on the bottom side is even, $B$ – the number is divisible by 3, and $C$ – the number is divisible by 5. The events are pairwise independent, but in totality, they are dependent.

The algebra is simple here, but what is the intuition? The "trick" is that events $AB$, $AC$, $BC$, and $ABC$ all coincide. In other words, $\mathbb{P}(A|BC) = 1$ even though $\mathbb{P}(A|B) = \mathbb{P}(A|C) = \mathbb{P}(A)$.

The concept of independence/dependence is not transitive. At first glance, it may seem incorrect. One may argue, "If $A$ depends on $B$, and $B$ depends on $C$, then $A$ should depend on $C$, right?" We can demonstrate that this reasoning is not correct with a simple example.

*Example 3.12.* Take a standard deck of 52 playing cards and replace the $Q\clubsuit$ with $Q\diamondsuit$. The deck still has 52 cards, two $Q\diamondsuit$ and no $Q\clubsuit$. From that deck draw a card at random and consider three events: $A$ – the card is a queen, $B$ – the card is red, and $C$ – the card is a $\heartsuit$. It is easy to see that $A$ and $B$ are dependent since $\mathbb{P}(AB) = 3/52$ does not equal $\mathbb{P}(A){\cdot}\mathbb{P}(B) = 4/52{\cdot}27/52$. Events $B$ and $C$ are dependent as well since event $C$ is contained in $B$, and $\mathbb{P}(BC) = \mathbb{P}(C)$ does not equal $\mathbb{P}(B){\cdot}\mathbb{P}(C)$. However, events $A$ and $C$ are independent since $\mathbb{P}(AC) = \mathbb{P}(Q\heartsuit) = \frac{1}{52} = \mathbb{P}(A)\mathbb{P}(C) = \frac{13}{52}\cdot\frac{4}{52}$.

## 3.7 Total Probability

The rule of total probability expresses the probability of an event $A$ as the weighted average of its conditional probabilities. The events that $A$ is conditioned upon need to be exclusive and should partition the sample space $\mathscr{S}$. Here are the definitions.

> Events $H_1, H_2, \ldots, H_n$ form a partition of the sample space $\mathscr{S}$ if
> (i) they are mutually exclusive ($H_i \cdot H_j = \emptyset$, $i \neq j$) and
> (ii) their union is the sample space $\mathscr{S}$, $\bigcup_{i=1}^{n} H_i = \mathscr{S}$.

The events $H_1, \ldots, H_n$ are usually called *hypotheses*. By this definition it follows that $\mathbb{P}(H_1) + \cdots + \mathbb{P}(H_n) = 1 \; (= \mathbb{P}(\mathscr{S}))$.

Let the event of interest $A$ happen under any of the hypotheses $H_i$ with a known (conditional) probability $\mathbb{P}(A|H_i)$. Assume, in addition, that the probabilities of hypotheses $H_1,\ldots,H_n$ are known. $\mathbb{P}(A)$ can then be calculated using the *rule of total probability*.

**Rule of Total Probability.**

$$\mathbb{P}(A) = \mathbb{P}(A|H_1)\mathbb{P}(H_1) + \cdots + \mathbb{P}(A|H_n)\mathbb{P}(H_n) \qquad (3.8)$$

Thus, the probability of $A$ is a weighted average of the conditional probabilities $\mathbb{P}(A|H_i)$ with weights given by $\mathbb{P}(H_i)$. Since $H_i$s partition the sample space, the sum of the weights is 1.

The proof is simple. From $\mathscr{S} = H_1 \cup H_2 \cup \cdots \cup H_n$ it follows that

$$A = A\mathscr{S} = A(H_1 \cup H_2 \cup \cdots \cup H_n) = AH_1 \cup AH_2 \cup \cdots \cup AH_n$$

(Fig. 3.12). Events $AH_i$ are all exclusive and by applying the additivity property,

$$\mathbb{P}(A) = \mathbb{P}(AH_1) + \mathbb{P}(AH_2) + \cdots + \mathbb{P}(AH_n).$$

Since each $\mathbb{P}(AH_i)$ is equal to $\mathbb{P}(A|H_i)\mathbb{P}(H_i)$ by the multiplication rule (3.6), the equality in (3.8) is true.

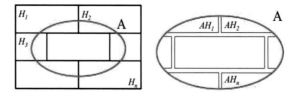

**Fig. 3.12** $A = A(H_1 \cup H_2 \cup \cdots \cup H_n) = AH_1 \cup AH_2 \cup \cdots \cup AH_n$, and the events $AH_i$ are exclusive.

*Example 3.13.* **Two-Headed Coin.** Out of 100 coins in a box, one has heads on both sides. The rest are standard fair coins. A coin is chosen at random from the box. Without inspecting whether it is fair or two-headed, the coin is flipped twice. What is the probability of getting two heads?

Let $A$ be the event that both flips resulted in heads. Let $H_1$ denote the event (hypothesis) that a fair coin was chosen. Then, $H_2 = H_1^c$ denotes the hypothesis that the two-headed coin was chosen.

$$\mathbb{P}(A) = \mathbb{P}(A|H_1)\mathbb{P}(H_1) + \mathbb{P}(A|H_2)\mathbb{P}(H_2)$$
$$= 1/4 \cdot 99/100 + 1 \cdot 1/100 = 103/400 = 0.2575.$$

The probability of the two flips resulting in tails is 0.2475 (check this!), which is slightly smaller than 0.2575. Is this an influence of the two-headed coin?

The next example is an interesting interplay between conditional and unconditional independence solved by the rule of total probability.

*Example 3.14.* **Accident Proneness.** Imagine a population with two types of individuals: $N$ normal, and $N^c$ accident prone. Suppose that 5/6 of these people are normal, so that if we randomly select a person from this population the probability that the chosen person will be normal is $\mathbb{P}(N) = 5/6$. Let $A_i$ be the event that an individual has an accident in year $i$. For each individual $A_i$ is independent of $A_j$ whenever $i \neq j$.

The accident probability is different for the two classes of individuals, $\mathbb{P}(A_i|N) = 0.01$ and $\mathbb{P}(A_i|N^c) = 0.1$. The chance of a randomly chosen individual having an accident in a given year is

$$\mathbb{P}(A_i) = \mathbb{P}(A_i|N)\mathbb{P}(N) + \mathbb{P}(A_i|N^c)\mathbb{P}(N^c) = 0.01 \times 5/6 + 0.1 \times 1/6 = 0.025.$$

The probability that a randomly chosen individual has an accident in both the first and second year follows from the rule of total probability and the fact that $A_1$ and $A_2$ are independent for a given individual

$$\begin{aligned}\mathbb{P}(A_1 \cap A_2) &= \mathbb{P}(A_1 \cap A_2|N)\mathbb{P}(N) + \mathbb{P}(A_1 \cap A_2|N^c)\mathbb{P}(N^c) \\ &= \mathbb{P}(A_1|N)\mathbb{P}(A_2|N)\mathbb{P}(N) + \mathbb{P}(A_1|N^c)\mathbb{P}(A_2|N^c)\mathbb{P}(N^c) \\ &= 0.01 \times 0.01 \times 5/6 + 0.1 \times 0.1 \times 1/6 = 0.00175.\end{aligned}$$

Note that

$$\mathbb{P}(A_2|A_1) = \mathbb{P}(A_1 \cap A_2)\mathbb{P}(A_2) = 0.00175/0.025 = 0.07 \neq 0.025 = \mathbb{P}(A_2).$$

Therefore $A_1$ and $A_2$ are not (unconditionally) independent!

## 3.8 Bayes' Rule

Bayes' rule is named after Thomas Bayes, a nonconformist priest from the eighteenth century who was among the first to use conditional probabilities. He first introduced "inverse" probabilities, which are the special case of what is now called *Bayes' rule* (Bayes, 1763). The general form was first used by Laplace (1774). Recall that the multiplication rule states

$$\mathbb{P}(AH) = \mathbb{P}(A)\mathbb{P}(H|A) = \mathbb{P}(H)\mathbb{P}(A|H).$$

This simple identity in association with the rule of total probability is the essence of Bayes' rule.

---

**Bayes' Rule.** Let the event of interest $A$ happen under any of the hypotheses $H_i$ with a known (conditional) probability $\mathbb{P}(A|H_i)$. Assume, in addition, that the probabilities of hypotheses $H_1,\ldots,H_n$ are known (*prior* probabilities). Then the conditional (*posterior*) probability of the hypothesis $H_i$, $i = 1,2,\ldots,n$, given that event $A$ happened, is

$$\mathbb{P}(H_i|A) = \frac{\mathbb{P}(A|H_i)\mathbb{P}(H_i)}{\mathbb{P}(A)},$$

where

$$\mathbb{P}(A) = \mathbb{P}(A|H_1)\mathbb{P}(H_1) + \cdots + \mathbb{P}(A|H_n)\mathbb{P}(H_n).$$

---

The proof is simple:

$$\mathbb{P}(H_i|A) = \frac{\mathbb{P}(AH_i)}{\mathbb{P}(A)} = \frac{\mathbb{P}(A|H_i)\mathbb{P}(H_i)}{\mathbb{P}(A)},$$

where $\mathbb{P}(A)$ is given by the rule of total probability.

Although Bayes' rule is a simple formula for finding conditional probabilities, it is a precursor for a coherent "statistical learning" that will be discussed in the following chapters. It concerns the transition from prior probabilities of hypotheses to the posterior probabilities once new information about the sample space is obtained.

---

$$\mathbb{P}(H) \quad \overset{\text{BAYES' RULE}}{\longrightarrow} \quad \mathbb{P}(H|A)$$

---

*Example 3.15.* **Many Flips of a Possibly Two-Headed Coin.** Assume that out of $N$ coins in a box, one has heads on both sides, and the remaining $N-1$ are fair. Assume that a coin is selected at random from the box and, without inspecting what kind of coin it was, flipped $k$ times. Every time the coin lands heads up. What is the probability that the two-headed coin was selected?

Let $A_k$ denote the event where a randomly selected coin lands heads up $k$ times. The hypotheses are $H_1$ – the coin is two-headed, and $H_2$ – the coin is fair. It is easy to see that $\mathbb{P}(H_1) = 1/N$ and $\mathbb{P}(H_2) = (N-1)/N$. The conditional probabilities are $\mathbb{P}(A_k|H_1) = 1$ for any $k$, and $\mathbb{P}(A_k|H_2) = 1/2^k$.

By the total probability rule,

$$\mathbb{P}(A_k) = \frac{2^k + N - 1}{2^k N},$$

and by Bayes' rule,

$$\mathbb{P}(H_1|A_k) = \frac{2^k}{2^k + N - 1}.$$

For $N = 1{,}000{,}000$ and $k = 1, 2, \ldots, 40$ the graph of posterior probabilities is given in Fig. 3.13.

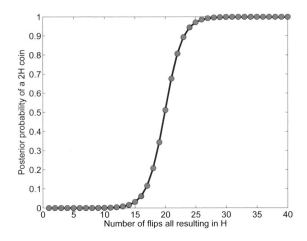

**Fig. 3.13** Posterior probability of a two-headed coin for $N = 1{,}000{,}000$ if in $k$ flips $k$ heads appeared. The *red dots* show the posterior probabilities for $k = 16$ and $k = 24$, equal to 0.0615 and 0.9437, respectively.

Note that our prior probability $\mathbb{P}(H_1) = 0.000001$ jumps to a posterior probability of 0.9991 after observing 30 heads in a row. The code ◀ twoheaded.m calculates the probabilities and plots the graph in Fig. 3.13. It is curious to observe how nonlinear the change of posterior probability is. This probability is quite stable for $k$ up to 15 and after 25. The most rapid change is in the range $16 \le k \le 24$, where it increases from 0.0615 to 0.9437. This illustrates the "learning" ability of Bayes' rule.

*Example 3.16.* **Prosecutor's Fallacy.** The prosecutor's fallacy is a fallacy commonly occurring in criminal trials but also in other various arguments involving rare events. It consists of a subtle exchange of $\mathbb{P}(A|B)$ for $\mathbb{P}(B|A)$. We will explain it in the context of Example 3.15. Assume that out of $N = 1{,}000{,}000$ coins in a box, one is two-headed and "guilty." Assume that a coin is selected at random from the box and, without inspection, flipped $k = 15$ times. All $k = 15$ times the coin lands heads up. Based on this evidence, the "prosecutor" claims the selected coin is guilty since if it were "innocent," the observed $k = 15$ heads in a row would be extremely unlikely, with a probability of $\left(\frac{1}{2}\right)^{15} \approx 0.00003$. But in reality, the probability that the "guilty" coin was selected and flipped is $\frac{1}{1+999999/2^{15}} \approx 0.03$ and the prosecutor is accusing an "innocent" coin with a probability of approx. 0.97.

Bayes' rule is even more revealing if expressed in terms of odds.

> The posterior odds of the hypothesis $H_i$ are equal to the product of its prior odds and Bayes' factor (likelihood ratio),
>
> $$\mathbb{O}dds(H_i|A) \; = \; \text{BF} \times \mathbb{O}dds(H_i),$$
>
> where BF $= \mathbb{P}(A|H_i)/\mathbb{P}(A|H_i^c)$.

Thus, the "updater" is Bayes' factor BF, which represents the ratio of probabilities of the evidence (event $A$) under $H_i$ and $H_i^c$. All available information from the experiment is contained in Bayes' factor, and Bayesian "learning" incorporates this information in a coherent way by transforming the prior odds to the posterior odds.

It is interesting to look at the log-odds equation

$$\log \mathbb{O}dds(H_i|A) = \log \text{BF} + \log \mathbb{O}dds(H_i).$$

If the prior log-odds $\log \mathbb{O}dds(H_i)$ increase/decrease for a constant $C$, then the posterior log-odds increase/decrease for the same constant, no matter what the Bayes' factor is. Likewise, if the log-Bayes factor increases/decreases for a constant $C$, then the log posterior odds increase/decrease for the same constant, no matter what the log prior odds are. This additivity property was used for constructing nomograms for fast approximate calculation of odds mainly in a medical context.

The log BF was also termed *weight of evidence* by Alan Turing, who used similar techniques during the Second World War when breaking German "Enigma Machine" codes.

*Example 3.17.* **A Bridge Connection.** Figure 3.14 shows a circuit $S$ that consists of components $e_i$, $i = 1,\dots,7$ which work (and fail) independently of each other. Note that the connection of component $e_7$ is neither parallel nor serial. The components are operational in some time interval $T$ with probabilities given in the following table.

| Component | $e_1$ | $e_2$ | $e_3$ | $e_4$ | $e_5$ | $e_6$ | $e_7$ |
|---|---|---|---|---|---|---|---|
| Probability of component working | 0.3 | 0.8 | 0.2 | 0.2 | 0.5 | 0.6 | 0.4 |

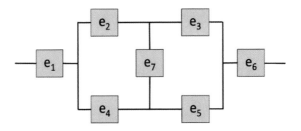

**Fig. 3.14** "Bridge" connection of $e_7$.

We will calculate the posterior odds of $e_7$ working, given the information that circuit $S$ is operational.

Assume two hypotheses, $H_1$ – the component $e_7$ is operational, and $H_2 = H_1^c$ – $e_7$ is not operational. Under hypothesis $H_1$, circuit $S$ can be expressed as

$$S|H_1 = e_1 \cap (e_2 \cup e_4) \cap (e_3 \cup e_5) \cap e_6,$$

while under $H_1^c$ the expression is

$$S|H_1^c = e_1 \cap ((e_2 \cap e_3) \cup (e_4 \cap e_5)) \cap e_6.$$

By calculations similar to that in Example 3.5 we find that $\mathbb{P}(S|H_1) = 0.09072$ and $\mathbb{P}(S|H_1^c) = 0.04392$.

Note that $\mathbb{P}(H_1) = \mathbb{P}(e_7 \text{ works}) = 0.4$ and $\mathbb{P}(H_2) = \mathbb{P}(H_1^c) = 0.6$, so by the total probability rule, $\mathbb{P}(S) = 0.06264$. The prior odds of $H_1$ are $\mathbb{O}dds(H_1) = 0.4/0.6 = 2/3$, and Bayes' factor is $BF = \mathbb{P}(S|H_1)/ \mathbb{P}(S|H_1^c) = 2.06557$. The posterior odds of $H_1$ are $\mathbb{O}dds(H_1|S) = BF \times \mathbb{O}dds(H_1) = 2.06557 \times 2/3 = 1.37705$.

Thus, the odds of $e_7$ working increased from 0.66667 to 1.30705, after learning that the circuit is operational. See also ◀ bridge.m.

*Example 3.18.* **Subsequent Transfers.** In each of $n$ boxes there are $a$ white and $b$ black balls. A ball is selected at random from the first box and placed into the second box. Then, from the second box another ball is selected at random and transferred to the third box, and so on. Finally, from the $(n-1)$th box a ball is selected at random and transferred to the $n$th box.

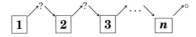

(a) After this series of consecutive transfers, a ball is selected from the $n$th box. What is the probability that this ball will be white?

(b) If the ball drawn from the fourth box was white, what is the probability that the first ball transferred had been white as well?

Let $A_i$ denote the event that in the $i$th transfer the white ball was selected. Then

$$\mathbb{P}(A_n) = \mathbb{P}(A_n|A_{n-1})\mathbb{P}(A_{n-1}) + \mathbb{P}(A_n|A_{n-1}^c)\mathbb{P}(A_{n-1}^c)$$
$$= \frac{a+1}{a+b+1}\mathbb{P}(A_{n-1}) + \frac{a}{a+b+1}\mathbb{P}(A_{n-1}^c)$$
$$= \frac{a+\mathbb{P}(A_{n-1})}{a+b+1}.$$

Since $\mathbb{P}(A_1) = \frac{a}{a+b}$, we find that

$$\mathbb{P}(A_2) = \mathbb{P}(A_3) = \cdots = \mathbb{P}(A_n) = \frac{a}{a+b}.$$

(b) By Bayes' rule, $\mathbb{P}(A_1|A_4) = \frac{\mathbb{P}(A_4|A_1)\mathbb{P}(A_1)}{\mathbb{P}(A_4)} = \mathbb{P}(A_4|A_1)$.
Since

$$\mathbb{P}(A_4|A_1) = \mathbb{P}(A_2A_3A_4|A_1) + \mathbb{P}(A_2^cA_3A_4|A_1) + \mathbb{P}(A_2A_3^cA_4|A_1) + \mathbb{P}(A_2^cA_3^cA_4|A_1)$$
$$= \frac{a+1}{a+b+1} \times \frac{a+1}{a+b+1} \times \frac{a+1}{a+b+1} + \frac{b}{a+b+1} \times \frac{a}{a+b+1} \times \frac{a+1}{a+b+1}$$
$$+ \frac{a+1}{a+b+1} \times \frac{b}{a+b+1} \times \frac{a}{a+b+1} + \frac{b}{a+b+1} \times \frac{b}{a+b+1} \times \frac{a}{a+b+1}$$
$$= \frac{(a+1)^3 + 2ab(a+1) + ab^2}{(a+b+1)^3}.$$

## 3.9 Bayesian Networks*

We will discuss simple Bayesian networks in which the nodes are events. Many events linked in a causal network form a Bayesian net. Graphically, Bayesian networks are directed acyclic graphs (DAGs) where the nodes represent events and directed edges capture their hierarchy and dependence. Consider a simple graph in Fig. 3.15.

**Fig. 3.15** $A \longrightarrow B$ graph. $A$ causes $B$ or $B$ is a consequence of $A$.

We would say that node $A$ is a parent of $B$, $B$ is a child of $A$, that $A$ influences, or causes $B$, and $B$ depends on $A$. This is captured by a directed edge (arrow) that leads from $A$ to $B$. The term *acyclic* in DAG relates to the fact that a closed loop of dependencies is not allowed. In other words, there does not exist a path consisting of nodes $A_1, \ldots, A_n$ such that

$$\boxed{A} \longrightarrow \boxed{A_1} \longrightarrow \cdots \longrightarrow \boxed{A_n} \longrightarrow \boxed{A}.$$

The independence of two nodes in a DAG depends on their relative position in the graph as well as on the knowledge of other nodes (conditioning) in the graph. The following simple example illustrates the influence of conditioning on independence.

*Example 3.19.* Let $A$ and $B$ be outcomes of flips of two fair coins and $C$ be an event that the two outcomes coincide. Thus, $\mathbb{P}(A = H) = \mathbb{P}(A = T) = 0.5$ and $\mathbb{P}(B = H) = \mathbb{P}(B = T) = 0.5$.

$$\boxed{A} \longrightarrow \boxed{C} \longleftarrow \boxed{B}$$

We are interested in $P(C|A,B)$. Nodes $A$ and $B$ are marginally independent (when we do not have evidence about $C$) but become dependent if the outcome of $C$ is known:

$$\frac{1}{2} = \mathbb{P}(A = T, B = T | C) \neq \mathbb{P}(A = T | C)\mathbb{P}(B = T | C) = \frac{1}{2} \cdot \frac{1}{2}.$$

*Hard evidence* for a node $A$ is evidence that the outcome of $A$ is known. Hard evidence about nodes is the information that we bring to the network, and it affects the probabilities of other nodes.

Bayesian networks possess a so-called Markov property. The conditional distribution of any node depends on its parental nodes. For instance, in the network

$$\boxed{A} \longrightarrow \boxed{B} \longrightarrow \boxed{C} \longrightarrow \boxed{D}$$

$\mathbb{P}(C|A,B) = \mathbb{P}(C|B)$ since $B$ is a parental node of $C$.

*Example 3.20.* **Alarm.** Your house has a security alarm system. The house is located in a seismically active area and the alarm system can be occasionally set off by an earthquake. You have two neighbors, Mary and John, who do not know each other. If they hear the alarm, they call you, but this is not guaranteed. They also call you from time to time just to chat.

Denote by $E, B, A, J,$ and $M$ the events earthquake, burglary, alarm, John's call, and Mary's call took place, and by $E^c, B^c, A^c, J^c,$ and $M^c$ the opposite events. The DAG of the network is shown in Fig. 3.16.

The known (or elicited) conditional probabilities are as follows:

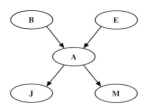

**Fig. 3.16** Alarm Bayesian network.

| $A^c$ | $A$ | Condition |
|---|---|---|
| 0.999 | 0.001 | $B^c\ E^c$ |
| 0.71 | 0.29 | $B^c\ E$ |
| 0.06 | 0.94 | $B\ E^c$ |
| 0.05 | 0.95 | $B\ E$ |

| $B^c$ | $B$ |
|---|---|
| 0.999 | 0.001 |

| $E^c$ | $E$ |
|---|---|
| 0.998 | 0.002 |

| $J^c$ | $J$ | condition |
|---|---|---|
| 0.95 | 0.05 | $A^c$ |
| 0.10 | 0.90 | $A$ |

| $M^c$ | $M$ | condition |
|---|---|---|
| 0.99 | 0.01 | $A^c$ |
| 0.30 | 0.70 | $A$ |

We are interested in $\mathbb{P}(J,M|B)$, i.e., the probability that both John and Mary call, given the burglary.

We will first calculate this probability exactly and then find an approximation using WinBUGS. If $E^*$ is either $E$ or $E^c$ and $A^*$ either $A$ or $A^c$, we have

$$\mathbb{P}(J,M|B) = \frac{1}{\mathbb{P}(B)} \times \mathbb{P}(B,J,M) = \frac{1}{\mathbb{P}(B)} \sum_{E^*,A^*} \mathbb{P}(B,E^*,A^*,J,M)$$

$$= \frac{1}{\mathbb{P}(B)}\{\mathbb{P}(B,E)\mathbb{P}(A|B,E)\mathbb{P}(J,M|A)$$

$$+\mathbb{P}(B,E^c)\mathbb{P}(A|B,E^c)\mathbb{P}(J,M|A)$$

$$+\mathbb{P}(B,E^c)\mathbb{P}(A^c|B,E^c)\mathbb{P}(J,M|A^c)$$

$$+\mathbb{P}(B,E)\mathbb{P}(A^c|B,E)\mathbb{P}(J,M|A^c)\}.$$

Given $A^*$ (either $A^c$ or $A$), $\mathbb{P}(J,M|A^*) = \mathbb{P}(J|A^*) \times \mathbb{P}(M|A^*)$, since John and Mary do not know each other, and their calls can be considered independent. After substituting the probabilities with their numerical values from the tables above, we obtain

$$\mathbb{P}(J,M|B) = \frac{1}{0.001}\,(0.001\cdot 0.002\cdot 0.95\cdot 0.90\cdot 0.70 + 0.001\cdot 0.998\cdot 0.94\cdot 0.90\cdot 0.70$$

$$+\ 0.001\cdot 0.998\cdot 0.06\cdot 0.05\cdot 0.01 + 0.001\cdot 0.002\cdot 0.05\cdot 0.05\cdot 0.01)$$

$$= 0.5922.$$

Thus, in the case of burglary, both John and Mary will call with probability of 0.5922.

This probability can be approximated in WinBUGS by simulation. Note that $\mathbb{P}(J,M|B) = \mathbb{P}(J|M,B)\mathbb{P}(M|B)$ by the chain rule. The probabilities $\mathbb{P}(J|M,B)$ and $\mathbb{P}(M|B)$ will be approximated separately.

First, we approximate $\mathbb{P}(M|B)$ by fixing hard evidence for a burglary. WinBUGS will use the code burglary =1 and burglary = 2 in the Data part to set the evidence that the burglary did not take place or that it took place, respectively. This is the only "hard evidence" here; all other nodes remain stochastic. The use of values 1,2 instead of the expected 0,1 is dictated by the categorical distribution dcat that takes only positive integers as realizations.

The WinBUGS code ( alarm.odc) is as follows:

```
model alarm
    {
    burglary ~ dcat(p.burglary[]);
    earthquake ~ dcat(p.earthquake[]);
    alarm ~ dcat(p.alarm[burglary, earthquake, ])
    john ~ dcat(p.john[alarm,]);
    mary ~ dcat(p.mary[alarm,]);
    }

DATA
list(
    p.earthquake=c(0.998, 0.002),
    p.alarm = structure(.Data = c(0.999, 0.001,
                                  0.71,0.29,
                                  0.06,0.94,
                                  0.05,0.95),
                         .Dim = c(2,2,2)),
p.john = structure(.Data = c(0.95,0.05,0.10,0.90),
                         .Dim = c(2,2)),
p.mary = structure(.Data = c(0.99,0.01,0.30,0.70),
                         .Dim = c(2,2)),
burglary = 2
)

INITS
list( earthquake = 1, alarm = 1, john = 1, mary = 1)
```

After 10,000 iterations, we obtain the mean value of $M$ as $\mathbb{E}M = 2 \cdot p_M + 1 \cdot (1 - p_M) = 1.661$, that is, $\mathbb{P}(M|B) = p_M = 1.661 - 1 = 0.661$. Any of the 1s in the initial values earthquake = 1, alarm = 1, john = 1, mary = 1 can be replaced by 2, as this would not influence the final approximation.

Next, to estimate $\mathbb{P}(J|M,B)$, we change WinBUGS' data by setting $B = M = 2$. This is hard evidence that a burglary occurred and Mary called.

```
DATA
list(
    p.earthquake=c(0.998, 0.002),
```

```
   p.alarm = structure(.Data = c(0.999, 0.001,
                                 0.71,0.29,
                                 0.06,0.94,
                                 0.05,0.95), .Dim = c(2,2,2)),
p.mary = structure(.Data = c(0.99,0.01,0.30,0.70), .Dim = c(2,2)),
burglary = 2,
mary=2
)
```

and change the initial values to

```
list(earthquake = 1, alarm = 1, john = 1)
```

After 10,000 iterations, the mean value of $J$ is obtained as $\mathbb{E}J = 2 \cdot p_J + (1_p J) = 1.899$. Then, $\mathbb{P}(J|M,B) = p_J = 0.899$. Thus, the final result is

$$\mathbb{P}(J,M|B) = \mathbb{P}(J|M,B)\mathbb{P}(M|B) = p_J \cdot p_M = 0.899 \cdot 0.661 = 0.5942$$

(which approximates 0.5922, from the exact probability calculation).

Bayesian networks can be useful in medical diagnostics if the conditional probabilities of the nodes are known. Here is the celebrated "Asia" example.

*Example 3.21.* **Asia.** Lauritzen and Spiegelhalter (1988) discuss a fictitious *expert system* for diagnosing a patient admitted to a chest clinic, who just returned from a trip to Asia and is experiencing dyspnoea.[2] A graphical model for the underlying process is shown in Fig. 3.17, where each variable is binary. The WinBUGS code is shown below with the conditional probabilities given as in Lauritzen and Spiegelhalter (1988).

```
model Asia;
    asia          ~ dcat(p.asia);
    smoking       ~ dcat(p.smoking[]);
    tuberculosis ~ dcat(p.tuberculosis[asia,]);
    lung.cancer  ~ dcat(p.lung.cancer[smoking,]);
    bronchitis   ~ dcat(p.bronchitis[smoking,]);
    either       <- max(tuberculosis,lung.cancer);
    xray          ~ dcat(p.xray[either,]);
    dyspnoea      ~ dcat(p.dyspnoea[either,bronchitis,])


DATA
list(asia = 2, dyspnoea = 2,
    p.tuberculosis = structure(.Data = c(0.99,0.01,0.95,0.05),
                               .Dim = c(2,2)),
    p.bronchitis = structure(.Data = c(0.70,0.30,0.40,0.60),
                             .Dim = c(2,2)),
    p.smoking = c(0.50,0.50),
```

---

[2] Difficulty in breathing, often associated with lung or heart disease and resulting in shortness of breath. Also called air hunger.

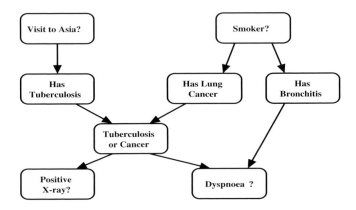

**Fig. 3.17** Lauritzen and Spiegelhalter's (1988) Asia Bayes net: a fictitious *expert system* representing the diagnosis of a patient having just returned from a trip to Asia and showing *dyspnoea*.

```
p.lung.cancer = structure(.Data = c(0.99,0.01,0.90,0.10),
                          .Dim = c(2,2)),
p.xray = structure(.Data = c(0.95,0.05,0.02,0.98),
                   .Dim = c(2,2)),
p.dyspnoea = structure(.Data = c(0.9,0.1,
                       0.2,0.8,
                       0.3,0.7,
                       0.1,0.9), .Dim = c(2,2,2)))

INITS
list(smoking = 1, tuberculosis = 1,
        lung.cancer = 1, bronchitis = 1, xray = 1)
```

| | mean | sd | MC error | val2.5pc | median | val97.5pc | start | sample |
|---|---|---|---|---|---|---|---|---|
| bronchitis | 1.812 | 0.3904 | 0.003988 | 1.0 | 2.0 | 2.0 | 2001 | 10000 |
| lung.cancer | 1.099 | 0.2985 | 0.003345 | 1.0 | 1.0 | 2.0 | 2001 | 10000 |
| smoking | 1.618 | 0.4859 | 0.004976 | 1.0 | 2.0 | 2.0 | 2001 | 10000 |
| tuberculosis | 1.095 | 0.2928 | 0.002706 | 1.0 | 1.0 | 2.0 | 2001 | 10000 |
| xray | 1.224 | 0.4171 | 0.004132 | 1.0 | 1.0 | 2.0 | 2001 | 10000 |

The results should be interpreted as follows. If the patient who visited Asia experiences dyspnoea (hard evidence in DATA: asia=2, dyspnoea = 2), then the probabilities of bronchitis, lung cancer, being a smoker, tuberculosis, and a positive xray, are 0.812, 0.099, 0.618, 0.095, and 0.224, respectively.

## 3.10 Exercises

3.1. **Event Differences.** Recall that the difference between events $A$ and $B$ was defined as $A \backslash B = A \cap B^c$. Using Venn diagrams demonstrate that
(a) $A \backslash (A \backslash B) = A \cap B$  and  $A \backslash (B \backslash A) = A$,
(b) $A \backslash (B \backslash C) = (A \cap C) \cup (A \cap B^c)$.

3.2. **Inclusion-Exclusion Principle in MATLAB.** From the set $\{1, 2, 3, \dots, 315\}$ a number is selected at random.
(a) Using MATLAB to count favorable outcomes, find the probability that the selected number is divisible by at least one of 3, 5, or 7.
(b) Compare this probability with a naïve solution $1/3 + 1/5 + 1/7 - 1/15 - 1/21 - 1/35 + 1/105 = 0.542857$, and show that the naïve solution is correct!
(c) Is the naïve solution correct for $\{1, 2, \dots, N\}$ if $N = 316$?
(d) Is the naïve solution correct for any other $N$ from $\{289, 290, \dots 340\}$? Plot this probability for $289 \le N \le 340$. Is there any symmetry in the plot?

3.3. **A Complex Circuit.** Figure 3.18 shows a circuit $S$ that consists of identical components $e_i$, $i = 1, \dots, 13$ that work (and fail) independently of each other. Any component is operational in some fixed time interval with probability 0.8.

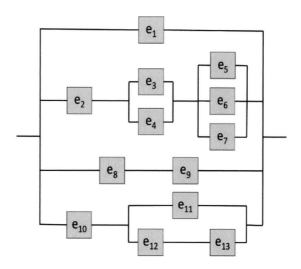

**Fig. 3.18** Each of 13 independent components in the circuit is operational with probability 0.8.

(a) Calculate the probability that circuit $S$ is operational.
(b) Write a MATLAB program that approximates the probability in (a) by simulation.

(c) Approximate the probability in (a) by WinBUGS simulations.

3.4. **De Mere Paradoxes.** In 1654 the Chevalier de Mere asked Blaise Pascal (1623–1662) the following two questions:

(a) Why would it be advantageous in a game of dice to bet on the occurrence of a 6 in 4 trials but not advantageous in a game involving two dice to bet on the occurrence of a double 6 in 24 trials?

(b) In playing a game with three dice, why is a sum of 11 more advantageous than a sum of 12 when both sums are the result of six configurations:
**11**: (1, 4, 6), (1, 5, 5), (2, 3, 6), (2, 4, 5), (3, 3, 5), (3, 4, 4);
**12**: (1, 5, 6), (2, 4, 6), (2, 5, 5), (3, 3, 6), (3, 4, 5), (4, 4, 4)?
How would you respond to the Chevalier?

3.5. **Probabilities of Some Composite Events.** Show that for arbitrary events $A, B$,
(a) $\mathbb{P}(A \Delta B) = \mathbb{P}(A \cup B) - \mathbb{P}(AB) = \mathbb{P}(A) + \mathbb{P}(B) - 2\mathbb{P}(AB)$;
(b) $\mathbb{P}(A \Delta B) \geq |\mathbb{P}(A) - \mathbb{P}(B)|$.
(c) For arbitrary event $C$, $\mathbb{P}(AC \Delta BC) \leq \mathbb{P}(A \Delta B)$ and
(d) $(\mathbb{P}(A) + \mathbb{P}(B)) \frac{1}{1 + 2\mathbb{P}(AB)/(\mathbb{P}(A) + \mathbb{P}(B))} \leq \mathbb{P}(A \cup B) \leq \mathbb{P}(A) + \mathbb{P}(B)$.

3.6. **Deighton's Novel.** In his World War II historical novel *Bomber* Len Dieghton argues that a pilot is "mathematically certain" to be shot down in 50 missions if the probability of being shot down on each mission is 0.02.
(a) Assuming independence of outcomes in each mission, is Deighton's reasoning correct?
(b) Find the probability of surviving all 50 missions without being shot down?

3.7. **Reliable System from Unreliable Components.** NASA is asking you to design a system that reliably performs a task on a space shuttle in the next 3 years with probability of $0.999999 = 1 - 10^{-6}$. In other words, the probability of failing during the next three years should not exceed one in a million. However, at your disposal you have components that in the next 3 years will fail with a probability of 0.2. Luckily, the weight and price of the components are not an issue and you can combine/link them to increase the system's reliability.
(a) Should you link the components in a serial or parallel fashion to increase the probability of reliable performance?
(b) What minimal number of components should be linked as in (a) to satisfy NASA's requirement of 0.999999 probability of reliable performance?

3.8. $k$-**out-of-**$n$ **Systems.** Suppose that $n$ independent components constitute an engineering system. The system is called a $k$-out-of-$n$ system if it works only when $k$ or more components are operational. This particular system has four components that are operational with probabilities $0.1, 0.8, 0.5$, and $0.4$. If the system is a 2-out-of-4, what is the probability that it works?

3.9. **Number of Dominos.** How many different dominos are a sample space if the number of dots on the dominos ranges between (a) 0 and 3, (b) 0 and 8, and (c) 0 and 16, inclusive?

3.10. **Counting Protocols.** Adel et al. (1993) applied various orders in drug combination sequence studies in search of a cure for human endometrial carcinoma. Four drugs A–D were evaluated for sequence-dependent inhibition of human tumor colony formation in soft agar.
(a) How many protocols are needed to evaluate all possible sequences of the four drugs?
(b) How many protocols are possible when only two drugs out of four are to be administered if the order of their administration is (i) important or (ii) not important?
(c) A fifth drug, E, is introduced. If drugs A and E cannot be given subsequent to each other because of cumulative toxicity concerns, how many protocols are possible if the order of drug administration is to be evaluated?

3.11. **Correlation Between Events.** The correlation between events $A$ and $B$ is defined as

$$\mathrm{Corr}(A,B) = \frac{\mathbb{P}(A \cap B) - \mathbb{P}(A)\mathbb{P}(B)}{\sqrt{\mathbb{P}(A)(1 - \mathbb{P}(A))}\,\sqrt{\mathbb{P}(B)(1 - \mathbb{P}(B))}}.$$

Show that $\mathrm{Corr}(A,A) = 1$ and $\mathrm{Corr}(A,B) = \mathrm{Corr}(A^c, B^c)$.

3.12. **A Fair Gamble with a Possibly Loaded Coin.** Suppose you have a coin for which you do not know the probability of its landing heads up. You suspect that the coin is loaded and that the probability of heads differs from 1/2.
(a) Can you emulate a fair coin by flipping the possibly biased one?
(b) Can you emulate the rolling of a fair die by flipping the possibly biased coin?
*Hint:* You may need to flip the coin more than once.

3.13. **Neural Signal.** A neuron will fire at random at any moment in $[0, T]$, with a probability of $p$. If up to time $t < T$ the neuron does not fire, what is the probability that it will fire in the remaining time, $(t, T]$?

3.14. **Guessing.** Subjects in an experiment are told that either a red or a green light will flash. Each subject is to guess which light will flash. The subject is told that the probability of a red light is 0.7, independently of guesses. Assume that the subject is a probability matcher, that is, guesses red with a probability of 0.7 and green with a probability of 0.3.
(a) What is the probability that the subject will guess correctly?
(b) Given that a subject guesses correctly, what is the probability that the light flashed red?

3.15. **Propagation of Genes.** The following example shows how the ideas of independence and conditional probability can be employed in studying genetic evolution. Consider a single gene that has two forms, *recessive (R)* and *dominant (D)*. Each individual in the population has two genes in his/her chromosomes and thus can be classified into the genotypes $DD$, $RD$, and $RR$. If an individual is drawn at random from the $n$th generation, then the probabilities of the three genotypes will be denoted by $p_n$, $2r_n$, and $q_n$, respectively. (Clearly, $p_n + q_n + 2\, r_n = 1$.)

The problem is expressing the probabilities $p_n$, $q_n$, and $r_n$ in terms of initial probabilities $p_0$, $q_0$, and $r_0$ and the method of reproduction. In *random Mendelian mating*, a single gene from each parent is selected at random and the selected pair determines the genotype of the offspring. These selections are carried independently of each other from generation to generation. Let $M_n$ be the event that $R$ is chosen from the male and $F_n$ be the event that $R$ is chosen from the counterpart female. Events $M_n$ and $F_n$ are independent and have the same probability. Thus:

$$\mathbb{P}(M_n) = \mathbb{P}(RR) \times \mathbb{P}(M_n|RR) + \mathbb{P}(RD) \times \mathbb{P}(M_n|RD) + \mathbb{P}(DD) \times \mathbb{P}(M_n|DD)$$
$$= \mathbb{P}(RR) \times 1 + \mathbb{P}(RD) \times 1/2 + \mathbb{P}(DD) \times 0$$
$$= q_n + 2r_n/2$$
$$= q_n + r_n$$

by the rule of total probability.
By the independence of $M_n$ and $F_n$,

$$q_{n+1} = \mathbb{P}(M_n \cap F_n) = \mathbb{P}(M_n) \cdot \mathbb{P}(F_n) = (q_n + r_n)^2.$$

Similarly,
$$p_{n+1} = (p_n + r_n)^2$$

and
$$2r_{n+1} = 1 - p_{n+1} - q_{n+1}.$$

The above equations govern the propagation of genotypes in this population.

Start with any initial probabilities $p_0$, $q_0$, and $r_0$. (Say, 0.3, 0.3, and 0.2; remember to check: $0.3 + 0.3 + 2 \cdot 0.2 = 1$.) Find iteratively $(p_1, q_1, r_1)$, $(p_2, q_2, r_2)$, and $(p_3, q_3, r_3)$, and demonstrate that $p_1 = p_2 = p_3$, $q_1 = q_2 = q_3$, and $r_1 = r_2 = r_3$. The fact that the probabilities remain the same is known as the *Hardy–Weinberg law*. It does not hold if other factors (mutation, selection, dependence) are introduced into the model.

3.16. **Easy Conditioning.** Assume $\mathbb{P}(\text{rain today}) = 40\%$, $\mathbb{P}(\text{rain tomorrow}) = 50\%$, and $\mathbb{P}(\text{rain today and tomorrow}) = 30\%$. Given that it is raining today, what is the chance that it will rain tomorrow?

3.17. **Eye Color.** The eye color of a child is determined by a pair of genes, one from each parent. If {b} and {B} denote blue- and brown-eyed genes, then a child can inherit the following pairs: {bb}, {bB}, {Bb}, and {BB}. The {B} gene is dominant, that is, the child will have brown eyes when the pairs are {Bb}, {bB}, or {BB} and blue eyes only for the {bb} combination. A parent passes to a child either gene from his/her pair with equal probability.[3]
Megan's parents are both brown-eyed, but Megan has blue eyes. Megan's brown-eyed sister is pregnant and her husband has blue eyes. What is the probability that the baby will have blue eyes?

3.18. **Dice.** In rolling ten fair dice we have information that at least one ⚃ appeared. What is the probability that there were at least two ⚃?

3.19. **Inflation and Unemployment.** Businesses commonly project revenues under alternative economic scenarios. For a stylized example, inflation could be high or low and unemployment could be high or low. There are four possible scenarios, with the following assumed probabilities:

| Scenario | Inflation | Unemployment | Probability |
|----------|-----------|--------------|-------------|
| 1 | High | High | 0.16 |
| 2 | High | Low | 0.24 |
| 3 | Low | High | 0.36 |
| 4 | Low | Low | 0.24 |

(a) What is the probability of high inflation?
(b) What is the probability of high inflation if unemployment is high?
(c) Are inflation and unemployment independent?

3.20. **Multiple Choice.** A student answers a multiple choice examination question that has four possible answers. Suppose that the probability that the student knows the answer to a question is 0.80 and the probability that the student guesses is 0.20. If the student guesses, the probability of guessing the correct answer is 0.25.
(a) What is the probability that the fixed question will be answered correctly?
(b) If it is answered correctly, what is the probability that the student really knew the correct answer?

3.21. **Manufacturing Bayes.** A factory has three types of machines producing an item. The probabilities that the item is conforming if it is produced on the $i$th machine are given in the following table:

---

[3] This description is simplified, and in fact there are several genes affecting eye color and the amount of yellow and black pigments in the iris, leading to shades of colors including green and hazel.

| Type of machine | Probability of item conforming |
|:---:|:---:|
| 1 | 0.94 |
| 2 | 0.95 |
| 3 | 0.97 |

The total production is distributed among the machines as follows: 30% is done on type 1, 50% on type 2, and 20% on type 3 machines. One item is selected at random from the production.
(a) What is the probability that it is conforming?
(b) If it is conforming, what is the probability that it was produced on a type 1 machine?

3.22. **Stanley.** Stanley takes an oral exam in statistics with several other students. He needs to answer the questions from an examination card drawn at random from the set of 20 cards. There are exactly 8 favorable cards among the 20 to which Stanley knows the answers. Stanley will get a grade of $A$ if he knows the answers, that is, if he draws a favorable card. What is the probability that Stanley will get an $A$ if he draws the card standing in line (a) first, (b) second, and (c) third?

3.23. **Kokomo, Indiana.** In Kokomo, IN, 65% of the people are conservative, 20% are liberal, and 15% are independent. Records show that in a particular election, 82% of conservatives voted, 65% of liberals voted, and 50% of independents voted. If a person from the city is selected at random and it is learned that she did not vote, what is the probability that the person is liberal?

3.24. **Mysterious Transfer.** Of two bags, one contains four white balls and three black balls and the other contains three white balls and five black balls. One ball is randomly selected from the first bag and placed unseen in the second bag.
(a) What is the probability that a ball now drawn from the second bag will be black?
(b) If the second ball is black, what is the probability that a black ball was transferred?

3.25. **Two Masked Robbers.** Two masked robbers try to rob a crowded bank during the lunch hour, but the teller presses a button that sets off an alarm and locks the front door. The robbers, realizing they are trapped, throw away their masks and disappear into the chaotic crowd. Confronted with 40 people claiming they are innocent, the police give everyone a lie detector test. Suppose that guilty people are detected with a probability of 0.85 and innocent people appear to be guilty with a probability of 0.08. What is the probability that Mr. Smith was one of the robbers given that the lie detector says he is a robber?

3.26. **Information Channel.** One of the three words AAAA, BBBB, and CCCC is transmitted via an information channel. The probabilities of these words being transmitted are 0.3, 0.5, and 0.2, respectively. Each letter is transmitted and received correctly with a probability of 0.6, independently of other letters. Since the channel is not perfect, the transmitted letter can change to one of the other two letters with an equal probability of 0.2. What is the probability that the word AAAA was submitted if the word ABCA is received?

3.27. **Quality Control.** An automatic machine in a small factory produces metal parts. Most of the time (90% according to long-term records), it produces 95% good parts, while the remaining parts have to be scrapped. Other times, the machine slips into a less productive mode and only produces 70% good parts. The foreman observes the quality of parts that are produced by the machine and wants to stop and adjust the machine when she believes that the machine is not working well. Suppose that the first dozen parts produced are given by the sequence

| s | u | s | s | s | s | s | s | s | u | s | u |
|---|---|---|---|---|---|---|---|---|---|---|---|

where **s** is satisfactory and **u** is unsatisfactory. After observing this sequence, what is the probability that the machine is in its productive state? If the foreman wishes to stop the machine when the probability of "good state" is under 0.7, when should she stop it?

3.28. **Let's Make a Deal.** Monty Hall was the host of the once-popular television game show *Let's Make a Deal*. At certain times during the show, a contestant was allowed to choose one of three identical doors A, B, and C, behind only one of which was a valuable prize (a new car). After the contestant picked a door (say, door A), Monty opened another door and showed the contestant that there was no prize behind that door. (Monty knew where the prize was and always chose a door where there was no prize.) He then asked the contestant whether he wanted to stick with his choice of door or switch to the remaining unopened door. Should the contestant have switched doors? Did it matter?

3.29. **Ternary Channel.** A communication system transmits three signals, $s_1, s_2$, or $s_3$, with equal probabilities. The reception is corrupted by noise, causing the transmission to be changed according to the following table of conditional probabilities:

|  |  | Received |  |  |
|---|---|---|---|---|
|  |  | $s_1$ | $s_2$ | $s_3$ |
|  | $s_1$ | 0.75 | 0.1 | 0.15 |
| Sent | $s_2$ | 0.098 | 0.9 | 0.002 |
|  | $s_3$ | 0.02 | 0.08 | 0.9 |

The entries in this table list the probability that $s_j$ is received, given that $s_i$ is sent, for $i, j = 1, 2, 3$. For example, if $s_1$ is sent, the conditional probability of receiving $s_3$ is 0.15.

(a) Compute the probabilities that $s_1, s_2$, and $s_3$ are received.

(b) Compute the probabilities $\mathbb{P}(s_i$ sent $|s_j$ received $)$ for $i,j = 1,2,3$. (Complete the table.)

|  |  | Sent |  |  |
|---|---|---|---|---|
|  |  | $s_1$ | $s_2$ | $s_3$ |
|  | $s_1$ | 0.8641 |  |  |
| Received | $s_2$ |  |  | 0.0741 |
|  | $s_3$ |  |  |  |

3.30. **Sprinkler Bayes Net.** Suppose that a sprinkler ($S$) or rain ($R$) can make the grass in your yard wet ($W$). The probability that the sprinkler was on depends on whether the day was cloudy ($C$). The probability of rain also depends on whether the day was cloudy. The DAG for events $C, S, R$, and $W$ is shown in Fig. 3.19.

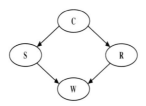

**Fig. 3.19** Sprinkler Bayes net.

The conditional probabilities of the nodes are given in the following tables.

| $C^c$ | $C$ |
|---|---|
| 0.5 | 0.5 |

| $S^c$ | $S$ | Condition |
|---|---|---|
| 0.50 | 0.50 | $C^c$ |
| 0.90 | 0.10 | $C$ |

| $R^c$ | $R$ | Condition |
|---|---|---|
| 0.80 | 0.20 | $C^c$ |
| 0.20 | 0.80 | $C$ |

| $W^c$ | $W$ | Condition |
|---|---|---|
| 1 | 0 | $S^c R^c$ |
| 0.10 | 0.90 | $S^c R$ |
| 0.10 | 0.90 | $S R^c$ |
| 0.01 | 0.99 | $S R$ |

Using WinBUGS, approximate the probabilities
(a) $\mathbb{P}(C|W)$, (b) $\mathbb{P}(S|W^c)$, and (c) $\mathbb{P}(C|R, W^c)$.

3.31. **Diabetes in Pima Indians.** The Pima Indians have the world's highest reported incidence of diabetes. Since 1965, this population has participated in a longitudinal epidemiological study of diabetes and its complications. The examinations have included a medical history for diabetes and other major health problems. A population of women who were at least 21 years old, of Pima Indian heritage, and living near Phoenix, AZ was tested for diabetes according to World Health Organization criteria.

The following conditions ("events"), constructed from the database, can be related to a randomly selected subject from this population.

| Event | Description |
|-------|-------------|
| P | Three or more pregnancies |
| A | Older than the database median age |
| O | Heavier than the database median weight |
| D | Diagnosis of diabetes |
| G | High plasma glucose concentration in an oral glucose tolerance test |
| I | High 2-h serum insulin ($\mu U/ml$) |
| B | High blood pressure |

The DAG in Fig. 3.20 simplifies the proposal of Tom Bylander from the University of Texas in San Antonio, who used Bayesian networks and the Pima Indians Diabetes Database in a machine learning example.[4]

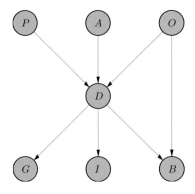

**Fig. 3.20** Pima Indians diabetes Bayes net.

From 768 complete records relative frequencies are used to approximate the conditional probabilities of the nodes. The probabilities are given in the following tables.

| $P^c$ | $P$ |
|-------|-----|
| 0.45 | 0.55 |

| $A^c$ | $A$ |
|-------|-----|
| 0.5 | 0.5 |

| $O^c$ | $O$ |
|-------|-----|
| 0.5 | 0.5 |

| $D^c$ | $D$ | condition |
|-------|-----|-----------|
| 0.95 | 0.05 | $P^c\ A^c\ O^c$ |
| 0.67 | 0.33 | $P^c\ A^c\ O$ |
| 0.59 | 0.41 | $P^c\ A\ O^c$ |
| 0.40 | 0.60 | $P^c\ A\ O$ |

| $D^c$ | $D$ | condition |
|-------|-----|-----------|
| 0.73 | 0.27 | $P\ A^c O^c$ |
| 0.66 | 0.34 | $P\ A^c\ O$ |
| 0.63 | 0.37 | $P\ A\ O^c$ |
| 0.41 | 0.59 | $P\ A\ O$ |

---

[4] http://www.cs.utsa.edu/~bylander/cs6243/bayes-example.pdf

| $G^c$ | $G$ | condition |
|---|---|---|
| 0.64 | 0.36 | $D^c$ |
| 0.21 | 0.79 | $D$ |

| $I^c$ | $I$ | condition |
|---|---|---|
| 0.49 | 0.51 | $D^c$ |
| 0.52 | 0.48 | $D$ |

| $B^c$ | $B$ | condition |
|---|---|---|
| 0.55 | 0.45 | $O^c$ $D^c$ |
| 0.58 | 0.42 | $O^c$ $D$ |
| 0.40 | 0.60 | $O$ $D^c$ |
| 0.49 | 0.51 | $O$ $D$ |

Suppose a female subject, older than 21, of Pima Indian heritage living near Phoenix, AZ, was selected at random. Using WinBUGS, approximate the probabilities (a) $\mathbb{P}(O|I)$, $\mathbb{P}(B|O^c,G)$, and $\mathbb{P}(G|B,A^c)$.

3.32. **A Simplified Probabilistic Model of Visual Pathway.** Our nervous system consists of specialized cells called neurons that are connected to one another in highly organized and specific ways. A neuron changes its membrane voltage in response to an external stimulus or a signal from an upstream neuron. For example, in the visual pathway, light excites photoreceptors in our eyes that are connected to more downstream neurons, and the activities of this neuronal network eventually elicit visual perceptions. The following diagram illustrates a simplified version of this connection ($P$ = photoreceptor cell; $B$ = bipolar cell; $R$ = retinal ganglion cell; $L$ = lateral geniculate nucleus ganglion cell, $V$ = primary visual cortex simple cell):

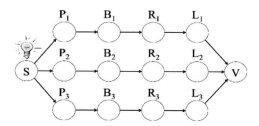

**Fig. 3.21** A simplified probabilistic model of visual pathway.

Let events $S$ and $S^c$ denote, respectively, the presence and absence of a light stimulus in a small time interval $\Delta t$. Let events $P, B, R$, and $L$ mean that the corresponding cells produce a response (fire) in the time interval $\Delta t$. Suppose that the conditional probabilities of firing for three parallel branches $i = 1, 2$, and 3 are the same:

$$\mathbb{P}(S) = 0.6$$
$$\mathbb{P}(P_i|S) = 0.95 \quad \mathbb{P}(P_i|S^c) = 0.1$$
$$\mathbb{P}(B_i|P_i) = 0.95 \quad \mathbb{P}(B_i|P_i^c) = 0.05$$
$$\mathbb{P}(R_i|B_i) = 0.9 \quad \mathbb{P}(R_i|B_i^c) = 0.02$$
$$\mathbb{P}(L_i|R_i) = 0.8 \quad \mathbb{P}(L_i|R_i^c) = 0.08$$
$$i = 1, 2, 3$$

In the schematic diagram in Fig. 3.21, three $L$ ganglion cells, $L_1$, $L_2$, and $L_3$, connect to one simple cell in the primary visual cortex, $V$. Assume that input from at least one of the $L$ cells is needed for the $V$ simple cell to respond with certainty. The $V$ simple cell will not respond if the input there from $L$ cells is absent.

The following questions may need the support of WinBUGS.

(a) What is the probability of the $V$ cell responding given that event $S$ (light stimulus present) has occurred?

(b) Assume that $L_1$ and $R_3$ have fired. What is the probability of $S$?

(c) Assume that $V$ responded. What is the probability of $S$?

---

| MATLAB AND WINBUGS FILES USED IN THIS CHAPTER |

http://springer.bme.gatech.edu/Ch3.Prob/

birthday.m, bridge.m, circuit.m, ComplexCircuit.m, demere.m, die.m, inclusionexclusion.m, mistery.m, rollingdie1.m, rollingdie2.m, sheriff.m, twoheaded.m,

alarm.odc, asia.odc, circuit1.odc, DeMere.odc, misterioustransfers.odc, pima.odc, sprinkler.odc,

---

# CHAPTER REFERENCES

Adel, A. L., Dorr, R. T., Liddil, J. D. (1993). The effect of anticancer drug sequence in experimental combination chemotherapy. *Cancer Invest.*, **11**, 1, 15–24.

Barbeau, E. (1993). The problem of the car and goats. *College Math. J.*, **24**, 2, 149–154.

Bayes, T. (1763). An essay towards solving a problem in the doctrine of chances. *Philos. Trans. R. Soc. Lond.*, *53*, 370–418.

Bonferroni, C. E. (1937). Teoria statistica delle classi e calcolo delle probabilita. In *Volume in Onore di Ricarrdo dalla Volta*, Universita di Firenze, 1–62.

Casscells, W., Schoenberger, A., and Grayboys, T. (1978). Interpretation by physicians of clinical laboratory results. *New Engl. J. Med.*, **299**, 999–1000.

Gillman, L. (1992). The car and the goats, *Am. Math. Mon.*, **99**, 1, 3–7.

Kolmogorov, A. N. (1933). *Grundbegriffe der Wahrscheinlichkeitsrechnung*, Springer, Berlin.

Laplace P. S. (1774). Mémoire sur la probabilité des causes par les évenements. *Mém. de l'Ac. R. des Sciences de Paris*, **6**, 621–656.

Lauritzen, S. L. and Spiegelhalter, D. L. (1988). Local computations with probabilities on graphical structures and their application to expert systems. *J. R. Stat. Soc. B*, **50**, 157–194.

Piaget, J. and Inhelder B. (1976). *The Origin of the Idea of Chance in Children*. W. W. Norton & Comp., NY, 276 pp.

Ross, S. (2009). *A First Course in Probability*, 8th Edition. Prentice Hall.

Selvin, S. (1975). A Problem in Probability. *Am. Stat.*, **29**, 1, 67.

Venn, J. (1880). On the employment of geometrical diagrams for the sensible representation of logical propositions. *Proc. Cambridge Philos. Soc.*, **4**, 47–59.

Wong, S. K. N., O'Connell, M., Wisdom, J. A., and Dai, H. (2005). Carbon nanotubes as multifunctional biological transporters and near-infrared agents for selective cancer cell destruction. *Proc. Natl. Acad. Sci.*, **102**, 11600–11605.

# Chapter 4
# Sensitivity, Specificity, and Relatives

*Poetry teaches us music, metaphor, condensation and specificity.*

– Walter Mosley

### WHAT IS COVERED IN THIS CHAPTER

• Definitions of Sensitivity, Specificity, Positive and Negative Predictive Values, Likelihood ratio Positive and Negative, Measure of Agreement.
• Combining Tests
• Performance of Tests: ROC Curves, Area Under ROC, Youden Index
• Two Examples: D-dimer and ADA

## 4.1 Introduction

This chapter introduces several notions fundamental for disease or device testing. The sensitivity, specificity, and positive and negative predictive values of a test are measures of the performance of a diagnostic test and are intimately connected with probability calculations (estimations) and Bayes' rule.

Although concepts such as "false positives," "true negatives," etc. are quite intuitive, many students and even health professionals have difficulties in assessing the associated probabilities. The following problem was posed by Casscells et al. (1978) to 60 students and staff at an elite medical school: *If a test to detect a disease whose prevalence is 1/1000 has a false positive rate of 5%, what is the chance that a person found to have a positive result actually has the disease, assuming you know nothing about the person's symptoms or signs?*

Assuming that the probability of a positive result given the disease is 1, the answer to this problem is approximately 2%. Casscells et al. found that only 18% of participants gave this answer. The most frequent response was 95%, presumably on the supposition that, because the error rate of the test is 5%, it must get 95% of results correct.

Examples of misconceptions of test precision measures, especially involving the sensitivity and positive predictive value, are abundant.

When a multiplicity of tests are possible and a researcher is to select the "best" test, the receiver operating characteristic (ROC) curve methodology is used. This methodology is especially useful in setting a threshold that separates positive and negative outcomes of a test.

## 4.2 Notation

Suppose that $n$ subjects are selected randomly from a given population. The population may not be very general; it could be a specific segment of subjects (patients in a hospital who were checked in 10+ days ago, subjects with a history of heart attack, etc.). Now suppose that the true disease status (disease present/absent) in all subjects is determined via a gold-standard assessment and that we are interested in evaluating a particular test for the disease assuming that the gold-standard results are always true. For example, in testing for breast cancer (BC), a mammogram is used as a test while the battery of numerous patient symptoms, medical history, and biopsy results are used as a gold standard.

A positive test would not necessarily mean the disease is present but rather would mean that the test says the disease is present. For instance, a patient's mammogram appears to show breast cancer. A true-positive test result not only means that the test says the disease is present, but that the disease *really is present*. In this case, a positive mammogram of a patient for which the gold standard indicates BC would be a true positive. In the same context, false positives, true negatives, and false negatives are defined in an obvious manner.

By classifying the patients with respect to the test results and the true disease status, the following table can be constructed:

|  | Disease (D) | No disease (C) | Total |
|---|---|---|---|
| Test positive (P) | TP | FP | nP = TP + FP |
| Test negative (N) | FN | TN | nN = FN + TN |
| Total | nD = TP + FN | nC = FP + TN | n = nD + nC = nP + nN |

where

| TP | True positive (test positive, disease present) |
|----|---|
| FP | False positive (test positive, disease absent) |
| FN | False negative (test negative, disease present) |
| TN | True negative (test negative, disease absent) |
| nP | Total number of positives (TP + FP) |
| nN | Total number of negatives (TN + FN) |
| nD | Total number with disease present (TP + FN) |
| nC | Total number without disease present (TN + FP) |
| n  | Total sample size (TP + FP + FN + TN) |

The test's effectiveness is measured by the number of true positives and true negatives relative to the total number of cases. It turns out that these two numbers are used to define two complementary measures of test performance: *sensitivity* and *specificity*.

We will illustrate the defined numbers with the following example.

*Example 4.1.* **BreastScreen Victoria.** This 1994 study involved women who participated in the BreastScreen Victoria initiative in Victoria, Australia, where free biennial screenings for BC are provided to women aged 40 and older. The data provided by Kavanagh at al. (2000) show that among 96,420 asymptomatic women, 5,401 had positive and 91,019 negative mammogram results. The mammograms were read independently by two radiologists. In the case of disagreement over whether to recall, a consensus was reached or a third reader made the decision. The women were then recommended for routine rescreen or referred for assessment. Assessment might include clinical examination, further radiographs, ultrasound, or biopsy. After assessment, women may have a cancer diagnosed, be recommended for routine rescreening, or be recommended for further assessment (early review). Out of 96,420 women, 665 were diagnosed with BC. Of those 665 diagnosed, the mammogram was positive for 495 and negative for 160 women. The table summarizing the data is below.

|  | BC diagnosed | BC not diagnosed | Total |
|---|---|---|---|
| Mammogram positive | 495 | 4906 | 5401 |
| Mammogram negative | 160 | 90859 | 91019 |
| Total | 665 | 95765 | 96420 |

Sensitivity is the ratio of the number of true positives and the number of subjects with a disease, while specificity is the ratio of the number of true negatives and the number of subjects without the disease. In the BC example sensitivity is 495/665 = 75.57%, and specificity is 90859/95765 = 94.88%.

Both measures have to be reported since reporting only sensitivity or only specificity reveals little information about the test. There are two extreme cases. Imagine a test that classifies *all* subjects as positive – trivially the sensitivity is 100%. Since there are no negatives, the specificity is zero. Likewise, a test that classifies all subjects as negative has a specificity of 100% and zero sensitivity.

The following table summarizes the key notions:

| | |
|---|---|
| Sensitivity (Se) | Se = TP/(TP + FN) = TP/nD |
| Specificity (Sp) | Sp = TN/(FP + TN) = TN/nC |
| Prevalence (Pre) | (TP + FN)/(TP + FP + FN + TN)= nD/n |
| Positive predictive value (PPV) | PPV = TP/(TP + FP) = TP/nP |
| Negative predictive value (NPV) | NPV = TN/(TN + FN) = TN/nN |
| Likelihood ratio positive (LRP) | LRP = Se/(1-Sp) |
| Likelihood ratio negative (LRN) | LRN = (1-Se)/Sp |
| Apparent prevalence (APre) | APre = nP/n |
| Concordance, agreement (Ag) | Ag =(TP + TN)/n |

The population prevalence of a disease is defined as the probability that a randomly selected person from this population will have the disease. As the table shows, the prevalence is estimated by (TP + FN)/(TP + FP + FN +TN) = nD/n. For the Victoria BC data the prevalence is 665/96420 = 0.0069. This is a valid estimator only if the table is a summary of a representative sample of the population under analysis. In other words, the sample should have been taken at random and the tabulation made subsequently. This is not the case in many studies. The prevalence of some diseases in a general population is often so small that insisting on a random sample would require huge sample sizes in order to obtain a nonzero TP or FN table entries. When the table is made from available cases and controls (convenience samples), the prevalence for the population cannot be estimated from it.

Related quantity is the incidence of a disease in a population. It is defined as the probability that a randomly selected person from the subset of people not affected by the disease will develop the disease in a fixed time window (week, month, year). While the prevalence relates to the magnitude, the incidence provides information about the progression and dynamics of the disease.

One of the most important measures is the positive predictive value (PPV). Based on the table it can be estimated as the proportion of true positives among all positives, TP/nP. This is correct only if the population prevalence is well estimated by nD/n, that is, if the table is representative of its population. This is approximately the case for the Victoria BC data; the PPV is well estimated by 495/5401 = 0.0916.

If the table is constructed from a convenience sample, the prevalence (Pre) would be external information, and the PPV is calculated as

$$PPV = \frac{Se \times Pre}{Se \times Pre + (1 - Sp) \times (1 - Pre)}.$$

This is simply the Bayes rule and will be discussed more in the next section.

Why is the PPV so important? Imagine an almost perfect test for a particular disease, with a sensitivity of 100% and specificity of 99%. If the prevalence of the disease in the population is 10%, then among ten positives there would be approx. one false positive. However, if the population prevalence is 1/10000, then for each true positive there would be approx. 100 false positives.

The *likelihood ratio positive*, or *Bayes factor positive*, represents the odds that a positive test result would be found in a patient with, vs. without, a disease. The *likelihood ratio negative*, or *Bayes factor negative*, represents the odds that a negative test result would be found in a patient without, vs. with, a disease. For example:

$$\text{Posttest disease odds } = \text{ LRP } \times \text{pretest disease odds;}$$
$$\text{Posttest no-disease odds } = \text{ LRN } \times \text{pretest no-disease odds.}$$

### *4.2.1 Conditional Probability Notation*

The definitions introduced in the previous section depend on the relative frequencies in the observed tables, and they are empirical. The theoretical counterparts are expressed in terms of probabilities. The analogy to this is the interplay of the probability of an event $A$, $\mathbb{P}(A)$, which is theoretical, and relative frequency of the event $n_A/n$, which is empirical.

Let $T$ be the event that a subject tests positive and $D, D^c$ the hypothesis that the subject does/does not have the disease. The sensitivity is the conditional probability $\mathbb{P}(T|D) = \mathbb{P}(T \cap D)/\mathbb{P}(D)$ [which is estimated by $(TP/n)/(nD/n)$ $= TP/nD$]. Analogously, the specificity is $\mathbb{P}(T^c|D^c) = \mathbb{P}(T^c \cap D^c)/\mathbb{P}(D^c)$, which is estimated by $(TN/n)/(nC/n) = TN/nC$. We have argued that $\mathbb{P}(D)$, the prevalence, cannot be estimated from the table unless the sample forming the table is representative of the population. In the case of "convenience" samples, the prevalence is evaluated separately or assumed known from other studies. If the sample is randomly obtained from the population, the prevalence can be estimated by $nD/n$. The probability of a positive test is in fact given by the rule of total probability, $\mathbb{P}(T) = \mathbb{P}(T|D)\mathbb{P}(D) + \mathbb{P}(T|D^c)\mathbb{P}(D^c)$. Note that $\mathbb{P}(T)$ depends on the prevalence and is estimated by $nP/n$ for a table from a random sample.

Finally, the PPV and NPV are determined by Bayes' rule. For example, the PPV is

$$\mathbb{P}(D|T) = \frac{\mathbb{P}(T|D)\mathbb{P}(D)}{\mathbb{P}(T)} = \frac{\mathbb{P}(T|D)\mathbb{P}(D)}{\mathbb{P}(T|D)\mathbb{P}(D) + \mathbb{P}(T|D^c)\mathbb{P}(D^c)}.$$

The posttest disease odds ratio is LRP times the pretest odds ratio. Because of this property, the LRP is in fact Bayes' factor in the terminology of Chaps. 3

and 8.

$$\frac{\mathbb{P}(D|T)}{\mathbb{P}(D^c|T)} = \frac{\mathbb{P}(T|D)\mathbb{P}(D)}{\mathbb{P}(T)} \bigg/ \frac{\mathbb{P}(T|D^c)\mathbb{P}(D^c)}{\mathbb{P}(T)} = \frac{\mathbb{P}(T|D)\mathbb{P}(D)}{\mathbb{P}(T|D^c)\mathbb{P}(D^c)} = \frac{\mathbb{P}(T|D)}{\mathbb{P}(T|D^c)} \times \frac{\mathbb{P}(D)}{\mathbb{P}(D^c)}.$$

Thus, posterior disease odds = LRP × prior disease odds.

The above definitions are illustrated on an example where researchers tested for acute pulmonary embolism.

*Example 4.2.* **D-Dimer.** When a vein or artery is injured and begins to leak blood, a sequence of clotting steps and factors (called the coagulation cascade) are activated by the body to limit the bleeding and create a blood clot to plug the hole. During this process, threads of a protein called fibrin are produced. These threads are cross-linked (chemically glued together) to form a fibrin net that catches platelets and helps hold the forming blood clot together at the site of the injury. Once the area has had time to heal, the body uses a protein called plasmin to break the clot (thrombus) into small pieces so that it can be removed. The fragments of the disintegrating fibrin in the clot are called fibrin degradation products (FDPs). One of the FDPs produced is D-dimer, which consists of variously sized pieces of cross-linked fibrin. D-dimer is normally undetectable in the blood and is produced only after a clot has formed and is in the process of being broken down. Measurement of D-dimer can indicate problems in the body's clotting mechanisms. The data below consist of quantitative plasma D-dimer levels among patients undergoing pulmonary angiography for suspected pulmonary embolism (PE). The patients who exceed the threshold of 500 ng/mL are classified as positive for a PE. The gold standard for PE is the pulmonary angiogram. Goldhaber et al. (1993), from Brigham and Women's Hospital at Harvard Medical School, considered a population of patients who are suspected of PE based on a battery of symptoms. The summarized data for 173 patients are provided in the table below.

|                                              | Acute PE | No PE present | Total |
|----------------------------------------------|:--------:|:-------------:|:-----:|
| Test positive (D-dimer ≥ 500 ng/mL)          |    42    |      96       |  138  |
| Test negative (D-dimer < 500 ng/mL)          |     3    |      32       |   35  |
| Total                                        |    45    |     128       |  173  |

A simple MATLAB file ◀ sesp.m will calculate the sensitivity, specificity, prevalence, positive and negative predictive values, and degree of agreement between the test and gold-standard results.

```
function [se sp pre ppv npv ag] = sesp(tp, fp, fn, tn)
  %D-dimer as a test for acute PE (Goldhaber et al, 1993)
  % [s1, s2, p1, p2, p3, a, yi] = sesp(42,96,3,32)
  %
  n = tp+tn+fn+fp; %total sample size
  np = tp + fp; %total positive
  nn = tn + fn; %total negative
```

```
nd = tp + fn; %total with disease
nc = tn + fp; %total control (without disease)
%--------------
se = tp/nd; %tp/(tp + fn):::sensitivity
sp = tn/nc; %tn/(tn + fp):::specificity
pre = nd/n; %(tp + fn)/(tp+tn+fn+fp):::prevalence
%only in for case when sample is random from the
%    population of interest. Otherwise, the prevalence
%    needed for calculating PPV and NPV is an input value
ppv = tp/np; %tp/(tp + fp):::positive predictive value
npv = tn/nn; %tn/(tn+fn):::negative predictive value
lrp = se/(1-sp); %:::likelihood ratio positive
lrn = (1-se)/sp; %:::likelihood ratio negative
ag = (tp+tn)/n;  %:::agreement
yi = (se + sp - 1)/sqrt(2); %:::youden index
%--------------
disp(' Se Sp Pre PPV NPV LRP Ag Yi')
disp([se, sp, pre, ppv, npv, lrp, ag yi])
        %spacing in disp depends on the font size.
```

For the D-dimer data, the result is

```
[a b c d e f g] = sesp(42,96,3,32);
  Se     Sp     Pre     PPV     NPV     LRP     Ag      Yi
0.9333 0.2500 0.2601 0.3043 0.9143 1.2444 0.4277 0.1296
```

Goldhaber et al. (1993) conclude: "The results of our study indicate that quantitative plasma D-dimer levels can be useful in screening patients with suspected PE who require pulmonary angiography. Plasma D-dimer values less than 500 ng/mL may obviate the need for pulmonary angiography, particularly among medical patients for whom the clinical suspicion of PE is low. The plasma D-dimer value, assayed using a commercially available enzyme-linked immunosorbent assay kit, is a sensitive but nonspecific test for the presence of acute PE."

## 4.3 Combining Two or More Tests

Suppose that $k$ independent tests for a particular condition are available and that their sensitivities and specificities are $Se_1$, $Sp_1$, $Se_2$, $Sp_2$, ..., $Se_k$, $Sp_k$. If these tests could be combined, what would be the sensitivity/specificity of the combined test?

First, it is important to define how the tests are going to be combined. There are two main strategies: *parallel* and *serial*. When the tests are assumed independent, the calculations are similar to those in circuit problems from Chap. 3, p. 68. Denote by Se and Sp the sensitivity and specificity of the combined test, respectively.

In the parallel strategy the combination is positive if at least one test is positive and negative if all tests are negative. Then the sensitivity is calculated as the probability of a union and the specificity as the probability of an intersection:

| Parallel combination (Positive if at least 1 positive) |
| --- |
| $Se = 1 - [(1 - Se_1) \times (1 - Se_2) \times \cdots \times (1 - Se_k)]$ |
| $Sp = Sp_1 \times Sp_2 \times \cdots \times Sp_k$ |

It is easy to see that in the parallel strategy the sensitivity is larger than any individual sensitivity and the specificity smaller than any individual specificity.

In the serial strategy, the combination is positive if all tests are positive and negative if at least one test is negative. Then the sensitivity is calculated as the probability of an intersection and the specificity as the probability of a union:

| Serial combination (Positive if all positive) |
| --- |
| $Se = Se_1 \times Se_2 \times \cdots \times Se_k$ |
| $Sp = 1 - [(1 - Sp_1) \times (1 - Sp_2) \times \cdots \times (1 - Sp_k)]$ |

Here, the overall sensitivity is smaller than any individual sensitivity, while the specificity is larger than any individual specificity.

There are other possible combinations as well as procedures that address bias and correlation among the individual tests.

*Example 4.3.* **Combining Two Tests for Sarcoidosis.** Parikh et al. (2008) provide an example of combining two tests for sarcoidosis. Sarcoidosis is an idiopathic multisystem granulomatous disease, where the diagnosis is made by a combination of clinical, radiological, and laboratory findings. The gold standard is a tissue biopsy showing noncaseating granuloma. Ocular sarcoidosis could present as anterior, intermediate, posterior, or panuveitis; but none of these is pathognomonic. Therefore, one has to rely on ancillary testing to confirm the diagnosis.

An angiotensin-converting enzyme (ACE) test has a sensitivity of 73% and a specificity of 83% to diagnose sarcoidosis. An abnormal gallium scan has a sensitivity of 91% and a specificity of 84%. Though individually the specificity of either test is not impressive, for the serial combination the specificity becomes

$$Sp = 1 - (1 - 0.84) \times (1 - 0.83) = 1 - (0.16 \times 0.17) = 0.97.$$

The combination sensitivity becomes $0.73 \times 0.91 = 0.66$. Note that the overall specificity drastically improves, but at the expense of overall sensitivity.

The independence of tests in the previous example is a quite limiting assumption. One may argue that two tests for the same disease are seldom independent by the very nature of the testing problem.

In the following example, we show how to handle more complex batteries of tests in which the tests could be dependent. The example considers two tests and a parallel combination strategy, but it could be extended to any number of tests and to more general combination strategies.

The approach is based on simulation since analytic solutions are typically computationally involved.

*Example 4.4.* **Simulation Approach.** Suppose a testing procedure consists of two tests given in a sequence. Test A has a sensitivity of 0.9 and a specificity of 0.8. Test B has a sensitivity of 0.7 and a specificity of 0.9 for subjects who tested negative in test A and a sensitivity of 0.95 and a specificity of 0.6 for subjects who tested positive in test A.

Clearly, test A and test B are dependent. If a subject is declared positive when the result of at least one of the two tests was positive (parallel link), what is the overall sensitivity/specificity of the described testing procedure? The population prevalence is considered known and is used in the simulation of a patient's status, but it does not affect the overall sensitivity/specificity.

Note that if a subject's status s is equal to 0/1 when the disease is absent/present, then the result of a test is s*(rand < se) + (1-s)*(rand > sp) for a known sensitivity and specificity, se, sp. The test outcome is binary, with 0/1 denoting a negative/positive test result.

The following MATLAB code ( ◢ simulatetesting2.m) considers 20,000 subjects from a population where the disease prevalence is 0.2. The estimated sensitivity/specificity was 0.97/0.72, but simulation results may vary slightly due to the random status of subjects.

```
nsubjects = 20000;
prevalence = 0.2;
se1  =0.9; sp1   = 0.8;  %se/sp of test1
se20 =0.7; sp20  = 0.9;  %se/sp of test2 if test1=0
se21 =0.95; sp21 = 0.6;  %se/sp of test2 if test1=1

tests = [];
ss=[]; tp=0; fp=0; fn=0; tn=0;
for i = 1:nsubjects
   %simulate a subject wp of disease equal to prevalence
   s = (rand < prevalence);
   %test the subject
   test1=s*(rand < se1) + (1-s)*(rand>sp1); %test is 0 or 1
   if (test1 == 0)
       test2=s*(rand < se20) + (1-s)*(rand>sp20);
   else
       test2=s*(rand < se21) + (1-s)*(rand>sp21);
   end
   %test = test1*test2;         %for serial
   test = (test1 + test2 > 0);  %for parallel
   ss=[ss s];                   %save subject's status
   tests = [tests test];        %save subject's test
   %building the test table
```

```
    tp = tp + test*s;              %true positives
    fp = fp + test*(1-s);          %false positives
    fn = fn + (1-test)*s;          %false negatives
    tn = tn + (1-test)*(1-s);      %true negatives
end
% estimate overall Se/Sp from the table
sens = tp/(tp+fn)
spec = tn/(tn+fp)
```

**Remark:** In the previous discussion we assumed that a true disease status was known and that a perfect gold standard test was available. In many cases an error-free assessment does not exist but a reference test, with known sensitivity $Se_R$ and specificity $Sp_R$, could be used. By taking this reference test as a gold standard and by not accounting for its errors would lead to biases in evaluating a new test. Staquet et al. (1981) provide a solution based on $Se_R$, $Sp_R$, and concordance of results between the two tests.

Another approach approach to this problem is "discrepant resolution," in which the subjects for whom the reference and new test disagreed were subjected to a third "resolver" test. Although commonly used, the resolver method can be biased and can overestimate sensitivity and specificity of a new test significantly (Hawkins et al., 2001; Qu and Hadgu, 1998).

## 4.4 ROC Curves

The receiver operating characteristic (ROC) curve was first used during World War II for the analysis of radar signals before it was employed in signal detection theory and, subsequently, in a range of fields where testing is critical. It is defined as a graphical plot of sensitivity vs. (1 - specificity) for a binary classifier system as its discrimination threshold (value that separates positives and negatives) varies.

Let us look at an ROC curve using the D-dimer example from the previous section. Mavromatis and Kessler (2001) report that in 18 publications (between 1988 and 1998) concerning D-dimer testing, the reported cut point for declaring the test positive ranged from 250 to 1000 ng/mL. What cut point should be recommended? To increase the apparently low specificity in the previous D-dimer analysis, suppose that the threshold for testing positive is increased from 500 to 650 ng/mL and that the data are distributed in the following way:

|  | Acute PE | No PE present | Total |
|---|---|---|---|
| Test positive (D-dimer $\geq$ 650 ng/mL) | 31 | 33 | 64 |
| Test negative (D-dimer $<$ 650 ng/mL) | 14 | 95 | 109 |
| Total | 45 | 128 | 173 |

This new table results in the following sesp output:

```
[a b c d e f] = sesp(31,33,14,95);
   Se      Sp     Pre    PPV    NPV    LRP     Ag      Yi
0.6889 0.7422 0.2601 0.4844 0.8716 2.6721 0.7283 0.3048
```

Combining this with the output of the 500-ng/mL threshold, we get the vectors `1-sp = [0 1-0.7422 1-0.25 1]` and `se = [0 0.6889 0.9333 1]`.

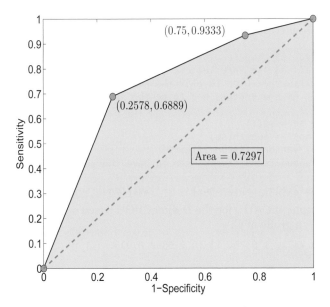

**Fig. 4.1** Rudimentary ROC curve for D-dimer data based on two thresholds.

The code ◀ RocDdimer.m plots this "rudimentary" ROC curve (Fig. 4.1). The curve is rudimentary since it is based on only two tests. Note that points (0,0) and (1,1) always belong to ROC curves. These two points correspond to the trivial tests in which all patients test negative or all patients test positive. The area under the ROC curve (AUC), is a well-accepted measure of test performance. The closer the area is to 1, the more unbalanced the ROC curve, implying that both sensitivity and specificity of the test are high. It is interesting that some researchers assign an academic scale to AUC as an informal measure of test performance.

| AUC | Performance |
|---------|:-----------:|
| 0.9–1.0 | A |
| 0.8–0.9 | B |
| 0.7–0.8 | C |
| 0.6–0.7 | D |
| 0.0–0.6 | F |

The following MATLAB program calculates AUC when the vectors `csp = 1 - specificity` and `sensitivity` are supplied.

```
function A = auc(csp, se)
%
% A = auc(csp,se) computes the area under the ROC curve
% where 'csp' and 'se' are vectors representing (1-specificity)
% and (sensitivity), used to plot the ROC curve
% The length of the vectors has to be the same

csp=csp(:); se = se(:);
if length(csp) ~= length(se)
error('Input vectors (1-specificity) ...
              and (sensitivity) should have the same length')
end
A = sum((csp(2:end)-csp(1:end-1)) .* (se(2:end)+se(1:end-1))/2 );
```

For example, the AUC for the D-dimer ROC based on the two thresholds is approx. 73%, a grade of C:

```
auc([0, 1-0.7422, 1-0.25, 1],[0 0.6889 0.9333 1])
  ans = 0.7297
```

To choose the best test out of a multiplicity of tests obtained by changing the threshold and generating the ROC curve, select the test corresponding to the point in the ROC curve most distant from the diagonal. This point corresponds to a Youden index

$$YI = max_i \frac{Se_i + Sp_i - 1}{\sqrt{2}},$$

where $Se_i$ and $Sp_i$ are, respectively, the sensitivity and specificity for the $i$th test. Thus, the Youden index is the distance of the most distant point $(1 - Sp, Se)$ on the ROC curve from the diagonal. It ranges between 0 and $\sqrt{2}/2$.

In the D-dimer example, the Youden index for the test with a 500-ng/mL threshold is 0.1296, compared to 0.3048 for the test with a 650-ng/mL threshold. Between the two tests, the test with the 650-ng/mL threshold is preferred.

*Example 4.5.* **ADA.** *Adenosine deaminase* (ADA) is an enzyme involved in the breakdown of adenosine to uric acid. ADA levels were found to be elevated in the pleural fluid of patients with tuberculosis (TB) pleural effusion. Pleural effusion is a very common clinical problem. It may occur in patients of pulmonary TB, pneumonia, malignancy, congestive cardiac failure, cirrhosis of the liver, nephrotic syndrome, pulmonary infarction, and connective tissue disorders. TB is one of the primary causes of pleural effusion. Numerous studies have evaluated the usefulness of ADA estimation in the diagnosis of TB pleural effusion. However, the sensitivity and specificity of ADA estimation and the cutoff level used for distinguishing TB pleural effusion from non-TB pleural effusion have varied between studies. The data (given in [image] ROCTBCA.XLS or ROC.mat) were collected by Dr. Mark Hopley of Chris-Hani Baragwanath Hospital (CHB, the largest hospital in the world), with the goal of critically evaluating the sensitivity and specificity of ADA estimation in the diagnosis of TB pleural effusion.

The data set consists of three columns:

Column 1 contains ADA levels.

Column 2 is an indicator of TB. The indicator is "1" if the patient had documented TB, zero otherwise.

Column 3 is an indicator of documented carcinoma. Six patients who had both carcinoma and TB have been excluded from the analysis.

To create an empirical ROC curve, the following four steps are applied:

(i) The data are sorted according to the ADA level, with the largest values first.

(ii) A column is created where each entry gives the total number of TB patients with ADA levels greater than or equal to the ADA value for that entry.

(iii) A column equivalent to that from step 2 is created for patients with cancer.

(iv) Two new columns are created, containing the true positive frequency (TPF) and false positive frequency (FPF) for each entry. The TPF is calculated by taking the number of TB cases identified at or above the ADA level for the current entry and dividing by the total number of TB cases. The FPF is determined by taking the number of "false TB alarms" (cancer patients) at or above that level and dividing by the total number of such non-TB patients.

This description can be simply coded in MATLAB thanks to the cumulative summation (cumsum) command:

```
disp('ROC Curve Example')
set(0, 'DefaultAxesFontSize', 16);
fs = 15;
     % data file ADA.mat should be on path
load 'ADA.mat'
     % columns in ADA.mat are:
     % 1. ADA level (ordered decreasingly)
     % 2. indicator of case TB
     % 3. indicator of non-case CA
cumultruepos  = cumsum(ada(:,2));
cumulfalsepos = cumsum(ada(:,3));
     % these are true positives/false positives if the
     % cut-level is from the sequence ada(:,1).
tpf = cumultruepos/cumultruepos(end);     %sensitivity
```

```
fpf = cumulfalsepos/cumulfalsepos(end);   %1-specificity
plot(fpf,tpf) %ROC, sensitivity against (1-specificity)
xlabel('1 - specificity')
ylabel('sensitivity')
```

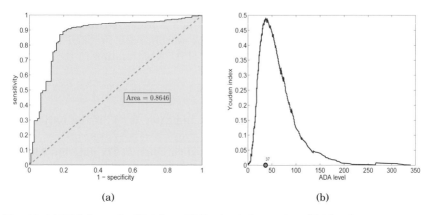

(a)                                    (b)

**Fig. 4.2** (a) ROC Curve for ADA data. (b) Youden index against ADA level.

Which ADA level should be recommended as a threshold? The Youden index for the ROC curve in Fig. 4.2a is 0.4910, which corresponds to ADA level of 37, Fig. 4.2b. For this particular threshold, the sensitivity and specificity are 0.8904 and 0.8039, respectively.

```
%youden index
yi = max((seth-cspth)/sqrt(2))        %0.4910
%ADA level corresponding to YI
ada((seth-cspth)/sqrt(2)== yi , 1)    %37
% sensitivity/specificity at YI
seth((seth-cspth)/sqrt(2)== yi)       %0.8904
1 - cspth((seth-cspth)/sqrt(2)== yi)  %0.8039
```

## 4.5 Exercises

4.1. **Stacked Auditory Brainstem Response.** The failure of standard auditory brainstem response (ABR) measures to detect small (<1 cm) acoustic tumors has led to the use of enhanced magnetic resonance imaging (MRI) as the standard to screen for small tumors. The study by Don et al. (2005) investigated the suitability of the stacked ABR as a sensitive screening alternative to MRI for small acoustic tumors (SATs). The objective of the study was to determine the sensitivity and specificity of the stacked ABR technique for detecting SATs. A total of 54 patients were studied who had MRI-identified acoustic tumors that were either <1 cm in size or undetected by standard ABR methods, irrespective of size. There were 78 nontumor normal-hearing subjects who tested as controls. The stacked ABR demonstrated 95% sensitivity and 88% specificity. Recover the testing table.

4.2. **Hypothyroidism.** Low values of a total thyroxine ($T4$) test can be indicative of *hypothyroidism* (Goldstein and Mushlin 1987). Hypothyroidism is a condition in which the body lacks sufficient thyroid hormone. Since the main purpose of the thyroid hormone is to "run the body's metabolism," it is understandable that people with this condition will have symptoms associated with a slow metabolism. Over five million Americans have this common medical condition.

A total of 195 patients, among which 59 have confirmed hypothyroidism, have been tested for the level of $T4$. If the patients with a $T4$ level $\leq 5$ are considered positive for hypothyroidism, the following table is obtained:

| $T4$ value | Hypothyroid | Euthyroid | Total |
|---|---|---|---|
| Positive, $T4 \leq 5$ | 35 | 5 | 40 |
| Negative, $T4 > 5$ | 24 | 131 | 155 |
| Total | 59 | 136 | 195 |

However, if the thresholds for $T4$ are 6, 7, 8, and 9, the following tables are obtained.

| $T4$ value | Hypothyroid | Euthyroid | Total |
|---|---|---|---|
| Positive, $T4 \leq 6$ | 39 | 10 | 49 |
| Negative, $T4 > 6$ | 20 | 126 | 146 |
| Total | 59 | 136 | 195 |

| $T4$ value | Hypothyroid | Euthyroid | Total |
|---|---|---|---|
| Positive, $T4 \leq 7$ | 46 | 29 | 75 |
| Negative, $T4 > 7$ | 13 | 107 | 120 |
| Total | 59 | 136 | 195 |

| $T4$ value | Hypothyroid | Euthyroid | Total |
|---|---|---|---|
| Positive, $T4 \leq 8$ | 51 | 61 | 112 |
| Negative, $T4 > 8$ | 8 | 75 | 83 |
| Total | 59 | 136 | 195 |

| $T4$ value | Hypothyroid | Euthyroid | Total |
|---|---|---|---|
| Positive, $T4 \leq 9$ | 57 | 96 | 153 |
| Negative, $T4 > 9$ | 2 | 40 | 42 |
| Total | 59 | 136 | 195 |

Notice that you can improve the sensitivity by moving the threshold to a higher $T4$ value; that is, you can make the criterion for a positive test less strict. You can improve the specificity by moving the threshold to a lower $T4$ value; that is, you can make the criterion for a positive test more strict. Thus, there is a tradeoff between sensitivity and specificity.

(a) For the test that uses $T4 = 7$ as the threshold, find the sensitivity, specificity, positive and negative predictive values, likelihood ratio, and degree of agreement. You can use the code ◢ sesp.m.

(b) Using the given thresholds for the test to be positive, plot the ROC curve. What threshold would you recommend? Explain your choice.

(c) Find the area under the ROC curve. You can use the code ◢ auc.m.

4.3. **Alzheimer's.** A medical research team wished to evaluate a proposed screening test for Alzheimer's disease. The test was given to a random sample of 450 patients with Alzheimer's disease and to an independent sample of 500 subjects without symptoms of the disease.

The two samples were drawn from a population of subjects who are 65 years old or older. The results are as follows:

| Test result  diagnosis | Diagnosed Alzheimer's, $D$ | No Alzheimer's symptoms, $D^c$ | Total |
|---|---|---|---|
| Positive test $T$ | 436 | 5 | 441 |
| Negative test $T^c$ | 14 | 495 | 509 |
| Total | 450 | 500 | 950 |

(a) Using the numbers from the table, estimate $\mathbb{P}(T|D)$ and $\mathbb{P}(T^c|D^c)$. Interpret these probabilities in terms of the problem.

The probability of $D$ (prevalence) is the rate of the disease in the relevant population ($\geq 65$ y.o.) and is estimated to be 11.3% (Evans 1990). Find $\mathbb{P}(D|T)$ (positive predicted value) using Bayes' rule. You cannot find $\mathbb{P}(D|T)$ using information from the table only – you need external info.

4.4. **Test for Being a Duchenne Muscular Dystrophy Carrier.** In Exercise 2.16 researchers used measures of pyruvate kinase and lactate dehydroginase to assess an individual's carrier status. The following table closely follows the authors' report.

|  | Woman carrier | Woman not carrier | Total |
|---|---|---|---|
| Test positive | 56 | 6 | 62 |
| Test negative | 11 | 121 | 132 |
| total | 67 | 127 | 194 |

(a) Find the sensitivity, specificity, and degree of agreement.

The sample is not representative of the general population for which the prevalence of carriers is 0.03%, or 3 in 10,000.

(b) With this information, find the PPV of the test, that is, the probability that a woman is a DMD carrier if she tested positive.

(c) What is the PPV if the table was constructed from a random sample of 194 subjects from a general population?

(d) Approximate the probability that among 15,000 women randomly selected from a general population, at least 2 are DMD carriers.

4.5. **Parkinson's Disease Statistical Excursions.** Parkinson's disease or, "shaking palsy," is a brain disorder that causes muscle tremor, stiffness, and weakness. Early symptoms of Parkinson's disease include muscular stiffness, a tendency to tire more easily than usual, and trembling that usually begins with a slight tremor in one hand, arm, or leg. This trembling is worse when the body is at rest but will generally stop when the body is in use, for example, when the hand becomes occupied by "pill rolling," or when the thumb and forefinger are rubbed together as if rolling a pill (Fig. 4.3).

**Fig. 4.3** "Pill rolling" stops muscle tremors in early Parkinson's disease.

In the later stages of Parkinson's disease, the affected person loses the ability to control his or her movements and the intellect begins to decline, making everyday activities hard to manage.

In a study by Schipper et al. (2008), 52 subjects, 20 with mild or moderate stages of Parkinson's disease and 32 age-matched controls, had whole blood samples analyzed using the near-infrared (NIR) spectroscopy and Raman spectroscopy methods. The data showed that the two independent biospectroscopy measurement techniques yielded similar and consistent results. In differentiating Parkinson's disease patients from the control group, Raman spectroscopy resulted in eight false positives and four false negatives. NIR spectroscopy resulted in four false positives and five false negatives.
(a) From the description above, construct tables for NIR spectroscopy and Raman spectroscopy containing TP, FP, FN and TN.
(b) For both methods find the sensitivity and specificity. Assume that the prevalence of Parkinson's disease in the age group matching this group is 1/120 for the general population. For both methods, also find the PPV, that is, the probability that a person who tested positive and was randomly selected from the same age group in the general population has the disease if no other clinical information is available.
(c) Mr. Smith is one of the 52 subjects in the study and he tested positive under a Raman spectroscopy test. What is the probability that Mr. Smith has the disease?

4.6. **Blood Tests in Diagnosis of Inflammatory Bowel Disease.** Cabrera-Abreu et al. (2004) explored the reliability of a panel of blood tests in screening for ulcerative colitis and Crohn's disease. The subjects were 153 children who were referred to a pediatric gastroenterology department with possible inflammatory bowel disease (IBD). Of these, 103 were found to have IBD (Crohn's disease 60, ulcerative colitis 37, indeterminate colitis 6). The 50 without IBD formed the controls. Blood tests evaluated several parameters including hemoglobin, platelet count, ESR, CRP, and albumin. The optimal screening strategy used a combination of hemoglobin and platelet count and "one of two abnormal" as the criterion for positivity. This was associated with a sensitivity of 90.3% and a specificity of 80.0%.
(a) Construct a table with TP, FP, FN and TN rounded to the nearest integer.
(b) Find the prevalence and PPV if the prevalence can be assessed from the table (the table is obtained from a random sample from the population of interest).

4.7. **Carpal Tunnel Syndrome Tests.** Three commonly used tests for carpal tunnel syndrome are Tinel's sign, Phalen's test, and the nerve conduction velocity test. Tinel's sign and Phalen's test are both highly sensitive (0.97 and 0.92, respectively) and specific (0.91 and 0.88, respectively). The sensitivity and specificity of the nerve conduction velocity test are 0.93 and 0.87, respectively. Assume that the tests are independent.
Calculate the sensitivity and specificity of a combined test if combining is done
(a) in a serial manner;
(b) in a parallel manner.
(c) Find PPV for tests from (a) and (b) if prevalence of carpal tunnel syndrome is approximately 50 cases per 1000 subjects in the general population.

4.8. **Hepatitic Scintigraphy.** A commonly used imaging procedure for detecting abnormalities in the liver is hepatitic scintigraphy. Drum and Christacopoulos (1972) reported data on 344 patients who underwent scintigraphy and were later examined by autopsy, biopsy, or surgical inspection for a gold-standard determination of the presence of liver pathology (parenchymal, focal, or infiltrative disease). The table summarizes the experimental results. Assume that this table is representative of the population of interest for this study.

|                          | Liver disease (D) | No liver disease (C) | Total |
|--------------------------|-------------------|----------------------|-------|
| Abnormal liver scan (P)  | 231               | 32                   | 263   |
| Normal liver scan (N)    | 27                | 54                   | 81    |
| total                    | 258               | 86                   | 344   |

Find the sensitivity, specificity, prevalence, PPV, NPV, LRP, LRN, and concordance. Interpret the meaning of a LRP.

4.9. **Apparent Prevalence.** When the disease status in a sample is not known, the prevalence cannot be estimated directly. It is estimated using apparent prevalence. There is a distinction between the true prevalence (Pre – the proportion of a population with the disease) and apparent prevalence (APre – the proportion of the population that tests positive for the disease). If the estimators of sensitivity, specificity, and apparent prevalence are available, show that the estimator of prevalence is

$$Pre = \frac{APre + Sp - 1}{Se + Sp - 1}.$$

4.10. **HAAH Improves the Test for Prostate Cancer.** A new procedure based on a protein called human aspartyl (asparaginyl) beta-hydroxylase, or HAAH, adds to the accuracy of standard prostate-specific antigen (PSA) testing for prostate cancer. The findings were presented at the 2008 Genitourinary Cancers Symposium (Keith et al. 2008).

The research involved 233 men with prostate cancer and 43 healthy men, all over 50 years old. Results showed that the HAAH test had an overall sensitivity of 95% and specificity of 93%.

Compared to the sensitivity and specificity of PSA (about 40%), this test may prove particularly useful for men with both low and high PSA scores. In men with high PSA scores (4 to 10), the addition of HAAH information could substantially decrease the number of unnecessary biopsies, according to the authors.

(a) From the reported percentages, construct a table with true positives, false positives, true negatives, and false negatives (tp, fp, tn, and fn). You will need to round to the nearest integer since the specificity and sensitivity were reported as integer percents.

(b) Suppose that for the men aged 50+ in the USA, the prevalence of prostate cancer is 7%. Suppose Jim Smith is randomly selected from this group and tested positive on the HAAH test. What is the probability that Jim has prostate cancer?

(c) Suppose that Bill Schneider is a randomly selected person from the sample of $n = 276$ ($= 233 + 43$) subjects involved in the HAAH study. What is the probability that Bill has prostate cancer if he tests positive and no other information is available? What do you call this probability? What is different here from (b)?

4.11. **Creatinine Kinase and Acute Myocardial Infraction.** In a study of 773 patients, Radack et al. (1986) used an elevated serum creatinine kinase concentration as a diagnostic test for acute myocardial infraction. The following thresholds of a diagnostic test have been suggested: 481, 361, 241, and 121 IU/l; if the creatine kinase concentration exceeds the selected threshold, the test for myocardial infraction is considered positive. The gold standard is dichotomized: myocardial infraction present (MIP) and myocardial infraction not present (MINP). Assume that the sample of 773 subjects is randomly selected from the population, so that the prevalence of the disease is estimated as 51/773.

| | MIP | MINP | Total |
|---|---|---|---|
| ≥ 481 IU/l | 9 | 14 | 23 |
| < 481 IU/l | 42 | 708 | 750 |
| ≥ 361 IU/l | 15 | 26 | 41 |
| < 361 IU/l | 36 | 696 | 732 |
| ≥ 241 IU/l | 22 | 50 | 72 |
| < 241 IU/l | 29 | 672 | 701 |
| ≥ 121 IU/l | 28 | 251 | 279 |
| < 121 IU/l | 23 | 471 | 494 |
| Total | 51 | 722 | 773 |

(a) For the test that uses 361 IU/l as a threshold, find the sensitivity, specificity, PPV, NPV, LRP, and degree of agreement.

(b) Using given thresholds plot the ROC curve. What threshold would you suggest?

(c) Find the area under the ROC curve.

4.12. **Asthma.** A medical research team wished to evaluate a proposed screening test for asthma. The test was given to a random sample of 100 patients with asthma and to an independent sample of 200 subjects without symptoms of the disease.

The two samples were drawn from a population of subjects who were 50 years old or older. The results are as follows.

| Test result | Asthma, $D$ | No asthma, $D^c$ | Total |
|---|---|---|---|
| Positive test $T$ | 92 | 13 | 105 |
| Negative test $T^c$ | 8 | 187 | 195 |
| Total | 100 | 200 | 300 |

(a) Using the numbers from the table, estimate the sensitivity and specificity. Interpret these proportions in terms of the problem, one sentence for each.

(b) The probability of $D$ (prevalence) as the rate of the disease in the relevant population ($\geq 50$ y.o.) is estimated to be 6.3%. Find the PPV using Bayes' rule.

---

**MATLAB FILES AND DATA SETS USED IN THIS CHAPTER**
http://springer.bme.gatech.edu/Ch4.ROC/

auc.m, hypothyroidism.m, Kinaseandmi.m rocada.m, RocDdimer.m, sesp.m, simulatetesting.m, simulatetseting2.m

ADA.mat, pasi.dat, roccreatine.vi, ROCTBCA.XLS

---

# CHAPTER REFERENCES

Cabrera-Abreu, J., Davies, P., Matek, Z., and Murphy, M. (2004). Performance of blood tests in diagnosis of inflammatory bowel disease in a specialist clinic. *Arch. Dis. Child.*, **89**, 69–71.

Casscells, W., Schoenberger, A., and Grayboys, T. (1978). Interpretation by physicians of clinical laboratory results. *New Engl. J. Med.*, **299**, 999–1000.

Don, M., Kwong, B., Tanaka, C., Brackmann, D., and Nelson, R. (2005). The stacked ABR: A sensitive and specific screening tool for detecting small acoustic tumors. *Audiol. Neurotol.*, **10**, 274–290.

Drum, D. E. and Christacopoulos, S. (1972). Hepatic seintigraphy in clinical decision making. *J. Nucl. Med.*, **13**, 908–915.

Evans, D. A. (1990). Estimated prevalence of Alzheimer's disease in the United States. *Milbank Q.*, **68**, 267–289.

Goldhaber, S. Z., Simons, G. R., Elliott, C. G., Haire, W. D., Toltzis, R., Blacklow, S. C., Doolittle M. H., and Weinberg, D. S. (1993). Quantitative plasma D-dimer levels among patients undergoing pulmonary angiography for suspected pulmonary embolism. *J. Am. Med. Assoc.*, **270**, 23, 819–822.

Goldstein, B. J. and Mushlin, A. I. (1987). Use of a single thyroxine test to evaluate ambulatory medical patients for suspected hypothyroidism. *J. Gen. Intern. Med.*, **2**, 20–24.

Hawkins, D. M., Garrett, J. A., and Stephenson, B. (2001). Some issues in resolution of diagnostic tests using an imperfect gold standard. *Stat. Med.*, **20**, 1987–2001.

Keith, S. N., Repoli, A. F., Semenuk, M., Harris, P. J., Ghanbari, H. A., Lebowitz, M. S. (2008). HAAH identifies prostate cancer, regardless of PSA level. Presentation at 2008 Genitourinary Cancers Symposium.

Mavromatis, B. H. and Kessler, C. M. (2001). D-Dimer testing: the role of the clinical laboratory in the diagnosis of pulmonary embolism. *J. Clin. Pathol.*, **54**, 664–668.

Parikh, R., Mathai, A., Parikh, S., Sekhar, G. C., Thomas, R. (2008). Understanding and using sensitivity, specificity and predictive values. *Indian J. Ophthalmol.*, **56**, 1, 45–50.

Qu, Y. and Hadgu, A. (1998). A model for evaluating sensitivity and specificity for correlated diagnostic tests in efficacy studies with an imperfect reference test. *J. Am. Stat. Assoc.*, **93**, 920–928.

Radack, K. L., Rouan, G., Hedges, J. (1986). The likelihood ratio: An improved measure for reporting and evaluating diagnostic test results. *Arch. Pathol. Lab. Med.*, **110**, 689–693.

Schipper, H., Kwok, C.-S., Rosendahl, S. M., Bandilla, D., Maes, O., Melmed, C., Rabinovitch, D., Burns, D. H. (2008). Spectroscopy of human plasma for diagnosis of idiopathic Parkinson's disease. *Biomark. Med.*, **2**, 3, 229–238.

Staquet, M., Rozencweig, M., Lee, Y.J., Muggia, F.M. (1981). Methodology for assessment of new dichotomous diagnostic tests. *J. Chronic. Dis.*, **34**, 599–610.

# Chapter 5
# Random Variables

*The generation of random numbers is too important to be left to chance.*

— Robert R. Coveyou

## 5.1 Introduction

Thus far we have been concerned with random experiments, events, and their probabilities. In this chapter we will discuss random variables and their probability distributions. The outcomes of an experiment can be associated with numerical values, and this association will help us arrive at the definition of a random variable.

A *random variable* is a variable whose numerical value is determined by the outcome of a random experiment.

Thus, a random variable is a mapping from the sample space of an experiment, $\mathscr{S}$, to a set of real numbers. In this respect, the term *random variable* is a misnomer. The more appropriate term would be *random function* or *random mapping*, given that $X$ maps a sample space $\mathscr{S}$ to real numbers. We generally denote random variables by capital letters $X, Y, Z, \ldots$.

*Example 5.1.* **Three Coin Tosses.** Suppose a fair coin is tossed three times. We can define several random variables connected with this experiment. For example, we can set $X$ to be the number of heads, $Y$ the difference between the number of heads and the number of tails, and $Z$ an indicator that heads appeared, etc.

Random variables $X$, $Y$, and $Z$ are fully described by their probability distributions, associated with the sample space on which they are defined.

For random variable $X$ the possible realizations are 0 (no heads in three flips), 1 (exactly one head), 2 (exactly two heads), and 3 (all heads). Fully describing random variable $X$ amounts to finding the probabilities of all possible realizations. For instance, the realization $\{X = 2\}$ corresponds to either outcome in the event $\{HHT, HTH, THH\}$. Thus, the probability of $X$ taking value 2 is equal to the probability of the event $\{HHT, HTH, THH\}$, which is equal to 3/8. After finding the probabilities for other outcomes, we determine the distribution of random variable $X$:

$$\begin{array}{c|cccc} X & 0 & 1 & 2 & 3 \\ \hline \text{Prob} & 1/8 & 3/8 & 3/8 & 1/8 \end{array}.$$

The *probability distribution* of a random variable $X$ is a table (assignment, rule, formula) that assigns probabilities to realizations of $X$, or sets of realizations.

Most random variables of interest to us will be the results of a random sampling. There is a general classification of random variables that is based on the nature of realizations they can take. Random variables that take values from a finite or countable set are called *discrete random variables*. Random variable $X$ from Example 5.1 is an example of a discrete random variable. Another type of random variable can take any value from an interval on a real line. These are called *continuous random variables*. The results of measurements are usually modeled by continuous random variables. Next, we will describe discrete and continuous random variables in a more structured manner.

## 5.2 Discrete Random Variables

Let random variable $X$ take discrete values $x_1, x_2, \ldots, x_n, \ldots$ with probabilities $p_1, p_2, \ldots, p_n, \ldots,$  $\sum_n p_n = 1$. The probability distribution function (PDF) is simply an assignment of probabilities to the realizations of $X$ and is given by the following table.

| $X$ | $x_1$ $x_2$ $\cdots$ $x_n$ $\cdots$ |
|-----|-------------------------------------|
| Prob | $p_1$ $p_2$ $\cdots$ $p_n$ $\cdots$ |

The probabilities $p_i$ sum up to 1: $\sum_i p_i = 1$. It is important to emphasize that discrete random variables can have an infinite number of realizations, as long as the infinite sum of the probabilities converges to 1. The PDF for discrete random variables is also called the probability mass function (PMF). The cumulative distribution function (CDF)

$$F(x) = P(X \leq x) = \sum_{n:x_n \leq x} p_n,$$

sums the probabilities of all realizations smaller than or equal to $x$. Figure 5.1a shows an example of a discrete random variable $X$ with four values and a CDF as the sum of probabilities in the range $X \leq x$ shown in yellow.

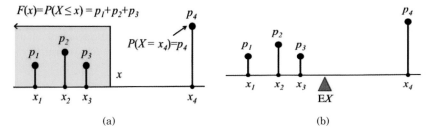

**Fig. 5.1** (a) An example of a cumulative distribution function for discrete random variable $X$. The CDF is the sum of probabilities in the region $X \leq x$ (*yellow*). (b) Expectation as a point of balance for "masses" $p_1, \ldots, p_4$ located at the points $x_1, \ldots, x_4$.

The expectation of $X$ is given by

$$\mathbb{E}X = x_1 p_1 + \cdots + x_n p_n + \cdots = \sum_n x_n p_n$$

and is a weighted average of all possible realizations with their probabilities as weights. Figure 5.1b illustrates the interpretation of the expectation as the point of balance for a system with weights $p_1,\ldots,p_4$ located at the locations $x_1,\ldots,x_4$.

The distribution and expectation of a function $g(X)$ are simple when $X$ is discrete: one applies function $g$ to realizations of $X$ and retains the probabilities:

$$\begin{array}{c|ccccc} g(X) & g(x_1) & g(x_2) & \cdots & g(x_n) & \cdots \\ \hline \text{Prob} & p_1 & p_2 & \cdots & p_n & \cdots \end{array}$$

and

$$\mathbb{E}g(X) = g(x_1)p_1 + \cdots + g(x_n)p_n + \cdots = \sum_n g(x_n)p_n.$$

The $k$th moment of a discrete random variable $X$ is defined as $\mathbb{E}X^k = \sum_n x_n^k p_n$, and the $k$th central moment is $\mathbb{E}(X - \mathbb{E}X)^k = \sum_n (x_n - \mathbb{E}X)^k p_n$. The first moment is the expectation and the second central moment is the variance, $\mathbb{V}\text{ar}(X) = \mathbb{E}(X - \mathbb{E}X)^2$. Thus, the variance for a discrete random variable is

$$\mathbb{V}\text{ar}(X) = \sum_n (x_n - \mathbb{E}X)^2 p_n.$$

The following properties are common for both discrete and continuous random variables.

For any set of random variables $X_1, X_2, \ldots, X_n$

$$\mathbb{E}(X_1 + X_2 + \cdots + X_n) = \mathbb{E}X_1 + \mathbb{E}X_2 + \cdots + \mathbb{E}X_n. \tag{5.1}$$

For any constant $c$, $\mathbb{E}(c) = c$ and $\mathbb{E}cX = c\mathbb{E}X$.

The independence of two random variables is defined via the independence of events. Two random variables $X$ and $Y$ are independent if for arbitrary intervals $A$ and $B$, the events $\{X \in A\}$ and $\{Y \in B\}$ are independent, that is, when

$$\mathbb{P}(X \in A, Y \in B) = \mathbb{P}(X \in A) \cdot \mathbb{P}(Y \in B),$$

holds.

If the random variables $X_1, X_2, \ldots, X_n$ are independent, then

$$\mathbb{E}(X_1 \cdot X_2 \cdots \cdot X_n) = \mathbb{E}X_1 \cdot \mathbb{E}X_2 \cdots \cdot \mathbb{E}X_n, \quad \text{and}$$

$$\mathbb{V}\text{ar}(X_1 + X_2 + \cdots + X_n) = \mathbb{V}\text{ar}\,X_1 + \mathbb{V}\text{ar}\,X_2 + \cdots + \mathbb{V}\text{ar}\,X_n. \tag{5.2}$$

For a constant $c$, $\mathbb{V}\text{ar}(c)=0$, and $\mathbb{V}\text{ar}(cX) = c^2 \mathbb{V}\text{ar}\,X$.

If $X_1, X_2, \ldots, X_n, \ldots$ are independent and identically distributed random variables, we will refer to them as i.i.d. random variables.

The arguments behind these properties involve the linearity of the sums (for discrete variables) and integrals (for continuous variables). The independence of the $X_i$s is critical for (5.2).

**Moment-Generating Function.** A particularly useful function for finding moments and for more advanced operations with random variables is the *moment-generating function*. For a random variable $X$, the moment-generating function is defined as

$$m_X(t) = \mathbb{E}e^{tX} = \sum_n p_n e^{tx_n}, \qquad (5.3)$$

which for discrete random variables has the form $m_X(t) = \sum_n p_n e^{tx_n}$. When the moment generating function exists, it uniquely determines the distribution. If $X$ has distribution $F_X$ and $Y$ has distribution $F_Y$, and if $m_X(t) = m_Y(t)$ for all $t$, then it follows that $F_X = F_Y$.

The name "moment generating" is motivated by the fact that the $k$th derivative of $m_X(t)$ evaluated at 0 results in the $k$th moment of $X$, that is, $m_X^{(k)}(t) = \sum_n p_n x_n^k e^{tx_n}$, and $\mathbb{E}X^k = m_X^{(k)}(0)$. For example, if

| $X$ | 0 | 1 | 3 |
|------|-----|-----|-----|
| Prob | 0.2 | 0.3 | 0.5 |

then $m_X(t) = 0.2 + 0.3\, e^t + 0.5\, e^{3t}$. Since $m_X'(t) = 0.3\, e^t + 1.5\, e^{3t}$, the first moment is $\mathbb{E}X = m'(0) = 0.3 + 1.5 = 1.8$. The second derivative is $m_X''(t) = 0.3\, e^t + 4.5\, e^{3t}$, the second moment is $\mathbb{E}X^2 = m''(0) = 0.3 + 4.5 = 4.8$, and so on.

In addition to generating the moments, moment generating functions satisfy

$$m_{X+Y}(t) = m_X(t)\, m_Y(t), \qquad (5.4)$$
$$m_{cX}(t) = m_X(ct),$$

which helps in identifying distributions of linear combinations of random variables whenever their moment-generating functions exist.

The properties in (5.4) follow from the properties of expectations. When $X$ and $Y$ are independent, $e^{tX}$ and $e^{tY}$ are independent as well, and by (5.2) $\mathbb{E}e^{t(X+Y)} = \mathbb{E}e^{tX}e^{tY} = \mathbb{E}e^{tX} \cdot \mathbb{E}e^{tY}$.

*Example 5.2.* **Apgar Score.** In the early 1950s, Dr. Virginia Apgar proposed a method to assess the health of a newborn child by assigning a grade referred to as the Apgar score (Apgar 1953). It is given twice for each newborn, once at 1 min after birth and again at 5 min after birth.

Possible values for the Apgar score are 0, 1, 2, $\cdots$, 9, and 10. A child's score is determined by five factors: muscle tone, skin color, respiratory effort, strength of heartbeat, and reflex, with a high score indicating a healthy infant. Let the random variable $X$ denote the Apgar score of a randomly selected newborn infant at a particular hospital. Suppose that $X$ has a given probability distribution:

| $X$ | 0 | 1 | 2 | 3 | 4 | 5 | 6 | 7 | 8 | 9 | 10 |
|------|------|------|------|------|-----|-----|-----|-----|-----|-----|-----|
| Prob | .002 | .001 | .002 | .005 | .02 | .04 | .17 | .38 | .25 | .12 | .01 |

The following MATLAB program calculates (a) $\mathbb{E}X$, (b) $\mathbb{V}ar(X)$, (c) $\mathbb{E}X^4$, (d) $F(x)$, (e) $\mathbb{P}(X < 4)$, and (f) $\mathbb{P}(2 < X \le 3)$:

```
X = 0:10;
p = [0.002  0.001  0.002  0.005  0.02  ...
        0.04  0.17 0.38  0.25  0.12  0.01];
EX = X * p'                    %(a) EX = 7.1600
VarX = (X-EX).^2 * p'          %(b) VarX = 1.5684
EX4 = X.^4 * p'                %(c) EX4 = 3.0746e+003
ps = [0 cumsum(p)];
Fx = @(x)   ps( min(max( floor(x)+2, 1),12) );  %handle
Fx(3.45)                       %(d) ans = 0.0100
sum(p(X < 4))                  %(e) ans = 0.0100
sum(p(X > 2 & X <= 3))         %(f) ans = 0.0050
```

Note that the CDF $F$ is expressed as function handle Fx to a custom-made function.

*Example 5.3.* **Cells.** Randomly observed circular cells on a plate have a diameter $D$ that is a random variable with the following PMF:

| $D$ | 8 | 12 | 16 |
|------|-----|-----|-----|
| Prob | 0.4 | 0.3 | 0.3 |

(a) Find the CDF for $D$.

(b) Find the PMF for the random variable $A = D^2\pi/4$ (the area of a cell). Show that $\mathbb{E}A \ne (\mathbb{E}D)^2\pi/4$. Explain.

(c) Find the variance $\mathbb{V}ar(A)$.

(d) Find the moment-generating functions $m_D(t)$ and $m_A(t)$. Find $\mathbb{V}ar(A)$ using its moment-generating function.

(e) It is known that a cell with $D > 8$ is observed. Find the probability of $D = 12$ taking into account this information.

**Solution:**

(a)

$$F_D(d) = \begin{cases} 0, & d < 8 \\ 0.4, & 8 \le d < 12 \\ 0.7, & 12 \le d < 16 \\ 1, & d \ge 16 \end{cases}$$

(b)

| $A$ | $8^2\pi/4$ | $12^2\pi/4$ | $16^2\pi/4$ |
|------|------|------|------|
| Prob | 0.4 | 0.3 | 0.3 |

| $A$ | $16\pi$ | $36\pi$ | $64\pi$ |
|------|------|------|------|
| Prob | 0.4 | 0.3 | 0.3 |

$\mathbb{E}A = 16\pi(\frac{4}{10}) + 36\pi(\frac{3}{10}) + 64\pi(\frac{3}{10}) = \frac{364\pi}{10} = 114.3540.$

$\mathbb{E}D = 8(\frac{4}{10}) + 12(\frac{3}{10}) + 16(\frac{3}{10}) = 116/10 = 11.6$

$\frac{(\mathbb{E}D)^2\pi}{4} = \frac{3364\pi}{100} \neq \frac{364\pi}{10}.$

The expectation is a linear operator, and such a "plug-in" operation would work only if the random variable $A$ were a linear function of $D$, i.e., if $A = \alpha D + \beta$, $\mathbb{E}A = \alpha\mathbb{E}D + \beta$. In our case, $A$ is quadratic in $D$, and "passing" the expectation through the equation is not valid.

(c)

$$\mathbb{V}\text{ar }A = \mathbb{E}A^2 - (\mathbb{E}A)^2 = 1720\pi^2 - 1324.96\pi^2 = 395.04\pi^2,$$

since

| $A^2$ | $16^2\pi^2$ | $36^2\pi^2$ | $64^2\pi^2$ |
|------|------|------|------|
| Prob | 0.4 | 0.3 | 0.3 |

and $\mathbb{E}A^2 = 1720\pi^2$.

(d) $m_D(t) = \mathbb{E}e^{tD} = 0.4e^{8t} + 0.3e^{12t} + 0.3e^{16t}$, and $m_A(t) = \mathbb{E}e^{tA} = 0.4e^{16\pi t} + 0.3e^{36\pi t} + 0.3e^{64\pi t}$.

From $m'_A(t) = 6.4e^{16\pi t} + 10.8e^{36\pi t} + 19.2e^{64\pi t}$, and $m''_A(t) = 6.4e^{16\pi t} + 10.8e^{36\pi t} + 19.2e^{64\pi t}$, we find $m'_A(0) = 36.4\pi$ and $m''_A(0) = 1720\pi$, leading to the result in (c).

(e) When $D > 8$ is true, only two values for $D$ are possible, 12 and 16. These values are equally likely. Thus, the distribution for $D|\{D > 8\}$ is

| $D|\{D > 8\}$ | 12 | 16 |
|------|------|------|
| Prob | 0.3/0.6 | 0.3/0.6 |

and $\mathbb{P}(D = 12|D > 8) = 1/2$. We divided 0.3 by 0.6 since $\mathbb{P}(D > 8) = 0.6$. From the definition of the conditional probability it follows that, $\mathbb{P}(D = 12|D > 8) = \mathbb{P}(D = 12, D > 8)/\mathbb{P}(D > 8) = \mathbb{P}(D = 12)/\mathbb{P}(D > 8) = 0.3/0.6 = 1/2.$

There are important properties of discrete distributions in which the realizations $x_1, x_2, \ldots, x_n$ are irrelevant and the focus is on the probabilities only, for example, the measure of *entropy*. For a discrete random variable where the probabilities are $\boldsymbol{p} = (p_1, p_2, \ldots, p_n)$ the (Shannon) entropy is defined as

$$\mathcal{H}(\boldsymbol{p}) = -\sum_i p_i \log(p_i).$$

Entropy is a measure of the uncertainty of a random variable and for finite discrete distributions achieves its maximum when the probabilities of realizations are equal, $\boldsymbol{p} = (1/n, 1/n, \ldots, 1/n)$.

For the distribution in Example 5.2, the entropy is 1.5812.

```
ps = [.002   .001   .002   .005   .02   .04   .17   .38  .25 .12 .01]
entropy = @(p)   -sum( p(p>0) .* log(p(p>0)))
entropy(ps) %1.5812
```

The maximum entropy for distributions with 11 possible realizations is 2.3979.

### 5.2.1 *Jointly Distributed Discrete Random Variables*

So far we have discussed probability distributions of a single random variable. As we delve deeper into this subject, a two-dimensional extension will be
needed.

When two or more random variables constitute the coordinates of a random
vector, their joint distribution is often of interest. For a random vector $(X,Y)$
the joint distribution function is defined via the probability of the event $\{X \le
x, Y \le y\}$,

$$F(x,y) = \mathbb{P}(X \le x, Y \le y).$$

The univariate case $\mathbb{P}(a \le X \le b) = F(b) - F(a)$ takes the bivariate form

$$\mathbb{P}(a_1 \le X \le a_2, b_1 \le Y \le b_2) = F(a_2, b_2) - F(a_1, b_2) - F(a_2, b_1) + F(a_1, b_1).$$

Marginal CDFs $F_X$ and $F_Y$ are defined as follows: for $X$, $F_X(x) = F(x, \infty)$
and for $Y$ as $F_Y(y) = F(\infty, y)$.
For a discrete bivariate random variable, the PMF is

$$p(x,y) = \mathbb{P}(X = x, Y = y), \quad \sum_{x,y} p(x,y) = 1,$$

while for marginal random variables $X$ and $Y$ the PMFs are

$$p_X(x) = \sum_y p(x,y), \quad p_Y(y) = \sum_x p(x,y).$$

The conditional distribution of $X$ given $Y = y$ is defined as

$$p_{X|Y=y}(x) = p(x,y)/p_Y(y),$$

and, similarly, the conditional distribution for $Y$ given $X = x$ is

$$p_{Y|X=x}(y) = p(x,y)/p_X(x).$$

When $X$ and $Y$ are independent, for any "cell" $(x,y)$, $p(x,y) = \mathbb{P}(X = x, Y =
y) = \mathbb{P}(X = x)\mathbb{P}(Y = y) = p_X(x)p_Y(y)$, that is, the joint probability of $(x,y)$
is equal to the product of the marginal probabilities. If, on the other hand,

$p(x,y) = p_X(x)p_Y(y)$ holds for every $(x,y)$, then $X$ and $Y$ are independent. The independence of two discrete random variables is fundamental for the inference in contingency tables (Chap. 14) and will be revisited later.

*Example 5.4.* The PMF of a two-dimensional discrete random variable is given by the following table:

|   |   | Y | |
|---|---|---|---|
|   | 5 | 10 | 15 |
| X   1 | 0.1 | 0.2 | 0.3 |
|     2 | 0.25 | 0.1 | 0.05 |

The marginal distributions for $X$ and $Y$ are

| X | 1 | 2 |
|---|---|---|
| Prob | 0.6 | 0.4 |

and

| Y | 5 | 10 | 15 |
|---|---|---|---|
| Prob | 0.35 | 0.3 | 0.35 |

while the conditional distribution for $X$ when $Y = 10$ and the conditional distribution for $Y$ when $X = 2$ are

| $X\vert Y = 10$ | 1 | 2 |
|---|---|---|
| Prob | 0.2/0.3 | 0.1/0.3 |

and

| $Y\vert X = 2$ | 5 | 10 | 15 |
|---|---|---|---|
| Prob | 0.25/0.4 | 0.1/0.4 | 0.05/0.4 |

respectively.

Here $X$ and $Y$ are not independent since

$$0.1 = \mathbb{P}(X = 1, Y = 5) \neq \mathbb{P}(X = 1)\mathbb{P}(Y = 5) = 0.6 \cdot 0.35 = 0.21.$$

For two independent random variables $X$ and $Y$, $\mathbb{E}XY = \mathbb{E}X \cdot \mathbb{E}Y$, that is, the expectation of a product of random variables is equal to the product of their expectations.

The *covariance* of two random variables $X$ and $Y$ is defined as

$$\mathrm{Cov}(X,Y) = \mathbb{E}((X - \mathbb{E}X) \cdot (Y - \mathbb{E}Y)) = \mathbb{E}XY - \mathbb{E}X \cdot \mathbb{E}Y.$$

For a discrete random vector $(X,Y)$, $\mathbb{E}XY = \sum_x \sum_y xy p(x,y)$, and the covariance is expressed as

$$\mathrm{Cov}(X,Y) = \sum_x \sum_y xy p(x,y) - \sum_x x p_X(x) \sum_y y p_Y(y).$$

It is easy to see that the covariance satisfies the following properties:

$$\mathbb{C}\mathrm{ov}(X,X) = \mathbb{V}\mathrm{ar}\,(X),$$
$$\mathbb{C}\mathrm{ov}(X,Y) = \mathbb{C}\mathrm{ov}(Y,X),\text{ and}$$
$$\mathbb{C}\mathrm{ov}(aX + bY,Z) = a\,\mathbb{C}\mathrm{ov}(X,Z) + b\,\mathbb{C}\mathrm{ov}(Y,Z).$$

For $(X,Y)$ from Example 5.4 the covariance between $X$ and $Y$ is $-1$. The calculation is provided in the following MATLAB code. Note that the distribution of the product $XY$ is found in order to calculate $\mathbb{E}XY$.

```
X=[1 2]; pX=[0.6 0.4]; EX = X * pX'
   %EX = 1.4000
Y=[5 10 15]; pY=[0.35  0.3  0.35]; EY = Y*pY'
   %EY =10
XY =[5 10 15 20 30];
pXY=[0.1   0.2+0.25   0.3  0.1  0.05];   EXY=XY * pXY'
   %EXY = 13
CovXY = EXY - EX * EY
   %CovXY = -1
```

The *correlation* between random variables $X$ and $Y$ is the covariance normalized by the standard deviations:

$$\mathbb{C}\mathrm{orr}(X,Y) = \frac{\mathbb{C}\mathrm{ov}(X,Y)}{\sqrt{\mathbb{V}\mathrm{ar}\,X \cdot \mathbb{V}\mathrm{ar}\,Y}}.$$

In Example 5.4, the variances of $X$ and $Y$ are $\mathbb{V}\mathrm{ar}\,X = 0.24$ and $\mathbb{V}\mathrm{ar}\,Y = 17.5$. Using these values, the correlation $\mathbb{C}\mathrm{orr}(X,Y)$ is $-1/\sqrt{0.24 \cdot 17.5} = -0.488$. Thus, the random components in $(X,Y)$ are negatively correlated.

## 5.3 Some Standard Discrete Distributions

### 5.3.1 Discrete Uniform Distribution

A random variable $X$ that takes values from 1 to $n$ with equal probabilities of $1/n$ is called a discrete uniform random variable. In MATLAB unidpdf and unidcdf are the PDF and CDF of $X$, while unidinv is its quantile. For example,

```
unidpdf(1:5, 5)
%ans =   0.2000    0.2000    0.2000    0.2000    0.2000

unidcdf(1:5, 5)
%ans =   0.2000    0.4000    0.6000    0.8000    1.0000
```

are the PDF and CDF of the discrete uniform distribution on $\{1,2,3,4,5\}$. From $\sum_{i=1}^{n} i = n(n+1)/2$, and $\sum_{i=1}^{n} i^2 = n(n+1)(2n+1)/6$ one can derive $\mathbb{E}X = (n+$

1)/2 and $\mathbb{V}\text{ar}\, X = (n^2 - 1)/12$. One of the important uses of discrete uniform distribution is in nonparametric statistics (p. 482).

*Example 5.5.* **Discrete Uniform: A Basis for Random Sampling.** Suppose that a population is finite and that we need a sample such that every subject in the population has an equal chance of being selected.

If the population size is $N$ and a sample of size $n$ is needed, then if replacement is allowed (each sampled object is recorded and then returned back to the population), there would be $N^n$ possible equally likely samples. If replacement is not allowed or possible (all subjects in the selected sample are to be different, that is, sampling is without replacement), then there would be $\binom{N}{n}$ different equally likely samples (see Sect. 3.5 for a definition of $\binom{N}{n}$).

The theoretical model for random sampling is the discrete uniform distribution. If replacement is allowed, each of $\{1, 2, \ldots, N\}$ has a probability of $1/N$ of being selected. In the case of no replacement, possible subsets of $n$ subjects can be indexed as $\{1, 2, \ldots, \binom{N}{n}\}$ and each subset has a probability of $1/\binom{N}{n}$ of being selected.

In MATLAB, random sampling is achieved by the function `randsample`. If the population has $n$ indexed subjects (from 1 to $n$), the indices in a random sample of size $k$ are found as `indices=randsample(n,k)`.

If it is possible to code the entire population as a vector `population`, then taking a sample of size $k$ is done by `y=randsample(population,k)`.

The default is set to sampling without replacement. For sampling with replacement, the flag for replacement should be `'true'`. If the sampling is done with replacement, it can be weighted with a nonnegative weight assigned to each subject in the population: `y=randsample(population,k,true,w)`. The size of weight vector `w` should be the same as that of `population`.

For instance,

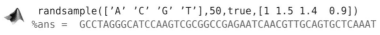

```
randsample(['A' 'C' 'G' 'T'],50,true,[1 1.5 1.4  0.9])
%ans =  GCCTAGGGCATCCAAGTCGCGGCCGAGAATCAACGTTGCAGTGCTCAAAT
```

## 5.3.2 *Bernoulli and Binomial Distributions*

A simple Bernoulli random variable $Y$ is dichotomous with $\mathbb{P}(Y = 1) = p$ and $\mathbb{P}(Y = 0) = 1 - p$ for some $0 \le p \le 1$ and is denoted as $Y \sim \mathscr{B}er(p)$. It is named after Jakob Bernoulli (1654–1705) a prominent Swiss mathematician and astronomer (Fig. 5.3a). Suppose that an experiment consists of $n$ independent trials $(Y_1, \ldots, Y_n)$ in which two outcomes are possible (e.g., success or failure), with $\mathbb{P}(\text{success}) = \mathbb{P}(Y = 1) = p$ for each trial. If $X = x$ is defined as the number of successes (out of $n$), then $X = Y_1 + Y_2 + \cdots + Y_n$ and there are $\binom{n}{x}$ arrangements of $x$ successes and $n - x$ failures, each having the same probability $p^x (1-p)^{n-x}$. $X$ is a *binomial* random variable with the PMF

$$p_X(x) = \binom{n}{x} p^x (1-p)^{n-x}, \; x = 0, 1, \ldots, n.$$

This is denoted by $X \sim \mathscr{B}in(n, p)$. From the moment-generating function $m_X(t) = (pe^t + (1-p))^n$ we obtain $\mu = \mathbb{E}X = np$ and $\sigma^2 = \mathbb{V}ar\,X = np(1-p)$.

The cumulative distribution for a binomial random variable is not simplified beyond the sum, i.e., $F(x) = \sum_{i \le x} p_X(i)$. However, interval probabilities can be computed in MATLAB using binocdf(x,n,p), which computes the CDF at value $x$. The PMF can also be computed in MATLAB using binopdf(x,n,p). In WinBUGS, the binomial distribution is denoted as dbin(p,n). Note the opposite order of parameters $n$ and $p$.

*Example 5.6.* **Left-Handed Families.** About 10% of the world's population is left-handed. Left-handedness is more prevalent in men (1/9) than in women (1/13). Studies have shown that left-handedness is linked to the gene LR-RTM1, which affects the symmetry of the brain. In addition to its genetic origins, left-handedness also has developmental origins. When both parents are left-handed, a child has a probability of 0.26 of being left-handed.

Ten families in which both parents are left-handed and have a single child are selected, and the ten children are inspected for left-handedness. Let $X$ be the number of left-handed among the inspected. What is the probability that $X$

    (a) Is equal to 3?

    (b) Falls anywhere between 3 and 6, inclusive?

    (c) Is at most 4?

    (d) Is not less than 4?

    (e) Would you be surprised if the number of left-handed children among the ten inspected was eight? Why or why not?

The solution is given by the following annotated MATLAB script.

```
% Solution
disp('(a) Bin(10, 0.26): P(X = 3)');
binopdf(3, 10, 0.26)
 % ans = 0.2563
disp('(b) Bin(10, 0.26): P(3 <= X <= 6)');
 % using binopdf(x, n, p)
disp('(b)-using PDF');    binopdf(3, 10, 0.26) + ...
binopdf(4, 10, 0.26) + binopdf(5, 10, 0.26)+ binopdf(6, 10, 0.26)
 % using binocdf(x, n, p)
disp('(b)-using CDF'); binocdf(6, 10, 0.26) - binocdf(2, 10, 0.26)
 % ans = 0.4998
%(c) at most four i.e., X <= 4
disp('(c) Bin(10, 0.26): P(X <= 4)'); binocdf(4, 10, 0.26)
 % ans = 0.9096
%(d) not less than 4 is 4,5,...,10, or complement of <=3
disp('(d)  Bin(12, 0.7): P(X >= 4)'); 1-binocdf(3, 10, 0.26)
 % ans = 0.2479
disp('(e) Bin(10, 0.26): P(X = 8)');
```

```
binopdf(8, 10, 0.26)
 % ans = 5.1459e-004
 % Yes, this is a surprising outcome since the probability
 % of this event is rather small, 0.0005.
```

Panels (a) and (b) in Fig. 5.2 show respectively the PMF and CDF for the binomial $\mathscr{B}in(10, 0.26)$ distribution.

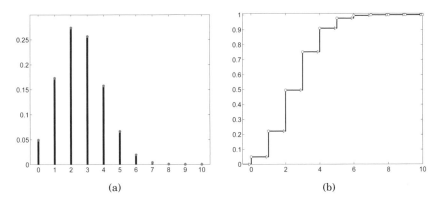

(a)                                          (b)

**Fig. 5.2** Binomial $\mathscr{B}in(10, 0.26)$ (a) PMF and (b) CDF.

How does one recognize that random variable $X$ has a binomial distribution?

(a) It allows an interpretation as the sum of "successes" in $n$ Bernoulli trials, for $n$ fixed.

(b) The Bernoulli trials are independent.

(c) The Bernoulli probability $p$ is constant for all $n$ trials.

Next we discuss how to deal with a binomial-like framework in which condition (c) is violated.

**Generalized Binomial Sampling*.** Suppose that $n$ independent experiments are performed and that an event $A$ has a probability of $p_i$ of appearing in the $i$th experiment.

We are interested in the probability that $A$ appeared exactly $k$ times in the $n$ experiments. The binomial setup is not directly applicable since the probabilities of $A$ differ from experiment to experiment. However, the binomial setup is useful as a hint on how to solve the general case. In the binomial setup the probability of $k$ events $A$ in $n$ experiments is equal to the coefficient of $z^k$ in the expansion of $G(z) = (pz + q)^n$. Indeed, $(pz + q)^n = p^n q^0 z^n + \cdots + \binom{n}{k} p^k q^{n-k} z^k + \cdots + npq^{n-1}z + p^0 q^n$.

The polynomial $G(z)$ is called the *probability-generating function.* If $X$ is a discrete integer-valued random variable such that $p_n = \mathbb{P}(X = n)$, then its probability-generating function is defined as

$$G_X(z) = \mathbb{E}z^X = \sum_n p_n z^n.$$

Note that in the polynomial $G_X(z)$, the probability $p_n = \mathbb{P}(X = n)$ is the coefficient of the power $z^n$. Also, $G_X(e^z)$ is the moment-generating function $m_X(z)$.

In the general binomial setup, the polynomial $(pz + q)^n$ becomes

$$G_X(z) = (p_1 z + q_1) \times (p_2 z + q_2) \times \cdots \times (p_n z + q_n) = \sum_{i=0}^{n} a_i z^i \qquad (5.5)$$

and the probability that there are $k$ events $A$ in $n$ experiments is equal to the coefficient $a_k$ of $z^k$ in the polynomial $G_X(z)$. This follows from the two properties of $G(z)$: (i) When $X$ and $Y$ are independent, $G_{X+Y}(z) = G_X(z) G_Y(z)$, and (ii) if $X$ is a Bernoulli $\mathcal{B}er(p)$, then $G_X(z) = pz + q$.

*Example 5.7.* **System with Unreliable Components.** Let $S$ be a system consisting of ten unreliable components that work and fail independently of each other. The components are operational in some fixed time interval $[0, T]$ with the probabilities

```
ps =[0.5 0.3 0.2 0.5 0.6 0.4 0.2 0.4 0.7 0.8];
```

Let a random variable $X$ represent the number of components that remain operational after time $T$.

Find (a) the distribution for $X$ and (b) $\mathbb{E}X$ and $\mathbb{V}\mathrm{ar}\, X$.

```
ps =[0.5 0.3 0.2 0.5 0.6 0.4 0.2 0.4 0.7 0.8];
qs = 1- ps;
all = [ps' qs'];
[m n]= size(all);
Gz = [1]; %initial
for i = 1:m
    Gz = conv(Gz, all(i,:) );
    % conv as polynomial multiplication
end
%at the end, Gz is the product of p_i x + q_i
%
sum(Gz) %the sum is 1
probs = Gz(end:-1:1);
k = 0:10
% probs=[0.0010  0.0117  0.0578  0.1547  0.2507 ...
% 0.2582  0.1716  0.0727  0.0188  0.0027  0.0002]
EX = k * probs' %expectation 4.6
EX2 = k.^2 * probs';
VX = EX2 - (EX)^2 %variance 2.12
```

Note that in the above script we used the convolution operation conv to multiply polynomials, as in

```
conv([2 -1],[1 3 2])
% ans =   2   5   1   -2,
```

which is interpreted as $(2z - 1) \cdot (z^2 + 3z + 2) = 2z^3 + 5z^2 + z - 2$.

From the MATLAB calculations we find that the probability-generating function $G(z)$ from (5.5) is

$$G(z) = 0.00016128z^{10} + 0.00268992z^9 + 0.01883264z^8 + 0.07273456z^7 +$$
$$0.17155808z^6 + 0.25816544z^5 + 0.25070848z^4 + 0.15470576z^3 +$$
$$0.05777184z^2 + 0.01170432z + 0.00096768,$$

and the random variable $X$, the number of operational items, has the following distribution (after rounding to four decimal places):

| $X$ | 0 | 1 | 2 | 3 | 4 | 5 | 6 | 7 | 8 | 9 | 10 |
|---|---|---|---|---|---|---|---|---|---|---|---|
| Prob | 0.0010 | 0.0117 | 0.0578 | 0.1547 | 0.2507 | 0.2582 | 0.1716 | 0.0727 | 0.0188 | 0.0027 | 0.0002 |

The answers to (b) are $\mathbb{E}X = 4.6$ and $\mathbb{V}\text{ar}\,X = 2.12$.

Note that a "solution" in which one finds the average of the component probabilities, ps, as $\bar{p} = \frac{1}{10}(0.5 + 0.3 + \cdots + 0.8) = 0.46$, and then applies the standard binomial calculation will lead to the correct expectation, 4.6, because of linearity. However, the variance and probabilities for $X$ would be different. For example, the probability $\mathbb{P}(X = 4)$ would be binopdf(4,10,0.46)=0.2331, while the correct value is 0.2507.

(a)                                          (b)

**Fig. 5.3** (a) Jacob Bernoulli (1654–1705), Swiss mathematician and astronomer. His monograph *Ars Conjectandi*, published posthumously in 1713, contains his explorations in probability theory, states a form of the law of large numbers, and describes experiments that we call now Bernoulli trials. (b) Siméon Denis Poisson (1781–1840), French mathematician and physicist. His book *Recherches sur la probabilité des jugements en matières criminelles et matière civile*, published in 1837, applies probability theory to the decisions of juries. It introduces a discrete probability distribution, now known as the Poisson distribution.

*Example 5.8.* **Surviving Pairs.** Daniel Bernoulli (1700–1782), a nephew of Jacob Bernoulli, posed and solved the following problem. If among $N$ married pairs there are $m$ random deaths, what is the expected number of intact marriages?

Suppose that there are $N$ pairs of balls denoted by 1,1, 2,2, $\ldots$, $N$,$N$. If $m$ balls are selected at random and removed, what is the expected number of intact pairs? Consider the pair $i$. Define a Bernoulli random variable $Y_i$ equal to 1 if pair $i$ remains intact after the removal of $m$ balls, and 0 otherwise. Then the number of unaffected pairs $N_m$ would be the sum of all $Y_i$, for $i = 1,\ldots,N$.

The probability that pair $i$ is not affected by the removal of $m$ balls is

$$\frac{\binom{2N-2}{m}}{\binom{2N}{m}} = \frac{\frac{(2N-2)(2N-3)\ldots(2N-2-m+2)(2N-2-m+1)}{m!}}{\frac{2N(2N-1)\ldots(2N-m+2)(2N-m+1)}{m!}} = \frac{(2N-m)(2N-m-1)}{2N(2N-1)},$$

and it is equal to $\mathbb{E}Y_i$. If $N_m$ is the number of unaffected pairs, then

$$N_m = Y_1 + Y_2 + \cdots + Y_N$$
$$\mathbb{E}N_m = \mathbb{E}Y_1 + \mathbb{E}Y_2 + \cdots + \mathbb{E}Y_N = N\mathbb{E}Y_i = \frac{(2N-m)(2N-m-1)}{2(2N-1)}.$$

For example, if among $N = 1000$ couples there are 100 random deaths, then the expected number of unaffected couples is 902.4762. If among $N = 1000$ couples there are 1936 deaths, then a single couple is expected to remain intact.

Even though $N_m$ is the sum of $N$ Bernoulli random variables $Y_i$, each with the same probability $p = \frac{(2N-m)(2N-m-1)}{2N(2N-1)}$, it does not have a binomial distribution due to the dependence among $Y_i$s.

### 5.3.3 Hypergeometric Distribution

Suppose a box contains $m$ balls, $k$ of which are white and $m - k$ of which are black. Suppose we randomly select and remove $n$ balls from the box *without replacement*, so that when sampling is finished, there are only $m - n$ balls left in the box. If $X$ is the number of white balls in $n$ selected, then the probability that $X = x$ is

$$p_X(x) = \frac{\binom{k}{x}\binom{m-k}{n-x}}{\binom{m}{n}}, \qquad x \in \{0, 1, \ldots, \min\{n, k\}\}.$$

Random variable $X$ is called hypergeometric and denoted by $X \sim \mathcal{HG}(m, k, n)$, where $m, k$, and $n$ are integer parameters.

This PMF can be deduced by counting rules. There are $\binom{m}{n}$ different ways of selecting the $n$ balls from a box with a total of $m$ balls. From these (each equally likely), there are $\binom{k}{x}$ ways of selecting $x$ white balls from the $k$ white balls in the box and, similarly, $\binom{m-k}{n-x}$ ways of choosing the black balls. The probability is the ratio of these two numbers. The PDF and CDF of $\mathscr{HG}(40, 15, 10)$ are shown in Fig. 5.4.

It can be shown that the mean and variance for the hypergeometric distribution are, respectively,

$$\mu = \frac{nk}{m} \quad \text{and} \quad \sigma^2 = \left(\frac{nk}{m}\right)\left(\frac{m-k}{m}\right)\left(\frac{m-n}{m-1}\right).$$

The MATLAB commands for hypergeometric CDF, PDF, quantile, and a random number are `hygecdf`, `hygepdf`, `hygeinv`, and `hygernd`. WinBUGS does not have a built-in command for a hypergeometric distribution.

*Example 5.9.* **CASES.** In a group of 40 people, 15 are "CASES" and 25 are "CONTROLS." A sample of 10 subjects is selected [(A) with replacement and (B) without replacement]. Find the probability $\mathbb{P}$(at least 2 subjects are CASES).

```
%Solution
%(A) - with replacement (binomial case);
%Let X be the number of CASES. The event
%X is at least 2 is the complement of X <= 1.
disp('(A) Bin(10, 15/40): P(X >= 2)');  1 - binocdf(1, 10, 15/40)
% ans = 0.9363
% or
 1 - binopdf(0, 10, 15/40) - binopdf(1, 10, 15/40)
% ans = 0.9363

%B - without replacement (hypergeometric case) hygecdf(x, m, k, n)
% where m size of population,
% k - number of cases among m,  and n sample size.
disp('(B) HyGe(40,15,10): P(X >=2)'); 1 - hygecdf(1, 40, 15, 10)
% ans = 0.9600,   or
 1 - hygepdf(0, 40, 15, 10)- hygepdf(1, 40, 15, 10)
% ans = 0.9600
```

*Example 5.10.* **Capture-Recapture Models.** Suppose that an unknown number $m$ of animals inhabit a particular region. To assess the population size, ecologists often apply the following capture-recapture scheme. They catch $k$ animals, tag them and release them back into the region. After some time, when the tagged animals are expected to be mixed well with the untagged, a second catch of size $n$ is made. Suppose that $x$ animals in the second sample are found to be tagged.

If catching any animal is assumed equally likely, the number $x$ of tagged animals in the second sample is hypergeometric $\mathscr{HG}(m, k, n)$. Ecologists use

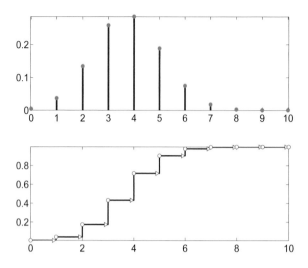

**Fig. 5.4** The PDF and CDF for hypergeometric distribution with $m = 40, k = 15$, and $n = 10$.

the observed ratio $x/n$ as an approximation to $k/m$, from which $m$ is estimated as

$$\hat{m} = \frac{k \times n}{x}.$$

A statistically better estimator of $m$ (known as the Schnabel formula) is given as

$$\hat{m} = \frac{(k + 1) \times (n + 1)}{(x + 1)} - 1.$$

In epidemiology and public health capture-recapture methods use multiple, routinely collected, computerized data sources to estimate various population indices.

For example, Gjini et al (2004) investigated the number of matching records of pneumococcal meningitis among adults in England by comparing data from Hospital Episode Statistics (HES) and the Public Health Laboratory Services reconciled laboratory records (RLR). The time period covered was April 1996 to December 1999. The authors found 646 records in RLR and 737 in HES, and matching based on demographic information was possible in 296 cases.

By the capture-recapture method the estimated incidence is $\hat{m} = 646 \cdot 737/296 = 1608.5 \approx 1609$. If Schnabel's formula is used, then $\hat{m} \approx 1607$. Thus, the total incidence of of pneumococcal meningitis in England between April 1996 to December 1999 is estimated to be 1607.

For large $m$, the hypergeometric distribution is close to binomial. More precisely, when $m \to \infty$ and $k/m \to p$, the hypergeometric distribution with parameters $(m, k, n)$ approaches a binomial with parameters $(n, p)$ for any value of $x$ between 0 and $n$.

```
format long
disp('(A)=(B) for large population');
1 - binocdf(1, 10, 150000/400000)   %ans = 0.936335370875895
1 - hygecdf(1, 400000, 150000, 10)  %ans = 0.936337703719839
```

We will use the hypergeometric distribution later in the book, (i) in the Fisher exact test (p. 546), and in Logrank test (p. 713).

### 5.3.4 Poisson Distribution

The PMF for the Poisson distribution is

$$p_X(x) = \frac{\lambda^x}{x!} e^{-\lambda}, \qquad x = 0, 1, 2, \ldots,$$

which is denoted by $X \sim \mathscr{P}oi(\lambda)$. This distribution is named after Simeon Denis Poisson (1781–1840), French mathematician, geometer and physicist (Fig. 5.3b).

From the moment-generating function $m_X(t) = \exp\{\lambda(e^t - 1)\}$ we have $\mathbb{E}X = \lambda$ and $\mathbb{V}\text{ar} X = \lambda$; the mean and the variance coincide.

The sum of a finite independent set of Poisson variables is also Poisson. Specifically, if $X_i \sim \mathscr{P}oi(\lambda_i)$, then $Y = X_1 + \cdots + X_k$ is distributed as $\mathscr{P}oi(\lambda_1 + \cdots + \lambda_k)$. If $X_1 \sim \mathscr{P}oi(\lambda_1)$ and $X_2 \sim \mathscr{P}oi(\lambda_2)$ are independent, then the distribution of $X_1$ given that $X_1 + X_2 = n$ is binomial $\mathscr{B}in\left(n, \frac{\lambda_1}{\lambda_1 + \lambda_2}\right)$ (Exercise 5.5).

Furthermore, the Poisson distribution is a limiting form for a binomial model, i.e.,

$$\lim_{n, np \to \infty, \lambda} \binom{n}{x} p^x (1 - p)^{n - x} = \frac{1}{x!} \lambda^x e^{-\lambda}. \tag{5.6}$$

The MATLAB commands for Poisson CDF, PDF, quantile, and random number are poisscdf, poisspdf, poissinv, and poissrnd. In WinBUGS the Poisson distribution is denoted as dpois(lambda).

*Example 5.11.* **Poisson Model for EBs.** After 7 days of aggregation, the microscopy images of 2000 embryonic bodies (EBs) are used to assess their surface area size. The probability that the area of a randomly selected EB exceeds the critical size $S_c$ is 0.001.

(a) Find the probability that the areas of exactly three EBs exceed the critical size.

(b) Find the probability that the number of EBs exceeding the critical size is between three and eight, inclusively.

We use a Poisson approximation to binomial probabilities since $n$ is large, $p$ is small, and product $np$ is moderate.

```
%Solution
disp('Poisson(2): P(X=3)'); poisspdf(3, 2)
  %ans= 0.1804
disp('Poisson(2): P(3 <= X <= 8)'); poisscdf(8, 2)-poisscdf(2, 2)
  %ans= 0.3231
```

Figure 5.5 shows the PMF and CDF of the $\mathscr{P}oi(2)$ distribution.

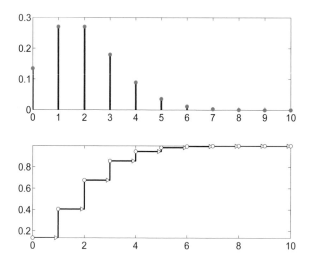

**Fig. 5.5** (Top) Poisson probability mass function. (Bottom) Cumulative distribution function for $\lambda = 2$.

In the binomial sampling scheme, when $n \to \infty$ and $p \to 0$, so that $np \to \lambda$, binomial probabilities converge to Poisson probabilities.

The following MATLAB simulation demonstrates the convergence. In the binomial distribution $n$ is increasing as 2,000, 200,000, 20,000,000 and $p$ is decreasing as 0.001, 0.00001, 0.0000001, so that $np$ remains constant and equal to 2. These binomial probabilities for $X = 3$ are compared to the Poisson distribution with the parameter $\lambda = 2$.

```
disp('P(X=3) for Bin(2000, 0.001), Bin(200000, 0.00001), ');
disp(' Bin(20000000, 0.0000001), and Poi(2) ');
format long
```

```
binopdf(3, 2000, 0.001)            % 0.180537328031786
binopdf(3, 200000, 0.00001)        % 0.180447946554779
binopdf(3, 20000000, 0.0000001)    % 0.180447058859339
poisspdf(3, 2)                     % 0.180447044315484
format short
```

*Example 5.12.* **Cold.** Suppose that the number of times during a year that an individual catches a cold can be modeled by a Poisson random variable with an expectation of 4. Further, suppose that a new drug based on vitamin C reduces this expectation to 3 (but the distribution still remains Poisson) for 90% of the population but has no effect on the remaining 10% of the population. We will calculate

(a) the probability that an individual taking the drug has two colds in a year if they are part of the population that benefits from the drug;

(b) the probability that a randomly chosen individual has two colds in a year if they take the drug; and

(c) the conditional probability that a randomly chosen individual is in that part of the population that benefits from the drug, given that they had two colds in the year during which they took the drug.

```
poisspdf(2,3)    %(Cold (a))
     %ans =   0.2240
poisspdf(2,3)*0.90 + poisspdf(2,4)*0.10    %(Cold (b))
     %ans = 0.2163
poisspdf(2,3)*0.90/(poisspdf(2,3)*0.90 + ...
                    poisspdf(2,4)*0.10) %(Cold (c))
     %ans = 0.9323
```

### 5.3.5 Geometric Distribution

Suppose that independent trials are repeated and that in each trial the probability of a success is equal to $0 < p < 1$. We are interested in the number of failures $X$ before the first success. The number of failures is a random variable with a geometric $\mathcal{G}e(p)$ distribution. Its PMF is given by

$$p_X(x) = p(1-p)^x, \qquad x = 0, 1, 2, \dots.$$

The expected value is $\mathbb{E}X = (1-p)/p$ and the variance is $\text{Var}\, X = (1-p)/p^2$. The moments can be found either directly or by the moment-generating function, which is

$$m_X(t) = \frac{p}{1 - (1-p)e^t}.$$

The geometric random variable possesses a "memoryless" property. That is, if we condition on the event $X \geq m$, for some nonnegative integer $m$, then for $n \geq m$, $\mathbb{P}(X \geq n | X \geq m) = \mathbb{P}(X \geq n - m)$ (Exercise 5.17). The MATLAB commands for geometric CDF, PDF, quantile, and random number are geocdf, geopdf, geoinv, and geornd. There are no special names for the geometric distribution in WinBUGS; the negative binomial can be used as dnegbin(p,1).

If instead of the number of failures before the first success ($X$) one is interested in the total number of experiments until the first success ($Y$), then the relationship is simple: $Y = X + 1$. In this formulation of the geometric distribution, $Y \sim \mathcal{G}eom(p)$, $\mathbb{E}Y = \mathbb{E}X + 1 = 1/p$, and $\mathbb{V}\mathrm{ar}\,Y = \mathbb{V}\mathrm{ar}\,X = (1 - p)/p^2$.

*Example 5.13.* **CASES I.** Let a subject constitute either a CASE or a CONTROL depending on whether the subject's level of LDL cholesterol is >160 mg/dL or ≤160 mg/dL, respectively. According to NHANES III, the prevalence of CASES among white male Americans aged 20 or older (target population) is $p = 20\%$. Subjects are sampled (when the population is large, it is unimportant if the sampling is done with or without replacement) until the first CASE is found. The number of CONTROLS sampled before finding the first CASE is a geometric random variable with parameter $p = 0.2$ (Fig. 5.6).

(a) Find the probability that seven CONTROLS will be sampled before we come across the first CASE.

(b) Find the probability that the number of CONTROLS before the first CASE will fall between four and eight, inclusively.

```
disp('X ~ Geometric(0.2):P(X=7)');
geopdf(7, 0.2)
     %ans=0.0419
disp('X ~ Geometric(0.2):P(4 <= X <= 8)');
geocdf(8, 0.2) - geocdf(3,0.2)
     %ans=0.2754
```

## 5.3.6 Negative Binomial Distribution

The negative binomial distribution was formulated by Montmort (1714). Suppose we are dealing with independent trials again. This time we count the number of failures observed until a fixed number of successes ($r \geq 1$) occur. Let $p$ be the probability of success in a single trial.

If we observe $r$ consecutive successes at the start of the experiment, then the count of failures is $X = 0$ and $\mathbb{P}(X = 0) = p^r$.

If $X = x$, then we have observed $x$ failures and $r$ successes in $x + r$ trials. There are $\binom{x+r}{x}$ different ways of arranging $x$ failures in those $x + r$ trials, but we can only be concerned with those arrangements in which the last

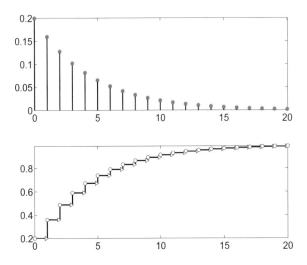

**Fig. 5.6** Geometric (Top) PMF and (Bottom) CDF for $p = 0.2$.

trial ended in a success. So there are really only $\binom{x+r-1}{x}$ equally likely arrangements. For any particular arrangement the probability is $p^r(1-p)^x$. Therefore, the PMF is

$$p_X(x) = \binom{r+x-1}{x} p^r (1-p)^x, \qquad x = 0, 1, 2, \ldots .$$

Sometimes this PMF is stated with $\binom{r+x-1}{r-1}$ in place of the equivalent $\binom{r+x-1}{x}$. This distribution is denoted as $X \sim \mathcal{NB}(r,p)$.

From its moment-generating function

$$m_X(t) = \left( \frac{p}{1 - (1-p)e^t} \right)^r,$$

the expectation of a negative binomial random variable is $\mathbb{E}X = r(1-p)/p$ and its variance is $\mathbb{V}\mathrm{ar}\,X = r(1-p)/p^2$.

Since the negative binomial $X \sim \mathcal{NB}(r,p)$ is a convolution (a sum) of $r$ independent geometric random variables, $X = Y_1 + Y_2 + \cdots + Y_r$, $Y_i \sim \mathcal{G}(p)$, the mean and variance of $X$ can be easily derived from the mean and variance of its geometric components $Y_i$, as in (5.1) and (5.2). Note also that $m_X(t) = (m_Y(t))^r$, where $m_Y(t) = \left( \frac{p}{1-(1-p)e^t} \right)$ is the moment-generating function of the component $Y_i$ in the sum. This is a consequence of the fact that a moment-generating function for a sum of independent random variables is the product of the moment-generating functions of the components, see (5.4).

The distribution remains valid if $r$ is not an integer, although an interpretation involving $r$ successes is lost. For an arbitrary nonnegative $r$, the distribution is called a generalized negative binomial or a Polya distribution. The constant $\binom{r+x-1}{x} = \frac{(r+x-1)!}{x!(r-1)!}$ is replaced by $\frac{\Gamma(r+x)}{x!\Gamma(r)}$, keeping in mind that $\Gamma(n) = (n-1)!$ when $n$ is an integer. The generalized negative binomial distribution is used in ecology for inference about the abundance of species in nature.

The MATLAB commands for negative binomial CDF, PDF, quantile, and random number are nbincdf, nbinpdf, nbininv, and nbinrnd. In WinBUGS the negative binomial distribution is denoted as dnegbin(p,r). Note the opposite order of parameters $r$ and $p$.

*Example 5.14.* **CASES II.** Assume as in Example 5.13 that the prevalence of "CASES" in a large population is $p = 20\%$. Subjects are sampled, one by one, until seven CASES are found and then the sampling is stopped.

(a) What is the probability that the number of CONTROLS among all sampled subjects will be 18?

(b) What is the probability of observing more than the "expected number" of CONTROLS?

The number of CONTROLS $X$ among all sampled subjects is a negative binomial, $X \sim \mathcal{NB}(7, 0.2)$.

$$P(X = 18) = \binom{25 + 7 - 1}{18} 0.2^7 (1 - 0.2)^1 8 = 0.0310.$$

Also, nbinpdf(18,7,0.2)=0.0310.

Thus, with a probability of 0.031 the number of CONTROLS sampled, before seven CASES are observed, is equal to 18.

(b) The expected number of CONTROLS is $\mathbb{E}X = 7\frac{0.8}{0.2} = 28$. The probability of $X > \mathbb{E}X$ is $P(X > 28) = 1 - P(X \le 28) = 1 - \sum_{x=0}^{28} \binom{7+x-1}{x} 0.8^x 0.2^7 = 0.4328$. In MATLAB $P(X > 28)$ is calculated as 1-nbincdf(28,7,0.20)=0.4328.

✎

The tail probabilities of a negative binomial distribution can be expressed by binomial probabilities. If $X \sim \mathcal{NB}(r, p)$, then

$$P(X > x) = P(Y < r),$$

where $Y \sim \mathcal{B}in(x + r, p)$. In words, if we have not seen $r$ successes after seing $x$ failures, then in $x + r$ experiments the number of successes will be less than $r$. In part (b) of the previous example, $r = 7, x = 28$, and $p = 0.20$, so

```
1 - nbincdf(28, 7, 0.20)    % 0.4328
binocdf(7-1, 28+7, 0.20)    % 0.4328
```

### 5.3.7 *Multinomial Distribution*

The binomial distribution was developed by counting the occurrences two complementary events, $A$ and $A^c$, in $n$ independent trials. Suppose, instead, that each trial results in one of $k > 2$ mutually exclusive events, $A_1, \ldots A_k$, so that $\mathscr{S} = A_1 \cup \cdots \cup A_k$. One can define the vector of random variables $(X_1, \ldots, X_k)$ where a component $X_i$ counts how many times $A_i$ appeared in $n$ trials. The defined random vector is called multinomial.

The probability mass function for $(X_1, \ldots, X_k)$ is

$$p_{X_1, \ldots, X_k}(x_1, \ldots, x_k) = \frac{n!}{x_1! \cdots x_k!} p_1^{x_1} \cdots p_k^{x_k},$$

where $p_1 + \cdots + p_k = 1$ and $x_1 + \cdots + x_k = n$. Since $p_k = 1 - p_1 - \cdots - p_{k-1}$, there are $k - 1$ free parameters to characterize the multinomial distribution, which is denoted by $X = (X_1, \ldots, X_k) \sim \mathcal{M}n(n, p_1, \ldots, p_k)$.

The mean and variance of the component $X_i$ are the same as in the binomial case. It is easy to see that the marginal distribution for a component $X_i$ is binomial since the events $A_1, \ldots, A_k$ can be grouped as $A_i, A_i^c$. Therefore, $\mathbb{E}(X_i) = np_i$, $\mathbb{V}\mathrm{ar}(X_i) = np_i(1 - p_i)$. The components $X_i$ are dependent since they sum up to $n$. For $i \neq j$, the covariance between $X_i$ and $X_j$ is

$$\mathbb{C}\mathrm{ov}(X_i, X_j) = \mathbb{E}X_i X_j - \mathbb{E}X_i \mathbb{E}X_j = -np_i p_j. \tag{5.7}$$

This is easy to verify if $X_i$ and $X_j$ are represented as the sums of Bernoullis $Y_{i1} + Y_{i2} + \cdots + Y_{ik}$ and $Y_{j1} + Y_{j2} + \cdots + Y_{jk}$, respectively. Since $Y_{im} Y_{jm} = 0$ (in a single trial $A_i$ and $A_j$ cannot occur simultaneously), it follows that

$$\mathbb{E}X_i X_j = (n^2 - n) p_i p_j.$$

Since $\mathbb{E}X_i \mathbb{E}X_j = n^2 p_i p_j$, the covariance in (5.7) follows.

If $X = (X_1, X_2, \ldots, X_k) \sim \mathcal{M}n(n, p_1, p_2, \ldots, p_k)$, then $X' = (X_1 + X_2, \ldots, X_k) \sim \mathcal{M}n(n, p_1 + p_2, \ldots, p_k)$. This is called the *fusing* property of the multinomial distribution.

If $X_1 \sim \mathcal{P}oi(\lambda_1)$, $X_2 \sim \mathcal{P}oi(\lambda_2)$, ..., $X_n \sim \mathcal{P}oi(\lambda_n)$ are $n$ independent Poisson random variables with parameters $\lambda_1, \ldots, \lambda_n$, then the conditional distribution of $X_1, X_2, \ldots, X_n$, given that $X_1 + X_2 + \cdots + X_n = n$, is $\mathcal{M}n(n, p_1, \ldots, p_k)$, where $p_i = \lambda_i/(\lambda_1 + \lambda_2 + \cdots + \lambda_n)$. This fact is used in modeling contingency tables with a fixed total and will be discussed in Chap. 14.

In MATLAB, the multinomial PMF is calculated by mnpdf(x,p), where $x$ is a $1 \times k$ vector of values, such that $\sum_{i=1}^{k} x_i = n$, and $p$ is a $1 \times k$ vector of probabilities, such that $\sum_{i=1}^{k} p_i = 1$.

For example,

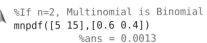

```
%If n=2, Multinomial is Binomial
mnpdf([5 15],[0.6 0.4])
      %ans = 0.0013
```

```
% is the same as
 binopdf(5, 5+15, 0.6)
          %ans = 0.0013
```

*Example 5.15.* Suppose that the probabilities of blood groups in a particular population are given as

| $O$ | $A$ | $B$ | $AB$ |
|------|------|------|------|
| 0.37 | 0.39 | 0.18 | 0.06 |

If eight subjects are selected at random from this population, what is the probability that

(a) $(O,A,B,AB) = (3,4,1,0)$?
(b) $O = 5$?

In (a), the probability is

```
   factorial(8) /(factorial(3) * ...
      factorial(4) * factorial(1) * factorial(0)) * ...
      0.37^3 * 0.39^4 * 0.18^1 * 0.06^0
   %ans = 0.0591
   %or
    mnpdf([3  4  1  0],[0.37 0.39 0.18 0.06])
   %ans = 0.0591.
```

In (b), $O \sim \mathscr{B}in(8, 0.37)$ and $\mathbb{P}(O = 3) = 0.2815$.

### 5.3.8 Quantiles

Quantiles of random variables are defined as follows. A $p$-quantile (or $100 \times p$ percentile) of random variable $X$ is the value $x$ for which $F(x) = p$, where $F$ is the cumulative distribution function for $X$. For discrete random variables this definition is not unique and modification is needed:

$$F(x) = \mathbb{P}(X \le x) \ge p \quad \text{and} \quad \mathbb{P}(X \ge x) \ge 1 - p.$$

For example, the 0.05 quantile of a binomial distribution with parameters $n = 12$ and $p = 0.7$ is $x = 6$ since $\mathbb{P}(X \le 6) = 0.1178 \ge 0.05$ and $\mathbb{P}(X \ge 6) = 1 - \mathbb{P}(X \le 5) = 1 - 0.0386 = 0.9614 \ge 0.95$. Binomial $\mathscr{B}in(12, 0.7)$ and geometric $\mathscr{G}(0.2)$ quantiles are shown in Fig. 5.7.

```
quab =[]; quag =[];
for p = 0.00:0.0001:1
    quab = [quab binoinv(p, 12, 0.7)];
    quag = [quag geoinv(p, 0.2)];
end
figure(1)
```

```
plot([0.00:0.0001:1],quab,'k-')
figure(2)
plot([0.00:0.0001:1],quag,'k-')
```

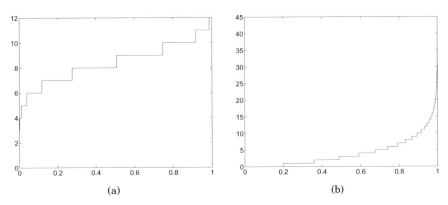

(a)                                                                (b)

**Fig. 5.7** (a) Binomial $\mathscr{B}in(12,0.7)$ and (b) geometric $\mathscr{G}(0.2)$ quantiles.

## 5.4 Continuous Random Variables

Continuous random variables take values within an interval $(a,b)$ on a real line **R**. The probability density function (PDF) $f(x)$ fully specifies the variable. The PDF is nonnegative, $f(x) \geq 0$, and integrates to 1, $\int_R f(x)\,dx = 1$. The probability that $X$ takes a value in an interval $(a,b)$ (and for continuous r.v. equivalently $[a,b), (a,b]$, or $[a,b]$) is $\mathbb{P}[X \in (a,b)] = \int_a^b f(x)dx$.

The CDF is

$$F(x) = \mathbb{P}(X \leq x) = \int_{-\infty}^x f(t)dt.$$

In terms of the CDF, $\mathbb{P}[X \in (a,b)] = F(b) - F(a)$.

The expectation of $X$ is given by

$$\mathbb{E}X = \int_R xf(x)dx.$$

The expectation of a function of a random variable $g(X)$ is

$$\mathbb{E}g(X) = \int_R g(x)f(x)dx.$$

The $k$th moment of a continuous random variable $X$ is defined as $\mathbb{E}X^k = \int_R x^k f(x)dx$, and the $k$th central moment is $\mathbb{E}(X - \mathbb{E}X)^k = \int_R (x - \mathbb{E}X)^k f(x)dx$. As in the discrete case, the first moment is the expectation and the second central moment is the variance, $\sigma^2(X) = \mathbb{E}(X - \mathbb{E}X)^2$.

The moment-generating function of a continuous random variable $X$ is

$$m(t) = \mathbb{E}e^{tX} = \int_{\mathbf{R}} e^{tx} f(x) dx.$$

Since $m^{(k)}(t) = \int_{\mathbf{R}} x^k e^{tx} f(x) dx$, $\mathbb{E}X^k = m^{(k)}(0)$.

Moment-generating functions are related to Laplace transforms of densities. Since the bilateral Laplace transform of $f(x)$ is defined as

$$\mathcal{L}(f) = \int_{\mathbf{R}} e^{-tx} f(x) dx,$$

it holds that $m(-t) = \mathcal{L}(f)$.

The entropy of a continuous random variable with a density $f(x)$ is defined as

$$\mathcal{H}(X) = -\int_{\mathbb{R}} f(x) \log f(x) dx$$

whenever this integral exists. Unlike the entropy for discrete random variables, $\mathcal{H}(X)$ can be negative and not necessarily invariant with respect to a transformation of $X$.

*Example 5.16.* **Markov's Inequality.** If $X$ is a random variable that takes only nonnegative values, then for any positive constant $a$

$$\mathbb{P}(X \geq a) \leq \frac{\mathbb{E}X}{a}. \tag{5.8}$$

Indeed,

$$\mathbb{E}X = \int_0^\infty x f(x) dx \geq \int_a^\infty x f(x) dx \geq \int_a^\infty a f(x) dx = a \int_a^\infty f(x) dx = a\mathbb{P}(X \geq a).$$

An average mass of a single cell of *E. coli* bacterium is 665 fg (femtogram, fg=$10^{-15}$g). If a particular cell of *E. coli* is inspected, what can be said about the probability that its weight will exceed 1000 fg? According to Markov's inequality, this probability does not exceed 665/1000 = 0.665.

## 5.4.1 Joint Distribution of Two Continuous Random Variables

Two random variables $X$ and $Y$ are jointly continuous if there exists a nonnegative function $f(x,y)$ so that for any two-dimensional domain $D$,

$$\mathbb{P}((X,Y) \in D) = \int\int_D f(x,y)dxdy.$$

When such a two-dimensional density $f(x,y)$ exists, it is a repeated partial derivative of the cumulative distribution function $F(x,y) = \mathbb{P}(X \le x, Y \le y)$,

$$f(x,y) = \frac{\partial^2 F(x,y)}{\partial x\, \partial y}.$$

The marginal densities for $X$ and $Y$ are, respectively, $f_X(x) = \int_{-\infty}^{\infty} f(x,y)dy$ and $f_Y(y) = \int_{-\infty}^{\infty} f(x,y)dx$. The conditional distributions of $X$ when $Y = y$ and of $Y$ when $X = x$ are

$$f(x|y) = f(x,y)/f_Y(y) \text{ and } f(y|x) = f(x,y)/f_X(x).$$

The distributional analogy of the multiplication probability rule $\mathbb{P}(AB) = \mathbb{P}(A|B)\mathbb{P}(B) = \mathbb{P}(B|A)\mathbb{P}(A)$ is

$$f(x,y) = f(x|y)f_Y(y) = f(y|x)f_X(x).$$

When $X$ and $Y$ are independent, the joint density is the product of marginal densities, $f(x,y) = f_X(x)f_Y(y)$. Conversely, if the joint density of $(X,Y)$ can be represented as a product of marginal densities, $X$ and $Y$ are independent.

The definition of covariance and the correlation for $X$ and $Y$ coincides with the discrete case equivalents:

$$\text{Cov}(X,Y) = \mathbb{E}XY - \mathbb{E}X \cdot \mathbb{E}Y \quad \text{and} \quad \text{Corr}(X,Y) = \frac{\text{Cov}(X,Y)}{\sqrt{\text{Var}(X) \cdot \text{Var}(Y)}}.$$

Here, $\mathbb{E}XY = \int_{\mathbb{R}^2} xyf(x,y)dxdy$.

*Example 5.17.* **Durability of the Starr–Edwards Valve.** The Starr–Edwards valve is one of the oldest cardiac valve prostheses in the world. The first aortic valve replacement (AVR) with a Starr–Edwards metal cage and silicone ball valve was performed in 1961. Follow-up studies have documented the excellent durability of the Starr–Edwards valve as an AVR. Suppose that the durability of the Starr–Edwards valve (in years) is a random variable $X$ with density

$$f(x) = \begin{cases} ax^2/100, & 0 < x < 10, \\ a(x-30)^2/400, & 10 \le x \le 30, \\ 0, & \text{otherwise.} \end{cases}$$

(a) Find the constant $a$.
(b) Find the CDF $F(x)$ and sketch graphs of $f$ and $F$.
(c) Find the mean and 60th percentile of $X$. Which is larger? Find the variance.

**Solution:** (a) Since $1 = \int_{\mathbb{R}} f(x)dx$,

$$1 = \int_0^{10} ax^2/100dx + \int_{10}^{30} a(x-30)^2/400dx = ax^3/300 \mid_0^{10} + a(x-30)^3/1200 \mid_{10}^{30}.$$

This gives $1000a/300 - 0 + 0 - (-20)^3a/1200 = 10a/3 + 20a/3 = 10a = 1$, that is, $a = 1/10$. The density is

$$f(x) = \begin{cases} x^2/1000, & 0 < x < 10, \\ (x-30)^2/4000, & 10 \le x \le 30, \\ 0, & \text{otherwise.} \end{cases}$$

(b) The CDF is

$$F(x) = \begin{cases} 0, & x < 0, \\ x^3/3000, & 0 < x < 10, \\ 1 + (x-30)^3/12000, & 10 \le x \le 30, \\ 1, & x \ge 30. \end{cases}$$

(c) The 60th percentile is a solution to the equation $1+(x-30)^3/12000 = 0.6$ and is $x = 13.131313....$ The mean is $\mathbb{E}X = 25/2$, and the 60th percentile exceeds the mean. $\mathbb{E}X^2 = 180$, thus the variance is $\text{Var}\,X = 180 - (25/2)^2 = 95/4 = 23.75$.
🖉

*Example 5.18.* A two-dimensional random variable $(X,Y)$ is defined by its density function, $f(x,y) = 2xe^{-x-2y}$, $x \ge 0, y \ge 0$.
 (a) Find the probability that random variable $(X,Y)$ falls in the rectangle $0 \le X \le 1, 1 \le Y \le 2$.
 (b) Find the marginal distributions of $X$ and $Y$.
 (c) Find the conditional distribution of $X|\{Y = y\}$ Does it depend on $y$?
 **Solution:** (a) The joint density separates variables $x$ and $y$, therefore

$$\mathbb{P}(0 \le X \le 1,\ 1 \le Y \le 2) = \int_0^1 xe^{-x}dx \times \int_1^2 2e^{-2y}dy.$$

Since

$$\int_0^1 xe^{-x}dx = -xe^{-x} \mid_0^1 + \int_0^1 e^{-x}dx = -e^{-1} - e^{-1} + 1 = 1 - 2/e,$$

and

$$\int_1^2 2e^{-2y}dy = -e^{-2y} \mid_1^2 = -e^{-4} + e^{-2} = \frac{e^2 - 1}{e^4}.$$

then

$$\mathbb{P}(0 \le X \le 1,\ 1 \le Y \le 2) = \frac{e-2}{e} \times \frac{e^2 - 1}{e^4} \approx 0.0309.$$

(b) Since the joint density separates the variables, it is a product of marginal densities $f(x,y) = f_X(x) \times f_Y(y)$. This is an analytic way to state that components $X$ and $Y$ are independent. Therefore, $f_X(x) = xe^{-x}$, $x \geq 0$ and $f_Y(y) = 2e^{2y}$, $y \geq 0$.

(c) The conditional densities for $X|\{Y = y\}$ and $Y|\{X = x\}$ are defined as

$$f(x|y) = f(x,y)/f_Y(y) \quad \text{and} \quad f(y|x) = f(x,y)/f_X(x).$$

Because of independence of $X$ and $Y$ the conditional densities coincide with the marginal densities. Thus, the conditional density for $X|\{Y = y\}$ does not depend on $y$.
✐

## 5.5 Some Standard Continuous Distributions

In this section we list some popular, commonly used continuous distributions: uniform, exponential, gamma, inverse gamma, beta, double exponential, logistic, Weibull, Pareto, and Dirichlet. The normal (Gaussian) distribution will be just briefly mentioned here. Due to its importance, a separate chapter will cover the details of the normal distribution and its close relatives: $\chi^2$, student $t$, Cauchy, $F$, and lognormal distributions. Some other continuous distributions will be featured in the examples, exercises, and other chapters, such as Maxwell and Rayleigh distributions.

### 5.5.1 Uniform Distribution

A random variable $X$ has a uniform $\mathscr{U}(a,b)$ distribution if its density is given by

$$f_X(x) = \begin{cases} \frac{1}{b-a}, & a \leq x \leq b, \\ 0, & \text{else}. \end{cases}$$

Sometimes, to simplify notation, the density can be written simply as

$$f_X(x) = \frac{1}{b-a} \mathbf{1}(a \leq x \leq b).$$

Here, $\mathbf{1}(A)$ is 1 if $A$ is a true statement and 0 if $A$ is false. Thus, for $x < a$ or $x > b$, $f_X(x) = 0$, since $\mathbf{1}(a \leq x \leq b) = 0$.

The CDF of $X$ is given by

$$F_X(x) = \begin{cases} 0, & x < a, \\ \frac{x-a}{b-a}, & a \le x \le b, \\ 1, & x > b. \end{cases}$$

The graphs of the PDF and CDF of a uniform $\mathcal{U}(-1,4)$ random variable are shown in Fig. 5.8.

The expectation of $X$ is $\mathbb{E}X = \frac{a+b}{2}$ and the variance is $\mathbb{V}\mathrm{ar}\, X = (b-a)^2/12$. The $n$th moment of $X$ is given by $\mathbb{E}X^n = \frac{1}{n+1} \sum_{i=0}^{n} a^i b^{n-i}$. The moment-generating function for the uniform distribution is $m(t) = \frac{e^{tb} - e^{ta}}{t(b-a)}$.

For $a = 0$ and $b = 1$, the distribution is called *standard uniform.*

If $U$ is $\mathcal{U}(0,1)$, then $X = -\lambda \log(U)$ is an exponential random variable with scale parameter $\lambda$.

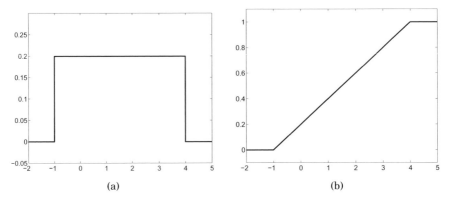

(a)                                                    (b)

**Fig. 5.8** (a) PDF and (b) CDF for uniform $\mathcal{U}(-1,4)$ distribution. The graphs are plotted as (a) `plot(-2:0.001:5, unifpdf(-2:0.001:5, -1, 4))` and (b) `plot(-2:0.001:5, unifcdf(-2:0.001:5, -1, 4))`.

The sum of two independent standard uniforms is triangular,

$$f_X(x) = \begin{cases} x, & 0 \le x \le 1, \\ 2 - x, & 1 \le x \le 2, \\ 0, & \text{else}. \end{cases}$$

This is sometimes called a "witch hat" distribution. The sum of $n$ independent standard uniforms is known as the Irwing–Hall distribution.

### 5.5.2 Exponential Distribution

The probability density function for an exponential random variable is

$$f_X(x) = \begin{cases} \lambda e^{-\lambda x}, & x \geq 0, \\ 0, & \text{else} , \end{cases}$$

where $\lambda > 0$ is called the *rate* parameter. An exponentially distributed random variable $X$ is denoted by $X \sim \mathcal{E}(\lambda)$. Its moment-generating function is $m(t) = \lambda/(\lambda - t)$ for $t < \lambda$, and the mean and variance are $1/\lambda$ and $1/\lambda^2$, respectively. The $n$th moment is $\mathbb{E}X^n = \frac{n!}{\lambda^n}$.

This distribution has several interesting features; for example, its *failure rate*, defined as

$$\lambda_X(t) = \frac{f_X(t)}{1 - F_X(t)},$$

is constant and equal to $\lambda$.

The exponential distribution has an important connection to the Poisson distribution. Suppose we observe i.i.d. exponential variates $(X_1, X_2, \ldots)$ and define $S_n = X_1 + \cdots + X_n$. For any positive value $t$, it can be shown that $\mathbb{P}(S_n < t < S_{n+1}) = p_Y(n)$, where $p_Y(n)$ is the probability mass function for a Poisson random variable $Y$ with parameter $\lambda t$.

Like a geometric random variable, an exponential random variable has the *memoryless property*, $\mathbb{P}(X \geq u + v | X \geq u) = \mathbb{P}(X \geq v)$ (Exercise 5.17).

The median value, representing a typical observation, is roughly 70% of the mean, showing how extreme values can affect the population mean. This is explicitly shown by the ease in computing the inverse CDF:

$$p = F(x) = 1 - e^{-\lambda x} \iff x = F^{-1}(p) = -\frac{1}{\lambda} \log(1 - p).$$

The MATLAB commands for exponential CDF, PDF, quantile, and random number are expcdf, exppdf, expinv, and exprnd. MATLAB uses the alternative parametrization with $1/\lambda$ in place of $\lambda$. Thus, the CDF of random variable $X$ with $\mathcal{E}(3)$ distribution evaluated at $x = 2$ is calculated in MATLAB as expcdf(2,1/3). In WinBUGS, the exponential distribution is coded as dexp(lambda).

*Example 5.19.* **Melanoma.** The 5-year cancer survival rate in the case of malignant melanoma of the skin at stage IIIA is 78%. Assume that the survival time $T$ can be modeled by an exponential random variable with unknown rate $\lambda$.

Using the given survival rate of 0.78, we first determine the parameter of the exponential distribution – the rate $\lambda$.

Since $\mathbb{P}(T > t) = \exp(-\lambda t)$, $\mathbb{P}(T > 5) = 0.78$ leads to $\exp\{-5\lambda\} = 0.78$, with solution $\lambda = -\frac{1}{5} \log(0.78)$, which can be rounded to $\lambda = 0.05$.

Next, we find the probability that the survival time exceeds 10 years, first directly using the CDF,

$$\mathbb{P}(T > 10) = 1 - F(10) = 1 - \left(1 - e^{-0.05 \cdot 10}\right) = \frac{1}{\sqrt{e}} = 0.6065,$$

and then by MATLAB. One should be careful when parameterizing the exponential distribution in MATLAB. MATLAB uses the scale parameter, a reciprocal of the rate $\lambda$.

```
1 - expcdf(10, 1/0.05)
   %ans =  0.6065
   %
   %Figures of PDF and CDF are produced by
time=0:0.001:30;
pdf = exppdf(time, 1/0.05);  plot(time, pdf, 'b-');
cdf = expcdf(time, 1/0.05);  plot(time, cdf, 'b-');
```

This is shown in Fig. 5.9.

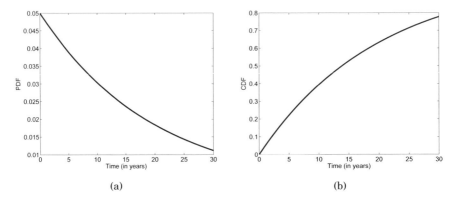

(a)                                                                                       (b)

**Fig. 5.9** Exponential (a) PDF and (b) CDF for rate $\lambda = 0.05$.

### 5.5.3 Normal Distribution

As indicated in the introduction, due to its importance, the normal distribution is covered in a separate chapter. Here we provide a definition and list a few important facts. The probability density function for a normal (Gaussian) random variable $X$ is given by

$$f_X(x) = \frac{1}{\sqrt{2\pi}\,\sigma}\,\exp\left\{-\frac{(x-\mu)^2}{2\sigma^2}\right\},$$

where $\mu$ is the mean and $\sigma^2$ is the variance of $X$. This will be denoted as $X \sim \mathcal{N}\left(\mu,\sigma^2\right)$.

For $\mu = 0$ and $\sigma = 1$, the distribution is called the standard normal distribution. The CDF of a normal distribution cannot be expressed in terms of elementary functions and defines a function of its own. For the standard normal distribution the CDF is

$$\Phi(x) = \int_{-\infty}^{x} \frac{1}{\sqrt{2\pi}} \exp\left\{-\frac{t^2}{2}\right\} dt.$$

The standard normal PDF and CDF are shown in Fig. 5.10a,b.

The moment-generating function is $m(t) = \exp\{\mu t + \sigma^2 t^2/2\}$. The odd central moments $\mathbb{E}(X - \mu)^{2k+1}$ are 0 because the normal distribution is symmetric about the mean. The even moments are

$$\mathbb{E}(X - \mu)^{2k} = \sigma^{2k} (2k - 1)!!,$$

where $(2k - 1)!! = (2k - 1) \cdot (2k - 3) \cdots 5 \cdot 3 \cdot 1$.

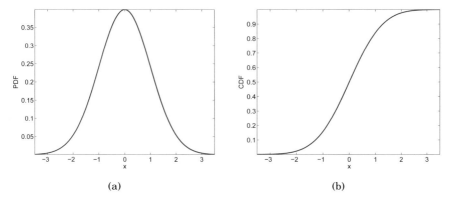

(a)            (b)

**Fig. 5.10** Standard normal (a) PDF and (b) CDF $\Phi(x)$.

The MATLAB commands for normal CDF, PDF, quantile, and a random number are `normcdf`, `normpdf`, `norminv`, and `normrnd`. In WinBUGS, the normal distribution is coded as `dnorm(mu,tau)`, where `tau` is a precision parameter, the reciprocal of variance.

### 5.5.4 Gamma Distribution

The gamma distribution is an extension of the exponential distribution. Prior to defining its density, we define the gamma function that is critical in normalizing the density. Function $\Gamma(x)$ defined via the integral $\int_0^\infty t^{x-1} e^{-t} dt, \quad x > 0$ is called the gamma function (Fig. 5.11a). If $n$ is a positive integer, then $\Gamma(n) = (n - 1)!$. In MATLAB: `gamma(x)`.

Random variable $X$ has a gamma $\mathcal{G}a(r, \lambda)$ distribution if its PDF is given by

$$f_X(x) = \begin{cases} \frac{\lambda^r}{\Gamma(r)} x^{r-1} e^{-\lambda x}, & x \geq 0, \\ 0, & \text{else} . \end{cases}$$

The parameter $r > 0$ is called the *shape* parameter, while $\lambda > 0$ is the *rate* parameter. Figure 5.11b shows gamma densities for $(r, \lambda) = (1, 1/3), (2, 2/3)$, and $(20, 2)$.

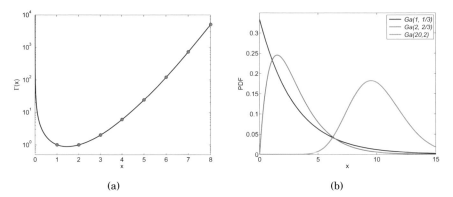

(a)                                          (b)

**Fig. 5.11** (a) Gamma function, $\Gamma(x)$. The *red dots* are values of the gamma function at integers, $\Gamma(n) = (n-1)!$; (b) Gamma densities: $\mathcal{G}a(1, 1/3)$, $\mathcal{G}a(2, 2/3)$, and $\mathcal{G}a(20, 2)$.

The moment-generating function is $m(t) = (\lambda/(\lambda - t))^r$, so in the case $r = 1$, the gamma distribution becomes the exponential distribution. From $m(t)$ we have $\mathbb{E}X = r/\lambda$ and $\mathbb{V}\text{ar}\, X = r/\lambda^2$.

If $X_1, \ldots, X_n$ are generated from an exponential distribution with (rate) parameter $\lambda$, it follows from $m(t)$ that $Y = X_1 + \cdots + X_n$ is distributed as gamma with parameters $\lambda$ and $n$; that is, $Y \sim \mathcal{G}a(n, \lambda)$. More generally, if $X_i \sim \mathcal{G}a(r_i, \lambda)$ are independent, then $Y = X_1 + \cdots + X_n$ is distributed as gamma with parameters $\lambda$ and $r = r_1 + r_2 + \cdots + r_n$,; that is, $Y \sim \mathcal{G}a(r, \lambda)$ (Exercise 5.16).

Often, the gamma distribution is parameterized with $1/\lambda$ in place of $\lambda$, and this alternative parametrization is used in MATLAB definitions. The CDF in MATLAB is `gamcdf(x,r,1/lambda)`, and the PDF is `gampdf(x,r,1/lambda)`. The function `gaminv(p,r,1/lambda)` computes the $p$th quantile of the $\mathcal{G}a(r, \lambda)$ random variable. In WinBUGS, $\mathcal{G}a(n, \lambda)$ is coded as `dgamma(n,lambda)`.

### 5.5.5 *Inverse Gamma Distribution*

Random variable $X$ is said to have an inverse gamma $\mathcal{IG}(r, \lambda)$ distribution with parameters $r > 0$ and $\lambda > 0$ if its density is given by

$$f_X(x) = \begin{cases} \frac{\lambda^r}{\Gamma(r)x^{r+1}} e^{-\lambda/x}, & x \geq 0, \\ 0, & \text{else}. \end{cases}$$

The mean and variance of $X$ are $\mathbb{E}X = \lambda/(r-1)$, $r > 1$, and $\mathbb{V}\text{ar}\,X = \lambda^2/[(r-1)^2(r-2)]$, $r > 2$, respectively. If $X \sim \mathcal{G}a(r,\lambda)$, then its reciprocal $X^{-1}$ is $\mathcal{I}\mathcal{G}(r,\lambda)$ distributed. We will see that in the Bayesian context, the inverse gamma is a natural prior distribution for a scale parameter.

### 5.5.6 Beta Distribution

We first define two special functions: beta and incomplete beta. The beta function is defined as $B(a,b) = \int_0^1 t^{a-1}(1-t)^{b-1}dt = \Gamma(a)\Gamma(b)/\Gamma(a+b)$. In MATLAB: beta(a,b). An incomplete beta is $B(x,a,b) = \int_0^x t^{a-1}(1-t)^{b-1}dt$, $0 \leq x \leq 1$. In MATLAB, betainc(x,a,b) represents the normalized incomplete beta defined as $I_x(a,b) = B(x,a,b)/B(a,b)$.

The density function for a beta random variable is

$$f_X(x) = \begin{cases} \frac{1}{B(a,b)}x^{a-1}(1-x)^{b-1}, & 0 \leq x \leq 1, \\ 0, & \text{else}, \end{cases}$$

and $B$ is the beta function. Because $X$ is defined only in the interval $[0,1]$, the beta distribution is useful in describing uncertainty or randomness in proportions or probabilities. A beta-distributed random variable is denoted by $X \sim \mathcal{B}e(a,b)$. The standard uniform distribution $\mathcal{U}(0,1)$ serves as a special case with $(a,b) = (1,1)$. The moments of beta distribution are

$$\mathbb{E}X^k = \frac{\Gamma(a+k)\Gamma(a+b)}{\Gamma(a)\Gamma(a+b+k)} = \frac{a(a+1)\ldots(a+k-1)}{(a+b)(a+b+1)\ldots(a+b+k-1)}$$

so that $\mathbb{E}(X) = a/(a+b)$ and $\mathbb{V}\text{ar}\,X = ab/[(a+b)^2(a+b+1)]$.

In MATLAB, the CDF for a beta random variable (at $x \in (0,1)$) is computed as betacdf(x,a,b), and the PDF is computed as betapdf(x,a,b). The $p$th percentile is betainv(p,a,b). In WinBUGS, the beta distribution is coded as dbeta(a,b).

To emphasize the modeling diversity of beta distributions, we depict densities for several choices of $(a,b)$, as in Fig. 5.12.

If $U_1, U_2, \ldots, U_n$ is a sample from a uniform $\mathcal{U}(0,1)$ distribution, then the distribution of the $k$th component in the ordered sample is beta, $U_{(k)} \sim \mathcal{B}e(k, n-k+1)$, for $1 \leq k \leq n$. Also, if $X \sim \mathcal{G}(m,\lambda)$ and $Y \sim \mathcal{G}(n,\lambda)$, then $X/(X+Y) \sim \mathcal{B}e(m,n)$.

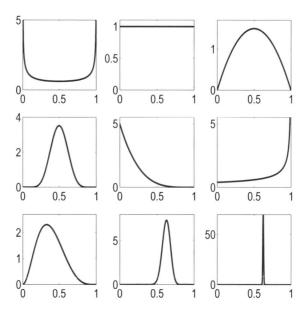

**Fig. 5.12** Beta densities for $(a,b)$ as (1/2, 1,2), (1,1), (2,2), (10,10), (1,5), (1, 0.4), (3,5), (50, 30), and (5000, 3000).

### 5.5.7 *Double Exponential Distribution*

A random variable $X$ has double exponential $\mathcal{DE}(\mu, \lambda)$ distribution if its density is given by

$$f_X(x) = \frac{\lambda}{2}\, e^{-\lambda|x-\mu|}, \quad -\infty < x < \infty, \lambda > 0.$$

The expectation of $X$ is $\mathbb{E}X = \mu$, and the variance is $\mathbb{V}\mathrm{ar}\,X = 2/\lambda^2$. The moment-generating function for the double exponential distribution is

$$m(t) = \frac{\lambda^2 e^{\mu t}}{\lambda^2 - t^2}, \quad |t| < \lambda.$$

The double exponential distribution is also known as the *Laplace distribution*. If $X_1$ and $X_2$ are independent exponential $\mathcal{E}(\lambda)$, then $X_1 - X_2$ is distributed as $\mathcal{DE}(0, \lambda)$. Also, if $X \sim \mathcal{DE}(0, \lambda)$, then $|X| \sim \mathcal{E}(\lambda)$. In MATLAB the double exponential distribution is not implemented since it can be readily obtained by folding the exponential distribution. In WinBUGS, $\mathcal{DE}(0, \lambda)$ is coded as ddexp(lambda).

## 5.5.8 Logistic Distribution

The logistic distribution is used for growth models in pharmacokinetics, regression with binary responses, and neural networks, to list just a few modeling applications.

The logistic random variable can be introduced by a property of its CDF expressed by a differential equation. Let $F(x) = P(X \leq x)$ be the CDF for which $F'(x) = F(x) \times (1 - F(x))$. One interpretation of this differential equation is as follows. For a Bernoulli random variable $\mathbf{1}(X \leq x) = \begin{cases} 1, & X \leq x \\ 0, & X > x \end{cases}$, the change in probability of 1 as a function of $x$ is equal to its variance. The solution in the class of CDFs is

$$F(x) = \frac{1}{1+e^{-x}} = \frac{e^x}{1+e^x},$$

which is called the logistic distribution. Its density is

$$f(x) = \frac{e^x}{(1+e^x)^2}.$$

Graphs of $f(x)$ and $F(x)$ are shown in Fig. 5.13. The mean of the distribution

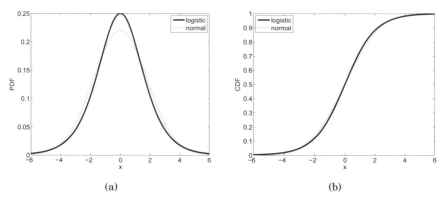

(a)                                    (b)

**Fig. 5.13** (a) Density and (b) CDF of logistic distribution. Superimposed (*dotted red*) is the normal distribution with matching mean and variance, 0 and $\pi^2/3$, respectively.

is 0 and the variance is $\pi^2/3$. For a more general logistic distribution given by the CDF

$$F(x) = \frac{1}{1+e^{-(x-\mu)/\sigma}},$$

the mean is $\mu$, variance $\pi^2\sigma^2/3$, skewness 0, and kurtosis 21/5. For the higher moments one can use the moment generating function

$$m(t) = \exp\{\mu t\} B (1 - \sigma t, 1 + \sigma t),$$

where $B$ is the beta function. In WinBUGS the logistic distribution is coded as `dlogis(mu,tau)`, where `tau` is the reciprocal of $\sigma$.

If $X$ has a logistic distribution, then $\log(X)$ has a log-logistic distribution (also known as the Fisk distribution). The log-logistic distribution is used in economics (population wealth distribution) and reliability.

The logistic distribution will be revisited in Chap. 17 where we deal with logistic regression.

### 5.5.9 *Weibull Distribution*

The Weibull distribution is one of the most important distributions in survival theory and engineering reliability. It is named after Swedish engineer and scientist Waloddi Weibull (Fig. 5.14) after his publication in the early 1950s (Weibull, 1951).

**Fig. 5.14** Ernst Hjalmar Waloddi Weibull (1887–1979).

The density of the two-parameter Weibull random variable $X \sim \mathscr{W}ei(r,\lambda)$ is given as

$$f(x) = \lambda r x^{r-1} e^{-\lambda x^r}, x > 0.$$

The CDF is given as $F(x) = 1 - e^{-\lambda x^r}$. Parameter $r$ is the shape parameter and $\lambda$ is the rate parameter. Both parameters are strictly positive.

In this form, Weibull $X \sim \mathscr{W}ei(r,\lambda)$ is a distribution of $Y^r$ for $Y$ exponential $\mathscr{E}(\lambda)$.

In MATLAB the Weibull distribution is parameterized by $a$ and $r$, as in $f(x) = a^{-r} r x^{r-1} e^{-(x/a)^r}$, $x > 0$. Note that in this parameterization, $a$ is the scale parameter and relates to $\lambda$ as $\lambda = a^{-r}$. So when $a = \lambda^{-1/r}$, the CDF in MATLAB is `wblcdf(x,a,r)`, and the PDF is `wblpdf(x,a,r)`. The function `wblinv(p,a,r)` computes the $p$th quantile of the $\mathscr{W}ei(r,\lambda)$ random variable. In WinBUGS, $\mathscr{W}ei(r,\lambda)$ is coded as `dweib(r,lambda)`.

The Weibull distribution generalizes the exponential distribution ($r = 1$) and Rayleigh distribution ($r = 2$).

The mean of a Weibull random variable $X$ is $\mathbb{E}X = \frac{\Gamma(1+1/r)}{\lambda^{1/r}} = a\Gamma\left(1+\frac{1}{r}\right)$, and the variance is $\mathbb{V}\mathrm{ar}\,X = \frac{\Gamma(1+2/r)-\Gamma^2(1+1/r)}{\lambda^{2/r}} = a^2(\Gamma\left(1+\frac{2}{r}\right)-\Gamma^2\left(1+\frac{1}{r}\right))$. The $k$th moment is $\mathbb{E}X^k = \frac{\Gamma(1+k/r)}{\lambda^{k/r}} = a^k\Gamma\left(1+\frac{k}{r}\right)$.

Figure 5.15 shows the densities of the Weibull distribution for $r = 2$ (blue), $r = 1$ (red), and $r = 1/2$ (black). In all three cases, $\lambda = 1/2$. The values for the scale parameter $a = \lambda^{-1/r}$ are $\sqrt{2}, 2$, and $4$, respectively.

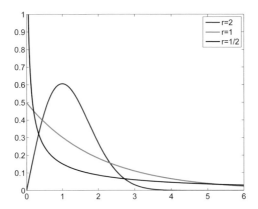

**Fig. 5.15** Densities of Weibull distribution with $r = 2$ (*blue*), $r = 1$ (*red*), and $r = 2$ (*black*). In all three cases, $\lambda = 1/2$.

### 5.5.10 Pareto Distribution

The Pareto distribution is named after the Italian economist Vilfredo Pareto. Some examples in which the Pareto distribution provides an exemplary model include wealth distribution in individuals, sizes of human settlements, visits to encyclopedia pages, and file size distribution of Internet traffic that uses the TCP protocol. A random variable $X$ has a Pareto $\mathscr{P}a(c,\alpha)$ distribution with parameters $0 < c < \infty$ and $\alpha > 0$ if its density is given by

$$f_X(x) = \begin{cases} \frac{\alpha}{c}\left(\frac{c}{x}\right)^{\alpha+1}, & x \geq c, \\ 0, & \text{else .} \end{cases}$$

The CDF is

$$F_X(x) = \begin{cases} 0, & x < c, \\ 1-\left(\frac{c}{x}\right)^{\alpha}, & x \geq c. \end{cases}$$

The mean and variance of $X$ are $\mathbb{E}X = \alpha c/(\alpha - 1)$ and $\mathbb{V}\mathrm{ar}\,X = \alpha c^2/[(\alpha - 1)^2(\alpha - 2)]$. The median is $m = c \cdot 2^{1/\alpha}$.

If $X_1, \ldots, X_n$ are independent $\mathscr{P}a(c, \alpha)$, then $Y = 2c\sum_{i=1}^{n}\ln(X_i) \sim \chi^2$ with $2n$ degrees of freedom.

In MATLAB one can specify the generalized Pareto distribution, which for some selection of its parameters is equivalent to the aforementioned Pareto distribution. In WinBUGS, the code is dpar(alpha,c) (note the permuted order of parameters).

### 5.5.11 Dirichlet Distribution

The Dirichlet distribution is a multivariate version of the beta distribution in the same way that the multinomial distribution is a multivariate extension of the binomial. A random variable $X = (X_1, \ldots, X_k)$ with a Dirichlet distribution of $(X \sim \mathscr{D}ir(a_1, \ldots, a_k))$ has a PDF of

$$f(x_1, \ldots, x_k) = \frac{\Gamma(A)}{\prod_{i=1}^{k}\Gamma(a_i)}\prod_{i=1}^{k}x_i^{a_i - 1},$$

where $A = \sum a_i$, and $x = (x_1, \ldots, x_k) \geq 0$ is defined on the simplex $x_1 + \cdots + x_k = 1$. Then

$$\mathbb{E}(X_i) = \frac{a_i}{A}, \quad \mathbb{V}\mathrm{ar}(X_i) = \frac{a_i(A - a_i)}{A^2(A + 1)}, \quad \text{and } \mathbb{C}\mathrm{ov}(X_i, X_j) = -\frac{a_i a_j}{A^2(A + 1)}.$$

The Dirichlet random variable can be generated from gamma random variables $Y_1, \ldots, Y_k \sim \mathscr{G}a(a, b)$ as $X_i = Y_i/S_Y$, $i = 1, \ldots, k$, where $S_Y = \sum_i Y_i$. Obviously, the marginal distribution of a component $X_i$ is $\mathscr{B}e(a_i, A - a_i)$. This is illustrated in the following MATLAB m-file that generates random Dirichlet vectors.

```
function drand = dirichletrnd(a,n)
% function drand = dirichletrnd(a,n)
% a - vector of parameters 1 x m
% n - number of random realizations
% drand - matrix m x n, each column one realization.
%-----------------------------------------------------
a=a(:);
m=size(a,1);
a1=zeros(m,n);
for i = 1:m
    a1(i,:) = gamrnd(a(i,1),1,1,n);
end
for i=1:m
drand(i, 1:n )= a1(i, 1:n ) ./ sum(a1);
end
```

## 5.6 Random Numbers and Probability Tables

In older introductory statistics texts, many backend pages have been devoted to various statistical tables. Several decades ago, many books of statistical tables were published. Also, the most respected of statistical journals occasionally published articles consisting of statistical tables.

In 1947 the RAND Corporation published the monograph *A Million Random Digits with 100,000 Normal Deviates*, which at the time was a state-of-the-art resource for simulation and Monte Carlo methods. The random digits can be found at http://www.rand.org/monograph_reports/MR1418.html.

These days, much larger tables of random numbers can be produced by a single line of code, resulting in a set of random numbers that can pass a battery of stringent randomness tests. With MATLAB and many other widely available software packages, statistical tables and tables of random numbers are now obsolete. For example, tables of binomial CDF and PDF for a specific $n$ and $p$ can be reproduced by

```
disp('binocdf(0:12, 12, 0.7)');
binocdf(0:12, 12, 0.7)
disp('binopdf(0:12, 12, 0.7)');
binopdf(0:12, 12, 0.7)
```

We will show how to sample and simulate from a few distributions in MATLAB and compare empirical means and variances with their theoretical counterparts. The following annotated MATLAB code simulates from binomial, Poisson, and geometric distributions and compares theoretical and empirical means and variances.

```
%various_simulations.m
simu = binornd(12, 0.7, [1,100000]);
  % simu is 10000 observations from Bin(12,0.7)
disp('simu = binornd(12, 0.7, [1,100000]); 12*0.7 - mean(simu)');

12*0.7 - mean(simu)   %0.001069
  %should be small since the theoretical mean is n*p
disp('simu = binornd(12, 0.7, [1,100000]); ...
                            12 * 0.7 * 0.3 - var(simu)');

12 * 0.7 * 0.3 - var(simu)   %-0.008350
  %should be small since the theoretical variance is n*p*(1-p)

%% Simulations from Poisson(2)
poi = poissrnd(2, [1, 100000]);
disp('poi = poissrnd(2, [1, 100000]); mean(poi)');
mean(poi)     %1.9976
disp('poi = poissrnd(2, [1, 100000]); var(poi)');
var(poi)      %2.01501
```

```
%%% Simulations from Geometric(0.2)
geo = geornd(0.2, [1, 100000]);
disp('geo = geornd(0.2, [1, 100000]); mean(geo)');
mean(geo)   %4.00281
disp('geo = geornd(0.2, [1, 100000]); var(geo)');
var(geo)    %20.11996
```

## 5.7 Transformations of Random Variables*

When a random variable with known density is transformed, the result is
a random variable as well. The question is how to find its distribution. The
general theory for distributions of functions of random variables is beyond the
scope of this text, and the reader can find comprehensive coverage in Ross
(2010a, b).

We have already seen that for a discrete random variable $X$, the PMF of a
function $Y = g(X)$ is simply the table

$$\begin{array}{c|ccccc} g(X) & g(x_1) & g(x_2) & \cdots & g(x_n) & \cdots \\ \hline \text{Prob} & p_1 & p_2 & \cdots & p_n & \cdots \end{array}$$

in which only realizations of $X$ are transformed while the probabilities are
kept unchanged.

For continuous random variables the distribution of a function is more complex. In some cases, however, looking at the CDF is sufficient.

In this section we will discuss two topics: (i) how to find the distribution for
a transformation of a single continuous random variable and (ii) how to approximate moments, in particular means and variances, of complex functions
of many random variables.

Suppose that a continuous random variable has a density $f_X(x)$ and that
a function $g$ is monotone on the domain of $f$, with the inverse function
$h$, $h = g^{-1}$. Then the random variable $Y = g(X)$ has a density

$$f_Y(y) = f(h(y))|h'(y)|. \tag{5.9}$$

If $g$ is not 1–1, but has $k$ 1–1 inverse branches, $h_1, h_2, \ldots, h_k$, then

$$f_Y(y) = \sum_{i=1}^{k} f(h_i(y))|h'_i(y)|. \tag{5.10}$$

An example of a non 1–1 function is $g(x) = x^2$, for which inverse branches
$h_1(y) = \sqrt{y}$ and $h_2(y) = -\sqrt{y}$ are 1–1.

*Example 5.20.* Let $X$ be a random variable with an exponential $\mathscr{E}(\lambda)$ distribution, where $\lambda > 0$ is the rate parameter. Find the distribution of the random variable $Y = \sqrt{X}$.

Here $g(x) = \sqrt{x}$ and $g^{-1}(y) = y^2$. The Jacobian is $|g^{-1}(y)'| = 2y,\ y \geq 0$. Thus,

$$f_Y(y) = \lambda e^{-\lambda y^2} \cdot 2y,\ y \geq 0, \lambda > 0,$$

which is known as the Rayleigh distribution.

An alternative approach to finding the distribution of $Y$ is to consider the CDF:

$$F_Y(y) = \mathbb{P}(Y \leq y) = \mathbb{P}(\sqrt{X} \leq y) = \mathbb{P}(X \leq y^2) = 1 - e^{-\lambda y^2}$$

since $X$ has the exponential distribution.

The density is now obtained by taking the derivative of $F_Y(y)$,

$$f_Y(y) = (F_Y(y))' = 2\lambda y e^{-\lambda y^2},\quad y \geq 0, \lambda > 0.$$

The distribution of a function of one or many random variables is an ultimate summary. However, the result could be quite messy and sometimes the distribution lacks a closed form. Moreover, not all facets of the resulting distribution may be of interest to researchers; sometimes only the mean and variance are needed.

If $X$ is a random variable with $\mathbb{E}X = \mu$ and $\operatorname{Var} X = \sigma^2$, then for a function $Y = g(X)$ the following approximation holds:

$$\mathbb{E}Y \approx g(\mu) + \frac{1}{2}g''(\mu)\sigma^2,$$

$$\operatorname{Var} Y \approx \left(g'(\mu)\right)^2 \sigma^2. \tag{5.11}$$

If $n$ independent random variables are transformed as $Y = g(X_1, X_2, \ldots, X_n)$, then

$$\mathbb{E}Y \approx g(\mu_1, \mu_2, \ldots, \mu_n) + \frac{1}{2}\sum_{i=1}^{n} \frac{\partial^2 g}{\partial x^2}(\mu_1, \mu_2, \ldots, \mu_n)\sigma_i^2,$$

$$\operatorname{Var} Y \approx \sum_{i=1}^{n}\left(\frac{\partial g}{\partial x_i}(\mu_1, \mu_2, \ldots, \mu_n)\right)^2 \sigma_i^2, \tag{5.12}$$

where $\mathbb{E}X_i = \mu_i$ and $\operatorname{Var} X_i = \sigma_i^2$.

The approximation for the mean $\mathbb{E}Y$ is obtained by the second-order Taylor expansion and is more precise than the approximation for the variance $\mathbb{V}\text{ar } Y$, which is of the first order ("linearization"). The second-order approximation for $\mathbb{V}\text{ar } Y$ is straightforward but involves third and fourth moments of $X$s. Also, when the variables $X_1,\ldots,X_n$ are correlated, the factor $2\sum_{1\le i<j\le n} \frac{\partial^2 g}{\partial x_i \partial x_j}(\mu_1,\ldots,\mu_n)\mathbb{C}\text{ov}(X_i,X_j)$ should be added to the expression for $\mathbb{V}\text{ar } Y$ in (5.12).

If $g$ is a complicated function, the mean $\mathbb{E}Y$ is often approximated by a first-order approximation, $\mathbb{E}Y \approx g(\mu_1,\mu_2,\ldots,\mu_n)$, that involves no derivatives.

*Example 5.21.* **String Vibrations.** In string vibration, the frequency of the fundamental harmonic is often of interest. The fundamental harmonic is produced by the vibration with nodes at the two ends of the string. In this case, the length of the string $L$ is half of the wavelength of the fundamental harmonic. The frequency $\omega$ (in Hz) depends also on the tension of the string $T$, and the string mass $M$,

$$\omega = \frac{1}{2}\sqrt{\frac{T}{ML}}.$$

Quantities $L, T$, and $M$ are measured imperfectly and are considered independent random variables. The means and variances are estimated as follows:

| Variable (unit) | Mean | Variance |
|---|---|---|
| $L$ (m) | 0.5 | 0.0001 |
| $T$ (N) | 70 | 0.16 |
| $M$ (kg/m) | 0.001 | $10^{-8}$ |

Approximate the mean $\mu_\omega$ and variance $\sigma_\omega^2$ of the resulting frequency $\omega$.

The partial derivatives

$$\frac{\partial\omega}{\partial T} = \frac{1}{4}\sqrt{\frac{1}{TML}}, \quad \frac{\partial^2\omega}{\partial T^2} = -\frac{1}{8}\sqrt{\frac{1}{T^3ML}},$$

$$\frac{\partial\omega}{\partial M} = -\frac{1}{4}\sqrt{\frac{T}{M^3L}}, \quad \frac{\partial^2\omega}{\partial M^2} = \frac{3}{8}\sqrt{\frac{T}{M^5L}},$$

$$\frac{\partial\omega}{\partial L} = -\frac{1}{4}\sqrt{\frac{T}{ML^3}}, \quad \frac{\partial^2\omega}{\partial L^2} = \frac{3}{8}\sqrt{\frac{T}{ML^5}},$$

evaluated at the means $\mu_L = 0.5$, $\mu_T = 70$, and $\mu_M = 0.001$, are

$$\frac{\partial \omega}{\partial T}(\mu_L, \mu_T, \mu_M) = 1.3363, \qquad \frac{\partial^2 \omega}{\partial T^2}(\mu_L, \mu_T, \mu_M) = -0.0095,$$

$$\frac{\partial \omega}{\partial M}(\mu_L, \mu_T, \mu_M) = -9.3541 \cdot 10^4, \qquad \frac{\partial^2 \omega}{\partial M^2}(\mu_L, \mu_T, \mu_M) = 1.4031 \cdot 10^8,$$

$$\frac{\partial \omega}{\partial L}(\mu_L, \mu_T, \mu_M) = -187.0829, \qquad \frac{\partial^2 \omega}{\partial L^2}(\mu_L, \mu_T, \mu_M) = 561.2486,$$

and the mean and variance of $\omega$ are

$$\boxed{\mu_\omega \approx 187.8117} \quad \text{and} \quad \boxed{\sigma_\omega^2 \approx 91.2857}.$$

The first-order approximation for $\mu_\omega$ is $\frac{1}{2}\sqrt{\frac{\mu_T}{\mu_M \mu_L}} = 187.0829$.

✎

## 5.8 Mixtures*

In modeling tasks it is sometimes necessary to combine two or more random variables in order to get a satisfactory model. There are two ways of combining random variables: by taking the linear combination $a_1 X_1 + a_2 X_2 + \ldots$ for which a density in the general case is often convoluted and difficult to express in a finite form, or by combining densities and PMFs directly.

For example, for two densities $f_1$ and $f_2$, the density $g(x) = \varepsilon f_1(x) + (1 - \varepsilon)f_2(x)$ is a mixture of $f_1$ and $f_2$ with weights $\varepsilon$ and $1 - \varepsilon$. It is important for the weights to be nonnegative and add up to 1 so that $g(x)$ remains a density.

Very popular mixtures are point mass mixture distributions that combine a density function $f(x)$ with a point mass (Dirac) function $\delta_{x_0}$ at a value $x_0$. The Dirac functions belong to a class of special functions. Informally, one may think of $\delta_{x_0}$ as a limiting function for a sequence of functions

$$f_{n,x_0} = \begin{cases} n, & x_0 - \frac{1}{2n} < x < x_0 + \frac{1}{2n}, \\ 0, & \text{else}, \end{cases}$$

when $n \to \infty$. It is easy to see that for any finite $n$, $f_{n,x_0}$ is a density since it integrates to 1; however, the function domain shrinks to a singleton $x_0$, while its value at $x_0$ goes to infinity.

For example, $f(x) = 0.3\delta_0 + 0.7 \times \frac{1}{\sqrt{2\pi}} \exp\{-\frac{x^2}{2}\}$ is a normal distribution contaminated by a point mass at zero with a weight 0.3.

## 5.9 Markov Chains*

You may have encountered statistical jargon containing the term "Markov chain." In Bayesian calculations the acronym MCMC stands for Markov chain Monte Carlo simulations, while in statistical models of genomes, hidden Markov chain models are popular. Here we give a basic definition and a few examples of Markov chains and provide references where Markov chains receive detailed coverage.

A sequence of random variables $X_0, X_1, \ldots, X_n, \ldots$, with values in the set of "states" $\mathscr{S} = \{1, 2, \ldots\}$, constitutes a Markov chain if the probability of transition to a future state, $X_{n+1} = j$, depends only on the value at the current state, $X_n = i$, and not on any previous values $X_{n-1}, X_{n-2}, \ldots, X_0$. A popular way of putting this is to say that in Markov chains the future depends on the present and not on the past. Formally,

$$\mathbb{P}(X_{n+1} = j | X_0 = i_0, X_1 = i_1, \ldots, X_{n-1} = i_{n-1}, X_n = i) = \mathbb{P}(X_{n+1} = j | X_n = i) = p_{ij},$$

where $i_0, i_1, \ldots, i_{n-1}, i, j$ are the states from $\mathscr{S}$. The probability $p_{ij}$ is independent of $n$ and represents the transition probability from state $i$ to state $j$. In our brief coverage of Markov chains, we will consider chains with a finite number of states, $N$.

For states $\mathscr{S} = \{1, 2, \ldots, N\}$, the transition probabilities form an $N \times N$ matrix $\mathbf{P} = (p_{ij})$. Each row of this matrix sums up to 1 since the probabilities of all possible moves from a particular state, including the probability of remaining in the same state, sum up to 1:

$$p_{i1} + p_{i2} + \cdots + p_{ii} + \cdots + p_{iN} = 1.$$

The matrix $\mathbf{P}$ describes the evolution and long-time behavior of the Markov chain it represents. In fact, if the distribution $\pi^{(0)}$ for the initial variable $X_0$ is specified, the pair $\pi^{(0)}, \mathbf{P}$ fully describes the Markov chain.

Matrix $\mathbf{P}^2$ gives the probabilities of transition in two steps. Its element $p_{ij}^{(2)}$ is $\mathbb{P}(X_{n+2} = j | X_n = i)$.

Likewise, the elements of matrix $\mathbf{P}^m$ are the probabilities of transition in $m$ steps,

$$p_{ij}^{(m)} = \mathbb{P}(X_{n+m} = j | X_n = i),$$

for any $n \geq 0$ and any $i, j \in \mathscr{S}$.

If the distribution for $X_0$ is $\pi^{(0)} = \left( \pi_1^{(0)}, \pi_2^{(0)}, \ldots, \pi_N^{(0)} \right)$, then the distribution for $X_n$ is

$$\pi^{(n)} = \pi^{(0)} \mathbf{P}^n. \tag{5.13}$$

Of course, if the state $X_0$ is known, $X_0 = i_0$, then $\pi^{(0)}$ is a vector of 0s except at position $i_0$, where the value is 1.

For $n$ large, the probability $\pi^{(n)}$ "forgets" the initial distribution at state $X_0$ and converges to $\pi = \lim_{n\to\infty}\pi^{(n)}$. This distribution is called the stationary distribution of a chain and satisfies

$$\pi = \pi\mathbf{P}.$$

**Result.** If for a finite state Markov chain one can find an integer $k$ so that all entries in $\mathbf{P}^k$ are strictly positive, then stationary distribution $\pi$ exist.

*Example 5.22.* **Point Accepted Mutation.** Point accepted mutation (PAM) implements a simple theoretical model for scoring the alignment of protein sequences. Specifically, at a fixed position the rate of mutation at each moment is assumed to be independent of previous events. Then the evolution of this fixed position in time can be treated as a Markov chain, where the PAM matrix represents its transition matrix. The original PAMs are $20 \times 20$ matrices describing the evolution of 20 standard amino acids (Dayhoff et al. 1978). As a simplified illustration, consider the case of a nucleotide sequence with only four states (A, T, G, and C). Assume that in a given time interval $\Delta T$ the probabilities that a given nucleotide mutates to each of the other three bases or remains unchanged can be represented by a $4 \times 4$ mutation matrix $M$:

$$M = \begin{pmatrix} & A & T & G & C \\ A & 0.98 & 0.01 & 0.005 & 0.005 \\ T & 0.01 & 0.96 & 0.02 & 0.01 \\ G & 0.01 & 0.01 & 0.97 & 0.01 \\ C & 0.02 & 0.03 & 0.01 & 0.94 \end{pmatrix}$$

Consider the fixed position with the letter T at $t = 0$:

$$s_0 = (0\ \ 1\ \ 0\ \ 0).$$

Then, at times $\Delta$, $2\Delta$, $10\Delta$, $100\Delta$, $1000\Delta$, and $10000\Delta$, by (5.13), the probabilities of the nucleotides (A, T, G, C) are $s_1 = s_0 M$, $s_2 = s_0 M^2$, $s_{10} = s_0 M^{10}$, $s_{100} = s_0 M^{100}$, $s_{1000} = s_0 M^{1000}$, and $s_{10000} = s_0 M^{10000}$, as given in the following table

|   | $\Delta$ | $2\Delta$ | $10\Delta$ | $100\Delta$ | $1000\Delta$ | $10000\Delta$ |
|---|---|---|---|---|---|---|
| A | 0.0100 | 0.0198 | 0.0909 | 0.3548 | 0.3721 | 0.3721 |
| T | 0.9600 | 0.9222 | 0.6854 | 0.2521 | 0.2465 | 0.2465 |
| G | 0.0200 | 0.0388 | 0.1517 | 0.2747 | 0.2651 | 0.2651 |
| C | 0.0100 | 0.0193 | 0.0719 | 0.1184 | 0.1163 | 0.1163 |

## 5.10 Exercises

5.1. **Phase I Clinical Trials and CTCAE Terminology.** In phase I clinical trials a safe dosage of a drug is assessed. In administering the drug doctors are grading subjects' toxicity responses on a scale from 0 to 5. In CTCAE (Common Terminology Criteria for Adverse Events, National Institute of Health) Grade refers to the severity of adverse events. Generally, Grade 0 represents no measurable adverse events (sometimes omitted as a grade); Grade 1 events are mild; Grade 2 are moderate; Grade 3 are severe; Grade 4 are life-threatening or disabling; Grade 5 are fatal. This grading system inherently places a value on the importance of an event, although there is not necessarily "proportionality" among grades (a "2" is not necessarily twice as bad as a "1"). Some adverse events are difficult to "fit" into this point schema, but altering the general guidelines of severity scaling would render the system useless for comparing results between trials, which is an important purpose of the system.

Assume that based on a large number of trials (administrations to patients with Renal cell carcinoma), the toxicity of drug PNU (a murine Fab fragment of the monoclonal antibody 5T4 fused to a mutated superantigen staphylococcal enterotoxin A) at a particular fixed dosage, is modeled by discrete random variable $X$,

| $X$ | 0 | 1 | 2 | 3 | 4 | 5 |
|------|------|------|------|------|------|------|
| Prob | 0.620 | 0.190 | 0.098 | 0.067 | 0.024 | 0.001 |

Plot the PMF and CDF and find $\mathbb{E}X$ and $\mathbb{V}\mathrm{ar}(X)$.

5.2. **Mendel and Dominance.** Suppose that a specific trait, such as eye color or left-handedness, in a person is dependent on a pair of genes and suppose that $D$ represents a dominant and $d$ a recessive gene. Thus a person having $DD$ is pure dominant and $dd$ is pure recessive while $Dd$ is a hybrid. The pure dominants and hybrids are alike in outward appearance. A child receives one gene from each parent.

Suppose two hybrid parents have 4 children. What is the probability that 3 out of 4 children have outward appearance of the dominant gene.

5.3. **Chronic Kidney Disease.** Chronic kidney disease (CKD) is a serious condition associated with premature mortality, decreased quality of life, and increased health-care expenditures. Untreated CKD can result in end-stage renal disease and necessitate dialysis or kidney transplantation. Risk factors for CKD include cardiovascular disease, diabetes, hypertension, and obesity. To estimate the prevalence of CKD in the United States (overall and by health risk factors and other characteristics), the CDC (CDC's MMWR Weekly, 2007; Coresh et al, 2003) analyzed the most recent data from the National Health and Nutrition Examination Survey (NHANES). The total crude (i.e., not age-standardized) CKD prevalence estimate for adults aged

> 20 years in the United States was 17%. By age group, CKD was more prevalent among persons aged > 60 years (40%) than among persons aged 40–59 years (13%) or 20–39 years (8%).

(a) From the population of adults aged > 20 years, 10 subjects are selected at random. Find the probability that 3 of the selected subjects have CKD.

(b) From the population of adults aged > 60, 5 subjects are selected at random. Find the probability that at least one of the selected have CKD.

(c) From the population of adults aged > 60, 16 subjects are selected at random and it was found that 6 of them had CKD. From this sample of 16, subjects are selected at random, *one-by-one with replacement,* and inspected. Find the probability that among 5 inspected (i) exactly 3 had CKD; (ii) at least one of the selected have CKD.

(d) From the population of adults aged > 60 subjects are selected at random until a subject is found to have CKD. What is the probability that exactly 3 subjects are sampled.

(e) Suppose that persons aged > 60 constitute 23% of the population of adults older than 20. For the other two age groups, 20–39 y.o., and 40–59 y.o., the percentages are 42% and 35%. Ten people are selected at random. What is the probability that 5 are from the > 60-group, 3 from the 20–39-group, and 2 from the 40-59 group.

5.4. **Ternary channel.** Refer to Exercise 3.29 in which a communication system was transmitting three signals, $s_1, s_2$ and $s_3$.

(a) If $s_1$ is sent $n = 1000$ times, find an approximation to the probability of the event that it was correctly received between 730 and 770 times, inclusive.

(b) If $s_2$ is sent $n = 1000$ times, find an approximation to the probability of the event that the channel did not switch to $s_3$ at all, i.e., if 1000 $s_2$ signals are sent not a single $s_3$ was received. Can you use the same approximation as in (a)?

5.5. **Conditioning a Poisson.** If $X_1 \sim \mathcal{P}oi(\lambda_1)$ and $X_2 \sim \mathcal{P}oi(\lambda_2)$ are independent, then the distribution of $X_1$, given $X_1 + X_2 = n$, is binomial $\mathcal{B}in(n, \lambda_1/(\lambda_1 + \lambda_2))$.

5.6. **Rh+ Plates.** Assume that there are 6 plates with red blood cells, three are Rh+ and three are Rh−.

Two plates are selected (a) with, (b) without replacement. Find the probability that one plate out of the 2 selected/inspected is of Rh+ type.

Now, increase the number of plates keeping the proportion of Rh+ fixed to 1/2. For example, if the total number of plates is 10000, 5000 of each type, what are the probabilities from (a) and (b)?

5.7. **Your Colleague's Misconceptions About Density and CDF.** Your colleague thinks that if $f$ is a probability density function for the continuous random variable $X$, then $f(10)$ is the probability that $X = 10$. (a) Explain to your colleague why his/her reasoning is false.

Your colleague is not satisfied with your explanation and challenges you by asking, "If $f(10)$ is not the probability that $X = 10$, then just what does $f(10)$ signify?" (b) How would you respond?

Your colleague now thinks that if $F$ is a cumulative probability density function for the continuous random variable $X$, then $F(5)$ is the probability that $X = 5$. (c) Explain why your colleague is wrong.

Your colleague then asks you, "If $F(5)$ is not the probability of $X = 5$, then just what does $F(5)$ represent?" (d) How would you respond?

5.8. **Falls among elderly.** Falls are the second leading cause of unintentional injury death for people of all ages and the leading cause for people 60 years and older in the United States. Falls are also the most costly injury among older persons in the United States.

One in three adults aged 65 years and older falls annually.

(a) Find the probability that 3 among 11 adults aged 65 years and older will fall in the following year.

(b) Find the probability that among 110,000 adults aged 65 years and older the number of falls will be between 36,100 and 36,700, inclusive. Find the exact probability by assuming a binomial distribution for the number of falls, and an approximation to this probability via deMoivre's Theorem.

5.9. **Cell clusters in 3-D Petri dishes.** The number of cell clusters in a 3-D Petri dish has a Poisson distribution with mean $\lambda = 5$. Find the percentage of Petri dishes that have (a) 0 clusters, (b) at least one cluster, (c) more than 8 clusters, and (d) between 4 and 6 clusters. Use MATLAB and `poisspdf`, `poisscdf` functions.

5.10. **Left-handed Twins.** The identical twin of a left-handed person has a 76 percent chance of being left-handed, implying that left-handedness has partly genetic and partly environmental causes. Ten identical twins of ten left-handed persons are inspected for left-handedness. Let $X$ be the number of left-handed among the inspected. What is the probability that $X$

(a) falls anywhere between 5 and 8, inclusive;

(b) is at most 6;

(c) is not less than 6.

(d) Would you be surprised if the number of left-handed among the 10 inspected was 3? Why or why not?

5.11. **Pot Smoking is Not Cool!** A nationwide survey of seniors by the University of Michigan reveals that almost 70% disapprove of daily pot smoking according to a report in "Parade," September 14, 1980. If 12 seniors are selected at random and asked their opinion, find the probability that the number who disapprove of smoking pot daily is

(a) anywhere from 7 to 9;

(b) at most 5;

(c) not less than 8.

5.12. **Emergency Help by Phone.** The emergency hotline in a hospital tries to answer questions to its patient support within 3 minutes. The probability is 0.9 that a given call is answered within 3 minutes and the calls are independent.

(a) What is the expected total number of calls that occur until the first call is answered late?

(b) What is the probability that exactly one of the next 10 calls is answered late?

5.13. **Min of Three.** Let $X_1, X_2$, and $X_3$ be three mutually independent random variables, with a discrete uniform distribution on $\{1, 2, 3\}$, given as $P(X_i = k) = 1/3$ for $k = 1, 2$ and 3.

(a) Let $M = \min\{X_1, X_2, X_3\}$. What is the distribution (probability mass function) and cumulative distribution function of $M$?

(b) What is the distribution (probability mass function) and cumulative distribution function of random variable $R = \max\{X_1, X_2, X_3\} - \min\{X_1, X_2, X_3\}$.

5.14. **Cystic Fibrosis in Japan.** Some rare diseases including those of genetic origin, are life-threatening or chronically debilitating diseases that are of such low prevalence that special combined efforts are needed to address them. An accepted definition of low prevalence is a prevalence of less than 5 in a population of 10,000. A rare disease has such a low prevalence in a population that a doctor in a busy general practice would not expect to see more than one case in a given year.

Assume that Cystic Fibrosis, which is a rare genetic disease in most parts of Asia, has a prevalence of 2 per 10000 in Japan. What is the probability that in a Japanese city of 15,000 there are

(a) exactly 3 incidences,

(b) at least one incidence,

of cystic fibrosis.

5.15. **Random Variables as Models.** Tubert-Bitter et al (1996) found that the number of serious gastrointestinal reactions reported to the British Committee on Safety of Medicines was 538 for 9,160,000 prescriptions of the anti-inflammatory drug *Piroxicam*.

(a) What is the rate of gastrointestinal reactions per 10,000 prescriptions?

(b) Using the Poisson model with the rate $\lambda$ as in (a), find the probability of exactly two gastrointestinal reactions per 10,000 prescriptions.

(c) Find the probability of finding at least two gastrointestinal reactions per 10,000 prescriptions.

5.16. **Additivity of Gammas.** If $X_i \sim \mathcal{G}a(r_i, \lambda)$ are independent, prove that $Y = X_1 + \cdots + X_n$ is distributed as gamma with parameters $r = r_1 + r_2 + \cdots + r_n$ and $\lambda$; that is, $Y \sim \mathcal{G}a(r, \lambda)$.

5.17. **Memoryless property.** Prove that the geometric distribution ($\mathbb{P}(X = x) =$
$(1 - p)^x p, x = 0, 1, 2, \ldots$ ) and the exponential distribution ($\mathbb{P}(X \le x) = 1 -$
$e^{-\lambda x}$, $x \ge 0, \lambda \ge 0$) both possess the *Memoryless Property*, that is, they satisfy

$$\mathbb{P}(X \ge u + v | X \ge u) = \mathbb{P}(X \ge v - u), u \le v.$$

5.18. **Rh System.** Rh antigens are transmembrane proteins with loops exposed
at the surface of red blood cells. They appear to be used for the transport
of carbon dioxide and/or ammonia across the plasma membrane. They are
named for the rhesus monkey in which they were first discovered. There
are a number of different Rh antigens. Red blood cells that are *Rh positive*
express the antigen designated as D. About 15% of the population do not
have RhD antigens and thus are *Rh negative*. The major importance of the
Rh system for human health is to avoid the danger of RhD incompatibility
between a mother and her fetus.
(a) From the general population 8 people are randomly selected and checked
for their Rh factor. Let $X$ be the number of Rh negative among the eight
selected. Find $\mathbb{P}(X = 2)$.
(b) In a group of 16 patients, three members are Rh negative. Eight patients
are selected at random. Let $Y$ be the number of Rh negative among the
eight selected. Find $\mathbb{P}(Y = 2)$.
(c) From the general population subjects are randomly selected and checked
for their Rh factor. Let $Z$ be the number of Rh positive subjects before the
first Rh negative subject is selected. Find $\mathbb{P}(Z = 2)$.
(d) Identify the distributions of the random variables in (a), (b), and (c)?
(e) What are the expectations and variances for the random variables in (a),
(b), and (c)?

5.19. **Blood Types.** The prevalence of blood types in the US population is O+:
37.4%, A+: 35.7%, B+: 8.5%, AB+: 3.4%, O–: 6.6%, A–: 6.3%, B–: 1.5%, and
AB–: 0.6%.
(a) A sample of 24 subjects is randomly selected from the US population.
What is the probability that 8 subjects are O+. Random variable $X$ de-
scribes the number of O+ subjects among 24 selected. Find $\mathbb{E}X$ and $\mathbb{V}\text{ar}\, X$.
(b) Among 16 subjects, eight are O+. From these 16 subjects, five are se-
lected at random as a group. What is the probability that among the five
selected at most two are O+.
(c) Use Poisson approximation to find the probability that among 500 randomly selected subjects the number of AB– subjects is at least 1.
(d) Random sampling from the population is performed until the first subject with B+ blood type is found. What is the expected number of subjects
sampled?

5.20. **Variance of the Exponential.** Show that for an exponential random variable $X$ with density $f(x) = \lambda e^{-\lambda x}$, $x \geq 0$, the variance is $1/\lambda^2$. (Hint: You can use the fact that $\mathbb{E}X = 1/\lambda$. To find $\mathbb{E}X^2$ you need to repeat the integration-by-parts twice.)

5.21. **Equipment Aging.** Suppose that the lifetime $T$ of a particular piece of laboratory equipment (in 1000 hour units) is an exponentially distributed random variable such that $\mathbb{P}(T > 10) = 0.8$.
(a) Find the "rate" parameter, $\lambda$.
(b) What are the mean and standard deviation of the random variable $T$?
(c) Find the median, the first and third quartiles, and the inter-quartile range of the lifetime $T$. Recall that for an exponential distribution, you can find any percentile exactly.

5.22. **A Simple Continuous Random Variable.** Assume that the measured responses in an experiment can be modeled as a continuous random variable with density

$$f(x) = \begin{cases} c - x, & 0 \leq x \leq c \\ 0, & \text{else} \end{cases}$$

(a) Find the constant $c$ and sketch the graph of the density $f(x)$.
(b) Find the CDF $F(x) = \mathbb{P}(X \leq x)$, and sketch its graph.
(c) Find $\mathbb{E}(X)$ and $\mathbb{V}\text{ar}(X)$.
(d) What is $\mathbb{P}(X \leq 1/2)$?

5.23. **2-D Continuous Random Variable Question.** A two dimensional random variable $(X, Y)$ is defined by its density function, $f(x, y) = Cxe^{-xy}$, $0 \leq x \leq 1$; $0 \leq y \leq 1$.
(a) Find the constant $C$.
(b) Find the marginal distributions of $X$ and $Y$.

5.24. **Insulin Sensitivity.** The insulin sensitivity (SI), obtained in a glucose tolerance test is one of patient responses used to diagnose type II diabetes. Leading a sedative lifestyle and being overweight are well-established risk factors for type II diabetes. Hence, body mass index (BMI) and hip to waist ratio (HWR = HIP/WAIST) may also predict an impaired insulin sensitivity. In an experiment, 106 males (coded 1) and 126 females (coded 2) had their SI measured and their BMI and HWR registered. Data (diabetes.xls) are available on the text web page. For this exercise you will need only the 8-th column of the data set, which corresponds to the SI measurements.
(a) Find the sample mean and sample variance of SI.
(b) A gamma distribution with parameters $\alpha$ and $\beta$ seems to be an appropriate model for SI. What $\alpha$, $\beta$ should be chosen so that the $\mathbb{E}X$ matches the sample mean of SI and $\mathbb{V}\text{ar}\,X$ matches the sample variance of SI.

(c) With $\alpha$ and $\beta$ selected as in (b), simulate a random sample from gamma distribution with a size equal to that of SI ($n = 232$). Use gamrnd. Compare two histograms, one with the simulated values from the gamma model and the second from the measurements of SI. Use 20 bins for the histograms. Comment on their similarities/differences.

(d) Produce a Q–Q plot to compare the measured SI values with the model. Suppose that you selected $\alpha = 3$ and $\beta = 3.3$, and that dia is your data set. Take $n = 232$ equally-spaced points between [0,1] and find their gamma quantiles using gaminv(points,alpha,beta). If the model fits the data, these theoretical quantiles should match the ordered sample.

*Hint:* The plot of theoretical quantiles against the ordered sample is called a Q–Q plot. An example of producing a Q–Q plot in MATLAB is as follows:

```
xx = 0.5/232: 1/232: 1;
yy=gaminv(xx, 3, 3.3);
plot(yy, sort(dia(:,8)),'*')
hold on
plot(yy, yy,'r-')
```

5.25. **Correlation Between a Uniform and its Power.** Suppose that $X$ has uniform $\mathcal{U}(-1,1)$ distribution and that $Y = X^k$.
(a) Show that for $k$ even, $\mathrm{Corr}(X,Y) = 0$.
(b) Show that for arbitrary $k$, $\mathrm{Corr}(X,Y) \to 0$, when $k \to \infty$.

5.26. **Precision of Lab Measurements.** The error $X$ in measuring the weight of a chemical sample is a random variable with PDF.

$$f(x) = \begin{cases} \frac{3x^2}{16}, & -2 < x < 2 \\ 0, & \text{otherwise} \end{cases}$$

(a) A measurement is considered to be *accurate* if $|X| < 0.5$. Find the probability that a randomly chosen measurement can be classified as accurate.
(b) Find and sketch the graph of the cumulative distribution function $F(x)$.
(c) The loss in dollars which is caused by measurement error is $Y = X^2$. Find the mean of $Y$ (expected loss).
(d) Compute the probability that the loss is less than \$3.
(e) Find the median of $Y$.

5.27. **Lifetime of Cells.** Cells in the human body have a wide variety of life spans. One cell may last a day; another a lifetime. Red blood cells (RBC) have a lifespan of several months and cannot replicate, which is the price RBCs pay for being specialized cells. The lifetime of a RBC can be modeled by an exponential distribution with density $f(t) = \frac{1}{\beta}e^{-t/\beta}$, where $\beta = 4$ (in units of months). For simplicity, assume that when a particular RBC dies it is instantly replaced by a newborn RBC of the same type. For example, a replacement RBC could be defined as any new cell born approximately at the time when the original cell died.

(a) Find the expected lifetime of a single RBC. Find the probability that the cell's life exceeds 150 days. *Hint:* Days have to be expressed in units of $\beta$.

(b) A single RBC and its replacements are monitored over the period of 1 year. How many deaths/replacements are observed on average? What is the probability that the number of deaths/replacements exceeds 5. *Hint:* Utilize a link between Exponential and Poisson distributions. In simple terms, if life- times are exponential with parameter $\beta$, then the number of deaths/replacements in the time interval $[0, t]$ is Poisson with parameter $\lambda = t/\beta$. Time units for $t$ and $\beta$ have to be the same.

(c) Suppose that a single RBC and two of its replacements are monitored. What is the distribution of their total lifetime? Find the probability that their total lifetime exceeds 1 year. *Hint:* Consult the gamma distribution. If $n$ random variables are exponential with parameter $\beta$ then their sum is gamma-distributed with parameters $\alpha = n$ and $\beta$.

(d) A particular RBC is observed $t = 2.2$ months after its birth and is found to still be alive. What is the probability that the total lifetime of this cell will exceed 7.2 months?

5.28. **Silver-coated Nylon Fiber.** Silver-coated nylon fiber is used in hospitals for its anti-static electricity properties, as well as for antibacterial and antimycotic effects. In the production of silver-coated nylon fibers, the extrusion process is interrupted from time to time by blockages occurring in the extrusion dyes. The time in hours between blockages, $T$, has an exponential $\mathscr{E}(1/10)$ distribution, where 1/10 is the rate parameter.

Find the probabilities that

(a) a run continues for at least 10 hours,

(b) a run lasts less than 15 hours, and

(c) a run continues for at least 20 hours, given that it has lasted 10 hours.

Use MATLAB and expcdf function. Be careful about the parametrization of exponentials in MATLAB.

5.29. **Xeroderma pigmentosum.** Xeroderma pigmentosum (XP) was first described in 1874 by Hebra et al. XP is the condition characterized as dry, pigmented skin. It is a hereditary condition with an incidence of 1:250000 live births (Robbin et al., 1974). In a city with a population of 1000000, find the distribution of the number of people with XP. What is the expected number? What is the probability that there are no XP-affected subjects?

5.30. **Failure Time.** Let $X$ model the time to failure (in years) of a Beckman Coulter TJ-6 laboratory centrifuge. Suppose the PDF of $X$ is $f(x) = c/(3+x)^3$ for $x \geq 0$.

(a) Find the value of $c$ such that $f$ is a legitimate PDF.

(b) Compute the mean and median time to failure of the centrifuge.

5.31. **Beta Fit.** Assume that the fraction of impurities in a certain chemical solution is modeled by a Beta $\mathscr{B}e(\alpha, \beta)$ distribution with known parameter $\alpha = 1$. The average fraction of impurities is 0.1.

(a) Find the parameter $\beta$.

(b) What is the standard deviation of the fraction of impurities?

(c) Find the probability that the fraction of impurities exceeds 0.25.

5.32. **Uncorrelated but Possibly Dependent.** Show that for any two random variables $X$ and $Y$ with equal second moments, the variables $Z = X + Y$ and $W = X - Y$ are uncorrelated. Note, that $Z$ and $W$ could be dependent.

5.33. **Nights of Mr. Jones.** If Mr. Jones had insomnia one night, the probability that he would sleep well the following night is 0.6, otherwise he would have insomnia. If he slept well one night, the probabilities of sleeping well or having insomnia the following night would be 0.5 each.

On Monday night Mr. Jones had insomnia. What is the probability that he had insomnia on the following Friday night.

5.34. **Stationary Distribution of MC.** Consider a Markov Chain with transition matrix

$$\mathbf{P} = \begin{pmatrix} 0 & 1/2 & 1/2 \\ 1/2 & 0 & 1/2 \\ 1/2 & 1/2 & 0 \end{pmatrix}.$$

(a) Show that all entries of $\mathbf{P}^2$ are strictly positive.

(b) Using MATLAB find $\mathbf{P}^{100}$ and guess what the stationary distribution $\pi = (\pi_1, \pi_2, \pi_3)$ would be. Confirm your guess by solving the equation $\pi = \pi\mathbf{P}$, which gives the exact stationary distribution. *Hint:* The system $\pi = \pi\mathbf{P}$ needs a closure equation $\pi_1 + \pi_2 + \pi_3 = 1$.

5.35. **Heat Production by a Resistor.** Joule's Law states that the amount of heat produced by a resistor is

$$Q = I^2 \, R \, T,$$

where
$Q$ is heat energy (in Joules),
$I$ is the current (in Amperes),
$R$ is the resistance (in Ohms), and
$T$ is duration of time (in seconds).
Suppose that in an experiment, $I$, $R$, and $T$ are independent random variables with means $\mu_I = 10A$, $\mu_R = 30\Omega$, and $\mu_T = 120sec$. Suppose that the variances are $\sigma_I^2 = 0.01A^2$, $\sigma_R^2 = 0.02\Omega^2$, and $\sigma_T^2 = 0.001sec^2$.
Estimate the mean $\mu_Q$ and the variance $\sigma_Q^2$ of the produced energy $Q$.

---

| MATLAB AND WINBUGS FILES AND DATA SETS USED IN THIS CHAPTER |
| --- |

http://springer.bme.gatech.edu/Ch5.RanVar/

bookplots.m, circuitgenbin.m, covcord2d.m, Discrete.m, empiricalcdf.m, histp.m, hyper.m, lefthanded.m, lifetimecells.m, markovchain.m, melanoma.m, plotbino.m, plotsdistributions.m, plotuniformdist.m, randdirichlet.m, stringerror.m

simulationc.odc, simulationd.odc

diabetes.xls

---

# CHAPTER REFERENCES

Apgar, V. (1953). A proposal for a new method of evaluation of the newborn infant. *Curr. Res. Anesth. Analg.*, **32**, (4), 260–267. PMID 13083014.

CDC (2007). Morbidity and mortality weekly report. **56**, 8, 161–165.

Coresh, J., Astor, B. C., Greene, T., Eknoyan, G., and Levey, A. S. (2003). Prevalence of chronic kidney disease and decreased kidney function in the adult US population: 3rd national health and nutrition examination survey. *Am. J. Kidney Dis.*, **41**, 1–12.

Dayhoff, M. O., Schwartz, R., and Orcutt, B. C. (1978). A model of evolutionary change in proteins. Atlas of protein sequence and structure (Volume 5, Supplement 3 ed.) *Nat. Biomed. Res. Found.*, 345–358, ISBN 0912466073.

Gjini, A., Stuart, J. M., George, R. C., Nichols, T., Heyderman, R. S. (2004). Capture-recapture analysis and pneumococcal meningitis estimates in England. *Emerg. Infect. Dis.*, **10**, 1, 87–93.

Hebra F. and Kaposi M. (1874). On diseases of the skin including exanthemata. *New Sydenham Soc.*, **61**, 252–258.

Montmort, P. R. (1714). *Essai d'Analyse sur les Jeux de Hazards*, 2ed., Jombert, Paris.

Robbin, J. H., Kraemer, K. H., Lutzner, M. A., Festoff, B. W., and Coon, H. P. (1974). Xeroderma pigmentosum: An inherited disease with sun sensitivity, multiple cutaneous neoplasms and abnormal DNA repair. *Ann. Intern. Med.*, **80**, 221–248.

Ross, M. S. (2010a). *A First Course in Probability*, Pearson Prentice-Hall, ISBN 013603313X.

Ross, M. S. (2010b) *Introduction to Probability Models*, 10th ed, Academic Press, Burlington.

Tubert-Bitter, P., Begaud, B., Moride, Y., Chaslerie, A., and Haramburu, F. (1996). Comparing the toxicity of two drugs in the framework of spontaneous reporting: A confidence interval approach. *J. Clin. Epidemiol.*, **49**, 121–123.

Weibull, W. (1951). A statistical distribution function of wide applicability. *J. Appl. Mech.*, **18**, 293–297.

# Chapter 6
# Normal Distribution

*The adjuration to be normal seems shockingly repellent to me.*

– Karl Menninger

---

**WHAT IS COVERED IN THIS CHAPTER**

• Definition of Normal Distribution. Bivariate Case
• Standardization. Quantiles of Normal Distribution. Sigma Rules
• Linear Combinations of Normal Random Variables
• Central Limit Theorem. de Moivre's Approximation
• Distributions Related to Normal: Chi-Square, Wishart, t, F, Lognormal, and Some Noncentral Distributions
• Transformations to Normality

---

## 6.1 Introduction

In Chaps. 2 and 5 we occasionally referred to a normal distribution either informally (bell-shaped distributions/histograms) or formally, as in Sect. 5.5.3, where the normal density and its moments were briefly introduced. This chapter is devoted to the normal distribution due to its importance in statistics. What makes the normal distribution so important? The normal distribution is the proper statistical model for many natural and social phenomena. But even if some measurements cannot be modeled by the normal distribution (it could be skewed, discrete, multimodal, etc.), their sample means would closely follow the normal law, under very mild conditions. The central limit theorem covered in this chapter makes it possible to use probabilities associated with the normal curve to answer questions about the sums and averages in sufficiently large samples. This translates to the ubiquity of normality – many estimators, test statistics, and nonparametric tests covered in later chapters of this text are approximately normal, when sample sizes are not small (typically larger than 20 to 30), and this asymptotic normality is used in a substantial way. Several other important distributions can be defined through a normal distribution. Also, normality is a quite stable property – an arbitrary linear combination of normal random variables remains normal. The property of linear combinations of random variables preserving the distribution of their components is not shared by any other probability law and is a characterizing property of a normal distribution.

## 6.2 Normal Distribution

In 1738, Abraham de Moivre (Fig. 6.1a) developed the normal distribution as an approximation to the binomial distribution, and it was subsequently used by Laplace in 1783 to study measurement errors and by Gauss (Fig. 6.1b) in 1809 in the analysis of astronomical data. The name *normal* came from Quetelet (Fig. 6.1c), who demonstrated that many human characteristics distributed themselves in a bell-shaped manner (centered about the "average man," *l'homme moyen*), including such measurements as chest girths of 5,738 Scottish soldiers, the heights of 100,000 French conscripts, and the body weight and height of people he measured. From his initial research on height and weight has evolved the internationally recognized measure of obesity called the Quetelet index (QI), or body mass index (BMI), QI = (weight in kilograms)/(squared height in meters).

Table 6.1 provides frequencies of chest sizes of 5738 Scottish soldiers as well as the relative frequencies. Using this now famous data set Quetelet argued that many human measurements distribute as normal. Figure 6.2 gives a normalized histogram of Quetelet's data set with superimposed normal density in which the mean and the variance are taken as the sample mean (39.8318) and sample variance ($2.0496^2$).

The PDF for a normal random variable with mean $\mu$ and variance $\sigma^2$ is

$$f(x) = \frac{1}{\sqrt{2\pi\sigma^2}} e^{-\frac{1}{2\sigma^2}(x-\mu)^2}, \qquad \infty < x < \infty.$$

The distribution function is computed using integral approximation because no closed form exists for the antiderivative of $f(x)$; this is generally not a problem for practitioners because most software packages will compute interval

(a)                              (b)                              (c)

**Fig. 6.1** (a) Abraham de Moivre (1667–1754), French mathematician; (b) Johann Carl Friedrich Gauss (1777–1855), German mathematician, astronomer, and physicist; (c) Lambert Adolphe Quetelet (1796–1874), Belgian astronomer, mathematician, and sociometrist

**Table 6.1** Chest sizes of 5738 Scottish soldiers, data compiled from the 13th edition of the Edinburgh Medical Journal (1817).

| Size | Frequency | Relative frequency (in %) |
|---|---|---|
| 33 | 3 | 0.05 |
| 34 | 18 | 0.31 |
| 35 | 81 | 1.41 |
| 36 | 185 | 3.22 |
| 37 | 420 | 7.32 |
| 38 | 749 | 13.05 |
| 39 | 1073 | 18.70 |
| 40 | 1079 | 18.80 |
| 41 | 934 | 16.28 |
| 42 | 658 | 11.47 |
| 43 | 370 | 6.45 |
| 44 | 92 | 1.60 |
| 45 | 50 | 0.87 |
| 46 | 21 | 0.37 |
| 47 | 4 | 0.07 |
| 48 | 1 | 0.02 |
| Total | 5738 | 99.99 |

probabilities numerically. In MATLAB, `normcdf(x, mu, sigma)` and `normpdf(x, mu, sigma)` calculate the CDF and PDF at $x$, and `norminv(p, mu, sigma)` computes the inverse CDF at given probability $p$, that is, the $p$-quantile. Equivalently, a normal CDF can be expressed in terms of a special function called the *error integral*:

$$\mathsf{erf}(\mathsf{x}) = \frac{2}{\sqrt{\pi}} \int_0^x e^{-t^2} dt.$$

It holds that `normcdf(x)= 1/2+1/2*erf(x/sqrt(2))`.

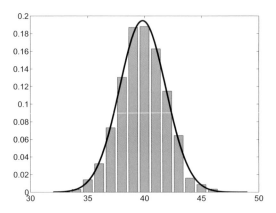

**Fig. 6.2** Normalized bar plot of Quetelet's data set. Superimposed is the normal density with mean $\mu = 39.8318$ and variance $\sigma^2 = 2.0496^2$

A random variable $X$ with a normal distribution will be denoted $X \sim \mathcal{N}(\mu, \sigma^2)$.

In addition to software, CDF values are often given in tables. Such tables contain only quantiles and CDF values for the *standard* normal distribution, $Z \sim \mathcal{N}(0, 1)$, for which $\mu = 0$ and $\sigma^2 = 1$. Such tables are sufficient since an arbitrary normal random variable $X$ can be *standardized* to $Z$ if its mean and variance are known:

$$X \sim \mathcal{N}(\mu, \sigma^2) \qquad \mapsto \qquad Z = \frac{X - \mu}{\sigma} \sim \mathcal{N}(0, 1).$$

For a standard normal random variable $Z$ the PDF is denoted by $\phi$, and CDF by $\Phi$,

$$\Phi(x) = \int_{-\infty}^{x} \phi(t)\, dt = \int_{-\infty}^{x} \frac{1}{\sqrt{2\pi}} e^{-t^2/2}\, dt. \quad \texttt{[normcdf(x)]}$$

Suppose we are interested in the probability that a random variable $X$ distributed as $\mathcal{N}(\mu, \sigma^2)$ falls between two bounds $a$ and $b$, $\mathbb{P}(a < X < b)$. It is irrelevant whether the bounds are included or not since the normal distribution is continuous and $\mathbb{P}(a < X < b) = \mathbb{P}(a \leq X \leq b)$. Also, any of the bounds can be infinite.

In terms of $\Phi$,

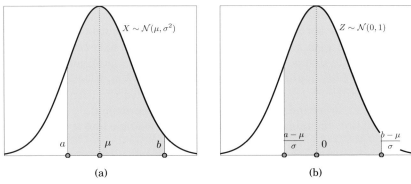

(a) (b)

**Fig. 6.3** Illustration of the relation $\mathbb{P}(a \leq X \leq b) = \mathbb{P}\left(\frac{a-\mu}{\sigma} \leq Z \leq \frac{b-\mu}{\sigma}\right)$

$X \sim \mathcal{N}(\mu, \sigma^2):$

$$\mathbb{P}(a \leq X \leq b) = \mathbb{P}\left(\frac{a-\mu}{\sigma} \leq Z \leq \frac{b-\mu}{\sigma}\right) = \Phi\left(\frac{b-\mu}{\sigma}\right) - \Phi\left(\frac{a-\mu}{\sigma}\right).$$

Figures 6.3 and 6.4 provide the illustration. In MATLAB:

```
normcdf((b-mu)/sigma) - normcdf((a - mu)/sigma)
        %or equivalently
normcdf(b, mu, sigma) - normcdf(a, mu, sigma)
```

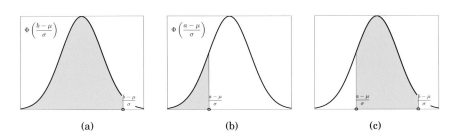

(a) (b) (c)

**Fig. 6.4** Calculation of $\mathbb{P}(a \leq X \leq b)$ for $X \sim \mathcal{N}(\mu, \sigma^2)$. (a) $\mathbb{P}(X \leq b) = \mathbb{P}(Z \leq \frac{b-\mu}{\sigma}) = \Phi\left(\frac{b-\mu}{\sigma}\right)$; (b) $\mathbb{P}(X \leq a) = \mathbb{P}(Z \leq \frac{a-\mu}{\sigma}) = \Phi\left(\frac{a-\mu}{\sigma}\right)$; (c) $\mathbb{P}(a \leq X \leq b)$ as the difference of the two probabilities in (a) and (b).

Note that when the bounds are infinite, since $\Phi$ is a CDF,

$$\Phi(-\infty) = 0, \text{ and } \Phi(\infty) = 1.$$

Traditional statistics textbooks provide tables of cumulative probabilities for the standard normal distribution, $p = \Phi(x)$, for values of $x$ typically between $-3$ and $3$ with an increment of $0.01$. The tables have been used in two ways: (i) directly, that is, for a given $x$ the user finds $p = \Phi(x)$; and (ii) inversely, given $p$, one finds approximately what $x$ gives $\Phi(x) = p$, which is of course a $p$-quantile of the standard normal. Given the limited precision of the tables, the results in direct and inverse uses have been approximate.

In MATLAB the tables can be reproduced by a single line of code:

```
x=(-3:0.01:3)'; tables=[x normcdf(x)]
```

Similarly, the normal $p$-quantiles $z_p$ defined as $p = \Phi(x_p)$ can be tabulated as

```
probs=(0.005:0.005:0.995)'; tables=[probs norminv(probs)]
```

There are several normal quantiles that are frequently used in the construction of confidence intervals and tests; these are the $0.9, 0.95, 0.975, 0.99, 0.995$, and $0.9975$ quantiles,

$$z_{0.9} = 1.28155 \approx 1.28 \quad z_{0.95} = 1.64485 \approx 1.64 \quad z_{0.975} = 1.95996 \approx 1.96$$
$$z_{0.99} = 2.32635 \approx 2.33 \quad z_{0.995} = 2.57583 \approx 2.58 \quad z_{0.9975} = 2.80703 \approx 2.81$$

For example, the $0.975$ quantile of the normal is $z_{0.975} = 1.96$. This is equivalent to saying that 95% of the area below the standard normal density $\phi(x) = \frac{1}{\sqrt{2\pi}} \exp\{-x^2/2\}$ lies between $-1.96$ and $1.96$. Note that the shortest interval containing $1 - \alpha$ probability is defined by quantiles $z_{\alpha/2}$ and $z_{1-\alpha/2}$ (see Fig. 6.5 as an illustration for $\alpha = 0.05$). Since the standard normal density is symmetric about 0, $z_p = -z_{1-p}$.

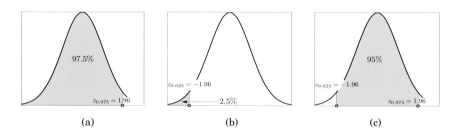

(a)                                      (b)                                      (c)

**Fig. 6.5** (a) Normal quantiles (a) $z_{0.975} = 1.96$; (b) $z_{0.025} = -1.96$; and (c) 95% area between quantiles $-1.96$ and $1.96$.

### 6.2.1 Sigma Rules

*Sigma rules* state that for any normal distribution, the probability that an observation will fall in the interval $\mu \pm k\sigma$ for $k = 1, 2$, and 3 is $68.27\%, 95.45\%$ and $99.73\%$, respectively. More precisely,

$$\mathbb{P}(\mu - \sigma < X < \mu + \sigma) = \mathbb{P}(-1 < Z < 1) = \Phi(1) - \Phi(-1) = 0.682689 \quad \approx 68.27\%$$
$$\mathbb{P}(\mu - 2\sigma < X < \mu + 2\sigma) = \mathbb{P}(-2 < Z < 2) = \Phi(2) - \Phi(-2) = 0.954500 \approx 95.45\%$$
$$\mathbb{P}(\mu - 3\sigma < X < \mu + 3\sigma) = \mathbb{P}(-3 < Z < 3) = \Phi(3) - \Phi(-3) = 0.997300 \approx 99.73\%$$

Did you ever wonder about the origin of the term *Six Sigma?* It does not involve $\mathbb{P}(\mu - 6\sigma < X < \mu + 6\sigma)$ as one may expect.

The Six Sigma doctrine is a standard according to which an item with measurement $X \sim \mathcal{N}(\mu, \sigma^2)$ should satisfy $X < 6\sigma$ to be conforming if $\mu$ is allowed to vary between $-1.5\sigma$ and $1.5\sigma$.

Thus effectively, accounting for the variability in the mean, the Six Sigma constraint becomes

$$\mathbb{P}(X < \mu + 4.5\sigma) = P(Z < 4.5) = \Phi(4.5) = 0.99999660.$$

This means that only 3.4 items per million produced are allowed to exceed $\mu + 4.5\sigma$ (be defective). Such standard of quality was set by the Motorola company in the 1980s, and it evolved into a doctrine for improving efficiency and quality in management.

### 6.2.2 Bivariate Normal Distribution*

When the components of a random vector have a normal distribution, we say that the vector has a multivariate normal distribution. For independent components, the density of a multivariate distribution is simply the product of the univariate densities. When components are correlated, the distribution involves the covariance matrix that describes the correlation. Next we discuss the bivariate normal distribution, which will be important later on, in the context of correlation and regression.

The pair $(X, Y)$ is distributed as bivariate normal $\mathcal{N}_2(\mu_X, \mu_Y, \sigma_X^2, \sigma_Y^2, \rho)$ if the joint density is

$$f(x, y) =$$

$$\frac{1}{2\pi\sigma_1\sigma_2\sqrt{1-\rho^2}} \exp\left\{ -\frac{1}{2(1-\rho^2)} \left[ \frac{(x-\mu_x)^2}{\sigma_1^2} - \frac{2\rho(x-\mu_x)(y-\mu_y)}{\sigma_1\sigma_2} + \frac{(y-\mu_y)^2}{\sigma_2^2} \right] \right\}. \quad (6.1)$$

The parameters $\mu_X, \mu_Y, \sigma_X^2, \sigma_Y^2$, and $\rho$ are

$$\mu_X = \mathbb{E}(X), \mu_Y = \mathbb{E}(Y), \sigma_X^2 = \mathbb{V}\mathrm{ar}(X), \sigma_Y^2 = \mathbb{V}\mathrm{ar}(Y), \text{ and } \rho = \mathbb{C}\mathrm{orr}(X, Y).$$

One can define bivariate normal distribution with a density as in (6.1) by transforming two independent, standard normal random variables $Z_1$ and $Z_2$,

$$X = \mu_1 + \sigma_X Z_1,$$

$$Y = \mu_2 + \rho\sigma_Y Z_1 + \sqrt{1-\rho^2}\sigma_Y Z_2.$$

The marginal distributions in (6.1) are $X \sim \mathcal{N}(\mu_X, \sigma_X^2)$ and $Y \sim \mathcal{N}(\mu_Y, \sigma_Y^2)$. The bivariate normal vector $(X,Y)$ has covariance a matrix

$$\Sigma = \begin{pmatrix} \sigma_X^2 & \sigma_X\sigma_Y\rho \\ \sigma_X\sigma_Y\rho & \sigma_Y^2 \end{pmatrix}. \tag{6.2}$$

The covariance matrix $\Sigma$ is nonnegative definite. A sufficient condition for non-negative definiteness in this case is $|\Sigma| \geq 0$ (see also Exercise 6.2).

Figure 6.6a shows the density of a bivariate normal distribution with mean

$$\mu = \begin{pmatrix} \mu_X \\ \mu_Y \end{pmatrix} = \begin{pmatrix} -1 \\ 2 \end{pmatrix}$$

and covariance matrix

$$\Sigma = \begin{pmatrix} 3 & -0.9 \\ -0.9 & 1 \end{pmatrix}.$$

Figure 6.6b shows contours of equal probability.

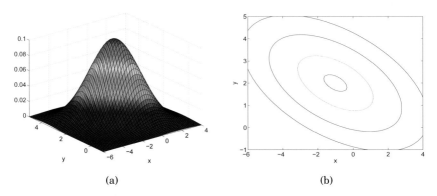

(a)                                                                            (b)

**Fig. 6.6** (a) Density of bivariate normal distribution with mean mu=[-1 2] and covariance matrix Sigma=[3 -0.9; -0.9 1]. (b) Contour plots of a density at levels [0.001 0.01 0.05 0.1]

Several properties of bivariate normal are listed below.

(i) If $(X,Y)$ is bivariate normal, then $aX + bY$ has a univariate normal distribution.

(ii) If $(X,Y)$ is bivariate normal, then $(aX + bY, cX + dY)$ is also bivariate normal.

(iii) If the components in $(X,Y)$ are such that $\mathrm{Cov}(X,Y) = \sigma_X\sigma_Y\rho = 0$, then $X$ and $Y$ are independent.

(iv) Any bivariate normal pair $(X,Y)$ can be transformed into a pair $(U,V) = (aX + bY, cX + dY)$ such that $U$ and $V$ are independent. If $\sigma_X^2 = \sigma_Y^2$, then one such transformation is $U = X + Y$, $V = X - Y$. For an arbitrary bivariate normal distribution, the rotation

$$U = X\cos\varphi - Y\sin\varphi$$
$$V = X\sin\varphi + Y\cos\varphi$$

makes components $(U,V)$ independent if the rotation angle $\varphi$ satisfies

$$\cot 2\varphi = \frac{\sigma_X^2 - \sigma_Y^2}{2\sigma_X\sigma_Y\rho}.$$

(v) If $(X,Y)$ is bivariate normal, then the conditional distribution of $Y$ when $X = x$ is normal with expectation and variance

$$\mu_X + \rho\frac{\sigma_Y}{\sigma_X}(x - \mu_X), \quad \text{and} \quad \sigma_Y^2(1 - \rho^2),$$

respectively. The linearity in $x$ of the conditional expectation of $Y$ will be the basis for linear regression, covered in Chap. 16. Also, the fact that $X = x$ is known decreases the variance of $Y$, indeed $\sigma_Y^2(1 - \rho^2) \le \sigma_Y^2$.

## 6.3 Examples with a Normal Distribution

We provide two examples with typical calculations involving normal distributions, with solutions in MATLAB and WinBUGS.

*Example 6.1.* **IgE Concentration.** Total serum IgE (immunoglobulin E) concentration allergy tests allow for the measurement of the total IgE level in a serum sample. Elevated levels of IgE are associated with the presence of an allergy. An example of testing for total serum IgE is the PRIST (paper radioimmunosorbent test). This test involves serum samples reacting with IgE that has been tagged with radioactive iodine. The bound radioactive iodine, calculated upon completion of the test procedure, is proportional to the amount of total IgE in the serum sample. The determination of normal IgE levels in a population of healthy nonallergic individuals varies by the fact that some individuals may have subclinical allergies and therefore have abnormal serum IgE levels. The log concentration of IgE (in IU/ml) in a cohort of healthy subjects is distributed as a normal $\mathcal{N}(9,(0.9)^2)$ random variable. What is the probability that in a randomly selected subject from the same cohort the log concentration will

- Exceed 10 IU/ml?
- Be between 8.1 and 9.9 IU/ml?

- Differ from the mean by no more than 1.8 IU/ml?
- Find the number $x_0$ such that the IgE log concentration in 90% of the subjects from the same cohort exceeds $x_0$.
- In what bounds (symmetric about the mean) does the IgE log concentration fall with a probability of 0.95?
- If the IgE log concentration is $\mathcal{N}(9,\sigma^2)$, find $\sigma$ so that

$$P(8 \le X \le 10) = 0.64.$$

Let $X$ be the IgE log concentration in a randomly selected subject. Then $X \sim \mathcal{N}(9, 0.9^2)$. The solution is given by the following MATLAB code ( ◀ ige.m):

```
%(1)
%P(X>10)= 1-P(X <= 10)
1-normcdf(10,9,0.9) %or    1-normcdf((10-9)/0.9)
%ans = 0.1333
%(2)
%P(8.1 <= X <= 9.9)
%P((8.1-9)/0.9 <= Z <= (9.9-9)/0.9)
%P(-1 <= Z <= 1) ::: Note 1-sigma rule.
normcdf(9.9, 9, 0.9) - normcdf(8.1, 9, 0.9)
%or, normcdf((9.9-9)/0.9)-normcdf((8.1-9)/0.9)
%ans = 0.6827
%(3)
%P(9-1.8 <= X <= 9+1.8) = P(-2 <= Z <= 2)
%Note 2-sigma rule.
normcdf(9+1.8, 9, 0.9) - normcdf(9-1.8, 9, 0.9)
% ans = 0.9545
%(4)
%0.90 = P(X > x_0)=1-P(X <= x0)
%that is P(Z <= (x0-9)/0.9)=0.1
norminv(1-0.9,   9, 0.9)
%ans = 7.8466
%(5)
%P(9-delta <= X <= 9+delta)=0.95
[9-0.9*norminv(1-0.05/2), 9+0.9*norminv(1-0.05/2)]
%ans =   7.2360   10.7640
%(6)
%P(-1/sigma) <= Z <= 1/sigma)=0.64
%note that 0.36/2 + 0.64 + 0.36/2 = 1
1/norminv( 1 - 0.36/2 )
%ans = 1.0925
```

*Example 6.2.* **Aplysia Nerves.** In this example, easily solved analytically and using MATLAB, we will show how to use WinBUGS and obtain an approximate solution. The analysis is not Bayesian; WinBUGS will simply serve as a random number generator and the required probability and quantile will be found approximately by simulation.

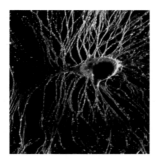

**Fig. 6.7** Sea slug (Aplysia) neuron stained with membrane dye. From the cover of the 28 January 2004 *Journal of Neuroscience*. Copyright 2004 Journal of Neuroscience

Characteristics of Aplysia nerves in response to extension were examined in Koike (1987). Only the Aplysia nerve (Fig. 6.7) was easily elongated up to about five times its resting or relaxing length without impairing propagation of the action potential along the axon in the nerve. The conduction velocity along the elongated nerve increased linearly in proportion to the nerve length in a range from the relaxing length to about 1 to 1.5 times extension. For an expansion factor of 1.5 the conducting velocity factors are normally distributed with a mean of 1.4 and a standard deviation of 0.1. Using WinBUGS, we are interested in finding

(a) the proportion of Aplysia nerves elongated by a factor of 1.5 for which the conduction velocity factor exceeds 1.5;

(b) the proportion of Aplysia nerves elongated by a factor of 1.5 for which the conduction velocity factor falls in the interval [1.35, 1.61]; and

(c) the velocity factor $x$ that is exceeded by 5% of Aplysia nerves elongated by a factor of 1.5.

```
#aplysia.odc
model{
mu <- 1.4
stdev <- 0.1
prec<- 1/(stdev * stdev)
y ~ dnorm(mu, prec)
#a
propexceed <- step(y - 1.5)
#b
propbetween <-  step(y-1.35)*step(1.61-y)
#c
#done in Sample Monitor Tool by
#selecting 95th percentile
}
```

There are no data to load; after the check model in Model>Specification go directly to compile, and then to gen inits. Update 10,000 iterations, and set in Sample Monitor Tool from Inference>Samples the nodes y, propexceed, and propbetween. For part (c) select the 95th percentile in Sample Monitor Tool under percentiles. Finally, run the Update Tool for 1,000,000 updates and check the results in Sample Monitor Tool by setting a star (*) in the node window and looking at stats.

| | mean | sd | MC error | val2.5pc | median | val97.5pc | start | sample |
|---|---|---|---|---|---|---|---|---|
| propbetween | 0.6729 | 0.4691 | 4.831E-4 | 0.0 | 1.0 | 1.0 | 10001 | 1000000 |
| propexceed | 0.1587 | 0.3654 | 3.575E-4 | 0.0 | 0.0 | 1.0 | 10001 | 1000000 |
| y | 1.4 | 0.1001 | 1.005E-4 | 1.204 | 1.4 | 1.565 | 10001 | 1000000 |

Here is the same computation in MATLAB.

```
1-normcdf(1.5, 1.4, 0.1)
 %ans = 0.1587
normcdf(1.61, 1.4, 0.1)-normcdf(1.35, 1.4, 0.1)
 %ans = 0.6736
norminv(1-0.05, 1.4, 0.1)
 %ans = 1.5645
```

## 6.4 Combining Normal Random Variables

Any linear combination of independent normal random variables is also normally distributed. Thus, we need only keep track of the mean and variance of the variables involved in the linear combination, since these two parameters completely characterize the distribution. Let $X_1, X_2, \ldots, X_n$ be independent normal random variables such that $X_i \sim \mathcal{N}(\mu_i, \sigma_i^2)$; then for any selection of constants $a_1, a_2, \ldots, a_n$

$$a_1 X_1 + a_2 X_2 + \cdots + a_n X_n = \sum_{i=1}^{n} a_i X_i \sim \mathcal{N}(\mu, \sigma^2),$$

where

$$\mu = a_1 \mu_1 + a_2 \mu_2 + \cdots + a_n \mu_n = \sum_{i=1}^{n} a_i \mu_i,$$

$$\sigma^2 = a_1^2 \sigma_1^2 + a_2^2 \sigma_2^2 + \ldots a_n^2 \sigma_n^2 = \sum_{i=1}^{n} a_i^2 \sigma_i^2.$$

Two special cases are important: (i) $a_1 = 1, a_2 = -1$ and (ii) $a_1 = \cdots = a_n = 1/n$. In case (i) we have a difference of two normals; its mean is the difference of the corresponding means and variance is a *sum* of two variances. Case (ii) corresponds to the arithmetic mean of normals, $\overline{X}$. For example, if $X_1,\ldots,X_n$ are i.i.d. $\mathcal{N}(\mu,\sigma^2)$, then the sample mean $\overline{X} = (X_1 + \cdots + X_n)/n$ has a normal $\mathcal{N}(\mu,\sigma^2/n)$ distribution. Thus, variances for $X_i$s and $\overline{X}$ are related as

$$\sigma_{\overline{X}}^2 = \frac{\sigma^2}{n}$$

or, equivalently, for standard deviations

$$\sigma_{\overline{X}} = \frac{\sigma}{\sqrt{n}}.$$

*Example 6.3.* **The Piston Production Error.** The profile of a piston comprises a ring in which inner and outer radii $X$ and $Y$ are normal random variables, $\mathcal{N}(20,0.01^2)$ and $\mathcal{N}(30,0.02^2)$, respectively. The thickness $D = Y - X$ is the random variable of interest.

(a) Find the distribution of $D$.

(b) For a randomly selected piston, what is the probability that $D$ will exceed 10.04?

(c) If $D$ is averaged over a batch of $n = 64$ pistons, what is the probability that $\overline{D}$ will exceed 10.04? Exceed 10.004?

```
sqrt(0.01^2 + 0.02^2)                               %0.0224
1-normcdf((10.04  - 10)/0.0224)                     %0.0371
1-normcdf((10.04  - 10)/(0.0224/sqrt(64)))  %0
1-normcdf((10.004 - 10)/(0.0224/sqrt(64)))  %0.0766
```

Compare the probabilities of events $\{D > 10.04\}$ and $\{\overline{D} > 10.04\}$. Why is the probability of $\{\overline{D} > 10.04\}$ essentially 0, when the analogous probability for an individual measure $D$ is 3.71%?

*Example 6.4.* **Diluting Acid.** In a laboratory, students are told to mix 100 ml of distilled water with 50 ml of sulfuric acid and 30 ml of $C_2H_5OH$. Of course, the measurements are not exact. The water is measured with a mean of 100 ml and a standard deviation of 4 ml, the acid with a mean of 50 ml and a standard deviation of 2 ml, and $C_2H_5OH$ with a mean of 30 ml and a standard deviation of 3 ml. The three measurements are normally distributed and independent.

(a) What is the probability of a given student measuring out at least 103 ml of water?

(b) What is the probability of a given student measuring out between 148 and 157 ml of water plus acid?

(c) What is the probability of a given student measuring out a total of between 175 and 180 ml of liquid?

```
1 - normcdf(103, 100, 4)                              %0.2266
normcdf(157, 150, sqrt(4^2 + 2^2)) ...
      - normcdf(148, 150, sqrt(4^2 + 2^2))            %0.6139
normcdf(180, 180, sqrt(4^2 + 2^2 + 3^2 )) ...
      - normcdf(175, 180, sqrt(4^2 + 2^2 + 3^2))      %0.3234
```

## 6.5 Central Limit Theorem

The central limit theorem (CLT) elevates the status of the normal distribution above other distributions. We have already seen that a linear combination of independent normals is the normal random variable itself. That is, if $X_1,\ldots,X_n \overset{\text{iid}}{\sim} \mathcal{N}(\mu,\sigma^2)$, then

$$\sum_{i=1}^{n} X_i \sim \mathcal{N}(n\mu, n\sigma^2), \quad \text{and} \quad \overline{X} = \frac{1}{n}\sum_{i=1}^{n} X_i \sim \mathcal{N}\left(\mu, \frac{\sigma^2}{n}\right).$$

The CLT states that $X_1,\ldots,X_n$ need not be normal in order for $\sum_{i=1}^{n} X_i$ or, equivalently, for $\overline{X}$ to be *approximately* normal. This approximation is quite good for $n$ as low as 30. As we said, variables $X_1, X_2,\ldots,X_n$ need not be normal but must satisfy some conditions. For CLT to hold, it is sufficient for $X_i$s to be independent, equally distributed, and have finite variances and, consequently, means. Other than that, the $X_i$s can be arbitrary – skewed, discrete, etc. The conditions of i.i.d. and finiteness of variances are sufficient – more precise formulations of the CLT are beyond the scope of this course.

**CLT.** Let $X_1, X_2,\ldots,X_n$ be i.i.d. random variables with finite means $\mu$ and variances $\sigma^2$. Then,

$$\sum_{i=1}^{n} X_i \overset{\text{approx}}{\sim} \mathcal{N}(n\mu, n\sigma^2), \quad \text{and} \quad \overline{X} = \frac{1}{n}\sum_{i=1}^{n} X_i \overset{\text{approx}}{\sim} \mathcal{N}\left(\mu, \frac{\sigma^2}{n}\right).$$

A special case of CLT involving Bernoulli random variables results in a normal approximation to binomials because the sum of many i.i.d. Bernoullis is at the same time exactly binomial and approximately normal. This approximation is handy when $n$ is very large.

**de Moivre (1738).** Let $X_1, X_2, \ldots, X_n$ be independent Bernoulli $\mathscr{B}er(p)$ random variables with parameter $p$.
   Then,

$$Y = \sum_{i=1}^{n} X_i \overset{\text{approx}}{\sim} \mathcal{N}(np, npq)$$

and

$$\mathbb{P}(k_1 \leq Y \leq k_2) = \Phi\left(\frac{k_2 + 1/2 - np}{\sqrt{npq}}\right) - \Phi\left(\frac{k_1 - 1/2 - np}{\sqrt{npq}}\right).$$

De Moivre's approximation is good if both $np$ and $nq$ exceed 10 and $n$ exceeds 30. If that is not the case, a Poisson approximation to binomial (p. 149) could be better.

The factors 1/2 in de Moivre's formula are continuity corrections. For example, $Y$, which is discrete, is approximated with a continuous distribution. $\mathbb{P}(Y \leq k_2 + 1)$ and $\mathbb{P}(Y < k_2 + 1)$ are the same for a normal but not for a binomial distribution for which $\mathbb{P}(Y < k_2 + 1) = \mathbb{P}(Y \leq k_2)$. Likewise, $\mathbb{P}(Y \geq k_1 - 1)$ and $\mathbb{P}(Y > k_1 - 1)$ are the same for a normal but not for a binomial distribution for which $\mathbb{P}(Y > k_1 - 1) = \mathbb{P}(Y \geq k_1)$. Thus, $\mathbb{P}(k_1 \leq Y \leq k_2)$ for a binomial distribution is better approximated by $\mathbb{P}(k_1 - 1/2 \leq Y \leq k_2 + 1/2)$.

All approximations used to be much more important in the era before modern computing power was available. MATLAB is capable of calculating exact binomial probabilities for huge values of $n$, and for practical reasons de Moivre's approximation is obsolete. For example,

```
format long
binocdf(1999988765, 4000000000, 1/2)
%ans = 0.361195130797824
format short
```

However, the theoretical value of de Moivre's approximation is significant since many estimators and tests based on a binomial distribution can use well-developed normal distribution machinery for an analysis beyond the computation.

The following MATLAB program exemplifies the CLT by averages of simulated uniform random variables.

```
% Central Limit Theorem Demo
figure;
subplot(3,2,1)
hist(rand(1, 10000),40)          %histogram of 10000 uniforms
subplot(3,2,2)
hist(mean(rand(2, 10000)),40)   %histogtam of 10000
```

```
                                    %averages of 2 uniforms
subplot(3,2,3)
hist(mean(rand(3, 10000)),40)   %histogtam of 10000
                                    %averages of 3 uniforms
subplot(3,2,4)
hist(mean(rand(5, 10000)),40)   %histogtam of 10000
                                    %averages of 5 uniforms
subplot(3,2,5)
hist(mean(rand(10, 10000)),40)  %histogtam of 10000
                                    %averages of 10 uniforms
subplot(3,2,6)
hist(mean(rand(100, 10000)),40) %histogtam of 10000
                                    %averages of 100 uniforms
```

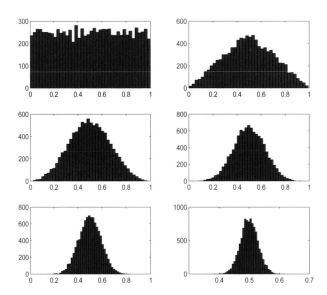

**Fig. 6.8** Convergence to normal distribution shown via averages of 1, 2, 3, 5, 10, and 100 independent uniform (0,1) random variables

Figure 6.8 shows the histograms of 10,000 simulations of averages of $k = 1,2,3,5,10$, and 100 uniform random variables. It is interesting to see the metamorphosis of a flat single uniform ($k = 1$), via a "witch hat distribution" ($k = 2$), into bell-shaped distributions close to the normal. For additional simulation experiments see the script ◀ cltdemo.m.

*Example 6.5.* **Is Grandpa's Genetic Theory Valid?** The domestic cat's wild appearance is increasingly overshadowed by color mutations, such as black, white spotting, maltesing (diluting), red and tortoiseshell, shading, and Siamese pointing. By favoring the odd or unusually colored and marked cats over the "plain" tabby, people have consciously and unconsciously enhanced these color mutations over the course of domestication. Today, "colored" cats outnumber the wild looking tabby cats, and pure tabbies are becoming rare. Some may not be quite as taken by the coat of our domestic feline friends as Jason's grandpa is. He has a genetic theory that asserts that three-fourths of cats with more than three colors in their fur are female. A total of $n = 300$ three-color cats (TCCs) are observed and 86 are found to be male. If Jason's grandpa's genetic theory is true, then the number of male TCCs is binomial $\mathcal{B}(300, 0.25)$, with an expectation of 75 and variance of $56.25 = 7.5^2$.

(a) What is the probability that, assuming Jason's grandpa's theory, one will observe 86 or more male cats? How does this finding support the theory?

(b) What is the probability that, assuming the independence of a cat's fur and gender, one will observe 86 or more male cats?

(c) What is the probability that one will observe exactly 75 male TCCs?

```
format long
1 - binocdf(85, 300, 0.25)
  %ans =   0.08221654140000

1 - normcdf(85, 75, 7.5)
  %ans =   0.09121121972587

1 - normcdf(86, 75, 7.5)
%ans =    0.07123337741399

1 - normcdf(85.5, 75, 7.5)
  %ans =   0.08075665923377

1 - binocdf(85, 300, 0.5)
  %ans =   0.99999999999998

binopdf(75, 300, 0.25)
  %ans =   0.05312831515720

normcdf(75.5, 75, 7.5)-normcdf(74.5, 75, 7.5)
  %ans =  0.05315292860073
```

*Example 6.6.* **Avio Company.** The Avio company sells 410 plane tickets for a 400-seater flight. Find the probability that the company overbooked the flight if a person who bought a ticket shows up at the gate with a probability of 0.96.

Each sold ticket can be thought of as an "experiment" where "success" means showing up at the gate for the flight. The number of people that show up $X$ is binomial $\mathcal{B}in(410, 0.96)$. The following MATLAB script calculates the normal approximation.

```
410*0.96
   %ans =   393.6000
```

```
sqrt(410*0.96*0.04)
   %ans =    3.9679

1-normcdf((400.5-393.6)/3.9679)
   %ans = 0.0410
```

Notice that in this case the normal approximation is not very good since the exact binomial probability is 0.0329:

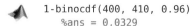

```
1-binocdf(400, 410, 0.96)
   %ans = 0.0329
```

The reason is that the normal approximation works well when the probabilities are not close to 0 or 1, and here 0.96 is quite close to 1 for a given sample size of 410.

The Poisson approximation to the binomial performs better. The probability of missing the flight is $1 - 0.96 = 0.04$, and overbooking will happen if 9 or fewer passengers miss the flight:

```
   %prob that 9 or less fail to show
poisscdf(9, 0.04*410)
   %ans =    0.0355
```

## 6.6 Distributions Related to Normal

Four distributions – chi-square $\chi^2$, Student's $t$, $F$, and lognormal – are specially related to the normal distribution. This relationship is described in terms of functions of independent standard normal variables. Let $Z_1, Z_2, \ldots, Z_n$ be $n$ independent standard normal (mean 0, variance 1) random variables. Then:

• The sum of squares $Z_1^2 + \cdots + Z_n^2$ is chi-square distributed with $n$ degrees of freedom, $\chi_n^2$:

$$\chi_n^2 \sim Z_1^2 + Z_2^2 + \cdots + Z_n^2.$$

• The ratio of a standard normal $Z$ and the square root of an independent chi-square $\chi^2$ random variable normalized by its number of degrees of freedom, has Student's $t$ distribution with $n$ degrees of freedom, $t_n$:

$$t_n \sim \frac{Z}{\sqrt{\frac{\chi_n^2}{n}}}.$$

• The ratio of two independent chi-squares normalized by their respective numbers of degrees of freedom is distributed as an $F$:

$$F_{m,n} \sim \frac{\chi_m^2/m}{\chi_n^2/n}.$$

The degrees of freedom for $F$ are $m$ – *numerator df* and $n$ – *denominator df*.

• As the name indicates, the lognormal ("log-is-normal") distribution is connected to a normal distribution via a logarithm function. If $X$ has a lognormal disrtibution, then the distribution of $Y = \log X$ is normal.

A more detailed description of these four distributions follows next.

### 6.6.1 Chi-square Distribution

The probability density function for a chi-square random variable with parameter $k$, called the *degrees of freedom*, is

$$f(x) = \frac{(1/2)^{k/2} \, x^{k/2-1}}{\Gamma(k/2)} e^{-x/2}, \; 0 \le x < \infty.$$

The chi-square distribution ($\chi^2$) is a special case of the gamma distribution with parameters $r = k/2$ and $\lambda = 1/2$. Its mean and variance are $\mu = k$ and $\sigma^2 = 2k$, respectively.

If $Z \sim \mathcal{N}(0,1)$, then $Z^2 \sim \chi_1^2$, that is, a chi-square random variable with one degree of freedom. Furthermore, if $U \sim \chi_m^2$ and $V \sim \chi_n^2$ are independent, then $U + V \sim \chi_{m+n}^2$.

From these results it can be shown that if $X_1, \ldots, X_n \sim \mathcal{N}(\mu, \sigma^2)$ and $\overline{X}$ is the sample mean, then the *sample variance* $s^2 = \sum_i (X_i - \overline{X})^2/(n-1)$ is proportional to a chi-square random variable with $n-1$ degrees of freedom:

$$\frac{(n-1)s^2}{\sigma^2} \sim \chi_{n-1}^2. \tag{6.3}$$

This result was proven first by German geodesist Helmert in 1875 (Fig. 6.11a). The $\chi^2$ distribution was previously defined by Abbe and Bienaymé in the mid-1800s.

The formal proof of (6.3) is beyond the scope of this text, but an intuition can be obtained by inspecting

$$\frac{(n-1)s^2}{\sigma^2} = \left(\frac{X_1 - \overline{X}}{\sigma}\right)^2 + \left(\frac{X_2 - \overline{X}}{\sigma}\right)^2 + \cdots + \left(\frac{X_n - \overline{X}}{\sigma}\right)^2$$
$$= (Y_1 - \overline{Y})^2 + (Y_2 - \overline{Y})^2 + \cdots + (Y_n - \overline{Y})^2,$$

where $Y_i$ are independent normal $\mathcal{N}(\mu/\sigma, 1)$.

$$(Y_1 - \overline{Y})^2 + (Y_2 - \overline{Y})^2 = \left(\frac{Y_1 - Y_2}{\sqrt{2}}\right)^2 = Z_1^2, \quad \text{for } \overline{Y} = \frac{Y_1 + Y_2}{2},$$

$$(Y_1 - \overline{Y})^2 + (Y_2 - \overline{Y})^2 + (Y_3 - \overline{Y})^2 = \left(\frac{Y_1 - Y_2}{\sqrt{2}}\right)^2 + \left(\frac{Y_1 + Y_2 - 2Y_3}{\sqrt{6}}\right)^2 = Z_1^2 + Z_2^2,$$

$$\text{for } \overline{Y} = \frac{Y_1 + Y_2 + Y_3}{3},$$

etc.

Note that the right-hand sides are sums of squares of uncorrelated standard normal variables.

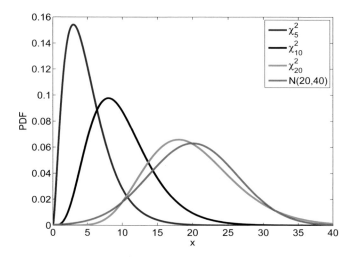

**Fig. 6.9** $\chi^2$ distribution with 5, 10, and 20 degrees of freedom. A normal $\mathcal{N}(20, 40)$ distribution is superimposed to illustrate a good approximation to $\chi_n^2$ by $\mathcal{N}(n, 2n)$ for $n$ large

In MATLAB, the CDF and PDF for a $\chi^2_k$ are `chi2cdf(x,k)` and `chi2pdf(x,k)`, respectively. The $p$th quantile of the $\chi^2_k$ distribution is `chi2inv(p,k)`.

*Example 6.7.* $\chi^2_{10}$ **as a Sum of Ten Standard Normals.** In this example we demonstrate by simulation that the sum of squares of standard normal random variates follows the $\chi^2$-distribution. In particular we compare $Z^2_1 + Z^2_2 + \cdots + Z^2_{10}$ with $\chi^2_{10}$.

Figure 6.10, produced by the code in ◀ nor2chi2.m, shows a normalized histogram of the sums of squares of ten standard normals with a superimposed $\chi^2_{10}$ density (above) and a Q–Q plot comparing the sorted generated sample with $\chi^2_{10}$ quantiles (below). As expected, the simulated empirical distribution is very close to the theoretical chi-square distribution.

```
figure;
subplot(2,1,1)
  %form a matrix of standard normals 10 x 10000
  %square the entries, sum up columnwise, to
  % get a vector of 10000 chi2 with 10 df.
  histn(sum(normrnd(0,1,[10, 10000]).^2),0, 1,30)
    hold on
  plot((0.1:0.1:30), chi2pdf((0.1:0.1:30),10),'r-','LineWidth',2)
  axis tight
subplot(2,1,2)
  %check the Q-Q plot
  xx = sum(normrnd(0,1,[10, 10000]).^2);
  tt = 0.5/10000:1/10000:1;
  yy = chi2inv(tt,10);
  plot(sort(xx), yy,'*')
    hold on
  plot(yy, yy,'r-')
```

*Example 6.8.* **Targeting Meristem Cells.** A gene transfer system for meristem cells can be developed on the basis of a ballistic approach (Sautter, 1993). Instead of a macroprojectile, microtargeting uses the law of Bernoulli for acceleration of highly uniform-sized gold particles. The particle is aimed at an area as small as 150 $\mu$m in diameter, which corresponds to the size of a meristem. Suppose that a particle is fired at a meristem at the origin of a plane coordinate system, with units in microns. The particle lands at $(X,Y)$, where $X$ and $Y$ are independent and each has a normal distribution with mean $\mu = 0$ and variance $\sigma^2 = 10^2$. The particle is successively delivered if it lands within $\sqrt{738}\ \mu$m of the target (origin). What is the probability of this event? The particle is successively delivered if $X^2 + Y^2 \le 738$, or $(X/10)^2 + (Y/10)^2 \le 7.38$. Since both $X/10$ and $Y/10$ have a standard normal distribution, random variable $(X/10)^2 + (Y/10)^2$ is $\chi^2_2$-distributed. Since `chi2cdf(7.38,2)=0.975`, we conclude that the particle is successfully delivered with a probability of 0.975.

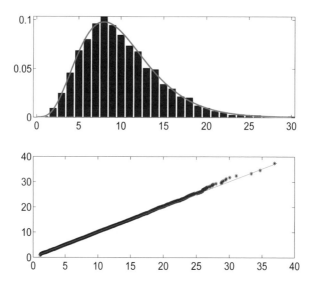

**Fig. 6.10** Sum of 10 squared standard normals compared to $\chi^2_{10}$ distribution. Above: Normalized histogram with superimposed $\chi^2_{10}$ density *(red)*; Below: Q–Q-plot of sorted sums against $\chi^2_{10}$ quantiles.

A multivariate version of the $\chi^2$ distribution is called a Wishart distribution. It is a distribution of random matrices that are symmetric and positive definite. As such it is a proper model for normal covariance matrices, and we will see later its use in Bayesian inference involving bivariate normal distributions.

A $p \times p$ random matrix $X$ has a Wishart distribution if its density is given by

$$f(X) = \frac{|X|^{(n-p-1)/2} \exp\{-\frac{1}{2} tr(\Sigma^{-1}X)\}}{2^{np/2} \pi^{p(p-1)/4} |\Sigma|^{n/2} \prod_{i=1}^{p} \Gamma\left(\frac{n+1-i}{2}\right)},$$

where $\Sigma$ is the scale matrix and $n$ is the number of degrees of freedom. Operator $tr$ is the trace of a matrix, that is, the sum of its diagonal elements, and $|\Sigma|$ and $|X|$ are determinants of $\Sigma$ and $X$, respectively.

For $p = 1$ and $\Sigma = 1$, the Wishart distribution is $\chi^2_n$. In MATLAB, it is possible to simulate from the Wishart distribution as wishrnd(Sigma,n). In WinBUGS, the Wishart distribution is coded as dwish(R[,],n), where the precision matrix $R$ is defined as $\Sigma^{-1}$.

### 6.6.2 *(Student's) t-Distribution*

Random variable $X$ has Student's $t$-distribution with $k$ degrees of freedom, $X \sim t_k$, if its PDF is

$$f_X(x) = \frac{\Gamma\left(\frac{k+1}{2}\right)}{\sqrt{k\pi}\,\Gamma(k/2)} \left(1 + \frac{x^2}{k}\right)^{-\frac{k+1}{2}}, \qquad -\infty < x < \infty.$$

The $t$ distribution is similar in shape to the standard normal distribution except for the fatter tails. If $X \sim t_k$, then $\mathbb{E}X = 0$, $k > 1$ and $\mathbb{V}\mathrm{ar}\,X = k/(k-2)$, $k > 2$. For $k = 1$, the $t$-distribution coincides with the Cauchy distribution.

(a)(b)(c)

**Fig. 6.11** (a) Friedrich Robert Helmert (1843–1917), (b) Jakob Lüroth (1844–1910), and (c) William Sealy Gosset (1876–1937).

The $t$-distribution has an important role to play in statistical inference. With a set of i.i.d. $X_1,\ldots,X_n \sim \mathcal{N}(\mu,\sigma^2)$, we can standardize the sample mean using the simple transformation of $Z = (\overline{X} - \mu)/\sigma_{\overline{X}} = \sqrt{n}(\overline{X} - \mu)/\sigma$. However, if the variance is unknown, by using the same transformation, except for substituting the sample standard deviation $s$ for $\sigma$, we arrive at a $t$-distribution with $n - 1$ degrees of freedom:

$$t = \frac{\overline{X} - \mu}{s/\sqrt{n}} \sim t_{n-1}.$$

More technically, if $Z \sim \mathcal{N}(0,1)$ and $Y \sim \chi_k^2$ are independent, then $t = Z/\sqrt{Y/k} \sim t_k$. In MATLAB, the CDF at $x$ for a $t$-distribution with $k$ degrees of freedom is calculated as tcdf(x,k), and the PDF is computed as tpdf(x,k). The $p$th percentile is computed with tinv(p,k). In WinBUGS, the $t$-distribution is coded as dt(mu,tau,k), where tau is a precision parameter and k is the number of degrees of freedom.

The $t$-distribution was originally found by German mathematician and astronomer Jacob Lüroth (Fig. 6.11b) in 1876. William Sealy Gosset (Fig. 6.11c) rediscovered the $t$-distribution in 1908 and published the results under the

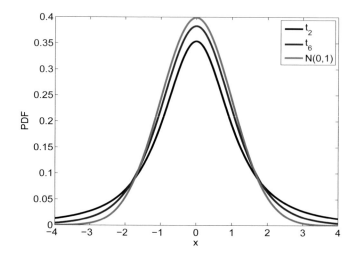

**Fig. 6.12** Student's $t$ with 3 and 6 degrees of freedom. A standard normal distribution is superimposed as the *dashed line*

pen name "Student." He was a researcher for Guinness Brewery, which reportedly forbade any of their employees from publishing "company secrets."

### 6.6.3 Cauchy Distribution

The Cauchy distribution a special case of the $t$-distribution; it is symmetric and bell-shaped like the normal distribution, but with much fatter tails. In fact, it is a popular distribution to use in nonparametric robust procedures and simulations because the distribution is so spread out; it has no mean and variance (none of the Cauchy moments exist). Physicists know this distribution as the *Lorentz distribution*. If $X \sim \mathscr{C}a(a,b)$, then $X$ has a density

$$f_X(x) = \frac{1}{\pi} \frac{b}{b^2 + (x-a)^2}, \quad -\infty < x < \infty.$$

The standard Cauchy $\mathscr{C}a(0,1)$ distribution coincides with the $t$-distribution with 1 degree of freedom.

The Cauchy distribution is also related to the normal distribution. If $Z_1$ and $Z_2$ are two independent $\mathscr{N}(0,1)$ random variables, then their ratio $C = Z_1/Z_2$ is Cauchy, $\mathscr{C}a(0,1)$. Finally, if $C_i \sim \mathscr{C}a(a_i,b_i)$ for $i = 1,\ldots,n$, then $S_n = C_1 + \cdots + C_n$ is Cauchy distributed with parameters $a_S = \sum_i a_i$ and $b_S = \sum_i b_i$. The consequence of this additivity is interesting. If one observes $n$ Cauchy $\mathscr{C}a(0,1)$ random variables $X_i, i = 1,\ldots,n$ and takes the average $\overline{X}$, the average is also Cauchy $\mathscr{C}a(0,1)$! This means that for Cauchy, a single measurement is as precise as the average of *any* number of measurements.

Here is a simple geometric example that leads to a Cauchy distribution.

*Example 6.9.* A ray passing through the point $(-1,0)$ in $R^2$ intersects the $y$-axis at the coordinate $(0,Y)$. If the angle $\alpha$ between the ray and the positive direction of the $x$-axis is uniform $\mathscr{U}(-\pi/2,\pi/2)$, what is the distribution for $Y$?

Here $Y = \tan\alpha$, $\alpha = h(Y) = \arctan(Y)$ and $h'(y) = \frac{1}{1+y^2}$.

The density for uniform $\mathscr{U}(-\pi/2,\pi/2)$ is constant $1/\pi$ if $\alpha \in (-\pi/2,\pi/2)$, and 0 else. From (5.9),

$$f_Y(y) = \frac{1}{\pi}|h'(y)| = \frac{1}{\pi}\frac{1}{1+y^2},$$

which is a density of the Cauchy $\mathscr{C}a(0,1)$ distribution.

### 6.6.4 F-Distribution

Random variable $X$ has an $F$-distribution with $m$ and $n$ degrees of freedom, denoted as $F_{m,n}$, if its density is given by

$$f_X(x) = \frac{m^{m/2}n^{n/2}}{B(m/2,n/2)}\,x^{m/2-1}(n+mx)^{-(m+n)/2}, \ x > 0.$$

The CDF of an $F$-distribution is not of closed form, but it can be expressed in terms of an incomplete beta function (p. 167) as

$$F(x) = 1 - I_v(n/2,m/2), \ \ v = n/(n+mx), \ x > 0.$$

The mean is given by $\mathbb{E}X = n/(n-2), n > 2$, and the variance by $\mathbb{V}ar\,X = \frac{2n^2(m+n-2)}{m(n-2)^2(n-4)}, n > 4$.

If $X \sim \chi^2_m$ and $Y \sim \chi^2_n$ are independent, then $(X/m)/(Y/n) \sim F_{m,n}$. Because of this representation, $m$ and $n$ are often called, respectively, the *numerator* and *denominator* degrees of freedom. $F$ and beta distributions are related. If $X \sim \mathscr{B}e(a,b)$, then $bX/[a(1-X)] \sim F_{2a,2b}$. Also, if $X \sim F_{m,n}$, then $mX/(n + mX) \sim \mathscr{B}e(m/2,n/2)$.

The $F$-distribution is one of the most important distributions for statistical inference; in introductory statistical courses, the test for equality of variances, ANOVA, and multivariate regression are based on the $F$-distribution. For example, if $s_1^2$ and $s_2^2$ are sample variances of two independent normal samples with variances $\sigma_1^2$ and $\sigma_2^2$ and sizes $m$ and $n$ respectively, the ratio $\frac{s_1^2/\sigma_1^2}{s_2^2/\sigma_2^2}$ is distributed as $F_{m-1,n-1}$. The $F$-distribution is named after Sir Ronald Fisher, who in fact tabulated not $F$ but $z = \frac{1}{2}\log F$. The $F$-distribution in its current

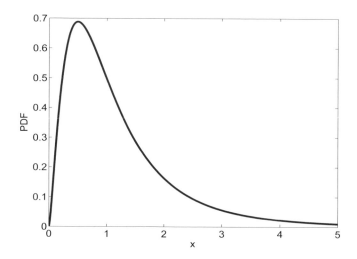

**Fig. 6.13** $F_{5,10}$ PDF. `t2 = 0:0.005:5; plot(t2, fpdf(t2, 5, 10))`

form was first tabulated and used by George W. Snedecor, and the distribution is sometimes called Snedecor's $F$, or the Fisher–Snedecor $F$.

In MATLAB, the CDF at $x$ for an $F$-distribution with $m,n$ degrees of freedom is calculated as `fcdf(x,m,n)`, and the PDF is computed as `fpdf(x,m,n)`. The $p$th percentile is computed with `finv(p,m,n)`. Figure 6.13 provides a plot of a $F_{5,10}$ PDF.

### 6.6.5 Noncentral $\chi^2$, $t$, and $F$ Distributions

Noncentral $\chi^2$, $t$, and $F$ distributions are generalizations of standard $\chi^2$, $t$, and $F$ distributions. They are used mainly in the power analysis of tests and sample size designs. For example, we will use noncentral $t$ for power analysis of one-sample and two-sample $t$ tests later in the text.

Random variable $\chi^2_{n,\delta}$ has a *noncentral $\chi^2$ distribution* with $n$ degrees of freedom and parameter of noncentrality $\delta$ if it can be represented as

$$\chi^2_{n,\delta} = Z_1 + Z_2 + \cdots + Z_{n-1} + X_n,$$

where $Z_1, Z_2, \ldots Z_{n-1}, X_n$ are independent random variables. Random variables $Z_1, \ldots, Z_{n-1}$ have a standard normal $\mathcal{N}(0,1)$ distribution while $X_n$ is distributed as $\mathcal{N}(\delta,1)$. In MATLAB the noncentral $\chi^2$ is denoted as `ncx2pdf`, `ncx2cdf`, `ncx2inv`, `ncx2stat`, and `ncx2rnd` for PDF, CDF, quantile, descriptive statistics, and random number generator.

Random variable $t_{n,\delta}$ has a *noncentral t distribution* with $n$ degrees of freedom and noncentrality parameter $\delta$ if it can be represented as

$$t_{n,\delta} = \frac{X}{\sqrt{\chi_n^2/n}},$$

where $X$ and $\chi_n^2$ are independent, $X \sim \mathcal{N}(\delta, 1)$, and $\chi_n^2$ has a (central) $\chi^2$ distribution with $n$ degrees of freedom. In MATLAB, functions `nctpdf`, `nctcdf`, `nctinv`, `nctstat`, and `nctrnd`, stand for PDF, CDF, quantile, descriptive statistics, and random number generator of the noncentral $t$.

Figure 6.14 plots the densities of noncentral $t$ for values of the noncentrality parameter $-1, 0$, and $2$. Noncentral $t$ for $\delta = 0$ is a standard $t$ distribution.

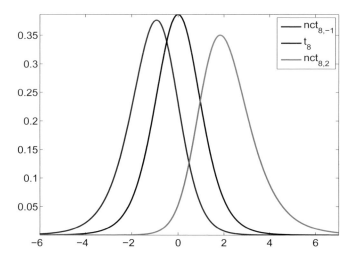

**Fig. 6.14** Densities of noncentral $t_{8,\delta}$ distribution for $\delta = -1, 0, 2$

Random variable $F_{m,n,\delta}$ has a *noncentral F-distribution* with $m, n$ degrees of freedom and parameter of noncentrality $\delta$ if it can be represented as

$$F_{m,n,\delta} = \frac{\chi_{m,\delta}^2/m}{\chi_n^2/n},$$

where $\chi_{m,\delta}^2$ and $\chi_n^2$ are independent, with noncentral ($\delta$) and standard $\chi^2$ distributions with $m$ and $n$ degrees of freedom, respectively. In MATLAB, functions `ncfpdf`, `ncfcdf`, `ncfinv`, `ncfstat`, and `ncfrnd`, stand for the PDF, CDF, quantile, descriptive statistics, and random number generator of the noncentral $F$.

The noncentral $F$ will be used in Chap. 11 for power calculations in several ANOVA designs.

### 6.6.6 Lognormal Distribution

A random variable $X$ has a lognormal distribution with parameters $\mu$ and $\sigma^2$, $X \sim \mathscr{LN}(\mu, \sigma^2)$, if its density function is given by

$$f(x) = \frac{1}{x\sqrt{2\pi}\sigma} \exp\left\{-\frac{(\log x - \mu)^2}{2\sigma^2}\right\}, \quad x > 0.$$

If $Y$ has a normal distribution, then $X = e^Y$ is lognormal. Parameter $\mu$ is the mean and $\sigma$ is the standard deviation of the distribution for the normal random variable $\log X$, not the lognormal random variable $X$, and this can sometimes be confusing.

The moments of the lognormal distribution can be computed from the moment-generating function of the normal distribution. The $n$th moment is $\mathbb{E}(X^n) = \exp\{n\mu + n^2\sigma^2/2\}$, from which the mean and variance of $X$ are

$$\mathbb{E}(X) = \exp\{\mu + \sigma^2/2\}, \quad \text{and} \quad \mathbb{V}\mathrm{ar}(X) = \exp\{2(\mu + \sigma^2)\} - \exp\{2\mu + \sigma^2\}.$$

The median is $\exp\{\mu\}$ and the mode is $\exp\{\mu - \sigma^2\}$.

The lognormality is preserved under multiplication and division, i.e., the products and quotients of lognormal random variables remain lognormally distributed. If $X_i \sim \mathscr{LN}(\mu_i, \sigma_i^2)$, then $\prod_{i=1}^n X_i \sim \mathscr{LN}(\sum_{i=1}^n \mu_i, \sum_{i=1}^n \sigma_i^2)$.

Several biomedical phenomena are well modeled by a lognormal distribution, for example, the age at onset of Alzheimer's disease, latent periods of infectious diseases, or survival time after diagnosis of cancer. For measurement errors that are multiplicative, the convenient model is lognormal. More applications and properties can be found in Crow and Shimizu (1988).

In MATLAB, the CDF of a lognormal distribution with parameters $m$ and $s$ is evaluated at $x$ as `logncdf(x,m,s)`, and the PDF is computed as `lognpdf(x,m,s)`. The $p$th percentile is computed with `logninv(p,m,s)`. Here the parameter `s` stands for $\sigma$, not $\sigma^2$. In WinBUGS, the lognormal distribution is coded as `dlnorm(mu,tau)`, where `tau` stands for the precision parameter $\frac{1}{\sigma^2}$.

*Example 6.10.* **Renner's Honey Data.** The content of hydroxymethylfurfurol (HMF, $\frac{mg}{kg}$) in 1573 honey samples (Renner, 1970) is well conforming to the lognormal distribution. The data set ▨ `renner.mat|dat` contains the interval midpoints (first column) and interval frequencies (second column). The parameter $\mu$ was estimated as $-0.6084$ and $\sigma$ as $1.0040$. The histogram and fitting density are shown in Fig. 6.15 and the code is given in ◀ `renner.m`.

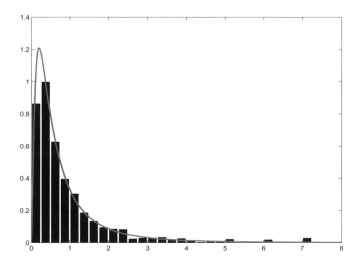

**Fig. 6.15** Normalized histogram of Renner's honey data and lognormal distribution with parameters $\mu = -0.6083$ and $\sigma^2 = 1.0040^2$ that fits data well

The goodness of such fitting procedures will be discussed in Chap. 13 more formally. Note that $\mu$ and $\sigma$ are the mean and standard deviation of the logarithms of observations, not the observations themselves.

```
load 'renner.dat'
% mid-intervals, int. length = 0.25
rennerx = renner(:,1);
% frequencies in the interval
rennerf = renner(:,2);
n = sum(renner(:,2)); % sample size (n=1573)
bar(rennerx,    rennerf./(0.25 * n))
hold on
    m = sum(log(rennerx) .* rennerf)/n    %m =-0.6083
    s = sqrt(   sum( rennerf .*(log(rennerx) - m).^2 )/n    )
    %s=1.0040
xx = 0:0.01:8;
yy = lognpdf(xx, m, s);
plot(xx, yy, 'r-','linewidth',2)
```

## 6.7 Delta Method and Variance Stabilizing Transformations*

The CLT states that for independent identically distributed random variables $X_1, \ldots, X_n$ with mean $\mu$ and finite variance $\sigma^2$,

$$\sqrt{n}(\overline{X} - \mu) \overset{approx}{\sim} \mathcal{N}(0, \sigma^2),$$

where the symbol $\overset{approx}{\sim}$ means *distributed approximately as.* Other than for a finite variance, there are no restrictions on the type, distribution, or any other feature of random variables $X_i$.

---

For a function $g$

$$\sqrt{n}\left(g(\overline{X}) - g(\mu)\right) \overset{approx}{\sim} \mathcal{N}(0, g'(\mu)^2 \sigma^2).$$

The only restriction on $g$ is that the derivative evaluated at $\mu$ must be finite and nonzero.

---

This result is called the *delta method* and the proof, which uses a simple Taylor expansion argument, will be omitted since it also uses facts concerning the convergence of random variables not covered in the text.

*Example 6.11.* For $n$ large

$$1/\overline{X} \overset{approx}{\sim} \mathcal{N}\left(\frac{1}{\mu}, \frac{\sigma^2}{\mu^4}\right)$$

$$\left(\overline{X}\right)^2 \overset{approx}{\sim} \mathcal{N}\left(\mu^2, 4\mu^2\sigma^2\right).$$

The delta method is useful for many asymptotic arguments. Now we focus on the selection of the transformation $g$ that stabilizes the variance.

Important statistical methodologies often assume that observations have variances that are constant for all possible values of the mean. Observations coming from a normal $\mathcal{N}(\mu, \sigma^2)$ distribution would satisfy this requirement since $\sigma^2$ does not depend on the mean $\mu$. However, constancy of variances with respect to the mean is rather an exception than the rule. For example, if random variates from the exponential $\mathcal{E}(\lambda)$ distribution are generated, then the variance $\sigma^2 = 1/\lambda^2$ depends on the mean $\mu = 1/\lambda$, as $\sigma^2 = \mu^2$.

For some important distributions we will find the transformation that will "free" the variance from the influence of the mean. This will prove beneficial for a range of inferential statistical procedures covered later in the text (confidence intervals, testing hypotheses).

Suppose that the variance $\mathbb{V}\mathrm{ar}\, X = \sigma_X^2(\mu)$ can be expressed as a function of the mean $\mu = \mathbb{E}X$. For $Y = g(X)$, $\mathbb{V}\mathrm{ar}\, Y \approx [g'(\mu)]^2 \sigma_X^2(\mu)$, see (5.12). The condition that the variance of $Y$ is constant leads to a simple differential equation

$$[g'(\mu)]^2 \sigma_X^2(\mu) = c^2$$

with the following solution:

$$g(x) = c \int \frac{dx}{\sigma_X(x)} \, dx. \tag{6.4}$$

This is the theoretical basis for many proposed variance stabilizing transforms. Note that $\sigma_X(x)$ in (6.4) is a function expressing the variance as a function of the mean.

*Example 6.12.* **Stabilizing Variance.**  Suppose data are sampled from (a) Poisson $\mathscr{P}oi(\lambda)$, (b) exponential $\mathscr{E}(\lambda)$, and (c) binomial $\mathscr{B}in(n,p)$ distributions. In (a), the mean and variance are equal, $\sigma^2(\mu) = \mu \, (= \lambda)$, and (6.4) becomes

$$g(x) = c \int \frac{dx}{\sqrt{x}} \, dx = 2c\sqrt{x} + d$$

for some constants $c$ and $d$. Thus, as the variance stabilizing transformation for Poisson observations one can take $g(x) = \sqrt{x}$.

In (b) and (c), $\sigma^2(\mu) = \mu^2$ and $\sigma^2(\mu) = \mu - \mu^2/n$, and, after solving the integral in (6.4), one finds that the transformations are $g(x) = \log(x)$ and $g(x) = \arcsin \sqrt{x/n}$ (Exercise 6.18).
✎

*Example 6.13.*  Box and Cox (1964) introduced a family of transformations, indexed by a parameter $\lambda$, applicable to positive data $X_1, \ldots, X_n$:

$$Y_i = \begin{cases} \frac{X_i^\lambda - 1}{\lambda}, & \lambda \neq 0 \\ \log X_i, & \lambda = 0. \end{cases} \tag{6.5}$$

This transformation is mostly applied to responses in linear models exhibiting nonnormality or heterogeneity of variances (heteroscedasticity). For properly selected $\lambda$, transformed data $Y_1, \ldots, Y_n$ may look "more normal" and amenable to standard modeling techniques. The parameter $\lambda$ is selected by maximizing,

$$(\lambda - 1) \sum_{i=1}^n \log X_i - \frac{n}{2} \log \left[ \frac{1}{n} \sum_{i=1}^n (Y_i - \overline{Y})^2 \right], \tag{6.6}$$

where $Y_i$ are as given in (6.5) and $\overline{Y} = \frac{1}{n} \sum_{i=1}^n Y_i$.

As an illustration, we apply the Box–Cox transformation to apparently skewed data of pyruvate kinase concentrations.

Exercise 2.16 featured a multivariate data set 🖫 dmd.dat in which the fourth column gives pyruvate kinase concentrations in 194 female relatives of boys with Duchenne muscular dystrophy (DMD). The distribution of this measurement is skewed to the right (Fig. 6.16a). We will find the Box–Cox transform to symmetrize the data (make it approximately normal). Panel (b) gives the values of likelihood in (6.6) for different values of $\lambda$. Note that (6.6) is maximized for $\lambda$ approximately equal to $-0.15$. Figure 6.16c gives the histogram for data transformed by the Box–Cox transformation with $\lambda = -0.15$. The histogram is notably symmetrized. For details see ◀ boxcox.m.

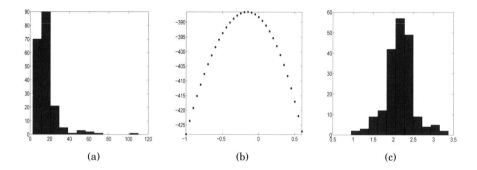

(a)                                    (b)                                    (c)

**Fig. 6.16** (a) Histogram of row data of pyruvate kinase concentrations; (b) Log-likelihood is maximized at $\lambda = -0.15$; and (c) Histogram of Box-Cox-transformed data.

## 6.8 Exercises

6.1. **Standard Normal Calculations.** Random variable $X$ has a standard normal distribution. What is larger, $\mathbb{P}(|X| \leq 0.7)$ or $\mathbb{P}(|X| \geq 0.7)$?

6.2. **Nonnegative Definiteness of $\Sigma$ Constrains $\rho$.** Show that condition $|\Sigma| \geq 0$ for $\Sigma$ in (6.2), implies $-1 \leq \rho \leq 1$.

6.3. **Herrings.** The alewife (*Pomolobus pseudoharengus*, Wilson 1811) grows to a length of about 15 in., but adults average only about 10.5 in. long and about 8 oz. in weight; 16,400,000 fish taken in New England in 1898 weighed about 8,800,000 lbs.

**Fig. 6.17** Alewife fish.

Assume that the length of an individual fish (Fig. 6.17) is normally distributed with mean 10.5 in. and standard deviation 1.6 in. and that the weight is distributed as $\chi^2$ with 8 degrees of freedom.
(a) What percentage of the fish are between 10.5 and 13 in. long?
(b) What percentage of the fish weigh more than 10 oz.?
(c) Ten percent of the fish are longer than $x$. Find $x$.

6.4. **Sea Urchins.** In a laboratory experiment, researchers at Barry University, (Miami Shores, FL) studied the rate at which sea urchins ingested turtle grass (*Florida Scientist*, Summer/Autumn 1991). The urchins starved for 48 h, were fed 5-cm blades of green turtle grass. The mean ingestion time was found to be 2.83 h and the standard deviation 0.79 h. Assume that green turtle grass ingestion time for the sea urchins has an approximately normal distribution.
(a) Find the probability that a sea urchin will require between 2.3 and 4 h to ingest a 5-cm blade of green turtle grass.
(b) Find the time $t^*$ (hours) so that 95% of sea urchins take more than $t^*$ hours to ingest a 5-cm blade of green turtle grass.

6.5. **Pyruvate Kinase for Controls Is Normal.** Refer to Exercise 2.16. The histogram for PK response for controls, $X$, is fairly bell-shaped (as much as 142 observations show), so you decided to fit it with a normal distribution, $\mathcal{N}(12, 4^2)$.
(a) How would you defend the choice of a normal model that allows for negative values when the measured level is always positive?
(b) Find the probability that $X$ falls between 4 and 20.
(c) Find the probability that $X$ exceeds 20.
(d) Find the value $x_0$ so that 93% of all PK measurements exceed $x_0$.

6.6. **Leptin.** Leptin (from the Greek word *leptos*, meaning thin) is a 16-kDa protein hormone that plays a key role in regulating energy intake and energy expenditure, including the regulation (decrease) of appetite and (increase) of metabolism. Serum leptin concentrations can be measured in several ways. One approach is by using a radioimmunoassay in venous blood samples (Linco Research Inc., St Charles, MO). Several studies have consistently found women to have higher serum leptin concentrations than do men. For example, among US adults across a broad age range, the mean serum leptin concentration in women is approximately normal $\mathcal{N}(12.7 \ \mu g/L, (1.3 \ \mu g/L)^2)$ and in men approximately normal $\mathcal{N}(4.6 \ \mu g/L, (0.5 \ \mu g/L)^2)$.
(a) What is the probability that the concentration of leptin in a randomly selected US adult male exceeds 6 $\mu$ g/L? (b) What proportion of US women have concentration of leptin in the interval $12.7 \pm 2 \ \mu$ g/L? (c) What interval, symmetric about the mean 12.7 $\mu g/L$, contains leptin concentrations of 95% of adult US women?

6.7. **Pulse Rate.** The pulse rate of 1-month-old infants has a mean of 115 beats per minute and a standard deviation of 16 beats per minute.
(a) Explain why the average pulse rate in a sample of 64 1-month-old infants is approximately normally distributed.
(b) Find the mean and the variance of the normal distribution in (a).
(c) Find the probability that the average pulse rate of a sample of 64 will exceed 120.

6.8. **Side Effects.** One of the side effects of flooding a lake in northern boreal forest areas[1] (e.g., for a hydroelectric project) is that mercury is leaked from the soil, enters the food chain, and eventually contaminates the fish. The concentration of mercury in fish will vary among individual fish because of differences in eating patterns, movements around the lake, etc. Suppose that the concentrations of mercury in individual fish follows an approximately normal distribution with a mean of 0.25 ppm and a standard deviation of 0.08 ppm. Fish are safe to eat if the mercury level is below 0.30 ppm. What proportion of fish are safe to eat?
a. 63.1%    b. 23.8%    c. 73.4%    d. 27.3%    e. 37.9%

6.9. **Macrolepiota Procera.** The size of mushroom caps varies. While many species of *Marasmius* and *Collybia* are only 12 to 20 mm (1/2 to 3/4 in.) in diameter, some fungi are nearly 200 mm (8 in.) across. The cap diameter of parasol mushroom (*Macrolepiota procera*, Fig. 6.18) is a normal random variable with parameters $\mu = 230$ mm and $\sigma = 25$ mm.

**Fig. 6.18** Parasol mushroom *Macrolepiota procera*.

(a) What proportion of parasol caps has a diameter between 200 and 250 mm?
(b) Five percent of parasol caps are larger than $x_0$ in diameter. Find $x_0$.

6.10. **Duration of Gestation in Humans.** Altman (1980) quotes the following incident from the UK: "In 1949 a divorce case was heard in which the sole

---

[1] The northern boreal forest, sometimes also called the taiga or northern coniferous forest, stretches unbroken from eastern Canada westward throughout the majority of Canada to the central region of Alaska.

evidence of adultery was that a baby was born 349 days after the husband had gone abroad on military service. The appeal judges agreed that medical evidence was unlikely but scientifically possible." So the appeal failed. "Most people think that the husband was hard done by," Altman adds.

So let us judge the judges. The reported mean duration of an uncomplicated human gestation is between 266 and 288 days, depending on many factors but mainly on the method of calculation. Assume that population mean and standard deviations are $\mu = 280$ and $\sigma = 10$ days, respectively. In fact, smaller standard deviations have been reported, so 10 days is a conservative choice. The normal model fits the data reasonably well if the samples are large.

Under the normal $\mathcal{N}(\mu, \sigma^2)$ model, find the probability that a gestation period will be equal to or greater than 349 days.

6.11. **Tolerance Design.** Eggert (2005) provides the following engineering design question. A 5-inch-diameter pin will be assembled into a 5.005 inch journal bearing. The pin manufacturing tolerance is specified to $t_{pin} = 0.003$ inch. A minimum clearance fit of 0.001 inch is needed.

Determine tolerance required of the hole, $t_{hole}$, such that 99.9% of the mates will exceed the minimum clearance. Assume that manufacturing variations are normally distributed. The tolerance is defined as 3 standard deviations.

6.12. **Ulnar Variance.** This exercise uses data reported in Jung et al. (2001), who studied radiographs of the wrists of 120 healthy volunteers in order to determine the normal range of ulnar variance, Fig. 6.19. The radiographs had been taken in various positions under both unloaded (static) and loaded (dynamic) conditions.

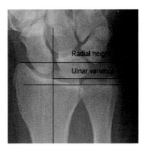

**Fig. 6.19** Ulnar variance (UV) is measured here using the method of perpendiculars, in which two lines are drawn perpendicular to the long axis of the radius. One line is drawn on the ulnar-side articular surface of the radius, and the other is drawn on the ulnar carpal surface. This image shows positive UV.

The ulnar variance in neutral rotation was modeled by normal distribution with a mean of $\mu = 0.74$ mm and standard deviation of $\sigma = 1.46$ mm.

(a) What is the probability that a radiogram of a normal person will show negative ulnar variance in neutral rotation (ulnar variance, unlike the statistical variance, can be negative)?

The researchers modeled the maximum ulnar variance ($UV_{max}$) as normal $\mathcal{N}(1.52, 1.56^2)$ when gripping in pronation and minimum ulnar variance ($UV_{min}$) as normal $\mathcal{N}(0.19, 1.43^2)$ when relaxed in supination.

(b) Find the probability that the mean dynamic range in ulnar variance, $C = UV_{max} - UV_{min}$, will exceed 1 mm.

6.13. **Independence of Sample Mean and Standard Deviation in Normal Samples.** Simulate 1000 samples from the standard normal distribution, each of size 100, and find their sample mean and standard deviation.

(a) Plot a scatterplot of sample means vs. the corresponding sample standard deviations. Are there any trends?

(b) Find the coefficient of correlation between sample means and standard deviations from (a) arranged as two vectors. Is the coefficient close to zero?

6.14. **Sonny and Multiple Choice Exam.** An instructor gives a 100-question multiple-choice final exam. Each question has 4 choices. In order to pass, a student has to have at least 35 correct answers. Sonny decides to guess at random on each question. What is the probability that Sonny will pass the exam?

6.15. **Amount of Liquid in a Bottle.** Suppose that the volume of liquid in a bottle of a certain chemical solution is normally distributed with a mean of 0.5 L and standard deviation of 0.01 L.

(a) Find the probability that a bottle will contain at least 0.48 L of liquid.

(b) Find the volume that corresponds to the 95th percentile.

6.16. **Meristem Cells in 3-D.** Suppose that a particle is fired at a cell sitting at the origin of a spatial coordinate system, with units in microns. The particle lands at $(X, Y, Z)$, where $X, Y$, and $Z$ are independent and each has a normal distribution with a mean of $\mu = 0$ and variance of $\sigma^2 = 250$. The particle is successfully delivered if it lands within 70 $\mu$m of the origin. Find the probability that the particle was not successfully delivered.

6.17. **Glossina morsitans.** *Glossina morsitans* (tsetse fly) is a large biting fly that inhabits most of midcontinental Africa. This fly is infamous as the primary biological vector (the meaning of vector here is epidemiological, not mathematical. A vector is any living carrier that transmits an infectious agent) of trypanosomes, which cause human sleeping sickness. The data in the table below are reported in Pearson (1914) and represent the frequencies of length in microns of trypanosomes found in *Glossina morsitans*.

| Microns | 15 | 16 | 17 | 18 | 19 | 20 | 21 | 22 | 23 | 24 | 25 |
|---|---|---|---|---|---|---|---|---|---|---|---|
| Frequency | 7 | 31 | 148 | 230 | 326 | 252 | 237 | 184 | 143 | 115 | 130 |

| Microns | 26 | 27 | 28 | 29 | 30 | 31 | 32 | 33 | 34 | 35 | Total |
|---|---|---|---|---|---|---|---|---|---|---|---|
| Frequency | 110 | 127 | 133 | 113 | 96 | 54 | 44 | 11 | 7 | 2 | 2500 |

The original data distinguished five different strains of trypanosomes, but it seems that the summary data set, as shown in the table, can be well approximated by a mixture of two normal distributions, $p_1 \mathcal{N}(\mu_1, \sigma_1^2) + p_2 \mathcal{N}(\mu_2, \sigma_2^2)$. Using MATLAB's `gmdistribution.fit` identify the means of the two normal components, as well as their weights in the mixture, $p_1$ and $p_2$. Plot the normalized histogram and superimpose the density of the mixture. Data can be found in 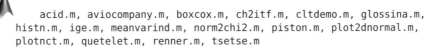 `glossina.mat`.

6.18. **Stabilizing Variance.** In Example 6.12 it was stated that the variance stabilizing transformations for exponential $\mathcal{E}(\lambda)$ and binomial $\mathcal{B}in(n, p)$ distributions are $g(x) = \log(x)$ and $g(x) = \arcsin\sqrt{\frac{x}{n}}$, respectively. Prove these statements.

6.19. **From Normal to Lognormal.** Derive the density of a lognormal distribution by transforming $X \sim \mathcal{N}(0, 1)$ into $Y = \exp\{X\}$.

6.20. **The Square of a Standard Normal.** If $X \sim \mathcal{N}(0, 1)$, show that $Y = X^2$ has a density of

$$f_Y(y) = \frac{1}{\sqrt{2}\,\Gamma\left(\frac{1}{2}\right)} y^{1/2-1} e^{-y/2}, \quad y \geq 0,$$

which is $\chi^2$ with 1 degree of freedom.

---

**MATLAB FILES AND DATA SETS USED IN THIS CHAPTER**

http://springer.bme.gatech.edu/Ch6.Norm/

    acid.m, aviocompany.m, boxcox.m, ch2itf.m, cltdemo.m, glossina.m,
  histn.m, ige.m, meanvarind.m, norm2chi2.m, piston.m, plot2dnormal.m,
  plotnct.m, quetelet.m, renner.m, tsetse.m

 aplysia.odc

 glossina.mat, renner.dat|mat

---

# CHAPTER REFERENCES

Altman, D. G. (1980). Statistics and ethics in medical research: misuse of statistics is uneth-
    ical. *Br. Med. J.*, **281**, 1182–1184.
Casella, G. and Berger, R. (2002). *Statistical Inference.* Duxbury Press, Belmont
Crow E. L. and Shimizu K. Eds., (1988). *Lognormal Distributions: Theory and Application.*
    Dekker, New York.
Eggert, R. J. (2005). *Engineering Design.* Pearson - Prentice Hall, Englewood Cliffs
Jung, J. M., Baek, G. H., Kim, J. H., Lee, Y. H., and Chung, M. S. (2001). Changes in ulnar
    variance in relation to forearm rotation and grip. *J. Bone Joint Surg. Br.*, **83**, 7, 1029–
    1033, PubMed PMID: 11603517.
Koike, H. (1987). The extensibility of Aplysia nerve and the determination of true axon
    length. *J. Physiol.*, **390**, 469–487.
Pearson, K. (1914). On the probability that two independent distributions of frequency are
    really samples of the same population, with special reference to recent work on the
    identity of trypanosome strains. *Biometrika*, **10**, 1, 85–143.
Renner E. (1970). *Mathematisch-statistische Methoden in der praktischen Anwendung.*
    Parey, Hamburg.
Sautter, C. (1993). Development of a microtargeting device for particle bombardment of
    plant meristems. *Plant Cell Tiss. Org.*, **33**, 251–257.

# Chapter 7
# Point and Interval Estimators

A *grade* is an inadequate report of an inaccurate judgment by a biased and variable judge of the extent to which a student has attained an undefined level of mastery of an unknown proportion of an indefinite amount of material. [Dressel, P. L. (1957). Facts and fancy in assigning grades. *Basic College Quarterly*, **2**, 6–12]

## WHAT IS COVERED IN THIS CHAPTER

- Moment Matching and Maximum Likelihood Estimators
- Unbiased and Consistent Estimators
- Estimation of Mean and Variance
- Confidence Intervals
- Estimation of Population Proportions
- Sample Size Design by Length of Confidence Intervals
- Prediction and Tolerance Intervals
- Intervals for the Poisson Rate

## 7.1 Introduction

One of the primary objectives of inferential statistics is estimation of population characteristics (descriptors) on the basis of limited information contained in a sample. The population descriptors are formalized by a statistical model, which can be postulated at various levels of specificity: a broad class of models, a parametric family, or a fully specific unique model. Often, a functional or distributional form is fully specified but dependent on one or more parameters. Such a model is called parametric. When the model is parametric, the task of estimation is to find the best possible sample counterparts as estimators for the parameters and to assess the accuracy of the estimators.

The estimation procedure follows standard rules. Usually, a sample is taken and a *statistic* (a function of observations) is calculated. The value of the statistic serves as a point estimator for the unknown population parameter. For example, responses in political pools observed as sample proportions are used to estimate the population proportion of voters in favor of a particular candidate. The associated model is binomial and the parameter of interest is the binomial proportion in the population.

The estimators for a parameter can be given as a single value – *point estimators* or as *interval estimators*. For example, the sample mean is a point estimator of the population mean. Confidence intervals and credible sets in a Bayesian context are examples of interval estimators.

In this chapter we first discuss general methods for finding estimators and then focus on estimation of specific population parameters: means, variances, proportions, rates, etc. Some estimators are universal, that is, they are not connected with any specific distribution. Universal estimators are a sample mean for the population mean and a sample variance for the population variance. However, for interval estimators and for Bayesian estimators, a knowledge of sampling distribution is critical.

In Chap. 2 we learned about many sample summaries that are good estimators for their population counterparts; these will be discussed further in this chapter. We have also seen some robust competitors based on order statistics and ranks; these will be discussed further in Chap. 12.

The methods for how to propose an estimator for a population parameter are discussed next. The methods will use knowledge of the form of population distribution or, equivalently, distribution of sample summaries treated as random variables.

## 7.2 Moment Matching and Maximum Likelihood Estimators

We describe two approaches for devising point estimators: moment matching and maximum likelihood.

**Matching Estimation.** Moment matching is a natural way to propose an estimator. Theoretical moments of a random variable $X$ with a density specified up to a parameter, $f(x|\theta)$, are functions of that parameter:

$$EX^k = h(\theta).$$

For example, if the measurements have a Poisson distribution $\mathscr{P}oi(\lambda)$, the second moment $\mathbb{E}X^2$ is $\lambda + \lambda^2$, which is a function of $\lambda$. Here, $h(x) = x + x^2$.

Suppose we obtained a sample $X_1, X_2, \ldots, X_n$ from $f(x|\theta)$. The empirical counterparts for theoretical moments $\mathbb{E}X^k$ are sample moments

$$\overline{X^k} = \frac{1}{n} \sum_{i=1}^{n} X_i^k.$$

By matching the theoretical and empirical moments, an estimator $\hat{\theta}$ is found as a solution of the equation

$$\overline{X^k} = h(\theta).$$

For example, for the exponential distribution $\mathscr{E}(\lambda)$, the first theoretical moment is $\mathbb{E}X = 1/\lambda$. An estimator for $\lambda$ is obtained by solving the moment-matching equation $\overline{X} = 1/\lambda$, resulting in $\hat{\lambda}_{mm} = 1/\overline{X}$. Moment-matching estimators are not unique; different theoretical and sample moments can be matched. In the context of an exponential model, the second theoretical moment is $\mathbb{E}X^2 = 2/\lambda^2$, leading to an alternative matching equation,

$$\overline{X^2} = 2/\lambda^2,$$

with the solution

$$\hat{\lambda}_{mm} = \sqrt{\frac{2}{\overline{X^2}}} = \sqrt{\frac{2n}{\sum_{i=1}^{n} X_i^2}}.$$

The following simple MATLAB code simulates a sample of size $10^6$ from an exponential distribution with rate parameter $\lambda = 3$ and then calculates moment-matching estimators based on the first two moments.

```
Y = exprnd(1/3, 10e6, 1);
%parametrization in MATLAB is 1/lambda
1/mean(Y) %matching the first moment
 ans = 2.9981
sqrt(2/mean(Y.^2)) %matching the second moment
 ans = 2.9984
```

*Example 7.1.* Consider a sample from a gamma distribution with parameters $r$ and $\lambda$. It is known that for $X \sim \mathscr{G}a(r,\lambda)$, $\mathbb{E}(X) = \frac{r}{\lambda}$, and $\mathbb{V}ar\,X = \mathbb{E}X^2 - (\mathbb{E}X)^2 = \frac{r}{\lambda^2}$. It is easy to see that

$$r = \frac{(\mathbb{E}X)^2}{\mathbb{E}X^2 - (\mathbb{E}X)^2} \quad \text{and} \quad \lambda = \frac{\mathbb{E}X}{\mathbb{E}X^2 - (\mathbb{E}X)^2}.$$

Thus, the moment matching estimators are

$$\hat{r}_{mm} = \frac{(\overline{X})^2}{\overline{X^2} - (\overline{X})^2} \quad \text{and} \quad \hat{\lambda}_{mm} = \frac{\overline{X}}{\overline{X^2} - (\overline{X})^2}.$$

Matching estimation uses mostly moments, but any other statistic that is (i) easily calculated from a sample and (ii) whose population counterpart depends on parameter(s) of interest can be used in matching. For example, the sample/population quantiles can be used.

*Example 7.2.* In one study, the highest 5-year survival rate (90%) for women was for malignant melanoma of the skin. Assume that survival time $T$ has an exponential distribution with an unknown rate parameter $\lambda$. Using quantiles, estimate $\lambda$.
    From

$$P(T > 5) = 0.90 \quad \Rightarrow \quad \exp\{-5 \cdot \lambda\} = 0.90$$

it follows that $\hat{\lambda} = 0.0211$.

**Maximum Likelihood.**  An alternative method, which uses a functional form for distributions of measurements, is the maximum likelihood estimation (MLE).
    The MLE was first proposed and used by R. A. Fisher (Fig. 7.1) in the 1920s and remains one of the most popular tools in estimation theory and broader statistical inference. The method can be formulated as an optimization problem involving the search for extrema when the model is considered as a function of parameters.
    Suppose that the sample $X_1, \ldots, X_n$ comes from a population with distribution $f(x|\theta)$ indexed by $\theta$ that could be a scalar or a vector of parameters. Elements of the sample are independent; thus the joint distribution of $X_1, \ldots, X_n$ is a product of individual densities:

$$f(x_1, \ldots, x_n|\theta) = \prod_{i=1}^{n} f(x_i|\theta).$$

When the sample is observed, the joint distribution remains dependent upon the parameter,

$$L(\theta|X_1, \ldots, X_n) = \prod_{i=1}^{n} f(X_i|\theta), \tag{7.1}$$

and, as a function of the parameter, $L$ is called the *likelihood*. The value of the parameter $\theta$ that maximizes the likelihood $L(\theta|X_1,\ldots,X_n)$ is the MLE, $\hat{\theta}_{mle}$.

**Fig. 7.1** Sir R. A. Fisher (1890–1962).

The problem of finding the maximum of $L$ and the value $\hat{\theta}_{mle}$ at which $L$ is maximized is an optimization problem. In some cases the maximum can be found directly or with the help of the log transformation of $L$. Other times the procedure must be iterative and the solution is an approximation. In some cases, depending on the model and sample size, the maximum is not unique or does not exist.

In the most common cases, maximizing the logarithm of likelihood, *log-likelihood*, is simpler than maximizing the likelihood directly. This is because the product in $L$ becomes the sum when a logarithm is applied:

$$\ell(\theta|X_1,\ldots,X_n) = \log L(\theta|X_1,\ldots,X_n) = \sum_{i=1}^{n} \log f(X_i|\theta),$$

and finding an extremum of a sum is simpler. Since the logarithm is a monotonically increasing function, the maxima of $L$ and $\ell$ are achieved at the same value $\hat{\theta}_{mle}$ (see Fig. 7.2 for an illustration).

Analytically,

$$\hat{\theta}_{mle} = \mathrm{argmax}_\theta\, \ell(\theta|X_1,\ldots,X_n),$$

and it can be found as a solution of

$$\frac{\partial \ell(\theta|X_1,\ldots,X_n)}{\partial \theta} = 0 \quad \text{subject to} \quad \frac{\partial^2 \ell(\theta|X_1,\ldots,X_n)}{\partial \theta^2} < 0.$$

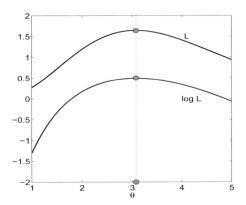

**Fig. 7.2** Likelihood and log-likelihood of exponential distribution with rate parameter $\lambda$ when the sample $X = [0.4, 0.3, 0.1, 0.5]$ is observed. The MLE is $1/\overline{X} = 3.077$.

In simple terms, the MLE makes the first derivative (with respect to $\theta$) of the log-likelihood equal to 0 and the second derivative negative, which is a condition for a maximum.

As an illustration, consider the MLE of $\lambda$ in the exponential model, $\mathscr{E}(\lambda)$. After $X_1, \ldots, X_n$ is observed, the likelihood becomes

$$L(\lambda | X_1, \ldots, X_n) = \prod_{i=1}^{n} \lambda e^{-\lambda X_i} = \lambda^n \exp\left\{ -\lambda \sum_{i=1}^{n} X_i \right\}.$$

The likelihood $L$ is obtained as a product of densities $f(x_i | \lambda)$ where the arguments $x_i$s are fixed observations $X_i$. The product is taken over all observations, as in (7.1). We can search for the maximum of $L$ directly, but since it is a product of two terms involving $\lambda$, it is beneficial to look at the log-likelihood instead.

The log-likelihood is

$$\ell(\lambda | X_1, \ldots, X_n) = n \log \lambda - \lambda \sum_{i=1}^{n} X_i.$$

The equation to be solved is

$$\frac{\partial \ell}{\partial \lambda} = \frac{n}{\lambda} - \sum_{i=1}^{n} X_i = 0,$$

and the solution is $\hat{\lambda}_{mle} = \frac{n}{\sum_{i=1}^{n} X_i} = 1/\overline{X}$. The second derivative of the log-likelihood, $\frac{\partial^2 \ell}{\partial \lambda^2} = -\frac{n}{\lambda^2}$, is always negative; thus the solution $\hat{\lambda}_{mle}$ maximizes $\ell$, and consequently $L$. Figure 7.2 shows the likelihood and log-likelihood as functions of $\lambda$. For sample $X = [0.4, 0.3, 0.1, 0.5]$ the maximizing $\lambda$ is $1/\overline{X} = 3.0769$.

Note that both the likelihood and log-likelihood are maximized at the same value.

For the alternative parametrization of exponentials via a scale parameter, as in MATLAB, $f(x|\lambda) = \frac{1}{\lambda}e^{-x/\lambda}$, the estimator is, of course, $\hat{\lambda}_{mle} = \overline{X}$.

An important property of MLE is their *invariance property*.

**Invariance Property of MLEs.** Let $\hat{\theta}_{mle}$ be an MLE of $\theta$ and let $\eta = g(\theta)$, where $g$ is an arbitrary function. Then $\hat{\eta}_{mle} = g(\hat{\theta}_{mle})$ is an MLE of $\eta$.

For example, if the MLE for $\lambda$ in the exponential distribution was $1/\overline{X}$, then for a function of the parameter $\eta = \lambda^2 - \sin(\lambda)$ the MLE is $(1/\overline{X})^2 - \sin\left(1/\overline{X}\right)$.

In MATLAB, the function mle finds the MLE when inputs are data and the name of a distribution with a list of options. The normal distribution is the default. For example, parhat = mle(data) calculates the MLE for $\mu$ and $\sigma$ of a normal distribution, evaluated at vector data. One of the outputs is the confidence interval. For example, [parhat, parci] = mle(data) returns MLEs and 95% confidence intervals for the parameters. The confidence intervals, as interval estimators, will be discussed later in this chapter. The command [...] = mle(data, 'distribution', dist) computes parameter estimations for the distribution specified by dist. Acceptable strings for dist are as follows:

| | | |
|---|---|---|
| 'beta' | 'bernoulli' | 'binomial' |
| 'discrete uniform' | 'exponential' | 'extreme value' |
| 'gamma' | 'generalized extreme value' | 'generalized pareto' |
| 'geometric' | 'lognormal' | 'negative binomial' |
| 'normal' | 'poisson' | 'rayleigh' |
| 'uniform' | 'weibull' | |

*Example 7.3.* **MLE of Beta in MATLAB.** The following MATLAB commands show how to estimate parameters $a$ and $b$ in a beta distribution. We will simulate a sample of size 1000 from a beta $\mathscr{B}e(2,3)$ distribution and then from the sample find MLEs of $a$ and $b$.

```
a = betarnd( 2, 3,[1, 1000]);
thetahat = mle(a,'distribution', 'beta')
%thetahat = 1.9991    3.0267
```

It is possible to find the MLE using MATLAB's mle command for distributions that are not on the list. The code is given at the end of Example 7.4in which moment-matching estimators and MLEs for parameters in a Maxwell distribution are compared.

*Example 7.4.* **Moment-Matching Estimators and MLEs in a Maxwell Distribution.** The Maxwell distribution models random speeds of molecules in thermal equilibrium as given by statistical mechanics. A random variable $X$ with a Maxwell distribution is given by the probability density function

$$f(x|\theta) = \sqrt{\frac{2}{\pi}} \, \theta^{3/2} \, x^2 \, e^{-\theta x^2/2}, \quad \theta > 0, x > 0.$$

Assume that we observed velocities $X_1, \ldots, X_n$ and want to estimate the unknown parameter $\theta$.

The following theoretical moments for the Maxwell distribution are available: the expectation $\mathbb{E}X = 2\sqrt{\frac{2}{\pi\theta}}$, the second moment $\mathbb{E}X^2 = 3/\theta$, and the fourth moment $\mathbb{E}X^4 = 15/\theta^2$. To find moment-matching estimators for $\theta$ the theoretical moments are "matched" with their empirical counterparts $\overline{X}$, $\overline{X^2} = \frac{1}{n}\sum_{i=1}^{n} X_i^2$, and $\overline{X^4} = \frac{1}{n}\sum_{i=1}^{n} X_i^4$, and the resulting equations are solved with respect to $\theta$:

$$\overline{X} = 2\sqrt{\frac{2}{\pi\theta}} \quad \Rightarrow \quad \hat{\theta}_1 = \frac{8}{\pi(\overline{X})^2},$$

$$\frac{1}{n}\sum_{i=1}^{n} X_i^2 = \frac{3}{\theta} \quad \Rightarrow \quad \hat{\theta}_2 = \frac{3n}{\sum_{i=1}^{n} X_i^2},$$

$$\frac{1}{n}\sum_{i=1}^{n} X_i^4 = \frac{15}{\theta^2} \quad \Rightarrow \quad \hat{\theta}_3 = \sqrt{\frac{15n}{\sum_{i=1}^{n} X_i^4}}.$$

To find the MLE of $\theta$ we show that the log-likelihood has the form $\frac{3n}{2}\log\theta - \frac{\theta}{2}\sum_{i=1}^{n} X_i^2 +$ part free of $\theta$. The maximum of the log-likelihood is achieved at $\hat{\theta}_{\text{MLE}} = \frac{3n}{\sum_{i=1}^{n} X_i^2}$, which is the same as the moment-matching estimator $\hat{\theta}_2$.

Specifically, if $X_1 = 1.4$, $X_2 = 3.1$, and $X_3 = 2.5$ are observed, the MLE of $\theta$ is $\hat{\theta}_{\text{MLE}} = \frac{9}{17.82} = 0.5051$. The other two moment-matching estimators are $\hat{\theta}_1 = 0.4677$ and $\hat{\theta}_3 = 0.5768$.

In MATLAB, the Maxwell distribution can be custom-defined using the function @:

```
maxwell = @(x,theta)  sqrt(2/pi)  *  ...
    theta.^(3/2) * x.^2   .* exp( - theta * x.^2/2);
mle([1.4 3.1 2.5], 'pdf', maxwell, 'start', rand)
    %ans = 0.5051
```

In most cases, taking the log of likelihood simplifies finding the MLE. Here is an example in which the maximization of likelihood was done without the use of derivatives.

*Example 7.5.* Suppose the observations $X_1 = 2$, $X_2 = 5$, $X_3 = 0.5$, and $X_4 = 3$ come from the uniform $\mathcal{U}(0,\theta)$ distribution. We are interested in estimating $\theta$. The density for the single observation $X$ is $f(x|\theta) = \frac{1}{\theta}\mathbf{1}(0 \leq x \leq \theta)$, and the likelihood, based on $n$ observations $X_1, \ldots, X_n$, is

$$L(\theta|X_1, \ldots, X_n) = \frac{1}{\theta^n} \cdot \mathbf{1}(0 \leq X_1 \leq \theta) \cdot \mathbf{1}(0 \leq X_2 \leq \theta) \cdot \ \ldots \ \cdot \mathbf{1}(0 \leq X_n \leq \theta).$$

The product in the above expression can be simplified: if all $X$s are less than or equal to $\theta$, then their maximum $X_{(n)}$ is less than $\theta$ as well. Thus,

$$\mathbf{1}(0 \leq X_1 \leq \theta) \cdot \mathbf{1}(0 \leq X_2 \leq \theta) \cdot \ \ldots \ \cdot \mathbf{1}(0 \leq X_n \leq \theta) = \mathbf{1}(X_{(n)} \leq \theta).$$

Maximizing the likelihood now can be performed by inspection. In order to maximize $\frac{1}{\theta^n}$, subject to $X_{(n)} \leq \theta$, one should take the smallest $\theta$ possible, and that $\theta$ is $X_{(n)} = \max X_i$. Therefore, $\hat{\theta}_{mle} = X_{(n)}$, and in this problem, the estimator is $X_{(4)} = X_2 = 5$.

An alternative estimator can be found by moment matching. One can show (the arguments are beyond the scope of this book) that in estimating $\theta$ in $\mathcal{U}(0,\theta)$, only $\max X_i$ should be used. What is the distribution of $\max X_i$?

We will find this distribution for general i.i.d. $X_i, i = 1, \ldots, n$, with CDF $F(x)$ and PDF $f(x) = F'(x)$.

The CDF is, by definition,

$$G(x) = \mathbb{P}(\max X_i \leq x) = \mathbb{P}(X_1 \leq x, X_2 \leq x, \ldots, X_n \leq x) = \prod_{i=1}^{n} \mathbb{P}(X_i \leq x) = (F(x))^n.$$

The reasoning in the above equation is as follows. If the maximum is $\leq x$, then all $X_i$ are $\leq x$, and vice versa. The density for $\max X_i$ is $g(x) = G'(x) = n F^{n-1}(x) f(x)$, and the first moment is

$$\mathbb{E}\max X_i = \int_{\mathbb{R}} x g(x) dx = \int_{\mathbb{R}} x n F^{n-1}(x) f(x) dx.$$

For the uniform distribution $\mathcal{U}(0,\theta)$,

$$\mathbb{E}\max X_i = \int_0^{\theta} x \cdot n (x/\theta)^{n-1} \cdot 1/\theta \, dx = \frac{n}{\theta^n} \int_0^{\theta} x^n \, dx = \frac{n}{n+1}\theta.$$

The expectation of the maximum $\mathbb{E}\max X_i$ is matched with the largest order statistic in the sample, $X_{(n)}$. Thus, by solving the moment-matching equation, one obtains an alternative estimator for $\theta$, $\hat{\theta}_{mm} = \frac{n+1}{n}X_{(n)}$. In this problem, $\hat{\theta}_{mm} = 25/4 = 6.25$. For a Bayesian estimator, see Example 8.6.

### 7.2.1 Unbiasedness and Consistency of Estimators

Based on a sample $X_1, \ldots, X_n$ from a population with distribution $f(x|\theta)$, let $\hat{\theta}_n = g(X_1, \ldots, X_n)$ be a statistic that estimates the parameter $\theta$. The statistic (estimator) $\hat{\theta}_n$ as a function of the sample is a random variable. As a random variable, the estimator would have an expectation of $\mathbb{E}\hat{\theta}_n$, a variance of $\mathbb{V}\mathrm{ar}\,\hat{\theta}_n$, and its own distribution called a *sampling distribution*.

*Example 7.6.* Suppose we are interested in finding the proportion of subjects with the AB blood group in a particular geographic region. This proportion, $\theta$, is to be estimated on the basis of the sample $Y_1, Y_2, \ldots, Y_n$, each having a Bernoulli $\mathcal{B}er(\theta)$ distribution taking values 1 and 0 with probabilities $\theta$ and $1-\theta$, respectively. The realization $Y_i = 1$ indicates the presence of the AB group in observation $i$. The sum $X = \sum_{i=1}^n Y_i$ is by definition binomial $\mathcal{B}in(n, \theta)$.

  The estimator for $\theta$ is $\hat{\theta}_n = \overline{Y} = \frac{X}{n}$. It is easy to check that this estimator is both moment matching ($\mathbb{E}Y_i = \theta$) and MLE (the likelihood is $\theta^{\sum Y_i}(1-\theta)^{n-\sum Y_i}$). Thus $\hat{\theta}_n$ has a binomial distribution with rescaled realizations $\{0, 1/n, 2/n, \ldots, (n-1)/n, 1\}$, that is,

$$\mathbb{P}\left(\hat{\theta}_n = \frac{k}{n}\right) = \binom{n}{k}\theta^k(1-\theta)^{n-k}, \quad k = 0, 1, \ldots, n,$$

which is the estimator's sampling distribution.

  It easy to show, by referring to a binomial distribution, that the expectation of $\hat{\theta}_n$ is $1/n$ times the expectation of the binomial, $n\theta$,

$$\mathbb{E}\hat{\theta}_n = \frac{1}{n} \times n\theta = \theta,$$

and that the variance is

$$\mathbb{V}\mathrm{ar}\,\hat{\theta}_n = \left(\frac{1}{n}\right)^2 \times n\theta(1-\theta) = \frac{\theta(1-\theta)}{n}.$$

If $\mathbb{E}\hat{\theta}_n = \theta$, then the estimator $\hat{\theta}$ is called *unbiased*. The expectation is taken with respect to the sampling distribution. The quantity

$$b(\theta) = \mathbb{E}\hat{\theta}_n - \theta$$

is called the *bias* of $\hat{\theta}$.

  The error in estimation can be assessed by various measures. The most common is the *mean squared error* (MSE).

The MSE is defined as

$$\text{MSE}(\hat{\theta}, \theta) = \mathbb{E}(\hat{\theta}_n - \theta)^2.$$

The MSE represents the expected squared deviation of the estimator from the parameter it estimates. This expectation is taken with respect to the sampling distribution of $\hat{\theta}_n$.

From the definition of MSE,

$$
\begin{aligned}
\mathbb{E}(\hat{\theta}_n - \theta)^2 &= \mathbb{E}(\hat{\theta}_n - \mathbb{E}\hat{\theta}_n + \mathbb{E}\hat{\theta}_n - \theta)^2 \\
&= \mathbb{E}(\hat{\theta}_n - \mathbb{E}\hat{\theta}_n)^2 - 2\mathbb{E}(\hat{\theta}_n - \mathbb{E}\hat{\theta}_n)(\mathbb{E}\hat{\theta}_n - \theta) + (\mathbb{E}\hat{\theta}_n - \theta)^2 \\
&= \mathbb{E}(\hat{\theta}_n - \mathbb{E}\hat{\theta}_n)^2 + (\mathbb{E}\hat{\theta}_n - \theta)^2.
\end{aligned}
$$

Consequently, the MSE can be represented as a sum of the variance of the estimator and its bias squared:

$$\text{MSE}(\hat{\theta}, \theta) = \mathbb{V}\text{ar}\,\hat{\theta} + b(\theta)^2.$$

The square root of the MSE is sometimes used; it is called the *root mean squared error* (RMSE). For example, in estimating the population proportion, the estimator $\hat{p} = X/n$, for the $X \sim \mathscr{B}in(n, p)$ model, is unbiased, $\mathbb{E}(\hat{p}) = p$. Then the MSE is $\mathbb{V}\text{ar}(\hat{p}) = pq/n$, and the RMSE is $\sqrt{pq/n}$. Note that the RMSE is a function of the parameter. If parameter $p$ is replaced by its estimator $\hat{p}$, then the RMSE becomes the *standard error*, s.e., of the estimator. For binomial $p$, the standard error of $\hat{p}$ is $s.e.(\hat{p}) = \sqrt{\hat{p}\hat{q}/n}$.

**Remark.** The *standard error* (s.e.) of any estimator usually refers to a sample counterpart of its RMSE (sample counterpart of standard deviation for unbiased estimators). For example, if $X_1, X_2, \ldots, X_n$ are $\mathscr{N}(\mu, \sigma^2)$, then $s.e.(\overline{X}) = s/\sqrt{n}$.

Inspecting the variance (when the sample size increases) of an unbiased estimator allows for checking its consistency. The consistency property is a desirable property of estimators. Informally, it is defined as the convergence of an estimator (in a stochastic sense) to the parameter it estimates.

If, for an unbiased estimator $\hat{\theta}_n$, $\mathbb{V}\text{ar}\,\hat{\theta}_n \to 0$ when the sample size $n \to \infty$, the estimator is called *consistent*.

More advanced definitions of convergences of random variables, which are beyond the scope of this text, are required in order to deduce more precise definitions of asymptotic unbiasedness and weak and strong consistency; therefore, these definitions will not be discussed here.

*Example 7.7.* Suppose that we are interested in estimating the parameter $\theta$ in a population with a distribution of $\mathcal{N}(0,\theta), \theta > 0$, and that the proposed estimator, when the sample $X_1, X_2, \ldots, X_n$ is observed, is $\hat{\theta} = \frac{1}{n} \sum_{i=1}^{n} X_i^2$.

It is easy to demonstrate that, when $X \sim \mathcal{N}(0,\theta)$, $\mathbb{E} X^2 = \theta$ and $\mathbb{E} X^4 = 3\theta^2$, by representing $X$ as $\sqrt{\theta} Z$ for $Z \sim \mathcal{N}(0,1)$ and using the fact that $\mathbb{E} Z^2 = 1$ and $\mathbb{E} Z^4 = 3$.

The estimator $\hat{\theta} = \frac{1}{n} \sum_{i=1}^{n} X_i^2 = \overline{X^2}$ is unbiased and consistent. Since $\mathbb{E} \hat{\theta} = \frac{1}{n} \sum_{i=1}^{n} \mathbb{E} X_i^2 = \frac{1}{n} n\theta = \theta$, the estimator is unbiased. To show consistency, it is sufficient to demonstrate that the variance tends to 0 as the sample size increases. This is evident from

$$\mathbb{V}\mathrm{ar}\, \hat{\theta} = \frac{1}{n^2} \sum_{i=1}^{n} \mathbb{V}\mathrm{ar}\, X_i^2 = \frac{1}{n^2} 3n\theta^2 = \frac{3\theta^2}{n} \to 0, \text{ when } n \to \infty.$$

Alternatively, one may use the fact that $\frac{1}{\theta} \sum_{i=1}^{n} X_i^2$ has a $\chi_n^2$ distribution, so that the sampling distribution of $\hat{\theta}$ is a scaled $\chi_n^2$, where the scaling factor is $\frac{1}{n\theta}$. The unbiasedness and consistency follow from $\mathbb{E} \chi_n^2 = n$ and $\mathbb{V}\mathrm{ar}\, \chi_n^2 = 2n$ by accounting for the scaling factor.
✐

Some important examples of unbiased and consistent estimators are provided next.

## 7.3 Estimation of a Mean, Variance, and Proportion

### 7.3.1 Point Estimation of Mean

For a sample $X_1, \ldots, X_n$ of size $n$ we have already discussed the sample mean $\overline{X} = \frac{1}{n} \sum_{i=1}^{n} X_i$ as an estimator of location. A natural estimator of the population mean $\mu$ is the sample mean $\hat{\mu} = \overline{X}$. The estimator $\overline{X}$ is an "optimal" estimator of a mean in many different models/distributions and for many different definitions of optimality.

The estimator $\overline{X}$ varies from sample to sample. More precisely, $\overline{X}$ is a random variable with a fixed distribution depending on the common distribution of observations, $X_i$.

The following is true for *any* distribution in the population as long as $\mathbb{E}X_i = \mu$ and $\mathbb{V}\mathrm{ar}(X_i) = \sigma^2$ exist:

$$\mathbb{E}\overline{X} = \mu, \quad \mathbb{V}\mathrm{ar}(\overline{X}) = \frac{\sigma^2}{n}. \tag{7.2}$$

The above equations are a direct consequence of independence in a sample and imply that $\overline{X}$ is an unbiased and consistent estimator of $\mu$.

If, in addition, we assume normality $X_i \sim \mathcal{N}(\mu, \sigma^2)$, then the sampling distribution of $\overline{X}$ is known exactly,

$$\overline{X} \sim \mathcal{N}\left(\mu, \frac{\sigma^2}{n}\right),$$

and the relations in (7.2) are apparent.

**Chebyshev's Inequality and Strong Law of Large Numbers\*.** There are two general results in probability that theoretically justify the use of the sample mean $\overline{X}$ to estimate the population mean, $\mu$. These are Chebyshev's inequality and strong law of large numbers (SLLN). We will discuss these results without mathematical rigor.

The Chebyshev inequality states that when $X_1, X_2, \ldots, X_n$ are i.i.d. random variables with mean $\mu$ and finite variance $\sigma^2$, the probability that $\overline{X}$ will deviate from $\mu$ is small,

$$\mathbb{P}(|\overline{X}_n - \mu| \geq \epsilon) \leq \frac{\sigma^2}{n\epsilon^2},$$

for any $\epsilon > 0$. The inequality is a direct consequence of (5.8) with $(\overline{X}_n - \mu)^2$ in place of $X$ and $\epsilon^2$ in place of $a$.

To translate this to specific numbers, choose $\epsilon$ small, say 0.000001. Assume that the $X_i$s have a variance of 1. The Chebyshev inequality states that with $n$ larger than the solution of $1/(n \times 0.000001^2) = 0.9999$, the distance between $\overline{X}_n$ and $\mu$ will be smaller than 0.000001 with a probability of 99.99%. Admittedly, $n$ here is an experimentally unfeasible number; however, for any small $\epsilon$, finite $\sigma^2$, and "confidence" close to 1, such $n$ is finite.

The laws of large numbers state that, as a numerical sequence, $\overline{X}_n$ converges to $\mu$. Care is needed here. The sequence $\overline{X}_n$ is not a sequence of numbers but a sequence of random variables, which are functions defined on sample spaces $\mathscr{S}$. Thus, direct application of a "calculus" type of convergence is not appropriate. However, for any fixed realization from sample space $\mathscr{S}$, the sequence $\overline{X}_n$ becomes numerical and a traditional convergence can be stated. Thus, a correct statement for the so-called SLLN is

$$\mathbb{P}(\overline{X}_n \to \mu) = 1,$$

that is, conceived as an event, $\left\{\overline{X}_n \to \mu\right\}$ is a sure event – it happens with a probability of 1.

### 7.3.2 Point Estimation of Variance

We will obtain an intuition starting, once again, with a finite population: $y_1, \ldots, y_N$. The population variance is $\sigma^2 = \frac{1}{N} \sum_{i=1}^{N} (y_i - \mu)^2$, where $\mu = \frac{1}{N} \sum_{i=1}^{N} y_i$ is the population mean.

> If a sample $X_1, X_2, \ldots, X_n$ is observed, an estimator of $\sigma^2$ is
>
> $$\hat{\sigma}^2 = \frac{1}{n} \sum_{i=1}^{n} (X_i - \mu)^2,$$
>
> for $\mu$ known, or
>
> $$\hat{\sigma}^2 = s^2 = \frac{1}{n-1} \sum_{i=1}^{n} (X_i - \overline{X})^2,$$
>
> for $\mu$ not known and estimated by $\overline{X}$.

In the expression for $s^2$ we divide by $n-1$ instead of the "expected" $n$ in order to ensure the unbiasedness of $s^2$, $\mathbb{E}s^2 = \sigma^2$. The proof is easy and does not require any distributional assumptions, except that the population variance $\sigma^2$ is finite.

Note that by the definition of variance, $\mathbb{E}(X_i - \mu)^2 = \sigma^2$ and $\mathbb{E}(\overline{X} - \mu)^2 = \sigma^2/n$.

$$
\begin{aligned}
(n-1)s^2 &= \sum_{i=1}^{n} (X_i - \overline{X})^2 \\
&= \sum_{i=1}^{n} [(X_i - \mu) - (\overline{X} - \mu)]^2 \\
&= \sum_{i=1}^{n} (X_i - \mu)^2 - 2n(\overline{X} - \mu) \sum_{i-1}^{n} (X_i - \mu) + n(\overline{X} - \mu)^2 \\
&= \sum_{i=1}^{n} (X_i - \mu)^2 - n(\overline{X} - \mu)^2, \quad \text{since } \sum_{i=1}^{n} (X_i - \mu) = n(\overline{X} - \mu).
\end{aligned}
$$

Then,

$$E(s^2) = \frac{1}{n-1} E(n-1)s^2$$

$$= \frac{1}{n-1} E[\sum_{i=1}^{n} (X_i - \mu)^2 - n(\overline{X} - \mu)^2]$$

$$= \frac{1}{n-1} (n\sigma^2 - n\frac{\sigma^2}{n})$$

$$= \frac{1}{n-1} (n-1)\sigma^2 = \sigma^2.$$

When, in addition, the population is normal $\mathcal{N}(\mu, \sigma^2)$, then

$$\frac{(n-1)s^2}{\sigma^2} \sim \chi^2_{n-1},$$

i.e., the statistic $\frac{(n-1)s^2}{\sigma^2} = \sum_{i=1}^{n} \left(\frac{X_i - \overline{X}}{\sigma}\right)^2$ has a $\chi^2$ distribution with $n-1$ degrees of freedom (see Eq. 6.3 and related discussion).

For a sample from a normal distribution, unbiasedness of $s^2$ is an easy consequence of the representation $s^2 \sim \frac{\sigma^2}{n-1} \chi^2_{n-1}$ and $E\chi^2_{n-1} = (n-1)$. The variance of $s^2$ is

$$\mathbb{V}\text{ar } s^2 = \left(\frac{\sigma^2}{n-1}\right)^2 \times \mathbb{V}\text{ar } \chi^2_{n-1} = \frac{2\sigma^4}{n-1} \qquad (7.3)$$

since $\mathbb{V}\text{ar } \chi^2_{n-1} = 2(n-1)$. Unlike the unbiasedness result, $E s^2 = \sigma^2$, which does not require a normality assumption, the result in (7.3) is valid only when observations come from a normal distribution.

Figure 7.3 indicates that the empirical distribution of normalized sample variances is close to a $\chi^2$ distribution. We generated $M = 100000$ samples of size $n = 8$ from a normal $\mathcal{N}(0, 5^2)$ distribution and found sample variances $s^2$ for each sample. The sample variances are multiplied by $n - 1 = 7$ and divided by $\sigma^2 = 25$. The histogram of such rescaled sample variances is plotted and the density of a $\chi^2$ distribution with 7 degrees of freedom is superimposed in red. The code generating Fig. 7.3 is given next.

```
M=100000;  n = 8;
   X = 5 * randn([n, M]);
   ch2 = (n-1) * var(X)/25;
 histn(ch2,0,0.4,30)
 hold on
plot( (0:0.1:30), chi2pdf((0:0.1:30), n-1),'r-')
```

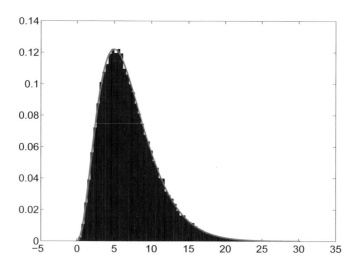

**Fig. 7.3** Histogram of normalized sample variances $(n-1)s^2/\sigma^2$ obtained from $M = 100000$ independent samples from $\mathcal{N}(0,5^2)$, each of size $n = 8$. The density of a $\chi^2$ distribution with 7 degrees of freedom is superimposed in *red*.

The code is quite efficient since a `for-end` loop is avoided. The simulated object X is an $n \times M$ matrix consisting of $M$ columns (samples) of length $n$. The operator `var(X)` acts on columns of X producing $M$ sample variances.

**Several Robust Estimators of the Standard Deviation\*.** Suppose that a sample $X_1, \ldots, X_n$ is observed and normality is not assumed. We discuss two estimators of the standard deviation that are calibrated by the normal distribution but quite robust with respect to outliers and deviations from normality.

Gini's mean difference is defined as

$$G = \frac{2}{n(n-1)} \sum_{1 \le i < j \le n} |X_i - X_j|.$$

The statistic $G\frac{\sqrt{\pi}}{2}$ is an estimator of the standard deviation and is more robust to outliers than the standard statistic $s$.

Another proposal by Croux and Rousseeuw (1992) involves absolute differences as in Gini's mean difference estimator but uses $k$th-order statistic rather than the average. The estimator of $\sigma$ is

$$Q = 2.2219 \, \{|X_i - X_j|, i < j\}_{(k)}, \quad \text{where } k = \binom{\lfloor n/2 \rfloor + 1}{2}.$$

The constant 2.2219 is needed for the calibration of the estimator, so that if the sample is a standard normal, then $Q = 1$. In calculating $Q$, all $\binom{n}{2}$ differences $|X_i - X_j|$ are ordered, and the $k$th in rank is selected and multiplied by 2.2219. This choice of $k$ requires an additional multiplicative correction factor $n/(n + 1.4)$ for $n$ odd, or $n/(n + 3.8)$ for $n$ even.

MATLAB scripts ◢ ginimd.m and crouxrouss.m evaluate the estimator. The algorithm is naïve and uses a double loop to evaluate $G$ and $Q$. The evaluation breaks down for sample sizes of less than 500 because of memory problems. A smarter algorithm that avoids looping is implemented in versions ginimd2.m and crouxrouss2.m. In these versions, the sample size can go up to 6000.

In the following MATLAB session we demonstrate the performance of robust estimators of the standard deviation. If 1000 standard normal random variates are generated and one value is replaced with a clear outlier, say $X_{1000} = 20$, we will explore the influence of this outlier to estimators of the standard deviation. Note that $s$ is quite sensitive and the outlier inflates the estimator by almost 20%. The robust estimators are affected as well, but not as much as $s$.

```
x =randn(1, 1000);
x(1000)=20;
std(x)
% ans = 1.1999
s1 = ginimd2(x)
%s1 =1.0555
s2 = crouxrouss2(x)
%s2 =1.0287
iqr(x)/1.349
%ans = 1.0172
```

There are many other robust estimators of the variance/standard deviation. Good references containing extensive material on robust estimation are Wilcox (2005) and Staudte and Sheater (1990).

### 7.3.3 Point Estimation of Population Proportion

It is natural to estimate the population proportion $p$ by a sample proportion. The sample proportion is the MLE and moment-matching estimator for $p$.

Sample proportions use a binomial distribution as the theoretical model. Let $X \sim \mathcal{B}in(n, p)$, where parameter $p$ is unknown. The MLE of $p$ based on a single observation $X$ is obtained by maximizing the likelihood

$$\binom{n}{X} p^X (1 - p)^{n - X}$$

or the log-likelihood

$$\text{factor free of } p + X\log(p) + (n - X)\log(1 - p).$$

The maximum is obtained by solving

$$(\text{factor free of } p + X\log(p) + (n - X)\log(1 - p))' = 0$$
$$\frac{X}{p} - \frac{n - X}{1 - p} = 0,$$

which after some algebra gives the solution $\hat{p}_{mle} = \frac{X}{n}$.

In Example 7.6 we argued that the exact distribution for $X/n$ is a rescaled binomial and that the statistic is unbiased, with the variance converging to 0 when the sample size increases. These two properties define a consistent estimator.

## 7.4 Confidence Intervals

Whenever the sampling distribution of a point estimator $\hat{\theta}_n$ is continuous, then necessarily $\mathbb{P}(\hat{\theta}_n = \theta) = 0$. In other words, the probability that the estimator exactly matches the parameter it estimates is 0.

Instead of the point estimator, one may report two estimators, $L = L(X_1, \ldots, X_n)$ and $U = U(X_1, \ldots, X_n)$, so that the interval $[L, U]$ covers $\theta$ with a probability of $1 - \alpha$, for small $\alpha$. In this case, the interval $[L, U]$ will be called a $(1 - \alpha)100\%$ confidence interval for $\theta$.

For the construction of a confidence interval for a parameter, one needs to know the sampling distribution of the associated point estimator. The lower and upper interval bounds $L$ and $U$ depend on the quantiles of this distribution. We will derive the confidence interval for the normal mean, normal variance, population proportion, and Poisson rate. Many other confidence intervals, including differences, ratios, and some functions of statistics, are tightly connected to testing methodology and will be discussed in subsequent chapters.

Note that when the population is normal and $X_1, \ldots, X_n$ is observed, the exact sampling distributions of

$$Z = \frac{\overline{X} - \mu}{\sigma/\sqrt{n}} \quad \text{and}$$

$$t = \frac{\overline{X} - \mu}{s/\sqrt{n}} = \frac{\overline{X} - \mu}{\sigma/\sqrt{n}} \times \frac{1}{\sqrt{\frac{(n-1)s^2}{\sigma^2}/(n-1)}}$$

are standard normal and Student $t_{n-1}$, respectively.

The expression for $t$ is shown as a product to emphasize the construction of a $t$-distribution from a standard normal (in *blue*) and $\chi^2$ (in *red*), as in p. 208).

When the population is not normal but $n$ is large, both statistics $Z$ and $t$ have an approximate standard normal distribution due to the CLT.

Wa saw that the point estimator for the population proportion (of "successes") is the sample proportion $\hat{p} = X/n$, where $X$ is the number of successes in $n$ trials. The statistic $X/n$ is based on a binomial sampling scheme in which $X$ has exactly a binomial $\mathscr{B}in(n,p)$ distribution. Using this exact distribution would lead to confidence intervals in which the bounds and confidence levels were discretized. The normal approximation to the binomial (CLT in the form of de Moivre's approximation) leads to

$$\hat{p} \overset{approx}{\sim} \mathscr{N}\left(p, \frac{p(1-p)}{n}\right), \tag{7.4}$$

and the confidence intervals for the population proportion $p$ would be based on normal quantiles.

### 7.4.1 Confidence Intervals for the Normal Mean

Let $X_1, \ldots, X_n$ be a sample from a $\mathscr{N}(\mu, \sigma^2)$ distribution where the parameter $\mu$ is to be estimated and $\sigma^2$ is known.

Starting from the identity

$$\mathbb{P}(-z_{1-\alpha/2} \leq Z \leq z_{1-\alpha/2}) = 1 - \alpha$$

and the fact that $\overline{X}$ has a $\mathscr{N}(\mu, \frac{\sigma^2}{n})$ distribution, we can write

$$\mathbb{P}\left(-z_{1-\alpha/2}\frac{\sigma}{\sqrt{n}} + \mu \leq \overline{X} \leq z_{1-\alpha/2}\frac{\sigma}{\sqrt{n}} + \mu\right) = 1 - \alpha;$$

see Fig. 7.4a for an illustration. Simple algebra gives

$$\overline{X} - z_{1-\alpha/2}\frac{\sigma}{\sqrt{n}} \leq \mu \leq \overline{X} + z_{1-\alpha/2}\frac{\sigma}{\sqrt{n}}, \tag{7.5}$$

which is a $(1-\alpha)100\%$ confidence interval.

If $\sigma^2$ is not known, then a confidence interval with the sample standard deviation $s$ in place of $\sigma$ can be used. The $z$ quantiles are valid for large $n$, but for small $n$ ($n < 40$) we use $t_{n-1}$ quantiles, since the sampling distribution for $\frac{\overline{X}-\mu}{s/\sqrt{n}}$ is $t_{n-1}$. Thus, for $\sigma^2$ unknown,

$$\overline{X} - t_{n-1,1-\alpha/2}\frac{s}{\sqrt{n}} \leq \mu \leq \overline{X} + t_{n-1,1-\alpha/2}\frac{s}{\sqrt{n}} \tag{7.6}$$

is the confidence interval for $\mu$ of level $1 - \alpha$.

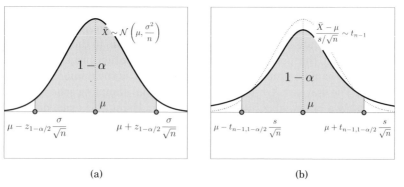

(a)                                                          (b)

**Fig. 7.4** (a) When $\sigma^2$ is known, $\overline{X}$ has a normal $\mathcal{N}(\mu, \sigma^2/n)$ distribution and $\mathbb{P}(\mu - z_{1-\alpha/2}\frac{\sigma}{\sqrt{n}} \leq \overline{X} \leq \mu + z_{1-\alpha/2}\frac{\sigma}{\sqrt{n}}) = 1 - \alpha$, leading to confidence interval (7.5). (b) If $\sigma^2$ is not known and $s^2$ is used instead, then $\frac{\overline{X}-\mu}{s/\sqrt{n}}$ is $t_{n-1}$, leading to the confidence interval in (7.6).

Below is a summary of the above-stated intervals.

> The $(1-\alpha)$ 100% confidence interval for an unknown normal mean $\mu$ on the basis of a sample of size $n$ is
>
> $$\left[\overline{X} - z_{1-\alpha/2}\frac{\sigma}{\sqrt{n}}, \overline{X} + z_{1-\alpha/2}\frac{\sigma}{\sqrt{n}}\right]$$
>
> when the variance $\sigma^2$ is known and
>
> $$\left[\overline{X} - t_{n-1,1-\alpha/2}\frac{s}{\sqrt{n}}, \overline{X} + t_{n-1,1-\alpha/2}\frac{s}{\sqrt{n}}\right]$$
>
> when the variance $\sigma^2$ is not known and $s^2$ is used instead.

**Interpretation of Confidence Intervals.** What does a "confidence of 95%" mean? A common misconception is that it means that the unknown mean falls in the calculated interval with a probability of 0.95. Such a probability statement is valid for credible sets in the Bayesian context, which will be discussed in Chap. 8.

The interpretation of the $(1 - \alpha)$ 100% confidence interval is as follows. If a random sample from a normal population is selected a large number of times and the confidence interval for the population mean $\mu$ is calculated, the proportion of such intervals covering $\mu$ approaches $1 - \alpha$.

The following MATLAB code illustrates this. The code samples $M = 10000$ times a random sample of size $n = 40$ from a normal population with a mean of $\mu = 10$ and a variance of $\sigma^2 = 4^2$ and calculates a 95% confidence interval. It then counts how many of the intervals cover the mean $\mu$, cover = 1, and, finally, finds their proportion, covers/M. The code was run consecutively several times and the following empirical confidences were obtained: 0.9461, 0.9484, 0.9469, 0.9487, 0.9502, 0.9482, 0.9502, 0.9482, 0.9530, 0.9517, 0.9503, 0.9514, 0.9496, 0.9515, etc., clearly scattering around 0.95. Figure 7.5a plots the behavior of the coverage proportion when simulations range from 1 to 10,000. Figure 7.5b plots the first 100 intervals in the simulation and their position with respect to $\mu = 10$. The confidence intervals in simulations 17, 37, 47, 58, 78, and 82 fail to cover $\mu$.

```
M=10000;           %simulate M times
n = 40;            % sample size
alpha = 0.05;      %1-alpha = confidence
tquantile = tinv(1-alpha/2, n-1);
covers =[];
for i = 1:M
    X = 10 + 4*randn(1,n); %sample, mean=10, var =16
    xbar = mean(X); s = std(X);
    LB = xbar - tquantile * s/sqrt(n);
    UB = xbar + tquantile * s/sqrt(n);
    % cover=1 if the interval covers population mean 10
    if UB < 10 | LB > 10
            cover = 0;
    else
            cover = 1;
    end
        covers =[covers cover]; %saves cover history
end
sum(covers)/M %proportion of intervals covering the mean
```

### 7.4.2 Confidence Interval for the Normal Variance

Earlier (p. 209) we argued that the sampling distribution of $\frac{(n-1)s^2}{\sigma^2}$ was $\chi^2$ with $n-1$ degrees of freedom. From the definition of $\chi^2_{n-1}$ quantiles,

$$1 - \alpha = \mathbb{P}(\chi^2_{n-1,\alpha/2} \leq \chi^2_{n-1} \leq \chi^2_{n-1,1-\alpha/2}),$$

as in Fig. 7.6. Replacing $\chi^2_{n-1}$ with $\frac{(n-1)s^2}{\sigma^2}$, we get

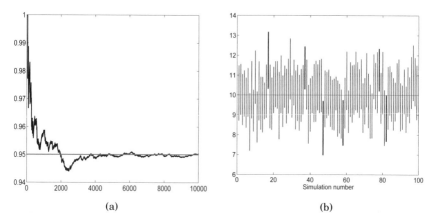

(a)                                                        (b)

**Fig. 7.5** (a) Proportion of intervals covering the mean plotted against the iteration number, as in `plot(cumsum(covers)./(1:length(covers))  )`. (b) First 100 simulated intervals. The intervals $17, 37, 47, 58, 78$, and $82$ fail to cover the true mean.

$$1 - \alpha = \mathbb{P}\left(\chi^2_{n-1,\alpha/2} \leq \frac{(n-1)s^2}{\sigma^2} \leq \chi^2_{n-1,1-\alpha/2}\right).$$

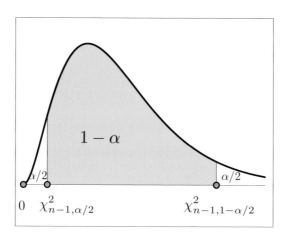

**Fig. 7.6** Confidence interval for normal variance $\sigma^2$ is derived from $\mathbb{P}(\chi^2_{n-1,\alpha/2} \leq (n-1)s^2/\sigma^2 \leq \chi^2_{n-1,1-\alpha/2}) = 1 - \alpha$.

Simple algebra with the above inequalities (taking the reciprocal of all three parts, being careful about the direction of the inequalities, and multiplying everything by $(n-1)s^2$) gives

$$\frac{(n-1)s^2}{\chi^2_{n-1,1-\alpha/2}} \le \sigma^2 \le \frac{(n-1)s^2}{\chi^2_{n-1,\alpha/2}}.$$

The $(1-\alpha)\,100\%$ confidence interval for an unknown normal variance is

$$\left[\frac{(n-1)s^2}{\chi^2_{n-1,1-\alpha/2}}, \frac{(n-1)s^2}{\chi^2_{n-1,\alpha/2}}\right]. \qquad (7.7)$$

**Remark.** If the population mean $\mu$ is known, then $s^2$ is calculated as $\frac{1}{n}\sum_{i=1}^{n}(X_i - \mu)^2$, and the $\chi^2$ quantiles gain one degree of freedom ($n$ instead of $n-1$). This makes the confidence interval a bit tighter.

*Example 7.8.* **Amanita muscaria.** With its bright red, sometimes dinner-plate-sized caps, the fly agaric (*Amanita muscaria*) is one of the most striking of all mushrooms (Fig. 7.7a). The white warts that adorn the cap, the white gills, a well-developed ring, and the distinctive volva of concentric rings distinguish the fly agaric from all other red mushrooms. The spores of the mushroom print white, are elliptical, and have a (maximal) diameter in the range of 7 to 13 $\mu$m (Fig. 7.7b).

 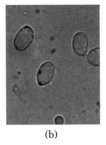

(a)                          (b)

**Fig. 7.7** *Amanita muscaria* and its spores. (a) Fly agaric or *Amanita muscaria*. (b) Spores of *Amanita muscaria*.

Measurements of the diameter $X$ of spores for $n = 51$ mushrooms are given in the following table:

|    |    |    |    |    |    |    |    |    |    |
|----|----|----|----|----|----|----|----|----|----|
| 10 | 11 | 12 | 9  | 10 | 11 | 13 | 12 | 10 | 11 |
| 11 | 13 | 9  | 10 | 9  | 10 | 8  | 12 | 10 | 11 |
| 9  | 10 | 7  | 11 | 8  | 9  | 11 | 11 | 10 | 12 |
| 10 | 8  | 7  | 11 | 12 | 10 | 9  | 10 | 11 | 10 |
| 8  | 10 | 10 | 8  | 9  | 10 | 13 | 9  | 12 | 9  |
| 9  |    |    |    |    |    |    |    |    |    |

Assume that the measurements are normally distributed with mean $\mu$ and variance $\sigma^2$, but both parameters are unknown. The sample mean and variances are $\overline{X} = 10.098$ , $s^2 = 2.1702$, and $s = 1.4732$. Also, the confidence interval would use an appropriate $t$-quantile, in this case `tinv(1-0.05/2, 51-1)` = `2.0086`.

The 95% confidence interval for the population mean, $\mu$, is

$$\left[ 10.098 - 2.0086 \times \frac{1.4732}{\sqrt{51}}, \quad 10.098 + 2.0086 \times \frac{1.4732}{\sqrt{51}} \right] = [9.6836, 10.5124].$$

Thus, the unknown mean $\mu$ belongs to the interval $[9.6836, 10.5124]$ with confidence 95%. That means that if the sample is obtained many times and for each sample the confidence interval is calculated, 95% of the intervals would contain $\mu$.

To find, say, the 90% confidence interval for the population variance, $\sigma^2$, we need $\chi^2$ quantiles, `chi2inv(1-0.10/2, 51-1)` = `67.5048`, and `chi2inv(0.10/2, 51-1)` = `34.7643`. According to (7.7), the interval is

$$[(51 - 1) \times 2.1702/67.5048, \quad (51 - 1) \times 2.1702/34.7643] = [1.6074, 3.1213].$$

Thus, the interval $[1.6074, 3.1213]$ covers the population variance $\sigma^2$ with a confidence of 90%.

✎

*Example 7.9.* An alternative confidence interval for the normal variance is possible. Since by the CLT $s^2 \overset{approx}{\sim} \mathcal{N}\left(\sigma^2, \frac{2\sigma^4}{n-1}\right)$ (Can you explain why?), when $n$ is not small, an approximate $(1 - \alpha)100\%$ confidence interval for $\sigma^2$ is

$$\left[ s^2 - z_{1-\alpha/2} \cdot \frac{\sqrt{2}\, s^2}{\sqrt{n-1}}, \quad s^2 + z_{1-\alpha/2} \cdot \frac{\sqrt{2}\, s^2}{\sqrt{n-1}} \right].$$

In Example 7.8, $s^2 = 2.1702$ and $n = 51$. A 90% confidence interval for the variance was $[1.6074, 3.1213]$. By normal approximation,

```
s2 = 2.1702; n=51; alpha = 0.1;
[s2 - norminv(1-alpha/2)*sqrt(2)* s2/sqrt(n-1), ...
 s2 + norminv(1-alpha/2)*sqrt(2)* s2/sqrt(n-1)]
%ans = 1.4563    2.8841
```

The interval $[1.4563, 2.8841]$ is shorter, compared to the standard confidence interval $[1.6074, 3.1213]$ obtained using $\chi^2$ quantiles, as $1.4278 < 1.5139$. Insisting on equal-probability tails does not lead to the shortest interval since the $\chi^2$ distribution is asymmetric. In addition, the approximate interval is centered at $s^2$. Why, then, is this interval not used? The coverage probability of a CLT-based interval is smaller than the nominal $1 - \alpha$, and unless $n$ is large ($>100$, say), this discrepancy can be significant (Exercise 7.26).

✎

### 7.4.3 Confidence Intervals for the Population Proportion

The sample proportion $\hat{p} = \frac{X}{n}$ has a range of optimality properties (unbiasedness, consistency); however, its realizations are discrete. For this reason confidence intervals for $p$ are obtained using the normal approximation, or connections of binomial with other continuous distributions, such as $F$.

Recall that for $n$ large and $np$ or $nq$ not small ($>10$), binomial $X$ can be approximated by a $\mathcal{N}(np, npq)$ distribution. This approximation leads to $\frac{X}{n} \overset{approx}{\sim} \mathcal{N}\left(p, \frac{pq}{n}\right)$.

Note, however, that the standard deviation of $\hat{p}$, $\sqrt{\frac{pq}{n}}$, is not known (it depends on $p$), and for the confidence interval one uses a plug-in estimator $\sqrt{\frac{\hat{p}\hat{q}}{n}}$ instead.

---

Let $p$ be the population proportion and $\hat{p}$ the observed sample proportion. Assume that the smaller number $\frac{np}{q}, \frac{nq}{p}$ is larger than 10. Then the $(1-\alpha)100\%$ confidence interval for unknown $p$ is

$$\left[ \hat{p} - z_{1-\alpha/2}\sqrt{\frac{\hat{p}\,\hat{q}}{n}}, \ \hat{p} + z_{1-\alpha/2}\sqrt{\frac{\hat{p}\,\hat{q}}{n}} \right].$$

This interval is known as the Wald interval (Wald and Wolfowitz, 1939).

---

The Wald interval is used most frequently but its performance is suboptimal and even poor when $p$ is close to 0 or 1. Figure 7.8a demonstrates the performance of Wald's 95% confidence interval for $n = 20$ and $p$ ranging from 0.05 to 0.95 with a step of 0.01. The plot is obtained by simulation ( ◀ waldsimulation.m). For each ("true") $p$, 100,000 binomial proportions are simulated, the Wald confidence intervals calculated, and the proportion of those intervals containing $p$ is plotted. Notice that for nominal 95% confidence, the actual coverage probability may be much smaller, depending on true $p$.

Unless the sample size $n$ is very large, the Wald interval should not be used. The performance of Wald's interval can be improved by continuity corrections:

$$\left[ \hat{p} - \frac{1}{2n} - z_{1-\alpha/2}\sqrt{\frac{\hat{p}\,\hat{q}}{n}}, \ \hat{p} + \frac{1}{2n} + z_{1-\alpha/2}\sqrt{\frac{\hat{p}\,\hat{q}}{n}} \right].$$

Figure 7.8b shows the coverage probability of Wald's corrected interval.

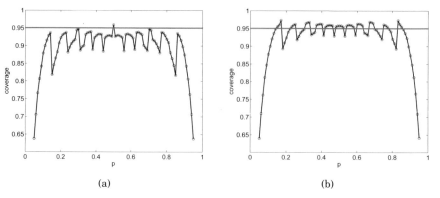

(a)                                                                  (b)

**Fig. 7.8** (a) Simulated coverage probability for Wald's confidence interval for the true binomial proportion $p$ ranging from 0.05 to 0.95, and $n = 20$. For each $p$, 100,000 binomial proportions are simulated, the Wald confidence intervals calculated, and the proportion of those containing $p$ plotted. (b) The same as (a), but for the corrected Wald interval.

There is a range of intervals that have a performance superior to Wald's interval. An overview of several alternatives is provided next.

**Adjusted Wald Interval.** The adjusted Wald interval (Agresti and Coull, 1998) uses $p^* = \frac{X+2}{n+4}$ as an estimator of the proportion. Adding "two successes and two failures" was proposed by Wilson (1927).

$$\left[ p^* - z_{1-\alpha/2} \sqrt{\frac{p^* q^*}{n+4}},\ p^* + z_{1-\alpha/2} \sqrt{\frac{p^* q^*}{n+4}} \right].$$

We will see in the next chapter that Wilson's proposal $p^*$ has a Bayesian justification (p. 289).

**Wilson Score Interval.** The Wilson score interval is another adjustment to the Wald interval based on the so-called Wilson-score test (Wilson, 1927; Hogg and Tanis, 2001):

$$\left[ \frac{1}{1+z^2/n} \left( \hat{p} + \frac{z^2}{2n} - z\sqrt{\frac{\hat{p}\,\hat{q}}{n} + \frac{z^2}{4n^2}} \right),\ \frac{1}{1+z^2/n} \left( \hat{p} + \frac{z^2}{2n} + z\sqrt{\frac{\hat{p}\,\hat{q}}{n} + \frac{z^2}{4n^2}} \right) \right],$$

where $z$ is $z_{1-\alpha/2}$. This interval can be easily obtained by solving the inequality

$$|\hat{p} - p| \leq z_{1-\alpha/2} \sqrt{\frac{p(1-p)}{n}}$$

with respect to $p$. After squaring the left- and right-hand sides and some algebra one gets the quadratic inequality

$$p^2 \left( 1 + \frac{z_{1-\alpha/2}^2}{n} \right) - p \left( 2\hat{p} + \frac{z_{1-\alpha/2}^2}{n} \right) + \hat{p}^2 \le 0,$$

for which the solution coincides with Wilson's score interval.

**Clopper–Pearson Interval.** The Clopper–Pearson confidence interval (Clopper and Pearson, 1934) does not use normal approximation but, rather, an exact link between binomial and $F$ distributions. For $0 < X < n$, the $(1-\alpha) \cdot 100\%$ Clopper–Pearson confidence interval is

$$\left[ \frac{X}{X + (n - X + 1)F^*}, \ \frac{(X+1)F^{**}}{n - X + (X+1)F^{**}} \right],$$

where $F^*$ is the $(1-\alpha/2)$-quantile of the $F_{v_1,v_2}$-distribution with $v_1 = 2(n-X+1)$ and $v_2 = 2X$ and $F^{**}$ is the $(1-\alpha/2)$-quantile of the $F_{v_1,v_2}$-distribution with $v_1 = 2(X+1)$ and $v_2 = 2(n-X)$. When $X = 0$, the interval is $[0, 1-(\alpha/2)^{1/n}]$ and for $X = n$, $[(\alpha/2)^{1/n}, 1]$.

**Anscombe's ArcSin Interval.** For $X \sim \mathcal{B}in(n,p)$ Anscombe (1948) showed that if $p^* = \frac{X+3/8}{n+3/4}$, then the quantity

$$2\sqrt{n}(\arcsin \sqrt{p^*} - \arcsin \sqrt{p})$$

has an approximately standard normal distribution. From this result it follows that

$$\left[ \sin^2 \left( \arcsin \sqrt{p^*} - \frac{z_{1-\alpha/2}}{2\sqrt{n}} \right), \ \sin^2 \left( \arcsin \sqrt{p^*} + \frac{z_{1-\alpha/2}}{2\sqrt{n}} \right) \right]$$

is the $(1-\alpha)100\%$ confidence interval for $p$.

The following example shows the comparative performance of different confidence intervals for the population proportion.

*Example 7.10.* **Cyclosporine Reversal Study.** An interesting case study involved research on the therapeutic benefits of cyclosporine on patients with chronic inflammatory bowel disease (Crohn's disease). In a double-blind clinical trial, researchers reported (Brynskov et al., 1989) that out of 37 patients with Crohn's disease resistant to standard therapies, 22 improved after a 3-month period. This proportion was significantly higher than that for the placebo group (11/34). The study was published in the *New England Journal of Medicine.*

However, at the 6-month follow-up, no significant differences were found between the treatment group and the control. In the cyclosporine group, 30 patients *did not* improve, compared to 23 out of 34 in the placebo group (Brynskov et al., 1991). Thus, the proportion of patients who benefited in the cyclosporine group dropped from $\hat{p}_1 = 22/37 = 59.46\%$ at the 3-month to $\hat{p}_2 = 7/37 = 18.92\%$ at the 6-month follow-up. The researchers state: "We conclude that a short course of cyclosporin treatment does not result in long-term improvement in active chronic Crohn's disease."

To illustrate the performance of several introduced confidence intervals for the population proportion, we will find Wald's, Wilson's, Wilson score, Clopper–Pearson's, and Arcsin 95% confidence intervals for the proportion of patients who benefited in the cyclosporine group at the 3-month and 6-month follow-ups. Calculations are performed in MATLAB.

```
%Cyclosporine Clinical Trials
%
n = 37; %number of subjects in cyclosporine group
%  three months
X1 = 22;       p1hat = X1/n;    q1hat = 1-p1hat;
%  six months
X2 = 7;        p2hat = X2/n;    q2hat = 1- p2hat;
%===============================
%Wald Intervals
W3 = [p1hat   - norminv(0.975) * sqrt( p1hat * q1hat / n), ...
    p1hat   + norminv(0.975) * sqrt( p1hat * q1hat / n)]
W6 = [p2hat   - norminv(0.975) * sqrt( p2hat * q2hat / n), ...
    p2hat   + norminv(0.975) * sqrt( p2hat * q2hat / n)]
%W3 = 0.4364          0.75279
%W6 = 0.06299         0.31539
%====================================
% Wilson Intervals
    p1hats = (X1+2)/(n+4);    q1hats = 1-p1hats;
    p2hats = (X2+2)/(n+4);    q2hats = 1- p2hats;
Wi3 = [p1hats - norminv(0.975)*sqrt( p1hats * q1hats/(n+4)), ...
    p1hats + norminv(0.975) * sqrt( p1hats * q1hats/(n+4))];
Wi6 = [p2hats - norminv(0.975)*sqrt( p2hats * q2hats/(n+4)), ...
    p2hats + norminv(0.975) * sqrt( p2hats * q2hats/(n+4))];
% Wi3 =       0.43457         0.73617
% Wi6 =       0.092815        0.34621
%==========================
%Wilson Score Intervals
z=norminv(0.975);
Wis3 = [  1/(1 + z^2/n) *  (p1hat   + z^2/(2 * n)   - ...
    z * sqrt( p1hat * q1hat / n  + z^2/(4 * n^2))), ...
    1/(1 + z^2/n) * (p1hat   + z^2/(2 * n)   +  ...
    z * sqrt( p1hat * q1hat / n  + z^2/(4 * n^2)))];
Wis6 = [  1/(1 + z^2/n) *  (p2hat   + z^2/(2 * n)   - ...
    z * sqrt( p2hat * q2hat / n  + z^2/(4 * n^2))), ...
    1/(1 + z^2/n) * (p2hat   + z^2/(2 * n)   +  ...
    z * sqrt( p2hat * q2hat / n  + z^2/(4 * n^2)))];
%Wis3 =     0.43486        0.73653
%Wis6 =     0.0948         0.34205
%==========================
```

```
% Clopper-Pearson Intervals
        Fs = finv(0.975, 2*(n-X1 + 1), 2*X1);
        Fss = finv(0.975, 2*(X1+1), 2*(n-X1));
        CP3 = [ X1/(X1 + (n-X1+1).*Fs),  ...
           (X1+1).*Fss./(n - X1 + (X1+1).*Fss)];

        Fs = finv(0.975, 2*(n-X2 + 1), 2*X2);
        Fss = finv(0.975, 2*(X2+1), 2*(n-X2));
        CP6 = [ X2/(X2 + (n-X2+1).*Fs), ...
           (X2+1).*Fss./(n - X2 + (X2+1).*Fss)];
%CP3 = 0.421            0.75246
%CP6 = 0.079621         0.35155
%==============================================
% Anscombe ARCSIN intervals
%
p1h = (X1 + 3/8)/(n + 3/4);   p2h = (X2 + 3/8)/(n + 3/4);

AA3 = [(sin(asin(sqrt(p1h))-norminv(0.975)/(2*sqrt(n))))^2, ...
       (sin(asin(sqrt(p1h))+norminv(0.975)/(2*sqrt(n))))^2];
AA6 = [(sin(asin(sqrt(p2h))-norminv(0.975)/(2*sqrt(n))))^2, ...
       (sin(asin(sqrt(p2h))+norminv(0.975)/(2*sqrt(n))))^2];

%AA3 = 0.43235          0.74353
%AA6 = 0.085489         0.3366
```

Figure 7.9 shows the pairs of confidence intervals at the 3- and 6-month follow-ups. Wald's intervals are in black, Wilson's in red, the Wilson score in green, Clopper–Pearson's in magenta, and ArcSin in blue. Notice that for all methods, the confidence intervals at the 3- and 6-month follow-ups are well separated, suggesting a significant change in the proportions. There are differences among the intervals, in their centers and lengths, for a particular time of follow-up. However, as Fig. 7.9 indicates, these differences are not large.
✐

How does one find the confidence interval for the probability of success if in $n$ trials no successes have been observed?

### 7.4.4 Confidence Intervals for Proportions When $X = 0$

When the binomial probability is small, then it is not unusual that out of $n$ trials no successes have been observed. How do we find a $(1 - \alpha)100\%$ confidence interval in such a case? The Clopper–Pearson interval is possible for $X = 0$, and it is given by $[0, 1 - (\alpha/2)^{1/n}]$.

An alternative interval can be established from the following consideration. Note that $(1 - p)^n$ is the probability of no success in $n$ trials, and this probability is at least $\alpha$:

$$(1 - p)^n \geq \alpha.$$

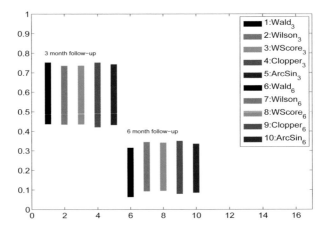

**Fig. 7.9** Confidence intervals at 3- and 6-month follow-ups. Wald's intervals are in *black*, Wilson's in *red*, the Wilson Score in *green*, Clopper–Pearson's in *magenta*, and ArcSin in *blue*.

Since $n \log(1 - p) \geq \log(\alpha)$ and $\log(1 - p) \approx -p$, then

$$p \leq -\log(\alpha)/n.$$

This is a basis for the so-called $3/n$ rule: 95% confidence interval for $p$ is $[0, 3/n]$ if no successes have been observed since $-\log(0.05) = 2.9957 \approx 3$. By symmetry, the 95% confidence interval for $p$ when $n$ successes are observed in $n$ experiments is $[1 - 3/n, 1]$. When $n$ is small, this rule leads to intervals that are too wide to be useful. See Exercise 7.29 for a comparison of the Clopper–Pearson and $3/n$-rule intervals. We will argue in the next chapter that in the case where no successes are observed, one should approach the inference in a Bayesian manner.

### 7.4.5 Designing the Sample Size with Confidence Intervals

In all previous examples it was assumed that we had data in hand. Thus, we looked at the data after the sampling procedure had been completed. It is often the case that we have control over what sample size to adopt before the sampling. How large should the sample be? A sample that is too small may affect the validity of our statistical conclusions. On the other hand, an unnecessarily large sample wastes money, time, and resources.

The length $L$ of the $(1 - \alpha)100\%$ confidence interval is $L = 2z_{1-\alpha/2}\sigma/\sqrt{n}$ for the normal mean and $L = 2z_{1-\alpha/2}\sqrt{\hat{p}(1 - \hat{p})/n}$ for the population proportion.

The sample size $n$ is determined by solving the above equations when $L$ is fixed.

(i) Sample size for estimating the mean: $\sigma^2$ is known:

$$n \geq \frac{4z_{1-\alpha/2}^2 \sigma^2}{L^2},$$

where $L$ is the length of the interval.

(ii) Sample size for estimating the proportion:

$$n \geq \frac{4z_{1-\alpha/2}^2 \hat{p}(1-\hat{p})}{L^2},$$

where $\hat{p}$ is the sample proportion.

Designing the sample size usually precedes the sampling. In the absence of data, $\hat{p}$ is our best guess. In the absence of any information, the most conservative choice is $\hat{p} = 0.5$.

It is possible to express $L^2$ in the units of variance of observations, $\sigma^2$, for the normal and $p(1-p)$ for the Bernoulli distribution. Therefore, it is sufficient to state that $L/\sigma$ is 1/2, for example, or that $L/\sqrt{p(1-p)}$ is 1/4, and the required sample size can be calculated.

*Example 7.11.* **Cholesterol Level.** Suppose you are designing a cholesterol study experiment and would like to estimate the mean cholesterol level of all students on a large metropolitan campus. You plan to take a random sample of $n$ students and measure their cholesterol levels. Previous studies have shown that the standard deviation is 25, and you intend to use this value in planning your study.

If a 99% confidence interval with a total length not exceeding 12 is desired, how many students should you include in your sample?

**Sol.** For a 99% confidence level, the normal 0.995 quantile is needed, $z_{0.995} = 2.58$. Then, $n \geq \frac{4 \cdot 2.5758^2 \cdot 25^2}{12^2} = 115.1892$, and the sample size is 116 since 115.1892 should be rounded to the closest larger integer.

📝

The *margin of error* is defined as half of the length of a 95% confidence interval for unknown proportion, location, scale, or some other population parameter of interest.

In popular use, however, margin of error is usually connected with public opinion polls and represents the quantifiable sampling error built into well-designed sampling schemes. For estimating the true proportion of voters favoring a particular candidate, an approximate 95% confidence interval is

$$\left[ \hat{p} - 1.96\sqrt{\hat{p}\,\hat{q}/n},\ \hat{p} + 1.96\sqrt{\hat{p}\,\hat{q}/n} \right],$$

where $\hat{p}$ is the sample proportion of voters favoring the candidate, $\hat{q} = 1 - \hat{p}$, 1.96 is the normal 97.5 percentile, and $n$ is the sample size. Since $\hat{p}\,\hat{q} \leq 1/4$, the margin of error, $1.96\sqrt{\hat{p}\,\hat{q}/n}$, is usually conservatively rounded to $\frac{1}{\sqrt{n}}$.

For example, if a survey of $n = 1600$ voters yields that 52% favor a particular candidate, then the margin of error can be estimated as $1/\sqrt{1600} = 1/40 = 0.025 = 2.5\%$ and is independent of the realized proportion of 52%.

However, if the true proportion is not close to 1/2, the precision of the margin of error can be improved by selecting a less conservative upper bound on $\hat{p}\,\hat{q}$. For example, if a survey of $n = 1600$ citizens yields that 16% of them favor policy $P$, the margin of error can be estimated as $1.96 \cdot \sqrt{0.2 \cdot 0.8/1600} \approx 1/50 = 0.02 = 2\%$ if we are sure that the true proportion of citizens supporting policy $P$ does not exceed 20%.

## 7.5 Prediction and Tolerance Intervals*

In addition to confidence intervals for the parameters, a researcher may be interested in predicting future observations. This leads to prediction intervals.

We will exemplify the prediction interval for predicting future observations from a normal population $\mathcal{N}(\mu,\sigma^2)$ once $X_1,\ldots,X_n$ have been observed. Denote the future observation by $X_{n+1}$.

Consider $\overline{X}$ and $X_{n+1}$. These two random variables are independent and their difference has a normal distribution,

$$\overline{X} - X_{n+1} \sim \mathcal{N}(0,\sigma^2/n + \sigma^2),$$

thus, $Z = \frac{\overline{X} - X_{n+1}}{\sigma\sqrt{1+1/n}}$ has a standard normal distribution. This leads to $(1-\alpha)100\%$ prediction intervals for $X_{n+1}$:

If $\sigma^2$ is known, then

$$\left[ \overline{X} - z_{1-\alpha/2}\,\sigma\sqrt{1+\frac{1}{n}},\ \overline{X} + z_{1-\alpha/2}\,\sigma\sqrt{1+\frac{1}{n}} \right].$$

If $\sigma^2$ is not known, then

$$\left[\overline{X} - t_{n-1,1-\alpha/2}\ s\ \sqrt{1 + \frac{1}{n}},\ \overline{X} + t_{n-1,1-\alpha/2}\ s\ \sqrt{1 + \frac{1}{n}}\right].$$

Note that prediction intervals contain the factor $\sqrt{1 + \frac{1}{n}}$ in place of $\sqrt{\frac{1}{n}}$ in matching confidence intervals for the normal population mean. When $n$ is large, the prediction interval can be substantially larger than the confidence interval. This is because the uncertainty about the future observation has two parts: (1) uncertainty about its mean plus (2) uncertainty about the individual response.

Tolerance intervals place bounds on fixed portions of population distributions with a specified confidence. For example, the question *What interval will contain 95% of the population measurements with 99% confidence?* is answered by a tolerance interval. The ends of a tolerance interval are called tolerance limits. A manufacturer of medical devices might be interested in the proportion of production for which a particular dimension falls within a given range. For normal populations, the two-sided interval is defined as

$$[\overline{X} - ks, \overline{X} + ks], \qquad k = \sqrt{\frac{(n^2 - 1)\ z^2_{1-\gamma/2}}{n\ \chi^2_{n-1,\alpha}}} \qquad (7.8)$$

and interpreted as follows. With a confidence of $1 - \alpha$, the proportion of $1 - \gamma$ of population measurements will fall between the lower and upper bounds in (7.8).

*Example 7.12.* For sample size $n = 20$, $\overline{X} = 12$, $s = 0.1$, confidence $1 - \alpha = 99\%$, and proportion $1 - \gamma = 95\%$, the tolerance factor $k$ is calculated as in the following MATLAB script:

```
n=20;
z = norminv(1-0.05/2)  %proportion of 1-0.05=0.95
  %z =  1.9600
xi = chi2inv(0.01, n-1)  %confidence 1-0.01=0.99
  %xi = 7.6327
k = sqrt( (n^2-1) * z^2/(n * xi) )
  %k = 3.1687
[12-k*0.1   12+k*0.1]
  %11.6831   12.3169
```

and the tolerance interval is [11.6831, 12.3169].

## 7.6 Confidence Intervals for Quantiles*

The confidence interval for a normal quantile is based on a noncentral $t$ distribution. Let $X_1, \dots, X_n$ be a sample of size $n$ with the sample mean $\overline{X}$ and sample standard deviation $s$.

It is of interest to find a confidence interval on the population's $p$th quantile, $\mu + z_p \times \sigma$, with a confidence level of $1 - \alpha$.

The confidence interval is given by $[L, U]$, where

$$L = \overline{X} + s \cdot nct(\alpha/2, n-1, \sqrt{n} \cdot z_p)/\sqrt{n},$$
$$U = \overline{X} + s \cdot nct(1 - \alpha/2, n-1\sqrt{n} \cdot z_p)/\sqrt{n},$$

and $nct(q, df, nc)$ is the $q$-quantile of the noncentral $t$ distribution (p. 217) with $df$ the degrees of freedom and noncentrality parameter $nc$.

The confidence intervals for quantiles can be based on order statistics when normality is not assumed. For example, instead of declaring the confidence interval for the mean, one should report the confidence interval on the median if the normality of the data is a concern. Let $X_{(1)}, X_{(2)}, \dots, X_{(n)}$ be the order statistics of the sample. Then a $(1 - \alpha)100\%$ confidence interval for the median $Me$ is

$$X_{(h)} \le Me \le X_{(n-h+1)}.$$

The value of $h$ is usually given by tables. For large $n$ $(n > 40)$ a good approximation for $h$ is an integer part of

$$\frac{n - z_{1-\alpha/2}\sqrt{n} - 1}{2}.$$

As an illustration, if $n = 300$, the 95% confidence interval for the median is $[X_{(132)}, X_{(169)}]$ since

```
n = 300;
h = floor( (n - 1.96 * sqrt(n) - 1)/2 )
% h = 132
n - h + 1
% ans =  169
```

## 7.7 Confidence Intervals for the Poisson Rate*

Recall that an observation $X$ coming from $\mathscr{P}oi(\lambda)$ has both a mean and a variance equal to the rate parameter, $\mathbb{E}X = \mathbb{V}\mathrm{ar}\,X = \lambda$, and that Poisson random variables are "additive in the rate parameter":

$$X_1, X_2, \ldots, X_n \sim \mathscr{P}oi(\lambda) \quad \Rightarrow n\overline{X} = \sum_{i=1}^{n} X_i \sim \mathscr{P}oi(n\lambda). \tag{7.9}$$

The asymptotically shortest Wald-type $(1-\alpha)100\%$ interval for $\lambda$ is obtained using the fact that $Z = \sqrt{\frac{n}{\lambda}}\,(\overline{X} - \lambda)$ is approximately the standard normal. The inequality

$$\sqrt{\frac{n}{\lambda}}\,|\overline{X} - \lambda| \leq z_{1-\alpha/2}$$

leads to

$$\lambda^2 - \lambda\left(2\overline{X} + \frac{z_{1-\alpha/2}^2}{n}\right) + (\overline{X})^2 \leq 0,$$

which solves to

$$\left[\overline{X} + \frac{z_{1-\alpha/2}^2}{2n} - z_{1-\alpha/2}\sqrt{\frac{\overline{X}}{n} + \frac{z_{1-\alpha/2}^2}{4n^2}}, \quad \overline{X} + \frac{z_{1-\alpha/2}^2}{2n} + z_{1-\alpha/2}\sqrt{\frac{\overline{X}}{n} + \frac{z_{1-\alpha/2}^2}{4n^2}}\right].$$

Other Wald-type intervals are derived from the fact that $\frac{\sqrt{\overline{X}} - \sqrt{\lambda}}{\sqrt{1/(4n)}}$ is approximately the standard normal. Variance stabilizing, modified variance stabilizing, and recentered variance stabilizing $(1-\alpha)100\%$ confidence intervals are given as (Barker, 2002)

$$\left[\overline{X} - z_{1-\alpha/2}\sqrt{\frac{\overline{X}}{n}}, \quad \overline{X} + z_{1-\alpha/2}\sqrt{\frac{\overline{X}}{n}}\right],$$

$$\left[\overline{X} + \frac{z_{1-\alpha/2}^2}{4n} - z_{1-\alpha/2}\sqrt{\frac{\overline{X}}{n}}, \quad \overline{X} + \frac{z_{1-\alpha/2}^2}{4n} + z_{1-\alpha/2}\sqrt{\frac{\overline{X}}{n}}\right],$$

$$\left[\overline{X} + \frac{z_{1-\alpha/2}^2}{4n} - z_{1-\alpha/2}\sqrt{\frac{\overline{X} + 3/8}{n}}, \quad \overline{X} + \frac{z_{1-\alpha/2}^2}{4n} + z_{1-\alpha/2}\sqrt{\frac{\overline{X} + 3/8}{n}}\right].$$

An alternative approach is based on the link between Poisson and $\chi^2$ distributions. Namely, if $X \sim \mathscr{P}oi(\lambda)$, then

$$\mathbb{P}(X > x) = \mathbb{P}(Y < 2\lambda), \text{ for } Y \sim \chi^2_{2x}$$

and the $(1 - \alpha)100\%$ confidence interval for $\lambda$ when $X$ is observed is

$$\left[ \frac{1}{2} \chi^2_{2X,\alpha/2}, \ \frac{1}{2} \chi^2_{2(X+1),1-\alpha/2} \right],$$

where $\chi^2_{2X,\alpha/2}$ and $\chi^2_{2(X+1),1-\alpha/2}$ are $\alpha/2$ and $1 - \alpha/2$ quantiles of the $\chi^2$ distribution with $2X$ and $2(X + 1)$ degrees of freedom, respectively. Due to the additivity property (7.9), the confidence interval changes slightly for the case of an observed sample of size $n$, $X_1, X_2, \ldots, X_n$. One finds $S = \sum_{i=1}^{n} X_i$, which is a Poisson with parameter $n\lambda$ and proceeds as in the single-observation case. The interval obtained is for $n\lambda$ and the bounds should be divided by $n$ to get the interval for $\lambda$:

$$\left[ \frac{1}{2n} \chi^2_{2S,\alpha/2}, \ \frac{1}{2n} \chi^2_{2(S+1),1-\alpha/2} \right].$$

*Example 7.13.* **Counts of $\alpha$-Particles.** Rutherford et al. (1930, pp. 171–172) provide descriptions and data on an experiment by Rutherford and Geiger (1910) on the collision of $\alpha$-particles emitted from a small bar of polonium with a small screen placed at a short distance from the bar. The number of such collisions in each of 2608 eight-minute intervals was recorded. The distance between the bar and screen was gradually decreased so as to compensate for the decay of radioactive substance.

| $X$ | 0 | 1 | 2 | 3 | 4 | 5 | 6 | 7 | 8 | 9 | 10 | 11 | $\geq 12$ |
|---|---|---|---|---|---|---|---|---|---|---|---|---|---|
| Frequency | 57 | 203 | 383 | 525 | 532 | 408 | 273 | 139 | 45 | 27 | 10 | 4 | 2 |

It is postulated that because of the large number of atoms in the bar and the small probability of any of them emitting a particle, the observed frequencies should be well modeled by a Poisson distribution.

```
%Rutherford.m
X=[ 0  1  2  3  4  5  6  7  8  9  10  11  12 ];
fr=[ 57 203  383  525  532  408  273  139  45  27  10  4  2];
n = sum(fr); %number of experiments//time intervals
rfr = fr./n;       %relative frequencies %n=2608
xbar = X * rfr' ;  %lambdahat = xbar =  3.8704
tc = X * fr';      %total number of counts  tc = 10094
%Recentered Variance Stabilizing
[xbar + (norminv(0.975))^2/(4*n)  - ...
        norminv(0.975) * sqrt(( xbar + 3/8)/n )...
 xbar + (norminv(0.975))^2/(4*n)  + ...
        norminv(0.975) * sqrt( (xbar+ 3/8)/n )]
             %    3.7917    3.9498
% Poisson/Chi2 link
[1/(2 *n) * chi2inv(0.025, 2 * tc)  ...
       1/(2 * n) * chi2inv(0.975, 2*(tc + 1))]
       %  3.7953    3.9467
```

The estimator for $\lambda$ is $\hat{\lambda} = \overline{X} = 3.8704$, the Wald-type recentered variance stabilizing interval is $[3.7917, 3.9498]$, and the Pearson/chi-square link confidence interval is $[3.7953, 3.9467]$. The intervals are very close to each other and quite tight due to the large sample size.

✎

**Sample Size for Estimating $\lambda$.** Assume that we are interested in estimating the Poisson rate parameter $\lambda$ from the sample $X_1, \ldots, X_n$. Chen (2008) provided a simple way of finding the sample size $n$ necessary for $\hat{\lambda} = \frac{1}{n} \sum_{i=1}^{n} X_i$ to satisfy

$$P(\{|\hat{\lambda} - \lambda| < \epsilon_a\} \text{ or } \{|\hat{\lambda}/\lambda - 1| < \epsilon_r\}) > 1 - \delta$$

for $\delta, \epsilon_r$, and $\epsilon_a$ small. The minimal sample size is

$$n > \frac{\epsilon_r}{\epsilon_a} \frac{\ln 2 - \ln \delta}{(1 + \epsilon_r) \ln(1 + \epsilon_r) - \epsilon_r}.$$

For example, the sample size needed to ensure that the absolute error is less than 0.5 or that the relative error is less than 5% with a probability of 0.99 is $n = 431$.

```
delta = 0.01; epsa =0.5; epsr = 0.05;
n = epsr/epsa * log(2/delta)/((1+epsr)*log(1+epsr)-epsr)
% n = 430.8723
```

Note that both tolerable relative and absolute errors, $\epsilon_r$ and $\epsilon_a$, need to be specified since the precision criterion involves the probability of a union.

## 7.8 Exercises

7.1. **Tricky Estimation.** A publisher gives the proofs of a new book to two different proofreaders, who read it separately and independently. The first proofreader found 60 misprints, the second proofreader found 70 misprints, and 50 misprints were found by both. Estimate how many misprints remain undetected in the book? *Hint:* Refer to Example 5.10.

7.2. **Laplace's Rule of Succession.** Laplace's Rule of Succession states that if an event appeared $X$ times out of $n$ trials, the probability that it will appear in a future trial is $\frac{X+1}{n+2}$.

(a) If $\frac{X+1}{n+2}$ is taken as an estimator for binomial $p$, compare the MSE of this estimator with the MSE of the traditional estimator, $\hat{p} = \frac{X}{n}$.

(b) Represent MSE from (a) as the sum of the estimator's variance and the bias squared.

7.3. **Neurons Fire in Potter's Lab.** The data set ⬛ neuronfires.mat was compiled by student Ravi Patel while working in Professor Steve Potter's lab at Georgia Tech. It consists of 989 firing times of a cell culture of neurons. The recorded firing times are time instances when a neuron sent a signal to another linked neuron (a spike). The cells, from the cortex of an embryonic rat brain, were cultured for 18 days on multielectrode arrays. The measurements were taken while the culture was stimulated at the rate of 1 Hz. It was postulated that firing times form a Poisson process; thus interspike intervals should have an exponential distribution.

(a) Calculate the interspike intervals $T$ using MATLAB's diff command. Check the histogram for $T$ and discuss its resemblance to the exponential density. By the moment matching estimator argue that exponential parameter $\lambda$ is close to 1.

(b) According to (a), the model for interspike intervals is $T \sim \mathcal{E}(1)$. You are interested in the proportion of intervals that are shorter than 3, $T \le 3$. Find this proportion from the theoretical model $\mathcal{E}(1)$ and compare it to the estimate from the data. For the theoretical model, use expcdf and for empirical data use sum(T <= 3)/length(T).

7.4. **The MLE in a Discrete Case.** A sample $-1,1,1,0,-1,1,1,1,0,1,1,0,-1,1,1$ was observed from a population with a probability mass function of

$$\begin{array}{c|ccc} X & -1 & 0 & 1 \\ \hline \text{Prob} & \theta & 2\theta & 1-3\theta \end{array}$$

(a) What is the possible range for $\theta$?

(b) What is the MLE for $\theta$.

(c) How would the MLE look like for a sample of size $n$?

7.5. **MLE for Two Continuous Distributions.** Find the MLE for parameter $\theta$ if the model for observations $X_1, X_2, \ldots, X_n$, is

$$\text{(a)} \quad f(x|\theta) = \frac{\theta}{x^2}, \quad 0 < \theta \le x;$$

$$\text{(b)} \quad f(x|\theta) = \frac{\theta-1}{x^\theta}, \quad x \ge 1, \ \theta > 1.$$

7.6. **Match the Moment.** The geometric distribution ($X$ is the number of failures before the first success) has a probability mass function of

$$f(x|p) = q^x p, \quad x = 0, 1, 2, \ldots.$$

Suppose $X_1, X_2, \ldots, X_n$ are observations from this distribution. It is known that $\mathbb{E}X_i = \frac{1-p}{p}$. What would you report as the moment-matching estimator if the sample $X_1 = 2, X_2 = 6, X_3 = 1$ were observed?
What is the MLE for $p$?

7.7. **Weibull Distribution.** The two-parameter Weibull distribution is given by the density

$$f(x) = a\lambda^a x^{a-1} e^{-(\lambda x)^a}, \ a > 0, \lambda > 0, x \geq 0,$$

with mean and variance

$$\mathbb{E}X = \frac{\Gamma(1 + 1/a)}{\lambda}, \quad \text{and } \mathbb{V}\text{ar}\, X = \frac{1}{\lambda^2}\left[\Gamma(1 + 2/a) - \Gamma(1 + 1/a)^2\right].$$

Assume that the "shape" parameter $a$ is known and equal to 1/2.
(a) Propose two moment-matching estimators for $\lambda$.
(b) If $X_1 = 1, X_2 = 3, X_3 = 2$, what are the values of the estimator?

*Hint:* Recall that $\Gamma(n) = (n-1)!$

7.8. **Rate Parameter of Gamma.** Let $X_1, \ldots, X_n$ be a sample from a gamma distribution given by the density

$$f(x) = \frac{\lambda^a x^{a-1}}{\Gamma(a)} e^{-\lambda x}, \ a > 0, \lambda > 0, x \geq 0,$$

where shape parameter $a$ is known and rate parameter $\lambda$ is unknown and of interest.
(a) Find the MLE of $\lambda$.
(b) Using the fact that $X_1 + X_2 + \cdots + X_n$ is also gamma distributed with parameters $na$ and $\lambda$, find the expected value of MLE from (a) and show that it is a biased estimator of $\lambda$.
(c) Modify the MLE so that it is unbiased. Compare MSEs for the MLE and the modified estimator.

7.9. **Estimating the Parameter of a Rayleigh Distribution.** If two random variables $X$ and $Y$ are independent of each other and normally distributed with variances equal to $\sigma^2$, then the variable $R = \sqrt{X^2 + Y^2}$ follows the Rayleigh distribution with scale parameter $\sigma$. An example of such a variable would be the distance of darts from the target in a dart-throwing game where the deviations in the two dimensions of the target plane are independent and normally distributed. The Rayleigh random variable $R$ has a density

$$f(r) = \frac{r}{\sigma^2} \exp\left\{-\frac{r^2}{2\sigma^2}\right\}, \quad r \geq 0,$$

$$\mathbb{E}R = \sigma\sqrt{\frac{\pi}{2}} \quad \mathbb{E}R^2 = 2\sigma^2.$$

(a) Find the two moment-matching estimators of $\sigma$.

(b) Find the MLE of $\sigma$.

(c) Assume that $R_1 = 3, R_2 = 4, R_3 = 2$, and $R_4 = 5$ are Rayleigh-distributed random observations representing the distance of a dart from the center. Estimate the variance of the horizontal error, which is theoretically a zero-mean normal.

(d) In Example 5.20, the distribution of a square root of an exponential random variable with a rate parameter $\lambda$ was Rayleigh with the following density:

$$f(r) = 2\lambda r \exp\{-\lambda r^2\}.$$

To find the MLE for $\lambda$, can you use the MLE for $\sigma$ from (b)?

7.10. **Monocytes Among Blood Cells.** Eisenhart and Wilson (1943) report the number of monocytes in 100 blood cells of a cow in 113 successive weeks.

| Monocytes | Frequency | Monocytes | Frequency |
|---|---|---|---|
| 0 | 0 | 7 | 12 |
| 1 | 3 | 8 | 10 |
| 2 | 5 | 9 | 11 |
| 3 | 13 | 10 | 7 |
| 4 | 19 | 11 | 3 |
| 5 | 13 | 12 | 2 |
| 6 | 15 | 13+ | 0 |

(a) If the underlying model is Poisson, what is the estimator of $\lambda$?

(b) If the underlying model is Binomial $\mathcal{B}in(100, p)$, what is the estimator of $p$?

(c) For the models specified in (a) and (b) find theoretical or "expected" frequencies.

*Hint:* Suppose the model predicts $\mathbb{P}(X = k) = p_k$, $k = 0, 1, \ldots, 13$. The expected frequency of $X = k$ is $113 \times p_k$.

7.11. **Estimation of $\theta$ in $\mathcal{U}(0, \theta)$.** Which of the two estimators in Example 7.5 is unbiased? Find the MSE of both estimators. Which one has a smaller MSE?

7.12. **Estimating the Rate Parameter in a Double Exponential Distribution.** Let $X_1, \ldots, X_n$ follow double exponential distribution with density

$$f(x|\theta) = \frac{\theta}{2} e^{-\theta|x|}, \quad -\infty < x < \infty, \, \theta > 0.$$

For this distribution, $\mathbb{E}X = 0$ and $\mathbb{V}\mathrm{ar}(X) = \mathbb{E}X^2 = 2/\theta^2$. The double exponential distribution, also known as Laplace's distribution, is a ubiquitous model in statistics. For example, it models the difference of two exponential variates, or absolute value of an exponential random variable, etc.

(a) Find a moment-matching estimator for $\theta$.

(b) Find the MLE of $\theta$.

(c) Evaluate the two estimators from (a) and (b) for a sample $X_1 = -2, X_2 = 3, X_3 = 2$ and $X_4 = -1$.

7.13. **Reaction Times I.** A sample of 20 students is randomly selected and given a test to determine their reaction time in response to a given stimulus. Assume that individual reaction times are normally distributed. If the mean reaction time is determined to be $\overline{X} = 0.9$ (in seconds) and the standard deviation is $s = 0.12$, find

(a) The 95% confidence interval for the unknown population mean $\mu$;

(b) The 98.5% confidence interval for the unknown population mean $\mu$;

(c) The 95% confidence interval for the unknown population variance $\sigma^2$.

7.14. **Reaction Times II.** Under the conditions in the previous problem, assume that the population standard deviation was known to be $\sigma = 0.12$. Find

(a) The 98.5% confidence interval for the unknown mean $\mu$;

(b) The sample size necessary to produce a 95% confidence interval for $\mu$ of length 0.07.

7.15. **Toxins.** An investigation on toxins produced by molds that infect corn crops was performed. A biochemist prepared extracts of the mold culture with organic solvents and then measured the amount of toxic substance per gram of solution. From 11 preparations of the mold culture the following measurements of the toxic substance (in milligrams) were obtained: 3, 2, 5, 3, 2, 6, 5, 4.5, 3, 3, and 4.

Compute a 99% confidence interval for the mean weight of toxic substance per gram of mold culture. State the assumption you make about the population.

7.16. **Bias of $s_*^2$.** For a sample $X_1, \ldots, X_n$ from a $\mathcal{N}(\mu, \sigma^2)$ population, find the bias of $s_*^2 = \frac{1}{n}\sum_i (X_i - \overline{X})^2$ as an estimator of variance $\sigma^2$.

7.17. **COPD Patients.** Acute exacerbations of disease symptoms in patients with chronic obstructive pulmonary disease (COPD) often lead to hospitalization and impose a great financial burden on the health care system. The study by Ghanei et al. (2007) aimed to determine factors that may predict rehospitalization in COPD patients.

A total of 157 COPD patients were randomly selected from all COPD patients admitted to the chest clinic of Baqiyatallah Hospital during the year 2006. Subjects were followed for 12 months to observe the occurrence of any disease exacerbation that might lead to hospitalization. Over the 12-month period, 87 patients experienced disease exacerbation. The authors found significant associations between COPD exacerbation and monthly income, comorbidity score, and depression using logistic regression tools. We are not interested in these associations in this exercise, but we are interested in the population proportion of all COPD patients that experienced disease exacerbation over a 12-month period, $p$.

(a) Find an estimator of $p$ based on the data available. What is an approximate distribution of this estimator?

(b) Find the 90% confidence interval for the unknown proportion $p$.

(c) How many patients should be sampled and monitored so that the 90% confidence interval as in (b) does not exceed 0.03 in length.

(d) The hospital challenges the claim by the local health system authorities that half of the COPD patients experience disease exacerbation in a 1-year period, claiming that the proportion is significantly higher. Can the hospital support their claim based on the data available? Use $\alpha = 0.05$. Would you reverse the decision if $\alpha$ were changed to 10%?

7.18. **Right to Die.** A Gallup Poll estimated the support among Americans for "right to die" laws. In the survey, 1528 adults were asked whether they favor voluntary withholding of life-support systems from the terminally ill. The results: 1238 said YES.

(a) Find the 99% confidence interval for the percentage of all adult Americans who are in favor of "right to die" laws.

(b) If the margin of error[1] is to be smaller than 0.01, what sample size is needed to achieve this requirement? Assume $\hat{p} = 0.8$.

7.19. **Exponentials Parameterized by the Scale.** A sample $X_1,\ldots,X_n$ was selected from a population that has an exponential $\mathcal{E}(\lambda)$ distribution with a density of $f(x|\lambda) = \frac{1}{\lambda}e^{-\frac{x}{\lambda}}$, $x \geq 0, \lambda > 0$. We are interested in estimating the parameter $\lambda$.

(a) What are the moment-matching and MLE estimators of $\lambda$ based on $X_1,\ldots,X_n$?

(b) Two independent observations $Y_1 \sim \mathcal{E}(\lambda/2)$ and $Y_2 \sim \mathcal{E}(2\lambda)$ are available. Combine them (make a specific linear combination) to obtain an unbiased estimator of $\lambda$. What is the variance of the proposed estimator?

(c) Two independent observations $Z_1 \sim \mathcal{E}(1.1\lambda)$ and $Z_2 \sim \mathcal{E}(0.9\lambda)$ are available. An estimator of $\lambda$ in the form $\hat{\lambda} = pZ_1 + (1-p)Z_2$, $0 \leq p \leq 1$ is proposed. What $p$ minimizes the magnitude of bias of $\hat{\lambda}$? What $p$ minimizes the variance of $\hat{\lambda}$?

7.20. **Bias in Estimator for Exponential $\lambda$ Distribution.** If the exponential distribution is parameterized with $\lambda$ as the scale parameter, $f(x|\lambda) = \frac{1}{\lambda}\exp\{-x/\lambda\}$, $x \geq 0, \lambda > 0$, (as in MATLAB), then $\hat{\lambda} = \overline{X}$ is an unbiased estimator of $\lambda$. However, if it is parameterized with $\lambda$ as a rate parameter, $f(x|\lambda) = \lambda\exp\{-\lambda x\}$, $x \geq 0, \lambda > 0$, then $\hat{\lambda} = 1/\overline{X}$ is biased. Find the bias of this

---

[1] There are several definitions for margin of error, the most common one is *half of the length of a 95% confidence interval*.

estimator. *Hint:* Argue that $1/\sum_{i=1}^{n} X_i$ has an inverse gamma distribution with parameters $n$ and $\lambda$ and take the expectation.

7.21. **Yucatan Miniature Pigs.** Ten adult male Yucatan miniature pigs were exposed to various durations of constant light ("Lighting"), then sacrificed after experimentally controlled time delay ("Survival"), as described in Dureau et al. (1996). Following the experimental protocol, entire eyes were fixed in Bouin's fixative for 3 days. The anterior segment (cornea, iris, lens, ciliary body) was then removed and the posterior segment divided into five regions: Posterior pole (including optic nerve head and macula) ("P"), nasal ("N"), temporal ("T"), superior ("S"), and inferior ("I"). Specimens were washed for 2 days, embedded in paraffin, and subjected to microtomy perpendicular to the retinal surface. Every 200 $\mu$m, a 10-$\mu$m-thick section was selected and 20 sections were kept for each retinal region. Sections were stained with hematoxylin. The outer nuclear layer (ONL) thickness was measured by an image-analyzing system (Biocom, Les Ulis, France), and three measures were performed for each section at regularly spaced intervals, so that 60 measures were made for each retinal region. The experimental protocol for 11 animals was as follows (Lighting and Survival times are in weeks).

| Animal | Lighting duration | Survival time |
|---|---|---|
| Control | 0 | 0 |
| 1 | 1 | 12 |
| 2 | 2 | 10 |
| 3 | 4 | 0 |
| 4 | 4 | 4 |
| 5 | 4 | 6 |
| 6 | 8 | 0 |
| 7 | 8 | 4 |
| 8 | 8 | 8 |
| 9 | 12 | 0 |
| 10 | 12 | 4 |

The data set 📀 pigs.mat contains the data structure pigs with pigs.pc, pigs.p1,...,pigs.p10, representing the posterior pole measurements for the 11 animals. This data set and complete data yucatanpigs.dat can be found on the book's Web site page.

Observe the data pigs.pc and argue that it deviates from normality by using MATLAB's qqplot. Transform pigs.pc as x = (pigs.pc - 14)/(33 -14), to confine $x$ between 0 and 1 and assume a beta $\mathcal{B}e(a,a)$ distribution. The MLE for $a$ is complex (involves a numerical solution of equations with digamma functions) but the moment matching estimator is straightforward.

Find a moment-matching estimator for $a$.

7.22. **Computer Games.** According to Hamilton (1990), certain computer games are thought to improve spatial skills. A mental rotations test, measuring spatial skills, was administered to a sample of school children after they had played one of two types of computer game. Construct 95% confidence intervals based on the following mean scores, assuming that the children were selected randomly and that the mental rotations test scores had a normal distribution in the population.
(a) After playing the "Factory" computer game: $\overline{X} = 22.47, s = 9.44, n = 19$.
(b) After playing the "Stellar" computer game: $\overline{X} = 22.68, s = 8.37, n = 19$.
(c) After playing no computer game (control group): $\overline{X} = 18.63, s = 11.13, n = 19$.

7.23. **Effectiveness in Treating Cerebral Vasospasm.** In a study on the effectiveness of hyperdynamic therapy in treating cerebral vasospasm, Pritz et al. (1996) reported on the therapy where success was defined as clinical improvement in terms of neurological deficits. The study reported 16 successes out of 17 patients. Using the methods discussed in the text find 95% confidence intervals for the success rate. Does any of the methods produce an upper bound larger than 1?

7.24. **Alcoholism and the Blyth–Still Confidence Interval.** Genetic markers were observed for a group of 50 Caucasian alcoholics in a study that aimed at determining whether alcoholism has (in part) a genetic basis. The antigen (marker) B15 was present in 5 alcoholics. Find the Blyth–Still 99% confidence interval for the proportion of Caucasian alcoholics having this antigen.

> If either $p$ or $q$ is close to 0, then a precise $(1 - \alpha)100\%$ confidence interval for the unknown proportion $p$ was proposed by Blyth and Still (1983). For $X \sim \mathcal{B}in(n, p)$,
>
> $$\left[ \frac{(X - 0.5) + \frac{z^2_{1-\alpha/2}}{2} - z_{1-\alpha/2}\sqrt{(X - 0.5) - \frac{(X-0.5)^2}{n} + \frac{z^2_{1-\alpha/2}}{4}}}{n + z^2_{1-\alpha/2}}, \right.$$
>
> $$\left. \frac{(X + 0.5) + \frac{z^2_{1-\alpha/2}}{2} + z_{1-\alpha/2}\sqrt{(X + 0.5) - \frac{(X+0.5)^2}{n} + \frac{z^2_{1-\alpha/2}}{4}}}{n + z^2_{1-\alpha/2}} \right].$$

7.25. **Spores of** *Amanita phalloides.* Exercise 2.4 provides measurements in $\mu$m of 28 spores of the mushroom *Amanita phalloides*.

Assuming a normality of measurements, find

(a) A point estimator for the unknown population variance $\sigma^2$. What is the sampling distribution of the point estimator?

(b) A 90% confidence interval for the population variance;

(c) (By MATLAB) the minimal sample size that ensures that the upper bound $U$ of the 90% confidence interval for the variance is at most 30% larger than the lower bound $L$, that is, $U/L \leq 1.3$.

(d) Miller (1991) showed that the coefficient of variation in a normal sample of size $n$ has an approximately normal distribution:

$$s/\overline{X} \overset{approx}{\sim} \mathcal{N}\left(\frac{\sigma}{\mu}, \frac{1}{n-1}\left(\frac{\sigma}{\mu}\right)^2\left[\frac{1}{2}+\left(\frac{\sigma}{\mu}\right)^2\right]\right).$$

Based on this asymptotic distribution, a $(1-\alpha)100\%$ confidence interval for the population coefficient of variation $\frac{\sigma}{\mu}$ is approximately

$$\left(\frac{s}{\overline{X}}-z_{1-\alpha/2}\frac{s}{\overline{X}}\sqrt{\frac{1}{n-1}\left[\frac{1}{2}+\left(\frac{s}{\overline{X}}\right)^2\right]}, \ \frac{s}{\overline{X}}+z_{1-\alpha/2}\frac{s}{\overline{X}}\sqrt{\frac{1}{n-1}\left[\frac{1}{2}+\left(\frac{s}{\overline{X}}\right)^2\right]}\right).$$

This approximation works well if $n$ exceeds 10 and the coefficient of variation is less than 0.7. Find the 95% confidence interval for the population coefficient of variation $\sigma/\mu$.

7.26. **CLT-Based Confidence Interval for Normal Variance.** Refer to Example 7.9. Using MATLAB, simulate a normal sample with mean 0 and variance 1 of size $n = 50$ and find if a 95% confidence interval for the population variance contains a 1 (the true population variance). Check this coverage for a standard confidence interval in (7.7) and for a CLT-based interval from Example 7.9. Repeat this simulation $M = 10000$ times, keeping track of the number of successful coverages. Show that the interval (7.7) achieves the nominal coverage, while the CLT-based interval has a smaller coverage of about 2%. Repeat the simulation for sample sizes of $n = 30$ and $n = 200$.

7.27. **Stent Quality Control.** A stent is a tube or mechanical scaffold used to counteract significant decreases in vessel or duct diameter by acutely propping open the conduit (Fig. 7.10). Stents are often used to alleviate diminished blood flow to organs and extremities beyond an obstruction in order to maintain an adequate delivery of oxygenated blood.

In the production of stents, the quality control procedure aims to identify defects in composition and coating. Precision $z$-axis measurements (10 nm and greater) are obtained along with surface roughness and topographic surface finish details using a laser confocal imaging system (an example is the Olympus LEXT OLS3000). Samples of 50 stents from a production process are selected every hour. Typically, 1% of stents are nonconforming. Let $X$ be the number of stents in the sample of 50 that are nonconforming.

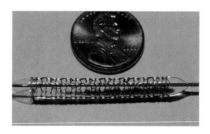

**Fig. 7.10** A stent used to restore blood flow following a heart attack, in a procedure called percutaneous coronary intervention.

A production problem is suspected if $X$ exceeds its mean by more than three standard deviations.

(a) Find the critical value for $X$ that will implicate the production problem.

(b) Find an approximation for the probability that in the next-hour batch of 50 stents, the number $X$ of nonconforming stents will be critical, i.e., will raise suspicions that the process might have a problem.

(c) Suppose now that the population proportion of nonconforming stents, $p$, is unknown. How would one estimate $p$ by taking a 50-stent sample? Is the proposed estimator unbiased?

(d) Suppose now that a batch of 50 stents produced $X = 1$. Find the 95% confidence interval for $p$.

7.28. **Right to Die.** A Gallup poll was taken to estimate the support among Americans for "right to die" laws. In the survey, 1528 adults were asked whether they favor voluntary withholding of life-support systems from the terminally ill; 1238 said yes. One is interested in conducting a similar study on a university campus.

If the margin of error is to be smaller than 2%, what sample size is needed to achieve that requirement. Assume $\hat{p} = 0.8$.

7.29. **Clopper–Pearson and $3/n$-Rule Confidence Intervals.** Using MAT-LAB compare the performance of Clopper–Pearson and $3/n$-rule confidence intervals when $X = 0$. Use $\alpha = 0.001, 0.005, 0.01, 0.05, 0.1$ and $n = 10 : 10 : 200$. Which interval is superior and under what conditions?

7.30. **Seventeen Pairs of Rats, Carbon Tetrachloride, and Vitamin B.** In a widely cited experiment by Sampford and Taylor (1959), 17 pairs of rats were formed by selecting pairs from the same litter. All rats were given carbon tetrachloride, and one rat from each pair was treated with vitamin $B_{12}$, while the other served as a control. In 7 of 17 pairs, the treated rat outlived the control rat.

(a) Based on this experiment, estimate the population proportion $p$ of pairs in which the treated rat would outlive the control rat.

(b) If the estimated proportion in (a) is the "true" population probability, what is the chance that in an independent replication of this experiment one will get exactly 7 pairs (out of 17) in which the treated rat outlives the control.

(c) Find the 95% confidence interval for the unknown $p$. Does the interval contain 1/2? What does $p = 1/2$ mean in the context of this experiment, and what do you conclude from the confidence interval.

Would the conclusion be the same if in 140 out of 340 pairs the treated rat outlived the control?

(d) The length of the 95% confidence interval based on $n = 17$ in (c) may be too large. What sample size (number of rat pairs) is needed so that the 95% confidence interval has a length of $\ell = 0.2$?

7.31. **Hemocytometer Counts.** A set of 1600 squares on a hemocytometer is inspected and the number of cells is counted in each square. The number of squares with a particular count is given in the table below:

| Count | 0 | 1 | 2 | 3 | 4 | 5 | 6 | 7 |
|---|---|---|---|---|---|---|---|---|
| # Squares | 5 | 24 | 77 | 139 | 217 | 262 | 251 | 210 |
| Count | 8 | 9 | 10 | 11 | 12 | 13 | 14 | 15 |
| # Squares | 175 | 108 | 63 | 36 | 20 | 9 | 2 | 1 |

Assume that the count has a Poisson distribution $\mathcal{P}oi(\lambda)$ and find an estimator of $\lambda$.

7.32. **Predicting Alkaline Phosphatase.** Refer to BUPA liver disorder data, BUPA.dat|mat|xlsx. The second column gives measurements of alkaline phosphatase among 345 male individuals affected by liver disorder. If variable $X$ represents the logarithm of this measurement, its distribution is symmetric and bell-shaped, so it can be assumed normal. From the data, $\overline{X} = 4.21$ and $s^2 = 0.0676$.

Suppose that a new patient with liver disorder just checked in. Find the 95% prediction interval for his log-level of alkaline phosphatase?

(a) Assume that the population variance is known and equal to 1/15.

(b) Assume that the population variance is not known.

(c) Compare the interval in (b) with a 95% confidence interval for the population mean. Why is the interval in (b) larger?

---

| MATLAB FILES AND DATA SETS USED IN THIS CHAPTER |
| http://springer.bme.gatech.edu/Ch7.Estim/ |

AmanitaCI.m, arcsinint.m, bickellehmann.m, clopperint.m, CLTvarCI.m, confintscatterpillar.m, crouxrouss.m, crouxrouss2.m, cyclosporine.m, dists2.m, ginimd.m, ginimd2.m, lfev.m, MaxwellMLE.m, MixtureModelExample.m, muscaria.m, plotlike.m, Rutherford.m, simuCI.m, tolerance.m, waldsimulation.m

neuronfires.mat

---

# CHAPTER REFERENCES

Agresti, A. and Coull, B. A. (1998). Approximate is better than "exact" for interval estimation of binomial proportions. *Am. Stat.*, **52**, 119–126.

Barker, L. (2002). A Comparison of nine confidence intervals for a Poisson parameter when the expected number of events is ≤ 5. *Am. Stat.*, **56**, 85–89.

Blyth, C. and Still, H. (1983). Binomial confidence intervals. *J. Am. Stat. Assoc.*, **78**, 108–116.

Brynskov, J., Freund, L., Rasmussen, S. N., et al. (1989). A placebo-controlled, double-blind, randomized trial of cyclosporine therapy in active chronic Crohn's disease. *New Engl. J. Med.*, **321**, 13, 845–850.

Brynskov, J., Freund, L., Rasmussen, S. N., et al. (1991). Final report on a placebo-controlled, double-blind, randomized, multicentre trial of cyclosporin treatment in active chronic Crohn's disease. *Scand. J. Gastroenterol.*, **26**, 7, 689–695.

Clopper, C. J. and Pearson, E. S. (1934). The use of confidence or fiducial limits illustrated in the case of the binomial. *Biometrika*, **26**, 404–413.

Croux, C. and Rousseeuw, P. J. (1992). Time-efficient algorithms for two highly robust estimators of scale. In Dodge, Y. and Whittaker, J.C. (eds), *Comput. Stat.*, **1**, 411–428, Physica-Verlag, Heidelberg.

Dureau, P., Jeanny, J.-C., Clerc, B., Dufier, J.-L., and Courtois, Y. (1996). Long term light-induced retinal degeneration in the miniature pig. *Mol. Vis.* **2**, 7.
http://www.emory.edu/molvis/v2/dureau.

Eisenhart, C. and Wilson, P. W. (1943). Statistical method and control in bacteriology. *Bact. Rev.*, **7**, 57–137,

Hamilton, L. C. (1990). *Modern Data Analysis, A First Course in Applied Statistics*. Brooks/Cole, Pacific Grove.

Hogg, R. V. and Tanis, E. A. (2001). *Probability and Statistical Inference*, 6th edn. Prentice-Hall, Upper Saddle River.

Miller, E. G. (1991). Asymptotic test statistics for coefficient of variation. *Comm. Stat. Theory Methods*, **20**, 10, 3351–3363.

Pritz, M. B., Zhou, X. H., and Brizendine, E. J. (1996). Hyperdynamic therapy for cerebral vasospasm: a meta-analysis of 14 studies. *J. Neurovasc. Dis.*, **1**, 6–8.

Rutherford, E., Chadwick, J., Ellis, C. D. (1930). *Radiations from Radioactive Substances*. Macmillan, London, pp. 171–172.

Rutherford, E. and Geiger, H. (1910). The probability variations in the distribution of $\alpha$-particles (with a note by H. Bateman). *Philos. Mag.*, **6**, 20, 697–707.

Sampford, M. R. and Taylor, J. (1959). Censored observations in randomized block experiments. *J. Roy. Stat. Soc. Ser. B*, **21**, 214–237.

Staudte, R. G. and Sheater, S. J. (1990). *Robust Estimation and Testing*, Wiley, New York.

Wald, A. and Wolfowitz, J. (1939). Confidence limits for continuous distribution functions. *Ann. Math. Stat.*, **10**, 105–118.

Wilcox, R. R. (2005). *Introduction to Robust Estimation and Hypothesis Testing*, 2nd edn. Academic, San Diego.

Wilson, E. B. (1927). Probable inference, the law of succession, and statistical inference. *J. Am. Stat. Assoc.*, **22**, 209–212.

# Chapter 8
# Bayesian Approach to Inference

*Bayesian: Breeding a statistician with a clergyman to produce the much sought honest statistician.*

– Anonymous

## WHAT IS COVERED IN THIS CHAPTER

- Bayesian Paradigm
- Likelihood, Prior, Marginal, Posterior, Predictive Distributions
- Conjugate Priors. Prior Elicitation
- Bayesian Computation
- Estimation, Credible Sets, Testing, Bayes Factor, Prediction

## 8.1 Introduction

Several paradigms provide a basis for statistical inference; the two most dominant are the *frequentist* (sometimes called classical, traditional, or Neyman–Pearsonian) and *Bayesian*. The term Bayesian refers to Reverend Thomas Bayes (Fig. 8.1), a nonconformist minister interested in mathematics whose posthumously published essay (Bayes, 1763) is fundamental for this kind of inference. According to the Bayesian paradigm, the unobservable parameters in a statistical model are treated as random. Before data are collected, a *prior distribution* is elicited to quantify our knowledge about the parameter. This knowledge comes from expert opinion, theoretical considerations, or previous similar experiments. When data are available, the prior distribution is updated to the *posterior distribution*. This is a conditional distribution that incorporates the observed data. The transition from the prior to the posterior is possible via Bayes' theorem.

The Bayesian approach is relatively modern in statistics; it became influential with advances in Bayesian computational methods in the 1980s and 1990s.

**Fig. 8.1** The Reverend Thomas Bayes (1702–1761), nonconformist priest and mathematician. His *Essay Towards Solving a Problem in the Doctrine of Chances*, presented to the Royal Society after his death, contains a special case of what is now known as Bayes' theorem.

Before launching into a formal exposition of Bayes' theorem, we revisit Bayes' rule for events (p. 86). Prior to observing whether an event $A$ has appeared or not, we set the probabilities of $n$ hypotheses, $H_1, H_2, \ldots, H_n$, under which event $A$ may appear. We called them *prior* probabilities of the hypotheses, $\mathbb{P}(H_1), \ldots, \mathbb{P}(H_n)$. Bayes' rule showed us how to update these prior probabilities to the posterior probabilities once we obtained information about event $A$. Recall that the posterior probability of the hypothesis $H_i$, given the evidence about $A$, was

$$\mathbb{P}(H_i|A) = \frac{\mathbb{P}(A|H_i)\mathbb{P}(H_i)}{\mathbb{P}(A)}.$$

Therefore, Bayes' rule gives a recipe for updating the prior probabilities of events to their posterior probabilities once additional information from the experiment becomes available. The focus of this chapter is on how to update prior knowledge about a model; however, this knowledge (or lack of it) is expressed in terms of probability distributions rather than by events.

Suppose that before the data are observed, a description of the population parameter $\theta$ is given by a probability density $\pi(\theta)$. The process of specifying the prior distribution is called *prior elicitation*. The data are modeled via the likelihood, which depends on $\theta$ and is denoted by $f(x|\theta)$. Bayes' theorem updates the prior $\pi(\theta)$ to the posterior $\pi(\theta|x)$ by incorporating observations $x$ summarized via the likelihood:

$$\pi(\theta|x) = \frac{f(x|\theta)\pi(\theta)}{m(x)}. \tag{8.1}$$

Here, $m(x)$ normalizes the product $f(x|\theta)\pi(\theta)$ to be a density and is a constant once the prior is specified and the data are observed. Given the data $x$ and the prior distribution, the posterior distribution $\pi(\theta|x)$ summarizes all available information about $\theta$.

Although the equation in (8.1) is referred to as a theorem, there is nothing to prove there. Recall that the probability of intersection of two events $A$ and $B$ was calculated as $\mathbb{P}(AB) = \mathbb{P}(A|B)\mathbb{P}(B) = \mathbb{P}(B|A)\mathbb{P}(B)$ [multiplication rule in (3.6)]. By analogy, the joint distribution of $X$ and $\theta$, $h(x,\theta)$, would have two representations depending on the order of conditioning:,

$$h(x,\theta) = f(x|\theta)\pi(\theta) = \pi(\theta|x)m(x),$$

and Bayes' theorem just solves this equation with respect to the posterior $\pi(\theta|x)$.

To summarize, Bayes' rule updates the probabilities of events when new evidence becomes available, while Bayes' theorem provides the recipe for updating prior distributions of model's parameters once experimental observations become available.

| $\mathbb{P}(\text{hypothesis})$ | BAYES' RULE $\longrightarrow$ | $\mathbb{P}(\text{hypothesis}|\text{evidence})$ |
|---|---|---|
| $\pi(\theta)$ | BAYES' THEOREM $\longrightarrow$ | $\pi(\theta|\text{data})$ |

The Bayesian paradigm has many advantages, but the two most important are that (i) the uncertainty is expressed via the probability distribution and the statistical inference can be automated; thus it follows a conceptually simple recipe embodied in Bayes' theorem; and (ii) available prior information is coherently incorporated into the statistical model describing the data.

The FDA guidelines document (FDA, 2010) recommends the use of a Bayesian methodology in the design and analysis of clinical trials for medical devices. This document eloquently outlines the reasons why a Bayesian methodology is recommended.

- Valuable prior information is often available for medical devices because of their mechanism of action and evolutionary development.
- The Bayesian approach, when correctly employed, may be less burdensome than a frequentist approach.
- In some instances, the use of prior information may alleviate the need for a larger sized trial. In some scenarios, when an adaptive Bayesian model is applicable, the size of a trial can be reduced by stopping the trial early when conditions warrant.
- The Bayesian approach can sometimes be used to obtain an exact analysis when the corresponding frequentist analysis is only approximate or is too difficult to implement.
- Bayesian approaches to multiplicity problems are different from frequentist ones and may be advantageous. Inferences on multiple endpoints and testing of multiple subgroups (e.g., race or sex) are examples of multiplicity.
- Bayesian methods allow for great flexibility in dealing with missing data.

In the context of clinical trials, an unlimited look at the accumulated data when sampling is of a sequential nature will not affect the inference. In the frequentist approach, interim data analyses affect type I errors. The ability to stop a clinical trial early is important from the moral and economic viewpoints. Trials should be stopped early due to both futility, to save resources or stop an ineffective treatment, and superiority, to provide patients with the best possible treatments as fast as possible.

Bayesian models facilitate meta-analysis. Meta-analysis is a methodology for the fusion of results of related experiments performed by different researchers, labs, etc. An example of a rudimentary meta-analysis is discussed in Sect. 8.10.

## 8.2 Ingredients for Bayesian Inference

A density function for a typical observation $X$ that depends on an unknown (possibly multivariate) parameter $\theta$ is called a model and denoted by $f(x|\theta)$. As a function of $\theta$, $f(x|\theta) = L(\theta)$ is called the *likelihood*. If a sample $x = (x_1, x_2, \ldots, x_n)$ is observed, the likelihood takes a familiar form, $L(\theta|x_1, \ldots, x_n) = \prod_{i=1}^{n} f(x_i|\theta)$. This form was used in Chap. 7 to produce MLEs for $\theta$.

Thus both terms model and likelihood are used to describe the distribution of observations. In the standard Bayesian inference the functional form of $f$ is given in the same manner as in the classical parametric approach; the functional form is fully specified up to a parameter $\theta$. According to the generally accepted *likelihood principle,* all information from the experimental data is summarized in the likelihood function, $f(x|\theta) = L(\theta|x_1, \ldots, x_n)$.

For example, if each datum $X|\theta$ were assumed to be exponential with the rate parameter $\theta$ and $X_1 = 2, X_2 = 3$, and $X_3 = 1$ were observed, then full information about the experiment would be given by the likelihood

$$\theta e^{-2\theta} \times \theta e^{-3\theta} \times \theta e^{-\theta} = \theta^3 e^{-6\theta}.$$

This model is $\theta^3 \exp\{-\theta \sum_{i=1}^{3} X_i\}$ if the data are kept unspecified, but in the likelihood function the expression $\sum_{i=1}^{3} X_i$ is treated as a constant term, as was done in the maximum likelihood estimation (p. 233).

The parameter $\theta$, with values in the parameter space $\Theta$, is not directly observable and is considered a random variable. This is the key difference between Bayesian and classical approaches. Classical statistics consider the parameter to be a fixed number or vector of numbers, while Bayesians express the uncertainty about $\theta$ by considering it as a random variable. This random variable has a distribution $\pi(\theta)$ called the prior distribution. The prior distribution not only quantifies available knowledge, but it also describes the uncertainty about a parameter before data are observed. If the prior distribution for $\theta$ is specified up to a parameter $\tau$, $\pi(\theta|\tau)$, then $\tau$ is called a *hyperparameter*. Hyperparameters are parameters of a prior distribution, and they are either specified or may have their own priors. This may lead to a hierarchical structure of the model where the priors are arranged in a hierarchy.

The previous discussion can be summarized as follows:

> The goal in Bayesian inference is to start with prior information on the parameter of interest, $\theta$, and update it using the observed data. This is achieved via Bayes' theorem, which gives a simple recipe for incorporating observations $x$ in the distribution of $\theta$, $\pi(\theta|x)$, called the *posterior* distribution. All information about $\theta$ coming from the prior distribution and the observations are contained in the posterior distribution. The posterior distribution is the ultimate summary of the parameter and serves as the basis for all Bayesian inferences.

According to Bayes' theorem, to find $\pi(\theta|x)$, we divide the *joint* distribution of $X$ and $\theta$ $(h(x,\theta) = f(x|\theta)\pi(\theta))$ by the *marginal* distribution for $X$, $m(x)$, which is obtained by integrating out $\theta$ from the joint distribution $h(x,\theta)$:

$$m(x) = \int_{\Theta} h(x,\theta)d\theta = \int_{\Theta} f(x|\theta)\pi(\theta)d\theta.$$

The marginal distribution is also called the *prior predictive* distribution. Thus, in terms of the likelihood and the prior distribution only, the Bayes theorem can be restated as

$$\pi(\theta|x) = \frac{f(x|\theta)\pi(\theta)}{\int_{\Theta} f(x|\theta)\pi(\theta)d\theta}.$$

The integral in the denominator is a main hurdle in Bayesian computation since for complex likelihoods and priors it could be intractable.

The following table summarizes the notation:

| | |
|---|---|
| Likelihood, model | $f(x|\theta)$ |
| Prior distribution | $\pi(\theta)$ |
| Joint distribution | $h(x,\theta) = f(x|\theta)\pi(\theta)$ |
| Marginal distribution | $m(x) = \int_{\Theta} f(x|\theta)\pi(\theta)d\theta$ |
| Posterior distribution | $\pi(\theta|x) = f(x|\theta)\pi(\theta)/m(x)$ |

We illustrate the above concepts by discussing a few examples in which one can find the posterior distribution explicitly. Note that the marginal distribution has the form of an integral, and in many cases these integrals cannot be found in a finite form. It is fair to say that the number of likelihood/prior combinations that lead to an explicit posterior is rather limited. However, in the general case, one can evaluate the posterior numerically or, as we will see later, simulate a sample from the posterior distribution. All of the above, admittedly abstract, concepts will be exemplified by several worked-out models. We start with the most important model in which both the likelihood and prior are normal.

*Example 8.1.* **Normal Likelihood with Normal Prior.** The normal likelihood and normal prior combination is important because it is frequently used in practice. Assume that an observation $X$ is normally distributed with mean $\theta$ and known variance $\sigma^2$. The parameter of interest, $\theta$, is normally distributed as well, with its parameters $\mu$ and $\tau^2$. Parameters $\mu$ and $\tau^2$ are hyperparameters, and we will consider them given. Starting with our Bayesian model of $X|\theta \sim \mathcal{N}(\theta,\sigma^2)$ and $\theta \sim \mathcal{N}(\mu,\tau^2)$, we will find the marginal and posterior distributions. Before we start with a derivation of the posterior and marginal, we need a simple algebraic identity:

$$A(x-a)^2 + B(x-b)^2 = (A+B)(x-c)^2 + \frac{AB}{A+B}(a-b)^2, \text{for } c = \frac{Aa+Bb}{A+B}. \quad (8.2)$$

We start with the joint distribution of $(X,\theta)$, which is the product of two distributions:

$$h(x,\theta) = \frac{1}{\sqrt{2\pi\sigma^2}} \exp\left\{-\frac{1}{2\sigma^2}(x-\theta)^2\right\} \times \frac{1}{\sqrt{2\pi\tau^2}} \exp\left\{-\frac{1}{2\tau^2}(\theta-\mu)^2\right\}.$$

The exponent in the joint distribution $h(x,\theta)$ is

$$-\frac{1}{2\sigma^2}(x-\theta)^2 - \frac{1}{2\tau^2}(\theta-\mu)^2,$$

which, after applying the identity in (8.2), can be expressed as

$$-\frac{\sigma^2+\tau^2}{2\sigma^2\tau^2}\left(\theta - \left(\frac{\tau^2}{\sigma^2+\tau^2}x + \frac{\sigma^2}{\sigma^2+\tau^2}\mu\right)\right)^2 - \frac{1}{2(\sigma^2+\tau^2)}(x-\mu)^2. \qquad (8.3)$$

Note that the exponent (8.3) splits into two parts, one containing $\theta$ and the other $\theta$-free. Accordingly the joint distribution $h(x,\theta)$ splits into the product of two densities. Since $h(x,\theta)$ can be represented in two ways, as $f(x|\theta)\pi(\theta)$ and as $\pi(\theta|x)m(x)$, by analogy to $P(AB) = P(A|B)P(B) = P(B|A)P(A)$, and since we started with $f(x|\theta)\pi(\theta)$, the exponent in (8.3) corresponds to $\pi(\theta|x)m(x)$. Thus, the marginal distribution simply resolves to $X \sim \mathcal{N}(\mu,\sigma^2+\tau^2)$ and the posterior distribution of $\theta$ comes out to be

$$\theta|X \sim \mathcal{N}\left(\frac{\tau^2}{\sigma^2+\tau^2}X + \frac{\sigma^2}{\sigma^2+\tau^2}\mu, \frac{\sigma^2\tau^2}{\sigma^2+\tau^2}\right).$$

Bellow is a specific example of our first Bayesian inference.

*Example 8.2.* **Jeremy's IQ.** Jeremy, an enthusiastic bioengineering student, posed a statistical model for his scores on a standard IQ test. He thinks that, in general, his scores are normally distributed with unknown mean $\theta$ (true IQ) and a variance of $\sigma^2 = 80$. Prior (and expert) opinion is that the IQ of bioengineering students in Jeremy's school, $\theta$, is a normal random variable, with mean $\mu = 110$ and variance $\tau^2 = 120$. Jeremy took the test and scored $X = 98$. The traditional estimator of $\theta$ would be $\hat{\theta} = X = 98$. The posterior is normal with a mean of $\frac{120}{80+120} \times 98 + \frac{80}{80+120} \times 110 = 102.8$ and a variance of $\frac{80 \times 120}{80+120} = 48$. We will see later that the mean of the posterior is Bayes' estimator of $\theta$, and a Bayesian would estimate Jeremy's IQ as 102.8.

If $n$ normal variates, $X_1, X_2, \ldots, X_n$, are observed instead of a single observation $X$, then the sample is summarized as $\overline{X}$ and the Bayesian model for $\theta$ is essentially the same as that for the single $X$, but with $\sigma^2/n$ in place of $\sigma^2$. In this case, the likelihood and the prior are

$$\overline{X}|\theta \sim \mathcal{N}\left(\theta, \frac{\sigma^2}{n}\right) \text{ and } \theta \sim \mathcal{N}(\mu,\tau^2),$$

producing

$$\theta|\overline{X} \sim \mathcal{N}\left(\frac{\tau^2}{\frac{\sigma^2}{n}+\tau^2}\overline{X} + \frac{\frac{\sigma^2}{n}}{\frac{\sigma^2}{n}+\tau^2}\mu, \frac{\frac{\sigma^2}{n}\tau^2}{\frac{\sigma^2}{n}+\tau^2}\right).$$

Notice that the posterior mean

$$\frac{\tau^2}{\frac{\sigma^2}{n}+\tau^2}\overline{X} + \frac{\frac{\sigma^2}{n}}{\frac{\sigma^2}{n}+\tau^2}\mu$$

is a weighted average of the MLE $\overline{X}$ and the prior mean $\mu$ with weights $w = n\tau^2/(\sigma^2+n\tau^2)$ and $1-w = \sigma^2/(\sigma^2+n\tau^2)$. When the sample size $n$ increases, the contribution of the prior mean to the estimator diminishes as $w \to 1$. On the other hand, when $n$ is small and our prior opinion about $\mu$ is strong (i.e., $\tau^2$ is small), the posterior mean remains close to the prior mean $\mu$. Later, we will explore several more cases in which the posterior mean is a weighted average of the MLE for the parameter and the prior mean.

*Example 8.3.* Suppose $n = 10$ observations are coming from $\mathcal{N}(\theta, 10^2)$. Assume that the prior on $\theta$ is $\mathcal{N}(20, 20)$. For the observations
$\{2.944, -13.361, 7.143, 16.235, -6.917, 8.580, 12.540, -15.937, -14.409, 5.711\}$
the posterior is $\mathcal{N}(6.835, 6.667)$. The three densities are shown in Fig. 8.2.

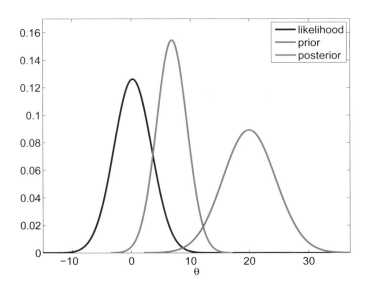

**Fig. 8.2** The likelihood centered at MLE $\overline{X} = 0.2529$, $\mathcal{N}(0.2529, 10^2/10)$ (*blue*), $\mathcal{N}(20, 20)$ prior (*red*), and posterior for data $\{2.9441, -13.3618, \ldots, 5.7115\}$ (*green*).

## 8.3 Conjugate Priors

A major technical difficulty in Bayesian analysis is finding an explicit posterior distribution, given the likelihood and prior. The posterior is proportional to the product of the likelihood and prior, but the normalizing constant, marginal $m(x)$, is often difficult to find since it involves integration.

In Examples 8.1 and 8.3, where the prior is normal, the posterior distribution remains normal. In such cases, the effect of likelihood is only to "update" the prior parameters and not to change the prior's functional form. We say that such priors are *conjugate* with the likelihood. Conjugacy is popular because of its mathematical convenience; once the conjugate pair likelihood/prior is identified, the posterior is found without integration. The normalizing marginal $m(x)$ is selected such that $f(x|\theta)\pi(\theta)$ is a density from the same class to which the prior belongs. Operationally, one multiplies "kernels" of likelihood and priors, ignoring all multiplicative terms that do not involve the parameter. For example, a kernel of gamma $\mathcal{G}a(r,\lambda)$ density $f(\theta|r,\lambda) = \frac{\lambda^r \theta^{r-1}}{\Gamma(r)} e^{-\lambda\theta}$ would be $\theta^{r-1}e^{-\lambda\theta}$. We would write: $f(\theta|r,\lambda) \propto \theta^{r-1}e^{-\lambda\theta}$, where the symbol $\propto$ stands for "proportional to." Several examples in this chapter involve conjugate pairs (Examples 8.4 and 8.6).

In the pre-Markov chain Monte Carlo era, conjugate priors were extensively used (and overused and misused) precisely because of this computational convenience. Today, the general agreement is that simple conjugate analysis is of limited practical value since, given the likelihood, the conjugate prior has limited modeling capability.

There are quite a few instances of conjugacy. The following table lists several important cases. For practice you may want to derive the posteriors in Table 8.1. It is recommended that you consult Chap. 5 on functional forms of densities involved in the Bayesian model.

*Example 8.4.* **Binomial Likelihood with Beta Prior.** An easy, yet important, example of a conjugate structure is the binomial likelihood and beta prior. Suppose that we observed $X$ from a binomial $\mathcal{B}in(n,p)$ distribution,

$$f(x|\theta) = \binom{n}{x} p^x (1-p)^{n-x},$$

and that the population proportion $p$ is the parameter of interest. If the prior on $p$ is beta $\mathcal{B}e(\alpha,\beta)$ with hyperparameters $\alpha$ and $\beta$ and density

$$\pi(p) = \frac{1}{B(\alpha,\beta)} p^{\alpha-1}(1-p)^{\beta-1},$$

the posterior is proportional to the product of the likelihood and the prior

$$\pi(p|x) = C \cdot p^x (1-p)^{n-x} \cdot p^{\alpha-1}(1-p)^{\beta-1} = C \cdot p^{x+\alpha-1}(1-p)^{n-x+\beta-1}$$

**Table 8.1** Some conjugate pairs. Here $\mathbf{X}$ stands for a sample of size $n$, $X_1,\ldots,X_n$. For functional expressions of the densities and their moments refer to Chap. 5

| Likelihood | Prior | Posterior |
|---|---|---|
| $X\|\theta \sim \mathcal{N}(\theta,\sigma^2)$ | $\theta \sim \mathcal{N}(\mu,\tau^2)$ | $\theta\|X \sim \mathcal{N}\left(\frac{\tau^2}{\sigma^2+\tau^2}X + \frac{\sigma^2}{\sigma^2+\tau^2}\mu, \frac{\sigma^2\tau^2}{\sigma^2+\tau^2}\right)$ |
| $X\|\theta \sim \mathcal{B}in(n,\theta)$ | $\theta \sim \mathcal{B}e(\alpha,\beta)$ | $\theta\|X \sim \mathcal{B}e(\alpha + x, n - x + \beta)$ |
| $X\|\theta \sim \mathcal{P}oi(\theta)$ | $\theta \sim \mathcal{G}a(\alpha,\beta)$ | $\theta\|\mathbf{X} \sim \mathcal{G}a(\sum_i X_i + \alpha, n + \beta)$. |
| $X\|\theta \sim \mathcal{N}\mathcal{B}(m,\theta)$ | $\theta \sim \mathcal{B}e(\alpha,\beta)$ | $\theta\|\mathbf{X} \sim \mathcal{B}e(\alpha + mn, \beta + \sum_{i=1}^{n} x_i)$ |
| $X \sim \mathcal{G}a(n/2, 1/(2\theta))$ | $\theta \sim \mathcal{I}\mathcal{G}(\alpha,\beta)$ | $\theta\|X \sim \mathcal{I}\mathcal{G}(n/2 + \alpha, x/2 + \beta)$ |
| $X\|\theta \sim \mathcal{U}(0,\theta)$ | $\theta \sim \mathcal{P}a(\theta_0,\alpha)$ | $\theta\|\mathbf{X} \sim \mathcal{P}a(\max\{\theta_0, X_1,\ldots,X_n\}, \alpha + n)$ |
| $X\|\theta \sim \mathcal{N}(\mu,\theta)$ | $\theta \sim \mathcal{I}\mathcal{G}(\alpha,\beta)$ | $\theta\|X \sim \mathcal{I}\mathcal{G}(\alpha + 1/2, \beta + (\mu - X)^2/2)$ |
| $X\|\theta \sim \mathcal{G}a(\nu,\theta)$ | $\theta \sim \mathcal{G}a(\alpha,\beta)$ | $\theta\|X \sim \mathcal{G}a(\alpha + \nu, \beta + x)$ |

for some constant $C$. The normalizing constant $C$ is free of $p$ and is equal to $\frac{\binom{n}{x}}{m(x)B(\alpha,\beta)}$, where $m(x)$ is the marginal distribution.

By inspecting the expression $p^{x+\alpha-1}(1-p)^{n-x+\beta-1}$, it is easy to see that the posterior density remains beta; it is $\mathcal{B}e(x + \alpha, n - x + \beta)$, and that normalizing constant resolves to $C = 1/B(x + \alpha, n - x + \beta)$. From the equality of constants, it follows that

$$\frac{\binom{n}{x}}{m(x)B(\alpha,\beta)} = \frac{1}{B(x + \alpha, n - x + \beta)},$$

and one can express the marginal

$$m(x) = \frac{\binom{n}{x}B(x + \alpha, n - x + \beta)}{B(\alpha,\beta)},$$

which is known as a *beta-binomial distribution*.

## 8.4 Point Estimation

The posterior is the ultimate experimental summary for a Bayesian. The posterior location measures (especially the mean) are of great importance. The posterior mean is the most frequently used Bayes estimator for a parameter. The posterior mode and median are alternative Bayes estimators.

The posterior mode maximizes the posterior density in the same way that the MLE maximizes the likelihood. When the posterior mode is used as an estimator, it is called the maximum posterior (MAP) estimator. The MAP estimator is popular in some Bayesian analyses in part because it is computationally less demanding than the posterior mean or median. The reason for this is simple; to find a MAP, the posterior does not need to be fully specified because $\text{argmax}_\theta \pi(\theta|x) = \text{argmax}_\theta f(x|\theta)\pi(\theta)$, that is, the product of the likelihood and the prior as well as the posterior are maximized at the same point.

*Example 8.5.* **Binomial-Beta Conjugate Pair.** In Example 8.4 we argued that for the likelihood $X|\theta \sim \mathcal{B}in(n,\theta)$ and the prior $\theta \sim \mathcal{B}e(\alpha,\beta)$, the posterior distribution is $\mathcal{B}e(x + \alpha, n - x + \beta)$. The Bayes estimator of $\theta$ is the expected value of the posterior

$$\hat{\theta}_B = \frac{\alpha + x}{(\alpha + x) + (\beta + n - x)} = \frac{\alpha + x}{\alpha + \beta + n}.$$

This is actually a weighted average of the MLE, $X/n$, and the prior mean $\alpha/(\alpha + \beta)$,

$$\hat{\theta}_B = \frac{n}{\alpha + \beta + n} \cdot \frac{X}{n} + \frac{\alpha + \beta}{\alpha + \beta + n} \cdot \frac{\alpha}{\alpha + \beta}.$$

Notice that, as $n$ becomes large, the posterior mean approaches the MLE, because the weight $\frac{n}{n+\alpha+\beta}$ tends to 1. On the other hand, when $\alpha$ or $\beta$ or both are large compared to $n$, the posterior mean is close to the prior mean. Because of this interplay between $n$ and prior parameters, the sum $\alpha + \beta$ is called the prior sample size, and it measures the influence of the prior as if additional experimentation was performed and $\alpha + \beta$ trials have been added. This is in the spirit of Wilson's proposal to "add two failures and two successes" to an estimator of proportion (p. 254). Wilson's estimator can be seen as a Bayes estimator with a beta $\mathcal{B}e(2,2)$ prior.

Large $\alpha$ indicates a small prior variance (for fixed $\beta$, the variance of $\mathcal{B}e(\alpha,\beta)$ is proportional to $1/\alpha^2$) and the prior is concentrated about its mean. ✎

In general, the posterior mean will fall between the MLE and the prior mean. This was demonstrated in Example 8.1. As another example, suppose we flipped a coin four times and tails showed up on all four occasions. We are interested in estimating the probability of showing heads, $\theta$, in a Bayesian fashion. If the prior is $\mathcal{U}(0,1)$, the posterior is proportional to $\theta^0(1-\theta)^4$, which is a beta $\mathcal{B}e(1,5)$. The posterior mean *shrinks* the MLE toward the expected value of the prior (1/2) to get $\hat{\theta}_B = 1/(1+5) = 1/6$, which is a more reasonable estimator of $\theta$ than the MLE. Note that the $3/n$ rule produces a confidence interval for $p$ of $[0, 3/4]$, which is too wide to be useful (Sect. 7.4.4).

*Example 8.6.* **Uniform/Pareto Model.** In Example 7.5 we had the observations $X_1 = 2$, $X_2 = 5$, $X_3 = 0.5$, and $X_4 = 3$ from a uniform $\mathcal{U}(0, \theta)$ distribution. We are interested in estimating $\theta$ in a Bayesian fashion. Let the prior on $\theta$ be Pareto $\mathcal{P}a(\theta_0, \alpha)$ for $\theta_0 = 6$ and $\alpha = 2$. Then the posterior is also Pareto $\mathcal{P}a(\theta^*, \alpha^*)$ with $\theta^* = \max\{\theta_0, X_{(n)}\} = \max\{6, 5\} = 6$, and $\alpha^* = \alpha + n = 2 + 4 = 6$. The posterior mean is $\frac{\alpha^* \theta^*}{\alpha^* - 1} = 36/5 = 7.2$, and the median is $\theta^* \cdot 2^{1/\alpha^*} = 6 \cdot 2^{1/6} = 6.7348$.

Figure 8.3 shows the prior (dashed red line) with the prior mean as a red dot. After observing $X_1, \ldots, X_4$, the posterior mode did not change since the elicited $\theta_0 = 6$ was larger than $\max X_i = 5$. However, the posterior has a smaller variance than the prior. The posterior mean is shown as a green dot, the posterior median as a black dot, and the posterior (and prior) mode as a blue dot.

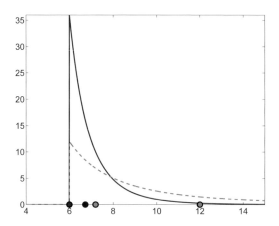

**Fig. 8.3** Pareto $\mathcal{P}a(6, 2)$ prior (*dashed red line*) and $\mathcal{P}a(6, 6)$ posterior (*solid blue line*). The *red dot* is the prior mean, the *green dot* is the posterior mean, the *black dot* is the posterior median, and the *blue dot* is the posterior (and prior) mode.

## 8.5 Prior Elicitation

Prior distributions are carriers of prior information that is coherently incorporated via Bayes' theorem into an inference. At the same time, parameters are unobservable, and prior specification is subjective in nature. The subjectivity of specifying the prior is a fundamental criticism of the Bayesian approach. Being subjective does not mean that the approach is nonscientific, as critics of Bayesian statistics often insinuate. On the contrary, vast amounts of scientific information coming from theoretical and physical models, previous experiments, and expert reports guides the specification of priors and merges such information with the data for better inference.

In arguing about the importance of priors in Bayesian inference, Garthwhite and Dickey (1991) state that "expert personal opinion is of great potential value and can be used more efficiently, communicated more accurately, and judged more critically if it is expressed as a probability distribution."

In the last several decades Bayesian research has also focused on priors that were noninformative and robust; this was in response to criticism that results of Bayesian inference could be sensitive to the choice of a prior.

For instance, in Examples 8.4 and 8.5 we saw that beta distributions are an appropriate family of priors for parameters supported in the interval $[0,1]$, such as a population proportion. It turns out that the beta family can express a wide range of prior information. For example, if the mean $\mu$ and variance $\sigma^2$ for a beta prior are elicited by an expert, then the parameters $(a,b)$ can be determined by solving $\mu = a/(a+b)$ and $\sigma^2 = ab/[(a+b)^2(a+b+1)]$ with respect to $a$ and $b$:

$$a = \mu\left(\frac{\mu(1-\mu)}{\sigma^2} - 1\right), \quad \text{and} \quad b = (1-\mu)\left(\frac{\mu(1-\mu)}{\sigma^2} - 1\right). \tag{8.4}$$

If $a$ and $b$ are not too small, the shape of a beta prior resembles a normal distribution and the bounds $[\mu-2\sigma, \mu+2\sigma]$ can be used to describe the range of likely parameters. For example, an expert's claim that a proportion is unlikely to be higher than 90% can be expressed as $\mu + 2\sigma = 0.9$.

In the same context of estimating the proportion, Berry and Stangl (1996) suggest a somewhat different procedure:

(i) Elicit the probability of success in the first trial, $p_1$, and match it to the prior mean $\alpha/(\alpha + \beta)$.

(ii) Given that the first trial results in success, the posterior mean is $\frac{\alpha+1}{\alpha+\beta+1}$. Match this ratio with the elicited probability of success in a second trial, $p_2$, conditional upon the first trial's resulting in success. Thus, a system

$$p_1 = \frac{\alpha}{\alpha + \beta} \quad \text{and} \quad p_2 = \frac{\alpha + 1}{\alpha + \beta + 1}$$

is obtained that solves to

$$\alpha = \frac{p_1(1-p_2)}{p_2 - p_1} \quad \text{and} \quad \beta = \frac{(1-p_1)(1-p_2)}{p_2 - p_1}. \tag{8.5}$$

See Exercise 8.12 for an application.

If one has no prior information, many noninformative choices are possible such as invariant priors, Jeffreys' priors, default priors, reference priors, and intrinsic priors, among others. Informally speaking, a noninformative prior is one which is dominated by the likelihood, or that is "flat" relative to the likelihood.

Popular noninformative choices are the flat prior $\pi(\theta) = C$ for the location parameter (mean) and $\pi(\theta) = 1/\theta$ for the scale/rate parameter. A vague prior for the proportion is $p^{-1}(1-p)^{-1}$, $0 < p < 1$. These priors are not proper probability distributions, that is, they are not densities because their integrals are not finite. However, Bayes' theorem usually leads to posterior distributions that are proper densities and on which Bayesian analysis can be carried out.

**Fig. 8.4** Sir Harold Jeffreys, (1891–1989).

Jeffreys' priors (named after Sir Harold Jeffreys, Fig. 8.4) are obtained from a particular functional of a density (called Fisher information), and they are also examples of vague and noninformative priors. For a binomial proportion Jeffreys' prior is proportional to $p^{-1/2}(1-p)^{-1/2}$, while for the rate of exponential distribution $\lambda$ Jeffreys' prior is proportional to $1/\lambda$. For a normal distribution, Jeffreys' prior on the mean is flat, while for the variance $\sigma^2$ it is proportional to $\frac{1}{\sigma^2}$.

*Example 8.7.* If $X_1 = 1.7$, $X_2 = 0.6$, and $X_3 = 5.2$ come from an exponential distribution with a rate parameter $\lambda$, find the Bayes estimator if the prior on $\lambda$ is $\frac{1}{\lambda}$.

The likelihood is $\lambda^3 e^{-\lambda \sum_{i=1}^3 X_i}$ and the posterior is proportional to

$$\frac{1}{\lambda} \times \lambda^3 e^{-\lambda \sum_{i=1}^3 X_i} = \lambda^{3-1} e^{-\lambda \sum X_i},$$

which is recognized as gamma $\mathcal{G}a\left(3, \sum_{i=1}^3 X_i\right)$. The Bayes estimator, as a mean of this posterior, coincides with the MLE, $\hat{\lambda} = \frac{3}{\sum_{i=1}^3 X_i} = \frac{1}{\bar{X}} = 1/2.5 = 0.4$.

An applied approach to prior selection was taken by Spiegelhalter et al. (1994) in the context of biomedical inference and clinical trials. They recommended a *community of priors* elicited from a large group of experts. A crude classification of community priors is as follows.

(i) Vague priors – noninformative priors, in many cases leading to posterior distributions proportional to the likelihood.

(ii) Skeptical priors – reflecting the opinion of a clinician unenthusiastic about the new therapy, drug, device, or procedure. This may be a prior of a regulatory agency.

(iii) Enthusiastic or clinical priors – reflecting the opinion of the proponents of the clinical trial, centered around the notion that a new therapy, drug, device, or procedure is superior. This may be the prior of the industry involved or clinicians running the trial.

For example, the use of a skeptical prior when testing for the superiority of a new treatment would be a conservative approach. In equivalence tests, both skeptical and enthusiastic priors may be used. The superiority of a new treatment should be judged by a skeptical prior, while the superiority of the old treatment should be judged by an enthusiastic prior.

## 8.6 Bayesian Computation and Use of WinBUGS

If the selection of an adequate prior is the major conceptual and modeling challenge of Bayesian analysis, the major implementational challenge is computation. When the model deviates from the conjugate structure, finding the posterior distribution and the Bayes rule is all but simple. A closed-form solution is more the exception than the rule, and even for such exceptions, lucky mathematical coincidences, convenient mixtures, and other tricks are needed to uncover the explicit expression.

If classical statistics relies on optimization, Bayesian statistics relies on integration. The marginal needed to normalize the product $f(x|\theta)\pi(\theta)$ is an integral

$$m(x) = \int_{\Theta} f(x|\theta)\pi(\theta)d\theta,$$

while the Bayes estimator of $h(\theta)$ is a ratio of integrals,

$$\delta_{\pi}(x) = \int_{\Theta} h(\theta)\pi(\theta|x)d\theta = \frac{\int_{\Theta} h(\theta)f(x|\theta)\pi(\theta)d\theta}{\int_{\Theta} f(x|\theta)\pi(\theta)d\theta}.$$

The difficulties in calculating the above Bayes rule derive from the facts that (i) the posterior may not be representable in a finite form and (ii) the integral of $h(\theta)$ does not have a closed form even when the posterior distribution is explicit.

The last two decades of research in Bayesian statistics has contributed to broadening the scope of Bayesian models. Models that could not be handled before by a computer are now routinely solved. This is done by *Markov chain*

*Monte Carlo* (MCMC) methods, and their introduction to the field of statistics revolutionized Bayesian statistics.

The MCMC methodology was first applied in statistical physics (Metropolis et al., 1953). Work by Gelfand and Smith (1990) focused on applications of MCMC to Bayesian models. The principle of MCMC is simple: One designs a Markov chain that samples from the target distribution. By simulating long runs of such a Markov chain, the target distribution can be well approximated. Various strategies for constructing appropriate Markov chains that simulate the desired distribution are possible: Metropolis–Hastings, Gibbs sampler, slice sampling, perfect sampling, and many specialized techniques. These are beyond the scope of this text, and the interested reader is directed to Robert (2001), Robert and Casella (2004), and Chen et al. (2000) for an overview and a comprehensive treatment.

In the examples that follow we will use WinBUGS for doing Bayesian inference when the models are not conjugate. Chapter 19 gives a brief introduction to the front end of WinBUGS. Three volumes of examples are a standard addition to the software; in the Examples menu of WinBUGS, see Spiegelhalter et al. (1996). It is recommended that you go over some of those examples in detail because they illustrate the functionality and modeling power of WinBUGS. A wealth of examples on Bayesian modeling strategies using WinBUGS can be found in the monographs of Congdon (2001, 2003, 2005) and Ntzoufras (2009).

The following example is a WinBUGS solution of Example 8.2.

*Example 8.8.* **Jeremy's IQ in WinBUGS.** We will calculate a Bayes estimator for Jeremy's true IQ, $\theta$, using simulations in WinBUGS. Recall that the model was $X \sim \mathcal{N}(\theta, 80)$ and $\theta \sim \mathcal{N}(100, 120)$. WinBUGS uses precision instead of variance to parameterize the normal distribution. Precision is simply the reciprocal of the variance, and in this example, the precisions are $1/120 = 0.00833$ for the prior and $1/80 = 0.0125$ for the likelihood. The WinBUGS code is as follows:

```
Jeremy in WinBUGS
model{
x ~ dnorm( theta, 0.0125)
theta ~ dnorm( 110, 0.008333333)
}
DATA
list(x=98)
INITS
list(theta=100)
```

Here is the summary of the MCMC output. The Bayes estimator for $\theta$ is rounded to 102.8. It is obtained as a mean of the simulated sample from the posterior.

|       | mean  | sd    | MC error | val2.5pc | median | val97.5pc | start | sample |
|-------|-------|-------|----------|----------|--------|-----------|-------|--------|
| theta | 102.8 | 6.943 | 0.01991  | 89.18    | 102.8  | 116.4     | 1001  | 100000 |

Because this is a conjugate normal/normal model, the exact posterior distribution, $\mathcal{N}(102.8,48)$, was easy to find, (Example 8.2). Note that in these simulations, the MCMC approximation, when rounded, coincides with the exact posterior mean. The MCMC variance of $\theta$ is $6.943^2 \approx 48.2$, which is close to the exact posterior variance of 48.

Another widely used conjugate pair is Poisson–gamma pair.

*Example 8.9.* **Poisson–Gamma Conjugate Pair.** Let $X_1,\dots,X_n$, given $\theta$ are Poisson $\mathcal{P}oi(\theta)$ with probability mass function

$$f(x_i|\theta) = \frac{\theta^{x_i}}{x_i!}e^{-\theta},$$

and $\theta \sim \mathcal{G}(\alpha,\beta)$ is given by $\pi(\theta) \propto \theta^{\alpha-1}e^{-\beta\theta}$. Then

$$\pi(\theta|X_1,\dots,X_n) = \pi(\theta|\textstyle\sum X_i) \propto \theta^{\sum X_i + \alpha - 1}e^{-(n+\beta)\theta},$$

which is $\mathcal{G}(\sum_i X_i + \alpha, n+\beta)$. The mean is $\mathbb{E}(\theta|X) = (\sum X_i + \alpha)/(n+\beta)$, and it can be represented as a weighted average of the MLE and the prior mean:

$$\mathbb{E}\theta|X = \frac{n}{n+\beta}\frac{\sum X_i}{n} + \frac{\beta}{n+\beta}\frac{\alpha}{\beta}.$$

Let us apply the above equation in a specific example. Let a rare disease have an incidence of $X$ cases per 100,000 people, where $X$ is modeled as Poisson, $X|\lambda \sim \mathcal{P}oi(\lambda)$, where $\lambda$ is the rate parameter. Assume that for different cohorts of 100,000 subjects, the following incidences are observed: $X_1 = 2$, $X_2 = 0$, $X_3 = 0$, $X_4 = 4$, $X_5 = 0$, $X_6 = 1$, $X_7 = 3$, and $X_8 = 2$. The experts indicate that $\lambda$ should be close to 2 and our prior is $\lambda \sim \mathcal{G}a(0.2,0.1)$. We matched the mean, since for a gamma distribution the mean is $0.2/0.1 = 2$ but the variance $0.2/0.1^2 = 20$ is quite large, thereby expressing our uncertainty. By setting the hyperparameters to 0.02 and 0.01, for example, the variance of the gamma prior would be even larger. The MLE of $\lambda$ is $\hat{\lambda}_{mle} = \overline{X} = 3/2$. The Bayes estimator is

$$\hat{\lambda}_B = \frac{8}{8+0.1}3/2 + \frac{0.1}{8+0.1}2 = 1.5062.$$

Note that since the prior was not informative, the Bayes estimator is quite close to the MLE.

*Example 8.10.* **Uniform/Pareto Model in WinBUGS.** In Example 8.6 we found that a posterior distribution of $\theta$, in a uniform $\mathcal{U}(0,\theta)$ model with a Pareto $\mathcal{P}a(6,2)$ prior, was Pareto $\mathcal{P}a(6,6)$. From the posterior we found the mean, median, and mode to be 7.2, 6.7348, and 6, respectively. These are reasonable estimators of $\theta$ as location measures of the posterior.

```
Uniform with Pareto in WinBUGS
model{
for (i  in 1:n){
   x[i] ~ dunif(0, theta);
   }
theta ~ dpar(2,6)
}
DATA
list(n=4, x = c(2, 5, 0.5, 3) )
INITS
list(theta= 7)
```

Here is the summary of the WinBUGS output. The posterior mean was found to be 7.196 and the median 6.736. Apparently, the mode of the posterior was 6, as is evident from Fig. 8.5. These approximations are close to the exact values found in Example 8.6.

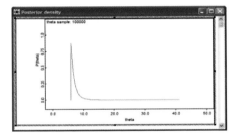

**Fig. 8.5** Output from Inference>Samples>density shows MCMC approximation to the posterior distribution.

|       | mean  | sd    | MC error | val2.5pc | median | val97.5pc | start | sample |
|-------|-------|-------|----------|----------|--------|-----------|-------|--------|
| theta | 7.196 | 1.454 | 0.004906 | 6.025    | 6.736  | 11.03     | 1001  | 100000 |

### 8.6.1 Zero Tricks in WinBUGS

Although the list of built-in distributions for specifying the likelihood or the prior in WinBUGS is rich (p. 742), sometimes we encounter densities that are not on the list.

How do we set the likelihood for a density that is not built into WinBUGS?

There are several ways, the most popular of which is the so-called *zero* trick. Let $f$ be an arbitrary model and $\ell_i = \log f(x_i|\theta)$ the log-likelihood for the $i$th observation. Then

$$\prod_{i=1}^{n} f(x_i|\theta) = \prod_{i=1}^{n} e^{\ell_i} = \prod_{i=1}^{n} \frac{(-\ell_i)^0 e^{-(-\ell_i)}}{0!} = \prod_{i=1}^{n} \mathscr{P}oi(0, -\ell_i).$$

The WinBUGS code for a zero trick can be written as follows.

```
for (i in 1:n){
zeros[i] <- 0
lambda[i] <- -llik[i] + 10000
  # Since lambda[i] needs to be positive as
  # a Poisson rate, an arbitrary constant C
  # can be added;  here we added C = 10000.
zeros[i] ~ dpois(lambda[i])
llik[i] <- ... write the log-likelihood function here
}
```

*Example 8.11.* This example finds the Bayes estimator of parameter $\theta$ in a Maxwell distribution with a density of $f(x|\theta) = \sqrt{\frac{2}{\pi}} \theta^{3/2} x^2 e^{-\theta x^2/2}$, $x \geq 0, \theta > 0$. The moment-matching estimator and the MLE have been discussed in Example 7.4. For a sample of size $n = 3$, $X_1 = 1.4$, $X_2 = 3.1$, and $X_3 = 2.5$ the MLE of $\theta$ was $\hat{\theta}_{\text{MLE}} = 0.5051$. The same estimator was found by moment matching when the second moment was matched. The Maxwell density is not implemented in WinBUGS and we will use a zero trick instead.

```
#Estimation of Maxwell's theta
#Using the zero trick
model{
   for (i in 1:n){
   zeros[i] <- 0
   lambda[i] <- -llik[i] + 10000
   zeros[i] ~ dpois(lambda[i])
   llik[i] <- 1.5 * log(theta)-0.5 * theta * pow(x[i],2)
}
      theta ~ dgamma(0.1, 0.1) #non-informative choice
}
DATA
list(n=3, x=c(1.4, 3.1, 2.5))
INITS
list(theta=1)
```

|  | mean | sd | MC error | val2.5pc | median | val97.5pc | start | sample |
|---|---|---|---|---|---|---|---|---|
| theta | 0.5115 | 0.2392 | 8.645E-4 | 0.1559 | 0.4748 | 1.079 | 1001 | 100000 |

Note that the Bayes estimator with respect to a noninformative prior dgamma(0.1, 0.1) is 0.5115.

## 8.7 Bayesian Interval Estimation: Credible Sets

The Bayesian term for an interval estimator of a parameter is *credible set*. Naturally, the measure used to assess the credibility of an interval estimator is the posterior distribution. Students learning concepts of classical confidence intervals often err by stating that "the probability that a particular confidence interval $[L, U]$ contains parameter $\theta$ is $1 - \alpha$." The correct statement seems more convoluted; one generates data from the underlying model many times and, for each generated data set, calculates the confidence interval. The proportion of confidence intervals covering the unknown parameter "tends to" $1 - \alpha$. The Bayesian interpretation of a credible set $C$ is arguably more natural: the probability of a parameter belonging to set $C$ is $1 - \alpha$. A formal definition follows.

Assume set $C$ is a subset of parameter space $\Theta$. Then $C$ is a *credible set* with credibility $(1 - \alpha)100\%$ if

$$\mathbb{P}(\theta \in C | X) = \mathbb{E}(I(\theta \in C) | X) = \int_C \pi(\theta | x) d\theta \geq 1 - \alpha.$$

If the posterior is discrete, then the integral is a sum (using the counting measure) and

$$\mathbb{P}(\theta \in C | X) = \sum_{\theta_i \in C} \pi(\theta_i | x) \geq 1 - \alpha.$$

This is the definition of a $(1 - \alpha)100\%$ credible set. For a fixed posterior distribution and a $(1 - \alpha)100\%$ "credibility," a credible set is not unique. We will consider two versions of credible sets: highest posterior density (HPD) and equal-tail credible sets.

**HPD Credible Sets.** For a given credibility level $(1 - \alpha)100\%$, the shortest credible set has obvious appeal. To minimize size, the sets should correspond to the highest posterior probability density areas.

**Definition 8.1.** The $(1 - \alpha)100\%$ HPD credible set for parameter $\theta$ is a set $C$, a subset of parameter space $\Theta$ of the form

$$C = \{\theta \in \Theta | \pi(\theta | x) \geq k(\alpha)\},$$

where $k(\alpha)$ is the largest constant for which

$$\mathbb{P}(\theta \in C | X) \geq 1 - \alpha.$$

Geometrically, if the posterior density is cut by a horizontal line at the height $k(\alpha)$, the credible set $C$ is the projection on the $\theta$-axis of the part of the line that lies below the density (Fig. 8.6).

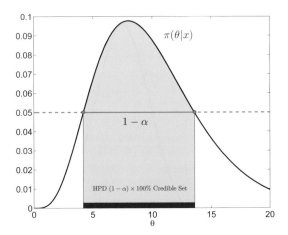

**Fig. 8.6** Highest posterior density (HPD) $(1-\alpha)100\%$ Credible Set (*blue*). The *area in yellow* is $1-\alpha$.

*Example 8.12.* **Jeremy's IQ, Continued.** Recall Jeremy, the enthusiastic bio-engineering student from Example 8.2 who used Bayesian inference in modeling his IQ test scores. For a score of $X$ he was using a $\mathcal{N}(\theta, 80)$ likelihood, while the prior on $\theta$ was $\mathcal{N}(110, 120)$. After the score of $X = 98$ was recorded, the resulting posterior was normal $\mathcal{N}(102.8, 48)$.

Here, the MLE is $\hat{\theta} = 98$, and a 95% confidence interval is $[98-1.96\sqrt{80},\ 98+1.96\sqrt{80}] = [80.4692, 115.5308]$. The length of this interval is approx. 35. The Bayesian counterparts are $\hat{\theta} = 102.8$, and $[102.8-1.96\sqrt{48},\ 102.8+1.96\sqrt{48}] = [89.2207, 116.3793]$. The length of the 95% credible set is approx. 27. The Bayesian interval is shorter because the posterior variance is smaller than the likelihood variance; this is a consequence of the presence of prior information. Figure 8.7 shows the credible set (in blue) and the confidence interval (in red).

From the WinBUGS output table in Jeremy's IQ estimation example (p. 294) the 95% credible set is $[89.18, 116.4]$.

|       | mean  | sd    | MC error | val2.5pc | median | val97.5pc | start | sample |
|-------|-------|-------|----------|----------|--------|-----------|-------|--------|
| theta | 102.8 | 6.943 | 0.01991  | **89.18** | 102.8 | **116.4** | 1001  | 100000 |

Other posterior quantiles that lead to credible sets of different "credibility" levels can be specified in `Sample Monitor Tool` under `Inference>Samples` in WinBUGS. The credible sets from WinBUGS are HPD only if the posterior is symmetric and unimodal.

**Equal-Tail Credible Sets.** HPD credible sets may be difficult to find for asymmetric posterior distributions, such as gamma, Weibull, etc. Much

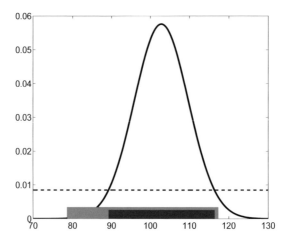

**Fig. 8.7** HPD 95% credible set based on a density of $\mathcal{N}(102.8, 48)$ (*blue*). The interval in *red* is a 95% confidence interval based on the observation $X = 98$ and likelihood variance $\sigma^2 = 80$.

simpler are *equal-tail credible sets* for which the tails have a probability of $\alpha/2$ each for a credibility of $1 - \alpha$. An equal-tail credible set may not be the shortest set, but to find it we need only $\alpha/2$ and $1 - \alpha/2$ quantiles of the posterior. These two quantiles are the lower and upper bounds $[L, U]$:

$$\int_{-\infty}^{L} \pi(\theta|x)\,d\theta = \alpha/2, \quad \int_{U}^{\infty} \pi(\theta|x)\,d\theta = 1 - \alpha/2.$$

Note that WinBUGS gives posterior quantiles from which one can directly establish several equal-tail credible sets (95%, 90%, 80%, and 50%) by selecting appropriate pairs of percentiles in the Sample Monitor Tool.

*Example 8.13.* **Bayesian** *Amanita muscaria.* Recall that in Example 7.8 (p. 251) observations were summarized by $\overline{X} = 10.098$ and $s^2 = 2.1702$, which are classical estimators of population parameters: mean $\mu$ and variance $\sigma^2$. We also obtained the 95% confidence interval for the population mean as $[9.6836, 10.5124]$ and the 90% confidence interval for the population variance as $[1.6074, 3.1213]$.

By assuming noninformative priors for the mean and variance, we use WinBUGS to find Bayesian counterparts of the estimators and confidence intervals. As we pointed out, the mean is a location parameter, and noninformative priors should be "flat." WinBUGS allows for flat priors, mu~dflat(), but any prior with a large variance (or small precision) is a possibility. We take a normal prior with a variance of 10,000. The inverse gamma distribution is traditionally used for a prior on variance; thus, for precision as a reciprocal of variance, the gamma prior is appropriate. As we discussed earlier, gamma distributions with small parameters will have a large variance, thereby making the prior vague/noninformative. We selected `prec~dgamma(0.001, 0.001)` as a noninformative choice. This prior is noninformative because it is essentially flat; its variance is $0.001/(0.001)^2 = 1000$ (p. 166). The WinBUGS program is simple:

```
model{
for ( i in 1:n ){
    amuscaria[i] ~ dnorm( mu, prec )
    }
    mu ~ dnorm(0, 0.00001)
    prec ~ dgamma(0.001, 0.001)
    sig2 <- 1/prec
}
DATA
list(n=51,amuscaria=c(10,11,12,9,10,11,13,12,10,11,11,13,9,10,
        9,10,8,12,10,11,9,10,7,11,8,9,11,11,10,12,10,8,7,11,12,
        10,9,10,11,10,8,10,10,8,9,10,13,9,12,9,9) )
INITS
list( mu =0, prec = 1 )
```

In WinBUGS' `Sample Monitor Tool` we asked for 2.5% and 97.5% posterior percentiles, which gives a 95% credible set and 5% and 95% posterior percentiles for the 90% credible set. The lower/upper bounds of the credible sets are given in boldface and the sets are $[9.684, 10.51]$ for the mean and $[1.607, 3.123]$ for the variance. The credible set for the mean is both HPD and equal-tail, but the credible set for the variance is only an equal-tail.

|      | mean   | sd      | MC error | val2.5pc | val5pc | val95pc | val97.5pc | start | sample |
|------|--------|---------|----------|----------|--------|---------|-----------|-------|--------|
| mu   | 10.1   | 0.2106  | 2.004E-4 | **9.684** | 9.752  | 10.44   | **10.51**  | 1001  | 100000 |
| prec | 0.4608 | 0.09228 | 9.263E-5 | 0.2983   | 0.3202 | 0.6224  | 0.6588    | 1001  | 100000 |
| sig2 | 2.261  | 0.472   | 4.716E-4 | 1.518    | **1.607** | **3.123** | 3.353     | 1001  | 100000 |

## 8.8 Learning by Bayes' Theorem

We start with an example.

*Example 8.14.* Freireich et al. (1963) conducted a remission maintenance therapy to compare 6-MP with placebo for prolonging the duration of remission in leukemia. From 42 patients affected with acute leukemia, but in a state of partial or complete remission, 21 pair was formed. One randomly selected patient from each pair was assigned the maintenance treatment 6-MP, while the other patient received a placebo. Investigators monitored which patient stayed in remission longer. If that was a patient from the 6-MP treatment arm, this was recorded as a "success" (S), otherwise it was a "failure" (F).

The results are given in the following table:

| Pair | 1 | 2 | 3 | 4 | 5 | 6 | 7 | 8 | 9 | 10 |
|---------|---|---|---|---|---|---|---|---|---|----|
| Outcome | S | F | S | S | S | F | S | S | S | S |

| 11 | 12 | 13 | 14 | 15 | 16 | 17 | 18 | 19 | 20 | 21 |
|----|----|----|----|----|----|----|----|----|----|----|
| S  | S  | S  | F  | S  | S  | S  | S  | S  | S  | S  |

The goal is to estimate $p$ – the probability of success. Suppose we got information only on 10 first subjects: 8 successes and 2 failures. When the prior on $p$ is uniform, and the likelihood binomial, the posterior is proportional to $p^8(1-p)^2 \times 1$, which is a beta $\mathcal{B}e(9,3)$.

Suppose now that the remaining 11 observations became available (10 successes and 1 failure). If the posterior from the first stage serves as a prior in the second stage, the updated posterior is proportional to $p^{10}(1-p)^1 \times p^8(1-p)^2$ which is a beta $\mathcal{B}e(19,4)$.

By sequentially updating prior we arrived to the same posterior as if all observations were available at the first place (18 successes and 3 failures). With a uniform prior, this would lead to the same beta $\mathcal{B}e(19,4)$ posterior. The final posterior would be the same even if the updating was done observation-by-observation.

This exemplified the *learning ability* of Bayes' theorem.

Suppose that observations $x_1, \ldots, x_n$ from the model $f(x|\theta)$ are available and that prior on $\theta$ is $\pi(\theta)$. Then the posterior is

$$\pi(\theta|\mathbf{x}) = \frac{f(\mathbf{x}|\theta)\pi(\theta)}{\int f(\mathbf{x}|\theta)\pi(\theta)d\theta},$$

where $\mathbf{x} = (x_1, \ldots, x_n)$ and $f(\mathbf{x}|\theta) = \prod_{i=1}^n f(x_i|\theta)$.

Suppose an additional observation $x_{n+1}$ was collected. Then

$$\pi(\theta|\mathbf{x}, x_{n+1}) = \frac{f(x_{n+1}|\theta)\pi(\theta|\mathbf{x})}{\int f(x_{n+1}|\theta)\pi(\theta|\mathbf{x})d\theta}.$$

Bayes' theorem updates inference in a natural way: the posterior based on previous observations serves as a new prior.

## 8.9 Bayesian Prediction

Up to now we have been concerned with Bayesian inference about population parameters. We are often faced with the problem of predicting a new observation $X_{n+1}$ after $X_1, \ldots, X_n$ from the same population have been observed. Assume that the prior for parameter $\theta$ is elicited. The new observation would have a likelihood of $f(x_{n+1}|\theta)$, while the observed sample $X_1,\ldots,X_n$ will lead to a posterior of $\theta$, $\pi(\theta|X_1,\ldots,X_n)$.

Then, the *posterior predictive distribution* for $X_{n+1}$ can be obtained from the likelihood after integrating out parameter $\theta$ using the posterior distribution,

$$f(x_{n+1}|X_1,\ldots,X_n) = \int_{\Theta} f(x_{n+1}|\theta)\pi(\theta|X_1,\ldots,X_n)d\theta,$$

where $\Theta$ is the domain for $\theta$. Note that the marginal distribution also integrates out the parameter, but using the prior instead of the posterior, $m(x) = \int_{\Theta} f(x|\theta)\pi(\theta)d\theta$. For this reason, the marginal distribution is sometimes called the *prior predictive* distribution.

The prediction for $X_{n+1}$ is the expectation $\mathbb{E}X_{n+1}$, taken with respect to the predictive distribution,

$$\hat{X}_{n+1} = \int_{\mathbb{R}} x_{n+1} f(x_{n+1}|X_1,\ldots,X_n)dx_{n+1},$$

while the *predictive variance*, $\int_{\mathbb{R}} (x_{n+1}-\hat{X}_{n+1})^2 f(x_{n+1}|X_1,\ldots,X_n)dx_{n+1}$, can be used to express the precision of the prediction.

*Example 8.15.* Consider the exponential distribution $\mathcal{E}(\lambda)$ for a random variable $X$ representing a survival time of patients affected by a particular disease. The density for $X$ is $f(x|\lambda) = \lambda\exp\{-\lambda x\}$, $x \geq 0$.

Suppose that the prior for $\lambda$ is gamma $\mathcal{Ga}(\alpha,\beta)$ with a density of $\pi(\lambda) = \frac{\beta^{\alpha}}{\Gamma(\alpha)}\lambda^{\alpha-1}\exp\{-\beta\lambda\}$, $\lambda \geq 0$.

The likelihood, after observing a sample $X_1,\ldots,X_n$ from $\mathcal{E}(\lambda)$ population, is

$$\lambda e^{-\lambda X_1}\cdot\ldots\cdot\lambda e^{-\lambda X_n} = \lambda^n \exp\left\{-\lambda\sum_{i=1}^{n} X_i\right\},$$

and the posterior is proportional to

$$\lambda^{n+\alpha-1}\exp\{-(\sum_{i=1}^{n} X_i + \beta)\lambda\},$$

which can be recognized as a gamma $\mathcal{Ga}(\alpha+n,\beta+\sum_{i=1}^{n} X_i)$ distribution and completed as

$$\pi(\lambda|X_1,\ldots,X_n) = \frac{(\sum_{i=1}^{n} X_i + \beta)^{n+\alpha}}{\Gamma(n+\alpha)}\lambda^{n+\alpha-1}\exp\{-(\sum_{i=1}^{n} X_i + \beta)\lambda\}, \ \lambda \geq 0.$$

The predictive distribution for a new $X_{n+1}$ is

$$f(x_{n+1}|X_1,\ldots,X_n) = \int_0^\infty \lambda \exp\{-\lambda x_{n+1}\}\pi(\lambda|X_1,\ldots,X_n)d\lambda$$

$$= \frac{(n+\alpha)(\sum_{i=1}^n X_i + \beta)^{n+\alpha}}{(\sum_{i=1}^n X_i + \beta + x_{n+1})^{n+\alpha+1}}, \ x_{n+1} > 0.$$

The expected value for a new observation (a Bayesian prediction) is

$$\hat{X}_{n+1} = \int_0^\infty x_{n+1} f(x_{n+1}|X_1,\ldots,X_n)dx_{n+1} = \frac{\sum_{i=1}^n X_i + \beta}{n+\alpha-1}.$$

One can show that the variance of the new observation is

$$\hat{\sigma}^2_{X_{n+1}} = \int_0^\infty (x_{n+1}-\hat{X}_{n+1})^2 f(x_{n+1}|X_1,\ldots,X_n)dx_{n+1} = \frac{(\sum_{i=1}^n X_i + \beta)^2(n+\alpha)}{(n+\alpha-1)^2(n+\alpha-2)}.$$

For example, if $X_1 = 2.1$, $X_2 = 5.5$, $X_3 = 6.4$, $X_4 = 8.7$, $X_5 = 4.9$, $X_6 = 5.1$, and $X_7 = 2.3$ are the observations, and $\alpha = 2$ and $\beta = 1$, then $\hat{X}_8 = 9/2$ and $\hat{\sigma}^2_{X_8} = 729/28 = 26.0357$. Figure 8.8 shows the posterior predictive distribution (solid blue line), observations (crosses), and prediction for the new observation (blue dot). The position of the mean of the data, $\overline{X} = 5$, is shown as a dotted red line.

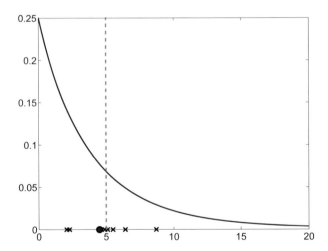

**Fig. 8.8** Bayesian prediction (*blue dot*) based on the sample (*black crosses*) X = [2.1, 5.5, 6.4, 8.7, 4.9, 5.1, 2.3] from the exponential distribution $\mathscr{E}(\lambda)$. The parameter $\lambda$ is given a gamma $\mathscr{G}a(2,1)$ distribution and the resulting posterior predictive distribution is shown as a *solid blue line*. The position of the sample mean is plotted as a *dotted red line*.

*Example 8.16.* To find the Bayesian prediction in WinBUGS, one simply samples a new observation from a likelihood that has updated parameters. In the
WinBUGS program below that implements Example 8.15, the observations
are read within the for loop. However, if a new variable is simulated from the
same likelihood, this is done for the current version of the parameter $\lambda$ and the
mean of simulations approximates the posterior mean of the new observation.

```
model{
 for (i in 1:7){
      X[i] ~ dexp(lambda)
 }
 lambda ~ dgamma(2,1)
 Xnew ~ dexp(lambda)
 }
DATA
list(X = c(2.1,  5.5,  6.4,  8.7,  4.9,  5.1,  2.3))
INITS
list(lambda=1, Xnew=1)
```

The output is

|  | mean | sd | MC error | val2.5pc | median | val97.5pc | start | sample |
|---|---|---|---|---|---|---|---|---|
| Xnew | 4.499 | 5.09 | 0.005284 | 0.1015 | 2.877 | 18.19 | 1001 | 100000 |
| lambda | 0.25 | 0.08323 | 8.343E-5 | 0.1142 | 0.2409 | 0.4378 | 1001 | 100000 |

Note that the posterior mean for Xnew is well approximated, $4.499 \approx 4.5$,
and that the standard deviation sd = 5.09 is close to $\sqrt{26.0357} = 5.1025$.

## 8.10 Consensus Means*

Suppose that several labs are reporting measurements of the same quantity
and that a consensus mean should be calculated. This problem appears in
interlaboratory studies, as well as in multicenter clinical trials and various
meta-analyses. In this section we provide a Bayesian solution to this problem
and compare it with some classical proposals.

Let $Y_{ij}, i = 1,\ldots,k;\ j = 1,\ldots,n_k$ be measurements made at $k$ laboratories,
where $n_i$ measurements come from lab $i$. Let $n = \sum_i n_i$ be the total sample
size.

We are interested in estimating the mean that would properly incorporate
information coming from all the labs, the so-called *consensus mean*. Why is
the solution not trivial and what is wrong with the average $\overline{Y} = 1/n \sum_i \sum_j Y_{ij}$?

There is nothing wrong, under the proper conditions: (a) variabilities
within the labs must be equal and (b) there must be no variability between
the labs.

When (a) is relaxed, proper pooling of the lab sample means is done via a
Graybill–Deal estimator:

$$\overline{Y}_{gd} = \frac{\sum_{i=1}^{k} \omega_i \overline{Y}_i}{\sum_{i=1}^{k} \omega_i}, \quad \omega_i = \frac{n_i}{s_i^2}.$$

When both conditions (a) and (b) are relaxed, there are many competing classical estimators. For example, the Schiller–Eberhardt estimator is given by

$$\overline{Y}_{se} = \frac{\sum_{i=1}^{k} \omega_i \overline{Y}_i}{\sum_{i=1}^{k} \omega_i}, \quad \omega_i = \frac{1}{s_i^2/n_i + s_b^2},$$

where $s_b^2$ is an estimator of the variance between the labs, $s_b^2 = \frac{(\bar{y}_{max} - \bar{y}_{min})^2}{12}$. The Mandel–Paule is the same as the Schiller–Eberhardt estimator but with $s_b^2$ obtained iteratively.

The Bayesian approach is conceptually simple. Individual means as random variables are generated from a single distribution. The mean of this distribution is the consensus mean. In somewhat convoluted wording, the consensus mean is the mean of a hyperprior placed on the individual means.

*Example 8.17.* **Selenium in Milk Powder.** The data on selenium in nonfat milk powder 🖳 selenium.dat are adapted from Witkovsky (2001). Four independent measurement methods are applied. The Bayes estimator of the consensus mean is 108.8.

In the WinBUGS program below, the individual means theta[i] have a $t$-prior with location mu, precision tau, and 5 degrees of freedom. The choice of $t$-prior, instead of the usual normal, is motivated by robustness considerations.

```
model{
for (i in 1:n)
    {
    sel[i] ~ dnorm( theta[lab[i]], prec[lab[i]])
    }
for (i in 1:k)
    {
    theta[i] ~ dt(mu, tau,5) #individual means
    prec[i] ~ dgamma(0.0001, 0.0001)
    sigma2[i] <- 1/prec[i]
    }
mu ~ dt(0,0.0001,5) #consensus mean
tau ~ dgamma(0.0001,0.0001)
si2   <-1/tau
}


DATA
list(lab=c(1,1,1,1,1,1,1,1,    2,2,2,2,2,2,2,2,2,2,2,2,
```

```
3,3,3,3,3,3,3,3,3,3,3,3,3,3,3,          4,4,4,4,4,4,4,4),
sel = c(
115.7,  113.5,  103.3,  119.1,  114.2,  107.3,   91.2,  104.4,
108.6,  109.1,  107.2,  111.5,  100.6,  106.3,  105.9,  109.7,
                                111.1,  107.9,  107.9,  107.9,
107.6,  107.26,109.7,  109.7,  108.5,  106.5,  110.2,  108.3,
                110.5,  108.5,  108.8,  110.1,  109.4,  112.4,
118.7,  109.7,  114.7,  105.4,  113.9,  106.3,  104.8,  106.3),
                                              k=4,  n=42)


INITS
list( mu=1, tau=1, prec=c(1,1,1,1), theta=c(1,1,1,1)  )
```

|          | mean   | sd    | MC error | val2.5pc  | median  | val97.5pc | start | sample |
|----------|--------|-------|----------|-----------|---------|-----------|-------|--------|
| mu       | 108.8  | 0.6499| 0.003674 | 107.6     | 108.9   | 110.0     | 5001  | 500000 |
| si2      | 0.7252 | 9.456 | 0.02088  | 1.024E-4  | 0.01973 | 4.875     | 5001  | 500000 |
| theta[1] | 108.8  | 0.8593| 0.003803 | 107.0     | 108.9   | 110.5     | 5001  | 500000 |
| theta[2] | 108.7  | 0.6184| 0.004188 | 107.2     | 108.7   | 109.7     | 5001  | 500000 |
| theta[3] | 108.9  | 0.4046| 0.00311  | 108.1     | 108.9   | 109.7     | 5001  | 500000 |
| theta[4] | 108.9  | 0.7505| 0.003705 | 107.6     | 108.9   | 110.7     | 5001  | 500000 |

Next, we compare the Bayesian estimator with the classical Graybill–Deal and Schiller–Eberhardt estimators, 108.8892 and 108.7703, respectively. The Bayesian estimator falls between the two classical ones. A 95% credible set for the consensus mean is [107.6, 110].

```
lab1=[115.7, 113.5, 103.3, 119.1, 114.2, 107.3,  91.2, 104.4];
lab2=[108.6, 109.1, 107.2, 111.5, 100.6, 106.3, 105.9, 109.7,...
                           111.1, 107.9, 107.9, 107.9];
lab3=[107.6, 107.26,109.7, 109.7, 108.5, 106.5, 110.2, 108.3,...
                    110.5, 108.5, 108.8, 110.1, 109.4, 112.4];
lab4=[118.7, 109.7, 114.7, 105.4, 113.9, 106.3, 104.8, 106.3];

m = [mean(lab1)  mean(lab2)  mean(lab3)  mean(lab4)];
s = [std(lab1)   std(lab2)   std(lab3)   std(lab4) ];
ni=[8 12 14 8]; k=length(m);

%Graybill-Deal Estimator
wei = ni./s.^2; %weights
m_gd = sum(m .* wei)/sum(wei)   %108.8892

%Schiller-Eberhardt Estimator
z = sort(m);
sb2 = (z(k)-z(1))^2/12;
wei = 1./(s.^2./ni + sb2);%weights
m_se = sum(m .* wei)/sum(wei)   %108.7703
```

**Borrowing Strength and Vague Priors.**  As popularly stated, this model allows for *borrowing strength* in the estimation of both the means $\theta_i$ and the variances $\sigma_i^2$. Even if some labs have extremely small sample sizes (as low as $n = 1$), the lab variances can be estimated through pooling via a hierarchical model structure. The prior distributions above are "vague," which is appropriate when prior information in the form of expert opinion or historic data is not available.

Analyses conducted using vague priors can be considered objective and are generally accepted by classical statisticians. When prior information is available in the form of a mean and variance of $\mu$, it can be included by simply changing the mean and variance of its prior, in our case the normal distribution. It is well known that Bayesian rules are sensitive with respect to changes in hyperparameters in light-tailed priors (e.g., normal priors). If more robustness is required, a $t$ distribution with a small number of degrees of freedom can be substituted for the normal prior.

Note that the consensus mean is the estimator of the mean of a hyperprior on $\theta_i$s, $\mu$. Via WinBUGS' MCMC sampling we get a full posterior distribution of $\mu$ as the ultimate summary information.

## 8.11 Exercises

8.1. **Exponential Lifetimes.** A lifetime $X$ (in years) of a particular device is modeled by an exponential distribution with unknown rate parameter $\theta$. The lifetimes of $X_1 = 5$, $X_2 = 6$, and $X_3 = 4$ are observed. Assume that an expert familiar with this type of device suggests that $\theta$ has an exponential distribution with a mean of 3.

(a) Write down the MLE of $\theta$ for those observations.

(b) Elicit a prior according to the expert assumptions.

(c) For the prior in (b), find the posterior. Is the problem conjugate?

(d) Find the Bayes estimator $\hat{\theta}_{Bayes}$, and compare it with the MLE from (a). Discuss.

(e) Check if the following WinBUGS program gives an estimator of $\lambda$ close to the Bayes estimator in (d):

```
model{
for (i in 1:n){
   X[i] ~ dexp(lambda)
 }
lambda ~ dexp(1/3)
 #note that dexp is parameterized
 #in WinBUGS by the rate parameter
}

DATA
list(n=3, X=c(5,6,4))
```

```
INITS
list(lambda=1)
```

8.2. **Uniform/Pareto.** Suppose $X = (X_1, \ldots, X_n)$ is a sample from $\mathcal{U}(0, \theta)$. Let $\theta$ have a Pareto $\mathcal{P}a(\theta_0, \alpha)$ distribution. Show that the posterior distribution is $\mathcal{P}a(\max\{\theta_0, x_1, \ldots, x_n\}, \alpha + n)$.

8.3. **Nylon Fibers.** Refer to Exercise 5.28, where times (in hours) between blockages of the extrusion process, $T$, had an exponential $\mathcal{E}(\lambda)$ distribution. Suppose that the rate parameter $\lambda$ is unknown, but there are three measurements of interblockage times, $T_1 = 3$, $T_2 = 13$, and $T_3 = 8$.

(a) Estimate parameter $\lambda$ using the moment-matching procedure. Write down the likelihood and find the MLE.

(b) What is the Bayes estimator of $\lambda$ if the prior is $\pi(\lambda) = \frac{1}{\sqrt{\lambda}}$, $\lambda > 0$.

(c) Using WinBUGS find the Bayes estimator and 95% credible set if the prior is lognormal with parameters $\mu = 10$ and $\tau = \frac{1}{\sigma^2} = 0.0001$.

*Hint:* In (b) the prior is not a proper distribution, but the posterior is. Identify the posterior from the product of the likelihood from (a) and the prior.

8.4. **Gamma–Inverse Gamma.** Let $X \sim \mathcal{G}a\left(\frac{n}{2}, \frac{1}{2\theta}\right)$, so that $X/\theta$ is $\chi_n^2$. Let $\theta \sim \mathcal{IG}(\alpha, \beta)$. Show that the posterior is $\mathcal{IG}(n/2 + \alpha, x/2 + \beta)$.

*Hint:* The likelihood is proportional to $\frac{x^{n/2-1}}{(2\theta)^{n/2}} e^{-x/(2\theta)}$ and the prior to $\frac{\beta^\alpha}{\theta^{\alpha+1}} e^{-\beta/\theta}$. Find their product and match the distribution for $\theta$. There is no need to find the marginal distribution and apply Bayes' theorem since the problem is conjugate.

8.5. **Negative Binomial–Beta.** If $X = (X_1, \ldots, X_n)$ is a sample from $\mathcal{NB}(m, \theta)$ and $\theta \sim \mathcal{B}e(\alpha, \beta)$, show that the posterior for $\theta$ is a beta $\mathcal{B}e(\alpha + mn, \beta + \sum_{i=1}^n x_i)$ distribution.

8.6. **Poisson–Gamma Marginal.** In Example 8.9 on p. 295, show that the marginal distribution is a generalized negative binomial.

8.7. **Exponential–Improper.** Find a Bayes estimator for $\theta$ if a single observation $X$ was obtained from a distribution with a density of $f(x|\theta) = \theta \exp\{-\theta x\}$, $x > 0, \theta > 0$. Assume priors (a) $\pi(\theta) = 1$ and (b) $\pi(\theta) = 1/\theta$.

8.8. **Normal Precision–Gamma.** Suppose $X = -2$ was observed from a population distributed as $\mathcal{N}\left(0, \frac{1}{\theta}\right)$ and one wishes to estimate the parameter $\theta$. (Here $\theta$ is the reciprocal of the variance $\sigma^2$ and is called a *precision parameter*. Precision parameters are used in WinBUGS to parameterize the normal distribution). An MLE of $\theta$ does exist, but one may be tempted to estimate $\theta$ as $1/\hat{\sigma}^2$, which is troublesome since there is a single observation. Suppose the analyst believes that the prior on $\theta$ is $\mathcal{G}a(1/2, 1)$.

(a) What is the MLE of $\theta$?

(b) Find the posterior distribution and the Bayes estimator of $\theta$. If the prior on $\theta$ is $\mathcal{G}a(r, \lambda)$, can you represent the Bayes estimator as the weighted average (sum of weights = 1) of the prior mean and the MLE?

(c) Find a 95% equal-tail credible set for $\theta$. Use MATLAB to evaluate the quantiles of the posterior distribution.

(d) Using WinBUGS, numerically find the Bayes estimator from (b) and credible set from (c).

*Hint.* The likelihod is proportional to $\theta^{1/2}e^{-\theta x^2/2}$ while the prior is proportional to $\theta^{r-1}e^{-\lambda\theta}$.

8.9. **Bayes Estimate in a Discrete Case.** Refer to the likelihood and data in Exercise 7.4, p. 266.

(i) If the prior for $\theta$ is

| $\theta$ | 1/12 | 1/6 | 1/4 |
|------|------|-----|-----|
| Prob | 0.3 | 0.3 | 0.4 |

find the posterior and the Bayes estimator.

(b) What would the Bayes estimator look like for a sample of size $n$?

8.10. **Histocompatibility.** A patient who is waiting for an organ transplant needs a histocompatible donor who matches the patient's human leukocyte antigen (HLA) type. For a given patient, the number of matching donors per 1000 National Blood Bank records is modeled as Poisson with an unknown rate $\lambda$. If a randomly selected group of 1000 records showed exactly one match, estimate $\lambda$ in a Bayesian fashion.

For $\lambda$ assume the following:

(a) Gamma $\mathcal{G}a(2, 1)$ prior;

(b) Flat prior $\lambda = 1$, for $\lambda > 0$;

(c) Invariance prior $\pi(\lambda) = \frac{1}{\lambda}$, for $\lambda > 0$;

(d) Jeffreys' prior $\pi(\lambda) = \frac{1}{\sqrt{\lambda}}$, for $\lambda > 0$.

Note that the priors in (b)–(d) are not proper densities (the integrals are not finite); nevertheless, the resulting posteriors are proper.

*Hint:* In all cases (a)–(d), the posterior is gamma. Write the product $\frac{\lambda^1}{1!}\exp\{-\lambda\} \times \pi(\lambda)$ and match the gamma parameters. The first part of the product is the likelihood when exactly one matching donor was observed.

8.11. **Neurons Fire in Potter's Lab 2.** Data set ⬛ neuronfires.mat consisting of 989 firing times in a cell culture of neurons was analyzed in Exercise 7.3. From this data set the count of firings in consecutive 20-ms time intervals was recorded:

| | | | | | | | | | |
|---|---|---|---|---|---|---|---|---|---|
| 20 | 19 | 26 | 20 | 24 | 21 | 24 | 29 | 21 | 17 |
| 23 | 21 | 19 | 23 | 17 | 30 | 20 | 20 | 18 | 16 |
| 14 | 17 | 15 | 25 | 21 | 16 | 14 | 18 | 22 | 25 |
| 17 | 25 | 24 | 18 | 13 | 12 | 19 | 17 | 19 | 19 |
| 19 | 23 | 17 | 17 | 21 | 15 | 19 | 15 | 23 | 22 |

It is believed that the counts are Poisson distributed with unknown parameter $\lambda$. An expert believes that the number of counts in the 20-ms interval
should be about 15.

(a) What is the likelihood function for these 50 observations?

(b) Using the information the expert provided elicit an appropriate gamma
prior. Is such a prior unique?

(c) For the prior suggested in (b) find the Bayes estimator of $\lambda$. How does
this estimator compare to the MLE?

(d) Suppose now that the prior is lognormal with a mean of 15 (one possible choice is $\mu = \log(15) - 1/2 = 2.2081$ and $\sigma^2 = 1$, for example). Using
WinBUGS, find the Bayes estimator for $\lambda$. Recall that WinBUGS uses the
precision parameter $\tau = 1/\sigma^2$ instead of $\sigma^2$.

8.12. **Eliciting a Beta Prior I.** This exercise is based on an example from Berry
and Stangl (1996). An important prognostic factor in the early detection of
breast cancer is the number of axillary lymph nodes. The surgeon will generally remove between 5 and 30 nodes during a traditional axillary dissection. We are interested in making an inference about the proportion of all
nodes affected by cancer and consult the surgeon in order to elicit a prior.
The surgeon indicates that the probability of a selected node testing positive
is 0.05. However, if the first node tested positive, the second will be found
positive with an increased probability of 0.2.

(a) Using Eqs. (8.5), elicit a beta prior that reflects the surgeon's opinion.

(b) If in a particular case two out of seven nodes tested positive, what is the
Bayes estimator of the proportion of affected nodes when the prior in (a) is
adopted.

8.13. **Eliciting a Beta Prior II.** A natural question for the practitioner in the
elicitation of a beta prior is to specify a particular quantile. For example,
we are interested in eliciting a beta prior with a mean of 0.8 such that the
probability of exceeding 0.9 is 5%. Find hyperparameters $a$ and $b$ for such
a prior. *Hint:* See file ◀ belicitor.m

8.14. **Eliciting a Weibull Prior.** Assume that the average recovery time for patients with a particular disease enters a statistical model as a parameter $\theta$
and that prior $\pi(\theta)$ needs to be elicited. Assume further that the functional
form of the prior is Weibull $\mathscr{W}ei(r, \lambda)$ so the elicitation amounts to specifying hyperparameters $r$ and $\lambda$. A clinician states that the first and third
quartiles for $\theta$ are $Q_1 = 10$ and $Q_3 = 20$ (in days). Elicit the prior.
*Hint:* The CDF for the prior is $\Pi(\theta) = 1 - e^{-\lambda \theta^r}$, which with conditions on $Q_1$
and $Q_3$ lead to two equations $- e^{-\lambda \theta^r} = 0.75$ and $e^{-\lambda \theta^r} = 0.25$. Take the log
twice to obtain a system of two equations with two unknowns $r$ and $\log \lambda$.

8.15. **Bayesian Yucatan Pigs.** Refer to Example 7.21 (Yucatan Pigs). Using
WinBUGS, find the Bayesian estimator of $a$ and plot its posterior distribution.

**8.16. Eliciting a Normal Prior.** We elicit a normal prior $\mathcal{N}(\mu, \sigma^2)$ from an expert who can specify percentiles. If the 20th and 70th percentiles are specified as 2.7 and 4.8, respectively, how should $\mu$ and $\sigma$ be elicited?
*Hint:* If $x_p$ is the $p$th quantile (100%$p$th percentile), then $x_p = \mu + z_p \sigma$. A system of two equations with two unknowns is formed with $z_p$s as `norminv(0.20)` = `-0.8416` and `norminv(0.70)` = `0.5244`.

**8.17. Is the Cloning of Humans Moral?** A recent Gallup poll estimates that about 88% of Americans oppose human cloning. Results are based on telephone interviews with a randomly selected national sample of $n = 1000$ adults, aged 18 and older. In these 1000 interviews, 882 adults opposed the cloning of humans.
(a) Write a WinBUGS program to estimate the proportion $p$ of people opposed to human cloning. Use a noninformative prior for $p$.
(b) Pretend that the original poll had $n = 1062$ adults, i.e., results for 62 adults are missing. Estimate the number of people opposed to cloning among the 62 missing in the poll.

**8.18. Poisson Observations with Truncated Normal Rate.** A sample average of $n = 15$ counting observations was found to be $\overline{X} = 12.45$. Assume that each count comes from a Poisson $\mathcal{P}oi(\lambda)$ distribution. Using WinBUGS find the Bayes estimator of $\lambda$ if the prior on $\lambda$ is a normal $\mathcal{N}(0, 10^2)$ constrained to $\lambda \geq 1$.
*Hint:* $n\overline{X} = \sum X_i$ is Poisson $\mathcal{P}oi(n\lambda)$.

**8.19. Counts of Alpha Particles.** In Example 7.13 we analyzed data from the experiment of Rutherford and Geiger on counting $\alpha$-particles.
The counts, given in the table below, can be well modeled by Poisson distribution.

| $X$ | 0 | 1 | 2 | 3 | 4 | 5 | 6 | 7 | 8 | 9 | 10 | 11 | $\geq 12$ |
|-----------|----|-----|-----|-----|-----|-----|-----|-----|----|----|----|---|----|
| Frequency | 57 | 203 | 383 | 525 | 532 | 408 | 273 | 139 | 45 | 27 | 10 | 4 | 2 |

(a) Find sample size $n$ and sample mean $\overline{X}$. In calculations for $\overline{X}$ take $\geq 12$ as 12.
(b) Elicit a gamma prior for $\lambda$ with rate parameter $\beta = 5$ and shape parameter $\alpha$ selected in such a way that the prior mean is 7.
(c) Find the Bayes estimator of $\lambda$ using the prior from (b). Is the problem conjugate? Use the fact that $\sum_{i=1}^{n} X_i \sim \mathcal{P}oi(n\lambda)$.
(d) Write a WinBUGS script that simulates the Bayes estimator for $\lambda$ and compare its output with the analytic solution from (c).

**8.20. Rayleigh Estimation by Zero Trick.** Referring to Exercise 7.9, find the Bayes estimator of $\sigma^2$ in a Rayleigh distribution using WinBUGS.
Since the Rayleigh distribution is not on the list of WinBUGS distributions, one may use a Poisson zero trick with a negative log-likelihood

as negloglik[i] <- C + log(sig2) + pow(r[i],2)/(2 * sig2), where sig2 is the parameter and r[i] are observations.

Since $\sigma$ is a scale parameter, it is customary to put an inverse gamma on $\sigma^2$. This can be achieved by putting a gamma prior on $1/\sigma^2$, as in

sig2 <- 1/isig2
isig2~dgamma(0.1, 0.1)

where the choice of dgamma(0.1, 0.1) is noninformative.

8.21. **Predictions in a Poisson/Gamma Model.** For a sample $X_1, \ldots, X_n$ from a Poisson $\mathcal{P}oi(\lambda)$ distribution and a gamma $\mathcal{G}a(\alpha, \beta)$ prior on $\lambda$,

(i) Prove that the marginal distribution is a generalized negative binomial, and identify its parameters.

(b) Show that the posterior predictive distribution for $X_{n+1}$ is also a generalized negative binomial. Identify its parameters and find the prediction $\hat{X}_4$ for $X_1 = 4$, $X_2 = 5$, and $X_3 = 4.2$, $\alpha = 2$ and $\beta = 1$. Support your findings with a WinBUGS simulation.

8.22. **Estimating Chemotherapy Response Rates.** An oncologist believes that 90% of cancer patients will respond to a new chemotherapy treatment and that it is unlikely that this proportion will be below 80%. Elicit a beta prior that models the oncologist's beliefs.

Hint: $\mu = 0.9$, $\mu - 2\sigma = 0.8$, and use Eqs. (8.4).

During the trial, in 30 patients treated, 22 responded. What are the likelihood and posterior distributions.

(a) Using MATLAB, plot the prior, likelihood, and posterior in a single figure.

(b) Using WinBUGS, find the Bayes estimator of the response rate and compare it to the posterior mean.

---

**MATLAB AND WINBUGS FILES AND DATA SETS USED IN THIS CHAPTER**

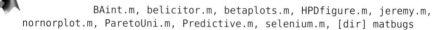

http://springer.bme.gatech.edu/Ch8.Bayes/

BAint.m, belicitor.m, betaplots.m, HPDfigure.m, jeremy.m, nornorplot.m, ParetoUni.m, Predictive.m, selenium.m, [dir] matbugs

copd.odc, ExeTransplant.odc, histocompatibility.odc, jeremy.odc|txt, jeremyminimal.odc, metalabs1.odc, metalabs2.odc, muscaria.odc, neurons.odc, pareto.odc, poistrunorm.odc, predictiveexample.odc, rayleigh.odc, rutherford.odc, selenium.odc, ztmaxwell.odc

selenium.dat

---

# CHAPTER REFERENCES

Anscombe, F. J. (1962). Tests of goodness of fit. *J. Roy. Stat. Soc. B*, **25**, 81–94.

Bayes, T. (1763). An essay towards solving a problem in the doctrine of chances. *Philos. Trans. R. Soc. Lond.*, **53**, 370–418.

Berger, J. O. (1985). *Statistical Decision Theory and Bayesian Analysis*, 2nd edn. Springer, Berlin Heidelberg New York.

Berger, J. O. and Delampady, M. (1987). Testing precise hypothesis. *Stat. Sci.*, **2**, 317–352.

Berger, J. O. and Selke, T. (1987). Testing a point null hypothesis: the irreconcilability of *p*-values and evidence (with discussion). *J. Am. Stat. Assoc.*, **82**, 112–122.

Berry, D. A. and Stangl, D. K. (1996). Bayesian methods in health-related research. In: Berry, D. A. and Stangl, D. K. (eds.). *Bayesian Biostatistics*, Dekker, New York.

Chen, M.-H., Shao, Q.-M., and Ibrahim, J. (2000). *Monte Carlo Methods in Bayesian Computation*. Springer, Berlin Heidelberg New York.

Congdon, P. (2001). *Bayesian Statistical Modelling*. Wiley, Hoboken.

Congdon, P. (2003). *Applied Bayesian Models*. Wiley, Hoboken.

Congdon, P. (2005). *Bayesian Models for Categorical Data*. Wiley, Hoboken.

FDA (2010). Guidance for the use of Bayesian statistics in medical device clinical trials. Center for Devices and Radiological Health Division of Biostatistics, Rockville, MD. http://www.fda.gov/downloads/MedicalDevices/ DeviceRegulationandGuidance/GuidanceDocuments/ucm071121.pdf

Finney, D. J. (1947). The estimation from individual records of the relationship between dose and quantal response. *Biometrika*, **34**, 320–334.

Freireich, E. J., Gehan, E., Frei, E., Schroeder, L. R., Wolman, I. J., Anbari, R., Burgert, E. O., Mills, S. D., Pinkel, D., Selawry, O. S., Moon, J. H., Gendel, B. R., Spurr, C. L., Storrs, R., Haurani, F., Hoogstraten, B. and Lee, S. (1963). The effect of 6-Mercaptopurine on the duration of steroid-induced remissions in acute leukemia: a model for evaluation of other potentially useful therapy. *Blood*, **21**, 699–716.

Garthwhite, P. H. and Dickey, J. M. (1991). An elicitation method for multiple linear regression models. *J. Behav. Decis. Mak.*, **4**, 17–31.

Gelfand, A. E. and Smith, A. F. M. (1990). Sampling-based approaches to calculating marginal densities. *J. Am. Stat. Assoc.*, **85**, 398–409.

Lindley, D. V. (1957). A statistical paradox. *Biometrika*, **44**, 187–192.

Martz, H. and Waller, R. (1985). *Bayesian Reliability Analysis*. Wiley, New York.

Metropolis, N., Rosenbluth, A., Rosenbluth, M., Teller, A., and Teller, E. (1953). Equation of state calculations by fast computing machines. *J. Chem. Phys.*, **21**, 1087–1092.

Ntzoufras, I. (2009). *Bayesian Modeling Using WinBUGS*. Wiley, Hoboken.

Robert, C. (2001). *The Bayesian Choice: From Decision-Theoretic Motivations to Computational Implementation*, 2nd edn. Springer, Berlin Heidelberg New York.

Robert, C. and Casella, G. (2004). *Monte Carlo Statistical Methods*, 2nd edn. Springer, Berlin Heidelberg New York.

Sellke, T., Bayarri, M. J., and Berger, J. O. (2001). Calibration of p values for testing precise null hypotheses. *Am. Stat.*, **55**, 1, 62–71.

Spiegelhalter, D. J., Thomas, A., Best, N. G., and Gilks, W. R. (1996). *BUGS Examples Volume 1*, v. 0.5. Medical Research Council Biostatistics Unit, Cambridge, UK (PDF document).

# Chapter 9
# Testing Statistical Hypotheses

*If one in twenty does not seem high enough odds, we may, if we prefer it, draw the line at one in fifty (the 2 percent point), or one in a hundred (the 1 percent point). Personally, the writer prefers to set a low standard of significance at the 5 percent point, and ignore entirely all results which fail to reach this level. A scientific fact should be regarded as experimentally established only if a properly designed experiment rarely fails to give this level of significance.*

– Ronald Aylmer Fisher

---

### WHAT IS COVERED IN THIS CHAPTER

- Basic Concepts in Testing: Hypotheses, Errors of the First and Second Kind, Rejection Regions, Significance Level, $p$-value, Power
- The Bayesian Approach to Testing
- Testing the Mean in a Normal Population: $z$ and $t$ Tests
- Testing the Variance in a Normal Population
- Testing the Population Proportion
- Multiple Testing, Bonferroni Correction, and False Discovery Rate

---

## 9.1 Introduction

The two main tasks of inferential statistics are parameter estimation and testing statistical hypotheses. In this chapter we will focus on the latter. Although the expositions on estimation and testing are separate, the two inference tasks are highly related, as it is possible to conduct testing by inspecting confidence intervals or credible sets. Both tasks can be unified via the so-called decision-theoretic approach in which both the estimator and the selection of a hypothesis represent an optimal action given the model, observations, and loss (utility) function.

Generally, any claim made about one or more populations of interest constitutes a *statistical hypothesis*. These hypotheses usually involve population parameters, the nature of the population, the relation between the populations, and so on. For example, we may hypothesize that:

- The mean of a population, $\mu$, is 2, or
- Two populations have the same variances, or
- A population is normally distributed, or
- The means in four populations are the same, or
- Two populations are independent, and so on.

Procedures leading to either the acceptance or rejection[1] of statistical hypotheses are called statistical tests.

We will discuss two approaches: the frequentist (classical) approach, which is based on the Neyman–Pearson lemma, and the Bayes approach, which assigns probabilities to hypotheses directly.

The Neyman–Pearson lemma is technical [details can be found in Casella and Berger (1990)], and the testing procedure based on it will be formulated as an algorithm (testing "recipe"). In fact, this recipe is a mix of Neyman–Pearson's and Fisherian approaches that takes the best from both: a framework for power analysis from the Neyman–Pearsonian approach and better sensitivity to the observations from the Fisherian method.

In the Bayesian framework, one simply finds and reports the probability that a particular hypothesis is true given the observations. The competing hypotheses are assigned probabilities and those with the larger probability are selected. Frequentist tests do not assign probabilities to hypotheses directly but rather to the statistic on which the test is based. This point will be emphasized later, since the $p$-values are often mistaken for probabilities of hypotheses.

We start by discussing the terminology and algorithm of the frequentist testing framework.

---

[1] Some authors recommend avoiding the use of jargon such as *accept* and *reject* in the testing context. The equivalent but conservative wording for *accept* would be: *there is not enough statistical evidence to reject*. We will use the terms "reject" and "do not reject" when appropriate, leaving the careful wording to practicing statisticians who could be liable for the undesirable consequences of their straightforward recommendations.

## 9.2 Classical Testing Problem

### 9.2.1 Choice of Null Hypothesis

The usual starting point in statistical testing is the formulation of statistical hypotheses. There will be at least (in most cases, exactly) two competing hypotheses. The hypothesis that reflects the current *state of nature*, adopted standard, or believed truth is denoted by $H_0$ (null hypothesis). The competing hypothesis $H_1$ is called the alternative or research hypothesis. Sometimes, the alternative hypothesis is denoted by $H_a$.

It is important which of the two hypotheses is assigned to be $H_0$ since the subsequent testing procedure depends on this assignment. The following "rule" describes the choice of $H_0$ and hints at the reason why it is termed the *null* hypothesis.

**Rule:** We want to establish an assertion about a population with substantive support obtained from the data. The negation of the assertion is taken to be the *null* hypothesis $H_0$, and the assertion itself is taken to be the research or alternative hypothesis $H_1$. In this context, the term *null* can be interpreted as a void research hypothesis.

The following example illustrates several hypothetical testing scenarios.

*Example 9.1.* (a) A biomedical engineer wants to determine whether a new chemical agent provides a faster reaction than the agent currently in use. The new agent is more expensive, so the engineer would not recommend it unless its faster reaction is supported by experimental evidence. The reaction times are observed in several experiments prepared with the new agent. If the reaction time is denoted by the parameter $\theta$, then the two hypotheses can be expressed in terms of that parameter. It is assumed that the reaction speed of the currently used agent is known, $\theta = \theta_0$. Null hypothesis $H_0$: The new agent is not faster ($\theta = \theta_0$). Alternative hypothesis $H_1$: The new agent is faster ($\theta > \theta_0$).

(b) A state labor department wants to determine if the current rate of unemployment varies significantly from the forecast of 8% made 2 months ago. Null hypothesis $H_0$: The current rate of unemployment is 8%. Alternative hypothesis $H_1$: The current rate of unemployment differs from 8%.

(c) A biomedical company claims that a new treatment is more effective than the standard treatment for prolonging the lives of terminal cancer patients. The standard treatment has been in use for a long time, and from reports in medical journals the mean survival period is known to be 5.2 years. Null hypothesis $H_0$: The new treatment is as effective as the standard one, that is, the survival time $\theta$ is equal to 5.2 years. Alternative hypothesis $H_1$: The new treatment is more effective than the standard one, i.e., $\theta > 5.2$.

(d) Katz et al. (1990) examined the performance of 28 students taking the SAT who answered multiple-choice questions without reading the referred passages. The mean score for the students was 46.6 (out of 100), with a standard deviation of 6.8. The expected score in random guessing is 20. Null hypothesis $H_0$: The mean score is 20. Alternative hypothesis $H_1$: The mean score is larger than 20.

(e) A pharmaceutical company claims that its best-selling painkiller has a mean effective period of at least 6 hours. Experimental data found that the average effective period was actually 5.3 hours. Null hypothesis $H_0$: The best-selling painkiller has a mean effective period of 6 hours. Alternative hypothesis $H_1$: The best-selling painkiller has a mean effective period of less than 6 hours.

(f) A pharmaceutical company claims that its generic drug has a mean AUC response equivalent to that of the innovative (brand name) drug. The regulatory agency considers two drugs bioequivalent if the population means in their AUC responses differ for no more than $\delta$.

Null hypothesis $H_0$: The difference in mean responses in AUC between the generic and innovative drugs is either smaller than $-\delta$ or larger than $\delta$. Alternative hypothesis $H_1$: The absolute difference in the mean responses is smaller than $\delta$, that is, the generic and innovative drugs are bioequivalent.

✎

When $H_0$ is stated as $H_0 : \theta = \theta_0$, the alternative hypothesis can be any of

$$\theta < \theta_0, \quad \theta \neq \theta_0, \quad \theta > \theta_0.$$

The first and third alternatives are one-sided while the middle one is two-sided. Usually the context of the problem indicates which one-sided alternative is appropriate. For example, if the pharmaceutical industry claims that the proportion of patients allergic to a particular drug is $p = 0.01$, then either $p \neq 0.01$ or $p > 0.01$ is a sensible alternative in this context, especially if the observed proportion $\hat{p}$ exceeds 0.01.

In the context of the bioequivalence trials, the research hypothesis $H_1$ states that the difference between the responses is "tolerable," as in (f). There, $H_0 : \mu_1 - \mu_2 < -\delta$ or $\mu_1 - \mu_2 > \delta$ and the alternative is $H_1 : -\delta \leq \mu_1 - \mu_2 \leq \delta$.

## 9.2.2 Test Statistic, Rejection Regions, Decisions, and Errors in Testing

Famous and controversial Cambridge astronomer Sir Fred Hoyle (1915–2001) once said: "I don't see the logic of rejecting data just because they seem incredible." The calibration of the credibility of data is done with respect to some theory or model; instead of rejecting data, the model should be questioned.

Suppose that a hypothesis $H_0$ and its alternative $H_1$ are specified, and a random sample from the population under research is obtained. As in the estimation context, an appropriate statistic is calculated from the random sample. Testing is carried out by evaluating the realization of this statistic. If the realization appears unlikely (under the assumption stipulated by $H_0$), $H_0$ is rejected, since the experimental support for $H_0$ is lacking.

If a null hypothesis is rejected when it is actually true, then a *type I error*, or *error of the first kind*, is committed. If, on the other hand, an incorrect null hypothesis is not rejected, then a *type II error*, or *error of the second kind*, is committed. It is customary to denote the probability of a type I error as $\alpha$ and the probability of a type II error as $\beta$.

This is summarized in the table below:

|  |  | decide $H_0$ | decide $H_1$ |
|---|---|---|---|
| true $H_0$ | | Correct action | Type I error |
| probability | | $1 - \alpha$ | $\alpha$ |
| true $H_1$ | | Type II error | Correct action |
| probability | | $\beta$ | power $= 1 - \beta$ |

We will also use the notation $\alpha = P(H_1|H_0)$ to denote the probability that hypothesis $H_1$ is decided when in fact $H_0$ is true. Analogously, $\beta = P(H_0|H_1)$.

A good testing procedure minimizes the probabilities of errors of the first and second kind. However, minimizing both errors at the same time for a fixed sample size is impossible. Controlling the errors is a tradeoff; when $\alpha$ decreases, $\beta$ increases, and vice versa. For this and other practical reasons, $\alpha$ is chosen from among several typical values: 0.01, 0.05, and 0.10.

Sometimes within testing problems there is no clear dichotomy: the *established truth* versus the *research hypothesis*, and both hypotheses may seem to be research hypotheses. For instance, the statements "The new drug is safe" and "The new drug is not safe" are both research hypotheses. In such cases $H_0$ is selected in such a way that the type I error is more severe than the type II error. If the hypothesis "The new drug is not safe" is chosen as $H_0$, then the type I error (rejection of a true $H_0$, "use unsafe drug") is more serious (at least for the patient) than the type II error (keeping a false $H_0$, "do not use a safe drug").

That is another reason why $\alpha$ is fixed as a small number; the probability of a more serious error should be controlled. The practical motivation for fixing a few values for $\alpha$ was originally the desire to keep the statistical tables needed to conduct a given test brief. This reason is now outdated since the "tables" are electronic and their brevity is not an issue.

## 9.2.3 Power of the Test

Recall that $\alpha = \mathbb{P}(\text{reject } H_0 | H_0 \text{ true})$ and $\beta = P(\text{reject } H_1 | H_1 \text{ true})$ are the probabilities of first- and second-type errors. For a specific alternative $H_1$, the probability $\mathbb{P}(\text{reject } H_0 | H_1 \text{true})$ is the *power* of the test.

Power $= 1 - \beta$  $( = \mathbb{P}(\text{reject } H_0 | H_1 \text{true}) )$

In plain terms, the power is measured by the probability that the test will reject a false $H_0$. To find the power, the alternative must be specific. For instance, in testing $H_0 : \theta = 0$ the alternative $H_1 : \theta = 2$ is specific but $H_1 : \theta > 0$ is not. A specific alternative is needed for the evaluation of the probability $\mathbb{P}(\text{reject } H_0 | H_1 \text{ true})$. The specific null and alternative hypotheses lead to the definition of *effect size*, a quantity that researchers want to set as a sensitivity threshold for a test.

Usually, the power analysis is prospective in nature. One plans the sample size and specifies the parameters in $H_0$ and $H_1$. This allows for the calculation of an error of the second kind, $\beta$, and the power as $1 - \beta$. This prospective power analysis is desirable and often required. In the real world of research and drug development, for example, no regulating agency will support a proposed clinical trial if the power analysis was not addressed.

The test protocols need sufficient sample sizes for the test to be sensitive enough to discrepancies from the null hypotheses. However, the sample sizes should not be unnecessarily excessive because of financial and ethical considerations (expensive sampling, experiments that involve laboratory animals). Also, overpowered tests may detect the effects of sizes irrelevant from a medical or engineering standpoint.

The calculation of the power after data are observed and the test was conducted (retrospective power) is controversial (Hoenig and Heisey, 2001). After the sampling is done, more information is available. If $H_0$ was not rejected, the researcher may be interested in knowing if the sampling protocol had enough power to detect effect sizes of interest. Inclusion of this new information in the power calculation and the perception that the goal of retrospective analysis is to justify the failure of a test to reject the null hypothesis lead to the controversy referred to earlier. Some researchers argue that retrospective power analysis should be conducted in cases where $H_0$ was rejected "in order not to declare $H_1$ true if the test was underpowered." However, this argument only emphasizes the need for the power analysis to be done beforehand. Calculating effect sizes from the collected data may also lead to a low retrospective power of well-powered studies.

### 9.2.4 Fisherian Approach: p-Values

A lot of information is lost by reporting only that the null hypothesis should or should not be rejected at some significance level. Reporting a measure of support for $H_0$ is much more desirable. For this measure of support, the p-value is routinely reported despite the controversy surrounding its meaning and use. The p-value approach was favored by Fisher, who criticized the Neyman–Pearsonian approach for reporting only a fixed probability of errors of the first kind, $\alpha$, no matter how strong the evidence against $H_0$ was. Fisher also criticized the Neyman–Pearsonian paradigm for its need for an alternative hypothesis and for a power calculation that depended on unknown parameters.

> A p-value is the probability of obtaining a value of the test statistic as extreme or more extreme (from the standpoint of the null hypothesis) than that actually obtained, given that the null hypothesis is true.

Equivalently, the p-value can be defined as the lowest significance level at which the observed statistic would be significant.

**Advantage of Reporting p-values.** When a researcher reports a p-value as part of their research findings, users can judge the findings according to the significance level of their choice.

> Decisions from a p-value:
> - If the p-value is less than $\alpha$: reject $H_0$.
> - If the p-value is greater than $\alpha$: do not reject $H_0$.

In the Fisherian approach, $\alpha$ is not connected to the error probability; it is a significance level against which the p-value is judged. The most frequently used value for $\alpha$ is 5%, though values of 1% or 10% are sometimes used as well. The recommendation of $\alpha = 0.05$ is attributed to Fisher (1926), whose "one-in-twenty" quote is provided at the beginning of this chapter.

A hypothesis may be rejected if the p-value is less than 0.05; however, a p-value of 0.049 is not the same evidence against $H_0$ as a p-value of 0.000001. Also, it would be incorrect to say that for any nonsmall p-value the null hypothesis is *accepted*. A large p-value indicates that the model stipulated under the null hypothesis is merely consistent with the observed data and that there could be many other such consistent models. Thus, the appropriate wording would be that the null hypothesis is *not rejected*.

Many researchers argue that the p-value is strongly biased against $H_0$ and that the evidence against $H_0$ derived from p-values not substantially smaller than 0.05 is rather weak. In Sect. 9.3.1 we discuss the calibration of $p$-values against Bayes factors and errors in testing.

The $p$-value is often confused with the probability of $H_0$, which it is not. It is the probability that the test statistic will be more extreme than observed when $H_0$ is true. If the $p$-value is small, then an unlikely statistic has been observed that casts doubt on the validity of $H_0$.

## 9.3 Bayesian Approach to Testing

In previous tests it was customary to formulate $H_0$ as $H_0 : \theta = 0$ versus $H_1 : \theta > 0$ instead of $H_0 : \theta \leq 0$ versus $H_1 : \theta > 0$, as one may expect. The reason was that we calculated the $p$-value under the assumption that $H_0$ is true, and this is why a precise null hypothesis was needed.

Bayesian testing is conceptually straightforward: The hypothesis with a higher posterior probability is selected.

Assume that $\Theta_0$ and $\Theta_1$ are two nonoverlapping sets for parameter $\theta$. We assume that $\Theta_1 = \Theta_0^c$, although arbitrary nonoverlapping sets $\Theta_0$ and $\Theta_1$ are easily handled. Let $\theta \in \Theta_0$ be the statement of the null hypothesis $H_0$ and let $\theta \in \Theta_1 = \Theta_0^c$ be the same for the alternative hypothesis $H_1$:

$$H_0 : \theta \in \Theta_0 \qquad H_1 : \theta \in \Theta_1.$$

Bayesian tests amount to a comparison of posterior probabilities of $\Theta_0$ and $\Theta_1$, the regions corresponding to the two competing hypotheses.

If $\pi(\theta|x)$ is the posterior probability, then the hypothesis corresponding to the larger of

$$p_0 = \mathbb{P}(H_0|X) = \int_{\Theta_0} \pi(\theta|x)d\theta,$$

$$p_1 = \mathbb{P}(H_1|X) = \int_{\Theta_1} \pi(\theta|x)d\theta$$

is accepted. Here $\mathbb{P}(H_i|X)$ is the notation for the posterior probability of hypothesis $H_i$, $i = 0, 1$.

Conceptually, this approach differs from frequentist testing where the $p$-value measures the agreement of data with the model postulated by $H_0$, but not the probability of $H_0$.

*Example 9.2.* We return to Jeremy (Examples 8.2 and 8.8) and consider the posterior for the parameter $\theta$, $\mathcal{N}(102.8, 48)$. Jeremy claims he had a bad day and his true IQ is at least 105. The posterior probability of $\theta \geq 105$ is

$$p_0 = \mathbb{P}(\theta \geq 105|X) = \mathbb{P}\left(Z \geq \frac{105 - 102.8}{\sqrt{48}}\right) = 1 - \Phi(0.3175) = 0.3754,$$

less than 1/2, so his claim is rejected.

We represent the prior and posterior odds in favor of the hypothesis $H_0$, respectively, as

$$\frac{\pi_0}{\pi_1} = \frac{\mathbb{P}(H_0)}{\mathbb{P}(H_1)} \quad \text{and} \quad \frac{p_0}{p_1} = \frac{\mathbb{P}(H_0|X)}{\mathbb{P}(H_1|X)}.$$

The *Bayes factor* in favor of $H_0$ is the ratio of the corresponding posterior to prior odds:

$$B_{01}^{\pi}(x) = \frac{\mathbb{P}(H_0|X)}{\mathbb{P}(H_1|X)} \bigg/ \frac{\mathbb{P}(H_0)}{\mathbb{P}(H_1)} = \frac{p_0/p_1}{\pi_0/\pi_1}. \tag{9.1}$$

In the context of Bayes' rule in Chap. 3 we discussed the Bayes factor (p. 88). Its meaning here is analogous: the Bayes factor updates the prior odds of hypotheses to their posterior odds, after an experiment was conducted.

*Example 9.3.* Jeremy continued: The posterior odds in favor of $H_0$ are $\frac{0.3754}{1-0.3754} = 0.4652$, less than 1.

When the hypotheses are simple (i.e., $H_0 : \theta = \theta_0$ versus $H_1 : \theta = \theta_1$) and the prior is just the two-point distribution $\pi(\theta_0) = \pi_0$ and $\pi(\theta_1) = \pi_1 = 1 - \pi_0$, then the Bayes factor in favor of $H_0$ becomes the likelihood ratio:

$$B_{01}^{\pi}(x) = \frac{\mathbb{P}(H_0|X)}{\mathbb{P}(H_1|X)} \bigg/ \frac{\mathbb{P}(H_0)}{\mathbb{P}(H_1)} = \frac{f(x|\theta_0)\pi_0}{f(x|\theta_1)\pi_1} \bigg/ \frac{\pi_0}{\pi_1} = \frac{f(x|\theta_0)}{f(x|\theta_1)}.$$

If the prior is a mixture of two priors, $\xi_0$ under $H_0$ and $\xi_1$ under $H_1$, then the Bayes factor is the ratio of two marginal (prior-predictive) distributions generated by $\xi_0$ and $\xi_1$. Thus, if $\pi(\theta) = \pi_0\xi_0(\theta) + \pi_1\xi_1(\theta)$, then

$$B_{01}^{\pi}(x) = \frac{\frac{\int_{\Theta_0} f(x|\theta)\pi_0\xi_0(\theta)d\theta}{\int_{\Theta_1} f(x|\theta)\pi_1\xi_1(\theta)d\theta}}{\frac{\pi_0}{\pi_1}} = \frac{m_0(x)}{m_1(x)}.$$

As noted earlier, the Bayes factor measures the relative change in prior odds once the evidence is collected. Table 9.1 offers practical guidelines for Bayesian testing of hypotheses depending on the value of the log-Bayes factor.

One could use $B_{01}^{\pi}(x)$, but then $a \leq \log B_{10}(x) \leq b$ becomes $-b \leq \log B_{01}(x) \leq -a$. Negative values of the log-Bayes factor are handled by using, in an obvious way, symmetry and changed wording.

**Table 9.1** Treatment of $H_0$ according to log-Bayes factor values.

| Value | Meaning |
|---|---|
| $0 \leq \log B_{10}(x) \leq 0.5$ | Evidence against $H_0$ is **poor** |
| $0.5 \leq \log B_{10}(x) \leq 1$ | Evidence against $H_0$ is **substantial** |
| $1 \leq \log B_{10}(x) \leq 2$ | Evidence against $H_0$ is **strong** |
| $\log B_{10}(x) > 2$ | Evidence against $H_0$ is **decisive** |

Suppose $X|\theta \sim f(x|\theta)$ is observed and we are interested in testing

$$H_0 : \theta = \theta_0 \quad v.s. \quad H_1 : \theta \neq \theta_0.$$

If the priors on $\theta$ are continuous distributions, Bayesian testing of precise hypotheses in the manner we just discussed is impossible. With continuous priors and subsequently continuous posteriors, the probability of a singleton $\theta = \theta_0$ is always 0, and the precise hypothesis is always rejected.

The Bayesian solution is to adopt a prior where $\theta_0$ has a probability (a distribution with point mass at $\theta_0$) of $\pi_0$ and the rest of the probability is spread by a distribution $\xi(\theta)$ that is the prior under $H_1$. Thus, the prior on $\theta$ is a mixture of the point mass at $\theta_0$ with a weight $\pi_0$ and a continuous density $\xi(\theta)$ with a weight of $\pi_1 = 1 - \pi_0$. One can show that the marginal density for $X$ is

$$m(x) = \pi_0 f(x|\theta_0) + \pi_1 m_1(x),$$

where $m_1(x) = \int f(x|\theta)\xi(\theta)d\theta$.

The posterior probability of the null hypothesis uses this marginal distribution and is equal to

$$\pi(\theta_0|x) = f(x|\theta_0)\pi_0/m(x) = \frac{f(x|\theta_0)\pi_0}{\pi_0 f(x|\theta_0) + \pi_1 m_1(x)} = \left(1 + \frac{\pi_1}{\pi_0} \cdot \frac{m_1(x)}{f(x|\theta_0)}\right)^{-1}.$$

There is an alternate way of testing the precise null hypothesis in a Bayesian fashion. One could test the hypothesis $H_0 : \theta = \theta_0$ against the two-sided alternative by credible sets for $\theta$. If $\theta_0$ belongs to a 95% credible set for $\theta$, then $H_0$ is not rejected. One-sided alternatives can be accommodated as well by one-sided credible sets. This approach is natural and mimics testing by confidence intervals; however, the posterior probabilities of hypotheses are not calculated.

**Testing Using WinBUGS.** WinBUGS generates samples from the posterior distribution. Testing hypotheses is equivalent to finding the relative frequencies of a posterior sample falling in competing regions $\Theta_0$ and $\Theta_1$. For example, if

$$H_0 : \theta \le 1 \quad \text{versus} \quad H_1 : \theta > 1$$

is tested in the WinBUGS program, the command ph1<-step(theta-1) will calculate the proportion of the simulated chain falling in $\Theta_1$, that is, satisfying $\theta > 1$. The step(x) is equal to 1 if $x \ge 0$ and 0 if $x < 0$.

### 9.3.1 Criticism and Calibration of p-Values*

In a provocative article, Ioannidis (2005) states that *many published research findings are false* because statistical significance by a particular team of researchers is found. Ioannidis lists several reasons: "...a research finding is less likely to be true when the studies conducted in a field are smaller; when effect sizes are smaller; when there is a greater number and lesser pre-selection of tested relationships; where there is greater flexibility in designs, definitions, outcomes, and analytical modes; when there is greater financial and other interest and prejudice; and when more teams are involved in a scientific field in case of statistical significance." Certainly great responsibility for an easy acceptance of research (alternative) hypotheses can be attributed to the *p*-values. There are many objections to the use of raw *p*-values for testing purposes.

Since the *p*-value is the probability of obtaining the statistic as large or more extreme than observed, when $H_0$ is true, the *p*-values measure how consistent the data are with $H_0$, and they may not be a measure of support for a particular $H_0$.

Misinterpretation of the *p*-value as the error probability leads to a strong bias against $H_0$. What is the posterior probability of $H_0$ in a test for which the reported *p*-value is *p*? Berger and Sellke (1987) and Sellke et al. (2001) show that the minimum Bayes factor (in favor of $H_0$) for a null hypothesis having a *p*-value of *p* is $-e\, p \log p$. The Bayes factor transforms the prior odds $\pi_0/\pi_1$ into the posterior odds $p_0/p_1$, and if the prior odds are 1 ($H_0$ and $H_1$ equally likely *a priori*, $\pi_0 = \pi_1 = 1/2$), then the posterior odds of $H_0$ are not smaller than $-e\, p \log p$ for $p < 1/e \approx 0.368$:

$$\frac{p_0}{p_1} \ge -ep \log p, \quad p < 1/e.$$

By solving this inequality with respect to $p_0$, we obtain a posterior probability of $H_0$ as

$$p_1 \geq \frac{1}{1+(-e\,p\log p)^{-1}},$$

which also has a frequentist interpretation as a type I error, $\alpha(p)$. Now, the effect of bias against $H_0$, when judged by the $p$-value, is clearly visible. The type I error, $\alpha$, always exceeds $(1+(-e\,p\log p)^{-1})^{-1}$.

It is instructive to look at specific numbers. Assume that a particular test yielded a $p$-value of 0.01, which led to the rejection of $H_0$ "with decisive evidence." However, if *a priori* we do not have a preference for either $H_0$ or $H_1$, the posterior odds of $H_0$ always exceed 12.53%. The frequentist type I error or, equivalently, the posterior probability of $H_0$ is never smaller than 11.13% – certainly not strong evidence against $H_0$.

Figure 9.1 (generated by ◢ SBB.m) compares a $p$-value (dotted line) with a lower bound on the Bayes factor (red line) and a lower bound on the probability of a type I error $\alpha$ (blue line).

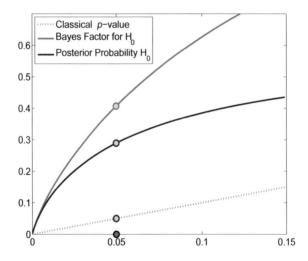

**Fig. 9.1** Calibration of $p$-values. A $p$-value (*dotted line*) is compared with a lower bound on the Bayes factor (*red line*) and a lower bound on the frequentist type I error $\alpha$ (*blue line*). The bound on $\alpha$ is also the lower bound on the posterior probability of $H_0$ when the prior probabilities for $H_0$ and $H_1$ are equal. For the $p$-value of 0.05, the type I error is never smaller than 0.2893, while the Bayes factor in favor of $H_0$ is never smaller than 0.4072.

```
sbb  = @(p)  - exp(1) * p .* log(p);
alph = @(p) 1./(1 + 1./(-exp(1)*p.*log(p)) );
%
```

```
pp = 0.0001:0.001:0.15
plot(pp, pp, ':', 'linewidth',lw)
hold on
plot(pp, sbb(pp), 'r-','linewidth',lw)
plot(pp, alph(pp), '-','linewidth',lw)
```

The interested reader is directed to Berger and Sellke (1987), Schervish (1996), and Goodman (1999a,b, 2001), among many others, for a constructive criticism of $p$-values.

## 9.4 Testing the Normal Mean

In testing the normal mean we will distinguish two cases depending on whether the population variance is known ($z$-test) or not known ($t$-test).

### 9.4.1 $z$-Test

Let us assume that we are interested in testing the null hypothesis $H_0 : \mu = \mu_0$ on the basis of a sample $X_1, \ldots, X_n$ from a normal distribution $\mathcal{N}(\mu, \sigma^2)$, where the variance $\sigma^2$ is assumed known. Situations in which the population mean is unknown but the population variance is known are rare, but not unrealistic. For example, a particular measuring equipment has well-known precision characteristics but might not be calibrated.

We know that $\overline{X} \sim \mathcal{N}(\mu, \sigma^2/n)$ and that $Z = \frac{\overline{X} - \mu_0}{\sigma/\sqrt{n}}$ is the standard normal distribution if the null hypothesis is true, that is, if $\mu = \mu_0$. This statistic, $Z$, is used to test $H_0$, and the test is called a $z$-test. Statistic $Z$ is compared to quantiles of the standard normal distribution.

The test can be performed using either (i) the rejection region or (ii) the $p$-value.

(i) The rejection region depends on the level $\alpha$ and the alternative hypothesis. For one-sided hypotheses the tail of the rejection region follows the direction of $H_1$. For example, if $H_1 : \mu > 2$ and the level $\alpha$ is fixed, the rejection region is $[z_{1-\alpha}, \infty)$. For the two-sided alternative hypothesis $H_1 : \mu \neq \mu_0$ and significance level of $\alpha$, the rejection region is two-sided, $(-\infty, z_{\alpha/2}] \cup [z_{1-\alpha/2}, \infty)$. Since the standard normal distribution is symmetric about 0 and $z_{\alpha/2} = -z_{1-\alpha/2}$, the two-sided rejection region is sometimes given as $(-\infty, -z_{1-\alpha/2}] \cup [z_{1-\alpha/2}, \infty)$.

The test is now straightforward. If statistic $Z$ is calculated from the observations $X_1, \ldots, X_n$ falls within the rejection region, the null hypothesis is rejected. Otherwise, we say that hypothesis $H_0$ is not rejected.

(ii) As discussed earlier, the $p$-value gives a more refined analysis in testing than the "reject–do not reject" decision rule. The $p$-value is the probability of the rejection-region-like area cut by the observed $Z$ (and, in the case of a two-sided alternative, by $-Z$ and $Z$) where the probability is calculated by the distribution specified by the null hypothesis.

The following table summarizes the $z$-test for $H_0 : \mu = \mu_0$ and $Z = \frac{\overline{X} - \mu_0}{\sigma/\sqrt{n}}$:

| Alternative | $\alpha$-level rejection region | $p$-value (MATLAB) |
|---|---|---|
| $H_1 : \mu > \mu_0$ | $[z_{1-\alpha}, \infty)$ | `1-normcdf(z)` |
| $H_1 : \mu \neq \mu_0$ | $(-\infty, z_{\alpha/2}] \cup [z_{1-\alpha/2}, \infty)$ | `2*normcdf(-abs(z))` |
| $H_1 : \mu < \mu_0$ | $(-\infty, z_\alpha]$ | `normcdf(z)` |

### 9.4.2 Power Analysis of a z-Test

The power of a test is found against a specific alternative, $H_1 : \mu = \mu_1$. In a $z$-test, the variance $\sigma^2$ is known and $\mu_0$ and $\mu_1$ are specified by their respective $H_0$ and $H_1$.

The power is the probability that a $z$-test of level $\alpha$ will detect the effect of size $e$ and thus, reject $H_0$. The effect size is defined as $e = \frac{|\mu_0 - \mu_1|}{\sigma}$.

Usually $\mu_1$ is selected such that effect $e$ has a medical or engineering relevance.

**Power of the $z$-test for $H_0 : \mu = \mu_0$ when $\mu_1$ is the actual mean.**
- One-sided test:

$$1 - \beta = \Phi\left(z_\alpha + \frac{|\mu_0 - \mu_1|}{\sigma/\sqrt{n}}\right) = \Phi\left(-z_{1-\alpha} + \frac{|\mu_0 - \mu_1|}{\sigma/\sqrt{n}}\right)$$

- Two-sided test:

$$1 - \beta = \Phi\left(-z_{1-\alpha/2} + \frac{(\mu_0 - \mu_1)}{\sigma/\sqrt{n}}\right) + \Phi\left(-z_{1-\alpha/2} + \frac{(\mu_1 - \mu_0)}{\sigma/\sqrt{n}}\right)$$

$$\approx \Phi\left(-z_{1-\alpha/2} + \frac{|\mu_0 - \mu_1|}{\sigma/\sqrt{n}}\right)$$

Typically the sample size is selected prior to the experiment. For example, it may be of interest to decide how many respondents to interview in a poll or how many tissue samples to process. We already selected sample sizes in the context of interval estimation to achieve a given interval size and confidence level.

In a testing setup, consider a problem of testing $H_0 : \mu = \mu_0$ using $\overline{X}$ from a sample of size $n$. Let the alternative have a specific value $\mu_1$, i.e., $H_1 : \mu = \mu_1(> \mu_0)$. Assume a significance level of $\alpha = 0.05$. How large should $n$ be so that the power $1 - \beta$ is 0.90?

Recall that the power of a test is the probability that a false null will be rejected, $\mathbb{P}(\text{reject } H_0 | H_0 \text{ false})$. The null is rejected when $\overline{X} > \mu_0 + 1.645 \cdot \frac{\sigma}{\sqrt{n}}$. We want the power of 0.90 leading to $P(\overline{X} > \mu_0 + 1.645 \cdot \frac{\sigma}{\sqrt{n}} | \mu = \mu_1) = 0.90$, that is,

$$P\left( \frac{\overline{X} - \mu_1}{\sigma/\sqrt{n}} > \frac{\mu_0 - \mu_1}{\sigma/\sqrt{n}} + 1.645 \right) = 0.9.$$

Since $P(Z > -1.282) = 0.9$, it follows that $\frac{\mu_0 - \mu_1}{\sigma/\sqrt{n}} = 1.282 - 1.645 \Rightarrow n = \frac{8.567 \cdot \sigma^2}{(\mu_1 - \mu_0)^2}$.

In general terms, if we want to achieve the power $1 - \beta$ within the significance level of $\alpha$ for the alternative $\mu = \mu_1$, we need $n \geq \frac{(z_{1-\alpha} + z_{1-\beta})^2 \sigma^2}{(\mu_0 - \mu_1)^2}$ observations. For two-sided alternatives $\alpha$ is replaced by $\alpha/2$.

The sample size for fixed $\alpha$, $\beta$, $\sigma$, $\mu_0$, and $\mu_1$ is

$$n = \frac{\sigma^2}{(\mu_0 - \mu_1)^2}(z_{1-\alpha} + z_{1-\beta})^2,$$

where $\sigma$ is either known or estimated from a pilot experiment. If the alternative is two-sided, then $z_{1-\alpha}$ is replaced by $z_{1-\alpha/2}$. In this case, the sample size is approximate.

If $\sigma$ is not known and no estimate exists, one can elicit the *effect size*, $e = |\mu_0 - \mu_1|/\sigma$, directly. This number is the distance between the competing means in units of $\sigma$. For example, for $e = 1/2$ we would like to find a sample size such that the difference between the true and postulated mean equal to $\sigma/2$ is detectable with a probability of $1 - \beta$.

### 9.4.3 Testing a Normal Mean When the Variance Is Not Known: t-Test

To test a normal mean when the population variance is unknown, we use the $t$-test. We are interested in testing the null hypothesis $H_0 : \mu = \mu_0$ against one of the alternatives $H_1 : \mu >, \neq, < \mu_0$ on the basis of a sample $X_1, \ldots, X_n$ from the normal distribution $\mathcal{N}(\mu, \sigma^2)$ where the variance $\sigma^2$ is unknown.

If $\overline{X}$ and $s$ are the sample mean and standard deviation, then under $H_0$ (which states that the true mean is $\mu_0$), the statistic $t = \frac{\overline{X}-\mu_0}{s/\sqrt{n}}$ has a $t$ distribution with $n-1$ degrees of freedom, see arguments on p. 247.

The test can be performed either using (i) the rejection region or (ii) the $p$-value. The following table summarizes the test.

| Alternative | $\alpha$-level rejection region | $p$-value (MATLAB) |
|---|---|---|
| $H_1 : \mu > \mu_0$ | $[t_{n-1,1-\alpha}, \infty)$ | `1-tcdf(t, n-1)` |
| $H_1 : \mu \neq \mu_0$ | $(-\infty, t_{n-1,\alpha/2}] \cup [t_{n-1,1-\alpha/2}, \infty)$ | `2*tcdf(-abs(t),n-1)` |
| $H_1 : \mu < \mu_0$ | $(-\infty, t_{n-1,\alpha}]$ | `tcdf(t, n-1)` |

It is sometimes argued that the $z$-test and the $t$-test are an unnecessary dichotomy and that only the $t$-test should be used. The population variance in the $z$-test is assumed "known," but this can be too strong an assumption. Most of the time when $\mu$ is not known, it is unlikely that the researcher would have definite knowledge about the population variance. Also, the $t$-test is more conservative and robust (to deviations from the normality) than the $z$-test. However, the $z$-test has an educational value since the testing process and power analysis are easily formulated and explained. Moreover, when the sample size is large, say, larger than 100, the $z$, and $t$-tests are practically indistinguishable, due to the CLT.

*Example 9.4.* **The Moon Illusion.** Kaufman and Rock (1962) stated that the commonly observed fact that the moon near the horizon appears larger than does the moon at its zenith (highest point overhead) could be explained on the basis of the greater *apparent* distance of the moon when at the horizon. The authors devised an apparatus that allowed them to present two artificial moons, one at the horizon and one at the zenith. Subjects were asked to adjust the variable horizon moon to match the size of the zenith moon or vice versa. For each subject the ratio of the perceived size of the horizon moon to the perceived size of the zenith moon was recorded. A ratio of 1.00 would indicate no illusion, whereas a ratio other than 1.00 would represent an illusion. (For example, a ratio of 1.50 would mean that the horizon moon appeared to have a diameter 1.50 times that of the zenith moon.) Evidence in support of an illusion would require that we reject $H_0 : \mu = 1.00$ in favor of $H_1 : \mu > 1.00$.

Obtained ratio: 1.73 1.06 2.03 1.40 0.95 1.13 1.41 1.73 1.63 1.56

For these data,

```
x = [1.73, 1.06, 2.03, 1.40, 0.95, 1.13, 1.41, 1.73, 1.63, 1.56];
n = length(x)
```

```
t = (mean(x)-1)/(std(x)/sqrt(n))
  % t= 4.2976
crit = tinv(1-0.05, n-1)
  % crit=1.8331. RR = (1.8331, infinity)
pval = 1-tcdf(t, n-1)
  % pval = 9.9885e-004 < 0.05
```

As evident from the MATLAB output, the data do not support $H_0$, and $H_0$ is rejected.

A Bayesian solution implemented in WinBUGS is provided next. Each parameter in a Bayesian model should be given a prior distribution. Here we have two parameters, the mean $\mu$, which is the population ratio, and $\sigma^2$, the unknown variance. The prior on $\mu$ is normal with mean 0 and variance $1/0.00001 = 100000$. We also restricted the prior to be on the nonnegative domain (since no negative ratios are possible) by WinBUGS option mu~dnorm(0,0.00001)I(0,). Such a large variance makes the normal prior essentially flat over $\mu \geq 0$. This means that our prior opinion on $\mu$ is vague, and the adopted prior is noninformative.

The prior on the precision, $1/\sigma^2$, is gamma with parameters 0.0001 and 0.0001. As we argued in Example 8.13, this selection of hyperparameters makes the gamma prior essentially flat, and we are not injecting any prior information about the variance.

```
model{
for (i in 1:n){
X[i] ~ dnorm(mu, prec)
}
  mu ~ dnorm(0, 0.00001) I(0, )
  prec ~ dgamma(0.0001, 0.0001)
  sigma <- 1/sqrt(prec)
  #TEST
  prH1 <- step(mu - 1)
  }
DATA
list(n=10, X=c(1.73, 1.06, 2.03, 1.40, 0.95,
               1.13, 1.41, 1.73, 1.63, 1.56) )

INITS
list(mu = 0, prec = 1)
```

|       | mean  | sd      | MC error | val2.5pc | median | val97.5pc | start | sample |
|-------|-------|---------|----------|----------|--------|-----------|-------|--------|
| mu    | 1.463 | 0.1219  | 1.26E-4  | 1.219    | 1.463  | 1.707     | 1001  | 100000 |
| prH1  | 0.999 | 0.03115 | 3.188E-5 | 1.0      | 1.0    | 1.0       | 1001  | 100000 |
| sigma | 0.3727| 0.101   | 1.14E-4  | 0.2344   | 0.354  | 0.6207    | 1001  | 100000 |

Note that the MCMC output produced $\mathbb{P}(H_0) = 0.001$ and $\mathbb{P}(H_1) = 0.999$ and the Bayesian solution agrees with the classical. Moreover, the posterior probability of hypothesis $H_0$ of 0.001 is quite close to the $p$-value of 0.000998, which is often the case when the priors in the Bayesian model are noninformative.

*Example 9.5.* **Hypersplenism and White Blood Cell Count.** Hypersplenism is a disorder that causes the spleen to rapidly and prematurely destroy blood cells. In the general population the count of white blood cells (per $mm^3$) is normal with a mean of 7200 and standard deviation of $\sigma = 1500$.

It is believed that hypersplenism decreases the leukocyte count. In a sample of 16 persons affected by hypersplenism, the mean white blood cell count was found to be $\overline{X} = 5213$. The sample standard deviation was $s = 1682$.

Using WinBUGS find the posterior probability of $H_1$ and estimate the mean and variance in the affected population. The program in WinBUGS will operate on the summaries $\overline{X}$ and $s$ since the original data are not available. The sample mean is normal and the precision (reciprocal of the variance) of the mean is $n$ times the precision of a single observation. In this case, knowledge of the population standard deviation $\sigma$ will guide the setting of an informative prior on the precision. To keep the numbers manageable, we will express the counts in 1000s, and $\overline{X}$ and $s$ will be coded as 5.213 and 1.682, respectively. Since $s = 1.682, s^2 = 2.8291, prec = 0.3535$, it is tempting to set the prior on the precision as precx~dgamma(0.3535,1) or precx~dgamma(3.535,10) since the mean of these priors will match the observed precision. However, this would be a "data-built" prior in the spirit of the empirical Bayes approach. We will use the fact that in the population $\sigma$ was 1.5 and we will elicit the prior precx~dgamma(4.444,10) since $1/1.5^2 = 0.4444$.

```
model {
precxbar <- n * precx
xbar ~ dnorm(mu, precxbar)
mu  ~    dnorm(0, 0.0001) I(0, )
    # sigma = 1.5, s^2 = 2.25, prec = 0.4444
    # X gamma(a,b) -> EX=a/b, Var X = a/b^2
precx ~ dgamma(4.444, 10 )
indh1 <- step(7.2 - mu)
sigx <- 1/sqrt(precx)
}
```

```
DATA
list(xbar = 5.213, n=16)
```

```
INITS
list(mu=1.000, precx=1.000)
```

|       | mean   | sd     | MC error  | val2.5pc | median | val97.5pc | start | sample |
|-------|--------|--------|-----------|----------|--------|-----------|-------|--------|
| indh1 | 0.9997 | 0.01643 | 3.727E-5 | 1.0      | 1.0    | 1.0       | 1001  | 200000 |
| mu    | 5.212  | 0.4263 | 9.842E-4  | 4.367    | 5.212  | 6.064     | 1001  | 200000 |
| sigx  | 1.644  | 0.4486 | 0.001081  | 1.032    | 1.561  | 2.749     | 1001  | 200000 |

Note that the posterior probability of $H_1$ is 0.9997 and this hypothesis is a clear winner.

### 9.4.4 Power Analysis of t-Test

When an experiment is planned, the data are not available. Even if the variance is unknown, as in the case of a $t$-test, it is elicited, or the difference $|\mu_0 - \mu_1|$ is expressed in units of standard deviation. Thus, in preexperimental planning the power analysis applicable to the $z$-test is also applicable to the $t$-test.

Once the data are available and the test is performed, the sample mean and sample variance are available and it becomes possible to assess the power retrospectively. We have already discussed controversies surrounding retrospective power analyses.

In a retrospective evaluation of the power it is not recommended to replace $|\mu_0 - \mu_1|$ by $|\mu_0 - \overline{X}|$, as is sometimes done, but to simply update the elicited $\sigma^2$ with the observed variance. When $\sigma$ is replaced by $s$, the expressions for finding the power involve $t$ and noncentral $t$ distributions. Here is an illustration.

*Example 9.6.* Suppose we are testing $H_0 : \mu = 10$ versus $H_1 : \mu > 10$, at a level $\alpha = 0.05$. A sample of size $n = 20$ gives $\overline{X} = 12$ and $s = 5$. We are interested in finding the power of the test against the alternative $H_1 : \mu = 13$.

The exact power is $\mathbb{P}(t \in RR | t \sim nct(df = n - 1, ncp = (\mu_1 - \mu_0)\sqrt{n}/\sigma))$, since under $H_1$, $t$ has a noncentral $t$ distribution with $n - 1$ degrees of freedom and a noncentrality parameter $\frac{(\mu_1 - \mu_0)\sqrt{n}}{\sigma}$. "RR" denotes the rejection region.

```
n=20; Xb=12; mu0 = 10; s=5; mu1= 13; alpha=0.05;
pow1 = nctcdf( -tinv(1-alpha, n-1), n-1,-abs(mu1-mu0)*sqrt(n)/s)
% or pow1=1-nctcdf(tinv(1-alpha, n-1),n-1,abs(mu1-mu0)*sqrt(n)/s)
% pow1 = 0.8266
%
pow = normcdf(-norminv(1-alpha) + abs(mu1-mu0)*sqrt(n)/s)
% or pow = 1-normcdf(norminv(1-alpha)-abs(mu1-mu0)*sqrt(n)/s)
% pow = 0.8505
```

For a large sample size the power calculated as in the $z$-test approximates the exact power, but from the "optimistic" side, that is, by always overestimating it. In this MATLAB script we find a power of approx. 85%, which in an exact calculation (as above) drops to 82.66%.

For the two-sided alternative $H_1 : \mu \neq 10$ the exact power decreases,

```
pow2 = nctcdf(tinv(1-alpha/2, n-1), n-1,-abs(mu1-mu0)*sqrt(n)/s)  ...
          -nctcdf(tinv(1-alpha/2, n-1), n-1, abs(mu1-mu0)*sqrt(n)/s)
%pow2 =0.7210
```

When easy calculation of the noncentral $t$ CDF is not available, a good approximation for the power is

$$1 - \Phi \left( \frac{t_{n-1,\alpha} - |\mu_1 - \mu_0| \sqrt{n}/s}{\sqrt{1 + \frac{t_{n-1,1-\alpha}^2}{2(n-1)}}} \right).$$

In our example,

```
1-normcdf((tinv(1-alpha,n-1)- ...
(mu1-mu0)/s * sqrt(n))/sqrt(1 + (tinv(1-alpha,n-1))^2/(2*n-2)))
%ans =    0.8209
```

Following is the summary of retrospective power calculations for the $t$-test.

---

**Power of the $t$-test for $H_0 : \mu = \mu_0$, when $\mu_1$ is the actual mean.**

• One-sided test:

$$1 - \beta = 1\text{-}nctcdf\left( t_{n-1,1-\alpha},\ n-1,\ \frac{|\mu_1 - \mu_0|}{s/\sqrt{n}} \right)$$

• Two-sided test:

$$1 - \beta = nctcdf\left( t_{n-1,1-\alpha/2},\ n-1,\ \frac{-|\mu_1 - \mu_0|}{s/\sqrt{n}} \right) -$$

$$nctcdf\left( t_{n-1,1-\alpha/2},\ n-1,\ \frac{|\mu_1 - \mu_0|}{s/\sqrt{n}} \right)$$

Here $nctcdf(x,df,\delta)$ is the CDF of a noncentral $t$ distribution, with $df$ degrees of freedom and noncentrality parameter $\delta$, evaluated at $x$. In MATLAB this function is nctcdf(x,df,delta).

---

## 9.5 Testing the Normal Variances

When we discussed the estimation of the normal variance (Sect.7.3.2), we argued that the statistic $(n-1)s^2/\sigma^2$ had a $\chi^2$ distribution with $n-1$ degrees of freedom. The test for the normal variance is based on this statistic and its distribution.

Suppose we want to test $H_0 : \sigma^2 = \sigma_0^2$ versus $H_1 : \sigma^2 \neq (<,>)\sigma_0^2$. The test statistic is

$$\chi^2 = \frac{(n-1)s^2}{\sigma_0^2}.$$

The testing procedure at the $\alpha$ level can be summarized by

| Alternative | $\alpha$-level rejection region | $p$-value (MATLAB) |
|---|---|---|
| $H_1 : \sigma > \sigma_0$ | $[\chi^2_{n-1,1-\alpha}, \infty)$ | `1-chi2cdf(chi2,n-1)` |
| $H_1 : \sigma \neq \sigma_0$ | $[0, \chi^2_{n-1,\alpha/2}] \cup [\chi^2_{n-1,1-\alpha/2}, \infty)$ | `2*chi2cdf(min(chi2,1/chi2),n-1)` |
| $H_1 : \sigma < \sigma_0$ | $[0, \chi^2_{n-1,\alpha}]$ | `chi2cdf(chi2,n-1)` |

The power of the test against the specific alternative is the probability of the rejection region evaluated as if $H_1$ were a true hypothesis. For example, if $H_1 : \sigma^2 > \sigma_0^2$, and specifically if $H_1 : \sigma^2 = \sigma_1^2$, $\sigma_1^2 > \sigma_0^2$, then the power is

$$
1 - \beta = \mathbb{P}\left( \frac{(n-1)s^2}{\sigma_0^2} \geq \chi^2_{1-\alpha,n-1} | H_1 \right) = \mathbb{P}\left( \frac{(n-1)s^2}{\sigma_1^2} \cdot \frac{\sigma_1^2}{\sigma_0^2} \geq \chi^2_{1-\alpha,n-1} | H_1 \right)
$$

$$
= \mathbb{P}\left( \chi^2 \geq \frac{\sigma_0^2}{\sigma_1^2} \chi^2_{1-\alpha,n-1} \right),
$$

or in MATLAB:

```
power=1-chi2cdf(sigmasq0/sigmasq1*chi2inv(1-alpha,n-1),n-1).
```

For the one-sided alternative in the opposite direction and for the two-sided alternative, finding the power is analogous. The sample size necessary to achieve a preassigned power can be found by trial and error or by using MATLAB's function `fzero`.

**LDL-C Levels.** A new handheld device for assessing cholesterol levels in the blood has been presented for approval to the FDA. The variability of measurements obtained by the device for people with normal levels of LDL cholesterol is one of the measures of interest. A calibrated sample of size $n = 224$ of serum specimens with a fixed 130-level of LDL-C is measured by the device. The variability of measurements is assessed.

(a) If $s^2 = 2.47$ was found, test the hypothesis that the population variance is 2 (as achieved by a clinical computerized Hitachi 717 analyzer, with enzymatic, colorimetric detection schemes) against the one-sided alternative. Use $\alpha = 0.05$.

(b) Find the power of this test against the specific alternative, $H_1 : \sigma^2 = 2.5$.

(c) What sample size ensures the power of 90% in detecting the effect $\sigma_0^2/\sigma_1^2 = 0.8$ significant.

```
n = 224; s2 = 2.47; sigmasq0 = 2; sigmasq1 = 2.5; alpha = 0.05;
%(a)
chisq = (n-1)*s2 /sigmasq0
  %test statistic chisq = 275.4050.
  %The alternative is H_1: sigma2 > 2
chi2crit = chi2inv( 1-alpha, n-1 )
  %one sided upper tail RR = [258.8365, infinity)
pvalue = 1 - chi2cdf(chisq, n-1) %pvalue = 0.0096
%(b)
power = 1-chi2cdf(sigmasq0/sigmasq1 * chi2inv(1-alpha, n-1), n-1 )
                            %power = 0.7708
%(c)
ratio = sigmasq0/sigmasq1   %0.8
pf = @(n) 1-chi2cdf( ratio * chi2inv(1-alpha, n-1), n-1 ) - 0.90;
ssize = fzero(pf, 300)      %342.5993 approx 343
```

## 9.6 Testing the Proportion

When discussing the CLT, in particular the de Moivre theorem, we saw that the binomial distribution could be well approximated with the normal if $n$ were large and $np(1-p) > 5$. The sample proportion $\hat{p}$ thus has an approximately normal distribution with mean $p$ and variance $p(1-p)/n$.

Suppose that we are interested in testing $H_0 : p = p_0$ versus one of the three possible alternatives. When $H_0$ is true, the test statistic

$$Z = \frac{\hat{p} - p_0}{\sqrt{p_0(1 - p_0)/n}}$$

has a standard normal distribution. Thus the testing procedure can be summarized in the following table:

| Alternative | $\alpha$-level rejection region | $p$-value (MATLAB) |
|---|---|---|
| $H_1 : p > p_0$ | $[z_{1-\alpha}, \infty)$ | `1-normcdf(z)` |
| $H_1 : p \neq p_0$ | $-\infty, z_{\alpha/2}] \cup [z_{1-\alpha/2}, \infty)$ | `2*normcdf(-abs(z))` |
| $H_1 : p < p_0$ | $(-\infty, z_\alpha]$ | `normcdf(z)` |

Using the normal approximation one can derive that the power against the specific alternative $H_1 : p = p_1$ is

$$\Phi\left[\sqrt{\frac{p_0(1 - p_0)}{p_1(1 - p_1)}}\left(z_\alpha + \frac{|p_1 - p_0|\sqrt{n}}{\sqrt{p_0(1 - p_0)}}\right)\right]$$

for the one-sided test. For the two-sided test $z_\alpha$ is replaced by $z_{\alpha/2}$. The sample size needed to find the effect $|p_0 - p_1|$ significant $(1-\beta)100\%$ of the time (the test would have a power of $1-\beta$) is

$$n = \frac{p_0(1-p_0)\left(z_{1-\alpha/2} + z_{1-\beta}\sqrt{\frac{p_1(1-p_1)}{p_0(1-p_0)}}\right)^2}{(p_0 - p_1)^2}.$$

*Example 9.7.* **Proportion of Hemorrhagic-type Strokes Among American Indians.** The study described in the American Heart Association news release of September 22, 2008 included 4,507 members of 13 American Indian tribes in Arizona, Oklahoma, and North and South Dakota. It found that American Indians have a stroke rate of 679 per 100,000, compared to 607 per 100,000 for African Americans and 306 per 100,000 for Caucasians. None of the participants, ages 45 to 74, had a history of stroke when they were recruited for the study from 1989 to 1992. Almost 60% of the volunteers were women.

During more than 13 years of follow-up, 306 participants suffered a first stroke, most of them in their mid-60s when it occurred. There were 263 strokes of the ischemic type – caused by a blockage that cuts off the blood supply to the brain – and 43 hemorrhagic (bleeding) strokes.

It is believed that in the general population one in five of all strokes is hemorrhagic.

(a) Test the hypothesis that the proportion of hemorrhagic strokes in the population of American Indians that suffered a stroke is lower than the national proportion of 0.2.

(b) What is the power of the test in (a) against the alternative $H_1 : p = 0.15$?

(c) What sample size ensures a power of 90% in detecting $p = 0.15$, if $H_0$ states $p = 0.2$?

Since $306 \times 0.2 > 10$, a normal approximation can be used.

```
z = (43/306 - 0.2)/sqrt(0.2 *(1- 0.2)/306)
% z = -2.6011

pval = normcdf(z)
% pval = 0.0046
```

Since the exact distribution is binomial and the $p$-value is the probability of observing a number of hemorrhagic strokes less than or equal to 43, we are able to calculate this $p$-value exactly:

```
pval = binocdf(43, 306, 0.2)
%pval = 0.0044
```

Note that the *p*-value of the normal approximation (0.0046) is quite close to the *p*-value of the exact test (0.0044).

```
%(b)
p0=0.2; p1=0.15; alpha=0.05; n=306;
 power = normcdf( sqrt(p0*(1-p0)/(p1*(1-p1))) * ...
     (norminv(alpha) + abs(p1-p0)*sqrt(n)/sqrt( p0*(1-p0)) ) )
%0.728

%(c)
beta = 0.1;
ssize = p0*(1-p0) * (norminv(1-alpha) + norminv(1-beta)*...
          sqrt( p1*(1-p1)/( p0 * (1-p0))))^2/(p0 - p1)^2
%497.7779   approx 498
```

The Bayesian test requires a prior on population proportion $p$. We select a beta prior with parameters 1 and 4 so that $\mathbb{E}p = 1/(1+4) = 0.2$ matches the mean under $H_0$.

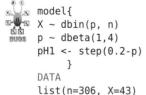

```
model{
 X ~ dbin(p, n)
 p ~ dbeta(1,4)
 pH1 <- step(0.2-p)
     }
DATA
list(n=306, X=43)
#Generate Inits
```

The output variable pH1 gives the posterior probability of the hypothesis $H_1 : p < 0.2$.

|      | mean   | sd      | MC error  | val2.5pc | median | val97.5pc | start | sample  |
|------|--------|---------|-----------|----------|--------|-----------|-------|---------|
| p    | 0.1415 | 0.01974 | 1.9158E-5 | 0.1051   | 0.1407 | 0.1823    | 1001  | 1000000 |
| pH1  | 0.9967 | 0.05694 | 5.675E-5  | 1.0      | 1.0    | 1.0       | 1001  | 1000000 |

If we wanted to be noninformative in eliciting a prior for $p$, several choices for such a prior are available. For example, the prior could be uniform, dbeta(1,1), but the differences in WinBUGS' output are minimal:

|      | mean    | sd      | MC error | val2.5pc | median | val97.5pc | start | sample |
|------|---------|---------|----------|----------|--------|-----------|-------|--------|
| p    | 0.1428  | 0.01987 | 6.337E-5 | 0.1061   | 0.1421 | 0.184     | 1001  | 100000 |
| test | 0.00402 | 0.06328 | 1.942E-4 | 0.0      | 0.0    | 0.0       | 1001  | 100000 |

Another noninformative choice is dbeta(0.5,0.5) (Jeffreys' prior, p. 292).

*Example 9.8.* **Savage's Disparity.** A Bayesian inference is conditional. It is based on data observed and not on data that could possibly be observed, or on the manner in which the sampling was conducted. This is not the case in classical testing, and the argument first put forth by Jimmie Savage at the Purdue Symposium in 1962 emphasizes the difference.

Suppose a coin is flipped 12 times and 9 heads and 3 tails are obtained. Let $p$ be the probability of heads. We are interested in testing whether the coin is fair against the alternative that it is more likely to come heads up, or

$$H_0 : p = 1/2 \quad \text{vs.} \quad H_1 : p > 1/2.$$

The $p$-value for this test is the probability that one observes 9 or more heads if the coin is fair (under $H_0$).

Consider the following two scenarios:

(A) Suppose that the number of flips $n = 12$ was decided a priori. Then the number of heads $X$ is binomial and under $H_0$ (fair coin) the $p$-value is

$$\mathbb{P}(X \geq 9) = 1 - \sum_{k=0}^{8} \binom{12}{k} p^k (1-p)^{12-k} = 1 - \texttt{binocdf}(8, 12, 0.5) = 0.0730.$$

At a 5% significance level $H_0$ is *not rejected*

(B) Suppose that the flipping is carried out until 3 tails have appeared. Let us call tails "success" and heads "failures." Then, under $H_0$, the number of failures (heads) $Y$ is a negative binomial $\mathcal{NB}(3, 1/2)$ and the $p$-value is

$$\mathbb{P}(Y \geq 9) = 1 - \sum_{k=0}^{8} \binom{3+k-1}{k} (1-p)^3 p^k = 1 - \texttt{nbincdf}(8, 3, 1/2) = 0.0327.$$

At a 5% significance level $H_0$ *is rejected.*

Thus, two Fisherian tests recommend opposite actions for the same data simply because of how the sampling was conducted.

Note that in (A) and (B) the likelihoods are proportional to $p^9 (1-p)^3$, and for a fixed prior on $p$ there is no difference in any Bayesian inference.

Edwards et al. (1963) note that "...the rules governing when data collection stops are irrelevant to data interpretation. It is entirely appropriate to collect data until a point has been proven or disproven, or until the data collector runs out of time, money, or patience."

✎

## 9.7 Multiplicity in Testing, Bonferroni Correction, and False Discovery Rate

Recall that when testing a single hypothesis $H_0$, a type I error is made if it is rejected, even if it is actually true. The probability of making a type I error in a test is usually controlled to be smaller than a certain level of $\alpha$, typically equal to 0.05.

When there are several null hypotheses, $H_{01}, H_{02}, \ldots, H_{0m}$, and all of them are tested simultaneously, one may want to control the type I error at some level $\alpha$ as well. In this scenario a type I error is then made if at least one true hypothesis in the family of hypotheses being tested is rejected. Because it pertains to the family of hypotheses, this significance level is called the familywise error rate (FWER).

If the hypotheses in the family are independent, then

$$\text{FWER} = 1 - (1 - \alpha_i)^m,$$

where FWER and $\alpha_i$ are overall and individual significance levels, respectively.

For arbitrary, possibly dependent, hypotheses, the Bonferroni inequality translates to

$$\text{FWER} \leq m\alpha_i.$$

Suppose $m = 15$ tests are conducted simultaneously. If an individual $\alpha_i = 0.05$, then FWER $= 1 - 0.95^{15} = 0.5367$. This means that the chance of claiming a significant result when there should not be one is larger than 1/2. For possibly dependent hypotheses the upper bound of FWER is 0.75.

> **Bonferroni:** To control FWER $\leq \alpha$ reject all $H_{0i}$ among $H_{01}, H_{02}, \ldots, H_{0m}$ for which the $p$-value is less than $\alpha/m$.

Thus, if for $n = 15$ arbitrary hypotheses we want an overall significance level of FWER $\leq 0.05$, then the individual test levels should be set to $0.05/15 = 0.0033$.

Testing for significance with gene expression data from DNA microarray experiments involves simultaneous comparisons of hundreds or thousands of genes and controlling the FWER by the Bonferroni method would require very small individual $\alpha_i$s. Setting such small $\alpha$ levels decreases the power of individual tests (many false $H_0$ are not rejected) and the Bonferroni correction is considered by many practitioners as overly conservative. Some call it a "panic approach."

**Remark.** If, in the context of interval estimation, $k$ simultaneous interval estimates are desired with an overall confidence level $(1 - \alpha)100\%$, one can construct each interval with confidence level $(1 - \alpha/k)100\%$, and the Bonferroni inequality insures that the overall confidence is at least $(1 - \alpha)100\%$.

The Bonferroni–Holm method is an iterative procedure in which individual significance levels are adjusted to increase power and still control the FWER. One starts by ordering the $p$-values of all tests for $H_{01}, H_{02}, \ldots, H_{0m}$ and then compares the smallest $p$-value to $\alpha/m$. If that $p$-value is smaller than $\alpha/m$, then reject that hypothesis and compare the second ranked $p$-value to $\alpha/(m-1)$. If this hypothesis is rejected, proceed to the third ranked $p$-value

**Fig. 9.2** Carlo Emilio Bonferroni (1892–1960).

and compare it with $\alpha/(m-2)$. Continue doing this until the hypothesis with the smallest remaining $p$-value cannot be rejected. At this point the procedure stops and all hypotheses that have not been rejected at previous steps are accepted.

For example, assume that five hypotheses are to be tested with a FWER of 0.05. The five $p$-values are 0.09, 0.01, 0.04, 0.012, and 0.004. The smallest of these is 0.004. Since this is less than 0.05/5, hypothesis four is rejected. The next smallest $p$-value is 0.01, which is also smaller than 0.05/4. So this hypothesis is also rejected. The next smallest $p$-value is 0.012, which is smaller than 0.05/3, and this hypothesis is rejected. The next smallest $p$-value is 0.04, which is not smaller than 0.05/2. Therefore, the hypotheses with $p$-values of 0.004, 0.01, and 0.012 are rejected while those with $p$-values of 0.04 and 0.09 are not rejected.

The false discovery rate paradigm (Benjamini and Hochberg, 1995) considers the proportion of falsely rejected null hypotheses (false discoveries) among the total number of rejections.

Controlling the expected value of this proportion, called the false discovery rate (FDR), provides a useful alternative that addresses low-power problems of the traditional FWER methods when the number of tested hypotheses is large. The test statistics in these multiple tests are assumed independent or positively correlated. Suppose that we are looking at the result of testing $m$ hypotheses among which $m_0$ are true. In terms of the table, $V$ denotes the number of false rejections, and the FWER is $\mathbb{P}(V \geq 1)$.

|            | $H_0$ not rejected | $H_0$ rejected | Total |
|------------|--------------------|----------------|-------|
| $H_0$ true | $U$                | $V$            | $m_0$ |
| $H_1$ true | $T$                | $S$            | $m_1$ |
| Total      | $W$                | $R$            | $m$   |

If $R$ denotes the number of rejections (declared significant genes, discoveries) then $V/R$, for $R > 0$, is the proportion of false rejected hypotheses. The FDR is

$$\mathbb{E}\left(\frac{V}{R}\Big| R > 0\right)\mathbb{P}(R > 0).$$

Let $p_{(1)} \leq p_{(2)} \leq \cdots \leq p_{(m)}$ be the ordered, observed $p$-values for the $m$ hypotheses to be tested. Algorithmically, the FDR method finds $k$ such that

$$k = \max\{i \,|\, p_{(i)} \leq (i/m)\alpha\}. \tag{9.2}$$

The FWER is controlled at the $\alpha$ level if the hypotheses corresponding to $p_{(1)}, \ldots, p_{(k)}$ are rejected. If no such $k$ exists, no hypothesis from the family is rejected. When the test statistics in the multiple tests are possibly negatively correlated as well, the FDR is modified by replacing $\alpha$ in (9.2) with $\alpha/(1 + 1/2 + \cdots + 1/m)$. The following MATLAB script ( ◢ FDR.m) finds the critical $p$-value $p_{(k)}$. If $p_{(k)} = 0$, then no hypothesis is rejected.

```
function pk = FDR(p,alpha)
%Critical p-value pk for FDR <= alpha.
%All hypotheses with p-value less than or equal
%to pk are rejected.
%if pk = 0 no hypothesis is to be rejected
m = length(p);      %number of hypotheses
po = sort(p(:));    %ordered p-values
i = (1:m)';         %index
pk = po(max(find(  po < i./m * alpha)));
%critical p-value
if ( isempty(pk)==1 )
    pk=0;
end
```

Suppose that we have 1000 hypotheses and all hypotheses are true. Then their $p$-values look like a random sample from the uniform $\mathcal{U}(0,1)$ distribution. About 50 hypotheses would have a $p$-value of less than 0.05. However, for all reasonable FDR levels (0.05–0.2) $p_{(k)} = 0$, as it should be since we do not want false discoveries.

```
p = rand(1000,1);
[FDR(p, 0.05), FDR(p, 0.2), FDR(p, 0.6), FDR(p, 0.92)]
%ans =      0       0     0.0022     0.0179
```

## 9.8 Exercises

9.1. **Public Health.** A manager of public health services in an area downwind of a nuclear test site wants to test the hypothesis that the mean amount of radiation in the form of strontium-90 in the bone marrow (measured in picocuries) for citizens who live downwind of the site does exceed that of citizens who live upwind from the site. It is known that "upwinders" have a mean level of strontium-90 of 1 picocurie. Measurements of strontium-90 radiation for a sample of $n = 16$ citizens who live downwind of the site were taken, giving $\overline{X} = 3$. The population standard deviation is $\sigma = 4$.
Test the (research, alternative) hypothesis that downwinders have a higher strontium-90 level than upwinders. Assume normality and use a significance level of $\alpha = 0.05$.

(a) State $H_0$ and $H_1$.

(b) Calculate the appropriate test statistic.

(c) Determine the critical region of the test.

(d) State your decision.

(e) What would constitute a type II error in this setup? *Describe in one sentence.*

9.2. **Testing IQ.** We wish to test the hypothesis that the mean IQ of the students in a school system is 100. Using $\sigma = 15$, $\alpha = 0.05$, and a sample of 25 students the sample value $\overline{X}$ is computed. For a two-sided test find:

(a) The range of $\overline{X}$ for which we would accept the hypothesis.

(b) If the true mean IQ of the students is 105, find the probability of falsely accepting $H_0 : \mu = 100$.

(c) What are the answers in (a) and (b) if the alternative is one-sided, $H_1 : \mu > 100$?

9.3. **Bricks.** A purchaser of bricks suspects that the quality of bricks is deteriorating. From past experience, the mean crushing strength of such bricks is 400 pounds. A sample of $n = 100$ bricks yielded a mean of 395 pounds and standard deviation of 20 pounds.

(a) Test the hypothesis that the mean quality has not changed against the alternative that it has deteriorated. Choose $\alpha = .05$.

(b) What is the $p$-value for the test in (a).

(c) Assume that the producer of the bricks contested your findings in (a) and (b). Their company suggested constructing the 95% confidence interval for $\mu$ with a total length of no more than 4. What sample size is needed to construct such a confidence interval?

9.4. **Soybeans.** According to advertisements, a strain of soybeans planted on soil prepared with a specific fertilizer treatment has a mean yield of 500 bushels per acre. Fifty farmers who belong to a cooperative plant the soybeans. Each uses a 40-acre plot and records the mean yield per acre. The mean and variance for the sample of 50 farms are $\bar{x} = 485$ and $s^2 = 10045$. Use the $p$-value for this test to determine whether the data provide sufficient evidence to indicate that the mean yield for the soybeans is different from that advertised.

9.5. **Great White Shark.** One of the most feared predators in the ocean is the great white shark *Carcharodon carcharias*. Although it is known that the white shark grows to a mean length of 14 ft. (record: 23 ft.), a marine biologist believes that the great white sharks off the Bermuda coast grow significantly longer due to unusual feeding habits. To test this claim a number of full-grown great white sharks are captured off the Bermuda coast, measured, and then set free. However, because the capture of sharks is difficult, costly, and very dangerous, only five are sampled. Their lengths are 16, 18, 17, 13, and 20 ft.

(a) What assumptions must be made in order to carry out the test?

(b) Do the data provide sufficient evidence to support the marine biologist's claim? Formulate the hypotheses and test at a significance level of $\alpha = 0.05$. Provide solutions using both a rejection-region approach and a $p$-value approach.

(c) Find the power of the test against the alternative $H_1 : \mu = 17$.

(d) What sample size is needed to achieve a power of 0.95 in testing the above hypothesis if $\mu_1 - \mu_0 = 3$ and $\alpha = 0.05$. Assume that the previous experiment was a pilot study to assess the variability in data and adopt $\sigma = 2.5$.

(e) Provide a Bayesian solution using WinBUGS with noninformative priors on $\mu$ and $\sigma^2$. Compare with classical results and discuss.

9.6. **Serum Sodium Levels.** A data set concerning the National Quality Control Scheme, Queen Elizabeth Hospital, Birmingham, referenced in Andrews and Herzberg (1985), provides the results of analysis of 20 samples of serum measured for their sodium content. The average value for the method of analysis used is 140 ppm.

$$140 \ 143 \ 141 \ 137 \ 132 \ 157 \ 143 \ 149 \ 118 \ 145$$
$$138 \ 144 \ 144 \ 139 \ 133 \ 159 \ 141 \ 124 \ 145 \ 139$$

Is there evidence that the mean level of sodium in this serum is different from 140 ppm?

9.7. **Weight of Quarters.** The US Department of Treasury claims that the procedure it uses to mint quarters yields a mean weight of 5.67 g with a standard deviation of 0.068 g. A random sample of 30 quarters yielded a mean of 5.643 g. Use a 0.05 significance to test the claim that the mean weight is 5.67 g.

(a) What alternatives make sense in this setup? Choose one sensible alternative and perform the test.

(b) State your decision in terms of accept–reject $H_0$.

(c) Find the $p$-value and confirm your decision from the previous bullet in terms of the $p$-value.

(d) Would you change the decision if $\alpha$ were 0.01?

9.8. **Dwarf Plants.** A genetic model suggests that three-fourths of the plants grown from a cross between two given strains of seeds will be of the dwarf variety. After breeding 200 of these plants, 136 were of the dwarf variety.

(a) Does this observation strongly contradict the genetic model?

(b) Construct a 95% confidence interval for the true proportion of dwarf plants obtained from the given cross.

(c) Answer (a) and (b) using Bayesian arguments and WinBUGS.

9.9. **Eggs in a Nest.** The average number of eggs laid per nest per season for the Eastern Phoebe bird is a parameter of interest. A random sample of

70 nests was examined and the following results were obtained (Hamilton, 1990),

| Number of eggs/nest | 1 | 2 | 3 | 4 | 5 | 6 |
|---|---|---|---|---|---|---|
| Frequency $f$ | 3 | 2 | 2 | 14 | 46 | 3 |

Test the hypothesis that the true average number of eggs laid per nest by the Eastern Phoebe bird is equal to five versus the two-sided alternative. Use $\alpha = 0.05$.

9.10. **Penguins.** Penguins are popular birds, and the Emperor penguin (*Aptenodytes forsteri*) is the most popular penguin of all. A researcher is interested in testing that the mean height of Emperor penguins from a small island is less than $\mu = 45$ in., which is believed to be the average height for the whole Emperor penguin population. The measurements of height of 14 randomly selected adult birds from the island are

$$41\ 44\ 43\ 47\ 43\ 46\ 45\ 42\ 45\ 45\ 43\ 45\ 47\ 40$$

State the hypotheses and perform the test at the level $\alpha = 0.05$.

9.11. **Hypersplenism and White Blood Cell Count.** In Example 9.5, the belief was expressed that hypersplenism decreased the leukocyte count and a Bayesian test was conducted. In a sample of 16 persons affected by hypersplenism, the mean white blood cell count (per $mm^3$) was found to be $\overline{X} = 5213$. The sample standard deviation was $s = 1682$.
(a) With this information test $H_0 : \mu = 7200$ versus the alternative $H_1 : \mu < 7200$ using both rejection region and $p$-value. Compare the results with WinBUGS output.
(b) Find the power of the test against the alternative $H_1 : \mu = 5800$.
(c) What sample size is needed if in a repeated study a difference of $|\mu_1 - \mu_0| = 600$ is to be detected with a power of 80%? Use the estimate $s = 1682$.

9.12. **Jigsaw.** An experiment with a sample of 18 nursery-school children involved the elapsed time required to put together a small jigsaw puzzle. The times were as follows:

| 3.1 | 3.2 | 3.4 | 3.6 | 3.7 | 4.2 | 4.3 | 4.5 | 4.7 |
|---|---|---|---|---|---|---|---|---|
| 5.2 | 5.6 | 6.0 | 6.1 | 6.6 | 7.3 | 8.2 | 10.8 | 13.6 |

(a) Calculate the 95% confidence interval for the population mean.
(b) Test the hypothesis $H_0 : \mu = 5$ against the two-sided alternative. Take $\alpha = 10\%$.

9.13. **Anxiety.** A psychologist has developed a questionnaire for assessing levels of anxiety. The scores on the questionnaire range from 0 to 100. People who obtain scores of 75 and greater are classified as *anxious*. The questionnaire has been given to a large sample of people who have been diagnosed with an anxiety disorder, and scores are well described by a normal model with a mean of 80 and a standard deviation of 5. When given to a large sample of people who do not suffer from an anxiety disorder, scores on the questionnaire can also be modeled as normal with a mean of 60 and a standard deviation of 10.

(a) What is the probability that the psychologist will misclassify a nonanxious person as anxious?

(b) What is the probability that the psychologist will erroneously label a truly anxious person as nonanxious?

9.14. **Aptitude Test.** An aptitude test should produce scores with a large amount of variation so that an administrator can distinguish between persons with low aptitude and those with high aptitude. The standard test used by a certain university has been producing scores with a standard deviation of 5. A new test given to 20 prospective students produced a sample standard deviation of 8. Are the scores from the new test significantly more variable than scores from the standard? Use $\alpha = 0.05$.

9.15. **Rats and Mazes.** Eighty rats selected at random were taught to run a new maze. All of them finally succeeded in learning the maze, and the number of trials to perfect the performance was normally distributed with a sample mean of 15.4 and sample standard deviation of 2. Long experience with populations of rats trained to run a similar maze shows that the number of trials to attain success is normally distributed with a mean of 15.

(a) Is the new maze harder for rats to learn than the older one? Formulate the hypotheses and perform the test at $\alpha = 0.01$.

(b) Report the $p$-value. Would the decision in (a) be different if $\alpha = 0.05$?

(c) Find the power of this test for the alternative $H_1 : \mu = 15.6$.

(d) Assume that the above experiment was conducted to assess the standard deviation, and the result was 2. You wish to design a sample size for a new experiment that will detect the difference $|\mu_0 - \mu_1| = 0.6$ with a power of 90%. Here $\alpha = 0.01$, and $\mu_0$ and $\mu_1$ are postulated means under $H_0$ and $H_1$, respectively.

9.16. **Hemopexin in DMD Cases I.** Refer to data set 📄 dmd.dat|mat|xls from Exercise 2.16. The measurements of hemopexin are assumed normal.

(a) Form a 95% confidence interval for the mean response of hemopexin h in a population of all female DMD carriers (carrier=1).

Although the level of pyruvate kinase seems to be the strongest single predictor of DMD, it is an expensive measure. Instead, we will explore the level of hemopexin, a protein that protects the body from oxidative damage. The level of hemopexin in a general population of women of the same age range as that in the study is believed to be 85.

(b) Test the hypothesis that the mean level of hemopexin in the population of women DMD carriers significantly exceeds 85. Use $\alpha = 5\%$. Report the $p$-value as well.

(c) What is the power of the test in (b) against the alternative $H_1 : \mu_1 = 89$.

(d) The data for this exercise come from a study conducted in Canada. If you wanted to replicate the test in the USA, what sample size would guarantee a power of 99% if $H_0$ were to be rejected whenever the difference from the true mean was 4, ($|\mu_0 - \mu_1| = 4$)? A small pilot study conducted to assess the variability of hemopexin level estimated the standard deviation as $s = 12$.

(e) Find the posterior probability of the hypothesis $H_1 : \mu > 85$ using Win-BUGS. Use noninformative priors. Also, compare the 95% credible set for $\mu$ that you obtained with the confidence interval in (a).

*Hint:* The commands

```
%file dmd.mat should be on path
load 'dmd.mat';  hemo = dmd( dmd(:,6)==1, 3);
```

will distill the levels of hemopexin in carrier cases.

9.17. **Retinol and a Copper-Deficient Diet.** The liver is the main storage site of vitamin A and copper. Inverse relationships between copper and vitamin A liver concentrations have been suggested. In Rachman et al. (1987) the consequences of a copper-deficient diet on liver and blood vitamin A storage in Wistar rats was investigated. Nine animals were fed a copper-deficient diet for 45 days from weaning. Concentrations of vitamin A were determined by isocratic high-performance liquid chromatography using UV detection. Rachman et al. (1987) observed in the liver of the rats fed a copper-deficient diet a mean level of retinol [in micrograms/g of liver] was $\overline{X} = 3.3$ and $s = 1.4$. It is known that the normal level of retinol in a rat liver is $\mu_0 = 1.6$.

(a) Find the 95% confidence interval for the mean level of liver retinol in the population of copper-deficient rats. Recall that the sample size was $n = 9$.

(b) Test the hypothesis that the mean level of retinol in the population of copper-deficient rats is $\mu_0 = 1.6$ versus a sensible alternative (either one-sided or two-sided), at the significance level $\alpha = 0.05$. Use both rejection region and $p$-value approaches.

(c) What is the power of the test in (b) against the alternative $H_1 : \mu = \mu_1 = 2.4$. Comment.

(d) Suppose that you are designing a new, larger study in which you are going to assume that the variance of observations is $\sigma^2 = 1.4^2$, as the limited nine-animal study indicated. Find the sample size so that the power of rejecting $H_0$ when an alternative $H_1 : \mu = 2.1$ is true is 0.80. Use $\alpha = 0.05$.

(e) Provide a Bayesian solution using WinBUGS.

9.18. **Aniline.** Organic chemists often purify organic compounds by a method known as *fractional crystallization*. An experimenter wanted to prepare and purify 5 grams of aniline. It is postulated that 5 grams of aniline would yield 4 grams of acetanilide. Ten 5-gram quantities of aniline were individually prepared and purified.

(a) Test the hypothesis that the mean dry yield differs from 4 grams if the mean yield observed in a sample was $\overline{X} = 4.21$. The population is assumed normal with known variance $\sigma^2 = 0.08$. The significance level is set to $\alpha = 0.05$.

(b) Report the $p$-value.

(c) For what values of $\overline{X}$ will the null hypothesis be rejected at the level $\alpha = 0.05$?

(d) What is the power of the test for the alternative $H_1 : \mu = 3.6$ at $\alpha = 0.05$.

(e) If you are to design a similar experiment but would like to achieve a power of 90% versus the alternative $H_1 : \mu = 3.6$ at $\alpha = 0.05$, what sample size would you recommended?

9.19. **DNA Random Walks.** DNA random walks are numerical transcriptions of a sequence of nucleotides. The imaginary walker starts at 0 and goes one step up ($s = +1$) if a purine nucleotide (A, G) is encountered, and one step down ($s = -1$) if a pyramidine nucleotide (C, T) is encountered. Peng et al. (1992) proposed identifying coding/noncoding regions by measuring the irregularity of associated DNA random walks. A standard irregularity measure is the Hurst exponent $H$, an index that ranges from 0 to 1. Numerical sequences with $H$ close to 0 are irregular, while the sequences with $H$ close to 1 appear more smooth.

Figure 9.3 shows a DNA random walk in the DNA of a spider monkey (*Ateles geoffroyi*). The sequence is formed from a noncoding region and has a Hurst exponent of $H = 0.61$.

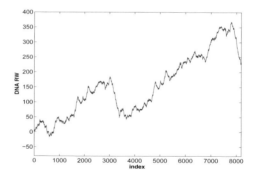

**Fig. 9.3** A DNA random walk formed by a noncoding region from the DNA of a spider monkey. The Hurst exponent is 0.61.

A researcher wishes to design an experiment in which $n$ nonoverlapping DNA random walks of a fixed length will be constructed, with the goal of testing to see if the Hurst exponent for noncoding regions is 0.6.

The researcher would like to develop a test so that an effect $e = |\mu_0 - \mu_1|/\sigma$ will be detected with a probability of $1 - \beta = 0.9$. The test should be two-sided with a significance level of $\alpha = 0.05$. Previous analyses of noncoding regions in the DNA of various species suggest that exponent $H$ is approximately normally distributed with a variance of approximately $\sigma^2 = 0.03^2$. The researcher believes that $|\mu_0 - \mu_1| = 0.02$ is a biologically meaningful difference. In statistical terms, a 5%-level test for $H_0 : \mu = 0.6$ versus the alternative $H_1 : \mu = 0.6 \pm 0.02$ should have a power of 90%. The preexperimentally assessed variance $\sigma^2 = 0.03^2$ leads to an effect size of $e = 2/3$.

(a) Argue that a sample size of $n = 24$ satisfies the power requirements. The experiment is conducted and the following 24 values for the Hurst exponent are obtained:

```
H =[0.56  0.61   0.62   0.53   0.54  0.60   0.56   0.59  ...
    0.60  0.60   0.62   0.60   0.58  0.57   0.61   0.64  ...
    0.60  0.61   0.58   0.59   0.55  0.59   0.60   0.65 ];
%   [mean(H)  std(H)]  %%% 0.5917   0.0293
```

(b) Using the $t$-test, test $H_0$ against the two-sided alternative at the level $\alpha = 0.05$ using both the rejection-region approach and the $p$-value approach.
(c) What is the retrospective power of your test? Use the formula with a noncentral $t$-distribution and $s$ found from the sample.

9.20. **Binding of Propofol.** Serum protein binding is a limiting factor in the access of drugs to the central nervous system. Disease-induced modifications of the degree of binding may influence the effect of anesthetic drugs.
The protein binding of *propofol*, an intravenous anaesthetic agent that is highly bound to serum albumin, has been investigated in patients with chronic renal failure. Protein binding was determined by the ultrafiltration technique using an Amicon Micropartition System, MPS-1.
The mean proportion of unbound propofol in healthy individuals is 0.96, and it is assumed that individual proportions follow a beta distribution, $\mathscr{B}e(96,4)$. Based on a sample of size $n = 87$ of patients with chronic renal failure, the average proportion of unbound propofol was found to be 0.93 with a sample standard deviation of 0.12.
(a) Test the hypothesis that the mean proportion of unbound propofol in a population of patients with chronic renal failure is 0.96 versus the one-sided alternative. Use $\alpha = 0.05$ and perform the test using both the rejection-region approach and the $p$-value approach. Would you change the decision if $\alpha = 0.01$?
(b) Even though the individual measurements (proportions) follow a beta distribution, one can use the normal theory in (a). Why?

9.21. **Improvement of Surgical Procedure.** In a disease in which the postoperative mortality is usually 10%, a surgeon devises a new surgical technique. He tries the technique on 15 patients and has no fatalities.
(a) What is the probability of the surgeon having no fatalities in treating 15 patients if the mortality rate is 10%.
(b) The surgeon claims that his new surgical technique significantly improves the survival rate. Is his claim justified?
(c) What is the minimum number of patients the surgeon needs to treat without a single fatality in order to convince you that his procedure is a

significant improvement over the old technique? Specify your criteria and justify your answer.

9.22. **Cancer Therapy.** Researchers in cancer therapy often report only the number of patients who survive for a specified period of time after treatment rather than the patients' actual survival times. Suppose that 40% of the patients who undergo the standard treatment are known to survive 5 years. A new treatment is administered to 200 patients, and 92 of them are still alive after a period of 5 years.

(a) Formulate the hypotheses for testing the validity of the claim that the new treatment is more effective than the standard therapy.

(b) Test with $\alpha = 0.05$ and state your conclusion; use the rejection-region method.

(c) Perform the test by finding the $p$-value.

9.23. **Is the Cloning of Humans Moral?** Gallup Poll estimates that 88% Americans believe that cloning humans is morally unacceptable. Results are based on telephone interviews with a randomly selected national sample of $n = 1000$ adults, aged 18 and older.

(a) Test the hypothesis that the true proportion is 0.9, versus the two-sided alternative, based on the Gallup data. Use $\alpha = 0.05$.

(b) Does 0.9 fall in the 95% confidence interval for the proportion.

(c) What is the power of this test against the specific alternative $p = 0.85$?

9.24. **Smoking Illegal?** In a recent Gallup poll of Americans, less than a third of respondents thought smoking in public places should be made illegal, a significant decrease from the 39% who thought so in 2001.

The question used in the poll was: *Should smoking in all public places be made totally illegal?* In the poll, 497 people responded and 154 answered yes. Let $p$ be the proportion of people in the US voting population supporting the idea that smoking in public places should be made illegal.

(a) Test the hypothesis $H_0 : p = 0.39$ versus the alternative $H_1 : p < 0.39$ at the level $\alpha = 0.05$.

(b) What is the 90% confidence interval for the unknown population proportion $p$? In terms of the Gallup pollsters, what is the "margin of error"?

9.25. **Spider Monkey DNA.** An 8192-long nucleotide sequence segment taken from the DNA of a spider monkey (*Ateles geoffroyi*) is provided in the file ◢ dnatest.m.

Find the relative frequency of adenine $\hat{p}_A$ as an estimator of the overall population proportion, $p_A$.

Find a 99% confidence interval for $p_A$ and test the hypothesis $H_0 : p_A = 0.2$ versus the alternative $H_1 : p_A > 0.2$. Use $\alpha = 0.05$.

---

| MATLAB AND WINBUGS FILES AND DATA SETS USED IN THIS CHAPTER |

http://springer.bme.gatech.edu/Ch9.Testing/

chi2test1.m, dnarw.m, dnatest.m FDR.m, hemopexin1.m, hemoragic.m, moon.m, powers.m, SBB.m, testexa.m

hemopexin.odc, hemorrhagic.odc, hypersplenism.odc, moonillusion.odc, retinol.odc, spikes.odc, systolic.odc

dnadat.mat|txt, spid.dat

---

# CHAPTER REFERENCES

Andrews, D. F. and Herzberg, A. M. (1985). *Data. A Collection of Problems from Many Fields for the Student and Research Worker*. Springer, Berlin Heidelberg New York.

Benjamini, Y. and Hochberg, Y. (1995) Controlling the false discovery rate: a practical and powerful approach to multiple testing. *J. R. Stat. Soc. B*, **57**, 289–300.

Berger, J. O. and Sellke, T. (1987). Testing a point null hypothesis: the irreconcilability of p-values and evidence (with discussion). *J. Am. Stat. Assoc.*, **82**, 112–122.

Casella, G. and Berger, R. (1990). *Statistical Inference*. Duxbury, Belmont.

Edwards, W., Lindman, H., and Savage, L. J. (1963). Bayesian statistical inference for psychological research. *Psychol. Rev.*, **70**, 193–242.

Fisher, R. A. (1925). *Statistical Methods for Research Workers*. Oliver and Boyd, Edinburgh.

Fisher, R. A. (1926). The arrangement of field experiments. *J. Ministry Agricult.*, **33**, 503–513.

Goodman, S. (1999a). Toward evidence-based medical statistics. 1: The p-value fallacy. *Ann. Intern. Med.*, **130**, 995–1004.

Goodman, S. (1999b). Toward evidence-based medical statistics. 2: The Bayes factor. *Ann. Intern. Med.*, **130**, 1005–1013.

Goodman, S. (2001). Of p-values and Bayes: a modest proposal. *Epidemiology*, **12**, 3, 295–297

Hamilton, L. C. (1990). *Modern Data Analysis: A First Course in Applied Statistics*. Brooks/Cole, Pacific Grove.

Hoenig, J. M. and Heisey, D. M. (2001). Abuse of power: the pervasive fallacy of power calculations for data analysis. *Am. Statist.*, **55**, 1, 19–24.

Ioannidis, J. P. (2005). Why most published research findings are false. PLoS Med 2(8): e124. doi:10.1371/journal.pmed.0020124.

Kaufman, L. and Rock, I. (1962). The moon illusion, I. *Science*, **136**, 953–961.

Katz, S., Lautenschlager, G. J., Blackburn, A. B., and Harris, F. H. (1990). Answering reading comprehension items without passages on the SAT. *Psychol. Sci.*, **1**, 122–127.

Peng, C. K., Buldyrev, S. V., Goldberger, A. L., Goldberg, Z. D., Havlin, S., Sciortino, E., Simons, M., and Stanley, H. E. (1992). Long-range correlations in nucleotide sequences. *Nature*, **356**, 168–170.

Rachman, F., Conjat, F., Carreau, J. P., Bleiberg-Daniel, F., and Amedee-Maneseme, O. (1987). Modification of vitamin A metabolism in rats fed a copper-deficient diet. *Int. J. Vitamin Nutr. Res.*, **57**, 247–252.

Sampford, M. R. and Taylor, J. (1959). [not cited?] Censored observations in randomized block experiments. *J. R. Stat. Soc. Ser. B*, **21**, 214–237.

Schervish, M. (1996). P-values: What they are and what they are not. *Am. Stat.*, **50**, 203–206.

Sellke, T., Bayarri, M. J., and Berger, J. O. (2001). Calibration of p values for testing precise null hypotheses. *Am. Stat.*, **55**, 62–71.

# Chapter 10
# Two Samples

*Given a choice between two theories, take the one which is funnier.*

– Blore's Razor

---

**WHAT IS COVERED IN THIS CHAPTER**

- Testing the Equality of Normal Means and Variances
- A Bayesian Approach to a Two-Sample Problem
- Paired $t$-Test
- Testing the Equality of Two Proportions
- Risk Jargon: Risk Differences, Ratios, and Odds Ratios
- Two Poisson Means
- Equivalence Testing

---

## 10.1 Introduction

A two-sample inference is one of the most common statistical procedures used in practice. For example, a colloquial use of "$t$-test" usually refers to the comparison of means from two independent normal populations rather than a single-sample $t$-test. In this chapter we will test the equality of two normal means for independent and dependent (paired) populations as well as the equality of two variances and proportions. In the context of comparing proportions, we will discuss the risk and odds ratios. In testing the equality of means in independent normal populations, we will distinguish two cases: (i) when the underlying population variances are the same and (ii) when no assumption about the variances is made. In this second case the population variances may be different, or even equal, but simply no assumption about their equality enters the test. Each of the tests involves the difference or ratio of the parameters (means, proportions, variances), and for each difference/ratio we provide the $(1 - \alpha)100\%$ confidence interval. For selected tests we will include the power analysis. This chapter is intertwined with parallel Bayesian solutions whenever appropriate.

It is important to emphasize that the normality of populations and large samples for the proportions (for the CLT to hold) are critical for some tests. Later in the text, in Chap. 12, we discuss distribution-free counterpart tests that relax the assumption of normality (sign test, Wilcoxon signed rank test, Wilcoxon Mann Whitney test) at the expense of efficiency if the normality holds.

## 10.2 Means and Variances in Two Independent Normal Populations

We start with an example that motivates the testing of two population means.

*Example 10.1.* **Lead Exposure.**  It is hypothesized that blood levels of lead tend to be higher for children whose parents work in a factory that uses lead in the manufacturing process. Researchers examined lead levels in the blood of 12 children whose parents worked in a battery manufacturing factory. The results for the "case children" $X_{11}, X_{12}, \ldots, X_{1,12}$ were compared to those of the "control" sample $X_{21}, X_{22}, \ldots, X_{2,15}$ consisting of 15 children selected randomly from families where the parents did not work in a factory that used lead. It is assumed that the measurements are independent and come from normal populations. The resulting sample means and sample standard deviations were $\overline{X}_1 = 0.015$, $s_1 = 0.004$, $\overline{X}_2 = 0.006$, and $s_2 = 0.006$.

Obviously, the sample mean for the "case" children is higher than the sample mean in the control sample. But is this difference significant?

To state the problem in more general terms, we assume that two samples $X_{11}, X_{12}, \ldots, X_{1,n_1}$ and $X_{21}, X_{22}, \ldots, X_{2,n_2}$ are observed from populations with normal $\mathcal{N}(\mu_1, \sigma_1^2)$ and $\mathcal{N}(\mu_2, \sigma_2^2)$ distributions, respectively.

We are interested in testing the hypothesis $H_0 : \mu_1 = \mu_2$ versus the alternative $H_1 : \mu_1 >, \neq, < \mu_2$ at a significance level $\alpha$.

For the lead exposure example, the null hypothesis being tested is that the parents' workplace has no effect on their children's lead concentration, that is, the two population means will be the same:

$$H_0 : \mu_1 = \mu_2.$$

Here the populations are defined as all children that are exposed or nonexposed. The alternative hypothesis $H_1$ may be either one- or two-sided. The two-sided alternative is simply $H_1 : \mu_1 \neq \mu_2$, the population means are not equal, and the difference can go either way. The choice of one-sided hypothesis should be guided by the problem setup, and sometimes by the observations. In the context of this example, it would not make sense to take the one-sided alternative as $H_1 : \mu_1 < \mu_2$ stating that the concentration in the exposed group is smaller than that in the control. In addition, $\overline{X}_1 = .015$ and $\overline{X}_2 = .006$ are observed. Thus, the sensible one-sided hypothesis in this context is $H_1 : \mu_1 > \mu_2$.

There are two testing scenarios of population means that depend on an assumption about associated population variances.

**Scenario 1: Variances unknown but assumed equal.** In this case, the joint $\sigma^2$ is estimated by both $s_1^2$ and $s_2^2$. The weighted average of $s_1^2$ and $s_2^2$ with weights $w$ and $1 - w$, depending on group sample sizes $n_1$ and $n_2$,

$$s_p^2 = \frac{(n_1 - 1)s_1^2 + (n_2 - 1)s_2^2}{n_1 + n_2 - 2} = \frac{n_1 - 1}{n_1 + n_2 - 2}s_1^2 + \frac{n_2 - 1}{n_1 + n_2 - 2}s_2^2 = ws_1^2 + (1 - w)s_2^2,$$

is called the pooled sample variance and it better estimates the population variance than any individual $s^2$. The square root of $s_p^2$ is called the pooled sample standard deviation and is denoted by $s_p$. One can show that when $H_0$ is true, that is, when $\mu_1 = \mu_2$, the statistic

$$t = \frac{\overline{X}_1 - \overline{X}_2}{s_p \sqrt{1/n_1 + 1/n_2}} \tag{10.1}$$

has Student's $t$ distribution with $df = n_1 + n_2 - 2$ degrees of freedom.

**Scenario 2: No assumption about the variances.** In this case, when $H_0$ is true, i.e., when $\mu_1 = \mu_2$, the statistic

$$t = \frac{\overline{X}_1 - \overline{X}_2}{\sqrt{s_1^2/n_1 + s_2^2/n_2}}$$

has a $t$ distribution with approximately

$$df = \frac{(s_1^2/n_1 + s_2^2/n_2)^2}{(s_1^2/n_1)^2/(n_1 - 1) + (s_2^2/n_2)^2/(n_2 - 1)} \tag{10.2}$$

degrees of freedom. This is a special case of the so-called Welch–Satterthwaite formula, which approximates the degrees of freedom for a linear combination of chi-square distributions (Satterthwaite 1946, Welch 1948).

For both scenarios:

| Alternative | $\alpha$-level rejection region | $p$-value |
|---|---|---|
| $H_1 : \mu_1 > \mu_2$ | $[t_{df,1-\alpha}, \infty)$ | 1-tcdf(t, df) |
| $H_1 : \mu_1 \neq \mu_2$ | $(-\infty, t_{df,\alpha/2}] \cup [t_{df,1-\alpha/2}, \infty)$ | 2*tcdf(-abs(t), df) |
| $H_1 : \mu_1 < \mu_2$ | $(-\infty, t_{df,\alpha}]$ | tcdf(t, df) |

When the population variances are known, the proper statistic is $Z$:

$$Z = \frac{\overline{X}_1 - \overline{X}_2}{\sqrt{\sigma_1^2/n_1 + \sigma_2^2/n_2}},$$

with a normal $\mathcal{N}(0,1)$ distribution, and the proper statistical analysis involves normal quantiles as in the $z$-test. Given the fact that in realistic examples the variances are not known when the means are tested, the $z$-test is mainly used as an asymptotic test.

When sample sizes $n_1$ and $n_2$ are large, the $z$-statistic can be used instead of $t$ even if the variances are not known, due to the CLT. This approximation was more interesting in the past when computing was expensive, but these days one should always use a $t$-test for any sample size, as long as the population variances are not known.

In Example 10.1, the variances are not known. We may assume that they are either equal or possibly not equal based on the nature of the experiment, sampling, and some other nonexperimental factors. However, we may also formally test whether the population variances are equal prior to deciding on the testing scenario.

We briefly interrupt our discussion of testing the equality of means with an exposition of how to test the equality of variances in two normal populations.

**Testing the Equality of Two Normal Variances.** Selecting the "scenario" for testing the equality of normal means requires an assumption about the associated variances. This assumption can be guided by an additional test for the equality of two normal variances prior to testing the means. The variance-before-the-means testing is criticized mainly on the grounds that tests for variances are not as robust (with respect to deviations from normality) compared to the test for means, especially when the sample sizes are small or unbalanced. Although we agree with this criticism, it should be noted that choosing the scenario seldom influences the resulting inference, except maybe in the borderline cases. See also discussion on p. 416. When in doubt, one should not make the assumption about variances and should use a more conservative scenario 2.

Suppose the samples $X_{11}, X_{12}, \ldots, X_{1,n_1}$ and $X_{21}, X_{22}, \ldots, X_{2,n_2}$ come from normal populations with distributions of $\mathcal{N}(\mu_1, \sigma_1^2)$ and $\mathcal{N}(\mu_2, \sigma_2^2)$, respectively. To choose the strategy for testing the means, the following test of variances with the two-sided alternative is helpful:

$$H_0 : \sigma_1^2 = \sigma_2^2 \text{ versus } H_1 : \sigma_1^2 \neq \sigma_2^2.$$

The testing statistic is the ratio of sample variances, $F = s_1^2/s_2^2$, that has an $F$ distribution with $n_1 - 1$ and $n_2 - 1$ degrees of freedom when $H_0$ is true. The decision can be made based on a $p$-value that is equal to

```
p = 2 * min( fcdf(F, n1-1, n2-1), 1-fcdf(F, n1-1, n2-1) )
```

**Remark.** We note that the most popular method (recommended in many texts) for calculating the $p$-value in two-sided testing uses either the expression 2*fcdf(F,n1-1,n2-1) or 2*(1-fcdf(F,n1-1,n2-1)), depending on whether F<1 or F>1. Although this approach leads to a correct $p$-value most of the time, it could lead to a $p$-value that exceeds 1 when the observed values of $F$ are close to 1. This is clearly wrong since the $p$-value is a probability. Exercise 10.4 describes such a case.

MATLAB has a built-in function, vartest2, for testing the equality of two normal variances.

Guided by the outcome of this test, either we assume that the population variances are the same and for testing the equality of means use a $t$-statistic with a pooled standard deviation and $df = n_1 + n_2 - 2$ degrees of freedom, or we use the $t$-test without making an assumption about the population variances and the degrees of freedom determined by the Welch–Satterthwaite formula in (10.2).

Next we summarize the test for both the one- and two-sided alternatives. When $H_0 : \sigma_1^2 = \sigma_2^2$ and $F = s_1^2/s_2^2$, the following table summarizes the test of the equality of normal variances against the one- or two-sided alternatives. Let $df_1 = n_1 - 1$ and $df_2 = n_2 - 1$.

| Alternative | $\alpha$-level rejection region | $p$-value |
|---|---|---|
| $H_1 : \sigma_1^2 > \sigma_2^2$ | $[F_{df_1,df_2,1-\alpha}, \infty)$ | 1-fcdf(F,df1,df2) |
| $H_1 : \sigma_1^2 \neq \sigma_2^2$ | $[0, F_{df_1,df_2,\alpha/2}] \cup [F_{df_1,df_2,1-\alpha/2}, \infty)$ | 2*min( fcdf(F,df1,df2), (1-fcdf(F,df1,df2)) ) |
| $H_1 : \sigma_1^2 < \sigma_2^2$ | $[0, F_{df_1,df_2,\alpha}]$ | fcdf(F,df1,df2) |

The $F$-test for testing the equality of variances assumes independent samples. Glass and Hopkins (1984, Sect. 13.9) gave a test statistic for testing the equality of variances obtained from paired samples with a correlation coefficient $r$. The test statistic has a $t$ distribution with $n - 2$ degrees of freedom,

$$t = \frac{s_1^2 - s_2^2}{2 s_1 s_2 \sqrt{(1 - r^2)/(n - 2)}},$$

where $s_1^2, s_2^2$ are the two sample variances, $n$ is the number of pairs of observations, and $r$ is the correlation between the two samples. This test is first
discussed in Pitman (1939).

*Example 10.2.* In Example 10.1, the $F$-statistic for testing the equality of variances is 0.4444 and the hypothesis of equality of variances is not rejected at a
significance level of $\alpha = 0.05$; the $p$-value is 0.18.

```
n1 = 12; X1bar = 0.010; s1 = 0.004; %exposed
n2 = 15; X2bar = 0.006; s2 = 0.006; %non-exposed
Fstat = s1^2/s2^2
    % Fstat = 0.4444
%The p-value is
pval = 2*min(fcdf(Fstat,n1-1,n2-1), 1-fcdf(Fstat,n1-1,n2-1))
    % pval =  0.1825
```

**Back to Testing Two Normal Means.** Guided by the previous test we assume that the population variances are the same, and for the original problem
of testing the means we use the $t$-statistic normalized by the pooled standard
deviation. The test statistic is

$$t = \frac{\overline{X}_1 - \overline{X}_2}{s_p \sqrt{1/n_1 + 1/n_2}}, \quad \text{where } s_p = \sqrt{\frac{(n_1 - 1)s_1^2 + (n_2 - 1)s_2^2}{n_1 + n_2 - 2}},$$

and it is $t$-distributed with $n_1 + n_2 - 2$ degrees of freedom.

```
sp = sqrt( ((n1-1)*s1^2 + (n2-1)*s2^2 )/(n1 + n2 - 2))
    % sp =0.0052
df= n1 + n2 - 2  %%df = 25
tstat = (X1bar - X2bar)/(sp * sqrt(1/n1 + 1/n2))
    % tstat=1.9803
pvalue = 1 - tcdf(tstat, n1+n2-2)
    % pvalue = 0.0294 approx 3%
```

The null hypothesis is rejected at the 5% level since $0.0294 < 0.05$.
Suppose that one wants to test $H_0$ using rejection regions. Since the alternative hypothesis is one-sided and right-tailed, as $\mu_1 - \mu_2 > 0$, the rejection
region is $RR = [t_{n_1+n_2-2, 1-\alpha}, \infty)$.

```
tinv(1-0.05, df)  %ans =1.7081
```

By rejection-region arguments, the hypothesis $H_0$ is rejected since $t > t_{n_1+n_2-2, 1-\alpha}$, that is, the observed value of statistic $t = 1.9803$ exceeds the critical value 1.7081.

### 10.2.1 Confidence Interval for the Difference of Means

Sometimes we might be interested in the $(1 - \alpha)100\%$ confidence interval for the difference of the population means. Such confidence intervals are easy to obtain and they depend, as do the tests, on the assumption about population variances. In general, the interval is

$$\left[ \overline{X}_1 - \overline{X}_2 - t_{df,1-\alpha/2} \, s^*, \ \overline{X}_1 - \overline{X}_2 + t_{df,1-\alpha/2} \, s^* \right],$$

where for the equal variance case $df = n_1 + n_2 - 2$ and $s^* = s_p \sqrt{1/n_1 + 1/n_2}$ and for no assumption about the population variances case $df$ is the Welch–Satterthwaite value in (10.2) and $s^* = \sqrt{s_1^2/n_1 + s_2^2/n_2}$.

For the lead exposure example, the 95% confidence interval for $\mu_1 - \mu_2$ is $[-0.00016, 0.0082]$:

```
sp=sqrt((((n1-1)*s1^2 + (n2-1)*s2^2 )/(n1+n2-2))    % sp = 0.0052
df = n1 + n2 - 2                                     % df = 25
LB=X1bar-X2bar-tinv(0.975,df)*sp*sqrt(1/n1+1/n2)     % LB =-0.00016
UB=X1bar-X2bar+tinv(0.975,df)*sp*sqrt(1/n1+1/n2)     % UB = 0.0082
```

Note that this interval barely covers 0. A test for the equality of two means against the two-sided alternative can be conducted by inspecting the confidence interval for their difference. For a two-sided test of level $\alpha$ one finds the $(1 - \alpha)100\%$ confidence interval, and if this interval contains 0, the null hypothesis is not rejected. What may be concluded from the interval $[-0.00016, 0.0082]$ is the following. If instead of the one-sided alternative that was found to be significant at the 5% level (the $p$-value was about 3%) one carried out the test against the two-sided alternative, the test of the same level would fail to reject $H_0$.

MATLAB's toolbox "stats" has a built-in function, ttest2, that performs two sample $t$-tests.

### 10.2.2 Power Analysis for Testing Two Means

In testing $H_0 : \mu_1 = \mu_2$ against the two-sided alternative $H_1 : \mu_1 \neq \mu_2$, for the specific alternative $|\mu_1 - \mu_2| = \Delta$, an approximation of power is

$$1 - \beta = \Phi \left( z_{\alpha/2} + \frac{\Delta}{\sqrt{\dfrac{\sigma_1^2}{n_1} + \dfrac{\sigma_2^2}{n_2}}} \right) + 1 - \Phi \left( z_{1-\alpha/2} + \frac{\Delta}{\sqrt{\dfrac{\sigma_1^2}{n_1} + \dfrac{\sigma_2^2}{n_2}}} \right). \tag{10.3}$$

If the alternative is one-sided, say $H_1 : \mu_1 > \mu_2$, then $\Delta = \mu_1 - \mu_2$ and the power is

$$1 - \beta = 1 - \Phi \left( z_{1-\alpha} - \frac{\Delta}{\sqrt{\frac{\sigma_1^2}{n_1} + \frac{\sigma_2^2}{n_2}}} \right) = \Phi \left( z_\alpha + \frac{\Delta}{\sqrt{\frac{\sigma_1^2}{n_1} + \frac{\sigma_2^2}{n_2}}} \right). \qquad (10.4)$$

The approximation is good if $n_1$ and $n_2$ are large, but for small to moderate values of $n_1$ and $n_2$ it tends to overestimate the power.

Equations (10.3) and (10.4) are standardly used but are somewhat obsolete since the noncentral $t$ distribution (p. 217) needed for an exact power is readily available.

We state the formulas in terms of MATLAB code:

```
1 - nctcdf(tinv(1-alpha/2,n1+n2-2), ...
                 n1+n2-2,Delta/(sp*sqrt(1/n1+1/n2)))...
  + nctcdf(tinv(alpha/2,n1+n2-2), ...
                 n1+n2-2,Delta/(sp*sqrt(1/n1+1/n2)))
```

For the one-sided alternative $H_1 : \mu_1 > \mu_2$ the code is

```
1 - nctcdf(tinv(1-alpha,n1+n2-2), ...
          n1+n2-2,Delta/(sp*sqrt(1/n1+1/n2)))
```

In testing the equality of means in two normal populations using independent samples, $H_0 : \mu_1 = \mu_2$, versus the one-sided alternative, the group sample size for fixed $\alpha, \beta$ is

$$n \geq \frac{2\sigma^2}{|\mu_1 - \mu_2|^2} (z_{1-\alpha} + z_{1-\beta})^2,$$

where $\sigma^2$ is the common population variance. If the alternative is two-sided, then $z_{1-\alpha}$ is replaced by $z_{1-\alpha/2}$. In that case, the sample size is approximate.

It is assumed that the group sample sizes are equal, i.e., that the total sample size is $N = 2n$. If the variances are not the same, then

$$n \geq \frac{\sigma_1^2 + \sigma_2^2}{|\mu_1 - \mu_2|^2} (z_{1-\alpha} + z_{1-\beta})^2.$$

In the context of Example 10.1 let us find the power of the test against the alternative $H_1 : \mu_1 - \mu_2 = 0.005$.

The power for a one-sided $\alpha$-level test against the alternative $H_1 : \mu_1 - \mu_2 = 0.005 (= \Delta)$ is given in (10.4). The normal approximation is used and $s_1^2$ and $s_2^2$ are plugged into the place of $\sigma_1^2$ and $\sigma_2^2$.

```
power = 1-normcdf(norminv(1-0.05)-0.005/sqrt(s1^2/n1+s2^2/n2) )
  %power=  0.8271
power = normcdf(norminv(0.05)+0.005/sqrt(s1^2/n1+s2^2/n2) )
  %power=  0.8271
```

Thus the power is about 83%. This is an approximation and normal approximation tends to overestimate the power. The exact power is about 81%,

```
power=1-nctcdf(tinv(1-0.05,n1+n2-2),...
                  n1+n2-2,0.005/sqrt(s1^2/n1+s2^2/n2))
  %power = 0.8084
```

We plan to design a future experiment to test the same phenomenon. When data are collected and analyzed, we would like for the $\alpha = 5\%$ test to achieve a power of $1 - \beta = 90\%$ against the specific alternative $H_1 : \mu_1 - \mu_2 = 0.005$. What sample size will be necessary? Since determining the sample size to meet a preassigned power and precision is prospective in nature, we assume that the previous data were obtained in a pilot study and that $\sigma_1^2$ and $\sigma_2^2$ are "known" and equal to the observed $s_1^2$ and $s_2^2$. The sample size formula is

$$n = \frac{(\sigma_1^2 + \sigma_2^2)(z_{1-\alpha} + z_{1-\beta})^2}{\Delta^2},$$

which in MATLAB gives

```
ssize = (s1^2 + s2^2)*(norminv(0.95)+norminv(0.9))^2/(0.005^2)
% ssize = 17.8128 approx 18 each
```

The number of children is 18 per group if one wishes for the sample sizes to be the same, $n_1 = n_2$. In the following section we discuss the design with $n_2 = k \times n_1$, for some $k$. Such designs can be justified by different costs of sampling, where the meaning of "cost" may be more general than only the financial one.

### 10.2.3 More Complex Two-Sample Designs

Suppose that we are interested in testing the equality of normal population means when the underlying variances in the two populations are $\sigma_1^2$ and $\sigma_2^2$, and not necessarily equal. Also assume that the desired proportion of sample sizes to be determined is $k = n_2/n_1$, that is, $n_2 = k \times n_1$. This proportion may be dictated by the cost of sampling or by the abundance of the populations. When equal group samples are desired, then $k = 1$,

$$n_1 = \frac{(\sigma_1^2 + \sigma_2^2/k)(z_{1-\alpha/2} + z_{1-\beta})^2}{|\mu_1 - \mu_2|^2}, \qquad n_2 = k \times n_1. \tag{10.5}$$

As before, $\mu_1$, $\mu_2$, $\sigma_1^2$, and $\sigma_2^2$ are unknown, and in the absence of any data, one can express $|\mu_1 - \mu_2|^2$ in units of $\sigma_1^2 + \sigma_2^2/k$ to elicit the effect size, $d^2$.

However, if preliminary or historic samples are available, then $\mu_1$, $\mu_2$, $\sigma_1^2$, and $\sigma_2^2$ can be estimated by $\overline{X}_1$, $\overline{X}_2$, $s_1^2$, and $s_2^2$, respectively, and plugged into formula (10.6).

*Example 10.3.* **Two Amanitas.** Suppose that two independent samples of $m = 12$ and $n = 15$ spores of A. *pantherina* ("Panther") and A. *rubescens* ("Blusher"), respectively, are only a pilot study. It was found that the means are $\overline{X}_1 = 6.3$ and $\overline{X}_2 = 7.5$ with standard deviations of $s_1 = 2.12$ and $s_2 = 1.94$. All measures are in $\mu$m. Suppose that Blushers are twice as common as Panthers.

Determine the sample sizes for future study that will find the difference obtained in the preliminary samples to be significant at the level $\alpha = 0.05$ with a power of $1 - \beta = 0.90$.

Here, based on the abundance of mushrooms, $2n_1 = n_2$ and $k = 2$. Substituting $\overline{X}_1$, $\overline{X}_2$, $s_1^2$, and $s_2^2$ into (10.6), one gets

$$n_1 = \frac{(2.12^2 + 1.94^2/2)(z_{0.975} + z_{0.9})^2}{|6.3 - 7.5|^2} = 46.5260 \approx 47.$$

Here, the effect size was $|6.3 - 7.5|/\sqrt{2.12^2 + 1.94^2/2} = 0.4752$, which corresponds to $d = 0.4752\sqrt{2}$.

The "plug-in" principle applied in the above example is controversial. Proponents argue that in the absence of any information on $\mu_1$, $\mu_2$, $\sigma_1^2$, and $\sigma_2^2$, the most "natural" procedure is to use their MLEs, $\overline{X}_1$, $\overline{X}_2$, $s_1^2$, and $s_2^2$. Opponents say that one is looking for a sample size that will find the pilot difference to be significant at a level $\alpha$ with a preassigned power. They further argue that, due to routinely small sample sizes in pilot studies, the estimators for population means and variances can be quite unreliable. This unreliability is further compounded by taking the ratios and powers in calculating the sample size.

**Remark.** Cochran and Cox (1957) proposed a method of testing two normal means with unequal variances according to which the rejection region is based on a linear combination of $t$ quantiles,

$$f_{1-\alpha} = \frac{(s_1^2/n_1)t_{n_1-1,1-\alpha} + (s_2^2/n_2)t_{n_2-1,1-\alpha}}{s_1^2/n_1 + s_2^2/n_2},$$

for one-sided alternatives. For two-sided alternative $1-\alpha$ is replaced by $1-\alpha/2$. This test is conservative, with the achieved level of significance smaller than the stated $\alpha$.

## 10.2.4 Bayesian Test of Two Normal Means

Bayesian testing of two means simply analyzes the posterior distribution of the means difference, given the priors and the data. We will illustrate a Bayesian approach for a simple noninformative prior structure.

Let $X_{11}, X_{12}, \ldots, X_{1,n_1}$ and $X_{21}, X_{22}, \ldots, X_{2,n_2}$ be samples from normal $\mathcal{N}(\mu_1, \sigma_1^2)$ and $\mathcal{N}(\mu_2, \sigma_2^2)$ distributions, respectively. We are interested in the posterior distribution of $\theta = \mu_2 - \mu_1$ when $\sigma_1^2 = \sigma_2^2 = \sigma^2$. If the priors on $\mu_1$ and $\mu_2$ are flat, $\pi(\mu_1) = \pi(\mu_2) = 1$, and the prior on the common $\sigma^2$ is noninformative, $\pi(\sigma^2) = 1/\sigma^2$, then the posterior of $\theta$, after integrating out $\sigma^2$, is a $t$ distribution. That is to say, if $\overline{X}_1$ and $\overline{X}_2$ are the sample means and $s_p$ is the pooled standard deviation, then

$$t = \frac{\theta - (\overline{X}_2 - \overline{X}_1)}{s_p \sqrt{1/n_1 + 1/n_2}} \tag{10.6}$$

has a $t$ distribution with $n_1 + n_2 - 2$ degrees of freedom (Box and Tiao, 1992, p. 103). Compare (10.6) with the distribution in (10.1). Although the two distributions coincide, they are conceptually different; (10.1) is the sampling distribution for the difference of sample means, while (10.6) gives the distribution for the difference of parameters.

In this case, the results of Bayesian inference coincide with frequentist results on the estimation of $\theta$, the confidence/credible intervals for $\theta$ and testing, as is usually the case when the priors are noninformative.

When $\sigma_1^2$ and $\sigma_2^2$ are not assumed equal and each has its own noninformative prior, finding the posterior in the previous model coincides with the Behrens–Fisher problem and the posterior is usually approximated (Patil's approximation, Box and Tiao, 1992, p. 107; Lee, 2004, p. 145).

When MCMC and WinBUGS are used in testing two normal means, we may entertain more flexible models. The testing becomes quite straightforward. Next, we provide an example.

*Example 10.4.* **Microdamage in Bones.** Bone is a hierarchical composite material that provides our bodies with mechanical support and facilitates mobility, among other functions. Figure 10.1 shows the structure of bone tubecules. Damage in bone, in the form of microcracks, occurs naturally during daily physiological loading. The normal bone remodeling process repairs this microdamage, restoring, if not improving, biomechanical properties. Numerous studies have shown that microdamage accumulates as we age due to impaired bone remodeling. This accumulation contributes to a reduction in bone biomechanical properties such as strength and stiffness by disrupting the local tissue matrix.

In order to better understand the role of microdamage in bone tissue matrix properties as we age, a study was conducted in the lab of Dr. Robert Guldberg at Georgia Institute of Technology. The interest was in changes in microdamage progression in human bone between young and old female donors.

**Fig. 10.1** Bone tubecules.

The data showing the score of normalized damage events are shown in the table below. There were $n_1 = 13$ donors classified as young ($\leq 45$ years old and $n_2 = 17$ classified as old (>45 years old). To calculate the microdamage progression score, the counts of damage events (extensions, surface originations, widenings, and combinations) are normalized to the bone area and summed up.

| Young | | Old | |
|---|---|---|---|
| 0.790 | 1.264 | 1.374 | 1.327 |
| 0.944 | 1.410 | 0.601 | 1.325 |
| 0.958 | 1.160 | 1.029 | 2.012 |
| 1.011 | 0.179 | 1.264 | 1.026 |
| 0.714 | | 1.183 | 1.130 |
| 0.256 | | 1.856 | 0.605 |
| 0.406 | | 1.899 | 0.870 |
| 0.135 | | 0.486 | 0.820 |
| 0.316 | | 0.813 | |

Assuming that the microdamage scores are normally distributed, test the hypothesis of equality of population means (young and old) against the one-sided alternative.

```
#microdamage.odc
model{
  for (i in 1:n){
    score[i] ~ dnorm(mu[age[i]], prec[age[i]])
  }
  mu[1] ~ dnorm(0, 0.00001)
  mu[2] ~ dnorm(0, 0.00001)
  prec[1] ~ dgamma(0.001, 0.001)
  prec[2] ~ dgamma(0.001, 0.001)
  d <- mu[1] - mu[2]
  r <- prec[1]/prec[2]
  ph1 <- step(-d)    #ph1=1 if d < 0
  ph0 <- 1-ph1
}

DATA
list(n=30,score=c(0.790, 0.944, 0.958, 1.011, 0.714, 0.256, 0.406,
      0.135, 0.316, 0.179, 1.264, 1.410, 1.160, 1.374,
      0.601, 1.029, 1.264, 1.183, 1.856, 1.899, 0.486,
      0.813, 0.820, 1.327, 1.325, 2.012, 1.026, 1.130,
      0.605, 0.870),
age = c(1,1,1,1,1,1,1,1,1,1,1,1,1,
      2,2,2,2,2,2,2,2,2,2,2,2,2,2,2,2))

INITS
list( mu = c(1,1), prec=c(1,1))
```

| | mean | sd | MC error | val2.5pc | median | val97.5pc | start | sample |
|---|---|---|---|---|---|---|---|---|
| d | −0.4194 | 0.1771 | 5.438E-4 | −0.7688 | −0.4192 | −0.071 | 1001 | 100000 |
| mu[1] | 0.7339 | 0.1326 | 4.099E-4 | 0.4703 | 0.7338 | 0.9968 | 1001 | 100000 |
| mu[2] | 1.153 | 0.118 | 3.685E-4 | 0.9188 | 1.153 | 1.389 | 1001 | 100000 |
| ph0 | 0.01011 | 0.1 | 3.112E-4 | 0.0 | 0.0 | 0.0 | 1001 | 100000 |
| ph1 | 0.9899 | 0.1 | 3.112E-4 | 1.0 | 1.0 | 1.0 | 1001 | 100000 |
| r | 1.244 | 0.7517 | 0.002553 | 0.3456 | 1.071 | 3.155 | 1001 | 100000 |

The posterior probability of $H_1$ is 0.9899, thus $H_0$ is rejected. Note that the credible interval for the difference $d$ is all negative, suggesting that the two-sided test would be significant (in Bayesian terms). The ratio of precisions (and variances) $r$ has a credible set that contains 1; thus the variances could be assumed equal. This assumption has no bearing on the Bayesian procedure (unlike the classical approach).

## 10.3 Testing the Equality of Normal Means When Samples Are Paired

When comparing two treatments it is desirable that the experimental units be as alike as possible so that the difference in responses can be attributed chiefly to the treatment. If many relevant factors (age, gender, body mass index, presence of risk factors, and so on) vary in an uncontrolled manner, a large portion of variability in the response can be attributed to these factors rather than the treatments.

The concept of pairing, matching, or blocking is critical to eliminate nuisance variability and obtain better experimental designs. Consider a sample consisting of paired elements, so that every element from population 1 has its match from population 2. A sample from population 1, $X_{11}, X_{12}, \ldots, X_{1,n}$, is thus paired with a sample from population 2, $X_{21}, X_{22}, \ldots, X_{2,n}$, so that a pair $(X_{1i}, X_{2i})$ represents the $i$th observation. Usually, observations in a pair are taken on the same subject (pretest–posttest, placebo–treatment) or on dependent subjects (brother–sister, two subjects with matching demographic characteristics, etc.). The examples are numerous, but most applications involve subjects with measurements taken at two different time points, during two different treatments, and so on. Sometimes this matching is called "blocking."

It is usually assumed that the samples come from normal populations with possibly different means $\mu_1$ and $\mu_2$ (subject to test) and with unknown variances $\sigma_1^2$ and $\sigma_2^2$. As linear combinations of normals, the differences $d_i = X_{1i} - X_{2i}$ are also normal,

$$d_i \sim \mathcal{N}(\mu_1 - \mu_2, \sigma_1^2 + \sigma_2^2 - 2 \cdot \sigma_{12}),$$

where $\sigma_{12}$ is the covariance, $\mathbb{E}[(X_1 - \mathbb{E}X_1)(X_2 - \mathbb{E}X_2)]$. Define

$$t = \frac{\overline{d}}{s_d/\sqrt{n}}, \tag{10.7}$$

where $\overline{d}$ is an average of the differences $d_i$ and $s_d$ is the sample standard deviation of the differences. Here, the sample size $n$ relates to the number of pairs and not the total number of observations, which is $2n$.

Note that one can express $s_d$ as $\sqrt{s_1^2 + s_2^2 - 2s_{12}}$, where $s_{12}$ is the estimator of covariance between the samples:

$$s_{12} = \frac{1}{n-1} \sum_{i=1}^{n} (X_{1i} - \overline{X}_1)(X_{2i} - \overline{X}_2).$$

For example, in MATLAB,

```
x1 = [2 4 5 6 5 7 8];
x2 = [6 8 6 5 3 4 2];
d = x1 - x2;
sd = std(d)
  %ans =3.6904

co=cov(x1, x2)
  %co =    3.9048   -2.7857
  %       -2.7857    4.1429
```

```
sqrt(co(1,1) + co(2,2) - 2 * co(1,2))
  %ans =3.6904
```

We remark that the assumption of equality of population variances is not necessary since we operate with the differences, and the population variance of the differences, $\sigma_d^2 = \sigma_1^2 + \sigma_2^2 - 2\sigma_{12}$, is unknown.

We are interested in testing the means, $H_0 : \mu_1 = \mu_2$, versus one of the three alternatives $H_1 : \mu_1 >, \neq, < \mu_2$. Under $H_0$ the test statistic $t$ has a $t$ distribution with $n-1$ degrees of freedom. Thus, the test coincides with the one-sample $t$-test, where the sample consists of all differences and where $H_0$ is the hypothesis that the mean in the population of differences is equal to 0. The popular name for this test is the *paired t-test*, which can be summarized as follows:

| Alternative | $\alpha$-level rejection region | $p$-value |
|---|---|---|
| $H_1 : \mu_1 > \mu_2$ | $[t_{n-1,1-\alpha}, \infty)$ | 1-tcdf(t, n-1) |
| $H_1 : \mu_1 \neq \mu_2$ | $(-\infty, t_{n-1,\alpha/2}] \cup [t_{n-1,1-\alpha/2}, \infty)$ | 2*tcdf(-abs(t), n-1) |
| $H_1 : \mu_1 < \mu_2$ | $(-\infty, t_{n-1,\alpha}]$ | tcdf(t, n-1) |

One can generalize this test to testing $H_0 : \mu_1 - \mu_2 = d_0$ versus the appropriate one- or two-sided alternative. The only modification needed is in the $t$-statistic (10.7), which now takes the form

$$t = \frac{\overline{d} - d_0}{s_d/\sqrt{n}},$$

which under $H_0$ has a $t$ distribution with $n-1$ degrees of freedom.

If $\delta = \mu_1 - \mu_2$ is the difference between the population means, then the $(1 - \alpha)100\%$ confidence interval for $\delta$ is

$$\left[ \overline{d} - t_{n-1,1-\alpha/2} \frac{s_d}{\sqrt{n}}, \overline{d} + t_{n-1,1-\alpha/2} \frac{s_d}{\sqrt{n}} \right].$$

Since matching (blocking) the observations eliminates the variability between the subjects entering the inference, the paired $t$-test is preferred to a two-sample $t$-test whenever such a design is possible. For example, the case-control study design selects one observation or experimental unit as the "case" variable and obtains as "matched controls" one or more additional observations or experimental units that are similar to the case, except for the variable(s) under study. When the treatments are assigned after subjects have been paired, this assignment should be random to avoid any potential systematic influences.

*Example 10.5.* **Psoriasis.** Woo and McKenna (2003) investigated the effect of broadband ultraviolet B (UVB) therapy and topical calcipotriol cream used together on areas of psoriasis. One of the outcome variables is the Psoriasis Area and Severity Index (PASI), where a lower score is better. The following table gives PASI scores for 20 subjects measured at baseline and after 8 treatments. Do these data provide sufficient evidence, at a 0.05 level of significance, to indicate that the combination therapy reduces PASI scores?

| Subject | Baseline | After 8 treatments | Subject | Baseline | After 8 treatments |
|---------|----------|--------------------|---------|----------|--------------------|
| 1  | 5.9  | 5.2  | 11 | 11.1 | 11.1 |
| 2  | 7.6  | 12.2 | 12 | 15.6 | 8.4  |
| 3  | 12.8 | 4.6  | 13 | 9.6  | 5.8  |
| 4  | 16.5 | 4.0  | 14 | 15.2 | 5.0  |
| 5  | 6.1  | 0.4  | 15 | 21.0 | 6.4  |
| 6  | 14.4 | 3.8  | 16 | 5.9  | 0.0  |
| 7  | 6.6  | 1.2  | 17 | 10.0 | 2.7  |
| 8  | 5.4  | 3.1  | 18 | 12.2 | 5.1  |
| 9  | 9.6  | 3.5  | 19 | 20.2 | 4.8  |
| 10 | 11.6 | 4.9  | 20 | 6.2  | 4.2  |

The data set is available as 🖳 pasi.dat|xls|mat.

We will import the data into MATLAB and test the hypothesis that the PASI significantly decreased after treatment at a significance of $\alpha = 0.05$. We will also find a 95% confidence interval for the difference between the population means $\delta = \mu_1 - \mu_2$.

```
baseline = [5.9 7.6 12.8 16.5 6.1 14.4 6.6 5.4 ...
    9.6 11.6 11.1 15.6 9.6 15.2 21 5.9 10 12.2 20.2 6.2];
after = [5.2 12.2 4.6 4 0.4 3.8 1.2 3.1 3.5 4.9 ...
    11.1 8.4 5.8 5 6.4 0 2.7 5.1 4.8 4.2];

d = baseline - after;
n = length(d);
dbar = mean(d)              %dbar = 6.3550
sdd= sqrt(var(d))           %sdd = 4.9309
tstat = dbar/(sdd/sqrt(n))  %tstat = 5.7637

    % Test using RR
critpt = tinv(0.95, n-1) %critpt = 1.7291
    % Rejection region (1.7291, infinity). Reject H_0 since
    % tstat=5.7637 falls in the rejection region.
    % Test using the p-value
p_value = 1-tcdf(tstat, n-1)  %p_value = 7.4398e-006
    % Reject H_0 at the level alpha=0.05
    % since the p_value = 0.00000744 < 0.05.
alpha = 0.05
LB =  dbar - tinv(1-alpha/2, n-1)*(sdd/sqrt(n))
    % LB = 4.0472
UB =  dbar + tinv(1-alpha/2, n-1)*(sdd/sqrt(n))
```

```
% UB =  8.6628
% 95% CI is  [4.0472, 8.6628]
```

Alternatively, if one uses d1=after-baseline, then care must be taken about the "direction" of $H_1$ and $p$-value calculations. The rejection region will be $(-\infty, -1.7291)$ and the 95% confidence interval $[-8.6628, -4.0472]$. A Bayesian solution is given next.

```
model{
for(i in 1:n){
d[i] <- baseline[i] - after[i]
d[i] ~ dnorm(mu, prec)
}
mu ~ dnorm(0, 0.00001)
pH1 <- step(mu-0)
prec ~ dgamma(0.001, 0.001)
sigma2  <- 1/prec;
sigma <- 1/sqrt(prec)
}

DATA
list(n=20,
baseline = c(5.9, 7.6, 12.8, 16.5, 6.1, 14.4, 6.6,
   5.4,  9.6, 11.6 ,11.1, 15.6, 9.6, 15.2, 21, 5.9,
   10, 12.2, 20.2, 6.2),
after = c(5.2, 12.2, 4.6, 4, 0.4 , 3.8, 1.2, 3.1, 3.5,
   4.9, 11.1, 8.4, 5.8, 5, 6.4, 0, 2.7, 5.1, 4.8, 4.2))

INITS
list(mu=0, prec=1)
```

| | mean | sd | MC error | val2.5pc | median | val97.5pc | start | sample |
|---|---|---|---|---|---|---|---|---|
| pH1 | 1.0 | 0.0 | 3.162E-13 | 1.0 | 1.0 | 1.0 | 1001 | 100000 |
| mu | 6.352 | 1.169 | 0.003657 | 4.043 | 6.351 | 8.666 | 1001 | 100000 |
| prec | 0.04108 | 0.01339 | 4.498E-5 | 0.01927 | 0.03959 | 0.07149 | 1001 | 100000 |
| sigma | 5.142 | 0.8912 | 0.003126 | 3.74 | 5.026 | 7.203 | 1001 | 100000 |
| sigma2 | 27.23 | 10.0 | 0.03528 | 13.99 | 25.26 | 51.88 | 1001 | 100000 |

Let us compare the classical and Bayesian solutions. The estimator for the difference between population means is 6.3550 in the classical case and 6.352 in the Bayesian case. The standard deviations of the difference are close as well: the classical is $4.9309/\sqrt{20} = 1.1026$ and the Bayesian is 1.169.

The 95% confidence interval for the difference is $[4.0472, 8.6628]$, while the 95% credible set is $[4.043, 8.666]$.

The posterior probability of $H_1$ is approx. 1, while the classical $p$-value (support for $H_0$) is 0.000007439.

This closeness of results is expected given that the priors mu~dnorm(0,0.00001) and prec~dgamma(0.001,0.001) are noninformative.

Next, we provide an example in which the measurements are taken on different subjects, but the subjects are matched with respect to some characteristics that may influence the response. Because of this matching, the sample is considered paired. This is often done when the application of both treatments to a single subject is either impossible or leads to biased responses.

*Example 10.6.* **IQ-Test Pairing.** In a study concerning memorizing verbal sentences, children were first given an IQ test. The two lowest-scoring children were randomly assigned, one to a "noun-first" task, the other to a "noun-last" task. The two next-lowest IQ children were similarly assigned, one to a "noun-first" task, the other to a "noun-last" task, and so on until all children were assigned. The data (scores on a word-recall task) are shown here, listed in order from lowest to highest IQ score:

| Noun-first | 12 21 12 16 20 39 26 29 30 35 38 34 |
|---|---|
| Noun-last | 10 12 23 14 16  8  16 22 32 13 32 35 |

If $\mu_1$ and $\mu_2$ are population means corresponding to "noun-first" and "noun-last" tasks, we will test the hypothesis $H_0 : \mu_1 - \mu_2 = 0$ against the two-sided alternative. The significance level is set to $\alpha = 5\%$.

Note that the two samples are not independent since the pairing is based on an ordered joint attribute, children's IQ scores. Thus, even though the subjects in the two groups are different, the paired $t$-test is appropriate.

```
% Noun First Example
disp('Noun First Example')
nounfirst = [12   21   12   16   20   39   26   29   30   35   38   34];
nounlast  = [10   12   23   14   16    8   16   22   32   13   32   35];
d=nounfirst - nounlast;
dbar = mean(d)
        %dbar = 6.5833
sd = std(d)
        %sd = 11.0409
n = length(d)
        %n = 12
t = dbar/(sd/sqrt(n))
        %t = 2.0655
pval = 2 * (1-tcdf(t, n-1))
        %pval = 0.0633
```

The null hypothesis is not rejected against the two-sided alternative $H_1$ : $\mu_1 \neq \mu_2$ at the 5% level. However, for the one-sided alternative, in this case $H_1$ : $\mu_1 > \mu_2$, the $p$-value would be less than 5% and the null would be rejected. The testing is equivalent to a one-sample $t$-test against the alternative $\mu_1 - \mu_2 > 0$.

```
pval = 1-tcdf(t, n-1)
        %pval = 0.0316
```

## 10.3.1 Sample Size in Paired t-Test

If in the paired $t$-test context the variance of the differences, $\sigma_d^2$, were known, then

$$Z = \frac{\overline{d} - d^*}{\sigma_d/\sqrt{n}}$$

would have a standard normal distribution. Thus, to achieve a power of $1 - \beta$ by an $\alpha$-level test against the alternative $H_1 : \mu_1 - \mu_2 = d^*$, the one-sided test would require

$$n \geq \frac{\sigma_d^2 \, (z_{1-\alpha} + z_{1-\beta})^2}{(d^*)^2} \tag{10.8}$$

observations. For the two sided alternative, the standard normal quantile $z_{1-\alpha}$ should be replaced by $z_{1-\alpha/2}$.

*Example 10.7.* Suppose that the study is to be designed for assessing the effect of a blood-pressure-lowering drug in middle-aged men. Each subject will have his systolic blood pressure (SBP) taken at the onset of the trial and after a 14-day regimen with the drug. In previous studies of related drugs, the variance of difference between the two measurements was found to be 300 (mm Hg)$^2$. The new drug would be of interest if it reduced the SBP by 5 mm Hg or more.

What sample size is needed so that a 5% level test detects a difference of 5 mm Hg at least 90% of time?

Direct application of (10.8) with $\sigma_d^2 = 300$, $d^* = 5$, $\alpha = 0.05$, and $\beta = 0.1$ gives a sample size of 103 subjects.

```
n = (300 * (norminv(0.95) + norminv(0.9))^2 )/(5^2)
%  102.7662
```

## 10.4 Two Variances

We have already seen the test for equality of variances from two normal populations when we discussed testing the equality of two independent normal means.

We will not repeat the summary table (p. 359) but will talk about how to find a confidence interval for the ratio of population variances, conduct power analyses, and provide some Bayesian considerations.

Let $s_1^2$ and $s_2^2$ be sample variances based on samples $X_{11}, X_{12}, \ldots, X_{1,n_1}$ and $X_{21}, X_{22}, \ldots, X_{2,n_2}$ from normal populations $\mathcal{N}(\mu, \sigma_1^2)$ and $\mathcal{N}(\mu_2, \sigma_2^2)$, respectively. The fact that the sampling distribution of $\frac{s_1^2}{\sigma_1^2}/\frac{s_2^2}{\sigma_2^2}$ is $F$ with $df_1 = n_1 - 1$ and $df_2 = n_2 - 1$ degrees of freedom was used in testing the equality of variances. The same statistic and its sampling distribution lead to a $(1 - \alpha)100\%$ confidence interval for $\sigma_1^2/\sigma_2^2$, as

$$\left[ \frac{s_1^2/s_2^2}{F_{n_1-1,n_2-1,1-\alpha/2}}, \frac{s_1^2/s_2^2}{F_{n_1-1,n_2-1,\alpha/2}} \right].$$

This follows from

$$\mathbb{P}\left( F_{n_2-1,n_1-1,\alpha/2} \leq \frac{s_2^2}{\sigma_2^2} / \frac{s_1^2}{\sigma_1^2} \leq F_{n_2-1,n_1-1,1-\alpha/2} \right) = 1 - \alpha \qquad (10.9)$$

and the property of $F$ quantiles,

$$F_{m,n,\alpha} = \frac{1}{F_{n,m,1-\alpha}}.$$

**Power Analysis for the Test of Two Variances**[*]. If $F = s_1^2/s_2^2$ is observed, and $n_1$ and $n_2$ are sample sizes, Desu and Raghavarao (1990) provide an approximation of the power of the test for the variance ratio,

$$1 - \beta \approx \Phi^{-1}\left( \sqrt{\frac{2(n_1-1)(n_2-2)}{n_1+n_2-2}} |\log(F)| - z_{1-\alpha} \right), \qquad (10.10)$$

where $z_{1-\alpha}$ is the $1 - \alpha$-quantile of the standard normal distribution. If the alternative is two-sided, the quantile $z_{1-\alpha/2}$ is used instead of $z_{1-\alpha}$.

The sample size necessary to achieve a power of $1 - \beta$ if the effect *eff* = $\sigma_1^2/\sigma_2^2 \neq 1$ is to be detected by a test of level $\alpha$ is

$$n = \left( \frac{z_{1-\alpha} + z_{1-\beta}}{\log(\text{eff})} \right)^2 + 2. \qquad (10.11)$$

This size is for each sample, so the total number of observations is $2n$. If unequal sample sizes are desired, the reader is referred to Zar (1996) and Desu and Raghavarao (1990).

*Example 10.8.* In the context of Example 10.1, we approximate the power of the two-sided, 5% level test of equality of variances against the specific alternative $H_1 : \sigma_1^2/\sigma_2^2 = 1.8$. Recall that the group sample sizes were $n_1 = 12$ and $n_2 = 15$. According to (10.10),

```
n1 = 12; n2=15; alpha = 0.05; F = 1.8;
normcdf(   sqrt(2 * (n1-1)*(n2 -2)/(n1 + n2 -2  )) ...
              * abs(log(F)) - norminv(1-alpha/2) )
% 0.5112
```

and the power is about 51%.

Next we find the retrospective power of the test in Example 10.2.

```
n1 = 12; n2=15; alpha = 0.05;
s12 = 0.004^2;   s22 = 0.006^2;
F=s12^2/s22^2; %0.1975
normcdf(   sqrt(2 * (n1-1)*(n2 -2)/(n1 + n2 -2  )) ...
              * abs(log(F)) - norminv(1-alpha/2) )
%   0.9998
```

The retrospective power of the test is quite high, 99.98%.

If the lead exposure trial from Example 10.1 is to be repeated, we will find the sample size that would guarantee that an effect of size 1.5 would be detected with a power of 90% in a two-sided 5% level test. Here the effect is defined as the ratio of population variances that is different than 1 and of interest to detect.

```
alpha = 0.05; beta = 0.1; eff = 1.5;
n = ( (norminv(1-alpha/2) + norminv(1-beta))/log(eff))^2 + 2
% n=66 (65.9130)
```

Therefore, each group will need 66 children.

**A Noninformative Bayesian Solution.** If the priors on the parameters are noninformative $\pi(\mu_1) = \pi(\mu_2) = 1, \pi(\sigma_1^2) = 1/\sigma_1^2$, and $\pi(\sigma_2^2) = 1/\sigma_2^2$, one can show that the posterior distribution of $(\sigma_1^2/\sigma_2^2)/(s_1^2/s_2^2)$ is $F$ with $n_2-1$ and $n_1-1$ degrees of freedom. Since

$$\frac{\sigma_1^2/\sigma_2^2}{s_1^2/s_2^2} = \frac{s_2^2}{\sigma_2^2} \Big/ \frac{s_1^2}{\sigma_1^2},$$

the posterior distribution for $\frac{\sigma_1^2/\sigma_2^2}{s_1^2/s_2^2}$ ($\sigma_1^2, \sigma_2^2$ random variables and $s_1^2, s_2^2$ constants) and a sampling distribution of $\frac{s_2^2}{\sigma_2^2} \Big/ \frac{s_1^2}{\sigma_1^2}$ ($s_1^2, s_2^2$ random variables and $\sigma_1^2, \sigma_2^2$ constants) coincide. Thus, a credible set based on this posterior coincides with a confidence interval by taking into account relation (10.9).

If the priors on the parameters are more general, then the MCMC method can be used.

*Example 10.9.* **The Discovery of Argon.** Lord Rayleigh, following an observation by Henry Cavendish, performed a series of experiments measuring the density of nitrogen and realized that atmospheric measurements gave consistently higher results than chemical measurements (that is, measurements from ammonia, oxides of nitrogen, etc.). This discrepancy of the order of 1/100 g was too large to be explained by the measurement error, which was of the order of approx. 2/10,000 g. Rayleigh postulated that atmospheric nitrogen contains a heavier constituent, which led to the discovery of argon in 1895 (Ramsay and Rayleigh). Rayleigh's data, published in the *Proceedings of the Royal Society* in 1893 and 1894, are given in the table below and shown as back-to-back histograms (Fig. 10.2).

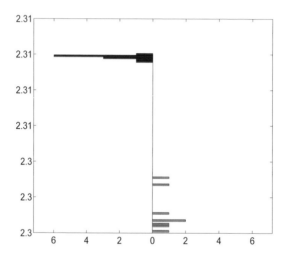

**Fig. 10.2** Back-to-back histogram of Rayleigh's measurements. Measurements from the air are given on the *left-hand side*, while the measurements obtained from chemicals are on the *right*.

| From air | 2.31035 | 2.31026 | 2.31024 | 2.31012 |
|---|---|---|---|---|
| | 2.31027 | 2.31017 | 2.30986 | 2.31010 |
| | 2.31001 | 2.31024 | 2.31010 | 2.31028 |
| From chemicals | 2.30143 | 2.29890 | 2.29816 | 2.30182 |
| | 2.29869 | 2.29940 | 2.29849 | 2.29889 |

Assume that measurements are normal with means $\mu_1$ and $\mu_2$ and variances $\sigma_1^2$ and $\sigma_2^2$, respectively.

Using WinBUGS and noninformative priors on the normal parameters, find 95% credible sets for

(a) $\theta = \mu_1 - \mu_2 - 0.01$ and

(b) $\rho = \frac{\sigma_2^2}{\sigma_1^2}$.

(c) Estimate the posterior probability of $\rho > 1$.

We are particularly interested in (b) and (c) since Fig. 10.2 indicates that the variability in the measurements obtained from chemicals is much higher than in measurements obtained from the air.

The WinBUGS file argon.odc solves (a)–(c).

```
#Discovery of Argon
model{
  for(i in 1:n1) {
    fromair[i] ~ dnorm(mu1, prec1)
    }
  for (j in 1:n2){
    fromchem[j] ~ dnorm(mu2, prec2)
    }
  mu1 ~ dflat()
  mu2 ~ dflat()
  prec1 ~ dgamma(0.0001, 0.0001)
  prec2 ~ dgamma(0.0001, 0.0001)
  theta <- mu1 - mu2 - 0.01
  sig2air <- 1/prec1
  sig2chem <- 1/prec2
  rho <- sig2chem/sig2air
  ph1 <- step(rho - 1)
}

DATA
list(n1=12, n2 = 8,
     fromair = c(2.31035, 2.31026, 2.31024, 2.31012, 2.31027,
                 2.31017, 2.30986, 2.31010, 2.31001,
                 2.31024, 2.31010, 2.31028),
     fromchem = c(2.30143, 2.29890, 2.29816, 2.30182,
                  2.29869, 2.29940,
                  2.29849, 2.29889) )

INITS
list(mu1 = 0, mu2 = 0, prec1 = 10, prec2 = 10)
```

| | mean | sd | MC error | val2.5pc | median | val97.5pc | start | sample |
|---|---|---|---|---|---|---|---|---|
| ph1 | 0.786 | 0.4101 | 4.473E-4 | 0.0 | 1.0 | 1.0 | 1001 | 1000000 |
| rho | 2.347 | 2.333 | 0.002684 | 0.4456 | 1.736 | 7.925 | 1001 | 1000000 |
| theta | 6.971E-4 | 0.002683 | 2.584E-6 | −0.004626 | 6.983E-4 | 0.006017 | 1001 | 1000000 |

It is instructive to compare a frequentist test for the variance ratio with WinBUGS output. The statistic $F = s_1^2/s_2^2 = 0.0099$ is strongly significant with a $p$-value of the order $10^{-9}$. At the same time, the posterior probability of $\rho > 1$, a Bayesian equivalent to the $F$ test, is only 0.786. The posterior probability of $\rho \leq 1$ is 0.214. Also, the 95% credible set for $\rho$ contains 1. Even though a Bayesian would favor the hypothesis $\rho > 1$, the evidence against $\rho \leq 0$ is not as strong as in the classical approach.

## 10.5 Comparing Two Proportions

Comparing two population proportions is arguably one of the most important tasks in statistical practice. For example, statistical support in clinical trials for new drugs, procedures, or medical devices almost always contains tests and confidence intervals involving two proportions: proportions of positive outcomes in control and treatment groups, proportions of readings within tolerance limits for proposed and currently approved medical devices, or proportions of cancer patients for which new and old treatment regimes manifested drug toxicity, to list just a few.

Sample proportions involve binomial distributions, and if sample sizes are not too small, the CLT implies their approximate normality.

Let $X_1 \sim \mathscr{B}in(n_1, p_1)$ and $X_2 \sim \mathscr{B}in(n_2, p_2)$ be the observed numbers of "events" and $\hat{p}_1$ and $\hat{p}_2$ be the sample proportions. Then the difference $\hat{p}_1 - \hat{p}_2 = X_1/n_1 - X_2/n_2$ has an approximately normal distribution (when $n_1$ and $n_2$ are not too small, say, >20) with mean $p_1 - p_2$ and variance $p_1(1-p_1)/n_1 + p_2(1-p_2)/n_2$.

A Wald-type confidence interval can be constructed using this normal approximation. Specifically, the $(1-\alpha)100\%$ confidence interval for the population proportion difference $p_1 - p_2$ is

$$
\left[ \hat{p}_1 - \hat{p}_2 - z_{1-\alpha/2} \sqrt{\frac{\hat{p}_1(1-\hat{p}_1)}{n_1} + \frac{\hat{p}_2(1-\hat{p}_2)}{n_2}}, \right.
$$

$$
\left. \hat{p}_1 - \hat{p}_2 + z_{1-\alpha/2} \sqrt{\frac{\hat{p}_1(1-\hat{p}_1)}{n_1} + \frac{\hat{p}_2(1-\hat{p}_2)}{n_2}} \right].
$$

In testing $H_0 : p_1 = p_2$ against one of the alternatives, the test statistic is

$$
Z = \frac{\hat{p}_1 - \hat{p}_2}{\sqrt{\frac{\hat{p}(1-\hat{p})}{n_1} + \frac{\hat{p}(1-\hat{p})}{n_2}}} = \frac{\hat{p}_1 - \hat{p}_2}{\sqrt{\hat{p}(1-\hat{p})}\sqrt{\frac{1}{n_1} + \frac{1}{n_2}}},
$$

where

$$
\hat{p} = \frac{X_1 + X_2}{n_1 + n_2} = \frac{n_1}{n_1 + n_2}\hat{p}_1 + \frac{n_2}{n_1 + n_2}\hat{p}_2
$$

is the pooled sample proportion. The pooled sample proportion is used since under $H_0$ the population proportions coincide and this common parameter should be estimated by all available data. By the CLT, the statistic $Z$ has an approximately standard normal $\mathscr{N}(0,1)$ distribution.

| Alternative | $\alpha$-level rejection region | $p$-value |
|---|---|---|
| $H_1 : p_1 > p_2$ | $[z_{1-\alpha}, \infty)$ | `1-normcdf(z)` |
| $H_1 : p_1 \neq p_2$ | $(-\infty, z_{\alpha/2}] \cup [z_{1-\alpha/2}, \infty)$ | `2*normcdf(-abs(z))` |
| $H_1 : p_1 < p_2$ | $(-\infty, z_\alpha]$ | `normcdf(z)` |

*Example 10.10.* **Vasectomies and Prostate Cancer.** Several studies have been conducted to analyze the relationship between vasectomy and prostate cancer. The study by Giovannucci et al. (1993) states that of 21,300 men who had not had a vasectomy, 69 were found to have prostate cancer, while of 22,000 men who had a vasectomy, 113 were found to have prostate cancer. Formulate hypotheses and perform a test at the 1% level.

```
x1=69; x2 = 113; n1 = 21300; n2 = 22000;
p1hat = x1/n1; p2hat  = x2/n2; phat = (x1 + x2)/(n1 + n2);
z=(p1hat - p2hat)/(sqrt(phat*(1-phat))*sqrt(1/n1 + 1/n2))
 % z = -3.0502
pval = normcdf(-3.0502)
 % pval = 0.0011
```

We tested $H_0 : p_1 = p_2$ versus $H_1 : p_1 < p_2$, where $p_1$ is the proportion of subjects with prostate cancer in the population of all subjects who had a vasectomy, while $p_2$ is the proportion of subjects with prostate cancer in the population of all subjects who did not have a vasectomy. Since the $p$-value was 0.0011, we concluded that vasectomy is a significant risk factor for prostate cancer.

## 10.5.1 The Sample Size

The sample size required for a two-sided $\alpha$-level test to detect the difference $\delta = |p_1 - p_2|$, with a power of $1 - \beta$, is

$$n \geq \frac{(z_{1-\alpha/2} + z_{1-\beta})^2 \times 2\overline{p}(1-\overline{p})}{\delta^2},$$

where $\overline{p} = (p_1 + p_2)/2$. The sample size $n$ is for each group, so that the total number of observations is $2n$. If the alternative is one-sided, $z_{1-\alpha/2}$ is replaced by $z_{1-\alpha}$. This formula requires some preliminary knowledge about $\overline{p}$. In the absence of any information about the proportions, the most conservative choice for $\overline{p}$ is 1/2.

*Example 10.11.* An investigator believes that a control group would have an annual event rate of 30% and that the treatment would reduce this rate to 20%. She wants to design a study to be one-sided with a significance level of $\alpha = 0.05$ and a power of $1 - \beta = 0.85$. The necessary sample size per group is

```
n = 2 *(norminv(1-0.05)+norminv(1-0.15))^2 *0.25*(1-0.25)/0.1^2
     % n = 269.5988
```

This number is rounded to 270, so that the total sample size is $2 \times 270 = 540$.
    More precise sample sizes are

$$
n' = \frac{\left( z_{1-\alpha/2} \sqrt{2\overline{p}(1-\overline{p})} + z_{1-\beta} \sqrt{p_1(1-p_1) + p_2(1-p_2)} \right)^2}{\delta^2},
$$

with $z_{1-\alpha/2}$ replaced by $z_{1-\alpha}$ for the one-sided alternative.
    Casagrande et al. (1978) propose a correction to $n'$ as

$$
n'' = n'/4 \times \left( 1 + \sqrt{1 + \frac{4}{n'\delta}} \right)^2,
$$

while Fleiss et al. (1980) suggest $n''' = n' + \frac{2}{\delta}$.
    In all three scenarios, however, preliminary knowledge about $p_1$ and $p_2$ is needed.

```
%Sample Size for Each Group:
n1 = (norminv(1-0.05) * sqrt(2 * 0.25 * 0.75)+ ...
      norminv(1-0.15) * sqrt(0.3*0.7+0.2*0.8))^2/0.1^2
%n1 = 268.2064
%Casagrande et al. 1978
n2 = n1/4 * (1 + sqrt(1 + 4/(n1 * 0.1)))^2
%n2 =287.8590
%Fleiss et al. 1980
n3 = n1 + 2/0.1
%n3 = 288.2064
```

The outputs $n'$, $n''$, and $n'''$ are rounded to 267, 288, and 289, so that the total sample sizes are 534, 576, and 578, respectively.

## 10.6 Risks: Differences, Ratios, and Odds Ratios

In epidemiological and population disease studies it is often the case that the findings are summarized as a table

| | Disease present (D) | No disease present (C) | Total |
|---|---|---|---|
| Exposed ($E$) | $a$ | $b$ | $n_1 = a + b$ |
| Nonexposed ($E^c$) | $c$ | $d$ | $n_2 = c + d$ |
| Total | $m_1 = a + c$ | $m_2 = b + d$ | $n = a + b + c + d$ |

In clinical trial studies, the risk factor status ($E/E^c$) can be replaced by a treatment/control or new treatment/old treatment, while the disease status ($D/D^c$) can be replaced by a improvement/nonimprovement.

**Remark.** In the context of epidemiology, the studies leading to tabulated data can be *prospective* and *retrospective*. In a prospective study, a group of $n$ disease-free individuals is identified and followed over a period of time. At the end of the study, the group, typically called the *cohort*, is assessed and tabulated with respect to disease development and exposure to the risk factor of interest.

In a retrospective study, groups of $m_1$ individuals with the disease (cases) and $m_2$ disease-free individuals (controls) are identified and their prior exposure histories are assessed. In this case, the table summarizes the numbers of exposure to the risk factor under consideration among the cases and controls.

## 10.6.1 Risk Differences

Let $p_1$ and $p_2$ be the population risks of a disease for exposed and nonexposed (control) subjects. These are probabilities that the subjects will develop the disease during the fixed interval of time for the two groups, exposed and nonexposed.

Let $\hat{p}_1 = a/n_1$ be an estimator of the risk of a disease for exposed subjects and $\hat{p}_2 = c/n_2$ be an estimator of the risk of that disease for control subjects.

The $(1 - \alpha)100\%$ confidence interval for the risk difference coincides with the confidence interval for the difference of proportions from p. 378:

$$\hat{p}_1 - \hat{p}_2 \ \pm \ z_{1-\alpha/2} \sqrt{\frac{\hat{p}_1(1 - \hat{p}_1)}{n_1} + \frac{\hat{p}_2(1 - \hat{p}_2)}{n_2}}.$$

Sometimes, better precision is achieved by a confidence interval with continuity corrections:

$$\left[ \hat{p}_1 - \hat{p}_2 \pm (1/(2n_1) + 1/(2n_2)) - z_{1-\alpha/2} \sqrt{\frac{\hat{p}_1(1-\hat{p}_1)}{n_1} + \frac{\hat{p}_2(1-\hat{p}_2)}{n_2}}, \right.$$

$$\left. \hat{p}_1 - \hat{p}_2 \pm (1/(2n_1) + 1/(2n_2)) + z_{1-\alpha/2} \sqrt{\frac{\hat{p}_1(1-\hat{p}_1)}{n_1} + \frac{\hat{p}_2(1-\hat{p}_2)}{n_2}} \right],$$

where the sign of the correction factor $1/(2n_1) + 1/(2n_2)$ is taken as "+" if $\hat{p}_1 - \hat{p}_2 < 0$ and as "−" if $\hat{p}_1 - \hat{p}_2 > 0$. The recommended sample sizes for the validity of the interval should satisfy $\min\{n_1 p_1(1 - p_1), n_2 p_2(1 - p_2)\} \geq 10$. For large sample sizes, the difference between "continuity-corrected" and uncorrected intervals is negligible.

### 10.6.2 Risk Ratio

The risk ratio in a population is the quantity $R = p_1/p_2$. It is estimated by $r = \hat{p}_1/\hat{p}_2$. The empirical distribution of $r$ does not have a simple form, and, moreover, it is typically skewed (Fig. 10.3b). If the logarithm is taken, the risk ratio is "symmetrized," the log ratio is equivalent to the difference between logarithms, and, given the independence of populations, the CLT applies. It is evident in Fig. 10.3c that the log risk ratios are well approximated by a normal distribution.

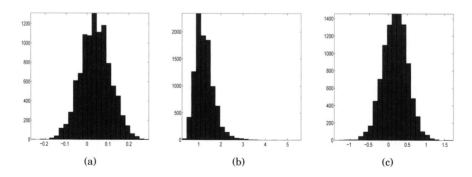

(a)                                          (b)                                          (c)

**Fig. 10.3** Two samples of size 10,000 are generated from $\mathcal{B}in(80, 0.21)$ and $\mathcal{B}in(60, 0.25)$ populations and risks $\hat{p}_1$ and $\hat{p}_2$ are estimated for each pair. The panels show histograms of (a) risk differences, (b) risk ratios, and (c) log risk ratios.

The following MATLAB code ( simulrisks.m) simulates 10,000 pairs from $\mathcal{B}in(80, 0.21)$ and $\mathcal{B}in(60, 0.25)$ populations representing exposed and nonexposed subjects. From each pair risks are assessed and histograms of risk differences, risk ratios, and log risk ratios are shown in Fig. 10.3a–c.

```
disexposed = binornd(60, 0.25, [1 10000]);
disnonexposed = binornd(80, 0.21, [1 10000]);
p1s = disexposed/60; p2s =disnonexposed/80;
figure; hist(p1s - p2s, 25)
figure; hist(p1s./p2s, 25)
figure; hist( log( p1s./p2s ), 25 )
```

### 10.6.3 Odds Ratios

For a particular proportion, $p$, the odds are defined as $\frac{p}{1-p}$. For two proportions $p_1$ and $p_2$, the odds ratio is defined as $O = \frac{p_1/(1-p_1)}{p_2/(1-p_2)}$, and its sample counterpart is $o = \frac{\hat{p}_1/(1-\hat{p}_1)}{\hat{p}_2/(1-\hat{p}_2)}$.

As evident in Fig. 10.4, the odds ratio is symmetrized by the log transformation, and it is the log domain where the normal approximations are used. The sample standard deviation for $\log o$ is $s_{\log o} = \sqrt{\frac{1}{a} + \frac{1}{b} + \frac{1}{c} + \frac{1}{d}}$, and the $(1-\alpha)100\%$ confidence interval for the log odds ratio is

$$\left[ \log o - z_{1-\alpha/2}\sqrt{\frac{1}{a} + \frac{1}{b} + \frac{1}{c} + \frac{1}{d}}, \ \log o + z_{1-\alpha/2}\sqrt{\frac{1}{a} + \frac{1}{b} + \frac{1}{c} + \frac{1}{d}} \right].$$

Of course, the confidence interval for the odds ratio is obtained by taking the exponents of the bounds:

$$\left[ \exp\left\{ \log o - z_{1-\alpha/2}\sqrt{\frac{1}{a} + \frac{1}{b} + \frac{1}{c} + \frac{1}{d}} \right\}, \ \exp\left\{ \log o + z_{1-\alpha/2}\sqrt{\frac{1}{a} + \frac{1}{b} + \frac{1}{c} + \frac{1}{d}} \right\} \right].$$

Many authors argue that only odds ratios should be reported and used because of their superior properties over risk differences and risk ratios (Edwards, 1963; Mosteller, 1968). For small sample sizes replacing counts $a, b, c,$ and $d$ by $a + 1/2, b + 1/2, c + 1/2,$ and $d + 1/2$ leads to a more stable inference.

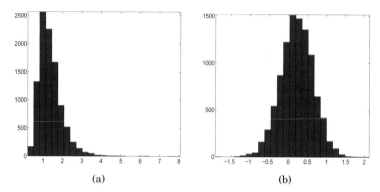

(a)                                                          (b)

**Fig. 10.4** For the data leading to Fig. 10.3, the histograms of (a) odds ratios and (b) log odds ratios are shown.

| | Risk difference | Relative risk | Odds ratio |
|---|---|---|---|
| Parameter | $D = p_1 - p_2$ | $R = p_1/p_2$ | $O = \frac{p_1/(1-p_1)}{p_2/(1-p_2)}$ |
| Estimator | $d = \hat{p}_1 - \hat{p}_2$ | $r = \hat{p}_1/\hat{p}_2$ | $o = \frac{\hat{p}_1/(1-\hat{p}_1)}{\hat{p}_2/(1-\hat{p}_2)}$ |
| St. deviation | $s_d = \sqrt{\frac{p_1 q_1}{n_1} + \frac{p_2 q_2}{n_2}}$ | $s_{\log r} = \sqrt{\frac{q_1}{n_1 p_1} + \frac{q_2}{n_2 p_2}}$ | $s_{\log o} = \sqrt{\frac{1}{a} + \frac{1}{b} + \frac{1}{c} + \frac{1}{d}}$ |

Interpretation of values for $RR$ and $OR$ are provided in the following table:

| Value in | Effect of exposure |
|---|---|
| $[0, 0.4)$ | Strong benefit |
| $[0.4, 0.6)$ | Moderate benefit |
| $[0.6, 0.9)$ | Weak benefit |
| $[0.9, 1.1]$ | No effect |
| $(1.1, 1.6]$ | Weak hazard |
| $(1.6, 2.5]$ | Moderate hazard |
| $> 2.5$ | Strong hazard |

*Example 10.12.* **Framingham Data.**  The table below gives the coronary heart disease status after 18 years, by level of systolic blood pressure (SBP). The levels of SBP $\geq 165$ are considered as an exposure to a risk factor.

| SBP (mmHg) | Coronary disease | No coronary disease | Total |
|---|---|---|---|
| $\geq 165$ | 95 | 201 | 296 |
| $< 165$ | 173 | 894 | 1067 |
| Total | 268 | 1095 | 1363 |

Find 95% confidence intervals for the risk difference, risk ratio, and odds ratio. The function ◀ risk.m calculates confidence intervals for risk differences, risk ratios, and odds ratios and will be used in this example.

```
function [rd rdl rdu rr rrl rru or orl oru] = risk(a, b, c, d, alpha)
%--------
%                  |   Disease    No disease      Total
%    --------------------------------------------------------
% Exposed          |      a           b         |   n1
% Nonexposed       |      c           d         |   n2
%    --------------------------------------------------------
if nargin < 5
alpha=0.05;
end
%--------
n1 = a + b;
n2 = c + d;
hatp1 = a/n1; hatp2 = c/n2;
%-------risk difference (rd) and CI [rdl, rdu] ---------
rd = hatp1 - hatp2;
stdrd = sqrt(hatp1 * (1-hatp1)/n1 + hatp2 * (1- hatp2)/n2 );
rdl = rd - norminv(1-alpha/2) * stdrd;
rdu = rd + norminv(1-alpha/2) * stdrd;
%----------risk ratio (rr) and CI [rrl, rru] -----------
rr = hatp1/hatp2;
lrr = log(rr);
stdlrr = sqrt(b/(a * n1) + d/(c*n2));
lrrl = lrr - norminv(1-alpha/2)*stdlrr;
rrl = exp(lrrl);
lrru = lrr + norminv(1-alpha/2)*stdlrr;
rru = exp(lrru);
%---------odds ratio (or) and CI [orl, oru] ------------
or = ( hatp1/(1-hatp1) )/(hatp2/(1-hatp2))
lor = log(or);
stdlor = sqrt(1/a + 1/b + 1/c + 1/d);
lorl = lor - norminv(1-alpha/2)*stdlor;
orl = exp(lorl);
loru = lor + norminv(1-alpha/2)*stdlor;
oru = exp(loru);
```

The solution is:

```
[rd rdl rdu rr rrl rru or orl oru] = risk(95,201,173,894)

%rd = 0.1588
%[rdl, rdu] = [0.1012, 0.2164]

%rr =   1.9795
%[rrl, rru]= [1.5971, 2.4534]

%or = 2.4424
%[orl, oru] = [1.8215,  3.2750]
```

*Example 10.13*. **Retrospective Analysis of Smoking Habits.** This example is adopted from Johnson and Albert (1999), who use data collected in a

study by Dorn (1954). A sample of 86 lung-cancer patients and a sample of
86 controls were questioned about their smoking habits. The two groups were
chosen to represent random samples from a subpopulation of lung-cancer patients and an otherwise similar population of cancer-free individuals. Of the
cancer patients, 83 out of 86 were smokers; among the control group, 72 out
of 86 were smokers. The scientific question of interest was to assess the difference between the smoking habits in the two groups. Uniform priors on the
population proportions were used as a noninformative choice.

```
model{
for(i in 1:2){
   r[i] ~ dbin(p[i],n[i])
   p[i] ~ dunif(0,1)
          }
RD <- p[1] - p[2]
RD.gt0 <- step(RD)

    RR <- p[1]/p[2]
    RR.gt1 <- step(RR - 1)

OR <-  (p[1]/(1-p[1]))/(p[2]/(1-p[2]))
OR.gt1 <- step(OR - 1)
}

DATA
list(r=c(83,72),n=c(86,86))
INITS
  #Generate Inits
```

| | mean | sd | MC error | val2.5pc | median | val97.5pc | start | sample |
|---|---|---|---|---|---|---|---|---|
| OR | 5.818 | 4.556 | 0.01398 | 1.556 | 4.613 | 17.29 | 1001 | 100000 |
| OR.gt1 | 0.9978 | 0.04675 | 1.469E-4 | 1.0 | 1.0 | 1.0 | 1001 | 100000 |
| RD | 0.125 | 0.0455 | 1.478E-4 | 0.0385 | 0.1237 | 0.2179 | 1001 | 100000 |
| RD.gt0 | 0.9978 | 0.04675 | 1.469E-4 | 1.0 | 1.0 | 1.0 | 1001 | 100000 |
| RR | 1.153 | 0.06276 | 2.038E-4 | 1.044 | 1.148 | 1.291 | 1001 | 100000 |
| RR.gt1 | 0.9978 | 0.04675 | 1.469E-4 | 1.0 | 1.0 | 1.0 | 1001 | 100000 |
| p[1] | 0.9546 | 0.02209 | 7.06E-5 | 0.9022 | 0.958 | 0.9873 | 1001 | 100000 |
| p[2] | 0.8296 | 0.03991 | 1.26E-4 | 0.7444 | 0.8322 | 0.9002 | 1001 | 100000 |

Note that 95% credible sets for the risk ratio and odds ratio are above 1,
and that the set for the risk difference does not contain 0. By all three measures the proportion of smokers among subjects with cancer is significantly
larger than the proportion among the controls. In Bayesian testing the hy-
potheses $H'_1 : p_1 > p_2$, $H''_1 : p_1/p_2 > 1$, and $H'''_1 : \frac{p_1}{1-p_1} \Big/ \frac{p_2}{1-p_2} > 1$ have posterior
probabilities of 0.9978 each. Therefore, in this retrospective study, smoking
status is indicated as a significant risk factor for lung cancer.

## 10.7 Two Poisson Rates*

There are several methods for devising confidence intervals on differences or the ratios of two Poisson rates. We will focus on the method for the ratio that modifies well-known binomial confidence intervals.

Let $X_1 \sim \mathcal{P}oi(\lambda_1 t_1)$ and $X_2 \sim \mathcal{P}oi(\lambda_2 t_2)$ be two Poisson counts with rates $\lambda_1$ and $\lambda_2$ observed during time intervals of length $t_1$ and $t_2$.

We are interested the confidence interval for the ratio $\lambda = \lambda_1/\lambda_2$.

Since $X_1$, given the sum $X_1 + X_2 = n$, is binomial $\mathcal{B}in(n, p)$ with $p = \frac{\lambda_1 t_1}{\lambda_1 t_1 + \lambda_2 t_2}$ (Exercise 5.5), the strategy is to find the confidence interval for $p$ and, from its confidence bounds $LB_p$ and $UB_p$, work out the bounds for the ratio $\lambda$.

$$LB_\lambda = \frac{LB_p}{1 - LB_p} \frac{t_2}{t_1} \qquad UB_\lambda = \frac{UB_p}{1 - UB_p} \frac{t_2}{t_1}.$$

For finding the $LB_p$ and $UB_p$ several methods are covered in Chap. 7. Note that there $\hat{p} = X_1/n$ and $n = X_1 + X_2$.

The design question can be addressed as well, but the "sample size" formulation needs to be expressed in terms of sampling durations $t_1$ and $t_2$. The sampling time frames $t_1$ and $t_2$, if assumed equal, can be determined on the basis of elicited precision for the confidence interval and preliminary estimates of the rates. Let $\bar{\lambda}_1$ and $\bar{\lambda}_2$ be preexperimental assessments of the rates and let the precision be elicited in the form of (a) the length of the interval $UB_\lambda - LB_\lambda = w$ or (b) the ratio of the bounds $UB_\lambda/LB_\lambda = w$.

Then, for achieving $(1 - \alpha)100\%$ confidence with an interval of length $w$, the sampling time frame required is

(a)

$$t \quad (= t_1 = t_2) = \frac{z_{1-\alpha/2}^2 \left(1/\bar{\lambda}_1 + 1/\bar{\lambda}_2\right)}{\arcsin\left(\frac{\bar{\lambda}_2}{\bar{\lambda}_1} \times \frac{w}{2}\right)}$$

and

(b)

$$t \quad (= t_1 = t_2) = \frac{4z_{1-\alpha/2}^2 \left(1/\bar{\lambda}_1 + 1/\bar{\lambda}_2\right)}{\log^2(w)}.$$

*Example 10.14.* **Wire Failures.** Price and Bonett (2000) provide an example with data from Gardner and Ringlee (1968), who found that bare wire had $X_1 = 69$ failures in a sample of $t_1 = 1079.6$ thousand foot-years, and a polyethylene-covered tree wire had $X_2 = 12$ failures in a sample of $t_2 = 467.9$ thousand foot-years. We are interested in a 95% confidence interval for the ratio of population failure rates.

The associated MATLAB file ◀ ratiopoissons.m calculates the 95% confidence interval for the ratio $\lambda = \lambda_1/\lambda_2$ using Wilson's proposal ("add two successes and two failures"). There, $\hat{p} = (X_1 + 2)/(n + 4)$, and the interval for $p$ is $[0.7564, 0.9141]$. After transforming the bounds to the $\lambda$ domain, the final interval is $[1.3461, 4.6147]$.

Suppose we want to replicate this study using a new shipment of each type of wire. We want to estimate the failure rate ratio with 99% confidence and $UB_\lambda/LB_\lambda = 2$. Using $\bar{\lambda}_1 = 69/1079.6 = 0.0691$ and $\bar{\lambda}_2 = 12/467.9 = 0.0833$ as our planning estimates of $\lambda_1$ and $\lambda_2$, we would sample $t = \dfrac{4(2.5758)^2\ (1/0.0691 + 1/0.0833)}{\log^2(2)} = $ 3018 foot-years from each shipment. If we want to complete the study in $k$ years, then we would sample $3018/k$ linear feet of wire from each shipment.

✐

```
%CI for Ratio of Two Poissons
    X1=69; t1 = 1079.6;
    X2=12; t2=467.9;
    n=X1 + X2;
    phat = X1/n; %0.8519
    phat1 = (X1 +2)/(n + 4); %0.8353
    qhat1 = 1 - phat1;        %0.1647
% Agresti-Coull CI for prop was selected.
LBp=phat1-norminv(0.975)*sqrt(phat1*qhat1/(n+4)) %0.7564
UBp=phat1+norminv(0.975)*sqrt(phat1*qhat1/(n+4)) %0.9141
LBlam = LBp/(1 - LBp) * t2/t1; %back to lambda
UBlam = UBp/(1 - UBp) * t2/t1;
[LBlam, UBlam]        %[1.3461    4.6147]
%Frame size in Poisson Sampling
    lambar1 = 69/1079.6;  %0.0639
    lambar2 = 12/467.9;   %0.0256
    w = 2;
  td =4* norminv(0.995)^2 *(1/lambar1+1/lambar2)/...
        (asin(lambar2/lambar1 * w/2));      %3511.8
  tr =  4 * norminv(0.995)^2 *...
  ( 1/lambar1 + 1/lambar2 )/(log(w))^2;   %3018.1
```

Cox (1953) gives an approximate test and confidence interval for the ratio that uses an $F$ distribution. He shows that the statistic

$$F = \frac{t_1\lambda_1}{t_2\lambda_2}\frac{X_2 + 1/2}{X_1 + 1/2}$$

has an approximate $F$ distribution with $2X_1 + 1$ and $2X_2 + 1$ degrees of freedom. From this, an approximate $(1 - \alpha)100\%$ confidence interval for $\lambda_1/\lambda_2$ is

$$\left[\frac{t_2}{t_1}\frac{X_1 + 1/2}{X_2 + 1/2}F_{2X_1+1,2X_2+1,\alpha/2},\ \frac{t_2}{t_1}\frac{X_1 + 1/2}{X_2 + 1/2}F_{2X_1+1,2X_2+1,1-\alpha/2}\right].$$

In the context of Example 10.14, the 95% confidence interval for the ratio $\lambda_1/\lambda_2$ is $[1.3932, 4.7497]$.

```
%Cox
LBlamc= t2/t1*(X1+1/2)/(X2+1/2)*finv(0.025, 2*X1+1, 2*X2+1);
UBlamc= t2/t1*(X1+1/2)/(X2+1/2)*finv(0.975, 2*X1+1, 2*X2+1);
[LBlamc, UBlamc]    %1.3932    4.7497
```

Note that this interval does not contain 1, which is equivalent to a rejection of $H_0 : \lambda_1 = \lambda_2$ in a test against the two-sided alternative, at the level $\alpha = 0.05$.
The test of $H_0 : \lambda_1 = \lambda_2$ can be conducted using the statistic

$$F = \frac{t_1}{t_2} \frac{X_2 + 1/2}{X_1 + 1/2},$$

which under $H_0$ has an $F$ distribution with $df_1 = 2X_1 + 1$ and $df_2 = 2X_2 + 1$ degrees of freedom.

| Alternative | $\alpha$-level rejection region | $p$-value |
|---|---|---|
| $H_1 : \lambda_1 < \lambda_2$ | $[F_{df_1,df_2,1-\alpha}, \infty)$ | 1-fcdf(F,df1,df2) |
| $H_1 : \lambda_1 \neq \lambda_2$ | $[0,F_{df_1,df_2,\alpha/2}] \cup [F_{df_1,df_2,1-\alpha/2}, \infty)$ | 2*fcdf(min(F,1/F),df1,df2) |
| $H_1 : \lambda_1 > \lambda_2$ | $[0,F_{df_1,df_2,\alpha}]$ | fcdf(F,df1,df2) |

In Example 10.14, the failure rate $\lambda_1$ for the bare wire is found to be significantly larger ($p$-value of 0.00066) than that of polyethylene-covered wire, $\lambda_2$.

```
%test against H_1: lambda1 > lambda2
pval =fcd(t1/t2*(X2+1/2)/(X1+1/2), 2*X1 + 1, 2*X2 + 1)
%6.6417e-004
```

## 10.8 Equivalence Tests*

In standard testing of two means, the goal is to show that one population mean is significantly smaller, larger, or different than the other. The null hypothesis is that there is no difference between the means. By not rejecting the null, the equality of means is not established – the test simply did not find enough statistical evidence for the alternative hypothesis. Absence of evidence is not evidence of absence.

In many situations (drug and medical procedure testing, device performance, etc.), one wishes to test the equivalence hypothesis, which states that the population means or population proportions differ for no more than a small tolerance value preset by a regulatory agency. If, for example, manufacturers of a generic drug are able to demonstrate bioequivalence to the brand-name product, they do not need to conduct costly clinical trials in order to demonstrate the safety and efficacy of their generic product. More importantly, established bioequivalence protects the public from unsafe or ineffective drugs.

In this kind of inference it is desired that "no difference" constitutes the research hypothesis $H_1$ and that significance level $\alpha$ relates to the probability of falsely rejecting the hypothesis that there is a difference when in fact the means are equivalent. In other words, we want to control the type I error and design the power properly in this context.

In drug equivalence testing typical measurements are the area under the concentration curve (AUC) or maximum concentration ($C_{max}$). The two drugs are bioequivalent if the population means of the AUC and $C_{max}$ are sufficiently close.

Let $\eta_T$ denote the population mean AUC for the generic (test) drug and let $\eta_R$ denote the population mean for the brand-name (reference) drug.

We are interested in testing

$$H_0 : \eta_T/\eta_R < \delta_L \quad \text{or} \quad \eta_T/\eta_R > \delta_U \qquad \text{versus} \qquad H_1 : \delta_L \leq \eta_T/\eta_R \leq \delta_U,$$

where $\delta_L$ and $\delta_U$ are the lower and upper tolerance limits, respectively. The FDA recommends $\delta_L = 4/5$ and $\delta_U = 5/4$ (FDA, 2001).

This hypothesis can be tested in the domain of original measurements (Berger and Hsu, 1996) or after taking the logarithm. This second approach is more common in practice since (i) AUC and Cmax measurements are consistent with the lognormal distribution (the pharmacokinetic rationale based on multiplicative compartmental models) and (ii) normal theory can be applied to logarithms of observations. The FDA also recommends a log-transformation of data by providing three rationales: clinical, pharmacokinetic, and statistical (FDA, 2001, Appendix D).

Since for lognormal distributions the mean $\eta$ is connected with the parameters of the associated normal distribution, $\mu$ and $\sigma^2$ (p. 218), by assuming equal variances we get $\eta_T = \exp\{\mu_T + \sigma^2/2\}$ and $\eta_R = \exp\{\mu_R + \sigma^2/2\}$. The equivalence hypotheses for the log-transformed data now take the form

$$H_0 : \mu_T - \mu_R \leq \theta_L \quad \text{or} \quad \mu_T - \mu_R \geq \theta_U, \qquad \text{versus} \qquad H_1 : \theta_L < \mu_T - \mu_R < \theta_U,$$

where $\theta_L = \log(\delta_L)$ and $\theta_U = \log(\delta_U)$ are known constants. Note that if $\delta_U = 1/\delta_L$, then the bounds $\theta_L$ and $\theta_U$ are symmetric about zero, $\theta_L = -\theta_U$.

Equivalence testing is an active research area and many classical and Bayesian solutions exist, as dictated by experimental designs in practice. The monograph by Wellek (2010) provides comprehensive coverage. We focus only on the case of testing the equivalence of two population means when unknown population variances are the same.

**TOST.** Schuirmann (1981) proposed two one-sided tests (TOSTs) for testing bioequivalence. Two $t$-statistics are calculated:

$$t_L = \frac{\overline{X}_T - \overline{X}_R - \theta_L}{s_p \sqrt{1/n_1 + 1/n_2}} \quad \text{and} \quad t_U = \frac{\overline{X}_T - \overline{X}_R - \theta_U}{s_p \sqrt{1/n_1 + 1/n_2}},$$

where $\overline{X}_T$ and $\overline{X}_R$ are test and reference means, $n_1$ and $n_2$ are test and reference sample sizes, and $s_p$ is the pooled sample standard deviation, as on p. 357. Note that here, the test statistic involves the acceptable bounds $\theta_L$ and $\theta_U$ in the numerator, unlike the standard two-sample $t$-test, where the numerator would be $\overline{X}_T - \overline{X}_R$.

The TOST is now carried out as follows.

(i) Using the statistic $t_L$, test $H_0' : \mu_T - \mu_R = \theta_L$ versus $H_1' : \mu_T - \mu_R > \theta_L$.

(ii) Using the statistic $t_U$, test $H_0'' : \mu_T - \mu_R = \theta_U$ versus $H_1'' : \mu_T - \mu_R < \theta_U$.

(iii) Reject $H_0$ at level $\alpha$, that is, declare the drugs equivalent if *both* hypotheses $H_0'$ and $H_0''$ are rejected at level $\alpha$, that is, if

$$t_L > t_{n_1+n_2-2,1-\alpha} \quad \text{and} \quad t_U < t_{n_1+n_2-2,\alpha}.$$

Equivalently, if $p_L$ and $p_U$ are the $p$-values associated with statistics $t_L$ and $t_U$, $H_0$ is rejected when $\max\{p_L, p_U\} < \alpha$.

**Westlake's Confidence Interval.** An equivalent methodology to test for equivalence is Westlake's confidence interval (Westlake, 1976). Bioequivalence is established at significance level $\alpha$ if a $t$-interval of confidence $(1 - 2\alpha)100\%$ is contained in the interval $(\theta_L, \theta_U)$:

$$\left[ \overline{X}_T - \overline{X}_R - t_{n_1+n_2-2,1-\alpha} \, s_p \, \sqrt{1/n_1 + 1/n_2}, \right.$$
$$\left. \overline{X}_T - \overline{X}_R + t_{n_1+n_2-2,1-\alpha} \, s_p \, \sqrt{1/n_1 + 1/n_2} \right] \in (\theta_L, \theta_U).$$

Here, the usual $t_{n_1+n_2-2,1-\alpha/2}$ is replaced by $t_{n_1+n_2-2,1-\alpha}$, and Westlake's interval coincides with the standard $(1 - 2\alpha)100\%$ confidence interval for a difference of normal means.

*Example 10.15.* **Equivalence of Generic and Brand-Name Drugs.** A manufacturer wishes to demonstrate that their generic drug for a particular metabolic disorder is equivalent to a brand-name drug. One indication of the disorder is an abnormally low concentration of levocarnitine, an amino acid derivative, in the plasma. Treatment with the brand-name drug substantially increases this concentration.

A small clinical trial is conducted with 43 patients, 18 in the brand-name drug arm and 25 in the generic drug arm. The following increases in the log-concentration of levocarnitine are reported.

| Increase for | 7 8 4 6 10 10 5 7 9 8 |
|---|---|
| brand-name drug | 6 7 8 4 6 10 8 9 |
| Increase for | 6 7 5 9 5 5 3 7 5 10 |
| generic drug | 2 5 8 4 4 8 6 11 7 5 |
| | 5 5 7 4 6 |

The FDA declares that bioequivalence among the two drugs can be established if the difference in response to the two drugs is within two units of the log-concentration. Assuming that the log-concentration measurements follow normal distributions with equal population variance, can these two drugs be declared bioequivalent within a tolerance of ±2 units?

```
brandname = [7    8    4    6   10   10    5    7    9 ...
             8    6    7    8    4    6   10    8    9 ];
generic   = [6    7    5    9    5    5    3    7    5 ...
            10    8    5    8    4    4    8    6   11 ...
             7    5    5    5    7    4    6            ];

          xbar1 = mean(brandname)    %7.3333
          xbar2 = mean(generic)      %6.2000

          s1 = std(brandname)        %1.9097
          s2 = std(generic)          %1.9791

          n1 = length(brandname)     %18
          n2 = length(generic)       %25
%
sp = sqrt( ((n1-1)*s1^2 + (n2-1)*s2^2)/(n1 + n2 - 2)) % 1.9506
tL = (xbar1 - xbar2 -(-2))/(sp * sqrt( 1/n1 + 1/n2 )) % 5.1965
tU = (xbar1 - xbar2 - 2 )/(sp * sqrt( 1/n1 + 1/n2 )) %-1.4373
pL = 1-tcdf(tL, n1+ n2  - 2)    %2.9745e-006
pU = tcdf(tU, n1 + n2 - 2)       %0.0791
max(pL, pU)                      %0.0791  > 0.05, no equivalence

alpha = 0.05;
[xbar1-xbar2 - tinv(1-alpha, n1+n2-2)*sp*sqrt(1/n1+1/n2),...
    xbar1-xbar2 + tinv(1-alpha, n1+n2-2)*sp*sqrt(1/n1+1/n2)]
%  0.1186    2.1481
%  (0.1186, 2.1481) is not contained in (-2,2), no equivalence
```

Note that the equivalence of the two drugs was not established. The TOST did not simultaneously reject null hypotheses $H_0'$ and $H_0''$, or, equivalently, Westlake's interval failed to be fully included in the preset tolerance interval $(-2, 2)$.

Bayesian solution is conceptually straightforward. One finds the posterior distribution for the difference of the means, and evaluates the probability of this difference falling in the interval $(-2, 2)$. The posterior probability of $(-2, 2)$ should be close to 1 (say, 0.95) in order to declare equivalence.

```
model{
for(i in 1:n) {
  increase[i] ~ dnorm(mu[type[i]],  prec)
  }
mu[1] ~ dnorm( 10, 0.00001)
mu[2] ~ dnorm( 10, 0.00001)
mudiff <- mu[1]-mu[2]
prec ~ dgamma(0.001, 0.001)
probint <- step( mudiff + 2) * step(2 - mudiff)
}

DATA
list( n=43, increase = c(7, 8, 4, 6, 10, 10, 5, 7, 9,
   8, 6, 7, 8, 4, 6, 10, 8, 9, 6, 7, 5, 9, 5, 5, 3, 7, 5,
   10, 8, 5, 8, 4, 4, 8, 6, 11, 7, 5, 5, 5, 7, 4, 6 ),
type = c(1, 1, 1, 1, 1, 1, 1, 1, 1, 1, 1, 1, 1, 1, 1, 1, 1, 1,
   2, 2, 2, 2, 2, 2, 2, 2, 2, 2, 2, 2, 2, 2, 2, 2, 2, 2,
   2, 2, 2, 2, 2, 2, 2))

INITS
list( mu = c(10, 10), prec = 1)
```

|  | mean | sd | MC error | val5.0pc | median | val95.0pc | start | sample |
|---|---|---|---|---|---|---|---|---|
| mudiff | 1.133 | 0.6179 | 6.238E-4 | 0.117 | 1.133 | 2.147 | 10001 | 1000000 |
| probint | 0.9213 | 0.2693 | 2.766E-4 | 0.0 | 1.0 | 1.0 | 10001 | 1000000 |

The Bayesian analysis closely matches the findings by TOST and Westlake's interval. Note that the posterior probability of the tolerance interval $(-2, 2)$ is 0.9213, short of 0.95. Also, the 90% credible set $(0.117, 2.147)$ is close to Westlake's interval $(0.1186, 2.1481)$. This closeness is a consequence of non-informative priors on the means and precision.

# 10.9 Exercises

10.1. **Testing Piaget.** Two groups of elementary school students are taught mathematics by two different methods: traditional (group 1) and small group interactive teaching by discovery based on Piagetian theory (group 2). The results of a learning test are analyzed to test the difference in mean scores using the two methods. Group 1 had 16 students while group 2 had 14 students and the scores are given below.

| Groups | Scores |
|---|---|
| Traditional | 80, 69, 85, 87, 74, 85, 95, 84, 87, 86, 82, 91, 79, 100, 83, 85 |
| Piagetian | 100, 89, 87, 76, 93, 68, 99, 100, 78, 99, 100, 74, 76, 97 |

Test the hypothesis that the methods have no influence on test scores
against the alternative that the students in group 2 have significantly
higher scores. Take $\alpha = .05$.

10.2. **Smoking and COPD.**    It is well established that long-term cigarette
smoking is associated with the activation of a cascade of inflammatory
responses in the lungs that lead to tissue injury and dysfunction. This is
manifested clinically as chronic obstructive pulmonary disease (COPD). It
is believed that smoking causes approx. 80 to 90% of COPD cases.

Nine life-long nonsmoking healthy volunteers (5 men and 4 women; mean
age $22.0 \pm 1.9$ years $[\pm SD]$) and 11 healthy volunteers (5 men and 6 women;
mean age $23.4 \pm 0.9$ years) with a $2.0 \pm 1.2$ pack-year cigarette smoking history were recruited from students attending the University of Illinois at
Chicago. No participants had a history of chronic respiratory tract disorders, including asthma and COPD, and denied symptoms of acute respiratory illness within 4 weeks preceding the study.

The study by Garey et al. (2004) have found various summaries involving proteins, nitrite, and other inflammatory cascade signatures in exhaled
breath condensate (EBC). Average nitrite concentration in EBC of nonsmokers was found to be $\overline{X}_1 = 16156$ (nmol/L) and the sample standard
deviation was $s_1 = 7029$ (nmol/L). For smokers the mean nitrite concentration was $\overline{X}_2 = 24672$ (nmol/L) with sample standard deviation $s_2 = 7534$
(nmol/L).

Assuming that the population variances are the same, test the hypothesis
that the nitrite concentration in EBC for smokers and nonsmokers are the
same versus the one-sided alternative. Use $\alpha = 5\%$.

10.3. **Noradrenergic Activity.**    Although loss of noradrenergic neurons in the
locus ceruleus has been consistently demonstrated postmortem in Alzheimer's disease, several studies suggest that indices of central noradrenergic
activity increase with the severity of Alzheimer's disease in living patients.
The research by Elrod et al. (1997) estimated the effect of Alzheimer's disease severity on central noradrenergic activity by comparing the CSF norepinephrine concentrations of subjects with Alzheimer's disease in early
and advanced stages. Lumbar punctures were performed in 29 subjects
with Alzheimer's disease of mild or moderate severity and 17 subjects with
advanced Alzheimer's disease. Advanced Alzheimer's disease was defined
prospectively by a mini-mental state score of less than 12. Norepinephrine
was measured by radioenzymatic assay, and it is assumed that the measurements followed a normal distribution.

The CSF norepinephrine concentration for Alzheimer's-free subjects has a
mean of 170 pg/ml. The patients with advanced Alzheimer's disease had a
mean CSF norepinephrine concentration of 279 pg/ml, with a sample standard deviation of 122, while in those with mild to moderate severity the
mean was 198 pg/ml with a standard deviation of 89. In both cases normality is assumed.

(a) Let $\mu_1$ and $\mu_2$ be the population means for CSF norepinephrine concentration for patients with advanced and mild-to-moderate severity, respectively. Test the hypotheses $H_0' : \mu_1 = 170$ and $H_0'' : \mu_2 = 170$ based on the available information.

(b) Test the hypothesis that the population variances are the same, $H_0 :$ $\sigma_1^2 = \sigma_2^2$, versus the two-sided alternative.

(c) Test the hypothesis $H_0 : \mu_1 = \mu_2$. Choose the type of $t$-test based on the decision in (b).

In all cases use $\alpha = 0.05$.

10.4. **Testing Variances.**  Consider the following annotated MATLAB file.

```
%The two samples x and y are:
x = [0.34   0.52   -0.67   -0.98   -2.46    0.05    1.12 ...
       1.80 -0.51    0.88    0.29    0.29    0.66    0.06  -0.29];
y = [0.70   -0.61   -1.09   1.68 -0.62   -0.57   0.41];

%Test the equality of population variances
%against the two sided alternative.
%The built-in MATLAB function
[h p]=vartest2(x, y)
gives: %h=0 (choose H0) and p-value p=0.9727.
```

(a) Find the $p$-value using MATLAB code in the table on p. 359. Does it coincide with the vartest2 result?

(b) Show that $F = s_1^2/s_2^2 > 1$. In such a case the standard recommendation is to calculate the two-sided $p$-value as p = 2 * (1 - fcdf(F,n1-1, n2-1)), where n1=length(x) and n2=length(y). Show that for the samples from this exercise, this "$p$-value" exceeds 1. Can you explain what the problem is?

10.5. **Mating Calls.**   In a study of mating calls in the gray treefrogs *Hyla hrysoscelis* and *Hyla versicolor*, Gerhart (1994) reports that in a location in Lousiana the following data on the length of male advertisement calls have been collected:

| | Sample size | Average duration | SD of duration | Duration range |
|---|---|---|---|---|
| *Hyla chrysoscelis* | 43 | 0.65 | 0.18 | 0.36–1.27 |
| *Hyla versicolor* | 12 | 0.54 | 0.14 | 0.36–0.75 |

The two species cannot be distinguished by external morphology, but *H. chrysoscelis* are diploids while *H. versicolor* are tetraploids. The triploid crosses exhibit high mortality in larval stages, and if they attain sexual maturity, they are sterile. Females responding to the mating calls try to avoid mismatches.

Based on the data summaries provided, test whether the length of call is a discriminatory characteristic? Use $\alpha = 0.05$.

10.6. **Fatigue.** According to the article "Practice and fatigue effects on the programming of a coincident timing response," published in the *Journal of Human Movement Studies* in 1976, practice under fatigued conditions distorts
mechanisms that govern performance. An experiment was conducted using
15 college males who were trained to make a continuous horizontal right-
to-left arm movement from a microswitch to a barrier, knocking over the
barrier coincident with the arrival of a clock's second hand to the 6 o'clock
position. The absolute value of the difference between the time, in milliseconds, that it took to knock over the barrier and the time for the second hand
to reach the 6 o'clock position (500 ms) was recorded. Each participant performed the task five times under prefatigue and postfatigue conditions, and
the sums of the absolute differences for the five performances were recorded
as follows:

|         | Absolute time differences (ms) | |
|---------|------------|-------------|
| Subject | Prefatigue | Postfatigue |
| 1       | 158        | 91          |
| 2       | 92         | 59          |
| 3       | 65         | 215         |
| 4       | 98         | 226         |
| 5       | 33         | 223         |
| 6       | 89         | 91          |
| 7       | 148        | 92          |
| 8       | 58         | 177         |
| 9       | 142        | 134         |
| 10      | 117        | 116         |
| 11      | 74         | 153         |
| 12      | 66         | 219         |
| 13      | 109        | 143         |
| 14      | 57         | 164         |
| 15      | 85         | 100         |

An increase in the mean absolute time differences when the task is performed under postfatigue conditions would support the claim that practice
under fatigued conditions distorts mechanisms that govern performance.
Assuming the populations to be normally distributed, test this claim at
level $\alpha = 0.01$.

10.7. **Mosaic Virus.** A single leaf is taken from each of 11 different tobacco
plants. Each leaf is then divided in half and given one of two preparations
of *mosaic* virus. Researchers wanted to examine if there was a difference in
the mean number of lesions from the two preparations.
Here are the raw data:

| Plant | Prep 1 | Prep 2 |
|-------|--------|--------|
| 1 | 18 | 14 |
| 2 | 20 | 15 |
| 3 | 6 | 9 |
| 4 | 13 | 11 |

(a) Is this experiment in accordance with a paired $t$-test setup?

(b) Test the hypothesis that the difference between the two population means, $\mu_1 - \mu_2$, is significantly positive. Use $\alpha = 0.05$.

10.8. **Dopamine $\beta$-hydroxylase Activity.** Postmortem brain specimens from nine chronic schizophrenic patients and nine controls were assayed for activity of dopamine $\beta$-hydroxylase (DBH), the enzyme responsible for the conversion of dopamine to norepinephrine (Wyatt et al., 1975).

The means and standard deviations of DBH activity in the hippocampus part of the brain are provided in the table. Assume that the data come from two normally distributed and independent populations.

|  | Schizophrenic subjects | Control subjects |
|--|------------------------|------------------|
| Sample size | $n_1 = 9$ | $n_2 = 9$ |
| Sample mean | $\overline{X}_1 = 35.5$ | $\overline{X}_2 = 39.8$ |
| Sample standard deviation | $s_1 = 6.93$ | $s_2 = 8.16$ |

(a) Test to determine if the mean activity is significantly lower for the schizophrenic subjects than for the control subjects. Use $\alpha = 0.05$.

(b) Construct a 99% confidence interval for the mean difference in enzyme activity between the two groups.

Solve the above in two ways: (i) by assuming that $\sigma_1 = \sigma_2$ and (ii) without such an assumption.

Wyatt et al. (1975) report that one of the control subjects with low DBH activity had unusually long death-to-morgue time (27 hours) and suggested excluding the subject from the study. The data for controls after exclusion were $n_2 = 8$, $\overline{X}_2 = 41.2$, and $s_2 = 7.52$. Repeat the test in (a) and (b) with these control data.

10.9. **5-HIAA Levels.** A rare and slow-growing form of cancer, carcinoid tumors, may develop anywhere in the body where neuroendocrine (hormone-producing) cells exist. Serotonin is one of the key body chemicals released by carcinoid tumors that are associated with carcinoid syndrome. The 5-hydroxyindoleacetic acid (5-HIAA) test is a 24-hour urine test that is specific to carcinoid tumors. Elevated levels of 5-HIAA, a byproduct of serotonin decomposition, can be detected from a urine sample.

Ross and Roberts (1985) provide results of a case/control study of a morphologically specific type of carcinoid disorder that involves the mural and valvular endocardium on the right side of the heart, known as carcinoid heart disease. Out of a total of 36 subjects they investigate, urinary excretion of 5-hydroxyindoleacetic acid (5-HIAA) was measured on 28 subjects, 16 cases with carcinoid hearth disease and 12 controls. The data (level of 5-HIAA in milligrams per 24 hours), also discussed in Dawson–Saunders and Trapp (1994), are provided in the table below.

| Patients | 263 | 288 | 432 | 890 |
|---|---|---|---|---|
| | 450 | 1270 | 220 | 350 |
| | 283 | 274 | 580 | 285 |
| | 524 | 135 | 500 | 120 |
| Controls | 60 | 119 | 153 | 588 |
| | 124 | 196 | 14 | 23 |
| | 43 | 854 | 400 | 73 |

Assuming that the data come from respective normal distributions, compare the means of the two populations in both a classical and a Bayesian fashion. For the Bayes model use noninformative priors.

10.10. **Stress, Diet, and Acids.** In the study "Interrelationships Between Stress, Dietary Intake, and Plasma Ascorbic Acid During Pregnancy," discussed by Walpole et al. (2007, p. 359), the plasma ascorbic acid levels of pregnant women were compared for smokers versus nonsmokers. Thirty-two healthy women, between 15 and 32 years old, in the last 3 months of pregnancy were selected for the study. Eight of the women were smokers. Prior to the lab tests, the participants were told to avoid food and vitamin supplements. From the blood samples, the following plasma ascorbic acid values of each subject were determined in milligrams per 100 ml:

| Plasma ascorbic acid values | | |
|---|---|---|
| Nonsmokers | | Smokers |
| 0.97 | 1.06 | 0.48 |
| 0.72 | 0.86 | 0.71 |
| 1.00 | 0.85 | 0.98 |
| 0.81 | 0.58 | 0.68 |
| 0.62 | 0.57 | 1.18 |
| 1.32 | 0.64 | 1.36 |
| 1.24 | 0.98 | 0.78 |
| 0.99 | 1.09 | 1.64 |
| 0.90 | 0.92 | |
| 0.74 | 0.78 | |
| 0.88 | 1.24 | |
| 0.94 | 1.18 | |

(a) Using WinBUGS, test the hypothesis of equality of levels of plasma ascorbic acid for the two populations. Use noninformative priors on population means and variances (precisions).

(b) Compare the results in (a) with a classical two-sample $t$-test with no assumption of equality of variances.

10.11. **A. pantherina and A. rubescens.** Making spore prints is an enormous
help in identifying genera and species of mushrooms. To make a spore print,
mushroom fans take a fresh, mature cap and lay it on a clean piece of glass.
Left overnight or possibly longer the cap should give you a good print.
The Amanitas Family is one that has the most poisonous (A. *phalloides,*
A. *verna,* A. *virosa,* A. *pantherina,* etc.), as well as the most delicious (A. *ce-*
*sarea,* A. *rubescens*) species. Two independent samples of $m = 12$ and $n = 15$
spores of A. *pantherina* ("Panther") and A. *rubescens* ("Blusher"), respectively, were analyzed. In both species of mushrooms the spores are smooth
and elliptical and the largest possible measurement was taken (great axis
of the ellipse). It was found that the means were $\overline{X}_1 = 6.3$ microns and
$\overline{X}_2 = 7.5$ microns with standard deviations of $s_1 = 2.12$ $\mu$m and $s_2 = 1.94$
$\mu$m.
(a) A researcher is interested in testing the hypothesis that the population
mean sizes of spores for these two mushrooms, $\mu_1$ and $\mu_2$, are the same,
versus the two-sided alternative. Use $\alpha = 0.05$.
(b) What sample sizes are needed so that the researcher should be able to
reject the null hypothesis of no effect (no difference between the population
means) with power $1 - \beta = 0.9$ versus the "medium" standardized effect size
of $d = 0.5$? The significance $\alpha$ is, as in (a), 0.05.

10.12. **Blood Volume in Infants.** The total blood volume of normal newborn infants was estimated by Schücking (1879) who took into account the addition
of placental blood to the circulation of the newborn infant when clamping
of he umbilical cord is delayed. Demarsh et al. (1942) further studied the
importance of early and late clamping. For 16 babies in whom the cord was
clamped early the total blood (as a percentage of weight) on the third day
was

| | | | | | | |
|---|---|---|---|---|---|---|
| 13.8 | 8.0 | 8.4 | 8.8 | 9.6 9.8 | 8.2 | 8.0 |
| 10.3 | 8.5 | 11.5 | 8.2 | 8.9 9.4 | 10.3 | 12.6 |

For 16 babies in whom the cord was not clamped until the placenta began
to descend, the corresponding figures were

| | | | | | | |
|---|---|---|---|---|---|---|
| 13.8 | 8.0 | 8.4 | 8.8 | 9.6 9.8 | 8.2 | 8.0 |
| 10.3 | 8.5 | 11.5 | 8.2 | 8.9 9.4 | 10.3 | 12.6 |

(a) Do these two samples provide evidence of a significant difference between the blood volumes? Perform the test at $\alpha = 0.05$.
(b) Using WinBUGS find the posterior probability of the hypothesis that
there is no difference in blood volumes. Use noninformative priors.

10.13. **Biofeedback.** In the past, many bodily functions were thought to be beyond conscious control. However, recent experimentation suggests that it

may be possible for a person to control certain bodily functions if that person is trained in a program of *biofeedback* exercises. An experiment is conducted to show that blood pressure levels can be consciously reduced in people trained in this program. The blood pressure measurements (in millimeters of mercury) listed in the table represent readings before and after the biofeedback training of five subjects.

| Subject | Before | After |
|---------|--------|-------|
| 1 | 137 | 130 |
| 2 | 201 | 180 |
| 3 | 167 | 150 |
| 4 | 150 | 153 |
| 5 | 173 | 162 |

(a) If we want to test whether the mean blood pressure decreases after the training, what are the appropriate null and alternative hypotheses?
(b) Perform the test in (a) with $\alpha = 0.05$.
(c) What assumptions are needed to assure the validity of the results?

10.14. **Hypertension.** Dernellis and Panaretou (2002) examined a small number of subjects with hypertension and healthy control subjects. One of the variables of interest was the aortic stiffness index. Measures of this variable were calculated from the aortic diameter evaluated by M-mode echocardiography and blood pressure measured by a sphygmomanometer. Generally, physicians wish to reduce aortic stiffness. From $n_1 = 15$ patients with hypertension (group 1), the mean aortic stiffness index was $\overline{X}_1 = 16.16$ with a standard deviation of $s_1 = 4.29$. In the $n_2 = 16$ control subjects (group 2), the mean aortic stiffness index was $\overline{X}_2 = 10.53$ with a standard deviation of $s_2 = 3.33$.
(a) Test the hypothesis that the population mean aortic stiffness indices for the two groups differ (use the two-sided alternative). Perform the test by both the rejection-region and $p$-value methods. Take $\alpha = 0.05$ and assume that the population variances are the same.
(b) What is the (observed, retrospective)[1] power of this test versus the specific alternative $\Delta = |\mu_2 - \mu_1| = 3$. Keep $\alpha = 0.05$.
(c) Pretend that this study is only a pilot study needed to design more elaborate clinical trials of the same type. What group sample sizes are needed to assure a 99% power versus the fixed alternative $\Delta = |\mu_2 - \mu_1| = 3$?

10.15. **Hemopexin in DMD Cases II.** Refer to Exercises 2.16 and 9.16 and data set ▣ dmd.dat|mat|xls.
(a) Form a 95% confidence interval for the difference in mean responses of hemopexin h in two populations: population 1 consisting of all women

---

[1] Attributes observed and retrospective stand for the power analysis of a test that has already been performed. The procedure is the same as for the *prospective* analysis but elicited $\sigma$s are replaced by observed estimators $s$.

who are not DMD carriers (`carrier=0`) and population 2 consisting of all
women who are DMD carriers (`carrier=1`).

(b) It is believed that the mean level of hemopexin in the population of
women DMD carriers exceeds the mean level in women noncarriers by 10.
Test the one-sided alternative that the mean levels of hemopexin for the
two populations differ by more than 10, that is, $H_0 : \mu_1 - \mu_2 = -10$ versus
$H_1 : \mu_1 - \mu_2 < -10$. Use $\alpha = 5\%$. Report the $p$-value as well. Compare the
population variances prior to deciding how to test the difference of means.
*Hint:* In testing the hypothesis $H_0 : \mu_1 - \mu_2 = C$, the associated $t$-statistic
has as its numerator $\overline{X}_1 - \overline{X}_2 - C$ instead of $\overline{X}_1 - \overline{X}_2$.

10.16. **Risk of Stroke.**    Abbott et al. (1986) evaluate the risk of stroke among
smokers and nonsmokers in a 12-year prospective Honolulu Hearth Program study. The data are given in the table below.

|  | Stroke yes | Stroke no | total |
|---|---|---|---|
| Smoker | 171 | 3264 | 3435 |
| Nonsmoker | 117 | 4320 | 4437 |
| Total | 288 | 7584 | 7872 |

Estimate

(a) the risk difference and find the 95% confidence interval for the population risk difference,

(b) the risk ratio and find the 95% confidence interval for the population
risk ratio, and

(c) the odds ratio and find the 95% confidence interval for the population
odds ratio.

10.17. **Cell Counts.**   Refer to Example 1.2 and assume that data is approximately
normal.

(a) Do automated and manual counts significantly differ for a fixed confluency level? Use a paired $t$-test with a two-sided alternative at significance
level $\alpha = 0.05$.

(b) What are 95% confidence intervals for the population differences?

(c) If the difference between automated and manual counts constitutes an
error, are the errors comparable for the two confluency levels? This is a
DoD (difference-of-differences) test and is equivalent to a two-sample $t$-test
when measurements are the differences. Use the two-sided alternative and
$\alpha = 0.05$.

10.18. **Impulses from Crayfish.**    The crayfish is utilized in numerous neuroscience labs in order to explore the role of the central pattern generator
in locomotion. In an experiment with crayfish, you design glass electrodes
to record from the abdominal ganglia nerves that contribute to the return
and power strokes of the swimmerets. What this means is that the nerves
you are recording from contain the motor output signals that tell the swimmerets to move water around their abdomen anterior or posterior. In the

experimental setup, glass extracellular electrodes are used to measure the action potentials or voltage waveform. Because this is a CPG, you normally expect bursting activity, but when you suction onto the nerve cord, you notice what seems to be regular spiking. A spike detector was applied on the voltage time trace and the interspike intervals recorded. The data were obtained under two treatments. (1) Drug: *Carbachol* (also known as *carbamylcholine*), a drug that binds and activates the acetylcholine receptor, was added to possibly induce faster spiking; and (2) control group. The data files ▣ carbachol.dat and ▣ control.dat containing the respective interspike times can be found at the book's Web site. Also, you can load both files by importing ▣ frankdata.mat or ▣ frankdata.xls. All interspike time measurements are given in seconds. (Thanks to Dr. Frank Lin for the data and description.)

(a) Find fundamental descriptive statistics for both samples carbachol and control. Is Gaussianity a reasonable assumption for the two populations?

(b) Find the 95% confidence interval for the population mean interspike time in the control case.

(c) It is believed that the spiking in the control population has a frequency of 5 Hz. Test this hypothesis by testing that the mean population interspike time is 0.2, versus the proper one-sided alternative. Use both RR and $p$-value approaches; $\alpha = 0.05$.

(d) Test the hypothesis that the mean interspike time in the control and carbachol populations are the same versus the one-sided alternative that states that carbachol increases the frequency of spikes (decreases the interspike times); $\alpha = 0.05$.

(e) Do (b) and (c) in a Bayesian fashion building on ▦ spikes.odc, which contains control data and hints on how to set the priors. Compare and discuss the 95% confidence interval from (b) and the 95% credible set from BUGS. Compare tests, the $p$-value from (c), and the probability of $H_0$ from WinBUGS.

10.19. **Aerobic Capacity.** The peak oxygen intake per unit of body weight, called the *aerobic capacity* of an individual performing a strenuous activity, is a measure of work capacity. For comparative study, measurements of aerobic capacities are recorded (Frisancho, 1975) for a group of 20 Peruvian Highland natives and for a group of 10 Peruvian lowlanders acclimatized as adults to high altitudes. The measurements are taken on a bicycle ergometer at high altitude (ml $kg^{-1}$ $min^{-1}$).

|  | Peruvian highland natives | Peruvian Lowlanders acclimatized as adults |
|---|---|---|
| Sample mean | 46.3 | 38.0 |
| Sample st. deviation | 5.0 | 5.2 |

(a) Test the hypothesis that the population mean aerobic capacities are the same versus the one-sided alternative. Assume an equality of population variances and take $\alpha = 0.05$.

(b) If you were to repeat this experiment, what sample size (per group) would give you a power of 90% to detect the difference between the means of magnitude 4, if you assumed that the common population variance was $\sigma^2 = 5^2 = 25$? The level of the test, $\alpha$, is to be kept at 5%.

10.20. **Cataract and Diabetes.** Hiller and Kahn (1976) consider diabetes as a risk factor for cataracts and provide the results of a case/control study.

| Diabetes | Cataract cases | No cataract | Total |
|----------|----------------|-------------|-------|
| Present | 56 | 84 | 140 |
| Absent | 552 | 1927 | 2479 |
| Total | 608 | 2011 | 2619 |

Find 95% confidence intervals for the risk difference, risk ratio, and odds ratio.

10.21. **Beginnings of Antiseptic Surgeries.** Sir Joseph Lister (1827–1912), Professor of Surgery at Glasgow University, influenced by Pasteur's ideas, found that a wound wrapped in bandages treated by Carbolic acid (phenol) would often not become infected. Here are Lister's data on treating open fractures and amputations:

| Period | Carbolic acid | Results |
|--------|---------------|---------|
| 1864–1866 | No | Treated 34 patients, 19 recovered and 15 died |
| 1867–1870 | Yes | Treated 40 patients, 34 recovered and 6 died |

(a) Find and interpret the risk difference, risk ratio, and odds ratio.
(b) Find a 95% confidence intervals for the parameters estimated in (a).

10.22. **Reaction Times.** Researchers are interested in reactions times to different color light stimuli, specifically, green and red. A randomized design is proposed – each subject is given a number of trials, such as GGRRGRRG, etc. As a measurement of reaction time for a subject we report an average speed of reaction to each color. The table below contains the measurements.

| Subject | X (red) | Y (green) |
|---------|---------|-----------|
| 1 | 18 | 22 |
| 2 | 16 | 20 |
| 3 | 23 | 29 |
| 4 | 30 | 35 |
| 5 | 32 | 27 |
| 6 | 30 | 29 |
| 7 | 31 | 33 |
| 8 | 25 | 29 |
| 9 | 27 | 31 |
| 10 | 21 | 24 |

Are the reaction times to red and green lights the same? Use a two-sided alternative and $\alpha = 0.05$.

10.23. **Gamma Globulin and Aspirin.** A clinical trial of gamma globulin in the treatment of children with Kawasaki syndrome randomized approximately half of the patients to receive gamma globulin. The standard treatment for Kawasaki syndrome was a regimen of aspirin; nevertheless, about one quarter of these patients developed coronary abnormalities, even under the standard treatment. The outcome of interest was the development of coronary abnormalities (CA) over a 7-week period. The following $2 \times 2$ table summarizes the results of the trial.

| Treatment group | $CA = 1$ | $CA = 0$ | Total |
|---|---|---|---|
| $GG = 1, A = 0$ | 5 | 78 | 83 |
| $GG = 0, A = 1$ | 21 | 63 | 84 |
| Total | 26 | 141 | 167 |

Find the 95% confidence interval for the (a) risk difference, (b) risk ratio, and (c) odds ratio.

10.24. **High/Low Protein Diet in Rats.** Armitage and Berry (1994, p. 111) report data on the weight gain of 19 female rats between 28 and 84 days after birth. The rats were placed on diets with high (12 animals) and low (7 animals) protein content.

| High protein | Low protein |
|---|---|
| 134 | 70 |
| 146 | 118 |
| 104 | 101 |
| 119 | 85 |
| 124 | 107 |
| 161 | 132 |
| 107 | 94 |
| 83 | |
| 113 | |
| 129 | |
| 97 | |
| 123 | |

We want to test the hypothesis on dietary effect. Did a low protein diet result in significantly lower weight gain?
(a) What test should be used?
(b) Perform the test at an $\alpha = 0.05$ level.
(c) What sample size is needed (per diet) if $\sigma_1^2 = \sigma_2^2 = 450$ and we are interested in detecting the difference $\Delta = |\mu_1 - \mu_2| = 20$ as a significant deviation from $H_0$ 95% of the time (power = 0.95). Use $\alpha = 0.05$.

10.25. **Spider Monkey DNA.** In the context of Exercise 9.25
(a) Test that proportion $p_A$ is significantly smaller than $p_T$.

(b) Demonstrate that proportions $p_G$ and $p_C$ are not significantly different. Use $\alpha = 0.05$.

*Hint:* Since this is a multinomial model, the variance for difference $\hat{p}_A - \hat{p}_T$ needed for $Z$ statistic is $\mathbb{V}\mathrm{ar}(\hat{p}_A - \hat{p}_T) = p_A(1 - p_A)/n + p_T(1 - p_T)/n + 2p_A p_T/n$. Under $H_0$ from (a) both $p_A$ and $p_T$ are estimated by $(\hat{p}_A + \hat{p}_T)/2$ which leads to $Z = \sqrt{n}(\hat{p}_A - \hat{p}_T)/\sqrt{\hat{p}_A + \hat{p}_T}$.

10.26. **PBSC versus BM for Unrelated Donor Allogeneic Transplants.** We are interested in determining whether bone marrow (BM) is equivalent to peripheral blood stem cells (PBSC) in myeloablative unrelated donor transplantation, using the data from Eapen et al. (2007). The greater graft-versus-host disease burden of PBSC might make clinicians less likely to use PBSC in this context.

By using equivalence margins of $\pm 10\%$, and proportions of relapse within 6 months, test whether PBSC and BM are equivalent at level $\alpha = 0.05$.

| Method | Number of patients | Relapsed after 6 months |
|--------|--------------------|-----------------------|
| BM | 583 | 93 (16%) |
| PBSC | 328 | 58 (18%) |

*Hint:* By mimicking Example 10.15 devise a TOST using a normal approximation to the binomial:

$$z_L = \frac{\hat{p}_{BM} - \hat{p}_{PB} - (-\delta)}{\sqrt{\frac{\hat{p}_{BM}(1 - \hat{p}_{BM})}{n_1} + \frac{\hat{p}_{PB}(1 - \hat{p}_{PB})}{n_2}}}, \quad p_L = 1 - \Phi(z_L);$$

$$z_U = \frac{\hat{p}_{BM} - \hat{p}_{PB} - \delta}{\sqrt{\frac{\hat{p}_{BM}(1 - \hat{p}_{BM})}{n_1} + \frac{\hat{p}_{PB}(1 - \hat{p}_{PB})}{n_2}}}, \quad p_U = \Phi(z_U);$$

$$p - \text{value} = \max\{p_L, p_U\}.$$

10.27. **Hydrogels.** It has been demonstrated that poly(ethylene glycol) diacrylate (PEG-DA)-based hydrogels could serve as direct cell carriers for engineering soft orthopedic tissues. Recent studies have also shown that PEG-DA can be rendered degradable by introducing an enzyme-sensitive peptide sequence into the polymer chains.

The data from Dr. Temenoff's lab at Georgia Tech represent proportions of surviving cells after gel degradation by the enzyme, a bacterial collagenase, in durations of 30 and 60 minutes.

It is postulated that the proportions of live cells after 30 minutes and 60 minutes of exposure to the enzyme are equivalent

The data collected in eight independent experiments represent the number of cells alive among the total number of cells recovered, as in the table below.

| 30 minutes | | 60 minutes | |
| --- | --- | --- | --- |
| Alive | Total | Alive | Total |
| 20250 | 44250 | 40500 | 69750 |
| 51000 | 126000 | 42750 | 76500 |
| 77250 | 100500 | 78750 | 155250 |
| 39000 | 58500 | 42750 | 67500 |

Form ratios alive/total for the two durations. The ratios could be assumed as approximately normal given the large number of cells recovered. Show that $H_0 : \mu_1 = \mu_2$ is not rejected against one- or two-sided alternatives. This, however, is not an evidence of equivalence. Test the hypothesis of equivalence with equivalence margins $\theta_U = -\theta_L = 0.1$, that is, $H_0 : \mu_1 - \mu_2 < -0.1$ or $\mu_1 - \mu_2 > 0.1$   versus   $H_1 : -0.1 < \mu_1 - \mu_2 < 0.1$. Use a Westlake interval and $\alpha = 0.05$.

---

**MATLAB AND WINBUGS FILES AND DATA SETS USED IN THIS CHAPTER**

http://springer.bme.gatech.edu/Ch10.Two/

            argon.m, dnatest2.m, equivalence1.m, hemoc.m, hemopexin2.m, leadexposure.m, miammt.m, microdamage.m, neanderthal.m, nounfirst.m, piaget.m, plazmaacid.m, ratiopoissons.m, risk.m, schizo.m, simulrisks.m, temenoff.m

            argon.odc, braintissue.odc, cancerprop.odc, equivalence.odc, microdamage.odc, plasma1.odc, plasma2.odc, psoriasis.odc, spikes.odc, stressacids.odc,

    carbachol.dat, lice.xls, pasi.dat|mat|xls

---

# CHAPTER REFERENCES

Abbott, R. D., Yin, Y., Reed, D. M., and Yano, K. (1986). Risk of stroke in male cigarette smokers. *N. Engl. J. Med.*, **315**, 717–720.

Armitage, P. and Berry, G. (1994). *Statistical Methods in Medical Research*, 3rd edn. Blackwell Science, Malden.

Berger, R. L. and Hsu, J. C. (1996). Bioequivalence trials, intersection-union tests and equivalence confidence sets. *Stat. Sci.*, **11**, 4, 283–302.

Casagrande, J. T., Pike M. C., and Smith, P. G. (1978). The power function of the exact test for comparing two binomial distributions. *Appl. Stat.*, **27**, 176–180.

Cochran, W. G. and Cox, G. M. (1957). *Experimental Designs*. Wiley, New York.

Cox, D. R. (1953). Some simple tests for Poisson variates. *Biometrika*, **40**, 3–4, 354–360.

Dawson-Saunders, B. and Trapp, R. G. (1994). *Basic and Clinical Biostatistics*, 2nd edn. Appleton & Lange, Norwalk.

Demarsh, Q. B., Windle, W. F., and Alt, H. L. (1942). Blood volume of newborn infant in relation to early and late clamping of umbilical cord. *Am. J. Dis. Child.*, **63**, 6, 1123–1129.

Dernellis, J. and Panaretou, M. (2002). Effect of thyroid replacement therapy on arterial blood pressure in patients with hypertension and hyperthyroidism. *Am. Heart J.*, **143**, 718–724.

Desu, M. M. and Raghavarao, D. (1990). *Sample Size Methodology*. Academic, San Diego.

Dorn, H. F. (1954). The relationship of cancer of the lung and the use of tobacco. *Am. Stat.*, **8**, 7–13.

Eapen, M., Logan, B., Confer, D. L., Haagenson, M., Wagner, J. E., Weisdorf, D. J., Wingard, J. R., Rowley, S. D., Stroncek, D., Gee, A. P., Horowitz, M. M., and Anasetti, C. (2007). Peripheral blood grafts from unrelated donors are associated with increased acute and chronic graft-versus-host disease without improved survival. *Biol. Blood Marrow Transplant.*, **13**, 1461–1468.

Edwards, A. W. F. (1963). The measure of association in a 2x2 table. *J. Roy. Stat. Soc., A*, **126**, 109–114.

Elrod R., Peskind, E. R., DiGiacomo, L., Brodkin, K., Veith, R. C., and Raskind M. A. (1997). Effects of Alzheimer's disease severity on cerebrospinal fluid norepinephrine. *Am. J. Psychiatry*, **154**, 25–30.

FDA. (2001). Guidance for Industry. Statistical Approaches to Establishing Bioequivalence. FDA-CDER. `http://www.fda.gov/downloads/Drugs/GuidanceCompliance RegulatoryInformation/Guidances/ucm070244.pdf`

Fleiss, J. L., Levin, B., and Paik, M. C. (2003). *Statistical Methods for Rates and Proportions*, 3rd edn. Wiley, New York.

Frisancho, A. R. (1975). Functional adaptation to high altitude hypoxia. *Science*, **187**, 313–319.

Gardner, E. S. and Ringlee, R. J. (1968). Line stormproofing receives critical evaluation. *Trans. Distrib.*, **20**, 59–61.

Garey, K. W., Neuhauser, M. M., Robbins, R. A., Danziger, L. H., and Rubinstein, I. (2004). Markers of inflammation in exhaled breath condensate of young healthy smokers. *Chest*, **125**, 1, 22–26. `10.1378/chest.125.1.22`.

Gerhardt, H. C. (1994). Reproductive character displacement of female mate choice in the grey treefrog, *Hyla chrysoscelis*. *Anim. Behav.*, **47**, 959–969.

Giovannucci, E., Tosteson, T. D., Speizer, F. E., Ascherio, A., Vessey, M. P., Colditz, G. A. (1993). A retrospective cohort study of vasectomy and prostate cancer in US men. *J. Am. Med. Assoc.*, **269**, 7, 878–882.

Glass, G. and Hopkins, K. (1984). *Statistical Methods in Education and Psychology*, 2nd edn. Allyn and Bacon, Boston.

Hiller, R. and Kahn, H. A. (1976). Senile cataract extraction and diabetes. *Br. J. Ophthalmol.*, **60**, 4, 283–286.

Johnson, V. and Albert, J. (1999). *Ordinal Data Modeling*. Statistics for Social Science and Public Policy. Springer, Berlin Heidelberg New York.

Mosteller, F. (1968). Association and estimation in contingency tables. *J. Am. Stat. Assoc.*, **63**, 1–28.

Pitman, E. (1939). A note on normal correlation. *Biometrika*, **31**, 9–12.

Price, R. M. and Bonett, D. G. (2000). Estimating the ratio of two Poisson rates. *Comput. Stat. Data Anal.*, **34**, 3, 345–356.

Ross, E. M. and Roberts, W. C. (1985). The carcinoid syndrome: comparison of 21 necropsy subjects with carcinoid heart disease to 15 necropsy subjects without carcinoid heart disease. *Am. J. Med.*, **79**, 3, 339–354.

Satterthwaite, F. E. (1946). An approximate distribution of estimates of variance components. *Biom. Bull.*, **2**, 110–114.

Schücking, A. (1879). Blutmenge des neugeborenen. *Berl. Klin. Wochenschr.*, **16**, 581–583.

Schuirmann, D. J. (1981). On hypothesis testing to determine if the mean of a normal distribution is contained in a known interval. *Biometrics*, **37**, 617.

Welch, B. L. (1947). The generalization of "Student's" problem when several different population variances are involved. *Biometrika*, **34**, 28–35.

Wellek, S. (2010). *Testing Statistical Hypotheses of Equivalence and Noninferiority*, 2nd edn. CRC Press, Boca Raton.

Walpole, R. E., Myers, R. H., Myers, S. L., and Ye, K. (2007). *Probability and Statistics for Engineers and Scientists,* 8th edn. Pearson-Prentice Hall, Englewood Cliffs.

Westlake, W. J. (1976). Symmetric confidence intervals for bioequivalence trials. *Biometrics*, **32**, 741–744.

Woo, W. K. and McKenna, K. E. (2003). Combination TL01 ultraviolet B phototherapy and topical calcipotriol for psoriasis: a prospective randomized placebo-controlled clinical trial. *Br. J. Dermatol.*, **149**, 146–150.

Wyatt, R. J., Schwarz, M. A., Erdelyi, E., and Barchas, J. D. (1975). Dopamine-hydroxylase activity in brains of chronic schizophrenic patients. *Science*, **187**, 368–370.

Zar, J. H. (1996). *Biostatistical Analysis*. Prentice-Hall, Englewood Cliffs.

# Chapter 11
# ANOVA and Elements of Experimental Design

*Design is art optimized to meet objectives.*

– Shimon Shmueli

## 11.1 Introduction

In Chap. 10 we discussed the test of equality of means from two populations, $H_0 : \mu_1 = \mu_2$. Under standard assumptions (normality and independence) the proper test statistic had $t$-distribution, and several tests (equal/different variances, paired/unpaired samples) shared the common name "two-sample $t$-tests."

Many experimental protocols involve more than two populations. For example, an experimenter may be interested in comparing the sizes of cells grown under several experimental conditions. At first glance it seems that we can apply the $t$-test on all possible pairs of means. This "solution" would not be satisfactory since the probability of type I error for such a procedure is unduly large. For example, if the equality of four means is tested by testing the equality of $\binom{4}{2} = 6$ different pairs, each at the level of 5%, then the probability of finding a significant difference when in fact all means are equal is about 26.5%. We already discussed in Sect. 9.7 the problems of controlling type I error in multiple tests.

An appropriate procedure for testing hypotheses of equality of several means is the analysis of variance (ANOVA). ANOVA is probably one of the most frequently used statistical procedures, and the reasoning behind it is applicable to several other, seemingly different, problems.

## 11.2 One-Way ANOVA

ANOVA is an acronym for Analysis of Variance. Even though we are testing for differences among population means, the variability among the group means and the variability of the data within the treatment group are compared. This technique of separating the total variability in data to *between* and *within* variabilities is due to Fisher (1918, 1921).

Suppose we are interested in testing the equality of $k$ means $\mu_1,\ldots,\mu_k$ characterizing $k$ independent populations $\mathscr{P}_1,\ldots,\mathscr{P}_k$. From the $i$th population $\mathscr{P}_i$ a sample

$$y_{i1}, y_{i2}, \ldots, y_{in_i}$$

of size $n_i$ is taken. Let $N = \sum_{i=1}^{k} n_i$ be the total sample size. The responses $y_{ij}$ are modeled as

$$y_{ij} = \mu_i + \epsilon_{ij}, \quad 1 \le j \le n_i; \ 1 \le i \le k, \tag{11.1}$$

where $\epsilon_{ij}$ represents the "error" and quantifies the stochastic variability of difference between observation $y_{ij}$ and the mean of the corresponding population, $\mu_i$. Sometimes the $\mu_i$s are called *treatment means*, with index $i$ representing one of the *treatments*.

We are interested in testing whether all population means $\mu_i$ are equal, and the procedure is called one-way ANOVA. The assumptions underlying one-way ANOVA are as follows:

(i) All populations are normally distributed;

(ii) The variances in all populations are the same and constant (assumption of homoscedasticity);

(iii) The samples are mutually independent.

That can be expressed by the requirement that all $y_{ij}$ in (11.1) must be i.i.d. normal $\mathcal{N}(\mu_i, \sigma^2)$ or, equivalently, all $\epsilon_{ij}$s must be i.i.d. normal $\mathcal{N}(0, \sigma^2)$.

|  | Normal population | | | |
|---|---|---|---|---|
|  | 1 | 2 | ... | $k$ |
| Population mean | $\mu_1$ | $\mu_2$ | ... | $\mu_k$ |
| Common variance | $\sigma^2$ | $\sigma^2$ | ... | $\sigma^2$ |

If the sample sizes are the same, i.e., $n_1 = n_2 = \cdots = n_k = N/k$, then the ANOVA is called *balanced*. It is often the case that many experiments are designed as a balanced ANOVA. During an experiment it may happen that a particular measurement is missing due to a variety of reasons, resulting in an *unbalanced* layout. Balanced designs are preferable since they lead to simpler computations and interpretations.

In terms of model (11.1), the null hypothesis to be tested is

$$H_0 : \mu_1 = \mu_2 = \cdots = \mu_k,$$

and the alternative is

$$H_1 : (H_0)^c \quad (\text{or} \quad \mu_i \neq \mu_j, \text{ for at least one pair } i, j.$$

Note that the alternative $H_1 \equiv (H_0)^c$ is any negation of $H_0$. Thus, for example, $\mu_1 > \mu_2 = \mu_3 = \cdots = \mu_k$ or $\mu_1 = \mu_2 \neq \mu_3 = \mu_4 = \cdots = \mu_k$ are valid alternatives. Later we will discuss how to assess the alternative if $H_0$ is rejected.

One can reparameterize the population mean $\mu_i$ as $\mu_i = \mu + \alpha_i$. Simply speaking, the treatment mean is equal to the mean common for all treatments (called grand mean $\mu$) plus the effect of the population group (treatment effect) $\alpha_i$. The hypotheses now can be restated in terms of treatment effects:

$$H_0 : \alpha_1 = \alpha_2 = \cdots = \alpha_k = 0.$$

The alternative is $H_1$: Not all $\alpha_i$s are equal to 0.

The representation $\mu_i = \mu + \alpha_i$ is not unique. We usually assume that $\sum_i \alpha_i = 0$. This is an identifiability assumption needed to ensure the uniqueness of the decomposition $\mu_i = \mu + \alpha_i$. Indeed, by adding and subtracting any number $c$, $\mu_i$ becomes $\mu + \alpha_i = (\mu + c) + (\alpha_i - c) = \mu' + \alpha'$, and uniqueness is assured by $\sum_i \alpha_i = 0$. This kind of constraint is sometimes called *sum-to-zero* (STZ) constraint. There are other ways to ensure uniqueness; a sufficient assumption is, for example, $\alpha_1 = 0$. In this case $\mu = \mu_1$ becomes the reference or baseline mean. Before providing the procedure for testing $H_0$, which is summarized in the ANOVA table below, we review the notation.

| $y_{ij}$ | $j$th observation from treatment $i$ |
|---|---|
| $\overline{y}_i$ | Sample mean from the treatment $i$ |
| $\overline{y}$ | Average of all observations |
| $\mu_i$ | Population treatment mean |
| $\mu$ | Population grand mean |
| $\alpha_i$ | $i$th treatment effect |
| $n_i$ | Size of $i$th sample |
| $k$ | Number of treatments |
| $N$ | Total sample size |

### 11.2.1 ANOVA Table and Rationale for F-Test

The ANOVA table displays the data summaries needed for inference about the ANOVA hypothesis. It also provides an estimator of the variance in measurements and assesses the goodness of fit of an ANOVA model.

The variability in data follows the so-called *fundamental ANOVA identity* in which the total sum of squares ($SST$) is represented as a sum of the treatment sum of squares ($SSTr$) and the sum of squares due to error ($SSE$):

$$SST = SSTr + SSE = SS_{\text{Between}} + SS_{\text{Within}},$$

$$\sum_{i=1}^{k}\sum_{j=1}^{n_i}(y_{ij} - \overline{y})^2 = \sum_{i=1}^{k} n_i(\overline{y}_i - \overline{y})^2 + \sum_{i=1}^{k}\sum_{j=1}^{n_i}(y_{ij} - \overline{y}_i)^2.$$

The standard output from most statistical software includes degrees of freedom (DF), mean sum of squares, $F$ ratio, and the corresponding $p$-value:

| Source | DF | Sum of squares | Mean squares | F | p |
|---|---|---|---|---|---|
| Treatment | $k-1$ | $SSTr$ | $MSTr = SSTr/(k-1)$ | $MSTr/MSE$ | $\mathbb{P}(F_{k-1,N-k} > F)$ |
| Error | $N-k$ | $SSE$ | $MSE = SSE/(N-k)$ | | |
| Total | $N-1$ | $SST$ | | | |

The null hypothesis is rejected if the $F$-statistic is large compared to the $(1-\alpha)$ quantile of an $F$ distribution with $k-1$ and $N-k$ degrees of freedom. A decision can also be made by looking at the $p$-value.

**Rationale.** The mean square error, $MSE$, is an unbiased estimator of $\sigma^2$. Indeed,

$$MSE = \frac{1}{N-k}\sum_{i=1}^{k}\sum_{j=1}^{n_i}(y_{ij}-\overline{y}_i)^2 = \frac{1}{N-k}\sum_{i=1}^{k}\left[(n_i-1)\frac{1}{n_i-1}\sum_{j=1}^{n_i}(y_{ij}-\overline{y}_i)^2\right]$$

$$= \frac{1}{N-k}\sum_{i=1}^{k}(n_i-1)s_i^2, \text{ and}$$

$$\mathbb{E}(MSE) = \frac{1}{N-k}\sum_{i=1}^{k}(n_i-1)\mathbb{E}(s_i^2) = \frac{1}{N-k}\sum_{i=1}^{k}(n_i-1)\sigma^2 = \sigma^2.$$

On the other hand, mean square error due to treatments,

$$MSTr = \frac{SSTr}{k-1} = \frac{1}{k-1}\sum_{i=1}^{k}n_i(\overline{y}_i-\overline{y})^2,$$

is an unbiased estimator of $\sigma^2$ only when $H_0$ is true, that is, when all $\mu_i$ are the same. This follows from the fact that

$$\mathbb{E}(MSTr) = \sigma^2 + \frac{\sum_i n_i(\alpha_i)^2}{k-1},$$

where $\alpha_i = \mu_i - \mu$ is the population effect of treatment $i$. Since under $H_0$ $\alpha_1 = \alpha_2 = \cdots = \alpha_k = 0$, the $\mathbb{E}(MSE)$ is equal to $\sigma^2$, and $MSTr$ is an unbiased estimator of variance. When $H_0$ is violated, not all $\alpha_i$ are 0, or, equivalently, $\sum\alpha_i^2 > 0$. Thus, the ratio $MSTr/MSE$ quantifies the departure from $H_0$, and large values of this ratio are critical.

*Example 11.1.* **Coagulation Times.** To illustrate the one-way ANOVA we work out an example involving coagulation times that is also considered by Box et al. (2005). Twenty-four animals are randomly allocated to 4 different diets, but the numbers of animals allocated to different diets are not the same. The blood coagulation time is measured for each animal. Does diet type significantly influence the coagulation time? The data and MATLAB solution are provided next.

```
times = [62, 60, 63, 59, 63, 67, 71, 64, 65, 66, 68, 66, ...
          71, 67, 68, 68, 56, 62, 60, 61, 63, 64, 63, 59];
diets = {'dietA','dietA','dietA','dietA','dietB','dietB',...
   'dietB','dietB','dietB','dietB','dietC','dietC','dietC',...
   'dietC','dietC','dietC','dietD','dietD','dietD','dietD',...
   'dietD','dietD','dietD','dietD'};
[p,table,stats] = anova1(times, diets,'on')

% p = 4.6585e-005
% table =
%'Source'    'SS'      'df'    'MS'        'F'          'Prob>F'
%'Groups'    [228]     [ 3]    [    76]    [13.5714]    [4.6585e-005]
%'Error'     [112]     [20]    [5.6000]           []               []
```

```
%'Total'      [340]    [23]          []          []                []
%
%stats =
%     gnames: 4x1 cell
%          n: [4 6 6 8]
%     source: 'anoval'
%      means: [61 66 68 61]
%         df: 20
%          s: 2.3664
```

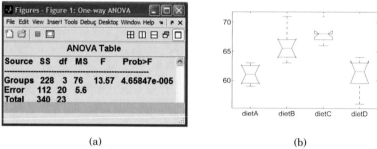

(a)                                                   (b)

**Fig. 11.1** (a) ANOVA table and (b) boxplots, as outputs of the anoval procedure on the coagulation times data.

From the ANOVA table we conclude that the null hypothesis is not tenable, the diets significantly affect the coagulation time. The $p$-value is smaller than $0.5 \cdot 10^{-4}$ which indicates strong support for $H_1$.

The ANOVA table featured in the output and Fig. 11.1a is a standard way of reporting the results of an ANOVA procedure. The $SS$ column in the ANOVA table restates the fundamental ANOVA identity $SSTr + SSE = SST$ as $228 + 112 = 340$. The degrees of freedom for treatments, $k - 1$, and the error, $n - k$, are additive and their sum is total number of degrees of freedom, $n - 1$. Here, $3 + 20 = 23$. The column with mean square errors is obtained when the sums of squares are divided by their corresponding degrees of freedom. The ratio $F = MSTr/MSE$ is the test statistic distributed as $F$ with $(3, 20)$ degrees of freedom. The observed $F = 13.5714$ exceeds the critical value finv(0.95, 3, 20)=3.0984, and $H_0$ is rejected. Recall that the rejection region is always right-tailed, in this case $[3.0984, \infty)$. The $p$-value is 1-fcdf(13.5714, 3, 20)= 4.6585e-005. Figure 11.1b shows boxplots of the coagulation times by the diet type.

Since the entries in ANOVA table are interrelated, it is possible to recover a table from only a few entries (e.g., Exercises 11.4, 11.7, and 11.14).

## 11.2.2 Testing Assumption of Equal Population Variances

There are several procedures that test for the fulfillment of ANOVA's condition of *homoscedasticity*, that is, the condition that the variances are the same and constant over all treatments.

A reasonably sensitive and simple procedure is Cochran's test.

### 11.2.2.1 Cochran's Test.

Cochran's test rejects the hypothesis that $k$ populations have the same variance if the statistic

$$C = \frac{s_{max}^2}{s_1^2 + \cdots + s_k^2}$$

is large. Here, $s_1^2, \ldots, s_k^2$ are sample variances in $k$ samples, and $s_{max}$ is the largest of $s_1^2, \ldots, s_k^2$. Cochran's test is implemented by ◀ Cochtest.m (Trujillo-Ortiz and Hernandez-Walls, MATLAB Central, File ID: #6431).

### 11.2.2.2 Levene's Test.

Levene's test (Levene, 1960) hinges on the statistic

$$L = \frac{(N-k)\sum_{i=1}^k n_i(\overline{Z}_{i\cdot} - \overline{Z}_{\cdot\cdot})^2}{(k-1)\sum_{i=1}^k \sum_{j=1}^{n_i}(Z_{ij} - \overline{Z}_{i\cdot})^2},$$

where $Z_{ij} = |y_{ij} - \overline{y}_{i\cdot}|$. To enhance the robustness of the procedure, the sample means could be replaced by the group medians, or trimmed means. The hypothesis $H_0 : \sigma_1 = \sigma_2 = \cdots = \sigma_k$ is rejected at level $\alpha$ if $L > F_{k-1,N-k,\alpha}$.

### 11.2.2.3 Bartlett's Test.

Another popular test for homoscedasticity is Bartlett's test. The statistic

$$B = \frac{(N-k)\log s_p^2 - \sum_{i=1}^k (n_i - 1)\log s_i^2}{1 + \frac{1}{3(k-1)} \times (\sum_{i=1}^k \frac{1}{n_i-1} - \frac{1}{N-k})},$$

where $s_p^2$ is the pooled sample variance $\frac{(n_1-1)s_1^2 + \cdots + (n_k-1)s_k^2}{N-k}$, has an approximately $\chi_{k-1}^2$ distribution. Large values of $B$ are critical, i.e., reject $H_0$ if $B > \chi_{k-1,\alpha}^2$, where $\alpha$ is the significance level.

In MATLAB, Bartlett's test is performed by the `vartestn(X)` command for samples formatted as columns of X, or as `vartestn(X, group)` for vector X, where group membership of $X$s is determined by the vector group. Bartlett's test is the default. Levene's test is invoked by optional argument, `vartestn(...,'robust')`. In the context of Example 11.1, Bartlett's and Levene's tests are performed in MATLAB as

```
% Coagulation Times: Testing Equality of Variances
[pval stats]=vartestn(times', diets','on')  %Bartlet
%   pval = 0.6441
%   chisqstat: 1.6680
%   df: 3

[pval stats]=vartestn(times', diets','on','robust') %Levene
%   pval = 0.6237
%   fstat: 0.5980
%   df: [3 20]
```

According to these tests the hypothesis of equal treatment variances is not rejected. Cochran's test agrees with Bartlett's and Levene's, giving a $p$-value of 0.6557.

**Remark.** As mentioned in the case of comparing the two means (p. 358), the variances-before-means type procedures are controversial. A celebrated statistician George E. P. Box criticized checking assumptions of equal variances before testing the equality means, arguing that comparisons of means are quite robust procedures compared to a non-robust variance comparison (Box, 1953). Aiming at Bartlett's test in particular, Box summarized his criticism as follows: "To make the preliminary test on variances is rather putting to sea in a rowing boat to find out whether conditions are sufficiently calm for an ocean liner to leave port!"

### 11.2.3 The Null Hypothesis Is Rejected. What Next?

When $H_0$ is rejected, the form for the alternative is not obvious as in the case of two means, and thus one must further explore relationships between the individual means. We will discuss two posttest ANOVA analyses: (i) tests for contrasts and (ii) pairwise comparisons. They both make sense only if the null hypothesis is rejected; if $H_0$ is not rejected, then both tests for contrasts and pairwise comparisons are trivial.

#### 11.2.3.1 Contrasts

A *contrast* is any linear combination of the population means,

$$C = c_1\mu_1 + c_2\mu_2 + \cdots + c_k\mu_k,$$

such that $\sum_{i=1}^{k} c_i = 0$.

For example, if $\mu_1, \ldots, \mu_5$ are means of $k = 5$ populations, the linear combinations (i) $2\mu_1 - \mu_2 - \mu_4$, (ii) $\mu_3 - \mu_2$, (iii) $\mu_1 + \mu_2 + \mu_3 - \mu_4 - 2\mu_5$, etc. are all contrasts since $2 - 1 + 0 - 1 + 0 = 0$, $0 - 1 + 1 + 0 + 0 = 0$, and $1 + 1 + 1 - 1 - 2 = 0$.

In the ANOVA model, $y_{ij} = \mu_i + \epsilon_{ij}$, $i = 1, \ldots, k; j = 1, \ldots, n_i$, sample treatment means $\bar{y}_i = \frac{1}{n_i} \sum_{j=1}^{n_i} y_{ij}$ are the estimators of the population treatment means $\mu_i$. Let $N = n_1 + \cdots + n_k$ be the total sample size and $s^2 = MSE$ the estimator of variance.

The test for a contrast

$$H_0 : \sum_{i=1}^{k} c_i\mu_i = 0 \quad \text{versus} \quad H_1 : \sum_{i=1}^{k} c_i\mu_i <, \neq, > 0 \qquad (11.2)$$

is based on the test statistic that involves sample contrast $c_i\bar{y}_i$

$$t = \frac{\sum_{i=1}^{k} c_i\bar{y}_i}{s\sqrt{\sum_{i=1}^{k} \frac{c_i^2}{n_i}}}$$

that has a $t$-distribution with $N - k$ degrees of freedom. Here, $\hat{C} = \sum_{i=1}^{k} c_i\bar{y}_i$ is an estimator of contrast $C$ and $s^2 \sum_{i=1}^{k} \frac{c_i^2}{n_i}$ is the sample variance of $\hat{C}$.

The $(1 - \alpha)100\%$ confidence interval for the contrast is

$$\left[ \sum_{i=1}^{k} c_i\bar{y}_i - t_{N-k,1-\alpha/2} \cdot s \cdot \sqrt{\sum_{i=1}^{k} \frac{c_i^2}{n_i}}, \ \sum_{i=1}^{k} c_i\bar{y}_i + t_{N-k,1-\alpha/2} \cdot s \cdot \sqrt{\sum_{i=1}^{k} \frac{c_i^2}{n_i}} \right].$$

Sometimes, contrast tests are called *single-degree F-tests* because of the link between $t$- and $F$-distributions. Recall that if random variable $X$ has a $t$-distribution with $n$ degrees of freedom, then $X^2$ has an $F$-distribution with 1 and $n$ degrees of freedom. Thus, the test of contrast in (11.2) against the two-sided alternative can equivalently be based on the statistic

$$F = \frac{\left(\sum_{i=1}^{k} c_i \bar{y}_i\right)^2}{s^2 \sum_{i=1}^{k} \frac{c_i^2}{n_i}},$$

which has an $F$-distribution with 1 and $N - k$ degrees of freedom. This $F$-test is good only for two-sided alternatives since the direction of deviation from $H_0$ is lost by squaring the $t$-statistic.

*Example 11.2.* As an illustration, let us test the hypothesis $H_0 : \mu_1 + \mu_2 = \mu_3 + \mu_4$ in the context of Example 11.1. The above hypothesis is a contrast since it can be written as $\sum_i c_i \mu_i = 0$ with $c = (1, 1, -1, -1)$. The following MATLAB code tests the contrast against the one-sided alternative and also finds the 95% confidence interval for $\sum_i c_i \mu_i$.

```
m = stats.means %[p,table,stats] = anova1(times, diets)
                %from Example Coagulation Times
c = [ 1  1 -1 -1 ];
L = c(1)*m(1) + c(2)*m(2)+c(3)*m(3) + c(4)*m(4) %L=-2
LL= m * c'   %LL=-2
stdL = stats.s * sqrt(c(1)^2/4+c(2)^2/6+c(3)^2/6+c(4)^2/8)
 %stdL = 1.9916
t = LL/stdL    %t =-1.0042

%test H_o: mu * c' = 0  H_1: mu * c' < 0
% p-value
tcdf(t, 23)  %0.1629

%or 95% confidence interval for population contrast
[LL -  tinv(0.975, 23)*stdL, LL +  tinv(0.975, 23)*stdL]
%   -6.1200    2.1200
```

The hypothesis $H_0 : \mu_1 + \mu_2 = \mu_3 + \mu_4$ is not rejected, and the $p$-value is 0.1629. Also, the confidence interval for $\mu_1 + \mu_2 - \mu_3 - \mu_4$ is $[-6.12, 2.12]$. ✏

**Orthogonal Contrasts\*.** Two or more contrasts are called orthogonal if their sample counterpart contrasts are uncorrelated. Operationally, two contrasts $c_1 \mu_1 + c_2 \mu_2 + \cdots + c_k \mu_k$ and $d_1 \mu_1 + d_2 \mu_2 + \cdots + d_k \mu_k$ are orthogonal if in addition to $\sum_i c_i = \sum_i d_i = 0$, the condition $c_1 d_1 + c_2 d_2 + \cdots + c_k d_k = \sum_i c_i d_i = 0$ holds. For unbalanced designs the condition is $\sum_i c_i d_i / n_i = 0$.

If there are $k$ treatments, only $k - 1$ mutually orthogonal contrasts can be constructed. Any additional contrast can be expressed as a linear combination of the original $k - 1$ contrasts. For example, if

| Contrast | Treatments | | | |
|---|---|---|---|---|
| | 1 | 2 | 3 | 4 |
| $C_1$ | 1 | −1 | −1 | 1 |
| $C_2$ | 1 | 0 | 0 | −1 |
| $C_3$ | 0 | 1 | −1 | 0 |

then the contrast $(1, -1, -3, 3)$ is $2C_1 - C_2 + C_3$. Any set of $k - 1$ orthogonal contrasts perfectly partitions the $SSTr$. If $SSC = \frac{(\sum c_i \bar{y}_i)^2}{\sum_i c_i^2/n_i}$ then

$$SSTr = SSC_1 + SSC_2 + \cdots + SSC_{k-1}.$$

This gives a possibility of simultaneous testing of any subset $\leq k - 1$ of orthogonal contrasts. Of particular interest are orthogonal contrasts sensitive to polynomial trends among the ordered and equally spaced levels of a factor. The requirement is that the design is balanced. For example, if $k = 4$,

| Contrast | Equispaced levels | | | |
|----------|----|----|----|----|
|          | 1  | 2  | 3  | 4  |
| $C_{\text{Linear}}$ | $-3$ | $-1$ | $1$ | $3$ |
| $C_{\text{Quadratic}}$ | $1$ | $-1$ | $-1$ | $1$ |
| $C_{\text{Cubic}}$ | $-1$ | $3$ | $-1$ | $3$ |

### 11.2.3.2 Pairwise Comparisons

After rejecting $H_0$, an assessment of $H_1$ can be conducted by pairwise comparisons. As the name suggests, this is a series of tests for all pairs of means in $k$ populations. Of course, there are $\binom{k}{2} = \frac{k(k-1)}{2}$ different tests.

A common error in doing pairwise comparisons is to perform $\frac{k(k-1)}{2}$ two-sample $t$-tests. The tests are dependent and the significance level $\alpha$ for simultaneous comparisons is difficult to control. This is equivalent in spirit to simultaneously testing multiple hypotheses and adjusting the significance level of each test to control overall significance level (p. 342). For example, for $k = 5$, the Bonferroni procedure will require $\alpha = 0.005$ for individual comparisons in order to control the overall significance level at 0.05, clearly a conservative approach.

Tukey (1952, unpubl. IMS address; 1953, unpubl. mimeograph) proposed a test designed specifically for pairwise comparisons sometimes called the "honestly significant difference test." The Tukey method is based on the so-called studentized range distribution with quantiles $q$. The quantile used in a test or a confidence interval is $q_{v,k,1-\alpha}$, with $\alpha$ being the overall significance level (or $(1 - \alpha)100\%$ overall confidence), $k$ the number of treatments, and $v$ the error degrees of freedom, $N - k$. The difference between two means $\mu_i$ and $\mu_j$ is significant if

$$|\bar{y}_i - \bar{y}_j| > q_{v,k,1-\alpha} \frac{s}{\sqrt{n}}, \tag{11.3}$$

where $s = \sqrt{MSE}$ and $n$ is the treatment sample size for a balanced design. If the design is not balanced, then replace $n$ in (11.3) by the harmonic mean of $n_i$s, $n_h = \frac{k}{\sum_{i=1}^{k} 1/n_i}$ (Tukey–Kramer procedure).

The function ◀ qtukey(v,k,p) (Trujillo-Ortiz and Hernandez-Walls, MATLAB Central #3469) approximates Tukey's quantiles for inputs $v = N - k$, $k$, and $p = 1 - \alpha$.

In biomedical experiments it is often the case that one treatment is considered a control and the only comparisons of interest are pairwise comparisons of all treatments with the control, forming a total of $k - 1$ comparisons. This is sometimes called the *many-to-one* procedure and it was developed by Dunnett (1955).

Let $\mu_1$ be the control mean. Then for $i = 2,\ldots,k$ the mean $\mu_i$ is different than $\mu_1$ if

$$|\overline{y}_i - \overline{y}_1| > d_{v,k-1,\alpha} \frac{s}{\sqrt{1/n_i + 1/n_1}},$$

where $\alpha$ is a joint significance test for $k-1$ tests, and $v = N-k$. The critical values $d_{v,k-1,\alpha}$ are available from the table at http://springer.bme.gatech.edu/dunnett.pdf.

It is recommended that the control treatment have more observations than other treatments. A discussion on the sample size necessary to perform Dunnett's comparisons can be found in Liu (1997).

*Example 11.3.* In the context of Example 11.1 (Coagulation Times), let us compare the means using the Tukey procedure. This is a default for MATLAB's command multcompare applied on the output stats in [p, table, stats] = anova1(times, diets). The multcompare command produces an interactive visual position for all means with their error bars and additionally gives an output with a confidence interval for each pair. If the confidence interval contains 0, then the means are not statistically different, according to Tukey's procedure. For example, the 95% Tukey confidence interval for $\mu_2 - \mu_3$ is $[-5.8241, 1.8241]$, and the means $\mu_2$ and $\mu_3$ are "statistically the same." On the other hand, Tukey's 95% interval for $\mu_1 - \mu_3$ is $[-11.2754, -2.7246]$, indicating that $\mu_1$ is significantly smaller than $\mu_3$.

```
multcompare(stats)   %[p,table,stats] = anova1(times, diets)
%Compares means: 1-2;  1-3;  1-4;  2-3;  2-4;  3-4
%ans =
%    1.0000    2.0000    -9.2754    -5.0000    -0.7246
%    1.0000    3.0000   -11.2754    -7.0000    -2.7246
%    1.0000    4.0000    -4.0560         0     4.0560
```

```
%    2.0000    3.0000    -5.8241    -2.0000    1.8241
%    2.0000    4.0000     1.4229     5.0000    8.5771
%    3.0000    4.0000     3.4229     7.0000   10.5771
```

We can also find the Tukey's confidence intervals by using ◀ qtukey.m. For example, the 95% confidence interval for $\mu_1 - \mu_2$ is

```
m=stats.means;
%1-2
[m(1)-m(2) - qtukey(20,4,0.95)*stats.s*sqrt(1/2 *(1/4+1/6)) ...
  m(1)-m(2) ...
  m(1)-m(2) + qtukey(20,4,0.95)*stats.s*sqrt(1/2*(1/4+1/6))]
%                  -9.3152    -5.0000    -0.6848
% Compare to: -9.2754    -5.0000    -0.7246 from multcompare
```

Although close to the output of multcompare, this interval differs due to a coarser approximation algorithm in qtukey.m.

In addition to Tukey and Dunnett, there is a range of other multiple comparison procedures. For example, Bonferroni is easy but too conservative. Since there are $\binom{k}{2}$ pairs among $k$ means, replacing $\alpha$ by $\alpha^* = \alpha/\binom{k}{2}$ would control all the comparisons at level $\alpha$. Scheffee's multiple comparison procedure provides a simultaneous $1 - \alpha$-level confidence interval for all linear combinations of population means and as a special case all pairwise differences. Scheffee's $(1 - \alpha)100\%$ confidence interval for $\mu_i - \mu_j$ is given by

$$|\bar{y}_i - \bar{y}_j| \pm s \sqrt{(k-1)F_{\alpha,k-1,n-k}\left(\frac{1}{n_i} + \frac{1}{n_j}\right)}.$$

Sidak's multiple comparison confidence intervals are

$$|\bar{y}_i - \bar{y}_j| \pm t_{n-k,\alpha^*/2} \, s \sqrt{\left(\frac{1}{n_i} + \frac{1}{n_j}\right)},$$

where $\alpha^* = 1-(1-\alpha)^{\frac{2}{k(k-1)}}$. Note that Sidak's comparisons are just slightly less conservative than Bonferroni's for which the $\alpha^* = \alpha/\binom{k}{2}$. Since $(1-\alpha)^m = 1 - m\alpha + \frac{m(m-1)}{2}\alpha^2 - \ldots$, Sidak's $1-(1-\alpha)^{\frac{2}{k(k-1)}}$ is approx. $\alpha/\binom{k}{2}$, which is Bonferroni's choice.

## 11.2.4 Bayesian Solution

Next we provide a Bayesian solution for the same problem. In the Win-BUGS code (✻anovacoagulation.odc) we stipulate that the data are normal with means equal to the grand mean, plus the effect of diet, mu[i]<-mu0 +

alpha[diets[i]]. The priors on alpha[i] are noninformative and depend on
the selection of identifiability constraint. Here the code uses *sum-to-zero*, a
STZ constraint that fixes one of the $\alpha$s, while the rest are given standard noninformative priors for the location. For example, $\alpha_1$ is fixed as $-(\alpha_2 + \cdots + \alpha_k)$,
which explains the term sum-to-zero. Another type of constraints that ensures
model identifiability is *corner* or CR constraint. In this case "corner" value $\alpha_1$
is set to 0. Then the treatment 1 is considered as a baseline category.

The grand mean mu0 is given a noninformative prior as well.

The parameter tau is a precision, that is, a reciprocal of variance. Traditionally, the noninformative prior on the precision is gamma with small parameters, in this case dgamma(0.001,0.001). From tau, the standard deviation
is calculated as sigma<-sqrt(1/tau). Thus, the highlights of the code are (i)
the indexing of alpha via diets[i], (ii) the identifiability constraints, and (iii)
the choice of noninformative priors.

```
model{
 for (i in 1:ntotal){
 times[i] ~ dnorm( mu[i], tau )
 mu[i] <-  mu0 + alpha[diets[i]]
 }
 #alpha[1] <- 0.0;       #CR Constraint
 alpha[1] <- -sum( alpha[2:a] ); #STZ Constraint

 mu0 ~ dnorm(0, 0.0001)
 alpha[2] ~ dnorm(0, 0.0001)
 alpha[3] ~ dnorm(0, 0.0001)
 alpha[4] ~ dnorm(0, 0.0001)
 tau ~ dgamma(0.001, 0.001)
 sigma <- sqrt(1/tau)
 }

DATA

list(ntotal = 24, a=4,
times =c(62, 60, 63, 59, 63, 67, 71, 64, 65, 66,
         68, 66, 71, 67, 68, 68, 56, 62, 60, 61, 63, 64, 63, 59),
diets = c(1,1,1,1, 2,2,2,2,2,2, 3,3,3,3,3,3,  4,4,4,4,4,4,4,4) )


INITS
list( mu0=0, alpha = c(NA,0,0,0), tau=1)
```

|          | mean   | sd      | MC error | val2.5pc | median  | val97.5pc | start | sample |
|----------|--------|---------|----------|----------|---------|-----------|-------|--------|
| alpha[1] | −3.001 | 1.03    | 0.002663 | −5.039   | −3.002  | −0.9566   | 1001  | 100000 |
| alpha[2] | 1.999  | 0.893   | 0.003573 | 0.2318   | 1.999   | 3.774     | 1001  | 100000 |
| alpha[3] | 4.001  | 0.8935  | 0.003453 | 2.232    | 4.002   | 5.779     | 1001  | 100000 |
| alpha[4] | −2.999 | 0.8178  | 0.003239 | −4.61    | −2.999  | −1.382    | 1001  | 100000 |
| mu0      | 64.0   | 0.5248  | 0.00176  | 62.96    | 64.0    | 65.03     | 1001  | 100000 |
| sigma    | 2.462  | 0.4121  | 0.001717 | 1.813    | 2.408   | 3.422     | 1001  | 100000 |
| tau      | 0.1783 | 0.05631 | 2.312E-4 | 0.08539  | 0.1724  | 0.3043    | 1001  | 100000 |

Figure 11.2a summarizes the posteriors of treatment effects, $\alpha_1, \ldots, \alpha_4$, as boxplots. Once the simulation for ANOVA is completed in WinBUGS, this graphical output becomes available under `Inference>Compare` tab.

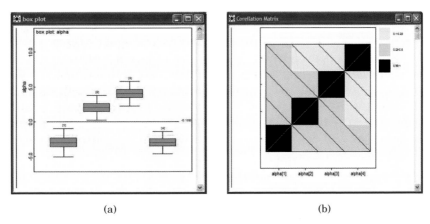

(a)                                               (b)

**Fig. 11.2** (a) WinBUGS output from `Inference>Compare`. Boxplots of posterior realizations of treatment effects `alpha`. (b) Matrix of correlations among components of `alpha`.

WinBUGS can also estimate the correlations between the treatment effects. The correlation matrix below and its graphical representation (Fig. 11.2b) are outputs from `Inference>Correlations`.

|          | alpha[1] | alpha[2] | alpha[3] | alpha[4] |
|----------|----------|----------|----------|----------|
| alpha[1] | 1.0      | −0.4095  | −0.4056  | −0.3694  |
| alpha[2] | −0.4095  | 1.0      | −0.3059  | −0.2418  |
| alpha[3] | −0.4056  | −0.3059  | 1.0      | −0.2476  |
| alpha[4] | −0.3694  | −0.2418  | −0.2476  | 1.0      |

Note that off-diagonal correlations are negative, as expected because of STZ constraint.

Expand the WinBUGS code 📄`anovacoagulation.odc` to accommodate the six differences `diff12 <- alpha[1]-alpha[2],...,diff34 <- alpha[3]-alpha[4]`. Compare credible sets for the differences with MATLAB's `multcompare` output.

## 11.2.5 Fixed- and Random-Effect ANOVA

In Example 11.1 (Coagulation Times) the levels of the factor are fixed: $dietA$, ..., $dietD$. Such ANOVAs are called fixed-effect ANOVAs. Sometimes, the number of factor levels is so large that a random subset is selected and serves as a set of levels. Then the inference is not concerned with these specific randomly selected levels but with the population of levels. For example, in measuring the response in animals to a particular chemical in food, a researcher may believe that the type of animal may be a factor. He/she would select a random but small number of different species as the levels of the factor. Inference about this factor would be translated to all potential species. In measuring the quality of healthcare, the researcher may have several randomly selected cities in the USA as the levels of the factor. For such models, the effects $\alpha_i$, $i = 1,\dots,k$ are assumed normal $\mathcal{N}(0,\sigma_\alpha^2)$, and the ANOVA hypothesis is equivalent to $H_0 : \sigma_\alpha^2 = 0$. Thus, for the random-effect model, $\mathbb{E}(MSTr) = \sigma^2 + \sigma_\alpha^2$, as opposed to the fixed-effect case, $\mathbb{E}(MSTr) = \sigma^2 + \frac{n}{k-1} \sum_{i=1}^{k} \alpha_i^2$.

If $s_1^2 = MSTr$, then in random effect ANOVA the *variance components* are estimated as:

$$\hat{\sigma}^2 = s^2 = MSE, \quad \text{and} \quad \hat{\sigma}_\alpha^2 = \frac{s_1^2 - s^2}{n}.$$

Thus testing $H_0 : \sigma_\alpha^2 = 0$ is based on $s_1^2/s^2 = MSTr/MSE$, which has an $F$-distribution with $k - 1, N - k$ degrees of freedom. Operationally, the random-effect and fixed-effect ANOVAs coincide, and the same ANOVA table can be used. The two differ mostly in the interpretation of the inference and the power analysis. Gauge R&R ANOVA in Sec. 11.10 is an example of a random-effect ANOVA.

## 11.3 Two-Way ANOVA and Factorial Designs

Many experiments involve two or more factors. For each combination of factor levels an experiment is performed and the response recorded. We will discuss only factorial designs with two factors; the interested reader is directed to Kutner et al. (2004) for a comprehensive treatment of multifactor designs and incomplete factorial designs.

Denote the two factors by $A$ and $B$ and assume that factor $A$ has $a$ levels and $B$ has $b$ levels. Then for each $a \times b$ combination of levels we perform the experiment $n \geq 1$ times. Measurements at fixed levels of $A$ and $B$ are called replicates. Such a design will be called a *factorial design*. When factors are arranged in a factorial design, they are called *crossed*. If the number of replicates is the same for each cell (fixed levels for $A$ and $B$), then the design is called balanced. We will be interested not only in how the factors influence the response, but also if the factors interact.

Suppose that the responses $y_{ijk}$ are obtained under the $i$th level of factor $A$ and the $j$th level of factor $B$. For each cell $(i,j)$ one obtains $n_{ij}$ replicates, and $y_{ijk}$ is the $k$th replicate. The model for $y_{ijk}$ is

$$y_{ijk} = \mu + \alpha_i + \beta_j + (\alpha\beta)_{ij} + \epsilon_{ijk}, i = 1,\dots,a; \; j = 1,\dots,b; \; k = 1,\dots,n_{ij}. \tag{11.4}$$

Thus, the observation $y_{ijk}$ is modeled as the grand mean $\mu$, plus the influence of factor $A$, $\alpha_i$, plus the influence of factor $B$, $\beta_j$, plus the interaction term $(\alpha\beta)_{ij}$, and, finally, plus the random error $\epsilon_{ijk}$. As in the one-way ANOVA, the errors $\epsilon_{ijk}$ are assumed independent normal with zero mean and constant variance $\sigma^2$ for all $i$, $j$, and $k$.

To ensure the identifiability of decomposition in (11.4) restrictions on $\alpha_i$s, $\beta_j$s, and $(\alpha\beta)_{ij}$s need to be imposed. Standardly, it is assumed that

$$\sum_{i=1}^{a} \alpha_i = 0, \quad \sum_{j=1}^{b} \beta_j = 0, \quad \sum_{i=1}^{a} (\alpha\beta)_{i,j} = 0, \quad \sum_{j=1}^{b} (\alpha\beta)_{i,j} = 0,$$

although different restrictions are possible, as we will see in the Bayesian models.

---

In two-factor factorial design there are three hypotheses to be tested: effects of factor $A$,

$$H_0' : \alpha_1 = \alpha_2 = \cdots = \alpha_a = 0 \quad \text{versus} \quad H_1' = (H_0')^c,$$

effects of factor $B$,

$$H_0'' : \beta_1 = \beta_2 = \cdots = \beta_b = 0 \quad \text{versus} \quad H_1'' = (H_0'')^c,$$

and the interaction of $A$ and $B$,

$$H_0''' : (\alpha\beta)_{11} = (\alpha\beta)_{12} = \cdots = (\alpha\beta)_{ab} = 0 \quad \text{versus} \quad H_1''' = (H_0''')^c.$$

---

The variability in observations follows the fundamental ANOVA identity in which the total sum of squares ($SST$) is represented as a sum of the $A$-treatment sum of squares ($SSA$), a sum of the $B$-treatment sum of squares ($SSB$), the interaction sum of squares ($SSAB$), and the sum of squares due to error ($SSE$). For a balanced design in which the number of replicates in all cells is $n$,

$$SST = SSA + SSB + SSAB + SSE$$
$$= \sum_{i=1}^{a} \sum_{j=1}^{b} \sum_{k=1}^{n} (y_{ijk} - \bar{y}_{...})^2$$
$$= bn \sum_{i=1}^{a} (\bar{y}_{i..} - \bar{y}_{...})^2 + an \sum_{j=1}^{b} (\bar{y}_{.j.} - \bar{y}_{...})^2 + n \sum_{i=1}^{a} \sum_{j=1}^{b} (\bar{y}_{ij.} - \bar{y}_{i..} - \bar{y}_{.j.} + \bar{y}_{...})^2$$
$$+ \sum_{i=1}^{a} \sum_{j=1}^{b} \sum_{k=1}^{n} (y_{ijk} - \bar{y}_{ij.})^2.$$

Here,

$$\overline{y}_{...} = \frac{1}{abn} \sum_{i=1}^{a} \sum_{j=1}^{b} \sum_{k=1}^{n} y_{ijk}, \ \overline{y}_{i..} = \frac{1}{bn} \sum_{j=1}^{b} \sum_{k=1}^{n} y_{ijk},$$

$$\overline{y}_{.j.} = \frac{1}{an} \sum_{i=1}^{a} \sum_{k=1}^{n} y_{ijk}, \text{ and } \overline{y}_{ij.} = \frac{1}{n} \sum_{k=1}^{n} y_{ijk}.$$

The point estimator for $\alpha_i$ effects is $\hat{\alpha}_i = \overline{y}_{i..} - \overline{y}_{...}$, for $\beta_j$ effects it is $\hat{\beta}_j = \overline{y}_{.j.} - \overline{y}_{...}$, and for the interaction $(\alpha\beta)_{ij}$ it is $\widehat{(\alpha\beta)}_{ij} = \overline{y}_{ij.} - \overline{y}_{i..} - \overline{y}_{.j.} + \overline{y}_{...}$.

The degrees of freedom are partitioned according to the ANOVA identity as $abn - 1 = (a-1)+(b-1)+(a-1)(b-1)+ab(n-1)$. Outputs in standard statistical packages include degrees of freedom (DF), mean sum of squares, $F$-ratios and their $p$-values.

| Source | DF | SS | MS | F | $p$-value |
|--------|-----|------|------|-----|-----------|
| Factor A | $a-1$ | SSA | $MSA = \frac{SSA}{a-1}$ | $F_A = \frac{MSA}{MSE}$ | $\mathbb{P}(F_{a-1,ab(n-1)} > F_A)$ |
| Factor B | $b-1$ | SSB | $MSB = \frac{SSB}{b-1}$ | $F_B = \frac{MSB}{MSE}$ | $\mathbb{P}(F_{b-1,ab(n-1)} > F_B)$ |
| A × B | $(a-1)(b-1)$ | SSAB | $MSAB = \frac{SSAB}{(a-1)(b-1)}$ | $F_{AB} = \frac{MSAB}{MSE}$ | $\mathbb{P}(F_{(a-1)(b-1),ab(n-1)} > F_{AB})$ |
| Error | $ab(n-1)$ | SSE | $MSE = \frac{SSE}{ab(n-1)}$ | | |
| Total | $abn-1$ | SST | | | |

$F_A$, $F_B$, and $F_{AB}$ are test statistics for $H_0'$, $H_0''$, and $H_0'''$, and their large values are critical. The rationale for these tests follows from the following expected values:

$$\mathbb{E}(MSE) = \sigma^2, \ \mathbb{E}(MSA) = \sigma^2 + \frac{nb}{a-1} \sum_{i=1}^{a} \alpha_i^2, \ \mathbb{E}(MSB) = \sigma^2 + \frac{na}{b-1} \sum_{j=1}^{b} \beta_j^2, \text{ and}$$

$$\mathbb{E}(MSAB) = \sigma^2 + \frac{n}{(a-1)(b-1)} \sum_{i=1}^{a} \sum_{j=1}^{b} (\alpha\beta)_{ij}^2.$$

*Example 11.4.* **Insulin Therapy.** Insulin has anti-inflammatory effects, as evaluated by its ability to reduce plasma concentrations of cytokines. The cytokine content in several organs after endotoxin (lipopolysaccharide, LPS) exposure and the effect of hyperinsulinaemia was examined in a porcine model (Brix-Christensen et al., 2005). All animals (35 to 40 kg) were subject to general anaesthesia and ventilated for 570 minutes. There were two possible interventions:

LPS: Lipopolysaccharide infusion for 180 minutes.

HEC: Hyperinsulinemic euglycemic clamp in 570 minutes (from start). Insulin was infused at a constant rate and plasma glucose was clamped at a certain level by infusion of glucose.

LPS induces a systemic inflammation (makes the animals sick) and HEC acts as a treatment. There were four experimental cells: (1) only anaesthesia (no HEC, no LPS), (2) HEC, (3) LPS, and (4) HEC and LPS.

The responses are levels of interleukin-10 (IL-10) in the kidney after 330 minutes have elapsed. The table corresponds to a balanced design $n = 8$, although the original experiment was unbalanced with ten animals in group 1, nine in group 2, ten in group 3, and nine in group 4.

|  | No HEC | | Yes HEC | |
|---|---|---|---|---|
|  | 7.0607 | 4.7510 | 3.0693 | 2.1102 |
| No LPS | 2.6168 | 2.9530 | 1.6489 | 3.1004 |
|  | 4.3489 | 3.6137 | 2.9160 | 4.1170 |
|  | 3.6356 | 5.6969 | 2.9149 | 3.0229 |
|  | 3.6911 | 4.5554 | 2.4159 | 1.8944 |
|  | 4.3933 | 3.8447 | 3.1493 | 3.5133 |
| Yes LPS | 6.0513 | 1.3590 | 4.4462 | 4.6254 |
|  | 4.2559 | 2.1449 | 2.8545 | 3.8967 |

```
%insulin.m
data2 = [...          %columns: IL10  LPS  HEC
7.0607  1 1;   2.6168  1 1;    4.3489  1 1;...
3.6356  1 1;   4.7510  1 1;    2.9530  1 1;...
3.6137  1 1;   5.6969  1 1;    3.0693  1 2;...
1.6489  1 2;   2.9160  1 2;    2.9149  1 2;...
2.1102  1 2;   3.1004  1 2;    4.1170  1 2;...
3.0229  1 2;   3.6911  2 1;    4.3933  2 1;...
6.0513  2 1;   4.2559  2 1;    4.5554  2 1;...
3.8447  2 1;   1.3590  2 1;    2.1449  2 1;...
2.4159  2 2;   3.1493  2 2;    4.4462  2 2;...
2.8545  2 2;   1.8944  2 2;    3.5133  2 2;...
4.6254  2 2;   3.8967  2 2];

IL10 = data2(:,1);   LPS=data2(:,2);   HEC=data2(:,3);
[p table stats terms] = anovan( IL10, {LPS,HEC}, ...
    'varnames',{'LPS','HEC'}, 'model','interaction')
```

The resulting ANOVA table provides the test for the two factors and their interaction.

| Source | DF | SS | MS | F | p-value |
|---|---|---|---|---|---|
| LPS | 1 | 0.0073 | 0.0073 | 0.0051 | 0.9436 |
| HEC | 1 | 7.2932 | 7.2932 | 5.0532 | 0.0326 |
| LPS*HEC | 1 | 2.1409 | 2.1409 | 1.4834 | 0.2334 |
| Error | 28 | 40.4124 | 1.4433 | | |
| Total | 31 | 49.8539 | | | |

Note that factor LPS is insignificant. The associated $F$ statistic is 0.0051 with $p$-value of 0.9436. The interaction (LPS*HEC) is insignificant as well ($p$-value of 0.2334), while HEC is significant ($p$-value of 0.0326).

Next, we generate the *interaction plots*.

```
%insulin.m continued
cell11 = mean(data2( 1: 8,1))    %L1 H1
cell12 = mean(data2( 9:16,1))    %L1 H2
cell21 = mean(data2(17:24,1))    %L2 H1
cell22 = mean(data2(25:32,1))    %L2 H2

figure;
plot([1 2],[cell11 cell12],'o','markersize',10, ...
    'MarkerEdgeColor','k','MarkerFaceColor','r')
hold on
plot([1 2],[cell11 cell12],'r-', 'linewidth',3)
plot([1 2],[cell21 cell22],'o','markersize',10, ...
    'MarkerEdgeColor','k','MarkerFaceColor','k')
plot([1 2],[cell21 cell22],'k-', 'linewidth',3)
title('Lines for LPS=1 (red) and LPS = 2 (black)')
xlabel('HEC')


figure;
plot([1 2],[cell11 cell21],'o','markersize',10, ...
    'MarkerEdgeColor','k','MarkerFaceColor','r')
hold on
plot([1 2],[cell11 cell21],'r-','linewidth',3)
plot([1 2],[cell12 cell22],'o','markersize',10, ...
    'MarkerEdgeColor','k','MarkerFaceColor','k')
plot([1 2],[cell12 cell22],'k-','linewidth',3)
title('Lines for HEC=1 (red) and HEC = 2 (black)')
xlabel('LPS')
```

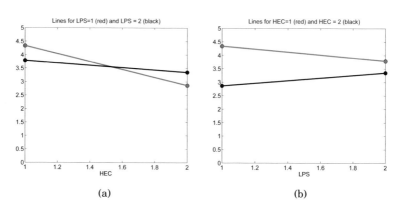

(a)                                              (b)

**Fig. 11.3** Interaction plots of LPS against HEC (*left*) and HEC against LPS (*right*) to explore the additivity of the model.

Figure 11.3 presents treatment mean plots, also known as interaction plots. The $x$-axis contains the levels of the factors, in this case both factors have levels 1 and 2. The $y$-axis contains the means of response (IL-10). The circles in both plots correspond to the cell means.

For example, in Fig. 11.3a $x$-axis has two levels of factor HEC. The means of IL-10 for LPS=1 are connected by the red line, while the means for LPS=2 are connected by the black line. When the lines on the plots are approximately parallel, the interaction between the factors is absent. Thus, the interaction plots serve as exploratory tools to check if the interaction term should be included in the ANOVA model. Note that some interaction between LPS and HEC is present (the lines are perfectly parallel); however, this interaction was found not statistically significant ($p$-value of 0.2334).

Next, we visualize the ANOVA fundamental identity.

```
%insulin.m continued
SSA = table{2,2}; SSB = table{3,2}; SSAB=table{4,2};
SSE=table{5,2}; SST = table{6,2};

%Display the budget of Sums of Squares
H=figure;
set(H,'Position',[400 400 400 400]);
y=[0 0 1 1];
hold on
h1=fill([0 SST SST 0],y,'c');
y=y+1;
h2=fill([0 SSA SSA 0],y,'y');
h3=fill([0 SSB SSB 0]+SSA,y,'r');
h4=fill([0 SSAB SSAB 0]+SSA+SSB,y,'g');
h5=fill([0 SSE SSE 0]+SSA+SSB+SSAB,y,'b');
y=y+1;
h6=fill([0 SST SST 0],y,'w');
hold off
legend([h1 h2 h3 h4 h5],'SST','SSA','SSB','SSAB','SSE',...
    'Location','NorthWest')
title('Sums of Squares')
```

Figure 11.4 shows the budget of sums of squares in this design. It is graphical representation of $SST = SSA + SSB + SSAB + SSE$. It shows the contributions to the total variability by the factors, their interaction and the error. Note that the sum of squares attributed to LPS (in yellow) is not visible in the plot. This is because its relative contribution to $SST$ is very small, $0.0073/49.8539 < 0.00015$.

The ANOVA table and both Figs. 11.3 and 11.4 are generated by file insulin.m. For a Bayesian solution of this example consult insulin.odc.

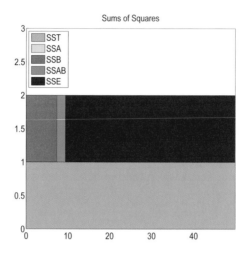

**Fig. 11.4** Budget of sums of squares for Insulin example.

## 11.4 Blocking

In many cases the design can account for the variability due to subjects or to experimental runs and focus on the variability induced by the treatments that constitute the factor of interest.

*Example 11.5.* **Blocking by Rats.** A researcher wishes to determine whether or not four different testing procedures produce different responses to the concentration of a particular poison in the blood of experimental rats. To minimize the influence of rat-to-rat variability, the biologist selects four rats from the same litter. Each of the four rats is given the same dose of poison per gram of body weight and then samples of their blood are tested by the four testing methods administered in random order. There are 4 different litters each containing 4 rats for a total of 16 animals involved.

```
concentration = [9.3 9.4 9.2 9.7 9.4 9.3 9.4 9.6 ...
   9.6 9.8 9.5 10.0 10.0 9.9 9.7 10.2];
procedure = [1 2 3 4 1 2 3 4 1 2 3 4 1 2 3 4];
litter = [1 1 1 1 2 2 2 2 3 3 3 3 4 4 4 4];
[p,table,stats,terms]=anovan(concentration,{procedure,litter},...
'varnames',char('Procedure','Rat'))

%p = 1.0e-003 *
%
%       0.8713
%       0.0452
%
%table =
```

```
%     'Source'     'Sum Sq.'   'd.f.'  'Singular?' 'Mean Sq.'
%     'Procedure' [0.3850]     [ 3]    [    0]     [ 0.1283]
%     'Rat'        [0.8250]    [ 3]    [    0]     [ 0.2750]
%     'Error'      [0.0800]    [ 9]    [    0]     [ 0.0089]
%     'Total'      [1.2900]    [15]    [    0]            []
%
%     'F'                'Prob>F'
%     [14.4375]          [8.7127e-004]
%     [30.9375]          [4.5233e-005]
%            []                 []
%            []                 []
```

In the above code, procedure is the factor of interest and litter is the blocking factor. Note that both the procedure and litter factors are highly significant at levels 0.0008713 and 0.0000452, respectively. We are interested in significant differences between the levels of procedure factor, but not between the litters or individual animals. However, it is desirable that the blocking factor turns out to be significant since in that case we will have accounted for significant variability attributed to blocks and separated it from the variability attributed to the test procedures. This makes the test more accurate.

To emphasize the benefits of blocking, we provide a nonsolution by treating this problem as a one-way ANOVA layout. This time, we fail to find any significant difference between the testing procedures ($p$-value 0.2196). Clearly, this approach is incorrect on other grounds: the condition of independence among the treatments, required for ANOVA, is violated.

```
   [p,table,stats] = anova1(concentration,procedure)

%p = 0.2196
%
%table =
%'Source'     'SS'        'df'     'MS'       'F'        'Prob>F'
%'Groups'    [0.3850]    [ 3]    [0.1283]   [1.7017]   [0.2196]
%'Error'     [0.9050]    [12]    [0.0754]        []         []
%'Total'     [1.2900]    [15]         []        []         []
```

In many cases the blocking is done by batches of material, animals from the same litter, by matching the demographic characteristics of the patients, etc. The repeated measures design is a form of block design where the blocking is done by subjects, individual animals, etc.

## 11.5 Repeated Measures Design

Repeated measures designs represent a generalization of the paired $t$-test to designs with more than two groups/treatments. In repeated measures designs the blocks are usually subjects, motivating the name "within-subject ANOVA" that is sometimes used. Every subject responds at all levels of the factor of interest, i.e., treatments. For example, in clinical trials, the subjects' responses could be taken at several time instances.

Such designs are sometimes necessary and have many advantages. The most important advantage is that the design controls for the variability between the subjects, which is usually not of interest. In simple words, subjects serve as their own controls, and the variability between them does not "leak" into the variability between the treatments. Another advantage is operational. Compared with factorial designs, repeated measures need fewer participants.

In the repeated measures design the independence between treatments (experimental conditions) is violated. Naturally, the responses of a subject are dependent on each other across treatments.

## ANOVA Table for Repeated Measures

| Treatment | Subject 1 | 2 | ... | $n$ | Treatment total |
|---|---|---|---|---|---|
| 1 | $y_{11}$ | $y_{12}$ | $\cdots$ | $y_{1n}$ | $y_{1\cdot}$ |
| 2 | $y_{21}$ | $y_{22}$ | $\cdots$ | $y_{2n}$ | $y_{2\cdot}$ |
| ... | | | ... | | ... |
| $k$ | $y_{k1}$ | $y_{k2}$ | $\cdots$ | $y_{kn}$ | $y_{k\cdot}$ |
| Subject total | $y_{\cdot 1}$ | $y_{\cdot 2}$ | $\cdots$ | $y_{\cdot n}$ | $y_{\cdot\cdot}$ |

The ANOVA model is

$$y_{ij} = \mu + \alpha_i + \beta_j + \epsilon_{ij}, \ i = 1,\dots,k; j = 1,\dots,n,$$

in which the hypothesis $H_0 : \alpha_i = 0$, $i = 1,\dots,k$, is of interest.

In the repeated measures design, total sum of squares $SST = \sum_{i=1}^{k}\sum_{j=1}^{n}(y_{ij} - \overline{y}_{..})^2$ partitions in the following way:

$$SST = SS_{BetweenSubjects} + SS_{WithinSubjects} = SSB + [SSA + SSE].$$

Equivalently,

$$\sum_{i=1}^{k}\sum_{j=1}^{n}(y_{ij} - \overline{y}_{..})^2 = k\sum_{j=1}^{n}(\overline{y}_{\cdot j} - \overline{y}_{..})^2 + \sum_{i=1}^{k}\sum_{j=1}^{n}(y_{ij} - \overline{y}_{\cdot j})^2$$

$$= k\sum_{j=1}^{n}(\overline{y}_{\cdot j} - \overline{y}_{..})^2 + \left[ n\sum_{i=1}^{k}(\overline{y}_{i\cdot} - \overline{y}_{..})^2 + \sum_{i=1}^{k}\sum_{j=1}^{n}(y_{ij} - \overline{y}_{i\cdot} - \overline{y}_{\cdot j} + \overline{y}_{..})^2 \right].$$

The degrees of freedom are split as

$$kn - 1 = (n - 1) + n(k - 1) = (n - 1) + [(k - 1) + (n - 1) \cdot (k - 1)].$$

The ANOVA table is

| Source | DF | SS | MS | F | p |
|--------|-----|-----|-----|-----|-----|
| Factor A | $k-1$ | SSA | $MSA = \frac{SSA}{k-1}$ | $F_A = \frac{MSA}{MSE}$ | $\mathbb{P}(F_{k-1,(k-1)(n-1)} > F_A)$ |
| Subjects B | $n-1$ | SSB | $MSB = \frac{SSB}{n-1}$ | $F_B = \frac{MSB}{MSE}$ | $\mathbb{P}(F_{n-1,(k-1)(n-1)} > F_B)$ |
| Error | $(k-1)(n-1)$ | SSE | $MSE = \frac{SSE}{(k-1)(n-1)}$ | | |
| Total | $kn-1$ | SST | | | |

The test statistic for Factor A ($H_0 : \alpha_i = 0, i = 1,\ldots,k$) is $F_A = MSA/MSE$ for which the p-value is $p = \mathbb{P}(F_{k-1,(k-1)(n-1)} > F_A)$. Usually we are not interested in $F_B = MSB/MSE$; however, its significance would mean that blocking by subjects is efficient in accounting for some variability thus making the inference about Factor A more precise.

*Example 11.6.* **Kidney Dialysis.** Eight patients each underwent three different methods of kidney dialysis (Daugridas and Ing, 1994). The following values were obtained for weight change in kilograms between dialysis sessions:

| Patient | Treatment 1 | Treatment 2 | Treatment 3 |
|---------|-------------|-------------|-------------|
| 1 | 2.90 | 2.97 | 2.67 |
| 2 | 2.56 | 2.45 | 2.62 |
| 3 | 2.88 | 2.76 | 1.84 |
| 4 | 1.73 | 1.20 | 1.33 |
| 5 | 2.50 | 2.16 | 1.27 |
| 6 | 3.18 | 2.89 | 2.39 |
| 7 | 2.83 | 2.87 | 2.39 |
| 8 | 1.92 | 2.01 | 1.66 |

Test the null hypothesis that there is no difference in mean weight change among treatments. Use $\alpha = 0.05$.

```
%dialysis.m
weich=[  2.90  2.97  2.67 ;...
         2.56  2.45  2.62 ;...
         2.88  2.76  1.84 ;...
         1.73  1.20  1.33 ;...
         2.50  2.16  1.27 ;...
         3.18  2.89  2.39 ;...
         2.83  2.87  2.39 ;...
         1.92  2.01  1.66 ];

subject=[1 2 3 4 5 6 7 8  ...
    1 2 3 4 5 6 7 8  1 2 3 4 5 6 7 8];
treatment = [1 1 1 1 1 1 1 1 ...
    2 2 2 2 2 2 2 2  3 3 3 3 3 3 3 3];
[p table stats terms] = ...
anovan(weich(:),{subject, treatment},'varnames',...
```

```
{'Subject' 'Treatment'} )

%'Source'     'Sum Sq.' 'd.f.'  'Mean Sq.'      'F'     'Prob>F'
%'Subject'    [5.6530]  [ 7]   [0.8076] [11.9341][6.0748e-005]
%'Treatment' [1.2510]   [ 2]   [0.6255] [ 9.2436][      0.0028]
%'Error'     [0.9474]   [14]   [0.0677]         []            []
%'Total'     [7.8515]   [23]        []          []            []


SST =   table{5,2}; SSE =   table{4,2};
SSTr = table{3,2}; SSBl = table{2,2};
SSW = SST - SSBl;
```

Since the hypothesis of equality of treatment means is rejected ($p$-val = 0.0028), one may look at the differences of treatment effects to find out which means are different. Command `multcompare(stats,'dimension',2)` will perform multiple comparisons along the second dimension, `Treatment`, and produces:

```
%1.0000    2.0000    -0.1917    0.1488    0.4892
%1.0000    3.0000     0.2008    0.5413    0.8817
%2.0000    3.0000     0.0521    0.3925    0.7329
```

This output is interpreted as $\alpha_1 - \alpha_2 \in [-0.1917, 0.4892]$, $\alpha_1 - \alpha_3 \in [0.2008, 0.8817]$ and $\alpha_2 - \alpha_3 \in [0.0521, 0.7329]$ with simultaneous confidence of 95%. Thus, by inspecting which interval contains 0 we conclude that treatment means 1 and 2 are not significantly different, while the mean for treatment 3 is significantly smaller from the means for treatments 1 or 2. For a Bayesian solution consult ⁂dialysis.odc.

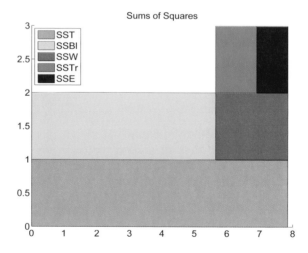

**Fig. 11.5** Budget of sums of squares for Kidney Dialysis example.

Figure 11.5 shows the budget of sums of squares. Note that the *SSTr* and *SSE* comprise *SSW* (Sum of Squares Within), while *SSBl* (variability due to subjects) is completely separated from *SSTr* and *SSE*. If the blocking by subjects is ignored and the problem is considered as a one-way ANOVA, the subject variability will be a part of *SSE* leading to wrong inference about the treatments (Exercise 11.16).

### 11.5.1 Sphericity Tests

Instead of the independence condition, as required in ANOVA, another condition is needed for repeated measures in order for an inference to be valid. This is the condition of *sphericity* or *circularity*. In simple terms, the sphericity assumption requires that all pairs of treatments be positively correlated in the same way. Another way to express sphericity is that variances of all pairwise differences between treatments are the same.

When an *F*-test in a repeated measures scenario is performed, many statistical packages (such as SAS, SPSS) automatically generate corrections for violations of sphericity. Examples are the Greenhouse–Geisser, the Huynh–Feldt, and lower-bound corrections. These packages correct for sphericity by altering the degrees of freedom, thereby altering the *p*-value for the observed *F*-ratio.

Opinions differ about which correction is the best, and this depends on what one wants to control in the analysis. An accepted universal choice is to use the Greenhouse–Geisser correction, $\epsilon_{GG}$. Violation of sphericity is more serious when $\epsilon_{GG}$ is smaller. When $\epsilon_{GG} > 0.75$ Huynh–Feldt correction $\epsilon_{HF}$ is recommended.

An application of Greenhouse–Geisser's and Huynh–Feldt's corrections, $\epsilon_{GG}$ and $\epsilon_{HF}$, respectively, is provided in the script bellow ( ◀ circularity.m) in the context of Example 11.16.

```
weich=[ 2.90  2.97  2.67 ;    2.56  2.45  2.62 ;...
        2.88  2.76  1.84 ;    1.73  1.20  1.33 ;...
        2.50  2.16  1.27 ;    3.18  2.89  2.39 ;...
        2.83  2.87  2.39 ;    1.92  2.01  1.66 ];
n = size(weich,1);      %number of subjects, n=8
k = size(weich,2);      %number of treatments, k=3
Sig = cov(weich);       %covariance matrix of weich
md = trace(Sig)/k;      %mean of diagonal entries of Sig
ma = mean(mean(Sig));   %mean of all components in Sig
mr = mean(Sig');        %row means of Sig
A = (k*(md-ma))^2;
B = (k-1)*(sum(sum(Sig.^2))-2*k*sum(mr.^2)+k^2*ma^2);
epsGG = A/B             %Greenhouse-Geisser epsilon 0.7038
epsHF = (n*(k-1)*epsGG-2)/((k-1)*((n-1)-(k-1)*epsGG))
                        %Huynh-Feldt epsilon 0.8281
%Corrections based on Trujillo-Ortiz et al. functions
```

```
%epsGG.m and epsHF.m available on MATLAB Central.
F = 9.2436; %F statistic for testing treatment differences
p = 1-fcdf(F,k-1,(n-1)*(k-1))  %original pvalue 0.0028
%
padjGG = 1-fcdf(F,epsGG*(k-1),epsGG*(n-1)*(k-1)) %0.0085
padjHF = 1-fcdf(F,epsHF*(k-1),epsHF*(n-1)*(k-1)) %0.0053
```

Note that both degrees of freedom in the $F$ statistic for testing the treatments are multiplied by correction factors, which increased the original $p$-value. In this example the corrections for circularity did not change the original decision of rejection of hypothesis $H_0$ stating the equality of treatment effects.

## 11.6 Nested Designs*

In the factorial design two-way ANOVA, two factors are *crossed*. This means that at each level of factor $A$ we get measurements under all levels of factor $B$, that is, all cells in the design are nonempty. In Example 11.4, two factors, HEC and LPS, given at two levels each ("yes" and "no"), form a $2 \times 2$ table with four cells. The factors are crossed, meaning that for each combination of levels we obtained observations. Sometimes this is impossible to achieve due to the nature of the experiment.

Suppose, for example, that four diets are given to mice and that we are interested in the mean concentration of a particular chemical in the tissue. Twelve experimental animals are randomly divided into four groups of three and each group put on a particular diet. After 2 weeks the animals are sacrificed and from each animal the tissue is sampled at five different random locations. The factors "diet" and "animal" cannot be crossed. After a single dietary regime, taking measurements on an animal requires its sacrifice, thus repeated measures designs are impossible.

The design is nested, and the responses are

$$y_{ijk} = \mu + \alpha_i + \beta_{j(i)} + \epsilon_{ijk}, i = 1, \ldots, 4; \; j = 1, \ldots, 3; \; k = 1, \ldots, 5,$$

where $\mu$ is the grand mean, $\alpha_i$ is the effect of the $i$th diet, and $\beta_{j(i)}$ is the effect of animal $j$, which is nested within treatment $i$.

For a general balanced two-factor nested design,

$$y_{ijk} = \mu + \alpha_i + \beta_{j(i)} + \epsilon_{ijk}, i = 1, \ldots, a; \; j = 1, \ldots, b; \; k = 1, \ldots, n,$$

the identifiability constraints are $\sum_{i=1}^{a} \alpha_i = 0$, and for each $i$, $\sum_{j=1}^{b} \beta_{j(i)} = 0$. The ANOVA identity $SST = SSA + SSB(A) + SSE$ is

$$\sum_{i=1}^{a}\sum_{j=1}^{b}\sum_{k=1}^{n}(y_{ijk}-\overline{y}_{...})^2$$

$$= bn\sum_{i=1}^{a}(\overline{y}_{i..}-\overline{y}_{...})^2 + n\sum_{i=1}^{a}\sum_{j=1}^{b}(\overline{y}_{ij.}-\overline{y}_{...})^2 + \sum_{i=1}^{a}\sum_{j=1}^{b}\sum_{k=1}^{n}(y_{ijk}-\overline{y}_{ij.})^2.$$

The degrees of freedom are partitioned according to the ANOVA identity as $abn - 1 = (a-1) + a(b-1) + ab(n-1)$. The ANOVA table is

| Source | DF | SS | MS |
|--------|-----|-----|-----|
| A | $a-1$ | SSA | $MSA = \frac{SSA}{a-1}$ |
| B(A) | $b-1$ | SSB | $MSB = \frac{SSB}{a(b-1)}$ |
| Error | $ab(n-1)$ | SSE | $MSE = \frac{SSE}{ab(n-1)}$ |
| Total | $abn-1$ | SST | |

Notice that the table does not provide the $F$-statistics and $p$-values. This is because the inferences differ depending on whether the factors are fixed or random.

The test for the main effect $H_0$ : all $\alpha_i = 0$ is based on $F = MSA/MSE$ if both factors $A$ and $B$ are fixed, and on $F = MSA/MSB(A)$ if at least one of the factors is random. The test $H_0$ : all $\beta_{j(i)} = 0$ is based on $F = MSB(A)/MSE$ in both cases. In the mouse-diet example, factor $B$ (animals) is random.

*Example 11.7.* Suppose that the data for the mouse-diet study are given as

| Diet | 1 | | | 2 | | | 3 | | | 4 | | |
|------|---|---|---|---|---|---|---|---|---|---|---|---|
| Animal | 1 | 2 | 3 | 1 | 2 | 3 | 1 | 2 | 3 | 1 | 2 | 3 |
| $k=1$ | 65 | 68 | 56 | 74 | 69 | 73 | 65 | 67 | 72 | 81 | 76 | 77 |
| $k=2$ | 71 | 70 | 55 | 76 | 70 | 77 | 74 | 59 | 63 | 75 | 72 | 69 |
| $k=3$ | 63 | 64 | 65 | 79 | 80 | 77 | 70 | 61 | 64 | 77 | 79 | 74 |
| $k=4$ | 69 | 71 | 68 | 81 | 79 | 79 | 69 | 66 | 69 | 75 | 82 | 79 |
| $k=5$ | 73 | 75 | 70 | 72 | 68 | 68 | 73 | 71 | 70 | 80 | 78 | 66 |

There are 12 mice in total, and the diets are assigned 3 mice each. On each mouse 5 measurements are taken. As we pointed out, this does not constitute a design with crossed factors since, for example, animal 1 under diet 1 differs from animal 1 under diet 2. Rather, the factor animal is nested within the factor diet. The following script is given in ⚓ nesta.m.

```
yijk =[...
        65  68  56     74  69  73     65  67  72     81  76  77 ;...
        71  70  55     76  70  77     74  59  63     75  72  69 ;...
        63  64  65     79  80  77     70  61  64     77  79  74 ;...
        69  71  68     81  79  79     69  66  69     75  82  79 ;...
        73  75  70     72  68  68     73  71  70     80  78  66  ];
```

```
a = 4;
b = 3;
n = 5;

%matrices of means (y..., y_i.., y_ij.)
yddd = mean(mean(yijk)) * ones(n, a*b)
yijd = repmat(mean(yijk), n, 1)
%yidd--------------------
    m=mean(yijk);
    mm=reshape(m', b, a);
    c=mean(mm);
    d=repmat(c',1,b);
    e=d';
yidd = repmat(e(:)',n,1)

SST = sum(sum((yijk - yddd).^2) )      %2.3166e+003
SSA = sum(sum((yidd - yddd).^2) )      %1.0227e+003
SSB_A = sum(sum((yijd - yidd).^2) ) %295.0667
SSE = sum(sum((yijk - yijd).^2) )      %998.8

MSA = SSA/(a-1)               %340.9111
MSB_A = SSB_A/(a * (b-1)) %36.8833
MSE = SSE/(a * b * (n-1)) %20.8083

%A fixed B(A) random //// 0r A random B(A) random
FArand = MSA/MSB_A      %9.2430
pa = 1- fcdf(FArand, a-1, a*(b-1))  %0.0056
%
FB_A = MSB_A/MSE  %1.7725
pb_a =   1- fcdf(FB_A,   a*(b-1), a*b*(n-1))   %0.1061
```

Figure 11.6 shows the budget of sums of squares for this example.

From the analysis we conclude that the effects of factor A (diet) were significant ($p = 0.0056$), while the (specific) effects of factor B (animal) were not significant ($p = 0.1061$).

✐

**Remarks.** (1) Note that the nested model does not have the interaction term. This is because the animals are nested within the diets. (2) If the design is not balanced, the calculations are substantially more complex and regression model should be used whenever possible.

## 11.7 Power Analysis in ANOVA

To design a sample size for an ANOVA test, one needs to specify the significance level, desired power, and a precision. Precision is defined in terms of ANOVA variance $\sigma^2$ and population treatment effects $\alpha_i$ coming from the null

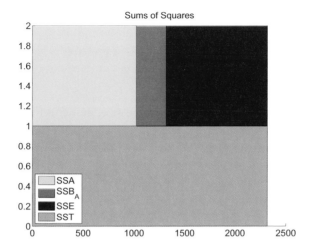

**Fig. 11.6** The budget of sums of squares for the mouse-diet study.

hypothesis $H_0 : \alpha_1 = \alpha_2 = \cdots = \alpha_k = 0$. It quantifies the extent of deviation from $H_0$ and usually is a function of the ratio $\frac{\sum n_i \alpha_i^2}{\sigma^2}$.

Under $H_0$, for a balanced design, the test statistic $F$ has an $F$-distribution with $k - 1$ and $N - k = k(n - 1)$ degrees of freedom, where $k$ is the number of treatments and $n$ the number of subjects at each level.

However, if $H_0$ is not true, the test statistic has a noncentral $F$-distribution with $k - 1$ and $k(n - 1)$ degrees of freedom and a noncentrality parameter $\lambda = n \frac{\sum_i \alpha_i^2}{\sigma^2}$.

An alternative way to set the precision is via Cohen's effect size, which for ANOVA takes the form $f^2 = \frac{1/k \sum_i \alpha_i^2}{\sigma^2}$. Note that Cohen's effect size and noncentrality parameter are connected via

$$\lambda = N f^2 = nk f^2.$$

Since determining the sample size is a prospective task, information about $\sigma^2$ and $\alpha_i$ may not be available. In the context of ANOVA Cohen (1988) recommends effect sizes of $f^2 = 0.1^2$ as small, $f^2 = 0.25^2$ as medium, and $f^2 = 0.4^2$ as large.

The power in ANOVA is, by definition,

$$1 - \beta = P(F^{nc}(k - 1, N - k, \lambda) > F^{-1}(1 - \alpha, k - 1, N - k)), \qquad (11.5)$$

where $F^{nc}(k - 1, N - k, \lambda)$ is a random variable with a noncentral $F$-distribution with $k - 1$ and $N - k$ degrees of freedom and noncentrality parameter $\lambda$. The quantity $F^{-1}(1 - \alpha, k - 1, N - k)$ is the $1 - \alpha$ percentile (a cut point for which the upper tail has a probability $\alpha$) of a standard $F$-distribution with $k - 1$ and $N - k$ degrees of freedom.

The interplay among the power, sample size, and effect size is illustrated by the following example.

*Example 11.8.* Suppose $k = 4$ treatment means are to be compared at a significance level of $\alpha = 0.05$. The experimenter is to decide how many replicates $n$ to run at each level, so that the null hypothesis is rejected with a probability of at least 0.9 if $f^2 = 0.0625$ or if $\sum_i \alpha_i^2$ is equal to $\sigma^2/4$.

For $n = 60$, i.e., $N = 240$, the power is calculated in a one-line command:

```
1-ncfcdf(finv(1-0.05, 4-1, 4*(60-1)), 4-1, 4*(60-1), 15)  %0.9122
```

Here we used $\lambda = nkf^2 = 60 \times 4 \times 0.0625 = 15$.

One can now try different values of $n$ (group sample sizes) to achieve the desired power; this change affects only two arguments in Eq. (11.5): $k(n - 1)$ and $\lambda = nkf^2$. Alternatively, one can use MATLAB's built in function `fzero(fun,x0)` which tries to find a zero of `fun` near some initial value x0.

```
k=4;  alpha = 0.05;  f2 = 0.0625;
f = @(n) 1-ncfcdf( finv(1-alpha, k-1,k*n-k),k-1,k*n-k,  n*k*f2 ) - 0.90;
ssize = fzero(f, 100) %57.6731
%Sample size of n=58 (per treatment) ensures the power of 90% for
%the effect size f^2=0.0625.
```

Thus, sample size of $n = 58$ will ensure the power of 90% for the specified effect size. The function `fzero` is quite robust with respect to specification of the initial value; in this case the initial value was $n = 100$.

If we wanted to plot the power for different sample sizes (Fig. 11.7), the following simple MATLAB script would do it. The inputs are the $k$ number of treatments and the significance level $\alpha$. The specific alternative $H_1$ is such that $\sum_i \alpha_i^2 = \sigma^2/4$, so that $\lambda = n/4$.

```
k=4;             %number of treatments
alpha = 0.05;    %significance level
y=[];            %set values of power for n
for n=2:100
y =[y 1-ncfcdf(finv(1-alpha, k-1, k*(n-1)), ...
                    k-1, k*(n-1), n/4)];
end
plot(2:100, y,'b-','linewidth',3)
xlabel('Group sample size n'); ylabel('Power')
```

In the above analysis, the total sample size is $N = k \times n = 240$.

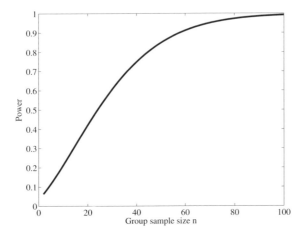

**Fig. 11.7** Power for $n \leq 100$ (size per group) in a fixed-effect ANOVA with $k = 4$ treatments and $\alpha = 0.05$. The alternative $H_1$ is defined as $\sum_i a_i^2 = \sigma^2/4$ so that the parameter of noncentrality $\lambda$ is equal to $n/4$.

**Sample Size for Multifactor ANOVA.** A power analysis for multifactor ANOVA is usually done by selecting the most important factor and evaluating the power for that factor. Operationally, this is the same as the previously discussed power analysis for one-way ANOVA, but with modified error degrees of freedom to account for the presence of other factors.

*Example 11.9.* Assume two-factor, fixed-effect ANOVA. The test for factor A at $a = 4$ levels is to be evaluated for its power. Factor B is present and has $b = 3$ levels. Assume a balanced design with $4 \times 3$ cells, with $n = 20$ subjects in each cell. The total number of subjects in the experiment is $N = 3 \times 4 \times 20 = 240$.

For $\alpha = 0.05$, a medium effect size $f^2 = 0.0625$, and $\lambda = Nf^2 = 20 \times 4 \times 3 \times 0.0625 = 15$, the power is 0.9121.

```
1-ncfcdf(finv(1-0.05, 4-1, 4*3*(20-1)), 4-1, 4*3*(20-1), 15)    %0.9121
```

Alternatively, if the cell sample size $n$ is required for a specified power we can use function fzero.

```
a=4; b=3; alpha = 0.05; f2=0.0625;
f = @(n) 1-ncfcdf( finv(1-alpha, a-1,a*b*(n-1)), ...
                   a-1, a*b*(n-1), a*b*n*f2) - 0.90;
ssize = fzero(f, 100) %19.2363
%sample size of 20 (by rounding 19.2363 up) ensures 90% power
%given the effect size and alpha
```

**Sample Size for Repeated Measures Design.** In a *repeated measures* design, each of the $k$ treatments is applied to every subject. Thus, the total sample size is equal to a treatment sample size, $N = n$. Suppose that $\rho$ is the correlation between scores for any two levels of the factor and it is assumed constant. Then the power is calculated by Eq. (11.5), where the noncentrality parameter is modified as

$$\lambda = \frac{n \sum_i \alpha_i^2}{(1-\rho)\sigma^2} = \frac{nkf^2}{1-\rho}.$$

*Example 11.10.* Suppose that $n = 25$ subjects go through $k = 3$ treatments and that a correlation between the treatments is $\rho = 0.6$. This correlation comes from the experimental design; the measures are repeated on the same subjects. Then for the medium effect size ($f = 0.25$ or $f^2 = 0.0625$) the achieved power is 0.8526.

```
n=25;   k=3;   alpha=0.05; rho=0.6;
f = 0.25;   %medium effect size f^2=0.0625
lambda = n * k * f^2/(1-rho)      %11.7188
power = 1-ncfcdf( finv(1-alpha, k-1, (n-1)*(k-1)),...
    k-1, (n-1)*(k-1), lambda)     %0.8526
```

If the power is specified at 85% level, then the number of subjects is obtained as

```
k=3;   alpha=0.05; rho=0.6;  f=0.25;
pf = @(n) 1-ncfcdf( finv(1-alpha, k-1, (n-1)*(k-1)),...
            k-1, (n-1)*(k-1), n*k*f^2/(1-rho) ) - 0.85;
ssize = fzero(pf, 100) %24.8342 (n=25 after rounding)
```

Often the sphericity condition is not met. This happens in longitudinal studies (repeated measures taken over time) where the correlation between measurements on days 1 and 2 may differ from the correlation between days 1 and 4, for example. In such a case, average correlation $\bar{\rho}$ is elicited and used; however, all degrees of freedom in central and noncentral F, as well as $\lambda$, are multiplied by sphericity parameter $\epsilon$.

*Example 11.11.* Suppose, as in Example 11.10, that $n = 25$ subjects go through $k = 3$ treatments, and that an average correlation between the treatments is $\bar{\rho} = 0.6$. If the sphericity is violated and parameter $\epsilon$ is estimated as $\epsilon = 0.7$, the power of 85.26% from Example 11.10 drops to 74.64%.

```
n=25;   k=3;   alpha=0.05; barrho=0.6; eps=0.7;
f = 0.25;
```

```
lambda =  n * k * f^2/(1-barrho)   %11.7188
power = 1-ncfcdf( finv(1-alpha, eps*(k-1), eps*(n-1)*(k-1)),...
     eps * (k-1), eps*(n-1)*(k-1), eps*lambda)    %0.7464
```

If the power is to remain at 85% level, then the number of subjects should increase from 25 to 32.

```
k=3;   alpha=0.05; barrho=0.6; eps=0.7;   f = 0.25;
pf = @(n) 1-ncfcdf( finv(1-alpha, eps*(k-1), eps*(n-1)*(k-1)),...
        eps*(k-1), eps*(n-1)*(k-1),  eps*n * k * f^2/(1-barrho) ) - 0.85;
ssize = fzero(pf, 100) %31.8147
```

## 11.8 Functional ANOVA*

Functional linear models have become popular recently since many responses are functional in nature. For example, in an experiment in neuroscience, observations could be functional responses, and rather than applying the experimental design on some summary of these functions, one could use the densely sampled functions as data.

We provide a definition for the one-way case, which is a "functionalized" version of standard one-way ANOVA.

Suppose that for any fixed $t \in T \subset \mathbb{R}$, the observations $\mathbf{y}$ are modeled by a fixed-effect ANOVA model:

$$y_{ij}(t) = \mu(t) + \alpha_i(t) + \epsilon_{ij}(t), \ i = 1, \dots, k, \ j = 1, \dots, n_i; \ \sum_{i=1}^{k} n_i = N, \quad (11.6)$$

where $\epsilon_{ij}(t)$ are independent $\mathcal{N}(0, \sigma^2)$ errors. When $i$ and $j$ are fixed, we assume that functions $\mu(t)$ and $\alpha_i(t)$ are square-integrable functions. To ensure the identifiability of treatment functions $\alpha_i$, one typically imposes

$$(\forall t) \sum_i \alpha_i(t) = 0. \quad (11.7)$$

In real life the measurements $\mathbf{y}$ are often taken at equidistant times $t_m$. The standard least square estimators for $\mu(t)$ and $\alpha_i(t)$

$$\hat{\mu}(t) = \overline{y}(t) = \frac{1}{n} \sum_{i,j} y_{ij}(t), \quad (11.8)$$

$$\hat{\alpha}_i(t) = \overline{y}_i(t) - \overline{y}(t), \quad (11.9)$$

where $\overline{y}_i(t) = \frac{1}{n_i} \sum_j y_{ij}(t)$, are obtained by minimizing the discrete version of LMSSE [for example, Ramsay and Silverman (1997) p. 141],

$$LMSSE = \sum_t \sum_{i,j} [y_{ij}(t) - (\mu(t) + \alpha_i(t))]^2, \quad (11.10)$$

subject to the constraint $(\forall t) \sum_i n_i \alpha_i(t) = 0$.

The fundamental ANOVA identity becomes a functional identity,

$$\text{SST}(t) = \text{SSTr}(t) + \text{SSE}(t), \tag{11.11}$$

with $\text{SST}(t) = \sum_{i,l} [y_{il}(t) - \overline{y}(t)]^2$, $\text{SSTr}(t) = \sum_i n_i [y_i(t) - \overline{y}(t)]^2$, and $\text{SSE}(t) = \sum_{i,l} [y_{il}(t) - \overline{y}_i(t)]^2$.

For each $t$, the function

$$F(t) = \frac{\text{SSTr}(t)/(k - 1)}{\text{SSE}(t)/(N - k)} \tag{11.12}$$

is distributed as noncentral $F_{k-1,N-k} \left( \frac{\sum_i n_i \alpha_i^2(t)}{\sigma^2} \right)$. Estimation of $\mu(t)$ and $\alpha_j(t)$ is straightforward, and estimators are given in Eqs. (11.8).

The testing of hypotheses involving functional components of the standard ANOVA method is hindered by dependence and dimensionality problems. Testing requires dimension reduction and this material is beyond the scope of this book. The interested reader can consult Ramsay and Silverman (1997) and Fan and Lin (1998).

*Example 11.12.* **FANOVA in Tumor Physiology.** Experiments carried out in vitro with tumor cell lines have demonstrated that tumor cells respond to radiation and anticancer drugs differently, depending on the environment. In particular, available oxygen is important. Efforts to increase the level of oxygen within tumor cells have included laboratory rats with implanted tumors breathing pure oxygen. Unfortunately, animals breathing pure oxygen may experience large drops in blood pressure, enough to make this intervention too risky for clinical use.

Mark Dewhirst, Department of Radiation Oncology at Duke University, sought to evaluate carbogen (95% pure oxygen and 5% carbon dioxide) as a breathing mixture that might improve tumor oxygenation without causing a drop in blood pressure. The protocol called for making measurements on each animal over 20 minutes of breathing room air, followed by 40 minutes of carbogen breathing. The experimenters took serial measurements of oxygen partial pressure ($PO_2$), tumor blood flow (LDF), mean arterial pressure (MAP), and heart rate. Microelectrodes, inserted into the tumors (one per animal), measured $PO_2$ at a particular location within the tumor throughout the study period. Two laser Doppler probes, inserted into each tumor, provided measurements of blood flow. An arterial line into the right femoral artery allowed measurement of MAP. Each animal wore a face mask for administration of breathing gases (room air or carbogen). [See Lanzen et al. (1998) for more information about these experiments.]

Nine rats had tumors transplanted within the quadriceps muscle (which we will denote by TM). For comparison, the studies also included eight rats with tumors transplanted subcutaneously (TS) and six rats without tumors

(N) in which measurements were made in the quadriceps muscle. The data
are provided in ▧ oxigen.dat.

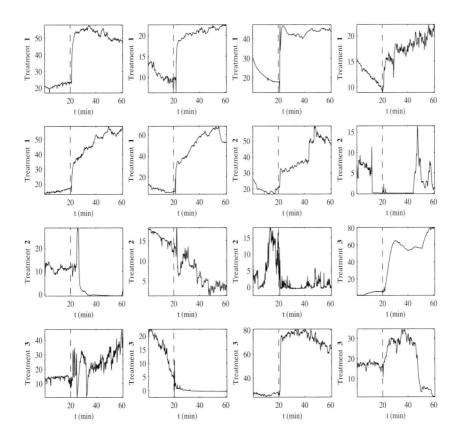

**Fig. 11.8** $PO_2$ measurements. Notice that despite a variety of functional responses and a
lack of a simple parametric model, at time $t^* = 20'$ the pattern generally changes.

Figure 11.8 show some of the data ($PO_2$). The plots show several features,
including an obvious rise in $PO_2$ at the 20-minute mark among some of the
animals. No physiologic model exists that would characterize the shapes of
these profiles mathematically. The primary study question concerned evaluating the effect of carbogen breathing on $PO_2$. The analysis was complicated by
the knowledge that there may be acute changes in $PO_2$ after carbogen breathing starts. The primary question of interest is whether the tumor tissue be-
haves differently than normal muscle tissue or whether a tumor implanted
subcutaneously responds to carbogen breathing differently than tumor tissue
implanted in muscle tissue in the presence of acute jumps in $PO_2$.

The analyses concern inference on changes in some physiologic measurements after an intervention. The problem for the data analysis is how best to define "change" to allow for the inference desired by the investigators.

From a statistical modeling point of view, the main issues concern building a flexible model for the multivariate time series $y_{ij}$ of responses and providing for formal inference on the occurrence of change at time $t^*$ and the "equality" of the $PO_2$ profiles. From the figures it is clear that the main challenge arises from the highly irregular behavior of responses. Neither physiologic considerations nor any exploratory data analysis motivates any parsimonious parametric form. Different individuals seem to exhibit widely varying response patterns. Still, it is clear from inspection of the data that for some response series a definite change takes place at time $t^*$.

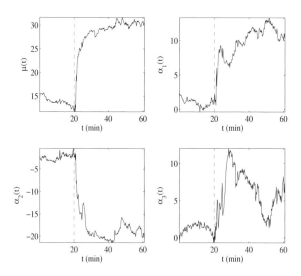

**Fig. 11.9** Functional ANOVA estimators.

Figure 11.9 shows the estimators of components in functional ANOVA. One can imagine from the figure that adding $\mu(t)$ and $\alpha_2(t)$ will lead to a relatively horizontal expected profile for group $i$, each fitted curve canceling the other to some extent. Files ◀ micePO2.m and fanova.m support this example.

# 11.9 Analysis of Means (ANOM)*

Some statistical applications, notably in the area of quality improvement, involve a comparison of treatment means to determine which are significantly different *from their overall average*. For example, a biomedical engineer might run an experiment to investigate which of six concentrations of an agent produces different output, in the sense that the average measurement for each concentration differs from the overall average.

Questions of this type are answered by the analysis of means (ANOM), which is a method for making multiple comparisons that is sometimes referred to as a "multiple comparison with the weighted mean". The ANOM answers different question than the ANOVA; however, it is related to ANOVA via multiple tests involving contrasts (Halperin et al., 1955; Ott, 1967)

The ANOM procedure consists of multiple testing of $k$ hypotheses $H_{0i} : \alpha_i = 0$, $i = 1, \ldots, k$ versus the two sided alternative. Since the testing is simultaneous, the Bonferroni adjustment to the type I error is used. Here $\alpha_i$ are, as in ANOVA, $k$ population treatment effects $\mu_i - \mu$, $i = 1, \ldots, k$ which are estimated as

$$\hat{\alpha}_i = \overline{y}_i - \overline{y},$$

in the usual ANOVA notation. Population effect $\alpha_i$ can be represented via the treatment means as

$$\alpha_i = \mu_i - \mu = \mu_i - \frac{\mu_1 + \cdots + \mu_k}{k}$$

$$= -\frac{1}{k}\mu_1 - \cdots - \frac{1}{k}\mu_{i-1} + \left(1 - \frac{1}{k}\right)\mu_i - \cdots - \frac{1}{k}\mu_k .$$

Since the constants $c_1 = -1/k$, ..., $c_i = 1 - 1/k$, ..., $c_k = -1/k$ sum up to 0, $\hat{\alpha}_i$ is an empirical contrast,

$$\hat{\alpha}_i = \overline{y}_i - \frac{\overline{y}_1 + \cdots + \overline{y}_k}{k}$$

$$= -\frac{1}{k}\overline{y}_1 - \cdots - \frac{1}{k}\overline{y}_{i-1} + \left(1 - \frac{1}{k}\right)\overline{y}_i - \cdots - \frac{1}{k}\overline{y}_k,$$

with standard deviation (as in p. 417),

$$s_{\hat{\alpha}_i} = s\sqrt{\sum_{j=1}^{k} \frac{c_j^2}{n_j}} = s\sqrt{\frac{1}{n_i}\left(1 - \frac{1}{k}\right)^2 + \frac{1}{k^2}\sum_{j\neq i}\frac{1}{n_j}}.$$

Here $n_i$ are treatment sample sizes, $N = \sum_{i=1}^{k} n_i$ is the total sample size, and $s = \sqrt{MSE}$.

All effects $\hat{\alpha}_i$ falling outside the interval

$$[-s_{\hat{\alpha}_i} \times t_{N-k,1-\alpha/(2k)}, \quad s_{\hat{\alpha}_i} \times t_{N-k,1-\alpha/(2k)}],$$

or equivalently, all treatment means $\overline{y}_i$ falling outside of the interval

$$[\overline{y} - s_{\hat{\alpha}_i} \times t_{N-k,1-\alpha/(2k)}, \quad \overline{y} + s_{\hat{\alpha}_i} \times t_{N-k,1-\alpha/(2k)}],$$

correspond to rejection of $H_{0i} : \alpha_i = \mu_i - \mu = 0$. The Bonferroni correction $1 - \alpha/(2k)$ in the $t$-quantile $t_{N-k,1-\alpha/(2k)}$ controls the significance of the procedure at level $\alpha$ (p. 342).

*Example 11.13.* We revisit Example 11.1 (Coagulation Times) and perform an ANOM analysis. In this example $\overline{y}_1 = 61$, $\overline{y}_2 = 66$, $\overline{y}_3 = 68$, and $\overline{y}_4 = 61$. The grand mean is $\overline{y} = 64$. Standard deviations for $\hat{\alpha}$ are $s_{\hat{\alpha}_1} = 0.9736$, $s_{\hat{\alpha}_2} = 0.8453$, $s_{\hat{\alpha}_3} = 0.8453$, and $s_{\hat{\alpha}_4} = 0.7733$, while the $t$-quantile corresponding to $\alpha = 0.05$ is $t_{1-0.05/8,20} = 2.7444$. This leads to ANOM bounds, $[61.328, 66.672]$, $[61.680, 66.320]$, $[61.680, 66.320]$, and $[61.878, 66.122]$.

Only the second treatment mean $\overline{y}_2 = 66$ falls in its ANOM interval $[61.680, 66.320]$, see Fig. 11.10 (generated by ◄ anomct.m). Consequently, the second population mean $\mu_2$ is not significantly different from the grand mean $\mu$.

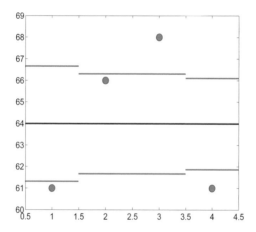

**Fig. 11.10** ANOM analysis in example Coagulation Times. *Blue* line is the overall mean, *red* lines are the ANOM interval bounds, while *green* dots are the treatment means. Note that only the second treatment mean falls in its interval.

## 11.10 The Capability of a Measurement System (Gauge R&R ANOVA)*

The gauge R&R methodology is concerned with the capability of measuring systems. Over time gauge R&R methods have evolved, and now two approaches have become widely accepted: (i) the average and range method, also known as the AIAG method, and (ii) the ANOVA method. We will focus on the ANOVA method and direct interested readers to Montgomery (2005) for comprehensive coverage of the topic.

In a typical gauge R&R study, several operators measure the same parts in random order. Most studies involve two to five operators and five to ten parts. There are usually several trial repetitions, which means that an operator performs multiple measurements on the same part.

The ANOVA method of analyzing measurement data provides not only the estimates for repeatability or equipment variation, reproducibility or appraiser variation, and part-to-part variation, but also accounts for possible interaction components.

We first define several key terms in gauge R&R context.

**Gauge or gage:** Any device that is used to obtain measurements.

**Part:** An item subjected to measurement. Typically a part is selected at random from the entire operating range of the process.

**Trial:** A set of measurements on a single part that is taken by an operator.

**Measurement system:** The complete measuring process involving gauges, operators, procedures, and operations. The system is evaluated as capable, acceptable, and not capable depending on the variabilities of its components. Notions of repeatability, reproducibility (R&R), and part variability as the main sources of variability in a measurement system are critical for its capability assessment.

**Repeatability:** The variance in measurements obtained with a measuring instrument when used several times by an appraiser while measuring the identical characteristic on the same part.

**Reproducibility:** The variation in measurements made by different appraisers using the same instrument when measuring an identical characteristic on the same part.

The measurements in a gauge R&R experiment are described as an ANOVA model

$$y_{ijk} = \mu + P_i + O_j + PO_{ij} + \epsilon_{ijk}, \quad i = 1,\ldots,p; \ j = 1,\ldots,o; \ k = 1,\ldots,n,$$

where $P_i$, $O_i$, $PO_{ij}$, and $\epsilon_{ijk}$ are independent random variables that represent the contributions of parts, operators, part–operator interaction, and random error to the measurement. We assume that the design is balanced, that there are $p$ parts and $o$ operators, and that for each part/operator combination there are $n$ trials.

This is an example of a random effect, two-way ANOVA, since the factors parts and operators are both randomly selected from the population consisting of many parts and operators. Except for the grand mean $\mu$, which is considered a constant, the random variables $P_i, O_j, PO_{ij}, \epsilon_{ijk}$ are considered independent zero-mean normal with variances $\sigma_P^2, \sigma_O^2, \sigma_{PO}^2$, and $\sigma^2$. Then the variability of measurement $y_{ijk}$ splits into the sum of four variances:

$$\mathbb{Var}(y_{ijk}) = \sigma_P^2 + \sigma_O^2 + \sigma_{PO}^2 + \sigma^2.$$

As in Sect. 11.3, the mean squares in the ANOVA table are obtained as

$$MSP = \frac{SSP}{p-1}, \ MSO = \frac{SSO}{o-1}, \ MSPO = \frac{SSPO}{(p-1)(o-1)}, \ MSE = \frac{SSE}{po(n-1)},$$

where the sums of squares are standardly defined (p. 425). Since this is a random-effect ANOVA, one can show that

$$\mathbb{E}(MSP) = \sigma^2 + n\sigma_{PO}^2 + on\sigma_P^2,$$

$$\mathbb{E}(MSO) = \sigma^2 + n\sigma_{PO}^2 + pn\sigma_O^2,$$

$$\mathbb{E}(MSPO) = \sigma^2 + n\sigma_{PO}^2,$$

$$\mathbb{E}(MSE) = \sigma^2.$$

The derivation of these expectations is beyond the scope of this text, but a thorough presentation of it can be found in Montgomery (1984) or Kutner et al. (2005). The above expectations, by moment matching, lead to the estimates

$$\hat{\sigma}_P^2 = \frac{MSP - MSPO}{on},$$

$$\hat{\sigma}_O^2 = \frac{MSO - MSPO}{pn},$$

$$\hat{\sigma}_{PO}^2 = \frac{MSPO - MSE}{n},$$

$$\hat{\sigma}^2 = MSE.$$

In the context of R&R analysis, the definitions of empirical variance components are as follows:

$$\hat{\sigma}^2_{Repeat} = \hat{\sigma}^2$$
$$\hat{\sigma}^2_{Reprod} = \hat{\sigma}^2_O + \hat{\sigma}^2_{PO}$$
$$\hat{\sigma}^2_{Gauge} = \hat{\sigma}^2_{Repeat} + \hat{\sigma}^2_{Reprod}$$
$$= \hat{\sigma}^2 + \hat{\sigma}^2_O + \hat{\sigma}^2_{PO}$$
$$\hat{\sigma}^2_{Total} = \hat{\sigma}^2_{Gauge} + \hat{\sigma}^2_{Part}$$

Next we provide several measures of the capability of a measurement system.

**Number of Distinct Categories.** The measure called signal-to-noise ratio, in the context of R&R, is defined as

$$SNR = \sqrt{\frac{2 \times \hat{\sigma}^2_{Part}}{\hat{\sigma}^2_{Gauge}}}.$$

The $SNR$ rounded to the closest integer defines the $NDC$ measure, which is a resolution of the measurement system.

The $NDC$ informally indicates how many "categories" the measurement system is able to differentiate. If $NDC = 0$ or 1, then the measurement system is useless. If $NDC = 2$, then the system can differentiate only between two categories ("small" and "large," in T-shirt terminology). If $NDC = 3$, then the system can distinguish "small," "medium," and "large," and so on. If $NDC \geq 5$, then the measurement system is capable. If $NDC \leq 1$, then the measurement system is not capable. Otherwise, the measurement system is evaluated as acceptable.

**Percent of R&R Variability.** The percent of R&R variability ($PRR$) measures the size of the R&R variation relative to the total data variation. It is defined as

$$PRR = \sqrt{\frac{\hat{\sigma}^2_{Gauge}}{\hat{\sigma}^2_{Total}}}.$$

If $PRR < 10\%$, then the measurement system is capable. If $PRR > 30\%$, the measurement system is not capable; the system is acceptable for $10\% \leq PRR \leq 30\%$. Also, the individual contribution of repeatability and reproducibility variances entering the summary measure $PRR$ are of interest. For instance, a large repeatability variation, relative to the reproducibility variation, indicates a need to improve the gauge. A high reproducibility variance relative to repeatability indicates a need for better operator training.

MATLAB's function ◀ gagerr(y,part,operator) performs a gauge repeatability and reproducibility analysis on measurements in vector y collected by operator on part. As in the anovan command, the number of elements in part and operator should be the same as in y. There are many important options in gagerr, and we recommend that the user carefully consult the function help.

*Example 11.14.* **Measurements of Thermal Impedance.** This example of a gauge R&R study comes from Houf and Berman (1988) and Montgomery (2005). The data, in the table below, represent measurements on thermal impedance (in °C per Watt × 100) on a power module for an induction motor starter. There are ten parts, three operators, and three measurements per part.

| Part | Operator 1 | | | Operator 2 | | | Operator 3 | | |
|---|---|---|---|---|---|---|---|---|---|
| number | Test 1 | Test 2 | Test 3 | Test 1 | Test 2 | Test 3 | Test 1 | Test 2 | Test 3 |
| 1 | 37 | 38 | 37 | 41 | 41 | 40 | 41 | 42 | 41 |
| 2 | 42 | 41 | 43 | 42 | 42 | 42 | 43 | 42 | 43 |
| 3 | 30 | 31 | 31 | 31 | 31 | 31 | 29 | 30 | 28 |
| 4 | 42 | 43 | 42 | 43 | 43 | 43 | 42 | 42 | 42 |
| 5 | 28 | 30 | 29 | 29 | 30 | 29 | 31 | 29 | 29 |
| 6 | 42 | 42 | 43 | 45 | 45 | 45 | 44 | 46 | 45 |
| 7 | 25 | 26 | 27 | 28 | 28 | 30 | 29 | 27 | 27 |
| 8 | 40 | 40 | 40 | 43 | 42 | 42 | 43 | 43 | 41 |
| 9 | 25 | 25 | 25 | 27 | 29 | 28 | 26 | 26 | 26 |
| 10 | 35 | 34 | 34 | 35 | 35 | 34 | 35 | 34 | 35 |

Using ANOVA R&R analysis evaluate the capability of the measurement system by assuming that parts and operators possibly interact. See Exercise 11.27 for the case where parts and operators do not interact (additive ANOVA model).

We will first analyze the problem as a two-way ANOVA ( ◀ RandR2.m) and then, for comparison, provide MATLAB's output from the function gagerr.

```
%model with interaction
impedance = [...
    37 38 37 41 41 40 41 42 41 42 41 43 42 42 42 43 42 43    ...
    30 31 31 31 31 31 29 30 28 42 43 42 43 43 43 42 42 42    ...
    28 30 29 29 30 29 31 29 29 42 42 43 45 45 45 44 46 45    ...
    25 26 27 28 28 30 29 27 27 40 40 40 43 42 42 43 43 41    ...
    25 25 25 27 29 28 26 26 26 35 34 34 35 35 34 35 34 35]'  ;

% forming part and operator vectors.
a = repmat([1:10],9,1);  part = a(:);
b = repmat([1:3], 3,1);  operator = repmat(b(:),10,1);

[p table stats terms] = anovan( impedance,{part, operator},...
  'model','interaction','varnames',{'part','operator'} )
```

```
MSE = table{5,5} %0.5111
MSPO =table{4,5} %2.6951
MSO = table{3,5} %19.6333
MSP = table{2,5} %437.3284
p = stats.nlevels(1); o = stats.nlevels(2);
n=length(impedance)/(p * o);%p=10, O=3, n=3

s2Part = (MSP - MSPO)/(o * n)   %48.2926
s2Oper = (MSO - MSPO)/(p * n)   %0.5646
s2PartOper = (MSPO - MSE)/n     %0.7280
s2 = MSE                        %0.5111

s2Repeat = s2                   %0.5111
s2Repro = s2Oper + s2PartOper   %1.2926
s2Gage = s2Repeat + s2Repro     %1.8037
s2Tot = s2Gage  + s2Part        %50.0963

%percent variation due to part
ps2Part = s2Part/s2Tot  %0.9640

% signal-to-noise ratio; > 5 measuring system is capable
snr = sqrt( 2 * ps2Part/(1-ps2Part))   %7.3177
% snr rounded is NDC (number of distinct categories).
ndc = round(snr)   %7


% percent of R&R variability
prr = sqrt(s2Gage/s2Tot)  %0.1897
```

Therefore, the measuring system is capable by $NDC = 7 \geq 5$ but falls in the "gray" zone $(10, 30)$ according to the $PRR$ measure.

MATLAB's built-in function gagerr produces a detailed output:

```
gagerr(impedance,{part, operator},'model','interaction')

% Source           Variance   %Variance   sigma 5.15*sigma 5.15*sigma
% ============================================================================
% Gage R&R             1.80       3.60     1.34      6.92       18.97
%    Repeatability     0.51       1.02     0.71      3.68       10.10
%    Reproducibility   1.29       2.58     1.14      5.86       16.06
%       Operator       0.56       1.13     0.75      3.87       10.62
%       Part*Operator  0.73       1.45     0.85      4.39       12.05
% Part                48.29      96.40     6.95     35.79       98.18
% Total               50.10     100.00     7.08     36.45
% ----------------------------------------------------------------------------
%            Number of distinct categories (NDC)     :   7
%            % of Gage R&R of total variations (PRR):  18.97
```

When operators and parts do not interact, an additive ANOVA model,

$$y_{ijk} = \mu + P_i + O_j + \epsilon_{ijk}, \quad i = 1,\ldots,p; \; j = 1,\ldots,o; \; k = 1,\ldots,n,$$

should be used. In this case, the estimators of variances are

$$\hat{\sigma}_P^2 = \frac{MSP - MSE}{on},$$

$$\hat{\sigma}_O^2 = \frac{MSO - MSE}{pn},$$

$$\hat{\sigma}^2 = MSE,$$

and $\hat{\sigma}_{Reprod}^2$ is simply reduced to $\hat{\sigma}_O^2$.

Exercise 11.27 solves the problem in Example 11.14 without the *PO*-interaction term and compares the analysis with the output from `gagerr`.

## 11.11  Testing Equality of Several Proportions

An ANOVA-type hypothesis can be considered in the context of several proportions. Consider $k$ independent populations from which $k$ samples of size $n_1, n_2, \ldots, n_k$ are taken. We record a binary attribute, say, 1 or 0. Let $X_1, X_2, \ldots, X_k$ be the observed number of 1s and $\hat{p}_1 = X_1/n_1, \ldots, \hat{p}_k = X_k/n_k$ the sample proportions.

To test $H_0 : p_1 = p_2 = \cdots = p_k$ against the general alternative $H_1 = H_0^c$, the statistic

$$\chi^2 = \sum_{i=1}^{k} \frac{(X_i - n_i \overline{p})^2}{n_i \overline{p}(1 - \overline{p})}$$

is formed. Here

$$\overline{p} = \frac{X_1 + X_2 + \cdots + X_k}{n_1 + n_2 + \cdots + n_k}$$

is a pooled sample proportion (under $H_0$ all proportions are the same and equal to $p$, and $\overline{p}$ is the best estimator). The statistic $\chi^2$ has a $\chi^2$-distribution with $k - 1$ degrees of freedom.

If $H_0$ is rejected, a Marascuillo procedure (Marascuillo and Serlin, 1988) can be applied to compare individual pairs of population proportions. For $k$ populations there are $k(k - 1)/2$ tests, and the two proportions $p_i$ and $p_j$ are different if

$$|\hat{p}_i - \hat{p}_j| > \sqrt{\chi_{k-1, 1-\alpha}^2} \times \sqrt{\hat{p}_i(1 - \hat{p}_i)/n_i + \hat{p}_j(1 - \hat{p}_j)/n_j}. \qquad (11.13)$$

*Example 11.15.* **Gender and Hair Color.** Zar (1996, p. 484) provides data on gender proportions in samples of subjects grouped by the color of their hair.

|        | Hair color |       |       |     |
|--------|------------|-------|-------|-----|
| Gender | Black      | Brown | Blond | Red |
| Male   | 32         | 43    | 16    | 9   |
| Female | 55         | 65    | 64    | 16  |
| Total  | 87         | 108   | 80    | 25  |

We will test the hypothesis that population proportions of males are the same for the four groups.

```
%Gender Proportions and Hair Color
Xi = [32  43  16  9];
ni = [87 108  80 25];
pi = Xi./ni      %0.3678    0.3981    0.2000    0.3600
pbar = sum(Xi)/sum(ni)      %0.3333
chi2 = sum(  (Xi - ni*pbar).^2./(ni * pbar * (1-pbar)) ) %8.9872
pval = 1-chi2cdf(chi2, 4-1)   %0.0295
```

Thus, with $p$-value of about 3%, the hypothesis of homogeneity of population proportions is rejected. Using the condition in (11.13) show that the difference between $\hat{p}_2$ and $\hat{p}_3$ is significant and responsible for rejecting $H_0$.

**Remark.** We will se later (Chap. 14) that this test and the test for homogeneity in $2 \times c$ contingency tables are equivalent. Here we assumed that the sampling design involved fixed totals $n_i$. If the sampling was fully random (i.e., no totals for hair color nor males/females were prespecified) then $H_0$ would be the hypothesis of independence between gender and hair color. ✐

For Tukey-type multiple comparisons and testing proportion trends, see Zar (1996) and Conover (1999).

## 11.12 Testing the Equality of Several Poisson Means*

When in each of $k$ treatments observations are multiple counts, the standard ANOVA is often inappropriately applied to test for the equality of means. For counting observations the Poisson model is more adequate than the normal, and using a standard ANOVA methodology may be problematic. For example, when an ANOVA hypothesis of equality of means is rejected, but the observations are in fact Poisson, multiple comparisons may be invalid due to unequal treatment variances (which for Poisson observations are equal to the means). Next, we describe an approach that is appropriate for Poisson observations.

Suppose that in treatment $i$ we observe $n_i$ Poisson counts,

$$X_{ij} \sim \mathcal{P}oi(\lambda_i), \ i = 1,\dots,k; \ j = 1,\dots,n_i.$$

Denote by $O_i$ the sum of all counts in treatment $i$, $O_i = \sum_{j=1}^{n_i} X_{ij}$, and by $O$ the sum of counts across all treatments, $O = \sum_{i=1}^{k} O_i$. The total number of observations is $N = \sum_{i=1}^{k} n_i$.

One can show that the distribution of the vector $(O_1, O_2, \ldots, O_k)$, given the total sum $O$, is multinomial (p. 155),

$$(O_1, O_2, \ldots, O_k | O) \sim \mathcal{M}n(O, \boldsymbol{p}),$$

where $\boldsymbol{p} = (p_1, p_2, \ldots, p_k)$ is defined via

$$p_i = \frac{n_i \lambda_i}{\sum_{i=1}^{k} n_i \lambda_i}.$$

Thus, when the hypothesis $H_0 : \lambda_1 = \lambda_2 = \cdots = \lambda_k$ is true, then $p_i = n_i/N$.

Suppose that counts $O_i$ are observed. Define expected counts $E_i$ as

$$E_i = p_i \times O = \frac{n_i}{N} \times O, \ i = 1, \ldots, k.$$

Then

$$\chi^2 = \sum_{i=1}^{k} \frac{(O_i - E_i)^2}{E_i}$$

is Pearson's test statistic for testing $H_0$. When $H_0$ is true, it has an approximately $\chi^2$-distribution with $k - 1$ degrees of freedom. Alternatively, one can use the likelihood statistic

$$G^2 = 2 \sum_{i=1}^{k} O_i \log \frac{O_i}{E_i},$$

which under $H_0$ has the same distribution as the $\chi^2$. Large values of $\chi^2$ or $G^2$ are critical for $H_0$.

*Example 11.16.* Assume that in an experiment three treatments are applied on 20 plates populated with a large number of cells, the first treatment on 7, the second on 5, and the third on 8 plates.

The following table gives the numbers of cells per plate that responded to the treatments.

| Treatment 1 | Treatment 2 | Treatment 3 |
|-------------|-------------|-------------|
| 1  6  4  2  3 | 5  5  9  2  7 | 2  3  0  2  1 |
| 8  2 |  | 4  5  2 |

We assume a Poisson model and wish to test the equality of mean counts, $H_0 : \lambda_1 = \lambda_2 = \lambda_3$.

The solution is provided in ⬦ poissonmeans.m. The hypothesis of the equality of means is rejected with a $p$-value of about 1.3%. Note that standard

ANOVA fails to reject $H_0$, for the $p$-value is 6.15%. This example demonstrates the inadequacy of standard ANOVA for this kind of data.

```
ncells =   [1 6 4 2 3 8 2  5 5 9 2 7  2 3 0 2 1 4 5 2]';
agent =    [1 1 1 1 1 1 1  2 2 2 2 2  3 3 3 3 3 3 3 3]';
ncells1=ncells(agent==1)
ncells2=ncells(agent==2)
ncells3=ncells(agent==3)
k = 3; %treatments
n1 = length(ncells1); n2 = length(ncells2); n3= length(ncells3);
ni=[n1 n2 n3]     %7     5     8
N = sum(ni) %20
O1 = sum(ncells1); O2 = sum(ncells2); O3 = sum(ncells3);
Oi=[O1  O2  O3]  %26     28     19
O = sum(Oi) %73
%expected
Ei = O .* ni/N        %25.55    18.25    29.2
%Poisson chi2
chi2 = sum( (Oi - Ei).^2 ./ Ei )      %8.7798
%Likelihood G2
G2 = 2 * sum( Oi .* log(Oi./Ei ) )    %8.5484
pvalchi2 = 1 - chi2cdf(chi2, k-1)     %0.0124
pvalG2 = 1 - chi2cdf(G2, k-1)         %0.0139

% If the problem is treated as ANOVA
[panova table] = anova1(ncells, agent)

% panova = 0.0615
% table =
%      'Source'   'SS'          'df'   'MS'       'F'        'Prob>F'
%      'Groups'   [ 32.0464]    [ 2]   [16.0232]  [3.3016]   [0.0615]
%      'Error'    [ 82.5036]    [17]   [ 4.8532]  []         []
%      'Total'    [114.5500]    [19]   []         []         []
```

## 11.13 Exercises

11.1. **Nematodes.** Some varieties of nematodes (roundworms that live in the soil and are frequently so small that they are invisible to the naked eye) feed on the roots of lawn grasses and crops such as strawberries and tomatoes. This pest, which is particularly troublesome in warm climates, can be treated by the application of nematocides. However, because of the size of the worms, it is very difficult to measure the effectiveness of these pesticides directly. To compare four nematocides, the yields of equal-size plots of one variety of tomatoes were collected. The data (yields in pounds per plot) are shown in the table.

| Nematocide A | Nematocide B | Nematocide C | Nematocide D |
|:---:|:---:|:---:|:---:|
| 18.6 | 18.7 | 19.4 | 19.0 |
| 18.4 | 19.0 | 18.9 | 18.8 |
| 18.4 | 18.9 | 19.5 | 18.6 |
| 18.5 | 18.5 | 19.1 | 18.7 |
| 17.9 |      | 18.5 |      |

(a) Write a statistical model for ANOVA and state $H_0$ and $H_1$ in terms of your model.
(b) What is your decision if $\alpha = 0.05$?
(c) For what values of $\alpha$ will your decision be different than that in (b)?

11.2. **Cell Folate Levels in Cardiac Bypass Surgery.** Altman (1991, p. 208) provides data on 22 patients undergoing cardiac bypass surgery. The patients were randomized to one of three groups receiving the following treatments:
**Treatment 1.** Patients received a 50% nitrous oxide and 50% oxygen mixture continuously for 24 hours.
**Treatment 2.** Patients received a 50% nitrous oxide and 50% oxygen mixture only during the operation.
**Treatment 3.** Patients received no nitrous oxide but received 35 to 50% oxygen for 24 hours.
The measured responses are red cell folate levels (ng/ml) for the three groups after 24 hours of ventilation.

| Treat1 | 243 251 275 291 347 354 380 392 |
|:---|:---|
| Treat2 | 206 210 226 249 255 273 285 295 309 |
| Treat3 | 241 258 270 293 328 |

The question of interest is whether the three ventilation methods result in a different mean red cell folate level.
If the hypothesis of equality of treatment means is rejected, which means are significantly different?

11.3. **Computer Games.** In Exercise 7.22, Mental Rotations Test scores were provided for three groups of children:
Group 1 ("Factory" computer game): $\overline{X}_1 = 22.47, s_1 = 9.44, n_1 = 19$.
Group 2 ("Stellar" computer game): $\overline{X}_2 = 22.68, s_2 = 8.37, n_2 = 19$.
Control (no computer game): $\overline{X}_3 = 18.63, s_3 = 11.13, n_3 = 19$.
Assuming a normal distribution of scores in the population and equal population variances, test the hypothesis that the population means are the same at a 5% significance level.

11.4. **MTHFR C677T Genotype and Levels of Homocysteine and Folate.** A study by Ozturk et al. (2005) considered the association of methylenetetrahydrofolate reductase (MTHFR) C677T polymorphisms with levels of homocysteine and folate. A total of $N = 815$ middle-aged and elderly subjects were stratified by MTHFR C677T genotype ($k = 3$) and their measurements summarized as in the table below.

| Characteristics | CC: $n_1 = 312$ $\overline{X}_1$ ($s_1$) | CT: $n_2 = 378$ $\overline{X}_2$ ($s_2$) | TT: $n_3 = 125$ $\overline{X}_3$ ($s_3$) |
|---|---|---|---|
| Homocysteine, nmol/l | 14.1 (1.9) | 14.2 (2.3) | 15.3 (3.0) |
| Red blood cell folate, nmol/l | 715 (258) | 661 (236) | 750 (348) |
| Serum folate, nmol/l | 13.1 (4.7) | 12.3 (4.2) | 11.4 (4.4) |

(a) Using one-way ANOVA, test the hypothesis that the population homocysteine levels for the three genotype groups are the same. Use $\alpha = 0.05$. *Hint:* Since the raw data are not given, calculate

$$MSTr = \frac{1}{k-1} \sum_{i=1}^{k} n_i (\overline{X}_i - \overline{X})^2 \quad \text{and} \quad MSE = \frac{1}{N-k} \sum_{i=1}^{k} (n_i - 1)s_i^2,$$

for

$$N = n_1 + \cdots + n_k \quad \text{and} \quad \overline{X} = \frac{n_1 \overline{X}_1 + \cdots + n_k \overline{X}_k}{N}.$$

Statistic $F$ is the ratio $MSTr/MSE$. It has $k-1$ and $N-k$ degrees of freedom.
(b) For red blood cell folate measurements complete the ANOVA table.

```
SS                   DF              MS                F          p
===============================================================================
SSTr=                DF1=            MSTr=469904.065   F=         p=
SSE =                DF2=            MSE = 69846.911
=============================
SST=                 DF=
```

11.5. **Beetles.** The following data were extracted from a more extensive study by Sokal and Karten (1964). The data represent mean dry weights (in milligrams) of three genotypes of beetles, *Tribolium castaneum*, reared at a density of 20 beetles per gram of flour. The four independent measurements for each genotype are recorded.

| | Genotypes | |
|---|---|---|
| ++ | +b | bb |
| 0.958 | 0.986 | 0.925 |
| 0.971 | 1.051 | 0.952 |
| 0.927 | 0.891 | 0.829 |
| 0.971 | 1.010 | 0.955 |

Using one-way ANOVA, test whether the genotypes differ in mean dry weight. Take $\alpha = 0.01$.

11.6. **ANOVA Table from Summary Statistics.** When accounts of one-way ANOVA designs are given in journal articles or technical reports, the data and the ANOVA table are often omitted. Instead, means and standard deviations for each treatment group are given (along with the $F$-statistic, its $p$-value, or the decision to reject or not). One can build the ANOVA table from this summary information (and thus verify the author's interpretation of the data).

The results below are from an experiment with $n_1 = n_2 = n_3 = n_4 = 10$ observations on each of $k = 4$ groups.

| Treatment $n_i$ | | $\overline{X}_i$ | $s_i$ |
|---|---|---|---|
| 1 | 10 | 100.40 | 11.68 |
| 2 | 10 | 103.00 | 11.58 |
| 3 | 10 | 107.10 | 10.05 |
| 4 | 10 | 114.80 | 10.61 |

(a) Write the model for this experiment and state the null hypothesis in terms of the parameters of the model.

(b) Use the information in the table above to show that $SST_r = 1186$ and $SSE = 4357$.

(c) Construct the ANOVA table for this experiment, with standard columns $SS$, $df$, $MS$, $F$, and $p$-value.

Hint: $SST_r = n_1(\overline{X}_1 - \overline{X})^2 + n_2(\overline{X}_2 - \overline{X})^2 + n_3(\overline{X}_3 - \overline{X})^2 + n_4(\overline{X}_4 - \overline{X})^2$ and $SSE = (n_1 - 1)s_1^2 + (n_2 - 1)s_2^2 + (n_3 - 1)s_3^2 + (n_4 - 1)s_4^2$.

Part (c) shows that it is possible to use treatment means and standard deviations to reconstruct the ANOVA table, find its $F$-statistic, and test the null hypothesis. However, without knowing the individual data values, one cannot carry out some other important statistical analyses (residual analysis, for example).

11.7. **Protein Content in Milk for Three Diets.** The data from Diggle et al. (1994) are used to compare the content of protein in milk in 79 cows randomly assigned to three different diets: barley, a barley-and-lupins mixture, and lupins alone.

The original data set is longitudinal in nature (protein monitored weekly for several months), but for the purpose of this exercise we took the protein concentration at 6 weeks following calving for each cow.

```
b=[3.7700    3.0000    2.9300    3.0500    3.6000    3.9200...
   3.6600    3.4700    3.2100    3.3400    3.5000    3.7300...
   3.4900    3.1600    3.4500    3.5200    3.1500    3.4200...
   3.6200    3.5700    3.6500    3.7100    3.5700    3.6300...
   3.6000];

m=[3.4000    3.8000    3.2900    3.7100    3.2800    3.3800...
   3.5700    2.9000    3.5500    3.5500    3.0400    3.4000...
   3.1500    3.1300    3.2500    3.1500    3.1000    3.1700...
   3.5000    3.5700    3.4700    3.4500    3.2600    3.2400...
   3.7000    3.0500    3.5400];
```

```
l=[3.0700    3.1200    2.8700    3.1100    3.0200    3.3800...
   3.0800    3.3000    3.0800    3.5200    3.2500    3.5700...
   3.3800    3.0000    3.0300    3.0600    3.9600    2.8300...
   2.7400    3.1300    3.0500    3.5500    3.6500    3.2700...
   3.2000    3.2700    3.7000];

resp = [b'; m'; l'];      %single column vector of all responses
class = [ones(25,1); 2*ones(27,1); 3*ones(27,1)]; %class vector
[pval, table, stats] = anoval(resp, class)
[cintsdiff,means] = multcompare(stats)    %default  comparisons
```

Partial output is given below.
(a) Fill in the empty spaces in the output.

```
%pval =
%     0.0053
%
%table
%'Source'    'SS'          'df'        'MS'        'F'        'Prob>F'
%'Groups'    [_____]    [_____]    [0.3735]    [_____]  [_____]
%'Error'     [_____]    [_____]    [_____]
%'Total'     [5.8056]     [_____]
%
%stats =
%     gnames: 3x1 cell
%          n: [25 27 27]
%     source: 'anoval'
%      means: [3.4688 3.3556 3.2293]
%         df: 76
%          s: 0.2580
%
% cintsdiff =
%    1.0000    2.0000    -0.0579    0.1132    0.2844
%    1.0000    3.0000     0.0684    0.2395    0.4107
%    2.0000    3.0000    -0.0416    0.1263    0.2941
%
% means =
%    3.4688    0.0516
%    3.3556    0.0497
%    3.2293    0.0497
```

(b) What is $H_0$ here and is it rejected? Use $\alpha = 5\%$.
(c) From the output cintsdiff discuss how the population means differ.

11.8. **Tasmanian Clouds.** The data [icon] clouds.txt provided by OzDASL were
collected in a cloud-seeding experiment in Tasmania between mid-1964 and
January 1971. Analysis of these data is discussed in Miller et al. (1979).
The rainfalls are period rainfalls in inches. Variables TE and TW are the
east and west target areas, respectively, while CN, CS, and CNW are the corresponding rainfalls in the north, south, and northwest control areas, respectively. S stands for seeded and U for unseeded. Variables C and T are
averages of control and target rainfalls. Variable DIFF is the difference T-C.

(a) Use additive two-way ANOVA to estimate and test treatment effects
Season and Seeded.

(b) Repeat the analysis from (a) after adding the interaction term.

11.9. **Clover Varieties.** Six plots each of five varieties of clover were planted
at the Danbury Experimental Station in North Carolina. Yields in tons per
acre were as follows:

| Variety | Yield |
|---------|-------|
| Spanish | 2.79, 2.26, 3.09, 3.01, 2.56, 2.82 |
| Evergreen | 1.93, 2.07, 2.45, 2.20, 1.86, 2.44 |
| Commercial Yellow | 2.76, 2.34, 1.87, 2.55, 2.80, 2.21 |
| Madrid | 2.31, 2.30, 2.49, 2.26, 2.69, 2.17 |
| Wisconsin A46 | 2.39,  2.05, 2.68, 2.96, 3.04, 2.60 |

(a) Test the hypothesis that the mean yields for the five clover varieties are
the same. Take $\alpha = 5\%$. What happens if your $\alpha$ is 1%.

(b) Which means are different at a 5% level?

(c) Is the hypothesis $H_0 : 3(\mu_1 + \mu_5) = 2(\mu_2 + \mu_3 + \mu_4)$ a contrast? Why? If yes,
test it against the two-sided alternative, at an $\alpha = 5\%$ level.

11.10. **Cochlear Implants.** A cochlear implant is a small, complex electronic device that can help to provide a sense of sound to a person who is profoundly
deaf or severely hard of hearing. The implant consists of an external portion
that sits behind the ear and a second portion that is surgically placed under
the skin (Fig. 11.11). Traditional hearing aids amplify sounds so they may
be detected by damaged ears. Cochlear implants bypass damaged portions
of the ear and directly stimulate the auditory nerve. Signals generated by
the implant are sent by way of the auditory nerve to the brain, which recog-
nizes the signals as sound. Hearing through a cochlear implant is different
from normal hearing and takes time to learn or relearn. However, it allows
many people to recognize warning signals, understand other sounds in the
environment, and enjoy a conversation in person or by telephone.

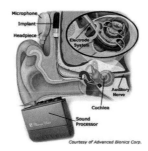

**Fig. 11.11** Cochlear implant.

Eighty-one profoundly deaf subjects in this experiment received one of three different brands of cochlear implant (A/B/G3). Brand G3 is a third-generation device, while brands A and B are second-generation devices. The research question was to determine whether the three brands differed in levels of speech recognition.

The data file is given in 📊 deaf.xlsx. The variables are as follows:

| Name   | Description |
|--------|-------------|
| ID     | Patient ID number |
| Age    | Patient's age at the time of implantation |
| Device | Type of cochlear implant (A/B/G3) |
| YrDeaf | Years of profound deafness prior to implantation |
| CS     | Consonant recognition score, sound only |
| CV     | Consonant recognition score, visual only (no sound) |
| CSV    | Consonant recognition score, sound and vision |
| SNT    | Sentence understanding score |
| VOW    | Vowel recognition score |
| WRD    | Word recognition score |
| PHN    | Phoneme recognition score |

Run three separate ANOVAs with three response variables, which are different tests of speech recognition:

(a) CSV, audiovisual consonant recognition (subjects hear triads like "ABA," "ATA," and "AFA" and have to pick the correct consonant on a touch screen);

(b) PHN, phoneme understanding (number of correct phonemes in random 5- to 7-word sentences like "The boy threw the ball."); and

(c) WRD, word recognition (number of words recognized in a list of random, unrelated monosyllabic words "ball," "dog," etc.).

11.11. **Bees.** The data for this problem are taken from Park (1932), who investigated changes in the concentration of nectar in the honey sac of the bee. Syrup of approx. 40% concentration was fed to the bees. The concentration in their honey sacs was determined upon their arrival at the hive. The decreases recorded in the table are classified according to date, both day (September 1931) and time of day being differentiated. The question to be answered is this: Were significant differences introduced by changes in the time of gathering the data, or may the six groups be considered random samples from a homogeneous population?

| 3 | 3 | 3 | 10 | 11 | 12 |
| 10:20 | 11:10 | 2:20 | 4:00 | 1:10 | 10:30 |
|---|---|---|---|---|---|
| 1.1 | 1.0 | 0.6 | −1.6 | 1.1 | 2.5 |
| 1.0 | 0.6 | 0.3 | 0.8 | 0.5 | 0.6 |
| 0.9 | 1.0 | −0.1 | 2.1 | 2.2 | 1.1 |
| 1.1 | 0.4 | 0.0 | 1.1 | 1.1 | 0.6 |
| 0.9 | 0.4 | 1.5 | 0.6 | 0.4 | 1.8 |
| 1.1 | 0.9 | 0.9 | 0.6 | −2.0 | 0.6 |
| 0.6 | 0.6 | 0.3 | 0.6 | 1.4 | 1.2 |
| 0.5 | 0.4 | 0.2 | 0.2 | −0.4 | 1.2 |
| 0.5 | 1.1 | 0.4 | 0.8 | 2.4 | 0.4 |
| 0.7 | 0.7 | 0.4 | 0.6 | 0.0 | 1.0 |

11.12. **SiRstv: NIST's Silicon Resistivity Data.** Measurements of bulk resistivity of silicon wafers were made at NIST with five probing instruments (columns in the data matrix) on each of 5 days (rows in the data matrix). The wafers were doped with phosphorous by neutron transmutation doping in order to have nominal resistibility of 200 ohm cm. Measurements were carried out with four-point DC probes according to ASTM Standard F84-93 and described in Ehrstein and Croarkin (1984).

| 1 | 2 | 3 | 4 | 5 |
|---|---|---|---|---|
| 196.3052 | 196.3042 | 196.1303 | 196.2795 | 196.2119 |
| 196.1240 | 196.3825 | 196.2005 | 196.1748 | 196.1051 |
| 196.1890 | 196.1669 | 196.2889 | 196.1494 | 196.1850 |
| 196.2569 | 196.3257 | 196.0343 | 196.1485 | 196.0052 |
| 196.3403 | 196.0422 | 196.1811 | 195.9885 | 196.2090 |

(a) Test the hypothesis that the population means of measurements produced by these five instruments are the same at $\alpha = 5\%$. Use MATLAB to produce an ANOVA table.

(b) Pretend now that some measurements for the first and fifth instruments are misrecorded (in italics) and that the modified table looks like this:

| 1 | 2 | 3 | 4 | 5 |
|---|---|---|---|---|
| 196.3052 | 196.3042 | 196.1303 | 196.2795 | *196.1119* |
| *196.2240* | 196.3825 | 196.2005 | 196.1748 | 196.1051 |
| *196.2890* | 196.1669 | 196.2889 | 196.1494 | 196.1850 |
| 196.2569 | 196.3257 | 196.0343 | 196.1485 | 196.0052 |
| 196.3403 | 196.0422 | 196.1811 | 195.9885 | *196.1090* |

Test now the same hypothesis as in (a). If $H_0$ is rejected, perform Tukey's multiple comparisons procedure (at Tukey's family error rate of 5%).

11.13. **Dorsal Spines of** *Gasterosteus aculeatus*. Bell and Foster (1994) were interested in the effect of predators on dorsal spine length evolution in *Gas-*

*terosteus aculeatus* (threespine stickleback, Fig. 11.12). Dorsal spines are thought to act as an antipredator mechanism.

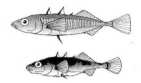

**Fig. 11.12** *Gasterosteus aculeatus.*

To examine this issue, researchers sampled eight sticklebacks from each of Benka Lake (no predators), Garden Bay Lake (some predators), and Big Lake (lots of predators). Their observations on spine length (in millimeters) are provided in the following table.

| Benka Lake | Garden Bay Lake | Big Lake |
|:---:|:---:|:---:|
| 4.2 | 4.4 | 4.9 |
| 4.1 | 4.6 | 4.6 |
| 4.2 | 4.5 | 4.3 |
| 4.3 | 4.2 | 4.9 |
| 4.5 | 4.4 | 4.7 |
| 4.4 | 4.2 | 4.4 |
| 4.5 | 4.5 | 4.5 |
| 4.3 | 4.7 | 4.4 |

They would like to know if spine lengths differ among these three populations and apply ANOVA with lakes as "treatments."

(a) Test the hypothesis that the mean lengths in the three populations are the same at level $\alpha = 0.05$. State your conclusion in terms of the problem.

(b) Would you change the decision in (a) if $\alpha = 0.01$? Explain why or why not.

11.14. **Incomplete ANOVA Table.** In the context of balanced two factor ANOVA recover entries $a$–$m$.

| Source of Variation | Sum of Squares | Degrees of Freedom | Mean Square | F | $p$-value |
|---|---|---|---|---|---|
| A | 256.12 | 2 | $a$ | 4.18 | $b$ |
| B | $c$ | 3 | 12.14 | $d$ | $e$ |
| A×B | 217.77 | $f$ | $g$ | $h$ | $i$ |
| Error | $j$ | $k$ | $l$ | | |
| Total | $m$ | 119 | | | |

11.15. **Maternal Behavior in Rats.** To investigate the maternal behavior of laboratory rats, researchers separated rat pups from their mother and recorded the time required for the mother to retrieve the pups. The study was run with 5-, 20-, and 35-day-old pups, six in each group. The pups were moved to a fixed distance from the mother and the time of retrieval (in seconds) was recorded (Montgomery, 1984):

| 5 days  | 15 10 25 15 20 18 |
|---------|-------------------|
| 20 days | 30 15 20 25 23 20 |
| 35 days | 40 35 50 43 45 40 |

State the inferential problem, pose the hypotheses, check for sphericity, and do the test at $\alpha = 0.05$. State your conclusions. .

11.16. **Comparing Dialysis Treatments.** In Example pretend that the three columns of measurements are independent, that is, that 24 independent patients were randomly assigned to one of the three treatments, 8 patients to each treatment. Test the null hypothesis that there is no difference in mean weight change among the treatments. Use $\alpha = 0.05$. Compare results with those in Example and comment.

11.17. **Materials Scientist and Assessing Tensile Strength.** A materials scientist wishes to test the effect of four chemical agents on the strength of a particular type of cloth. Because there might be variability from one bolt to another, the scientist decides to use a randomized block design, with the bolts of cloth considered as blocks. She selects five bolts and applies all four chemicals in random order to each bolt. The resulting tensile strengths follow. Analyze the data and draw appropriate conclusions.

|          | Bolt            |
|----------|-----------------|
| Chemical | 1  2  3  4  5   |
| 1        | 73 68 74 71 67  |
| 2        | 73 67 75 72 70  |
| 3        | 75 68 78 73 68  |
| 4        | 73 71 75 75 69  |

11.18. **Oscilloscope.** Montgomery (1984) discusses an experiment conducted to study the influence of operating temperature and three types of face-plate glass in the light output of an oscilloscope tube. The following data are collected.

|            | Temperature |      |      |
| Glass type | 100         | 125  | 150  |
| --- | --- | --- | --- |
|            | 580 | 1090 | 1392 |
| 1          | 568 | 1087 | 1380 |
|            | 570 | 1085 | 1386 |
|            |     |      |      |
|            | 550 | 1070 | 1328 |
| 2          | 530 | 1035 | 1312 |
|            | 579 | 1000 | 1299 |
|            |     |      |      |
|            | 546 | 1045 | 867  |
| 3          | 575 | 1053 | 904  |
|            | 599 | 1066 | 889  |

11.19. **Magnesium Ammonium Phosphate and Chrysanthemums.** Walpole et al. (2007) provide data on a study "Effect of magnesium ammonium phosphate on height of chrysanthemums" that was conducted at George Mason University to determine a possible optimum level of fertilization, based on the enhanced vertical growth response of the chrysanthemums. Forty chrysanthemums seedlings were assigned to 4 groups, each containing 10 plants. Each was planted in a similar pot containing a uniform growth medium. An increasing concentration of $MgNH_4PO_4$, measured in grams per bushel, was added to each plant. The 4 groups of plants were grown under uniform conditions in a greenhouse for a period of 4 weeks. The treatments and the respective changes in heights, measured in centimeters, are given in the following table:

| Treatment |          |          |          |
| 50 g/bu | 100 g/bu | 200 g/bu | 400 g/bu |
| --- | --- | --- | --- |
| 13.2 | 16.0 | 7.8  | 21.0 |
| 12.4 | 12.6 | 14.4 | 14.8 |
| 12.8 | 14.8 | 20.0 | 19.1 |
| 17.2 | 13.0 | 15.8 | 15.8 |
| 13.0 | 14.0 | 17.0 | 18.0 |
| 14.0 | 23.6 | 27.0 | 26.0 |
| 14.2 | 14.0 | 19.6 | 21.1 |
| 21.6 | 17.0 | 18.0 | 22.0 |
| 15.0 | 22.2 | 20.2 | 25.0 |
| 20.0 | 24.4 | 23.2 | 18.2 |

(a) Do different concentrations of $MgNH_4PO_4$ affect the average attained height of chrysanthemums? Test the hypothesis at the level $\alpha = 0.10$.
(b) For $\alpha = 10\%$ perform multiple comparisons using multcompare.
(c) Find the 90% confidence interval for the contrast $\mu_1 - \mu_2 - \mu_3 + \mu_4$.

11.20. **Color Attraction for** *Oulema melanopus.* Some colors are more attractive to insects than others. Wilson and Shade (1967) conducted an experiment aimed at determining the best color for attracting cereal leaf beetles
(*Oulema melanopus*). Six boards in each of four selected colors (lemon yellow, white, green, and blue) were placed in a field of oats in July. The following table (modified from Wilson and Shade, 1967) gives data on the number
of cereal leaf beetles trapped.

| Board color | Insects trapped |
|---|---|
| Lemon yellow | 45 59 48 46 38 47 |
| White | 21 12 14 17 13 17 |
| Green | 37 32 15 25 39 41 |
| Blue | 16 11 20 21 14  7 |

(a) Based on computer output, state your conclusions about the attractiveness of these colors to the beetles. See also Fig. 11.13a.
In MATLAB:

```
ntrap=[ 45,  59,  48,  46,  38,  47,  21,  12,  14,  17,...
        13,  17,  37,  32,  15,  25,  39,  41,  16,  11,...
        20,  21,  14,   7];
color={'ly','ly','ly','ly','ly','ly',...
       'wh','wh','wh','wh','wh','wh',...
       'gr','gr','gr','gr','gr','gr',...
       'bl','bl','bl','bl','bl','bl'};
[p, table, stats] = anova1(ntrap, color)
multcompare(stats)

%
%'Source'   'SS'            'df'        'MS'          'F'        'Prob>F'
%'Groups'   [4.2185e+003]   [ 3]   [1.4062e+003]  [30.5519]  [1.1510e-007]
%'Error'    [  920.5000]    [20]   [  46.0250]    []         []
%'Total'    [5.1390e+003]   [23]                  []         []
%
%Pairwise Comparisons
%    1.0000    2.0000    20.5370    31.5000    42.4630
%    1.0000    3.0000     4.7037    15.6667    26.6297
%    1.0000    4.0000    21.3703    32.3333    43.2963
%    2.0000    3.0000   -26.7963   -15.8333    -4.8703
%    2.0000    4.0000   -10.1297     0.8333    11.7963
%    3.0000    4.0000     5.7037    16.6667    27.6297
```

(b) The null hypothesis is rejected. Which means differ? See Fig. 11.13b.
(c) Perform an ANOM analysis. Which means are different from the (overall) grand mean?

11.21. **Raynaud's Phenomenon.** Raynaud's phenomenon is a condition resulting in a discoloration of the fingers or toes after exposure to temperature

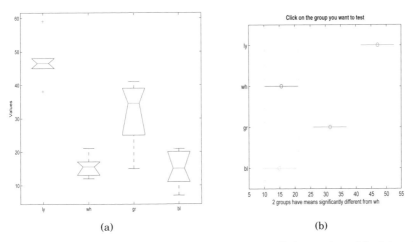

(a)                                                                    (b)

**Fig. 11.13** (a) Barplots of ntrap for the four treatments. (b) A snapshot of the interactive plot for multiple comparisons.

changes or emotional events. Skin discoloration is caused by an abnormal spasm of the blood vessels, diminishing blood supply to the local tissues. Kahan et al. (1987) investigated the efficacy of the calcium-channel blocker nicardipine in the treatment of Raynaud's phenomenon. This efficacy was assessed in a prospective, double-blind, randomized, crossover trial in 20 patients. Each patient received 20 mg nicardipine or placebo three times a day for 2 weeks and then was crossed over for 2 weeks. To suppress any carryover effect, there was a 1-week washout period between the two treatments. The researchers were interested in seeing if nicardipine significantly decreased the frequency and severity of Raynaud's phenomenon as compared with placebo. The data consist of a number of attacks in 2 weeks.

| | Period 1 | Period 2 | | Period 1 | Period 2 |
|---|---|---|---|---|---|
| Subject | Nicardipine | Placebo | Subject | Placebo | Nicardipine |
| 1 | 16 | 12 | 11 | 18 | 12 |
| 2 | 26 | 19 | 12 | 12 | 4 |
| 3 | 8 | 20 | 13 | 46 | 37 |
| 4 | 37 | 44 | 14 | 51 | 58 |
| 5 | 9 | 25 | 15 | 28 | 2 |
| 6 | 41 | 36 | 16 | 29 | 18 |
| 7 | 52 | 36 | 17 | 51 | 44 |
| 8 | 10 | 11 | 18 | 46 | 14 |
| 9 | 11 | 20 | 19 | 18 | 30 |
| 10 | 30 | 27 | 20 | 44 | 4 |

Download the MATLAB format of these data ( ◀ raynaud.m) and perform an ANOVA test in MATLAB. Select an additive model, with the number of attacks $y_{ijk}$ represented as

$$y_{ijk} = \mu + \alpha_i + \beta_j + \gamma_k + \epsilon_{ijk}.$$

Here $\mu$ is the grand mean, $\alpha_i$, $i = 1,\ldots,20$ is the subject effect, $\beta_j$, $j = 1,2$ is the drug effect, $\gamma_k$, $k = 1,2$ is the period effect, and $\epsilon_{ijk}$ is the zero-mean normal random error.

We are not interested in testing the null hypothesis about the subjects,

$$H_0 : \alpha_i = 0,$$

since the subjects are blocks here. We are interested in testing the effect of the drug,

$$H'_0 : \beta_1 = \beta_2 = 0,$$

and in testing the effect of the washout period,

$$H''_0 : \gamma_1 = \gamma_2 = 0.$$

(a) Provide an ANOVA table for this model using MATLAB. Test hypotheses $H'_0$ and $H''_0$. Describe your findings.

(b) What proportion of variability in the data $y_{ijk}$ is explained by the above ANOVA model?

(c) What happens if we believe that there are 40 subjects and they are randomized to drug/period treatments. Test $H'_0$ and $H''_0$ in this context and comment.

11.22. **Simvastatin.** In a quantitative physiology lab II at Georgia Tech, students were asked to find a therapeutic model to test on MC3T3-E1 cell line to enhance osteoblastic growth. The students found a drug called Simvastatin, a cholesterol lowering drug to test on these cells. Using a control and three different concentrations $10^{-9}$M, $10^{-8}$M, and $10^{-7}$M, cells were treated with the drug. These cells were plated on four, 24 well plates with each well plate having a different treatment. To test for osteoblastic differentiation an assay, pNPP, was used to test for alkaline phosphatase activity. The higher the alkaline phosphatase activity the better the cells are differentiating, and become more bone like. This assay was performed 6 times total within 11 days. Each time the assay was performed, four wells from each plate were used. The data ( simvastatin.dat) are provided in the following table:

| | Time | | | | | |
|---|---|---|---|---|---|---|
| Concentration | Day 1 | Day 3 | Day 5 | Day 7 | Day 9 | Day 11 |
| A (Control) | 0.062 | 0.055 | 0.055 | 1.028 | 0.607 | 0.067 |
| | 0.517 | 0.054 | 0.059 | 1.067 | 0.104 | 0.093 |
| | 0.261 | 0.056 | 0.062 | 1.128 | 0.163 | 0.165 |
| | 0.154 | 0.063 | 0.062 | 0.855 | 0.109 | 0.076 |
| B ($10^{-7}$M) | 0.071 | 0.055 | 0.067 | 0.075 | 0.068 | 0.347 |
| | 0.472 | 0.060 | 1.234 | 0.076 | 0.143 | 0.106 |
| | 0.903 | 0.057 | 1.086 | 0.090 | 0.108 | 0.170 |
| | 0.565 | 0.056 | 0.188 | 1.209 | 0.075 | 0.097 |
| C ($10^{-8}$M) | 0.068 | 0.059 | 0.092 | 0.091 | 0.098 | 0.115 |
| | 0.474 | 0.070 | 0.096 | 0.218 | 0.122 | 0.085 |
| | 0.063 | 0.090 | 0.123 | 0.618 | 0.837 | 0.076 |
| | 0.059 | 0.064 | 0.091 | 0.093 | 0.142 | 0.085 |
| D ($10^{-9}$M) | 0.066 | 0.447 | 0.086 | 0.248 | 0.108 | 0.290 |
| | 0.670 | 0.091 | 0.076 | 0.094 | 0.105 | 0.090 |
| | 0.076 | 0.079 | 0.082 | 0.215 | 0.093 | 0.518 |
| | 0.080 | 0.071 | 0.080 | 0.401 | 0.580 | 0.071 |

In this design there are two crossed factors Concentration and Time.
(a) Using the two-way ANOVA show that the interaction between factors Time and Concentration is significant.
(b) Since the presence of a significant interaction affects the tests for main effects, conduct a conditional test for the Concentration factor if Time is fixed at Day 3. Also, conduct the conditional test for Time factor if the Concentration level is C.

11.23. **Antitobacco Media Campaigns.** Since the early 1990s, the US population has been exposed to a growing number and variety of televised antitobacco advertisements. By 2002, 35 states had launched antitobacco media campaigns.

Four states (factor $A$) participated in an antitobacco awareness study. Each state independently devised antitobacco program. Three cities (factor $B$) within each state were selected for participation and 10 households within each city were randomly selected to evaluate effectiveness of the program. All members of the selected households were interviewed and a composite index was formed for each household measuring the extent of antitobacco awareness. The data are given in the table (larger the index, the greater the awareness)

| State $i$ | | 1 | | | 2 | | | 3 | | | 4 | | |
|---|---|---|---|---|---|---|---|---|---|---|---|---|---|
| City $j$ | | 1 | 2 | 3 | 1 | 2 | 3 | 1 | 2 | 3 | 1 | 2 | 3 |
| | 1 | 42 | 26 | 33 | 48 | 56 | 44 | 23 | 47 | 48 | 70 | 56 | 35 |
| | 2 | 56 | 38 | 51 | 54 | 65 | 34 | 31 | 39 | 40 | 61 | 49 | 30 |
| | 3 | 35 | 42 | 43 | 48 | 71 | 39 | 34 | 45 | 43 | 57 | 58 | 41 |
| | 4 | 40 | 29 | 49 | 57 | 66 | 40 | 28 | 51 | 48 | 69 | 55 | 52 |
| Household $k$ | 5 | 54 | 44 | 42 | 51 | 64 | 51 | 30 | 38 | 51 | 71 | 47 | 39 |
| | 6 | 49 | 43 | 39 | 51 | 51 | 47 | 36 | 47 | 57 | 49 | 50 | 44 |
| | 7 | 51 | 45 | 41 | 56 | 60 | 34 | 41 | 39 | 40 | 61 | 49 | 30 |
| | 8 | 48 | 39 | 47 | 50 | 72 | 41 | 35 | 44 | 41 | 57 | 61 | 40 |
| | 9 | 52 | 30 | 45 | 50 | 60 | 42 | 28 | 44 | 49 | 69 | 54 | 52 |
| | 10 | 54 | 40 | 51 | 49 | 63 | 49 | 33 | 46 | 49 | 67 | 52 | 45 |

(a) Discuss and propose the analyzing methodology.

(b) Assume that nested design with fixed factor effects is appropriate. Provide the ANOVA table.

(c) Test whether or not the mean awareness differ for the four states. Use $\alpha = 0.05$. State the alternatives and conclusion.

11.24. **Orthosis.** The data 🖳 Ortho.dat were acquired and computed by Dr. Amarantini David and Dr. Martin Luc (Laboratoire Sport et Performance Motrice, Grenoble University, France). The purpose of recording such data was the interest to better understand the processes underlying movement generation under various levels of an externally applied moment to the knee. In this experiment, stepping-in-place was a relevant task to investigate how muscle redundancy could be appropriately used to cope with an external perturbation while complying with the mechanical requirements related either to balance control and/or minimum energy expenditure. For this purpose, 7 young male volunteers wore a spring-loaded orthosis of adjustable stiffness under 4 experimental conditions: a control condition (without orthosis), an orthosis condition (with the orthosis only), and two conditions (Spring1, Spring2) in which stepping in place was perturbed by fitting a spring-loaded orthosis onto the right knee joint.

For each stepping-in-place replication, the resultant moment was computed at 256 time points equally spaced and scaled so that a time interval corresponds to an individual gait cycle. A typical moment observation is therefore a one-dimensional function of normalized time $t$ so that $t \in [0, 1]$. The data set consists in 280 separate runs and involves the $j = 7$ subjects over $i = 4$ described experimental conditions, replicated $k = 10$ times for each subject. Figure 11.14 shows the data set; typical moment plots over gait cycles. Model the data as arising from a fixed-effects FANOVA model with 2 qualitative factors (Subjects and Treatments), 1 quantitative factor (Time) and 10 replications for each level combination.

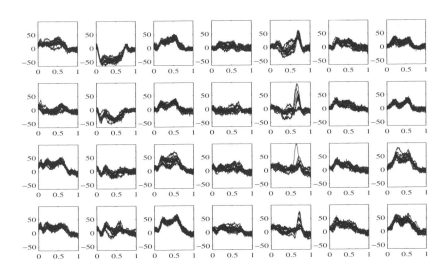

**Fig. 11.14** Orthosis data set: The panels in rows correspond to *Treatments* while the panels in columns correspond to *Subjects*; there are 10 repeated measurements in each panel.

If the functional ANOVA model is $\mathbb{E}y_{ijk}(t) = \mu(t) + \alpha_i(t) + \beta_j(t)$, find and plot functional estimators for the grand mean $\mu(t)$ and treatment effects $\alpha_1(t) - \alpha_4(t)$.

11.25. **Bone Screws.** Bone screws are the most commonly used type of orthopedic implant. Their precision is critical and several dimensions are rigorously checked for conformity (outside and root diameters, pitch, tread angle, etc.). The screws are produced in lots of 20000. To confirm the homogeneity of production, 6 lots were selected and from each lot a sample of size 100 was obtained. The following numbers of nonconforming screws are found in the six samples: 20, 17, 23, 9, 15, and 14.

(a) Test the hypothesis of homogeneity of proportions of nonconforming screws in the six lots.

(b) If the null hypothesis in (a) is rejected, find which lots differ. Assume $\alpha = 0.05$.

11.26. **R&R Study.** Five parts are measured by two appraisers using the same measuring instrument. Each appraiser measured each part three times. Find the relevant parameters of the measuring system and assess its capability (*PRR* and *NDC*).

```
measure =[ ...
217 220 217 214 216 ...
216 216 216 212 219 ...
216 218 216 212 220 ...
216 216 215 212 220 ...
```

```
219 216 215 212 220 ...
220 220 216 212 220]';

part = [ ...
    1 2 3 4 5 1 2 3 4 5 ...
    1 2 3 4 5 1 2 3 4 5 ...
    1 2 3 4 5 1 2 3 4 5]';

appraiser =[...
    1 1 1 1 1 ...
    1 1 1 1 1 ...
    1 1 1 1 1 ...
    2 2 2 2 2 ...
    2 2 2 2 2 ...
    2 2 2 2 2]';
```

11.27. **Additive R&R ANOVA for Measuring Impedance.** Repeat the analysis
in Example 11.14 by assuming that parts and operators do not interact,
that is, with an additive ANOVA model. Provide comparisons with results
from gagerr, where the option ('model','linear') is used.

---

| MATLAB AND WINBUGS FILES AND DATA SETS USED IN THIS CHAPTER |
| --- |

http://springer.bme.gatech.edu/Ch11.Anova/

anomct.m, anovarep.m, Barspher.m, C677T.m, cardiac.m,
chrysanthemum.m, circularity.m, coagulationtimes.m, cochcdf1.m,
Cochtest.m, dialysis.m, dorsalspines.m, dunn.m, epsGG.m, fanova.m,
fibers.m, insects.m, insulin.m, maternalbehavior.m, Mauspher.m,
Mausphercnst.m, micePO2.m, nest1.m, nest2.m, nesta.m, nestAOV.m,
nicardipine.m, orthosis.m, orthosisraw.m, pedometer1300.m,
poissonmeans.m, powerANOVA.m, qtukey.m, RandR.m, RandR2.m, RandR3.m,
ratsblocking.m, raynaud.m, screws.m, secchim.m, sphertest.m,
ssizerepeated.m, tmcomptest.m

anovacoagulation.odc, dialysis.odc, insulin.odc, vortex.odc

arthritis.dat|mat, clouds.txt, deaf.xls, Ortho.dat, oxygen.dat,
pmr1300.mat, porcinedata.xls, Proliferation.xls, secchi.mat|xls,
secchi.xls, silicone.dat, silicone1.dat, simvastatin.dat

---

# CHAPTER REFERENCES

Altman, D. G. (1991). *Practical Statistics for Medical Research*, Chapman and Hall, London.

Bell, M. A. and Foster, S. A. (1994). Introduction to the evolutionary biology of the three-spine stickleback. In: Bell and Foster (eds.) *The Evolutionary Biology of the Threespine Stickleback*. Oxford University Press, Oxford, pp. 277–296.

Box, G. E. P. (1953). Non-normality and tests on variances. *Biometrika*, **40**, 318–335.

Box, G. E. P, Hunter, J. S., and Hunter, W. G. (2005). *Statistics for Experimenters II*, 2nd edn. Wiley, Hoboken.

Brix-Christensen, V., Vestergaard, C., Andersen, S. K., Krog, J., Andersen, N. T., Larsson, A., Schmitz, O., and Tønnesen, E. (2005). Evidence that acute hyperinsulinaemia increases the cytokine content in essential organs after an endotoxin challenge in a porcine model. *Acta Anaesthesiol. Scand.*, **49**, 1429–1435.

Conover, W. J. (1999). *Practical Nonparametric Statistics*, 3rd edn. Wiley, Hoboken.

Daugridas, J. T. and Ing, T. (1994). *Handbook of Dialysis*. Little, Brown, Boston.

Diggle, P. J., Liang, K. Y., and Zeger, S. L. (1994). *Analysis of Longitudinal Data*. Oxford University Press, Oxford.

Dunnett, C. W. (1955). A multiple comparison procedure for comparing several treatments with a control. *J. Am. Stat. Assoc.*, **50**, 1096–1121.

Ehrstein, J. and Croarkin, M. C. (1984). Standard Method for Measuring Resistivity of Silicon Wafers with an In-Line Four-Point Probe. Unpublished NIST data set. In: http://www.itl.nist.gov/div898/strd/anova/anova.html.

Fan, J. and Lin, S-K. (1998). Test of significance when data are curves. *J. Am. Stat. Assoc.*, **93**, 1007–1021.

Fisher, R. A. (1918). Correlation between relatives on the supposition of Mendelian inheritance. *Trans. R. Soc. Edinb.*, **52**, 399–433.

Fisher, R. A. (1921). Studies in crop variation. I. An examination of the yield of dressed grain from Broadbalk. *J. Agric. Sci.*, **11**, 107–135.

Halperin, M., Greenhouse, S. W., Cornfield, J., and Zalokar, J. (1955). Tables of percentage points for the studentized maximum absolute deviation in normal samples. *J. Am. Stat. Assoc.*, **50**, 185–195.

Houf, R. E. and Burman, D. B. (1988). Statistical analysis of power module thermal test equipment performance. *IEEE Trans. Components Hybrids Manuf. Technol.*, **11**, 516–520.

Kahan, A., Amor, B., Menkès, C. J., Weber, S., Guérin, F., and Degeorges, M. (1987). Nicardipine in the treatment of Raynaud's phenomenon: a randomized double-blind trial. *Angiology*, **38**, 333–337.

Kutner, M. H., Nachtsheim, C. J., Neter, J., and Li, W. (2004). *Applied Linear Statistical Models*, 5th edn. McGraw-Hill, Boston.

Lanzen, J. L., Braun, R. D., Ong, A. L., and Dewhirst, M. W. (1998). Variability in blood flow and PO2 in tumors in response to carbogen breathing. *Int. J. Radiat. Oncol. Biol. Phys.*, **42**, 855–859.

Levene, H. (1960). In: Olkin et al. (eds.) *Contributions to Probability and Statistics: Essays in Honor of Harold Hotelling, I*. Stanford University Press, Palo Alto, pp. 278–292.

Liu, W. (1997). Sample size determination of Dunnett's procedure for comparing several treatments with a control. *J. Stat. Plann. Infer.*, **62**, 2, 255–261.

Miller, A. J., Shaw, D. E., Veitch, L. G., and Smith, E. J. (1979). Analyzing the results of a cloud-seeding experiment in Tasmania. *Commun. Stat. Theory Methods*, **8**, 10, 1017–1047.

Marascuillo, L. A. and Serlin, R. C. (1988). *Statistical Methods for the Social and Behavioral Sciences*. Freeman, New York.

Montgomery, D. C. (1984). *Design and Analysis of Experiments*, 2nd edn. Wiley, New York.

Montgomery, D. C. (2005). *Statistical Control Theory*, 5th edn. Wiley, Hoboken.

Ott, E. R. (1967). Analysis of means – A graphical procedure. *Industrial Quality Control*, **24**, 101–109.

Ozturk, H., Durga, J., van de Rest, O., and Verhoef, P. (2005). The MTHFR 677C→T genotype modifies the relation of folate intake and status with plasma homocysteine in middle-aged and elderly people. *Ned. Tijdschr. Klin. Chem. Labgeneesk*, **30**, 208–217.

Park, W. (1932). Studies on the change in nectar concentration produced by the honeybee, *Apis mellifera*. Part I: Changes that occur between the flower and the hive. *Iowa Agric. Exp. Stat. Res. Bull.*, **151**.

Ramsay, J. and Silverman, B. (1997). *Functional Data Analysis*. Springer, Berlin Heidelberg New York.

Sokal, R. and Karten, I. (1964). Competition among genotypes in *Tribolium castaneum* at varying densities and gene frequencies (the black locus). *Genetics*, **49**, 195–211.

Walpole, R. E., Myers, R. H., Myers, S. L., and Ye, K. (2007). *Probability and Statistics for Engineers and Scientists*, 8th edn. Pearson-Prentice Hall, Upper Saddle River.

Wilson, M. C. and Shade, R. E. (1967). Relative attractiveness of various luminescent colors to the cereal leaf beetle and the meadow spittle-bug. *J. Econ. Entomol.*, **60**, 578–580.

Zar, J. H. (1996). *Biostatistical Analysis*. Prentice-Hall, Englewood Cliffs.

# Chapter 12
# Distribution-Free Tests

*Assumptions are the termites of relationships.*

– Henry Winkler

---

**WHAT IS COVERED IN THIS CHAPTER**

- Sign Test
- Ranks
- Wilcoxon Signed-Rank Test and Wilcoxon Sum Rank Test
- Kruskal–Wallis and Friedman Test
- Walsh Test for Outliers

---

## 12.1 Introduction

Most of the methods we have covered until now are based on parametric assumptions; the data are assumed to follow some well-known family of distributions, such as normal, exponential, Poisson, and so on. Each of these distributions is indexed by one or more parameters (e.g., the normal distribution has $\mu$ and $\sigma^2$), and at least one is presumed unknown and must be inferred. However, with complex experiments and messy sampling plans, the generated data might not conform to any well-known distribution. In the case where the experimenter is not sure about the underlying distribution of the data, statistical techniques that can be applied regardless of the true distribution of the data are needed. These techniques are called *distribution-free* or *nonparametric*. To quote statistician J.V. Bradley (1968):

> The terms nonparametric and distribution-free are not synonymous.... Popular usage, however, has equated the terms....Roughly speaking, a nonparametric test is one which makes no hypothesis about the value of a parameter in a statistical density function, whereas a distribution-free test is one which makes no assumptions about the precise form of the sampled population.

Many basic statistical procedures, such as the $t$-test, test for Pearson's correlation coefficient, analysis of variance, etc., assume that the sampled data come from a normal distribution or that the sample size is sufficiently large for the CLT to make the normality assumption reasonable.

In this chapter, we will revisit the testing of means, the two-sample problem, ANOVA, and the block design, without making strict distributional assumptions about the data. The material in the next chapter dealing with goodness-of-fit tests is also considered a nonparametric methodology.

For a more comprehensive account on the theory and use of nonparametric methods in engineering research, we direct the reader to Kvam and Vidakovic (2007).

The following table gives nonparametric counterparts to a few of the most popular inferential procedures. The acronym WSiRT stands for Wilcoxon's Signed Rank Test, while WSuRT stands for Wilcoxon's Sum Rank Test.

| Parametric | Nonparametric |
|---|---|
| One-sample $t$-test for the location | Sign test, WSiRT |
| Paired $t$-test | Sign test, WSiRT |
| Two-sample $t$-test | WSuRT, Wilcoxon–Mann–Whitney |
| One-way ANOVA | Kruskal–Wallis test |
| Block design ANOVA | Friedman test |

## 12.2 Sign Test

We start with the simplest nonparametric procedure, the sign test. Suppose we are interested in testing the hypothesis $H_0$ that a population with a continuous CDF has a median $m_0$ against one of the alternatives $H_1 : med >\neq< m_0$.

We assign a + sign when $X_i > m_0$, i.e., when the difference $X_i - m_0$ is positive, and a − sign otherwise. The case $X_i = m_0$ (a tie) is theoretically impossible for continuous distributions, although when observations are coded with limited precision, ties are quite possible and most of the time present. Henceforth we will assume an ideal situation in which no ties occur.

If $m_0$ is the median and if $H_0$ is true, then by the definition of the median, $\mathbb{P}(X_i > m_0) = \mathbb{P}(X_i < m_0) = \frac{1}{2}$. Thus, the statistic $T$ representing the total number of + is equal to

$$T = \sum_{i=1}^{n} \mathbf{1}(X_i > m_0)$$

and has a binomial distribution with parameters $n$ and $1/2$.

Let the level of the test, $\alpha$, be specified. When the alternative is $H_1 : med > m_0$, the critical values of $T$ are integers greater than or equal to $k_\alpha$, which is defined as the smallest integer for which the relationship

$$\sum_{t=k_\alpha}^{n} \binom{n}{k} \left(\frac{1}{2}\right)^n < \alpha$$

holds.

Likewise, if the alternative is $H_1 : med < m_0$, the critical values of $T$ are integers less than or equal to $k'_\alpha$, which is defined as the largest integer for which the relationship

$$\sum_{t=0}^{k'_\alpha} \binom{n}{k} \left(\frac{1}{2}\right)^n < \alpha$$

holds.

If the alternative is two-sided, i.e., $H_1 : med \neq m_0$, the critical values of $T$ are integers less than or equal to $k'_{\alpha/2}$ and greater than or equal to $k_{\alpha/2}$, which are defined, respectively, as the largest and smallest integers for which the inequalities

$$\sum_{t=0}^{k'_{\alpha/2}} \binom{n}{k} \left(\frac{1}{2}\right)^n < \alpha/2, \text{ and } \sum_{t=k_{\alpha/2}}^{n} \binom{n}{k} \left(\frac{1}{2}\right)^n < \alpha/2$$

hold.

If the value $T$ is observed, then in testing against the alternative $H_1 : med > m_0$, large values of $T$ are critical and the $p$-value is $p = \sum_{i=T}^{n} \binom{n}{i} 2^{-n} = \sum_{i=0}^{n-T} \binom{n}{i} 2^{-n}$. When testing against the alternative $H_1 : med < m_0$, small values of $T$ are critical and the $p$-value is $p = \sum_{i=0}^{T} \binom{n}{i} 2^{-n}$. When the alternative is two-sided, the $p$-value is $p = 2 \sum_{i=0}^{T'} \binom{n}{i} 2^{-n}$ for $T' = \min\{T, n - T\}$.

| Alternative | $p$-value |
|---|---|
| $H_1 : med > m_0$ | $\sum_{i=T}^{n} \binom{n}{i} 2^{-n}$ |
| $H_1 : med \neq m_0$ | $2 \sum_{i=0}^{T'} \binom{n}{i} 2^{-n}$ for $T' = \min\{T, n - T\}$ |
| $H_1 : med < m_0$ | $\sum_{i=0}^{T} \binom{n}{i} 2^{-n}$ |

Consider now the two-paired-sample case, and suppose that the samples $X_1, \ldots, X_n$ and $Y_1, \ldots, Y_n$ are observed. We are interested in knowing whether the population median of the differences $X_1 - Y_1, X_2 - Y_2, \ldots, X_n - Y_n$ is equal to 0. In this case, $T = \sum_{i=1}^{n} \mathbf{1}(X_i > Y_i)$, is the total number of strictly positive differences.

Although it is true that the hypothesis of equality of means is equivalent to the hypothesis that the mean of the differences is 0, for the medians an analogous statement is not true in general. More precisely, if $D = X - Y$, then $med(D)$ may not be equal to $med(X) - med(Y)$. Thus, with the sign test we are not testing the equality of medians, but whether the median of the differences is 0.

Ties complicate the calculations but can be handled. Even when observations come from a continuous distribution, ties appear due to limited precision in the application part. There are several ways of dealing with ties:

(i) Ignore them. If there are $s$ ties, use only the "untied" observations. Of course, the sample size drops to $n - s$.

(ii) Assign the winning sign to tied pairs. For example, if there are two minuses, two ties, and six pluses, consider the two ties as pluses.

(iii) Randomize. If you have two ties, flip a coin twice and assign a plus if the coin lands heads and minus if the coin lands tails.

In script ◀ signtst.m conducting the sign test (not to be mixed with MATLAB's built in signtest), the options for handling ties are: I, C, and R, for policies in (i)-(iii).

*Example 12.1.* **TCDD Levels.** Many Vietnam veterans have dangerously high levels of the dioxin 2,3,7,8-TCDD in their blood and fat tissue as a result of their exposure to the defoliant Agent Orange. A study published in *Chemosphere* (vol. 20, 1990) reported on the TCDD levels of 20 Massachusetts Vietnam veterans who had possibly been exposed to Agent Orange. The amounts of TCDD (measured in parts per trillion) in blood plasma and fat tissue drawn from each veteran are shown in the table below.

| TCDD levels in plasma | TCDD levels in fat tissue |
|:---:|:---:|
| 2.5  3.1  2.1 | 4.9  5.9  4.4 |
| 3.5  3.1  1.8 | 6.9  7.0  4.2 |
| 6.8  3.0  36.0 | 10.0  5.5  41.0 |
| 4.7  6.9  3.3 | 4.4  7.0  2.9 |
| 4.6  1.6  7.2 | 4.6  1.4  7.7 |
| 1.8  20.0  2.0 | 1.1  11.0  2.5 |
| 2.5  4.1 | 2.3  2.5 |

Is there sufficient evidence of a difference between the distributions of TCDD levels in plasma and fat tissue for Vietnam veterans exposed to Agent Orange? Use the sign test and $\alpha = 0.10$.

```
tcddpla = [2.5 3.1 2.1 3.5 3.1 1.8 6.8 3.0 36.0 ...
   4.7 6.9 3.3 4.6 1.6 7.2 1.8 20.0 2.0 2.5 4.1];
tcddfat = [4.9 5.9 4.4 3.5 7.0 4.2 10.0 5.5 41.0 ...
   4.4 7.0 2.9 4.6 1.4 7.7 1.8 11.0 2.5 2.3 2.5];
% ignore ties
[pvae, pvaa, n, plusses, ties] = signtst(tcddpla, tcddfat)
%pvae =0.1662
%pvaa =0.1660
%n =17
%plusses =6
%ties =3

% randomize ties
[pvae, pvaa, n, plusses, ties] = signtst(tcddpla, tcddfat,'R')
% pvae =0.2517
% take the conservative, least favorable approach
[pvae, pvaa, n, plusses, ties] = signtst(tcddpla, tcddfat,'C')
% pvae =0.0577
```

Overall, the sign test failed to find significant differences between the distributions of TCDD levels. Only the conservative assignment of ties (least favorable to $H_0$) produced $p$-value of 0.0577, significant at $\alpha = 10\%$ level. Compare these results with MATLAB's built-in function signtest.

## 12.3 Ranks

Let $X_1, X_2, \ldots, X_n$ be a sample from a population with a continuous distribution $F$. Many distribution-free procedures are based on how observations within the sample are *ranked* compared to either a parameter $\theta$ or to another sample. The ranks of a sample $X_1, X_2, \ldots, X_n$ are defined as indices of ordered sample

$$r(X_1), r(X_2), \ldots, r(X_n).$$

For example,

```
ranks([10 20 25 7])
%ans = 2     3     4     1
```

The function ranks.m is

```
function   r = ranks(data, glob)
%-------------------------------------------
if nargin < 2
    glob = 1;
end
```

```
    shape = size(data);
if glob == 1
    data=data(:);
  end
% Ties ranked from UptoDown
  [ irrelevant , indud ]  =  sort(data);
  [ irrelevant , rUD ]    =  sort(indud);
% Ties ranked from RtoL
  [ irrelevant , inddu ]  =  sort(flipud(data));
  [ irrelevant , rDU ]    =  sort(inddu);
% Averages ranks of ties, keeping ranks
% of no-tie-observations the same
r = (rUD + flipud(rDU))./2;
r = reshape(r,shape);
```

For example, when the input is a matrix, the optional parameter glob = 1 produces global ranking, while for glob not equal to 1, columnwise ranking is performed.

```
%a =
%     0.8147     0.9134     0.2785     0.9649
%     0.9058     0.6324     0.5469     0.1576
%     0.1270     0.0975     0.9575     0.9706

  ranks(a)
% ans =
%     7      9      4     11
%     8      6      5      3
%     2      1     10     12

  ranks(a,2)
% ans =
%     2      3      1      2
%     3      2      2      1
%     1      1      3      3
```

In the case of ties, it is customary to average the tied rank values. The script ⊲ ranks.m does just that:

```
ranks([2 1 7 1 15 9])
%ans = 3.0000    1.5000    4.0000    1.5000    6.0000    5.0000
```

Here $r(2) = 3$, $r(1) = 1.5$, $r(7) = 4$, and so on. Note that 1 appears twice and ranks 1 and 2 are averaged. In the case

```
ranks([9 1 7 1 9 9])
%ans = 5.0000    1.5000    3.0000    1.5000    5.0000    5.0000
```

the ranks of three 9s are 4, 5, and 6, which are averaged to 5.

Suppose that a random sample from continuous distribution $X_1,\dots,X_n$ is ranked and that $R_i = r(X_i)$, $i = 1,\dots,n$ are the ranks. Ranks $R_i$ are random variables with discrete uniform distribution (p. 141). The properties of integer sums lead to the following properties for ranks:

$$E(R_i) = \sum_{j=1}^{n} \frac{j}{n} = \frac{n+1}{2},$$

$$E(R_i^2) = \sum_{j=1}^{n} \frac{j^2}{n} = \frac{n(n+1)(2n+1)}{6n} = \frac{(n+1)(2n+1)}{6}$$

$$Var(R_i) = \frac{n^2 - 1}{12}.$$

These relationships follow from the fact that for a random sample, ranks are distributed as discrete uniform, namely, for any $i$,

$$P(R_i = j) = \frac{1}{n}, \ 1 \le j \le n.$$

## 12.4 Wilcoxon Signed-Rank Test

More powerful than the sign test is Wilcoxon's signed-rank test (Wilcoxon, 1945), where, in addition to signs, particular ranks are taken into account.

Let the paired sample $(X_i, Y_i)$, $i = 1, \ldots, n$ be observed and let $D_i = X_i - Y_i$, $i = 1, \ldots, n$ be the differences. In a two-sample problem, we are interested in testing that the true mean of the differences is 0.

It is also possible to consider a one-sample scenario in which testing the hypothesis about the median *med* is of interest. Here $H_0 : med = m_0$ is tested versus the one- or two-sided alternative. Then observations $X_i$, $i = 1, \ldots, n$ are compared to $m_0$, and the differences are $D_i = X_i - m_0$, $i = 1, \ldots, n$.

The only assumption is that the distribution of the differences $D_i$, $i = 1, \ldots, n$ is symmetric about 0. This implies that positive and negative differences are equally likely. For this test, the absolute values of the differences $(|D_1|, |D_2|, \ldots, |D_n|)$ are ranked. Let $r(|D_1|), r(|D_2|), \ldots, r(|D_n|)$ be the ranks of the differences.

Under $H_0$, the expectations of the sum of positive differences and the sum of negative differences should be equal. Define

$$W^+ = \sum_{i=1}^{n} S_i \ r(|D_i|)$$

and

$$W^- = \sum_{i=1}^{n} (1 - S_i) \ r(|D_i|),$$

where $S_i = 1$ if $D_i > 0$ and $S_i = 0$ if $D_i < 0$. Cases where $D_i = 0$ are ties and are ignored. Thus, $W^+ + W^-$ is the sum of all ranks, and in the case of no ties is equal to $\sum_{i=1}^{n} i = n(n+1)/2$. The statistic for the WSiRT is the difference between the ranks of positive differences and ranks of negative differences:

$$W = W^+ - W^- = 2 \sum_{i=1}^{n} r(|D_i|)S_i - n(n+1)/2.$$

**Rule:** For the WSiRT, it is suggested that a large-sample approximation should be used for $W$. If the samples are from the same population, the differences should be well mixed, and the sum of the ranks of positive differences should be close to the sum of the ranks of negative differences. Thus, in this case, $\mathbb{E}(W) = 0$ and $\mathbb{V}\mathrm{ar}(W) = \sum_i (r(|D_i|)^2) = \sum_i i^2 = n(n+1)(2n+1)/6$ under $H_0$ and no ties in differences. The statistic

$$Z = \frac{W}{\sqrt{\mathbb{V}\mathrm{ar}(W)}}$$

has an approximately standard normal distribution, so normcdf can be used to evaluate the $p$-values of the observed statistic $W$ with respect to a particular alternative (see the m-file ◀ wsirt.m).

```
function [W, Z, p] = wsirt( data1, data2, alt )
% -----------------------------------------------------------
% WILCOXON SIGNED RANK TEST
% Input:  data1, data2 - first and second sample
%         alt - code for alternative hypothesis;
%             -1  mu1<mu2; 0 mu1 ne mu2; and 1 mu1>mu2
% Output: W - sum of all signed ranks
%         Z - standardized W but adjusted for the ties
%         p - p-value for testing equality of distribs
%             (equality of locations) against the
%             alternative specified by the input alt
% Example of use:
% >  dat1=[1 3 2 4 3 5 5 4 2 3 4 3 1 7 6 6 5 4 5 8 7];
% >  dat2=[2 5 4 3 4 3 2 2 1 2 3 2 3 4 3 2 3 4 4 3 5];
% >  [srs, tstat, pval] = wsirt(dat1, dat2, 1)
%
%  Needs: M-FILE ranks.m (ranking procedure)
%-----------------------------------------------------------
data1 = data1(:)' ;  % convert sample 1 to a row vector
data2 = data2(:)' ;  % convert sample 2 to a row vector
if length(data1) ~= length(data2)
    error('Sample sizes should coincide')
end
difs = data1 - data2;
difs = difs( difs ~= 0); % exclude ties
rank_all = ranks(abs(difs));
signs = 2.*(difs > 0)-1;
sig_ranks = signs .* rank_all;
   W = sum( sig_ranks );      %sum of all signed ranks
```

```
  W2 = sum( ( sig_ranks.^2 ) );
Z  = W/sqrt(W2);
cc = 1/sqrt(W2); %continuity correction
%------------------------- alternatives -------------
% alt == 0 for two sided;
% alt == -1 for mu1 < mu2; alt == 1 for mu1 > mu2.
if       alt == 0
             p = 2*normcdf(-abs( Z ) + cc);
elseif  alt == -1
             p = normcdf( Z  + cc);
elseif  alt == 1
             p = normcdf( -Z + cc);
else
      error('Input "alt" should be either 0,-1,or 1.')
end
```

*Example 12.2.* **Identical Twins.** This data set was discussed in Conover (1999). Twelve pairs of identical twins underwent psychological tests to measure the amount of aggressiveness in each person's personality. We are interested in comparing the twins to each other to see if the first-born twin tends to be more aggressive than the other. The results are as follows (the higher score indicates more aggressiveness).

| First-born, $X_i$ | 86 71 77 68 91 72 77 91 70 71 88 87 |
|---|---|
| Second twin, $Y_i$ | 88 77 76 64 96 72 65 90 65 80 81 72 |

The hypotheses are: $H_0$ : the mean aggressiveness scores for the two twins are the same, that is, $\mathbb{E}(X_i) = \mathbb{E}(Y_i)$, and $H_1$ : the first-born twin tends to be more aggressive than the other, i.e., $\mathbb{E}(X_i) > \mathbb{E}(Y_i)$. The WSiRT is appropriate if we assume that $D_i = X_i - Y_i$ are independent and symmetric. Below is the output of wsirt, where the $T$ statistic has been used.

```
fb = [86 71 77 68 91 72 77 91 70 71 88 87];
sb = [88 77 76 64 96 72 65 90 65 80 81 72];
[w1, z1, p] = wsirt(fb, sb, 1)
%w1 =   17              %value of T
%z1 =   0.7565          %value of Z
%p  =   0.2382          %p-value of the test
```

Note that the test failed to reject $H_0$.

The WSiRT can be used to test the hypothesis of location $H_0 : \mu = \mu_0$ using a single sample, as in a one-sample $t$-test. The differences in the WSiRT are $X_1 - \mu_0, X_2 - \mu_0, \ldots, X_n - \mu_0$ instead of $X_1 - Y_1, X_2 - Y_2, \ldots, X_n - Y_n$, as in the two-sample WSiRT.

*Example 12.3.* In the Moon Illusion (Example 9.4), we tested $H_0 : \mu = 1$ against $H_1 : \mu > 1$.

```
moon = [1.73 1.06 2.03 1.40 0.95 1.13 1.41 1.73 1.63 1.56];
mu0 = 1;
mu0vec = mu0 * ones(size(moon));
[w, z, p]=wsirt(moon, mu0vec, 1)
%w = 53
%z = 2.7029
%p = 0.0040
```

Compared to the $t$-test where the $p$-value was found to be pval = 9.9885e-04, the WSiRT still rejects $H_0$ even though the $p$-value is higher, p = 0.004.

Equivalently, the WSiRT can be based on the sum of the ranks of positive differences only (or, equivalently, the sum of the ranks of negative differences only). In that case, under $H_0$, $\mathbb{E}W^+ = n(n+1)/4$ and $\mathbb{V}\text{ar}(W^+) = n(n+1)(2n+1)/24$, leading to

$$Z = \frac{W^+ - n(n+1)/4}{\sqrt{n(n+1)(2n+1)/24}}.$$

## 12.5 Wilcoxon Sum Rank Test and Wilcoxon–Mann–Whitney Test

The Wilcoxon sum rank test (WSuRT) and Wilcoxon–Mann–Whitney test (WMW) are equivalent tests and we will discuss only the former, WSuRT. The tests are named after statisticians shown in Fig. 12.1a-c.

(a)                          (b)                          (c)

**Fig. 12.1** (a) Frank Wilcoxon (1892–1965), (b) Henry Berthold Mann (1905–2000), and (c) Donald Ransom Whitney (1915–2001).

The WSuRT is often used in place of a two-sample $t$-test when the populations being compared are independent, but possibly not normally distributed.

An example of the sort of data for which this test could be used is responses on a Likert scale (e.g., 1 = much worse, 2 = worse, 3 = no change, 4 = better, 5 = much better). It would be inappropriate to use the $t$-test for such data because for ordinal data the normality assumption does not hold. The WSuRT tells us more generally whether the groups are homogeneous or if one group is "better' than the other. More generally, the basic null hypothesis of the WSuRT is that the two populations are equal. That is, $H_0 : F_X(x) = F_Y(x)$. When stated in this way, this test assumes that the shapes of the distributions are similar, which is not a stringent assumption.

Let $\boldsymbol{X} = X_1,\dots,X_{n_1}$ and $\boldsymbol{Y} = Y_1,\dots,Y_{n_2}$ be two samples of sizes $n_1$ and $n_2$, respectively, from the populations that we want to compare. Assume that the samples are put together and that $n = n_1 + n_2$ ranks are assigned to their concatenation. The test statistic $W_n$ is the sum of ranks (1 to $n$) corresponding to the first sample, $\boldsymbol{X}$. For example, if $X_1 = 1, X_2 = 13, X_3 = 7, X_4 = 9$, and $Y_1 = 2, Y_2 = 0, Y_3 = 18$, then the value of $W_n$ is $2 + 4 + 5 + 6 = 17$.

If the two populations have the same distribution, then the sum of the ranks of the first sample and those in the second sample should be close, relative to their sample sizes. The WSuRT statistic is

$$W_n = \sum_{i=1}^{n} i S_i(\boldsymbol{X},\boldsymbol{Y}),$$

where $S_i(\boldsymbol{X},\boldsymbol{Y})$ is an indicator function defined as 1 if the $i$th ranked observation is from the first sample and as 0 if the observation is from the second sample.

For example, for $X_1 = 1, X_2 = 13, X_3 = 7, X_4 = 9$ and $Y_1 = 2, Y_2 = 0, Y_3 = 18$, $S_1 = 0, S_2 = 1, S_3 = 0, S_4 = 1, S_5 = 1, S_6 = 1, S_7 = 0$. Thus

$$W_n = 1 \times 0 + 2 \times 1 + 3 \times 0 + 4 \times 1 + 5 \times 1 + 6 \times 1 + 7 \times 0 = 2 + 4 + 5 + 6 = 17.$$

If there are no ties, then under $H_0$

$$\mathbb{E}(W_n) = \frac{n_1(n+1)}{2} \quad \text{and} \quad \mathbb{V}\text{ar}(W_n) = \frac{n_1 n_2(n+1)}{12}.$$

The statistic $W_n$ achieves its minimum when the first sample is entirely smaller than the second, and its maximum when the opposite occurs:

$$\min W_n = \sum_{i=1}^{n_1} i = \frac{n_1(n_1+1)}{2}, \quad \max W_n = \sum_{i=n-n_1+1}^{n} i = \frac{n_1(2n-n_1+1)}{2}.$$

For the statistic $W_n$ a normal approximation holds:

$$W_n \sim \mathcal{N}\left(\frac{n_1(n+1)}{2}, \frac{n_1 n_2(n+1)}{12}\right).$$

A better approximation is

$$\mathbb{P}(W_n \le w) \approx \Phi(x) + \phi(x)(x^3 - 3x)\frac{n_1^2 + n_2^2 + n_1 n_2 + n}{20 n_1 n_2 (n+1)},$$

where $\phi(x)$ and $\Phi(x)$ are the PDF and CDF of a standard normal distribution, respectively, and $x = (w - \mathbb{E}(W_n) + 0.5)/\sqrt{\mathbb{V}\text{ar}(W_n)}$. This approximation is satisfactory for $n_1 > 5$ and $n_2 > 5$ if there are no ties.

```
function  [W, Z, p] = wsurt( data1, data2, alt )
%  ------------------------------------------------------------
%  WILCOXON SUM RANK TEST
%  Input:  data1, data2 - first and second sample
%          alt - code for alternative hypothesis;
%               -1  mu1<m2; 0 mu1 ne m2; and 1 mu1>mu2
%  Output: W - sum of the ranks for the first sample. If
%              there is no ties, the standardization by ER &
%              Var R allows using standard normal quantiles
%              as long as sample sizes are larger than 15-20.
%          Z - standardized R but adjusted for the ties
%          p - p-value for testing equality of distributions
%              (equality of locations) against the alternative
%              specified by input "alt"
%  Example of use:
%  >  dat1=[1 3 2 4 3 5 5 4 2 3 4 3 1 7 6 6 5 4 5 8 7 3 3 4];
%  >  dat2=[2 5 4 3 4 3 2 2 1 2 3 2 3 4 3 2 3 4 4 3 5];
%  >  [sumranks1, tstat, pval] = wsurt(dat1, dat2, 1)
%
%  Needs: M-FILE ranks.m (ranking procedure)
%-------------------------------------------------------------
data1 = data1(:)' ;         %convert sample 1 to a row vector
  n1 = length( data1 );     %n1 - size of first sample, data1
data2 = data2(:)' ;         %convert sample 2 to a row
  n2 = length( data2 );     %n2 - size of second sample, data2
  n =n1+ n2;                 %n is the total sample size
mergeboth = [ data1 data2 ];
ranksall = ranks( mergeboth ); %ranks of merged observations
  W2 = sum( ( ranksall.^2 ) ); %sum of all ranks squared
% needed to make adjustment for the ties; if no ties are
% present, this sum is equal to the sum of squares of the
% first n integers: n(n+1)(2 n+1)/6.
ranksdata1 = ranksall( :, 1:n1); %ranks of first sample
  W = sum( ranksdata1 );    % statistic for WMW
%-------------------------------------------------------------
Z = (W - n1 * (n+1)/2 )/sqrt( n1*n2*W2/(n * (n-1)) ...
                   - n1*n2*(n+1)^2/(4*(n-1)));
% Z is approximately standard normal and approximation is
% quite good if n1,n2 > 15. Since W ranges over integers
% and half integers, a continuity correction, cc, may be
% used for improving the accuracy of p-values.
cc = 0.25/sqrt( n1*n2*W2/(n * (n-1)) - ...
                   n1*n2*(n+1)^2/(4*(n-1)));
```

```
%-------------------------- alternatives -----------------
% alt == 0 for two sided;
% alt == -1 for mu1 < mu2; alt == 1 for mu1 > mu2.
if      alt == 0
            p = 2*normcdf(- abs( Z ) - cc);
elseif  alt == -1
            p = normcdf( Z - cc );
elseif  alt == 1
            p = normcdf( - Z - cc );
else
        error('Input "alt" should be either 0, -1, or 1.')
end
%-----------------------------------------------------------
```

*Example 12.4.* **Nanoscale Probes.** The development of intracellular nanoscale probes against various biomolecules is very important in the furthering of basic studies of cellular biology and pathology. One way to improve the binding properties of these probes is to have multiple binding domains or ligands on the surface of the probes. The more ligands, the greater the chance of binding. One issue that comes up with such probes is whether, due to the multiple ligands, they will cause aggregation of target molecules within the cell. In order to show that new probes do not induce aggregation, researchers in the lab of Dr. Phil Santangelo at Georgia Tech compared the number of granules using a monovalent and a tetravalent probe in a cell plated at the same time and under the same biological conditions. The numbers of granules detected by each probe were recorded, and the researchers were interested to see if there were real differences between the numbers.

Using WSuRT we test the hypothesis of equality of distributions and, subsequently, all theoretical moments at the significance level $\alpha = 0.05$.

```
monovalent =  [117   92   84 213   89   76   96 104 114 142 ...
               122 154 124   65 129   67 100 127   63   82 ...
               114   93 117   83   82   83 111   78   92   91];

tetravalent = [103   78 155 107 113   75   74   80 120 112 ...
               158   72   81 124 110   90   64   74 110 149 ...
                97   70 105   94 110   93 115 114 110 95];

[sumranks1, tstat, pval] = wsurt(monovalent, tetravalent, 0)

% sumranks1 = 925.5000
%
% tstat = 0.1553
%
% pval =   0.8737
```

The statistical evidence that the two samples come from the same distribution is overwhelming since the *p*-value of 0.8737 exceeds the significance level $\alpha = 0.05$.

## 12.6 Kruskal–Wallis Test

The Kruskal–Wallis (KW) test is a generalization of the WSuRT. It is a non-parametric test used to compare three or more samples. It is used to test the null hypothesis that all populations have identical distribution functions against the alternative hypothesis that at least two of the samples differ only with respect to location (median), if at all.

The KW test is an analog to the $F$-test used in one-way ANOVA. While ANOVA tests depend on the assumption that all populations under comparison are independent and normally distributed, the KW test places no such restriction. Suppose the data consist of $k$ independent random samples with sample sizes $n_1, \ldots, n_k$. Let $n = n_1 + \cdots + n_k$.

| Sample 1 | $X_{11},$ | $X_{12},$ | $\ldots$ | $X_{1,n_1}$ |
|---|---|---|---|---|
| Sample 2 | $X_{21},$ | $X_{22},$ | $\ldots$ | $X_{2,n_2}$ |
| $\vdots$ | $\vdots$ | | | |
| Sample $k-1$ | $X_{k-1,1},$ | $X_{k-1,2},$ | $\ldots$ | $X_{k-1,n_{k-1}}$ |
| Sample $k$ | $X_{k1},$ | $X_{k2},$ | $\ldots$ | $X_{k,n_k}$ |

Under the null hypothesis, we can claim that all of the $k$ samples are from a common population.

The expected sum of ranks for sample $i$, $\mathbb{E}(R_i)$, would be $n_i$ times the expected rank for a single observation. That is, $n_i(n+1)/2$, and the variance can be calculated as $\mathbb{V}\mathrm{ar}\,(R_i) = n_i(n+1)(n-n_i)/12$. One way to test $H_0$ is to calculate $R_i = \sum_{j=1}^{n_i} r(X_{ij})$ – the total sum of ranks in sample $i$. The statistic

$$\sum_{i=1}^{k} \left[ R_i - \frac{n_i(n+1)}{2} \right]^2 \tag{12.1}$$

will be large if the samples differ, so the idea is to reject $H_0$ if (12.1) is "too large." However, its distribution is quite messy, even for small samples, so we can use the normal approximation

$$\frac{R_i - \mathbb{E}(R_i)}{\sqrt{\mathbb{V}\mathrm{ar}\,(R_i)}} \overset{\mathrm{appr}}{\sim} \mathcal{N}(0,1) \Rightarrow \sum_{i=1}^{k} \frac{(R_i - \mathbb{E}(R_i))^2}{\mathbb{V}\mathrm{ar}\,(R_i)} \overset{\mathrm{appr}}{\sim} \chi^2_{k-1},$$

where the $\chi^2$-statistic has only $k-1$ degrees of freedom due to the fact that only $k-1$ ranks are unique.

Based on this idea, Kruskal and Wallis (1952) proposed the test statistic

$$H' = \frac{1}{S^2} \left[ \sum_{i=1}^{k} \frac{R_i^2}{n_i} - \frac{n(n+1)^2}{4} \right], \tag{12.2}$$

where

$$S^2 = \frac{1}{n-1} \left[ \sum_{i=1}^{k} \sum_{j=1}^{n_i} r(X_{ij})^2 - \frac{n(n+1)^2}{4} \right].$$

If there are no ties in the data, (12.2) simplifies to

$$H = \frac{12}{n(n+1)} \sum_{i=1}^{k} \frac{1}{n_i} \left[ R_i - \frac{n_i(n+1)}{2} \right]^2. \tag{12.3}$$

Kruskal and Wallis (Fig. 12.2a,b) showed that this statistic has an approximate $\chi^2$-distribution with $k-1$ degrees of freedom.

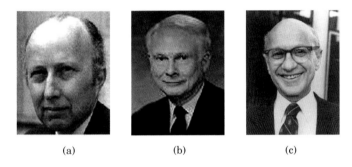

(a)                              (b)                              (c)

**Fig. 12.2** (a) William Henry Kruskal (1919–2005), (b) Wilson Allen Wallis (1912–1998), and (c) Milton Friedman (1912–2006).

The MATLAB routine ◀ kw.m implements the KW test using a vector to represent the responses and another to identify the population from which the response came. Suppose we have the following responses from three treatment groups:

$$(1,3,4), (3,4,5), (4,4,4,6,5).$$

The code for testing the equality of locations of the three populations computes a $p$-value of 0.1428.

```
data   = [ 1 3 4   3 4 5   4 4 4 6 5 ];
belong = [ 1 1 1   2 2 2   3 3 3 3 3 ];
[H, p] = kw(data, belong)

% [H, p] =   3.8923       0.1428
```

*Example 12.5.* The following data are from a classic agricultural experiment measuring crop yield in four different plots. For simplicity, we identify the treatment (plot) using the integers $\{1,2,3,4\}$. The third treatment mean measures far above the rest, and the null hypothesis (the treatment means are equal) is rejected with a $p$-value less than 0.0002.

```
data= [83 91 94 89 89 96 91 92 90 84 91 90 81 83 84 83 ...
   88 91 89 101 100 91 93 96 95 94 81 78  82  81 77 79 81 80];
```

```
belong = [1 1 1 1 1 1 1 1 1 1 2 2 2 2 2 2 2 2 2 2 ...
          3 3 3 3 3 3 3 3 4 4 4 4 4 4 4 4];
[H, p] = kw(data, belong)

%  H  =  20.3371
%  p = 1.4451e-004
```

**Kruskal–Wallis Pairwise Comparisons.** If the KW test detects treatment differences, we can determine if two particular treatment groups (say, $i$ and $j$) are different at level $\alpha$ if

$$\left| \frac{R_i}{n_i} - \frac{R_j}{n_j} \right| > t_{n-k,1-\alpha/2} \sqrt{\frac{S^2(n-1-H')}{n-k} \cdot \left( \frac{1}{n_i} + \frac{1}{n_j} \right)}. \tag{12.4}$$

*Example 12.6.* Since in Example 12.5 we found the four crop treatments significantly different, it would be natural to find out which ones seem better and which ones seem worse. In the table below, we compute the statistic

$$T = \frac{\left| \frac{R_i}{n_i} - \frac{R_j}{n_j} \right|}{\sqrt{\frac{S^2(n-1-H')}{n-k} \left( \frac{1}{n_i} + \frac{1}{n_j} \right)}}$$

for every combination of $1 \le i \ne j \le 4$ and compare it to $t_{30,0.975} = 2.042$.

| $(i,j)$ | 1 | 2 | 3 | 4 |
|---|---|---|---|---|
| 1 | 0 | 1.856 | 1.859 | 5.169 |
| 2 | 1.856 | 0 | 3.570 | 3.363 |
| 3 | 1.859 | 3.570 | 0 | 6.626 |
| 4 | 5.169 | 3.363 | 6.626 | 0 |

This shows that the third treatment is the best, but not significantly different from the first treatment, which is second best. Treatment 2, which is third best, is not significantly different from treatment 1, but is different from treatments 4 and 3.

## 12.7 Friedman's Test

*Friedman's test* is a nonparametric alternative to the randomized block design (RBD) in regular ANOVA. It replaces the RBD when the assumptions of normality are in question or when the variances are possibly different from population to population. This test uses the ranks of the data rather than their raw values to calculate the test statistic. Because the Friedman test does not make distribution assumptions, it is not as powerful as the standard test if the populations are indeed normal.

Milton Friedman (Fig. 12.2c) published the first results for this test, which was eventually named after him. He received the Nobel Prize in economics in 1976, and one of the listed breakthrough publications was his article "The use of ranks to avoid the assumption of normality implicit in the analysis of variance," published in 1937 (Friedman, 1937).

Recall that the RBD design requires repeated measures for each block at each level of treatment. Let $X_{ij}$ represent the experimental outcome of subject (or "block") $i$ with treatment $j$, where $i = 1, \ldots, b$, and $j = 1, \ldots, k$.

| Block | Treatment | | | |
|-------|-----------|-----------|-----|-----------|
|       | 1         | 2         | ... | $k$       |
| 1     | $X_{11}$  | $X_{12}$  | ... | $X_{1k}$  |
| 2     | $X_{21}$  | $X_{22}$  | ... | $X_{2k}$  |
| $\vdots$ | $\vdots$ | $\vdots$ |     | $\vdots$  |
| $b$   | $X_{b1}$  | $X_{b2}$  | ... | $X_{bk}$  |

To form the test statistic, we assign ranks $\{1, 2, \ldots, k\}$ to each *row* in the table of observations. Thus the expected rank of any observation under $H_0$ is $(k + 1)/2$. We next sum all the ranks by columns (by treatments) to obtain $R_j = \sum_{i=1}^{b} r(X_{ij})$, $1 \leq j \leq k$. If $H_0$ is true, the expected value for $R_j$ is $\mathbb{E}(R_j) = b(k + 1)/2$. The statistic

$$\sum_{j=1}^{k} \left( R_j - \frac{b(k + 1)}{2} \right)^2$$

is an intuitive formula to reveal treatment differences. It has expectation $bk(k^2 - 1)/12$ and variance $k^2 b(b - 1)(k - 1)(k + 1)^2/72$. Once normalized to

$$S = \frac{12}{bk(k + 1)} \sum_{j=1}^{k} \left( R_j - \frac{b(k + 1)}{2} \right)^2, \tag{12.5}$$

it has moments $\mathbb{E}(S) = k - 1$ and $\mathbb{V}\mathrm{ar}(S) = 2(k - 1)(b - 1)/b \approx 2(k - 1)$, which coincide with the first two moments of $\chi^2_{k-1}$. Higher moments of $S$ also approximate well those of $\chi^2_{k-1}$ when $b$ is large.

In the case of ties, a modification to $S$ is needed. Let $C = bk(k + 1)^2/4$ and $R^* = \sum_{i=1}^{b} \sum_{j=1}^{k} r(X_{ij})^2$. Then,

$$S' = \frac{k - 1}{R^* - bk(k + 1)^2/4} \left( \sum_{j=1}^{k} R_j^2 - bC \right) \tag{12.6}$$

is also approximately distributed as $\chi^2_{k-1}$.

Although the Friedman statistic makes for a sensible, intuitive test, it turns out there is a better one to use. As an alternative to $S$ (or $S'$), the test statistic

$$F = \frac{(b-1)S}{b(k-1)-S}$$

is approximately distributed as $F_{k-1,(b-1)(k-1)}$, and tests based on this approximation are generally superior to those based on chi-square tests that use $S$. For details on the comparison between $S$ and $F$, see Iman and Davenport (1980)

*Example 12.7.* In an evaluation of vehicle performance, six professional drivers (labeled I, II, III, IV, V, VI) evaluated three cars (A, B, and C) in a randomized order. Their grades concern only the performance of the vehicles and supposedly are not influenced by the vehicle brand name or similar exogenous information. Here are their rankings on a scale of 1 to 10:

| Car | I | II | III | IV | V | VI |
|-----|---|----|-----|----|---|----|
| A | 7 | 6 | 6 | 7 | 7 | 8 |
| B | 8 | 10 | 8 | 9 | 10 | 8 |
| C | 9 | 7 | 8 | 8 | 9 | 9 |

To use the MATLAB procedure ◢ friedmanGT the input data should be formatted as a matrix in which the rows are blocks and the columns are treatments.

```
data = [7  8  9;   6  10  7;   6  8  8; ...
        7  9  8;   7  10  9;   8  8  9];
[S,F,pS,pF] = friedmanGT(data)

%S = 8.2727
%F = 11.0976
%pS = 0.0160
%pF = 0.0029
% pF p-value is more reliable
```

**Friedman's Pairwise Comparisons.** If the $p$-value is small enough to warrant multiple comparisons of treatments, we consider two treatments $i$ and $j$ to be different at level $\alpha$ if

$$\left| R_i - R_j \right| > t_{(b-1)(k-1),1-\alpha/2} \sqrt{2 \cdot \frac{bR^* - \sum_{j=1}^{k} R_j^2}{(b-1)(k-1)}}. \tag{12.7}$$

*Example 12.8.* From Example 12.7, the three cars A, B, and C are considered significantly different at test level $\alpha = 0.01$ (if we use the $F$-statistic). We can use the function ◀ friedmanpairwise(x,i,j,alpha) to make a pairwise comparison between treatments $i$ and $j$ at level alpha. The output = 1 if treatments $i$ and $j$ are different; otherwise it is 0. The Friedman pairwise comparison reveals that car A is rated significantly lower than both cars B and C, but cars B and C are not considered to be different.

## 12.8 Walsh Nonparametric Test for Outliers*

Suppose that $r$ outliers are suspect, where $r \geq 1$ and fixed. Order observations $X_1, \ldots, X_n$ and obtain the order statistic $X_{(1)} \leq X_{(2)} \leq \cdots \leq X_{(n)}$, and set the significance level $\alpha$. The test is nonparametric and no assumptions about the underlying distribution are made.

We will explain the steps and provide a MATLAB implementation, but readers interested in details are directed to Walsh (1962). The Walsh recipe has the following steps, following Madansky (1988).

**STEP 1:** Calculate $c = \lfloor \sqrt{2n} \rfloor$, where $\lfloor x \rfloor$ is the largest integer $\leq x$, $b^2 = 1/\alpha$, $k = r + c$, and $a = (1 + b\sqrt{\frac{c-b^2}{c-1}})/(c - b^2 - 1)$.
**STEP 2:** The smallest $r$ values $X_{(1)} \leq \cdots \leq X_{(r)}$ are outliers if

$$rL = X_{(r)} - (1+a)X_{(r+1)} + aX_k < 0,$$

and the largest $r$ values $X_{(n-r+1)} \leq \cdots \leq X_{(n)}$ are outliers if

$$rU = X_{(n-r+1)} - (1+a)X_{(n-r)} + aX_{n-k+1} > 0.$$

The sample size has to satisfy $n \geq \frac{1}{2}(1 + \frac{1}{\alpha})^2$. To achieve $\alpha = 0.05$, a sample size of at least 221 is needed. As an outcome, one either rejects no outliers, rejects the smallest $r$, the largest $r$, or even both the smallest and largest $r$, thus potentially rejecting a total of $2r$ observations. Here is an example of Walsh's procedure implemented by m-function ◀ walshnp.m.

```
dat = normrnd(0, 2, [300 1]);
data = [-11.26; dat; 13.12];
walshnp(data) %default r=1, alpha=0.05
 %Lower outliers are: -11.26
 %Upper outliers are: 13.12
 %ans = -11.2600   13.1200
```

## 12.9 Exercises

12.1. **Friday the 13th.** The following data set is part of the larger study from Scanlon et al. (1993) titled "Is Friday the 13th bad for your health?" The data analysis in this paper addresses the issues of how superstitions regarding Friday the 13th affect human behavior. The authors reported and analyzed data on traffic accidents for Friday the 6th and Friday the 13th between October 1989 and November 1992. The data consist of the number of patients accepted in SWTRHA (South West Thames Regional Health Authority, London) hospital on the dates of Friday the 6th and Friday the 13th.

| Year, month | # of accidents Friday 6th | Friday 13th | Sign |
|---|---|---|---|
| 1989, October | 9 | 13 | − |
| 1990, July | 6 | 12 | − |
| 1991, September | 11 | 14 | − |
| 1991, December | 11 | 10 | + |
| 1992, March | 3 | 4 | − |
| 1992, November | 5 | 12 | − |

Use the sign test at the level $\alpha = 10\%$ to test the hypothesis that the "Friday the 13th effect" is present. The m-file ◀ signtst.m could be applied.

12.2. **Reaction Times.** In Exercise 10.22 the paired $t$-test was used to assess the differences between reaction times to red and green lights. Repeat this analysis using both sign test and WSiRT.

12.3. **Simulation.** To compare the $t$-test with the Wilcoxon signed-rank test, set up the following simulation in MATLAB:
(1) Generate $n = 20$ observations from $\mathcal{N}(0,1)$ as your first sample $X$.
(2) Form Y = X + randn(size(x)) + 0.5 as your second sample paired with the first.
(3) For the test of $H_0 : \mu_1 = \mu_2$ versus $H_1 : \mu_1 < \mu_2$, perform the $t$-test at $\alpha = 0.05$.
(4) Run the Wilcoxon signed-rank test.
(5) Repeat this simulation 1000 times and compare the powers of the tests by counting the number of times $H_0$ was rejected.

12.4. **Grippers.** Measurements of the left- and right-hand gripping strengths of ten left-handed writers are recorded.

| Person | 1 | 2 | 3 | 4 | 5 | 6 | 7 | 8 | 9 | 10 |
|---|---|---|---|---|---|---|---|---|---|---|
| Left hand ($X$) | 140 | 90 | 125 | 130 | 95 | 121 | 85 | 97 | 131 | 110 |
| Right hand ($Y$) | 138 | 87 | 110 | 132 | 96 | 120 | 86 | 90 | 129 | 100 |

(a) Does the data provide strong evidence that people who write with their left hand have a greater gripping strength in the left hand than they do in their right hand? Use the Wilcoxon signed-rank test and $\alpha = 0.05$.

(b) Would you change your opinion on the significance if you used the paired $t$-test?

12.5. **Iodide and Serum Concentration of Thyroxine.** The effect of iodide administration on serum concentration of thyroxine ($T_4$) was investigated in Vagenakis et al. (1974). Twelve normal volunteers (9 male and 3 female) were given 190 mg iodide for 10 days. The measurement $X$ is an average of $T_4$ in the last 3 days of administration while $Y$ is the mean value in three successive days after the administration stopped.

| Subject | 1 | 2 | 3 | 4 | 5 | 6 | 7 | 8 | 9 | 10 | 11 | 12 |
|---|---|---|---|---|---|---|---|---|---|---|---|---|
| Iodide ($X$) | 7.9 | 9.1 | 9.2 | 8.1 | 4.2 | 7.2 | 5.4 | 4.9 | 6.6 | 4.7 | 5.2 | 7.3 |
| Control ($Y$) | 10.2 | 10.2 | 11.5 | 8.0 | 6.6 | 7.4 | 7.7 | 7.2 | 8.2 | 6.2 | 6.0 | 8.7 |

Assume that the difference $D = X - Y$ has a symmetric distribution. Compare the $p$-values of WSiRT and the paired $t$-test in testing the hypothesis that the mean of control measurements exceed that of the iodide measurements.

12.6. **Weightlifters.** Blood lactate levels were determined in a group of amateur weightlifters following a competition of 10 repetitions of 5 different upper-body lifts, each at 70% of each lifter's single maximum ability. A sample of 17 randomly selected individuals was tested: 10 males and 7 females. The following table gives the blood lactate levels in female and male weightlifters, in units of mg/100 ml of blood.

| Gender | N | Blood lactate |
|---|---|---|
| Female | 7 | 7.9 8.2 8.7 12.3 12.5 16.7 20.2 |
| Male | 10 | 5.2 5.4 6.7  6.9  8.2  8.7 14.2 14.2 17.4 20.3 |

Test an $H_0$ that blood lactate levels were not different between the two genders at the level $\alpha = 0.05$. Assume that the data, although consisting of measurements of continuous variables, are **not** distributed normally or variances are possibly heteroscedastic (nonequal variances). In other words, the traditional $t$-test may not be appropriate.

12.7. **Cartilage Thickness in Two Osteoarthritis Models.** Osteoarthritis (OA), characterized by gradual degradation of the cartilage extracellular matrix, articular cartilage degradation, and subchondral bone remodeling, is the most common degenerative joint disease in humans. One of the most common ways for researchers to study the progression of OA in a controlled manner is to use small animal models. In a study conducted in the laboratory of Robert Guldberg at the Georgia Institute of Technology, a group of 14 rats was randomly divided into two subgroups, and each subgroup was subjected to an induced form of OA. One of the subgroups ($n_1 = 7$) was subjected to a chemically induced form of OA via an intra-articular injection of monosodium iodoacetate (MIA), while the other subgroup ($n_2 = 7$) was subjected to a surgically induced form of OA by transecting the medial meniscus (MMT). In this study, all rats had OA induced by either MIA or MMT on the left knee, with the right knee serving as a contralateral control.

The objectives of this study were to quantify changes in cartilage thickness of a selected area of the medial tibial plateau, compare thickness values between the treatments and the controls, and determine if OA induced by MIA produced different results from OA inducted by MMT. Cartilage thickness values were measured 3 weeks after the treatments using an ex vivo micro-CT scanner; the data are provided in the table below.

| Rat | MIA Treated | Control | Rat | MMT Treated | Control |
|-----|-------------|---------|-----|-------------|---------|
| 1 | 0.1334 | 0.2194 | 8 | 0.2569 | 0.2726 |
| 2 | 0.1214 | 0.1929 | 9 | 0.2101 | 0.2234 |
| 3 | 0.1276 | 0.1833 | 10 | 0.1852 | 0.2216 |
| 4 | 0.1152 | 0.1879 | 11 | 0.1798 | 0.1905 |
| 5 | 0.1047 | 0.2529 | 12 | 0.1049 | 0.1444 |
| 6 | 0.1312 | 0.2527 | 13 | 0.2649 | 0.2841 |
| 7 | 0.1222 | 0.2595 | 14 | 0.2383 | 0.2731 |

(a) Using the Wilcoxon signed-rank test, test the hypothesis that the difference between MIA-treated rats and its control thickness is significant.
(b) Find the difference in thickness (treatment–control) for MIA and MMT and test the hypothesis that they are the same. Apply a nonparametric version of a two-sample $t$-test on the differences.
Conduct both tests at a 5% significance level.

12.8. **A Claim.** Professor Scott claims that 50% of his students in a big class get a final score of 90 and higher.
A suspicious student asks 17 randomly selected students from Professor Scott's class and they report the following scores:

$$80 \ 81 \ 87 \ 94 \ 79 \ 78 \ 89 \ 90 \ 92 \ 88 \ 81 \ 79 \ 82 \ 79 \ 77 \ 89 \ 90$$

Test the hypothesis that Professor Scott's claim does not conform to the evidence, i.e., that the 50th percentile (0.5-quantile, median) is different than 90. Use $\alpha = 0.05$.

12.9. **Claustrophobia.** Sixty subjects seeking treatment for claustrophobia are independently sorted into two groups, the first of size $n = 40$ and the second of size $m = 20$. The members of the first group each individually receive treatment A over a period of 15 weeks, while those of the second group receive treatment B. The investigators' directional hypothesis is that treatment A will prove to be more effective. At the end of the experimental treatment period, the subjects are individually placed in a series of claustrophobia test situations, knowing that their reactions to these situations are being recorded on videotape. Subsequently three clinical experts, uninvolved in the experimental treatment and not knowing which subject received which treatment, independently view the videotapes and rate each subject according to the degree of claustrophobic tendency shown in the test situations. Each subject is rated by the experts on a scale of 0 (no claustrophobia) to 10 (an extreme claustrophobia). The following table shows the average ratings for each subject in each of the two groups.

| A |
| --- |
| 4.6 4.7 4.9 5.1 7.0 4.9 |
| 5.1 5.2 5.5 4.8 5.7 5.0 |
| 5.8 6.1 6.5 7.0 6.4 5.2 |
| 4.6 4.7 4.9 6.4 5.9 4.7 |
| 5.8 5.2 5.4 6.1 7.7 6.2 |
| 5.8 5.1 6.5 2.2 6.9 5.0 |
| 6.5 7.2 8.2 6.7 |

| B |
| --- |
| 5.2 5.3 5.4 7.7 8.1 4.9 |
| 5.6 6.2 6.3 7.0 7.0 7.8 |
| 6.8 7.7 8.0 6.6 5.5 8.2 |
| 8.1 5.0 |

The investigators expected treatment A to prove more effective, and sure enough it is group A that appears to show the lower mean level of claustrophobic tendency in the test situations.

Using Wilcoxon's sum rank test, test the hypothesis $H_0$ that treatments A and B are of the same effectiveness, versus the alternative that treatment A is more effective.

12.10. **Nonparametric Stats with Raynaud's Phenomenon.** (a) Refer to Raynaud's phenomenon data from Exercise 11.21. Using Wilcoxon's signed-rank test, compare the responses (number of attacks) for drug/placebo effect. Ignore the Period variable. Compare this result with the result from a paired $t$-test.

(b) Compare the drug/placebo effect using Bayesian inference. Write a WinBUGS program that will read the drug/placebo info, take the difference between the measurements, d, and model it as a normal, with noninformative priors on the mean mu and precision prec. Check if the credible set for mu contains 0 and draw the appropriate conclusion.

12.11. **Cotinine and Nicotine Oxide.** Garrod et al. (1974) measured the nicotine metabolites, cotinine, and nicotine-1'-$N$-oxide in 24-hour urine collections from normal healthy male smokers and smokers affected with cancer of the urinary bladder. The data, also discussed by Daniel (1978), show the ratio of cotinine to nicotine-1'-$N$-oxide in the two groups of subjects.

| | Ratio | | | | | | | | | | | |
| --- | --- | --- | --- | --- | --- | --- | --- | --- | --- | --- | --- | --- |
| Cancer | 5.0 | 8.3 | 6.7 | 3.0 | 2.5 | 12.5 | 2.4 | 5.5 | 5.2 | 21.3 | 5.1 | 1.6 |
| patients | 2.1 | 4.6 | 3.2 | 2.2 | 7.0 | 3.3 | 6.7 | 11.1 | 3.4 | 5.9 | 27.4 | |
| Control | 2.3 | 1.9 | 3.6 | 2.5 | 0.75 | 2.5 | 2.1 | 1.1 | 2.3 | 2.2 | 3.5 | 1.8 |
| subjects | 2.3 | 1.4 | 2.1 | 2.0 | 2.3 | 2.4 | 3.6 | 2.6 | 1.5 | | | |

Test the difference between the two populations using WSuRT, at $\alpha = 0.05$.

12.12. **Coagulation Times.** In Example 11.1 a standard one-way ANOVA gave $p$-value of 4.6585e-05. Repeat the test of equality of means using the Kruskal–Wallis procedure. If the test turns out significant, compare the means using the Kruskal–Wallis pairwise comparisons.

12.13. **Blocking by Rats.** In Example 11.5 factor `Procedure` was found significant with $p$-value of 8.7127e-04. Is this factor significant according to Friedman's test?

---

| **MATLAB FILES USED IN THIS CHAPTER** |
http://springer.bme.gatech.edu/Ch12.NP/

claustrophobia.m, friedmanGT.m, friedmanpairwise.m, grubbs.m, kw.m, miammt.m, NPdemo.m, npexamples.m, ranks.m, signtst.m, walshnp.m, wsirt.m, wsurt.m

---

# CHAPTER REFERENCES

Bradley, J. V. (1968). *Distribution Free Statistical Tests*. Prentice Hall, Englewood Cliffs.

Conover, W. J. (1999). *Practical Nonparametric Statistics*, 3rd edn. Wiley, New York.

Daniel, W. W. (1978). *Applied Nonparametric Statistics*, Houghton Mifflin, Boston.

Friedman, M. (1937). The use of ranks to avoid the assumption of normality implicit in the analysis of variance. *J. Am. Stat. Assoc.*, **32**, 675–701.

Garrod, J. W., Jenner, P., Keysell, G. R., Mikhael, B. R. (1974). Oxidative metabolism of nicotine by cigarette smokers with cancer of urinary bladder. *J. Nat. cancer Inst.*, **52**, 1421–1924.

Iman, R. L. and Davenport, J. M. (1980). Approximations of the critical region of the Friedman statistic. *Commun. Stat. A.*, **9**, 571–595.

Kruskal, W. H. and Wallis, W. A. (1952). Use of ranks in one-criterion variance analysis. *J. Am. Stat. Assoc.*, **47**, 583–621.

Kvam, P. and Vidakovic, B. (2007). *Nonparametric Statistics with Applications to Science and Engineering*. Wiley, Hoboken.

Madansky, A. (1988). *Prescriptions for Working Statisticians*. Springer, Berlin Heidelberg New York.

Scanlon, T. J., Luben, R. N., Scanlon, F. L., and Singleton, N. (1993). Is Friday the 13th bad for your health? *Br. Med. J.*, **307**, 1584–1586.

Vagenakis, A. G., Rapoport, B., Azizi, F., Portnay, G. I., Braverman, L. E., and Ingbar, S. H. (1974). Hyperresponse to thyrothropin-releasing hormone acompanying small decreases in serum thyroid concentrations. *J. Clin. Invest.*, **54**, 913–918.

Walsh, J. E. (1962). *Handbook of Nonparametric Statistics I and II*. Van Nostrand, Princeton.

Wilcoxon, F. (1945). Individual comparisons by ranking methods. *Biometrics*, **1**, 80–83.

# Chapter 13
# Goodness-of-Fit Tests

*When schemes are laid in advance, it is surprising how often the circumstances fit in with them.*

– William Osler, Canadian Physician (1849–1919)

---

### WHAT IS COVERED IN THIS CHAPTER

- Quantile–Quantile Plots
- Pearson's $\chi^2$-Test
- Kolmogorov–Smirnov Goodness-of-Fit Test
- Smirnov's Two-Sample Test
- Moran's Test
- Testing Departures from Normality

---

## 13.1 Introduction

Goodness-of-fit tests are batteries of tests that test that the distribution of a sample is equal to some fixed-in-advance distribution. We already saw Q–Q plots in Chap. 5 where the samples were compared to some theoretical distributions but in a descriptive fashion, without formal inference. In this chapter we discuss the celebrated Pearson's $\chi^2$-test and the Kolmogorov–Smirnov (KS) test.

## 13.2  Quantile–Quantile Plots

*Quantile–quantile*, or Q–Q, plots are a popular and informal diagnostic tool for assessing the distribution of data and comparing two distributions. There are two kinds of Q–Q plot, one that compares an empirical distribution with a theoretical one and another that compares two empirical distributions.

We will explain the plots with an example. Suppose that we generated some random sample $X$ from a uniform $\mathcal{U}(-10, 10)$ distribution of size $n = 200$. Suppose for a moment that we do not know the distribution of this data and want to check for normality by a Q–Q plot. We generate quantiles of a normal distribution for $n$ equally spaced probabilities between 0 and 1, starting at $0.5/n$ and ending at $1 - 0.5/n$. The empirical quantiles for $X$ are simply the elements of a sorted sample. If the empirical distribution matches the theoretical, the Q–Q plot is close to a straight line.

```
n=200;
X = unifrnd(-10, 10, [1 n]);
q=((0.5/n):1/n:(1-0.5/n));
qY = norminv(q);
qZ = unifinv(q);
qX = sort(X);
figure(1); plot(qX, qY, '*')
figure(2); plot(qX, qZ, '*')
```

The Q–Q plot is given in Fig.13.1a. Note that the ends of the Q–Q plot curve up and down from the straight line, indicating that the sample is not normal.

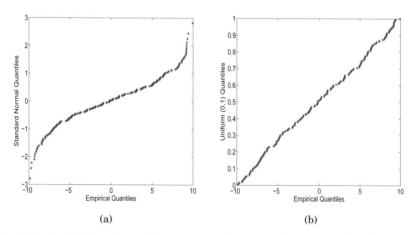

(a)                                                    (b)

**Fig. 13.1** (a) Q–Q plot of a uniform sample against normal quantiles. (b) Q–Q plot of a uniform sample against uniform quantiles.

If, on the other hand, the statement qY=norminv(q) is replaced by qY=unifinv(q), then the Q–Q plot indicates that the match is good and that the sample is uniform (Fig. 13.1b). Note that while the moments (mean and variance) of the sample and theoretical distributions may differ, matching the family of distributions is important. If the mean and variance differ, the straight line in the plot is shifted and has a different slope. Sometimes the straight line passing through the first and third quartiles is added to the Q–Q plot for reference.

MATLAB has built-in functions qqplot, normplot, probplot, and cdfplot for a variety of visualization and diagnostic tasks.

In our example we used $p_i$-quantiles for $p_i = (i - 0.5)/n$, $i = 1, \ldots, n$. In general, one can use $p_i = (i - c)/(n - 2c + 1)$ for any $c \in [0, 1]$, and popular choices are $c = 1/3$, $c = 1/2$ (our choice), and $c = 1$. The choice $c = 0.3175$ is recommended if the distributions are long-tailed since the extreme points on a Q–Q plot are more stable. This choice of $c$ corresponds to medians of order statistics from a uniform distribution on (0,1) and has invariance property, $F^{-1}(p_i)$ are medians of order statistics from $F$.

Multivariate extensions are possible. A simple procedure that can be used to assess multivariate normality is described next.

Given multivariate data $y_i = (y_{i1}, y_{i2}, \ldots, y_{ip})$, $i = 1, \ldots, n$ one finds

$$d_i^2 = (y_i - \bar{y}_i) \, S^{-1} \, (y_i - \bar{y}_i)', \tag{13.1}$$

where $\bar{y}_i = \frac{1}{n} \sum_{i=1}^n y_i$ is the sample mean and $S = \frac{1}{n-1} \sum_{i=1}^n (y_i - \bar{y}_i)'(y_i - \bar{y}_i)$ is the sample covariance matrix. If the $y_i$s are distributed as multivariate normal, then the $d_i^2$s have an approximately $\chi_p^2$-distribution, and plotting the empirical quantiles of $d_i^2$ against the corresponding quantiles of $\chi_p^2$ assesses the goodness of a multivariate normal fit.

*Example 13.1.* The data set ▨ pearson.dat compiled by K. Pearson contains the hereditary stature of 1078 fathers and their sons. This data set will be revisited in Chap. 16 in the context of regression. We will visualize the agreement of pairs of heights to a bivariate normal distribution by inspecting a Q–Q plot. We will calculate and sort $d^2$s in (13.1) and plot them against quantiles of a $\chi_2^2$-distribution (Fig. 13.2).

```
%qqbivariatenormal.m
load 'pearson.dat'
S = cov(pearson)
[m n] = size(pearson)    %[1078 2]
mp = repmat(mean(pearson), m, 1);
d2 = diag((pearson - mp) * inv(S)  * (pearson - mp)');

% chi-square (2 df) quantiles
p = 0.5/m:1/m:1;
quanch2 = chi2inv(p,2);
```

```
figure(1)
loglog(quanch2, sort(d2),'o','Markersize',msize,...
    'MarkerEdgeColor','k', 'MarkerFaceColor','g')
hold on
loglog([0.0002 50],[0.0002 50],'r','linewidth',lw)
axis([10^-4, 100, 10^-4, 100])
xlabel('Quantiles of chi^2_2 ')
ylabel('Sorted d^2')
```

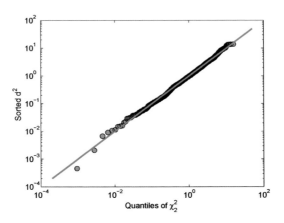

**Fig. 13.2** Sorted values of $d^2$ plotted against quantiles of a $\chi_2^2$-distribution. The linearity of the plot indicates a good fit of Pearson's data to a bivariate normal distribution.

*Example 13.2.* **Poissonness Plots.** Poissonness plots were developed by Hoaglin (1980) and consist of assessing the goodness of fit of observed frequencies to frequencies corresponding to Poisson-distributed data. Suppose that frequencies $n_i, i = 1, \ldots, k$ corresponding to realizations $0, 1, \ldots, k-1$ are observed and that for $i \geq k$ all $n_i = 0$. Let $N = \sum_{i=1}^{k} n_i$.

Then, the frequency $n_i$ and the theoretical counterpart $e_i = N \times \frac{\lambda^i}{i!} \exp\{-\lambda\}$ should be close, if the data are consistent with a Poisson distribution,

$$n_i \sim N \times \frac{\lambda^i}{i!} \exp\{-\lambda\}.$$

By taking the logarithms of both sides one gets

$$\log n_i + \log(i!) - \log N \sim \log \lambda \cdot i - \lambda.$$

Thus, the plot of $\log n_i + \log(i!) - \log N$ against integer $i$ should be linear with slope $\log \lambda$ if the frequencies are consistent with a Poisson $\mathscr{P}oi(\lambda)$ distribution. Such a plot is called a *poissonness plot*.

The script  poissonness.m illustrates this. $N = 50{,}000$ Poisson random variables with $\lambda = 4$ are generated, observed frequencies are calculated, and a Poissonness plot is made (Fig.13.3).

```
rand('state',2)
N = 50000;   xx = poissrnd(4, 1, N); %simulate N Poisson(4) rv's.
% observed frequencies
ni  =   [sum(xx==0) sum(xx==1) sum(xx==2) sum(xx==3)  sum(xx==4) ...
             sum(xx==5) sum(xx==6) sum(xx==7) sum(xx==8)  sum(xx==9) ...
        sum(xx==10) sum(xx==11) sum(xx==12) sum(xx==13)  sum(xx==14)]' ;

i=(0:14)'
poissonness = log(ni) + log(factorial(i)) - log(N);

plot(i, poissonness,'o')
xlabel('i')
ylabel('log(n_i) + log(i!) - log(N)')
hold on
plot(i, log(4)*i - 4,'r:') %theoretical line with lambda=4
[beta] = regress(poissonness, [ones(size(i)) i])

% beta =
%    -4.0591
%     1.4006
%
%     Slope in the linear fit is 1.4006
%     lambda can be estimated as  exp(1.4006) = 4.0576,
%     also, as negative intercept, 4.0509
```

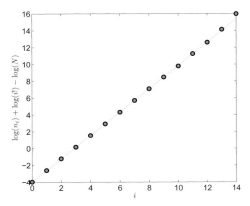

**Fig. 13.3** Poissonness plot for simulated example involving $N = 50{,}000$ Poisson $\mathscr{P}oi(4)$ random variates. The slope of this plot estimates $\log \lambda$ while the intercept estimates $-\lambda$. The theoretical line, $\log(4)i - 4$ (*red*), fits the points $(i, \log(n_i) + \log(i!) - \log(N))$ well.

## 13.3 Pearson's Chi-Square Test

Pearson's $\chi^2$-test (Pearson, 1900) is the first formally formulated testing procedure in the history of statistical inference. In his 1984 *Science* article entitled "Trial by number," Ian Hacking says that the goodness-of-fit chi-square test introduced by Karl Pearson (Fig. 13.4) *ushered in a new kind of decision making* and gives it a place among the top 20 discoveries since 1900 considering all branches of science and technology.

**Fig. 13.4** Karl Pearson (1857–1936).

Suppose that $X_1, X_2, \ldots, X_n$ is a sample from an unknown distribution. We are interested in testing that this distribution is equal to some specific distribution $F_0$, i.e., in testing the goodness-of-fit hypothesis $H_0 : F_X(x) = F_0(x)$.

Suppose that the domain $(a, b)$ of the distribution $F_0$ is split into $r$ nonoverlapping intervals, $I_1 = (a, x_1], I_2 = (x_1, x_2] \ldots I_r = (x_{r-1}, b)$. Such intervals have probabilities $p_1 = F_0(x_1) - F_0(a)$, $p_2 = F_0(x_2) - F_0(x_1)$, ..., $p_r = F_0(b) - F_0(x_{r-1})$, under the assumption that $H_0$ is true. Of course it is understood that the observations belong to the domain of $F_0$; if this is not the case, the null hypothesis is automatically rejected.

Let $n_1, n_2, \ldots, n_r$ be the observed frequencies of observations falling in the intervals $I_1, I_2, \ldots, I_r$. In this notation, $n_1$ is the count of observations from the sample $X_1, \ldots, X_n$ that fall in the interval $I_1$. Of course, $n_1 + \cdots + n_r = n$ since the intervals partition the domain of the sample.

The discrepancy between observed frequencies $n_i$ and frequencies under $F_0$, $np_i$ is the rationale for forming the statistic

$$\chi^2 = \sum_{i=1}^{r} \frac{(n_i - np_i)^2}{np_i},\tag{13.2}$$

which has a $\chi^2$-distribution with $r - 1$ degrees of freedom. Alternative representations include

$$\chi^2 = \sum_{i=1}^{r} \frac{n_i^2}{np_i} - n \quad \text{and} \quad \chi^2 = n \left[ \sum_{i=1}^{r} \left( \frac{\hat{p}_i}{p_i} \right) \hat{p}_i - 1 \right],$$

where $\hat{p}_i = n_i/n$.

In some cases, the distribution under $H_0$ is not *fully* specified; for example, one might conjecture that the data is exponentially distributed without knowing the exact value of $\lambda$. In this case, the unknown parameter can be estimated from the sample.

If $k$ parameters needed to fully specify $F_0$ are estimated from the sample, the $\chi^2$-statistic in (13.2) has a $\chi^2$-distribution with $r - k - 1$ degrees of freedom. A degree of freedom in the test statistic is lost for each estimated parameter.

Firm recommendations on how to select the intervals or even the number of intervals for a $\chi^2$-test do not exist. For a continuous distribution function one may make the intervals approximately equal in probability. Practitioners may want to arrange interval selection so that all $np_i > 1$ and that at least 80% of the $np_i$s exceed 5. The rule of thumb is $n \geq 10$, $r \geq 3$, $n^2/r \geq 10$, $np_i \geq 0.25$.

Some statisticians refer to the $\chi^2$-test as Pearson's "poorness-of-fit" test. This attribute is appropriate since the test can only conclude that the model does not fit the data well. If the model fits the data well, then it may not be unique and one could possibly find other models that would pass the test as well.

When some of the expected frequencies are less than 5, or when the number of classes is small, then the $\chi^2$-test with Yates (Yates, 1934) corrections is recommended,

$$\chi^2 = \sum_{i=1}^{r} \frac{(|n_i - np_i| - 0.5)^2}{np_i},$$

which, under $H_0$, has a $\chi^2$-distribution with $r - 1$ degrees of freedom.

*Example 13.3.* **Weldon's 26,306 Rolls of 12 Dice.** Pearson (1900) discusses Weldon's data (table below) as the main illustration for his test. Raphael Weldon, an evolutionary biologist and founder of biometry, rolled 12 dice simultaneously and recorded the number of times a 5 or a 6 was rolled. In his letter to Galton, dated 2 February 1894, Weldon provided the data and asked for Galton's opinion about their validity. Three contemporary British statisticians, Pearson, Edgeworth, and Yule, have also considered Weldon's data. An interesting historic account can be found in Kemp and Kemp (1991).

The results from 26,306 rolls of 12 dice are summarized in the following table:

| No. of dice resulting in $\boxed{\because}$ or $\boxed{::}$ when 12 dice are rolled | Observed frequency |
|:---:|:---:|
| 0 | 185 |
| 1 | 1149 |
| 2 | 3265 |
| 3 | 5475 |
| 4 | 6114 |
| 5 | 5194 |
| 6 | 3067 |
| 7 | 1331 |
| 8 | 403 |
| 9 | 105 |
| 10 | 14 |
| 11 | 4 |
| 12 | 0 |

In Pearson (1900), the value of $\chi^2$-statistic was quoted as 43.9, which is slightly different than the correct value. The discrepancy is probably due to the use of expected frequencies rounded to the nearest integer and due to the accumulation of rounding errors when calculating $\chi^2$.

To find the expected frequencies, recall the binomial distribution. The number of times $\boxed{\because}$ or $\boxed{::}$ was rolled with 12 dice is binomial: $\mathcal{B}in(12, 1/3)$. To find the expected frequencies, multiply the total number of rolls 26,306 by the expected (theoretical) probabilities obtained from $\mathcal{B}in(12, 1/3)$, as in the following MATLAB output:

```
obsfreq = [ 185  1149  3265  5475  6114  5194 ...
            3067  1331   403   105    14     4    0];

n = sum(obsfreq) %26306
expected = n * binopdf(0:12, 12, 1/3)

%expected =
%   1.0e+003 *
%
%   0.2027  1.2165  3.3454  5.5756  6.2726  5.0180  2.9272
%   1.2545  0.3920  0.0871  0.0131  0.0012  0.0000

chisqs = (obsfreq - expected).^2./expected
%chisqs =
%   1.5539  3.7450  1.9306  1.8155  4.0082  6.1695  6.6772
%   4.6635  0.3067  3.6701  0.0665  6.6562  0.0495

chi2 = sum(chisqs)
%chi2 =   41.3122

pval=1 - chi2cdf(chi2, 13-1)
       % pval = 4.3449e-005
```

```
crit = chi2inv(0.95, 13-1)
    % crit = 21.0261
```

As evident from the MATLAB output, the observations are not supporting the fact that the dice were fair ($p$-value of 4.3449e-05), see also Fig. 13.5.

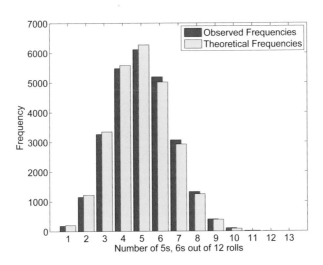

**Fig. 13.5** Bar plot of observed frequencies and theoretical frequencies. Although the *graphs* appear close, the large sample size makes the discrepancy of this size very unlikely. It is almost certain that one or more of Weldon's 12 dice were not well balanced.

*Example 13.4.* **Is the Process Poisson?** The Poisson process is one of the most important stochastic models. For example, random arrival times of patients to a clinic are well modeled by a Poisson process. This means that in any interval of time, say $[0, t]$, the number of arrivals is Poisson with parameter $\lambda t$. For such Poisson processes, the interarrival times follow an exponential distribution with density $f(t) = \lambda e^{-\lambda t}$, $t \geq 0, \lambda > 0$. It is often of interest to establish the Poissonity of a process since many theoretical results are available for such processes, which are ubiquitous in queueing theory and various engineering applications.

The interarrival times of the arrival process were recorded, and it was observed that $n = 109$ recorded times could be categorized as follows:

| Interval | $0 \leq T < 1$ | $1 \leq T < 2$ | $2 \leq T < 3$ | $3 \leq T < 4$ | $4 \leq T < 5$ | $5 \leq T < 6$ | $T \geq 6$ |
|---|---|---|---|---|---|---|---|
| Frequency | 34 | 20 | 16 | 15 | 9 | 7 | 8 |

A simple calculation determined that the sample mean was $\overline{T} = 5/2$. Test the hypothesis that the process described with the above interarrival times is Poisson, at level $\alpha = 0.05$.

Given this data, one should test the hypothesis that the interarrivals times are exponential. The density $f(t) = \lambda e^{-\lambda t}$, $t \geq 0, \lambda > 0$ corresponds to the CDF $F(t) = 1 - e^{-\lambda t}$, $t \geq 0, \lambda > 0$, and the theoretical probability of an interval $[a, b]$ is $F(b) - F(a)$.

But we first need to estimate the parameter $\lambda$ in order to calculate the (theoretical) probabilities. A standard estimator for the parameter $\lambda$ is $\hat{\lambda} = 1/\overline{T} = 1/(5/2) = 0.4$.

The theoretical frequencies of intervals $[3, 4]$ and $[4, 5]$ are $p_4 = 109 \cdot (F(4) - F(3)) = 109 \cdot (1 - e^{-0.4 \cdot 4} - (1 - e^{-0.4 \cdot 3})) = 109 \cdot (e^{-0.4 \cdot 3} - e^{-0.4 \cdot 3}) = 0.099298 \cdot 109 = 10.823$ and $0.066561 \cdot 109 = 7.255$, respectively, yielding $\chi^2$ equal to

$$\frac{(34 - 35.9335)^2}{35.935} + \frac{(20 - 24.089)^2}{24.089} + \frac{(16 - 16.147)^2}{16.147} + \frac{(15 - 10.823)^2}{10.823} +$$
$$\frac{(9 - 7.255)^2}{7.255} + \frac{(7 - 4.863)^2}{4.863} + \frac{(8 - 9.888)^2}{9.888} = 4.13.$$

The number of degrees of freedom is $df = 7 - 1 - 1 = 5$ and the 95% quantile for $\chi_5^2$ is chi2inv(0.95,5)=11.071. Thus, we do not reject the hypothesis that the interarrival times are exponential, i.e., the observed process is consistent with a Poisson process.

*Example 13.5.* A sample of n=1000 exponential $\mathcal{E}(1/2)$ random variates is generated. We pretend that the generating distribution is unknown. Using MATLAB's chi2gof function we test the consistency of the generated data with an exponential distribution with the rate $\lambda$ estimated from the sample.

```
X = exprnd(2, [1, 1000]);
[h,p,stats] = ...
chi2gof(X,'cdf',@(z)expcdf(z,mean(X)),'nparams',1,'nbins',7)
%
%h = 0
%p = 0.6220
%stats = chi2stat: 2.6271
%            df: 4
%         edges: [1x7 double]
%             O: [590 258 96 34 14 8]
%             E: [1x6 double]
```

The sample is consistent with the exponential distribution with p-value of 0.6220. Note that the number of intervals selected by MATLAB is 6, not the requested 7. This is because the upper tail intervals with a low expected count ($< 5$) are merged.

*Example 13.6.* **Wrinkled Peas.** Mendel crossed peas that were heterozygotes for smooth/wrinkled, where smooth is dominant. The expected ratio in the offspring is 3 smooth: 1 wrinkled. He observed 423 smooth and 133 wrinkled peas. The expected frequency of smooth is calculated by multiplying the sample size (556) by the expected proportion (3/4) to yield 417. The same is done for wrinkled to yield 139. The number of degrees of freedom when an extrinsic hypothesis is used is the number of values of the nominal variable minus one. In this case, there are two values (smooth and wrinkled), so there is one degree of freedom.

```
chisq = (556/4 - 133)^2/(556/4) + (556*3/4 - 423)^2/(556*3/4)
   %chisq = 0.3453
1 - chi2cdf(chisq, 2-1)
     %ans = 0.5568
chisqy = (abs(556/4 - 133)-0.5)^2/(556/4) + ...
   (abs(556*3/4 - 423)-0.5)^2/(556*3/4) %with Yates correction
   %chisq = 0.2902
1-chi2cdf(chisqy, 2-1)
     %ans = 0.5901
```

We conclude that the theoretical odds 3:1 in favor of smooth peas are consistent with the observations at level $\alpha = 0.05$.

*Example 13.7.* **Horse-Kick Fatalities.**  During the latter part of the nineteenth century, Prussian officials gathered information on the hazards that horses posed to cavalry soldiers. Fatal accidents for 10 cavalry corps were collected over a period of 20 years (Preussischen Statistik). The number of fatalities due to kicks, $x$, was recorded for each year and each corps. The table below shows the distribution of $x$ for these 200 "corps-years."

| $x$ = number of deaths | Observed number of corps-years in which $x$ fatalities occurred |
|---|---|
| 0 | 109 |
| 1 | 65 |
| 2 | 22 |
| 3 | 3 |
| $\geq 4$ | 1 |
| | 200 |

Altogether there were 122 fatalities $[109(0) +65(1) + 22(2) +3(3) + 1(4)]$, meaning that the observed fatality *rate* was 122/200, or 0.61 fatalities per corps-year. Von Bortkiewicz (1898) proposed a Poisson model for $X$ with a mean of $c = 0.61$. The table below shows the observed and expected frequencies corresponding to $x = 0, 1, 2, \ldots$, etc. The expected frequencies are

$$n \times \frac{0.61^i}{i!} \exp\{-0.61\},$$

for $n = 200$ and $i = 0, 1, 2$, and 3. We put together all values $\geq 4$ as a single class; this will ensure that the sum of the theoretical probabilities is equal to 1. For example, the expected frequencies $np_i$ are

```
npi = 200 * [poisspdf(0:3, 0.61)  1-poisscdf(3, 0.61)]
   %npi = 108.6702   66.2888   20.2181   4.1110   0.7119
```

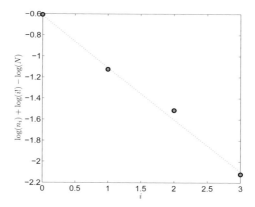

**Fig. 13.6** Poissonness plot for von Bortkiewicz's data. The theoretical line, $\log(0.61)i - 0.61$, is shown in *red*.

| i | # of fatalities | Observed number of corps-years, $n_i$ | Expected number of corps-years, $e_i$ |
|---|---|---|---|
| 1 | 0 | 109 | 108.6702 |
| 2 | 1 | 65 | 66.2888 |
| 3 | 2 | 22 | 20.2181 |
| 4 | 3 | 3 | 4.1110 |
| 5 | ≥ 4 | 1 | 0.7119 |
|   |   | 200 | 200 |

Now we calculate the statistic $\chi^2 = \sum_{i=1}^{5}(n_i - np_i)^2/(np_i)$ and find the *p*-value for the test and rejection region. Note that the number of degrees of freedom is $df = 5 - 1 - 1$ since $\lambda = 0.61$ was estimated from the data.

```
ni = [109 65 22 3 1]
   % ni = 109    65    22    3    1
ch2 = sum( (ni-npi).^2 ./npi )
   % ch2 = 0.5999
1-chi2cdf(0.5999, 5-1-1)
   % ans = 0.8965,  pvalue
chi2inv(0.95, 5-1-1)
   % ans = 7.8147, critical value
   % rejection region [7.8147, infinity)
```

The Poisson distribution in $H_0$ is consistent with the data at the level $\alpha =$ 0.05. Clearly the agreement between the observed and expected frequencies is remarkable, see also Poissonness plot in Fig. 13.6.

## 13.4 Kolmogorov–Smirnov Tests

The first measure of goodness of fit for general distributions was derived by Kolmogorov (1933). Andrei Nikolaevich Kolmogorov (Fig. 13.7a), the most accomplished and celebrated Russian mathematician of all time, made fundamental contributions to probability theory, including a test statistic for distribution functions, some of which are named after him. Nikolai Vasilyevich Smirnov (Fig. 13.7b), another Russian mathematician, extended Kolmogorov's results to two samples.

(a)                                        (b)

**Fig. 13.7** (a) Andrei Nikolaevich Kolmogorov (1905–1987); (b) Nikolai Vasilyevich Smirnov (1900–1966).

### 13.4.1 Kolmogorov's Test

Let $X_1, X_2, \ldots, X_n$ be a sample from a population with a continuous, but unknown, CDF $F$. As in (2.2), let $F_n(x)$ be the empirical CDF based on $X_1, X_2, \ldots, X_n$.

We are interested in testing the hypothesis

$$H_0 : F(x) = F_0(x), \; (\forall x) \quad \text{versus} \quad H_1 : F(x) \neq F_0(x), \; (\exists x),$$

where $F_0(x)$ is a fully specified continuous distribution.

The test statistic $D_n$ is calculated from the sample as

$$D_n = \max_{1 \le i \le n} \left\{ \frac{i}{n} - F_0(X_i), \; F_0(X_i) - \frac{i-1}{n} \right\}.$$

When hypothesis $H_0$ is true, Kolmogorov (1933) showed that the statistic $\sqrt{n}D_n$ is approximately distributed as

$$\lim_{n \to \infty} \mathbb{P}(\sqrt{n}D_n \le x) = K(x) = 1 - 2\sum_{k=1}^{\infty} (-1)^{k-1} e^{-2k^2 x^2}, \; x \ge 0,$$

which allows one to calculate critical regions and $p$-values for this test.

The MATLAB file ◀ kscdf.m calculates the CDF $K$. Since large values of $D_n$ are critical for $H_0$, the $p$-value of the test is approximated as

$$p \approx 1 - K(\sqrt{n}D_n),$$

or in MATLAB as 1-kscdf(sqrt(n)*Dn).

In practice, most KS-type tests are two-sided, testing whether $F$ is equal to $F_0$, the distribution postulated by $H_0$, or not. Alternatively, one might test to see if the distribution is larger or smaller than a hypothesized $F_0$ [see, for example, Kvam and Vidakovic (2007)].

*Example 13.8.* To illustrate the Kolmogorov test we simulate 30 observations from an exponential distribution with $\lambda = 1/2$.

```
% rand('state', 0);
% n = 30;   i = 1:n;
% x = exprnd(1/2,[1,n]); x = sort(x);
x = [...
    0.0256    0.0334    0.0407    0.0434    0.0562    0.0575...
    0.0984    0.1034    0.1166    0.1358    0.1518    0.2427...
    0.2497    0.2523    0.3608    0.3921    0.4052    0.4455...
    0.4511    0.5208    0.6506    0.7324    0.7979    0.8077...
    0.8079    0.8679    0.9870    1.4246    1.9949    2.309];
distances = [i./n - expcdf(x, 1/2); expcdf(x, 1/2) - (i-1)./n ];
Dn = max(max(distances))              %0.1048
pval = 1 - kscdf(sqrt(n)*Dn)          %0.8966
```

The $p$-value is 0.8966 and $H_0$ is not rejected. In other words, the sample is consistent with the hypothesis of the population's exponential distribution $\mathscr{E}(1/2)$.

The CDF $K(x)$ is an approximate distribution for $\sqrt{n}D_n$ and for small values of $n$ may not be precise. Better approximations use a continuity correction:

$$p \approx 1 - K\left(\sqrt{n}D_n + \frac{1}{6\sqrt{n}}\right).$$

This approximation is satisfactory for $n \geq 20$.

```
1 - kscdf(sqrt(n) * Dn + 1/(6 * sqrt(n)) ) %0.8582
```

MATLAB has a built-in function, kstest, that produces similar output.

```
% form theoretical cdf
t = 0:0.01:20;
y = expcdf(t,1/2);
cdf = [t' y'];
[decision, pval, KSstat, critValue] = kstest(x, cdf, 0.01, 0)
          %decision  =    0
          %pval   = 0.8626
          %KSstat = 0.1048
          %critValue = 0.2899
```

Figure 13.8 shows the empirical and theoretical distribution in this example and it is produced by the code below.

```
%Plot
xx = 0:0.01:4;
plo = cdfplot(x); set(plo,'LineWidth',2);
hold on
plot(xx,expcdf(xx, 1/2),'r-','LineWidth',2);
legend('Empirical','Theoretical Exponential',...
        'Location','SE')
```

Note that in the two calculations the values of $D_n$ statistics coincide but the $p$-values differ. This is because kstest uses a different approximation to the $p$-value.

The Kolmogorov test has advantages over exact tests based on the $\chi^2$ goodness-of-fit statistic, which depend on an adequate sample size and proper interval assignments for the approximations to be valid. A shortcoming of the KS test is that the $F_0$ distribution in $H_0$ must be fully specified. That is, if location, scale, or shape parameters are estimated from the data, the critical region of the KS test is no longer valid. An example is Lilliefors' test for departures from normality when the null distribution is not fully specified.

## 13.4.2 Smirnov's Test to Compare Two Distributions

Smirnov (1939) extended the Kolmogorov test to compare two distributions based on independent samples from each population. Let $X_1, X_2, \ldots, X_m$ and

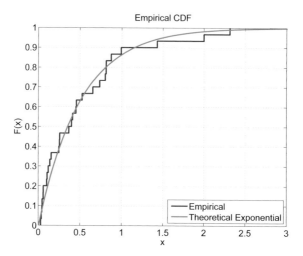

**Fig. 13.8** Empirical and theoretical distributions in the simulated example.

$Y_1, Y_2, \ldots, Y_n$ be two independent samples from populations with unknown CDFs $F_X$ and $G_Y$. Let $F_m(x)$ and $G_n(x)$ be the corresponding empirical distribution functions.

We would like to test

$$H_0 : F_X(x) = G_Y(x) \ \forall x \quad \text{versus} \quad H_1 : F_X(x) \neq G_Y(x) \text{ for some } x.$$

We will use the analog of the Kolmogorov statistic

$$D_{m,n} = \max\left\{ \max_{1 \leq i \leq m}\left\{\frac{i}{m} - G_n(X_i)\right\}, \ \max_{1 \leq j \leq n}\left\{\frac{j}{n} - F_m(Y_j)\right\} \right\}.$$

The limiting distribution for $D_{m,n}$ can be expressed by Kolmogorov's CDF $K$ as

$$\lim_{m,n \to \infty} \mathbb{P}\left( \sqrt{\frac{mn}{m+n}} D_{m,n} \leq x \right) = K(x),$$

and the $p$-value for the test is approximated as

$$p \approx 1 - K\left( \sqrt{\frac{mn}{m+n}} D_{m,n} \right). \tag{13.3}$$

This approximation is good when both $m$ and $n$ are large.

**Remark.** The approximation of the $p$-value in (13.3) can be improved by continuity corrections as

$$p \approx 1 - K\left(\sqrt{\frac{mn}{m+n}}\left(D_{m,n} + \frac{|m-n|}{6mn} + b_{m,n}\right)\right),$$

where

$$b_{m,n} = \frac{(m+n)\,(\min(m,n) - \gcd(m,n))}{2mn(m+n+\gcd(m,n))}$$

and $\gcd(m,n)$ is the greatest common divisor of $m$ and $n$. This approximation is satisfactory if $m,n > 20$.

If $m = n$ are not large, then exact $p$-values can be found by using ranks and combinatorial calculations.

*Example 13.9.* To illustrate Smirnov's test we simulate $m = 39$ observations from a normal $\mathcal{N}(-1, 2^2)$ distribution and $n = 35$ observations from Student's $t$-distribution with 10 degrees of freedom. The hypothesis of equality of distributions is rejected at the level $\alpha = 0.05$ since the $p$-values are approximately in the neighborhood of 2%.

Notice that the corrected $p$-value 0.0196 is close to that in kstest2, and the differences are due to different approximation formulas.

```
x =[...
    -5.75    -3.89    -3.69    -3.68    -3.54    -2.59 ...
    -2.53    -2.40    -2.39    -2.27    -2.02    -1.72 ...
    -1.64    -1.54    -1.27    -1.11    -1.08    -1.00 ...
    -0.88    -0.84    -0.47    -0.36    -0.29    -0.24 ...
    -0.20    -0.19     0.02     0.15     0.25     0.45 ...
     0.51     0.72     0.74     0.96     1.25     1.33 ...
     1.49     2.39     2.59 ];

y=[...
    -2.72    -2.18    -1.31    -1.17    -1.00    -0.94 ...
    -0.78    -0.65    -0.63    -0.52    -0.40    -0.37 ...
    -0.30    -0.21    -0.19    -0.11    -0.05     0.12 ...
     0.14     0.25     0.27     0.35     0.45     0.48 ...
     0.48     0.60     0.71     0.76     0.79     1.01 ...
     1.10     1.10     1.12     1.36     2.03 ];

m = length(x); n=length(y);
i = 1:m; j = 1:n;
distances = [i./m - empiricalcdf(x, y), j./n - empiricalcdf(y, x)];
Dmn = max(max(distances))    %0.3414
z = sqrt(m * n/(m + n))*Dmn %1.4662
pval1 = 1 - kscdf(z)    %0.0271
%
bmn = (m+n)/(2*m*n) * (min(m,n) - gcd(m, n))/(m + n + gcd(m,n)) %0.0123
zz =  sqrt(m * n/(m + n))*(Dmn + abs(m-n)/(6*m*n) +  bmn )    %1.5211
```

```
pval2 = 1 - kscdf(zz)   %0.0196
% MATLAB's built in function
[h,p,k] = kstest2(x,y)
    % 1
    % 0.0202
    % 0.3414
```

## 13.5 Moran's Test*

Before discussing Moran's test, we note that any goodness-of-fit null hypothesis $H_0 : F = F_0$, where $F_0$ is a fixed continuous distribution, can be restated in terms of a uniform distribution. This is a simple consequence of the following result:

> **Result.** Let random variable $X$ have a continuous distribution $F$. Then, $F(X)$ has a uniform distribution on $[0, 1]$.

This important fact has a simple proof, since the continuity of $F$ ensures the existence of $F^{-1}$:

$$\mathbb{P}(F(X) \le x) = \mathbb{P}(X \le F^{-1}(x)) = F(F^{-1}(x)) = x, \ 0 \le x \le 1.$$

Thus the hypothesis $H_0 : F = F_0$ for observed $x_1, \ldots, x_n$ is equivalent to $H_0$: the CDF is uniform on $[0, 1]$ for transformed values $y_1 = F_0(x_1), y_2 = F_0(x_2), \ldots, y_n = F_0(x_n)$.

Define spacings

$$d_i = y_{(i+1)} - y_{(i)}, \ i = 0, 1, 2, \ldots, n,$$

where $y_{(1)} \le y_{(2)} \le \cdots \le y_{(n)}$ is the order statistic of $y_1, \ldots, y_n$ and $y_{(0)} = 0$ and $y_{(n+1)} = 1$. Then Moran's statistic

$$M = \sqrt{n} \left( \frac{n}{2} \sum_{i=0}^{n} d_i^2 - 1 \right)$$

has an approximately standard normal distribution. This approximation is adequate if $n > 30$.

> The hypothesis $H_0 : F = F_0$ is rejected if $M > z_\alpha$. The $p$-value is
> `1-normcdf(M)`.

Since Moran's $M$ is minimized by identical spacing $d_i = 1/(n + 1)$, small values of $M$ may indicate nonrandom data.

Moran's test is complementary to other goodness-of-fit tests since it is sensitive to data clustering and anomalous spacing. The alternatives, consisting of long-tailed densities that remain undetected by Pearson's $\chi^2$ or KS tests, may be detected by Moran's test. The following example demonstrates that Moran's test is superior in detecting a long-tailed alternative compared to Kolmogorov's test.

*Example 13.10.* In the following MATLAB script we repeated 1000 times the following: (i) a sample of size 200 was generated from a $t_4$ distribution and (ii) the sample was tested for standard normality using Moran's and Kolmogorov's tests. Moran's test rejected $H_0$ 203 times while by Kolmogorov's test $H_0$ was rejected 122 times. The exact number of times $H_0$ was rejected varies slightly depending on the random number seed; in this example we used a combined multiple recursive generator: stream=RandStream('mrg32k3a');

```
n=200;
stream = RandStream('mrg32k3a');
RandStream.setDefaultStream(stream);
pvalmoran =[]; pvalks=[];
for i = 1:1000
    x =  trnd(4,[n 1]);              %simulate t_4
    y=normcdf(x,0,1);               %H0: F=N(0,1)
    yy =[0; sort(y); 1];
    Sd = sum(diff(yy,1).^2);
    M  =sqrt(n)*(n * Sd/2 - 1);   %Moran's Stat
    p = 1-normcdf(M);
    pvalmoran = [pvalmoran p];
    [h pv dn] = kstest(x, [x normcdf(x, 0, 1)]);
    pvalks = [pvalks pv];
end
%Number of times H_0 was rejected in 1000 runs
sum(pvalmoran < 0.05)          %203
sum(pvalks < 0.05)             %122
```

## 13.6 Departures from Normality

Several tests are available specifically for the normal distribution. The Jarque–Bera test (Jarque and Bera, 1980) is a goodness-of-fit measure of departure from normality, based on the sample kurtosis and skewness. The test statistic is defined as

$$\chi^2_{JB} = \frac{n}{6}\left(\gamma_n^2 + \frac{(\kappa_n - 3)^2}{4}\right),$$

where $n$ is the sample size, $\gamma_n$ is the sample skewness,

$$\gamma_n = \frac{\frac{1}{n}\sum_{i=1}^{n}(X_i - \overline{X})^3}{\left(\frac{1}{n}\sum_{i=1}^{n}(X_i - \overline{X})^2\right)^{3/2}},$$

and $\kappa_n$ is the sample kurtosis,

$$\kappa_n = \frac{\frac{1}{n}\sum_{i=1}^{n}(X_i - \overline{X})^4}{\left(\frac{1}{n}\sum_{i=1}^{n}(X_i - \overline{X})^2\right)^{2}},$$

as on p. 20.

Note that the first four moments are used jointly for the calculation of $\chi^2_{JB}$. The statistic $\chi^2_{JB}$ has an asymptotic chi-square distribution with 2 degrees of freedom and is a measure of deviation from the moments of normal distribution ($Sk = 0$ and $\kappa = 3$). In MATLAB, the Jarque–Bera test is called [h,p,jbstat,critval]=jbtest(x,alpha).

Lilliefors' test (Lilliefors, 1967) for departures from normality is a version of the KS test, where the theoretical distribution is normal but not fully specified, as the KS test requires. The test statistic is of the same type as in the KS test. The MATLAB command for Lilliefors' test is [h,p,ksstat,critval] = lillietest(x, alpha), where the data to be tested are in vector x.

*Example 13.11.* Suppose that we generate a random sample of size 30 from an $F$ distribution with 5 and 10 degrees of freedom and we want to test that the sample is normal.

```
x = frnd(5, 10, [1, 30]);
alpha=0.05;
[h1,p1] = jbtest(x, alpha)
 %h1 = 1
 %p1 = 1.0000e-003
[h2,p2] = lillietest(x, alpha)
 %h2 = 1
 %p2 = 0.0084
```

Both the Jarque–Bera and Lilliefors' tests reject $H_0$ that the sample is normally distributed. In this example, the Jarque–Bera test resulted in a smaller $p$-value.

## 13.7 Exercises

13.1. **Q–Q Plot for $\sqrt{2\chi^2}$.** Simulate $N = 10,000$ $\chi^2$ random variables with $k = 40$ degrees of freedom. Demonstrate empirically that

$$Z = \sqrt{2\chi^2} - \sqrt{2k - 1}$$

is approximately standard normal $\mathcal{N}(0, 1)$ by plotting a histogram and Q–Q plot.

*Hint:* `df=40; zs=sqrt(2*chi2rnd(df,[1 10000]))-sqrt(2*df-1);`

13.2. **Not at All Like Me.** You have a theory that if you ask subjects to sort one-sentence characteristics of people (e.g., "I like to get up early in the morning") into five groups ranging from *not at all like me* (group 1) to *very much like me* (group 5), the percentage falling in each of the five groups will be approximately 10, 20, 40, 20, and 10. You have one of your friends sort 50 statements, and you obtain the following data: 8, 9, 21, 8, and 4.
Do these data support your hypothesis at $\alpha = 5\%$?

13.3. **Cell Counts.** A student takes the blood cell count of five random blood samples from a larger volume of solution to determine if it is well mixed. She expects the cell counts to be distributed uniformly. The data is given below:

| Sample blood cell count | Expected blood cell count |
|:---:|:---:|
| 35 | 27.2 |
| 20 | 27.2 |
| 25 | 27.2 |
| 25 | 27.2 |
| 31 | 27.2 |
| 136 | 136 |

Can she depend on the results to be uniformly distributed? Do the chi-square test with $\alpha = 0.05$.

13.4. **GSS Data.** Below is part of the data from the 1984 and 1990 General Social Survey, conducted annually by the National Opinion Research Center. Random samples of 1473 persons in 1984 and 899 persons in 1990 were taken using multistage cluster sampling. One of the questions in a 67-question-long questionnaire was: *Do you think most people would try to take advantage of you if they got a chance, or would they try to be fair?*

|                              | 1984 survey | 1990 survey |
|------------------------------|-------------|-------------|
| 1. Would take advantage of you | 507       | 325         |
| 2. Would try to be fair      | 913         | 515         |
| 3. Depends                   | 47          | 53          |
| 4. No answer                 | 6           | 6           |

Assuming that 1984 frequencies are theoretical, using Pearson's $\chi^2$ test explore how the 1990 frequencies agree. Use $\alpha = 0.05$. State your findings clearly.

**13.5. Strokes on "Black Monday."** In a long-term study of heart disease, the day of the week on which 63 seemingly healthy men died was recorded. These men had no history of disease and died suddenly.

| Day of week  | Mon. | Tue. | Wed. | Thu. | Fri. | Sat. | Sun. |
|--------------|------|------|------|------|------|------|------|
| No. of deaths | 22  | 7    | 6    | 13   | 5    | 4    | 6    |

(a) Test the hypothesis that these men were just as likely to die on one day as on any other. Use $\alpha = 0.05$.

(b) Explain in words what constitutes the error of the second kind in the testing in (a).

**13.6. Benford's Law.** Benford's law (Benford, 1938; Hill, 1998) concerns the relative frequencies of leading digits of various data sets, numerical tables, accounting data, etc. Benford's law, also called the first digit law, states that in numbers from many sources, the leading digit is much more often a 1 than it is any other digit (specifically about 30% of the time). Furthermore, the higher the digit, the less likely it is to occur as the leading digit of a number. This applies to figures related to the natural world or figures of social significance, be it numbers taken from electricity bills, newspaper articles, street addresses, stock prices, population numbers, death rates, areas or lengths of rivers, or physical and mathematical constants.

**Fig. 13.9** Frank Benford (1883–1948).

More precisely, the Benford law states that the leading digit $n$, $(n = 1, \ldots, 9)$ occurs with probability $P(n) = \log_{10}(n+1) - \log_{10}(n)$, or as in the table below [probabilities $P(n)$ rounded to 4 decimal places]:

| Digit $n$ | 1 | 2 | 3 | 4 | 5 | 6 | 7 | 8 | 9 |
|-----------|---|---|---|---|---|---|---|---|---|
| $P(n)$ | 0.3010 | 0.1761 | 0.1249 | 0.0969 | 0.0792 | 0.0669 | 0.0580 | 0.0512 | 0.0458 |

An oft-cited data set is the distribution of the leading digit of all 309 numbers appearing in a particular issue of Reader's Digest.

| Digit | 1 | 2 | 3 | 4 | 5 | 6 | 7 | 8 | 9 |
|-------|---|---|---|---|---|---|---|---|---|
| Count | 103 | 57 | 38 | 23 | 20 | 23 | 17 | 15 | 13 |

At level $\alpha = 0.05$, test the hypothesis that the observed distribution of leading digits in Reader's Digest numbers follows Benford's law.

*Hint:* To speed up the calculation, some theoretical frequencies rounded to two decimal places are provided:

| 93.01 | 54.41 | • | 29.94 | 24.47 | • | • | 15.82 | 14.15 |
|-------|-------|---|-------|-------|---|---|-------|-------|

13.7. **Simulational Exercise.**

- Generate a sample of size $n = 216$ from a standard normal distribution.
- Select intervals by cutting the range of simulated values by points $-2.7$, $-2.2$, $-2$, $-1.7$, $-1.5$, $-1.2$, $-1$, $-0.8$, $-0.5$, $-0.3$, 0, 0.2, 0.4, 0.9, 1, 1.4, 1.6, 1.9, 2, 2.5, and 2.8.
- Using a $\chi^2$-test confirm the normality of the sample.
- Repeat the previous test if your sample is contaminated by a Cauchy $Ca(0,1)$ distribution in the following way: `0.95 * normal_sample + 0.05 * cauchy_sample`.

13.8. **Deathbed Scenes.** Can some people postpone their death until after a special event takes place? It is believed that famous people do so with respect to their birthdays to which they attach some importance. A study by Philips (1972) seems to be consistent with this notion. Philips obtained data[1] on the months of birth and death of 1251 famous Americans; the deaths were classified by the time period between the birth dates and death dates as shown in the following table:

| b | e | f | o | r | e | birth | a | f | t | e | r |
|---|---|---|---|---|---|-------|---|---|---|---|---|
| 6 | 5 | 4 | 3 | 2 | 1 | month | 1 | 2 | 3 | 4 | 5 |
| 90 | 100 | 87 | 96 | 101 | 86 | 119 | 118 | 121 | 114 | 113 | 106 |

(a) Clearly formulate the statistical question based on the above observations.

(b) Provide a solution for the question formulated in (a). Use $\alpha = 0.05$.

13.9. **Grouping in a Vervet Monkey Troop.** Struhsaker (1965) recorded, individual by individual, the composition of sleeping groups of wild verver

---

[1] 348 were people listed in *Four Hundred Notable Americans* and 903 are listed as the foremost families in three volumes of *Who Was Who* for the years 1951–1960, 1943–1950, and 1897–1942.

monkeys (*Cercopithecus aethiops*) in East Africa. A particular large troop was observed for 22 nights.

| Size | 1 | 2 | 3 | 4 | 5 6 7 8 9 | 10 | 11 | ≥12 |
|------|---|---|---|---|-----------|----|----|-----|
| Freq | 19 | 14 | 19 | 11 | 7 7 3 2 3 | 1 | 2 | 1 |

It is well known that the sizes of groups among humans in various social settings (shoppers, playgroups, pedestrians) follows a truncated Poisson distribution:

$$P(X = k) = \frac{\lambda^k \exp\{-\lambda\}}{k!(1 - \exp\{-\lambda\})}, \quad k = 1, 2, 3, \ldots.$$

Using a $\chi^2$ goodness-of-fit test, demonstrate that the truncated Poisson is not a good model for this data at the significance level $\alpha = 0.05$. To find the theoretical frequencies, $\lambda$ can be estimated by the method of moments using an average size of $\overline{X} = 3.74$ and the equation $\lambda/(1 - \exp\{-\lambda\}) = 3.74$. In MATLAB fzero(@(lam) lam - 3.74*(1-exp(-lam)), 3) gives $\hat{\lambda} = 3.642$. Struhsaker further argues that in this context the truncated negative binomial gives a reasonably good fit.

13.10. **Crossing Mushrooms.** In a botany experiment the results of crossing two hybrids of a species of mushrooms (*Agaricus bisporus*) gave observed frequencies of 120, 53, 36, and 15. Do these results disagree with theoretical frequencies that specify a 9:3:3:1 ratio? Use $\alpha = 0.05$.

13.11. **Renner's Honey Data Revisited.** In Example 6.10 we argued that a lognormal distribution with parameters $\mu =$ and $\sigma^2 = 1.0040^2$ provides a good fit for Renner's honey data. Find $\chi^2$-statistics and the *p*-value of the fit. State decision at level $\alpha = 0.05$. The midintervals in the first column of data set 🖭 renner.mat|dat, 0.1250, 0.3750, 0.6250, ..., 7.1250 correspond to the intervals (0, 0.25], (0.25, 0.5], (0.5, 0.75], ..., (7, 7.5].

13.12. **PCB in Yolks of Pelican Eggs.** A well-known data set for testing agreement with a normal distribution comes from Risebrough (1972) who explored concentrations of polychlorinated biphenyl (PCB) in yolk lipids of pelican eggs. For $n = 65$ pelicans from Anacapa Island (northwest of Los Angeles), the concentrations of PCB were as follows:

| | | | | | | | | | | | |
|---|---|---|---|---|---|---|---|---|---|---|---|
| 452 | 184 | 115 | 315 | 139 | 177 | 214 | 356 | 166 | 246 | 177 | 289 | 175 |
| 296 | 205 | 324 | 260 | 188 | 208 | 109 | 204 | 89 | 320 | 256 | 138 | 198 |
| 191 | 193 | 316 | 122 | 305 | 203 | 396 | 250 | 230 | 214 | 46 | 256 | 204 |
| 150 | 218 | 261 | 143 | 229 | 173 | 132 | 175 | 236 | 220 | 212 | 119 | 144 |
| 147 | 171 | 216 | 232 | 216 | 164 | 185 | 216 | 199 | 236 | 237 | 206 | 87 |

Test the hypothesis that PCB concentrations are consistent with the normal distribution at the level $\alpha = 0.05$.

13.13. **Number of Leaves per Whorl in** *Ceratophyllum demersum*. After spend-
ing a year with Karl Pearson in London, biometrician Raymond Pearl published his work on mathematical and statistical modeling of growth of an
aquatic plant, *Ceratophyllum demersum* (Pearl, 1907). Among several interesting data sets, Pearl gives the table of frequencies of numbers of leaves
per whorl.

| Leaves | 5 | 6 | 7 | 8 | 9 | 10 | 11 | 12 | Total |
|--------|---|---|---|---|---|----|----|----|-------|
| Whorls | 6 | 169 | 258 | 495 | 733 | 617 | 48 | 2 | 2328 |

Is the binomial distribution $\mathscr{B}in(12, p)$ a good fit for the data? *Hint:* The
probability $p$ can be estimated from the data; the average of *Leaves* should
be close to $12p$.

13.14. **From 1998–2002 U.S. National Health Interview Survey (NHIS).**
Stansfield and Carlton (2007) analyzed data in sibships of size 2 and 3 from
the 1998–2002 U.S. National Health Interview Survey (NHIS).
For the 25,468 sibships of size 2, among all 50,936 children, 10 years of age
and younger, in the NHIS data set (the youngest age cohort available), they
found that 51.38% were boys (B) and 48.62% were girls (G); the B/G ratio
was 1.0566, or about 106 boys for every 100 girls. The number of boys is
given in the following table:

| Number of boys | 0 | 1 | 2 |
|----------------|---|---|---|
| Observed sibships | 5,844 | 13,079 | 6,545 |

For the 7,541 sibships of size 3 from the same NHIS data set the number of
boys among the first two children is given in the following table:

| Number of boys | 0 | 1 | 2 |
|----------------|---|---|---|
| Observed sibships | 1,941 | 3,393 | 2,207 |

(a) If the number of boys among the first two children in sibships of size 3
is binomial $\mathscr{B}in(2, 0.515)$, find what theoretical frequencies are expected in
the table for sibships of size 3. Are the observed and theoretical frequencies
close? Comment. *Hint:* Find the binomial probabilities and multiply them
by $n = 7{,}541$.
(b) For the 25,468 sibships of size 2, test the hypothesis that the probability
of a boy is 1/2 versus the one-sided alternative, that is,

$$H_0 : p = 0.5 \ \ vs \ \ H_1 : p > 0.5,$$

at the level $\alpha = 0.05$. *Hint:* Be careful about the $n$ here. The rejection region
of this test is $RR = [1.645, \infty)$.

13.15. **Neuron Fires Revisited.** The 989 firing times of a cell culture of neurons
have been recorded. The records are time instances when a neuron sends a

signal to another linked neuron (a spike) and the largest time instance is 1001.

The count of firings in 200 consecutive time intervals of length 5 (= 1000/200) time units is as follows:

| | | | |
|---|---|---|---|
| 2 7 5 | 5 1 6 8 4 6 4 5 7 | 7 5 4 7 4 4 5 | 9 6 5 6 6 5 |
| 5 6 4 | 10 7 8 8 2 9 5 4 4 | 4 3 8 3 2 7 6 | 7 5 6 6 4 6 |
| 5 7 3 | 5 6 5 5 2 7 7 6 4 | 8 8 9 7 3 3 3 | 5 6 6 4 6 4 |
| 5 4 4 | 5 2 3 5 1 4 4 3 2 | 10 4 7 2 1 7 9 | 4 3 6 8 6 5 |
| 2 4 4 | 3 7 2 2 6 1 3 7 6 | 6 4 7 5 5 8 7 | 3 5 5 4 5 6 |
| 10 3 6 | 9 5 7 2 8 4 4 2 3 | 4 3 3 3 5 4 2 | 2 7 4 5 4 5 |
| 3 5 3 | 5 5 7 2 5 5 4 8 4 | 3 5 7 4 4 8 5 | 4 3 6 4 6 4 |
| 1 7 9 | 4 2 5 4 4 4 4 7 2 | 4 4 5 2 4 7 5 | 12 1 5 4 6 7 |

Are the data consistent with a Poisson distribution?

**13.16. Cloudiness in Greenwich.** Pearse (1928) cites data describing the degree of cloudiness in Greenwich, UK for days in the month of July between 1890 and 1904.

| Cloudiness | 0 | 1 | 2 | 3 | 4 | 5 | 6 | 7 | 8 | 9 | 10 | Total |
|---|---|---|---|---|---|---|---|---|---|---|---|---|
| Days | 320 | 129 | 74 | 68 | 45 | 45 | 55 | 65 | 90 | 148 | 676 | 1715 |

Assume that the measure of cloudiness can be modeled as a continuous random variable $C = 10 \cdot X$, where $X$ has a beta $\mathcal{B}e(\alpha, \beta)$ distribution. Find the best fitting parameters $\alpha$ and $\beta$ and assess the goodness of fit.

**13.17. Distance Between Spiral Reversals in Cotton Fibers.** Tippett (1941) discusses a frequency histogram for a data set consisting of intervals and counts for a distance between spiral reversals in cotton fiber (Fig. 13.10). The distances are given in units of $mm^{-2}$, and the sample size was 1117.

| Distance | $[0, 2.5)$ | $[2.5, 4.5)$ | $[4.5, 6.5)$ | $[6.5, 8.5)$ | $[8.5, 10.5)$ |
|---|---|---|---|---|---|
| Number | 7 | 48 | 100 | 106 | 84 |

| $[10.5, 12.5)$ | $[12.5, 16.5)$ | $[16.5, 20.5)$ | $[20.5, 24.5)$ | $[24.5, 28.5)$ |
|---|---|---|---|---|
| 72 | 136 | 94 | 78 | 69 |

| $[28.5, 32.5)$ | $[32.5, 36.5)$ | $[36.5, 40.5)$ | $[40.5, 50.5)$ | $[50.5, 60.5)$ |
|---|---|---|---|---|
| 53 | 45 | 36 | 69 | 40 |

| $[60.5, 70.5)$ | $[70.5, 80.5)$ | $[80.5, 90.5)$ | $[90.5, \infty)$ | Total |
|---|---|---|---|---|
| 31 | 21 | 17 | 11 | 1117 |

Using Pearson's $\chi^2$ criterion and significance level $\alpha = 0.05$, test the hypothesis that the data are consistent with the $\chi^2$-distribution. Use the midpoints of intervals to estimate the mean of a theoretical $\chi^2$-distribution. Recall that the mean in this case is equal to the number of degrees of freedom.

**Fig. 13.10** Structure of spiral reversals in cotton fiber.

---

| MATLAB FILES AND DATA SETS USED IN THIS CHAPTER |
|---|

http://springer.bme.gatech.edu/Ch13.Goodness/

anacapa.m, consolidator.m, empiricalcdf.m, examples.m, horsekicks.m, kolsm.m, ks2cdf.m, kscdf.m, moran.m, neuronfiresqq.m, neurongof.m, notatalllikeme.m, perk.m, plotsimulqq.m, poissonness.m, qqbivariatenormal.m, qqnorm.m, qqplotsGOF.m, RANDmillion.m, smirnov.m, smirnov2tests.m, tippett.m, weldon.m

neuronfires.mat, RAND1Millrandomdigits.txt

---

# CHAPTER REFERENCES

Benford, F. (1938). The law of anomalous numbers. *Proc. Am. Philos. Soc.*, **78**, 551–572.

Hill, T. (1998). The first digit phenomenon. *Am. Sci.*, **86**, 358–363.

Hoaglin, D. C. (1980). A poissonness plot. *Am. Stat.*, **34**, 146–149.

Jarque, C. and Bera, A. (1980). Efficient tests for normality, homoscedasticity and serial independence of regression residuals. *Econ. Lett.*, **6**, 3, 255–259.

Kemp, W. A. and Kemp, D. C. (1991). Weldon's dice data revisited. *Am. Stat.*, **45**, 3, 216–222.

Kolmogorov, A. N. (1933). Sulla determinazione empirica di una legge di distributione. *Gior. Ist. Ital. Attuari*, **4**, 83–91.

Kvam, P. and Vidakovic, B. (2007). *Nonparametric Statistics with Applications to Science and Engineering*. Wiley, Hoboken.

Lilliefors, H. W. (1967). On the Komogorov-Smirnov test for normality with mean and variance unknown. *J. Am. Stat. Assoc.*, **62**, 399–402.

Pearl, R. (1907). Variation and differentiation in *Ceratophyllum*. *Carnegie Inst. Wash. Publ.*, **58**, 1–136.

Pearse, G. E. (1928). On corrections for the moment-coefficients of frequency distributions. *Biometrika*, **20** A, 314–355.

Pearson, K. (1900). On the criterion that a given system of deviations from the probable in the case of a correlated system of variables is such that it can be reasonably supposed to have arisen from random sampling. *Philos. Mag. Ser.*, **5**, 50, 157–175.

Phillips, D. P. (1972). Deathday and birthday: an unexpected connection. In: Tanur, J. M. ed. *Statistics: a guide to the unknown*. Holden-Day, San Francisco, 52–65.

Risebrough, R. W. (1972). Effects of environmental pollutants upon animals other than man. Proc. Sixth Berkeley Symp. on Math. Statist. and Prob., vol. 6, 443–453.

Smirnov, N. (1939). On the estimation of the discrepancy between empirical curves of distribution for two independent samples. *Bull. Math. Univ. Moscou*, **2**, 3–14.

Stansfield, W. D. and Carlton, M. A. (2007). Human sex ratios and sex distribution in sibships of size 2. *Hum. Biol.*, **79**, 255–260.

Struhsaker, T. T. (1965). Behavior of the vervet monkey (*Cercopithecus aethiops*). Ph.D. dissertation, University of California-Berkeley.

von Bortkiewicz, L. (1898). Das Gesetz der kleinen Zahlen. Teubner, Leipzig.

Yates, F. (1934). Contingency table involving small numbers and the $\chi^2$ test. *J. R. Stat. Soc.*, **1** (suppl.), 2, 217–235.

# Chapter 14
# Models for Tables

*It seems to me that everything that happens to us is a disconcerting mix of choice and contingency.*

– Penelope Lively

### WHAT IS COVERED IN THIS CHAPTER

• Contingency Tables and Testing for Independence in Categorical Data
  • Measuring Association in Contingency Tables
  • Three-dimensional Tables
  • Tables with Fixed Marginals: Fisher's Exact Test
  • Combining Contingency Tables: Mantel–Haenszel Theory
  • Paired Tables: McNemar Test
  • Risk Differences, Risk Ratios, and Odds Ratios for Paired Experiments

## 14.1 Introduction

The focus of this chapter is the analysis of tabulated data. Although the measurements could be numerical, the tables summarize only the counts along the levels of two or more crossed factors according to which the data are tabulated. In this text, we go beyond the traditional coverage and discuss topics such as three-dimensional tables, multiple tables (Mantel–Haenszel theory), paired tables (McNemar test), and risk theory (risk differences, relative risk, and odds ratios) for paired tables. The risk theory for ordinary (unpaired) tables was already discussed in Chap. 10 in the context of comparing two population proportions.

The dominant statistical procedure in this chapter is testing for the independence of two cross-tabulated factors. In cases where the marginal counts are fixed before the sampling, the test for independence becomes the test for homogeneity of one factor across the levels of the other factor. Although the concepts of homogeneity and independence are different, the mechanics of the two tests is the same. Thus, it is important to know how an experiment was conducted and whether the table marginal counts were fixed in advance.

The paired tables, like the paired $t$-test or block designs, are preferred to ordinary (or parallel) tables whenever pairing is feasible. In paired tables, we would usually be interested in testing the agreement of proportions.

## 14.2 Contingency Tables: Testing for Independence

To formulate a test for the independence of two crossed factors, we need to recall the definition of independence of two events and two random variables. Two events $R$ and $C$ are independent if the probability of their intersection is equal to the product of their individual probabilities,

$$\mathbb{P}(R \cap C) = \mathbb{P}(R) \cdot \mathbb{P}(C).$$

For random variables independence is defined using the independence of events. For example, two random variables $X$ and $Y$ are independent if the events $\{X \in I_x\}$ and $\{Y \in I_y\}$, where $I_x$ and $I_y$ are arbitrary intervals, are independent.

To motivate inference for tabulated data we also need a brief review of two-dimensional discrete random variables, discussed in Chap. 4. A two-dimensional discrete random variable $(X, Y)$, where $X \in \{x_1, \ldots, x_r\}$ and $Y \in \{y_1, \ldots, y_c\}$, is fully specified by its probability distribution, which is given in the form of a table:

|          | $y_1$    | $y_2$    | $\cdots$ | $y_c$    | Marginal  |
|----------|----------|----------|----------|----------|-----------|
| $x_1$    | $p_{11}$ | $p_{12}$ |          | $p_{1c}$ | $p_{1\cdot}$ |
| $x_2$    | $p_{21}$ | $p_{22}$ |          | $p_{2c}$ | $p_{2\cdot}$ |
|          |          |          |          |          |           |
| $x_r$    | $p_{r1}$ | $p_{r2}$ |          | $p_{rc}$ | $p_{r\cdot}$ |
| Marginal | $p_{\cdot1}$ | $p_{\cdot2}$ |      | $p_{\cdot c}$ | 1         |

The two components (marginal variables) $X$ and $Y$ in $(X, Y)$ are independent if all cell probabilities are equal to the product of the associated marginal probabilities, that is, if $p_{ij} = P(X = x_i, Y = y_j) = P(X = x_i)P(Y = y_j) = p_{i\cdot} \times p_{\cdot j}$ for each $i,j$. If there exists a cell $(i,j)$ for which $p_{ij} \neq p_{i\cdot} \times p_{\cdot j}$, then $X$ and $Y$ are dependent. The marginal distributions for components $X$ and $Y$ are obtained by taking the sums of probabilities in the table, row-wise and column-wise, respectively: $\dfrac{X \mid x_1 \ x_2 \ \dots \ x_r}{p \mid p_1 \ p_2 \ \dots \ p_r}$ and $\dfrac{Y \mid y_1 \ y_2 \ \dots \ y_c}{p \mid p_{\cdot 1} \ p_{\cdot 2} \ \dots \ p_{\cdot c}}$ .

*Example 14.1.* If $(X,Y)$ is defined by

| $X \setminus Y$ | 10 | 20 | 30 | 40 |
|---|---|---|---|---|
| −1 | 0.1 | 0.2 | 0 | 0.05 |
| 0 | 0.2 | 0 | 0.05 | 0.1 |
| 1 | 0.1 | 0.1 | 0.05 | 0.05 |

then the marginals are $\dfrac{X \mid -1 \quad 0 \quad 1}{p \mid 0.35 \ 0.35 \ 0.3}$ and $\dfrac{Y \mid 10 \ 20 \ 30 \ 40}{p \mid 0.4 \ 0.3 \ 0.1 \ 0.2}$ , and $X$ and $Y$ are dependent since we found a cell, for example, $(2,1)$, such that $0.2 = P(X = 0, Y = 10) \neq P(X = 0) \cdot P(Y = 10) = 0.35 \cdot 0.4 = 0.14$. As we indicated, it is sufficient for one cell to violate the condition $p_{ij} = p_{i\cdot} \times p_{\cdot j}$ in order for $X$ and $Y$ to be dependent.

✐

Instead of random variables and cell probabilities, we will consider an empirical counterpart, a table of observed frequencies. The table is defined by the levels of two factors $R$ and $C$. The levels are not necessarily numerical but could be, and most often are, categorical, ordinal, or interval. For example, when assessing the possible dependence between gender (factor $R$) and personal income (factor $C$), the levels for $R$ are categorical {male, female}, and for the $C$ interval, say, $\{[0, 30K), [30K, 60K), [60K, 100K), \geq 100K\}$. In the table below, factor $R$ has $r$ levels coded as $1, \dots, r$ and factor $C$ has $c$ levels coded as $1, \dots, c$. A cell $(i,j)$ is an intersection of the $i$th row and the $j$th column and contains $n_{ij}$ observations. The sum of the $i$th row is denoted by $n_{i\cdot}$ while the sum of the $j$th column is denoted by $n_{\cdot j}$.

| | 1 | 2 | $\cdots$ | $c$ | Total |
|---|---|---|---|---|---|
| 1 | $n_{11}$ | $n_{12}$ | | $n_{1c}$ | $n_{1\cdot}$ |
| 2 | $n_{21}$ | $n_{22}$ | | $n_{2c}$ | $n_{2\cdot}$ |
| | | | | | |
| $r$ | $n_{r1}$ | $n_{r2}$ | | $n_{rc}$ | $n_{r\cdot}$ |
| Total | $n_{\cdot 1}$ | $n_{\cdot 2}$ | | $n_{\cdot c}$ | $n_{\cdot\cdot}$ |

Denote the total number of observations $n_{\cdot\cdot} = \sum_{i=1}^{r} n_{i\cdot} = \sum_{j=1}^{c} n_{\cdot j}$ simply by $n$. The empirical probability of the cell $(i,j)$ is $\frac{n_{ij}}{n}$, and the empirical marginal probabilities of levels $i$ and $j$ are $\frac{n_{i\cdot}}{n}$ and $\frac{n_{\cdot j}}{n}$, respectively.

When factors $R$ and $C$ are independent, the frequency in the cell $(i,j)$ is *expected* to be $n \cdot p_{i\cdot} \cdot p_{\cdot j}$. This can be estimated by empirical frequencies

$$e_{ij} = n \times \frac{n_{i\cdot}}{n} \times \frac{n_{\cdot j}}{n},$$

that is, as the product of the total number of observations $n$ and the corresponding empirical marginal probabilities. After simplification, the empirical frequency in the cell $(i,j)$ for independent factors $A$ and $B$ is

$$e_{ij} = \frac{n_{i\cdot} \times n_{\cdot j}}{n}.$$

By construction, the table containing "independence" frequencies $e_{ij}$ would have the same row and column totals as the table containing observed frequencies $n_{ij}$, that is,

|       | 1        | 2        | $\cdots$ | $c$      | Total       |
|-------|----------|----------|----------|----------|-------------|
| 1     | $e_{11}$ | $e_{12}$ |          | $e_{1c}$ | $n_{1\cdot}$ |
| 2     | $e_{21}$ | $e_{22}$ |          | $e_{2c}$ | $n_{2\cdot}$ |
|       |          |          |          |          |             |
| $r$   | $e_{r1}$ | $e_{r2}$ |          | $e_{rc}$ | $n_{r\cdot}$ |
| Total | $n_{\cdot 1}$ | $n_{\cdot 2}$ |     | $n_{\cdot c}$ | $n_{\cdot\cdot}$ |

To measure the deviation from independence of the factors, we compare $n_{ij}$ and $e_{ij}$ over all cells. There are several measures for discrepancy between observed and expected frequencies, the two most important being Pearson's $\chi^2$,

$$\chi^2 = \sum_{i=1}^{r} \sum_{j=1}^{c} \frac{(n_{ij} - e_{ij})^2}{e_{ij}}, \qquad (14.1)$$

and the *likelihood ratio* statistic $G^2$,

$$G^2 = 2 \sum_{i=1}^{r} \sum_{j=1}^{c} n_{ij} \log\left(\frac{n_{ij}}{e_{ij}}\right).$$

Both statistics $\chi^2$ and $G^2$ are approximately distributed as chi-square with $(r-1) \times (c-1)$ degrees of freedom and their large values are critical for $H_0$. The approximation is good when cell frequencies are not small. Informal requirements are that no empty cells should be present and that observed counts should not be less than 5 for at least 80% of the cells. We focus on $\chi^2$ since the inference using statistic $G^2$ is similar.

Thus, for testing $H_0$ : Factors $R$ and $C$ are independent, versus $H_1$ : Factors $R$ and $C$ are dependent, the test statistic is $\chi^2$ given in (14.1) and the test is summarized as

| Null | Alternative | $\alpha$-level rejection region | $p$-value (MATLAB) |
|---|---|---|---|
| $H_0 : R,C$ ind. | $H_1 : R,C$ dep. | $[\chi^2_{df,1-\alpha},\infty)$ | `1-chi2cdf(chi2, df)` |

where $df = (r-1)\cdot(c-1)$.

When some cells have expected frequencies $< 5$, the Yates correction for continuity is recommended, and statistic $\chi^2$ gets the form

$$\chi^2 = \sum_{i=1}^{r} \sum_{j=1}^{c} \frac{(|n_{ij} - e_{ij}| - 0.5)^2}{e_{ij}}.$$

We denote by $p_{ij}$ the population counterpart to $\hat{p}_{ij} = \frac{n_{ij}}{n}$ and by $p_{i\cdot}$ and $p_{\cdot j}$ the population counterparts of $\hat{p}_{i\cdot} = n_{i\cdot}/n$ and $\hat{p}_{\cdot j} = n_{\cdot j}/n$. In terms of $p$, the independence hypothesis takes the form

$H_0 : p_{ij} = p_{i\cdot} \times p_{\cdot j}$ for all $i,j$ versus $H_1 : p_{ij} \neq p_{i\cdot} \times p_{\cdot j}$ for at least one $i,j$.

The following MATLAB program, ◀ `tablerxc.m`, calculates expected frequencies, the value of the $\chi^2$ statistic, and the associated $p$-value.

```
function [chi2, pvalue, exp, assoc] = tablerxc(obs)
% Contingency Table r x c for testing the probabilities
%
% Input:
%    obs - r x c matrix of observations.
%
% Output:
%    chi2 - statistic (approx distributed as chi-square
%                      with (r-1)(c-1) degrees of freedom)
%    pvalue - p - value
%    exp - matrix of expected frequencies
%    asoc - structure containing association measures: phi,
%                C, and Cramer's V
% Example of use:
% [chi2,pvalue,exp,assoc]=tablerxc([6 14 17 9; 30 32 17 3])
%-------------------------------------------------------------
  [r c]=size(obs);                    %size of matrix
  n = sum(sum(obs));                  %total s. size
  columns = sum(obs);                 %col sums 1 x c
  rows = sum(obs')';                  %row sums r x 1
  exp = rows * columns ./ n;          %[r x c] matrix
  chi2 = sum(sum((exp - obs).^2./exp ));
  df=(r-1)*(c-1);
```

```
pvalue = 1- chi2cdf(chi2, df);
% measures of association
if (df == 1)
  assoc.phi = sqrt(chi2/n);
end
assoc.C = sqrt(chi2/(n + chi2));
assoc.V = sqrt(chi2/(n * (min(r,c)-1)));
```

*Example 14.2.* **PAS and Streptomycin Cures for Pulmonary Tuberculosis.** Data released by the British Medical Research Council in 1950 (analyzed also in Armitage and Berry, 1994) concern the efficiency of para-amino-salicylic acid (PAS), streptomycin, and their combination in the treatment of pulmonary tuberculosis. Outcomes of sputum culture test performed on patients after the treatment are categorized as "Positive Smear," "Negative Smear & Positive Culture," and "Negative Smear & Negative Culture."

The table below summarizes the findings on 273 treated patients with a TB diagnosis.

|                      | Smear (+) | Smear (−) Culture (+) | Smear (−) Culture (−) | Total |
|----------------------|-----------|-----------------------|-----------------------|-------|
| PAS                  | 56        | 30                    | 13                    | 99    |
| Streptomycin         | 46        | 18                    | 20                    | 84    |
| PAS & Streptomycin   | 37        | 18                    | 35                    | 90    |
| Total                | 139       | 66                    | 68                    | 273   |

There are two factors here: cure and sputum results. We will test the hypothesis of their independence at the level $\alpha = 0.05$.

```
D =[ ...
    56     30     13 ;   %99
    46     18     20 ;   %84
    37     18     35 ]; %90
%  139     66     68    273
 [chi2,pvalue,exp] = tablerxc(D)

%  chi2 = 17.6284
%  pvalue = 0.0015
%  exp =
%      50.4066   23.9341   24.6593
%      42.7692   20.3077   20.9231
%      45.8242   21.7582   22.4176
```

The *p*-value is 0.0015, which is significant. For $\alpha = 0.05$, the rejection region of the test comprise values larger than chi2inv(1-0.05, (3-1)*(3-1)), which is the interval $[9.4877, \infty)$. Since $17.6284 > 9.4877$, the hypothesis of independence is rejected.

## *14.2.1 Measuring Association in Contingency Tables*

In a contingency table, statistic $\chi^2$ tests the hypothesis of independence and in some sense measures the association between the two factors. However, as a measure, $\chi^2$ is not normalized and depends on the sample size, while its distribution depends on the number of rows and columns. There are several measures that are calibrated to the interval $[0,1]$ and measure the strength of association in a way similar to $R^2$ in a regression context. We discuss three measures of association: the $\phi$-coefficient, the contingency coefficient $C$, and Cramer's $V$ coefficient.

**$\phi$-Coefficient.** If a $2 \times 2$ contingency table classifying $n$ elements produces statistic $\chi^2$, then the $\phi$-coefficient is defined as

$$\phi = \sqrt{\frac{\chi^2}{n}}.$$

**Contingency Coefficient $C$.** If an $r \times c$ contingency table classifying $n$ elements produces statistic $\chi^2$, then the contingency coefficient $C$ is defined as

$$C = \sqrt{\frac{\chi^2}{\chi^2 + n}}.$$

**Cramer's $V$ Coefficient.** If an $r \times c$ contingency table classifying $n$ elements produces statistic $\chi^2$, then Cramer's $V$ coefficient is defined as

$$V = \sqrt{\frac{\chi^2}{n(k-1)}},$$

where $k$ is the smaller of $r$ and $c$, $k = \min\{r, c\}$. If the number of levels for any factor is 2, then Cramer's $V$ becomes the $\phi$-coefficient.

*Example 14.3.* **Thromboembolism and Contraceptive Use.** A data set, considered in more detail by Worchester (1971), contains a cross-classification of 174 subjects with respect to the presence of thromboembolism and contraceptive use as a risk factor.

| | Contraceptive use | No contraceptive use | Total |
|---|---|---|---|
| Thromboembolism | 26 | 32 | 58 |
| Control | 10 | 106 | 116 |
| Total | 36 | 138 | 174 |

Statistic $\chi^2$ is 30.8913 with p-value $= 2.7289 \cdot 10^{-8}$, and the hypothesis of independence of thromboembolism and contraceptive usage is strongly rejected. For this table:

$$\phi = \sqrt{\frac{\chi^2}{n}} = \sqrt{\frac{30.8913}{174}} = 0.4214,$$

$$C = \sqrt{\frac{\chi^2}{\chi^2 + n}} = \sqrt{\frac{30.8913}{30.8913 + 174}} = 0.3883.$$

For $2 \times 2$ tables $\phi$ and Cramer's $V$ coincide. The fourth component in the output of tablerxc.m is a data structure with measures of association.

```
[chi2, pvalue, exp, stat]=tablerxc([26,32; 10,106])

%chi2 =     30.8913
%pvalue =   2.7289e-008
%
%exp =
%     12      46
%     24      92
%
%stat =
%     phi: 0.4214
%       C: 0.3883
%       V: 0.4214
```

*Example 14.4.* **Galton and ALW Features in Fingerprints.** In his influential book *Finger Prints*, Sir Francis Galton details the description and distributions of arch-loop-whorl (ALW) features of human fingerprints (Fig. 14.1). Some of his findings from 1892 are still in use today.

Tables 14.1a,b are from Galton (1892) and tabulate ALW features on forefingers in pairs of school children.

In Table 14.1a subjects A and B are paired at random from a large population of subjects. In Table 14.1b, subjects A and B are two brothers with a random assignment of order (A, B).

Galton was interested in knowing if there was any influence of fraternity on the dependence of ALW features. Using a $\chi^2$-test, test for the dependence in both tables.

$H_0$: Individuals $A$ and $B$ are independent if characterized by their fingerprint features.

(a)          (b)          (c)

**Fig. 14.1** (a) Arch, (b) loop, and (c) whorl features in a fingerprint.

**Table 14.1** (a) Table with random pairing of children. (b) Table with fraternal pairing.

|   |   | A | | |
|---|---|---|---|---|
|   |   | Arch | Loop | Whorl |
|   | Arch | 5 | 12 | 8 |
| B | Loop | 8 | 18 | 8 |
|   | Whorl | 9 | 20 | 13 |

(a)

|   |   | A | | |
|---|---|---|---|---|
|   |   | Arch | Loop | Whorl |
|   | Arch | 5 | 12 | 2 |
| B | Loop | 4 | 42 | 15 |
|   | Whorl | 1 | 14 | 10 |

(b)

MATLAB output:

```
[chisq, p, expected,assoc]=tablerxc([5 12 8; 8 18 8; 9 20 13])
%(a) random pairing
%chisq = 0.6948
% p = 0.9520
% expected =
%     5.4455    12.3762     7.1782
%     7.4059    16.8317     9.7624
%     9.1485    20.7921    12.0594
% assoc =
%     C: 0.0827
%     V: 0.0586


[chisq, p, expected,assoc]=tablerxc([5 12 2; 4 42 15; 1 14 10])
%(b) fraternal pairing
% chisq = 11.1699
% p = 0.0247
% expected =
%     1.8095    12.3048     4.8857
%     5.8095    39.5048    15.6857
%     2.3810    16.1905     6.4286
% assoc =
%     C: 0.3101
%     V: 0.2306
```

It is evident that for random pairing, the hypothesis of independence $H_0$ is not rejected ($p$-value 0.9520), while in the case of fraternal pairing the fingerprint features are significantly dependent ($p$-value 0.0247).

**Power Analysis.** Power analysis for contingency tables involves evaluation of a noncentral $\chi^2$ in which the noncentrality parameter $\lambda$ depends on an appropriate effect size.

The traditional effect size in this context is Cohen's $w$, which is in fact equivalent to the $\phi$-coefficient, $\phi = \sqrt{\chi^2/n}$, where $n$ is the total table size. Unlike the $\phi$-coefficient which is used for $2 \times 2$ tables only, $w$ is used for arbitrary $r \times c$ tables and can exceed 1. Effects $w = 0.1, 0.3$ and $0.5$ correspond to a small, medium and large size, respectively.

For prospective analyses, the noncentrality parameter $\lambda$ is $nw^2$ while for retrospective analyses $\lambda$ is the observed $\chi^2$. The power is

$$1 - \beta = 1 - nc\chi^2(\chi^2_{k,1-\alpha}, k, \lambda),$$

where $k = (r-1) \times (c-1)$ is the number of degrees of freedom and $\chi^2_{k,1-\alpha}$ is the $(1-\alpha)$-quantile of a (central) $\chi^2_k$ distribution.

*Example 14.5.* (a) Find the power in a $2 \times 6$ contingency table, for $n = 180$ and $w = 0.3$ (medium effect).

```
w = 0.3; n = 180; k = (2-1)*(6-1); alpha = 0.05; lambda=n*w^2
pow = 1-ncx2cdf( chi2inv(1-alpha,k), k, lambda) %0.8945
```

(b) What sample size ensures the power of 95% in a contingency table $2 \times 6$, for an effect of $w = 0.3$ and $\alpha = 0.05$?

```
beta = 0.05; alpha = 0.05; k = (2-1)*(6-1); w=0.3;
pf = @(n)  ncx2cdf( chi2inv(1-alpha,k), k, n*w^2) - beta;
ssize = fzero(pf, 200)        %219.7793 approx 220
```

## 14.2.2 Cohen's Kappa

Cohen's kappa is an important descriptor of agreement between two testing procedures. This descriptor is motivated by calibrating the observed agreement by an agreement due to chance. If $p_c$ is the proportion of agreement due to chance and $p_o$ the proportion of observed agreement, then

$$\hat{\kappa} = \frac{p_o - p_c}{1 - p_c}. \tag{14.2}$$

For a paired table representing the results of $n$ tests by two devices or ratings by two raters

|   | $+$ | $-$ |  |
|---|-----|-----|-----|
| $+$ | $a$ | $b$ | $a+b$ |
| $-$ | $c$ | $d$ | $c+d$ |
|   | $a+c$ | $b+d$ | $n = a+b+c+d$ |

Cohen's kappa index is defined as

$$\hat{\kappa} = \frac{2(ad - bc)}{(a + b)(b + d) + (a + c)(c + d)}. \qquad (14.3)$$

The expression in Eq. (14.2) is equivalent to that in Eq. (14.3) for $p_o = a/n + d/n$ and $p_c = (a + b)(a + c)/n^2 + (b + d)(c + d)/n^2$, respectively. The former is the observed agreement equal to the proportion of $(+,+)$ and $(-,-)$ outcomes, while the latter is the proportion of agreement when the results are independent within fixed marginal proportions.

There are no formal rules for judging $\hat{\kappa}$, but here is the standard:

| $\hat{\kappa}$ | Degree of agreement |
|---|---|
| <0.20 | Poor |
| 0.20 – 0.40 | Fair |
| 0.40 – 0.60 | Moderate |
| 0.60 – 0.80 | Good |
| 0.8 – 1 | Very good |

The MLE of $\kappa$ is

$$\hat{\kappa}_{mle} = \frac{4(ad - bc) - (b - c)^2}{(2a + b + c)(2d + b + c)},$$

and it is obtained from (14.2) by taking $p_o = a/n + d/n$ and $p_c = P^2 + (P')^2$, where $P = (2a + b + c)/(2n)$ and $P' = (2d + b + c)/(2n)$ are, respectively, the MLEs of the prevalence of $+$ and $-$ in the population.

There are several expressions for the variance of $\hat{\kappa}$. Two standard estimators are the Block–Kraemer (BK) and Garner (G) approximations:

(1) (BK):

$$\mathbb{V}\mathrm{ar}\,\hat{\kappa} \approx \frac{1 - \hat{\kappa}}{n}\left[(1 - \hat{\kappa})(1 - 2\hat{\kappa}) + \frac{\hat{\kappa}(2 - \hat{\kappa})}{2P(1 - P')}\right];$$

(2) (G)

$$\mathbb{V}\mathrm{ar}\,\hat{\kappa} \approx \frac{4}{(1 - p_c)^2 n^2 (\frac{1}{a+1} + \frac{1}{b+1} + \frac{1}{c+1} + \frac{1}{d+1})}.$$

The sampling distribution of $\hat{\kappa}$ is asymptotically normal and the approximation is satisfactory if $n$ is large and $\kappa$ not too close to 1. This leads, in a standard manner, to approximate confidence intervals for $\kappa$.

*Example 14.6.* A company producing a medical sensor A is applying for FDA approval of a new version B. Both sensors A and B are prone to errors, and a gold standard is absent. The FDA is requesting that the new sensor be comparable to the currently used one, and the company decides to include Cohen's $\hat{\kappa}$ statistic in the report.

The experiment consisted of $n = 2803$ trials and resulted in a paired table [97 11; 6 2689], where 97 was the number of (+,+) outcomes and 2689 the number of (−,−) outcomes. The code ◢ cohen.m finds Cohen's $\hat{\kappa}$ and 95% confidence intervals for the population $\kappa$, based on the two estimators of variance, Block-Kraemer and Garner.

```
data =[97 11;   6 2689]
%data =
% 97            11
% 6           2689
a=data(1,1); b=data(1,2); c=data(2,1); d=data(2,2);
apb = a+b; cpd = c+d; apc = a+c; bpd = b+d;
n = a + b + c + d;
%------------
p0 = (a + d)/n;                    %Observed agreement
pc = (apb*apc + cpd*bpd)/n^2; %Chance agreement
Pre  = (2*a + b + c)/(2 * n); %Prevalence of +
pcc = Pre^2 + (1-Pre)^2;           %MLE of chance agreement
%------------
kappa = 2 * (a * d - b * c)/(apb*bpd + apc*cpd) %0.9163
%or kappa=(p0-pc)/(1-pc)

kappamle = (4 * (a*d - b*c) - (b -c)^2)/...
   ((2 * a + b + c) * (2 * d + b + c))  %0.9163
%or kappamle=(p0-pcc)/(1-pcc)
%------------
%Block-Kraemer variance estimator
varbk = (1- kappa)/n * ( (1- kappa)*(1-2* kappa) + ...
   (kappa * (2 -  kappa)/(2 * Pre * (1-Pre)))) %4.0731e-004
%Garner variance estimator
vargarner = 4/( (1- pc)^2 * n^2 * (1/(a+1) + 1/(b+1) + ...
    1/(c+1) + 1/(d + 1) ) )  %4.0971e-004
%------------
%Confidence intervals
[ kappa - 1.96 * sqrt(varbk) ...
        kappa + 1.96 * sqrt(varbk)]  %0.8767   0.9558
[ kappa - 1.96 * sqrt(vargarner) ...
    kappa + 1.96 * sqrt(vargarner)]  %0.8766   0.9560
```

Cohen's $\kappa$ is estimated to be 91.63%, which represents very good agreement.

## 14.3 Three-Way Tables

A natural extension of two-way tables for testing the independence of two factors are $n$-dimensional tables for testing the independence of $n$ factors. We will discuss a three-dimensional extension; the interested reader can consult Zar (2007), Agresti (2002), or Fienberg (2000) for more detailed coverage. In three-dimensional tables the counts constitute three-dimensional arrays characterized by rows, columns, and pages. We associate factors with these three dimensions and consider row, column, and page factors ($R$, $C$, and $P$). In the cell $(i,j,k)$ there are $n_{ijk}$ observations, $i = 1,\ldots,r$; $j = 1,\ldots,c$; and $k = 1,\ldots,p$.

Denote the total number of observations by $n$. The empirical probability of the cell $(i,j,k)$ is $n_{ijk}/n$, and the empirical marginal probabilities of row $i$, column $j$, and page $k$ are $n_{i..}/n$, $n_{.j.}/n$, and $n_{..k}/n$. The numerators are calculated as the sums over all indices replaced by dots. For example, $n_{.j.} = \sum_{i=1}^{r} \sum_{k=1}^{p} n_{ijk}$, $j = 1,\ldots,c$.

We are interested in testing the hypothesis $H_0$ that the factors $R$, $C$, and $P$ are independent. The alternative $H_1$ would be that the factors are dependent. Under $H_0$, the frequency in the cell $(i,j,k)$ is *expected* to be

$$e_{ijk} = n \times \frac{n_{i..}}{n} \times \frac{n_{.j.}}{n} \times \frac{n_{..k}}{n},$$

as the product of the total number of observations $n$ and the corresponding empirical marginal probabilities. After simplification, the expected frequency in the cell $(i,j,k)$ becomes

$$e_{ijk} = \frac{n_{i..} \times n_{.j.} \times n_{..k}}{n^2}.$$

The test statistic is

$$\chi^2 = \sum_{i=1}^{r} \sum_{j=1}^{c} \sum_{k=1}^{p} \frac{(n_{ijk} - e_{ijk})^2}{e_{ijk}}, \tag{14.4}$$

and the *likelihood ratio* statistic is

$$G^2 = 2 \sum_{i=1}^{r} \sum_{j=1}^{c} \sum_{k=1}^{p} n_{ijk} \log\left(\frac{n_{ijk}}{e_{ijk}}\right).$$

When $H_0$ is true, both statistics $\chi^2$ and $G^2$ follow approximately a $\chi^2$ distribution with $rcp - r - c - p + 2$ degrees of freedom. Large values of $\chi^2$ are critical for $H_0$.

If $H_0$ is rejected, then multiple alternatives are possible. For example, all three factors $R$, $C$, and $P$ are mutually dependent, factors $R$ and $C$ are dependent but both are independent of factor $P$, and so on. To specify why $H_0$ is rejected, one needs to test the hypothesis whether each single factor is independent of the other two. There are three such tests, and they are summarized in the following table:

| Factor | $e_{ijk}$ | $df$ |
|---|---|---|
| $R$ vs. $C,P$ | $\frac{n_{i\cdot\cdot} \times n_{\cdot jk}}{n}$ | $rcp - cp - r + 1$ |
| $C$ vs. $R,P$ | $\frac{n_{\cdot j\cdot} \times n_{\cdot\cdot k}}{n}$ | $rcp - rp - c + 1$ |
| $P$ vs. $R,C$ | $\frac{n_{\cdot\cdot k} \times n_{\cdot jk}}{n}$ | $rcp - rc - p + 1$ |

Here $n_{ij\cdot} = \sum_k n_{ijk}$, $n_{\cdot jk} = \sum_i n_{ijk}$, and $n_{i\cdot k} = \sum_j n_{ijk}$. Also, as before $n_{i\cdot\cdot} = \sum_j \sum_k n_{ijk}$, $n_{\cdot j\cdot} = \sum_i \sum_k n_{ijk}$, and $n_{\cdot\cdot k} = \sum_i \sum_j n_{ijk}$. The $\chi^2$ statistic is calculated as in (14.4) and the degrees of freedom are given in the table.

*Example 14.7.* **Anolis Lizards of Bimini.** This well-known data set comes from the paper of Schoener (1968) and is also used in Fienberg (1970). The researcher was interested in structural habitat categories for Anolis lizards of Bimini: *sagrei* (brown anole) adult males versus *distichus* (trunk anole) adult and subadult males. The brown anole and trunk anole are medium-sized, fairly robust, "trunk-ground" lizards. They generally prefer the fairly open vegetation of disturbed sites, where they adopt a head-down, sit-and-wait posture and perch low on large trunks or fenceposts (Fig. 14.2).

**Fig. 14.2** Brown anole (*Anolis sagrei*) in typical position at its perch.

The researcher was interested in the preferences of these two species with respect to the perch height and diameter.

| | A. sagrei | | A. distichus | |
|---|---|---|---|---|
| | Perch Diameter | | Perch Diameter | |
| | ≤ 4 | > 4 | ≤ 4 | > 4 |
| Perch Height > 4.75 | 32 | 11 | 61 | 41 |
| (in feet)    ≤ 4.75 | 86 | 35 | 73 | 70 |

The data is classified with respect to three dichotomous factors: Height at levels Low ($\leq 4.75$) and High ($> 4.75$), Diameter at levels Small ($\leq 4$) and Large ($> 4$), and Species with levels *A. sagrei* and *A. distichus*. Are these three factors independent? The null hypothesis is that the factors are independent and the alternative is that they are not. The MATLAB program ◀ tablerxcxp.m calculates the expected frequencies (under mutual independence hypothesis) and provides the $\chi^2$ statistic, its degrees of freedom, and the $p$-value. This program also outputs the expected frequencies in the format that matches the input data. Here, Height is the row factor $R$, Diameter is the column factor $C$, and Species is the page factor $P$.

```
anolis = [32 11; 86 35];
anolis(:,:,2)=[61 41; 73 70];
[ch2 df pv exp]=tablerxcxp(anolis)
%ch2 = 23.9055
%df = 4
%pv =8.3434e-005
%exp(:,:,1) =
%   35.8233 22.3185
%   65.2231 40.6351
%exp(:,:,2) =
%   53.5165 33.3417
%   97.4370 60.7048
```

The hypothesis $H_0$ is rejected (with $p$-value $< 5\%$), and we infer that the three factors are not independent. However, this analysis does not fully explain the dependencies responsible for rejecting $H_0$. Are all three factors dependent, or maybe two of the factors are mutually dependent and the third is independent of both? When $H_0$ is rejected, we need a partial independence test, similar to pairwise comparisons in the case where the ANOVA hypothesis is rejected. The MATLAB program for this test is ◀ partialrxcxp.m.

```
[ch2 df pv exp]=partialrxcxp(anolis,'r_cp')
   %ch2 = 12.3028; df=3; pv = 0.0064, exp=...
[ch2 df pv exp]=partialrxcxp(anolis,'c_pr')
   %ch2 = 14.4498; df=3; pv = 0.0024, exp=...
[ch2 df pv exp]=partialrxcxp(anolis,'p_rc')
   %ch2 = 23.9792; df=3; pv =2.5231e-005, exp=...
```

Three tests are performed: (i) the row factor independent of column/page factors (partial = 'r_cp'), (ii) the column factor independent of row/page factors (partial = 'c_rp'), and (iii) the page factor independent of row/column factors (partial = 'p_rc'). It is evident from the output that all three tests produced significant $\chi^2$, that is, each factor depends on the two others. Since Height was the row factor $R$, Diameter the column factor $C$, and Species the page factor $P$, we conclude that the strongest dependence is that of Species factor on the height and diameter of the perch, partialrxcxp(anolis,'p_rc'), with a $p$-value of 0.000025.

✎

## 14.4  Contingency Tables with Fixed Marginals: Fisher's Exact Test

In his book, Fisher (1935) provides an example of a small $2 \times 2$ contingency table related to a tea-tasting experiment, namely, a woman claimed to be able to judge whether tea or milk was poured in a cup first. The woman was given eight cups of tea, in four of which tea was poured first, and was told to guess which four had tea poured first. The contingency table for this design is

|              | Guess milk first | Guess tea first | Total |
|--------------|:----------------:|:---------------:|:-----:|
| Milk first   | $x$              | $4 - x$         | 4     |
| Tea first    | $4 - x$          | $x$             | 4     |
| Column total | 4                | 4               | 8     |

The number of correct guesses "Milk first" in the cell $(1,1)$, $x$, can take values 0, 1, 2, 3, or 4 with the probabilities $\frac{\binom{4}{x}\binom{4}{4-x}}{\binom{8}{4}}$, as in

```
hygepdf(0:4, 8,4,4)
% ans =   0.0143   0.2286   0.5143   0.2286   0.0143
```

These are hypergeometric probabilities, applicable here since the marginal counts are fixed.

For $x = 4$ the probability of obtaining this table by chance is 0.0143, while for $x = 3$ the probability of getting this or a more extreme table by chance is $0.2286 + 0.0143 = 0.2429$, and so on. Thus the probability of the woman's guessing correctly, i.e., if $x = 4$, would be less than 5%.

Suppose that in the table

|              | Column 1 | Column 2 | Row total             |
|--------------|:--------:|:--------:|:---------------------:|
| Row 1        | $a$      | $b$      | $a + b$               |
| Row 2        | $c$      | $d$      | $c + d$               |
| Column total | $a + c$  | $b + d$  | $n = a + b + c + d$   |

marginal counts $a + b$, $c + d$, $a + c$, and $b + d$ are fixed. Then $a$, $b$, $c$, and $d$ are constrained by these marginals. We are interested if the probabilities that an observation will be in column 1 are the same for rows 1 and 2. Denote these probabilities as $p_1$ and $p_2$. The null hypothesis here is not the hypothesis of independence but the hypothesis of homogeneity, $H_0 : p_1 = p_2$. The test is close to the two-sample problem considered in Chap. 10; however, in this case the samples are dependent because of marginal constraints.

The statistic $T$ to test $H_0$ is simply the number of observations in the cell (1,1):

$$T = a.$$

If $H_0$ is true, then $T$ has a hypergeometric distribution $\mathscr{HG}(n, a+c, a+b)$, that is

$$\mathbb{P}(T = x) = \frac{\binom{a+c}{x} \cdot \binom{b+d}{a+b-x}}{\binom{n}{a+b}}, \quad x = 0, 1, \ldots, \min\{a+b, a+c\}.$$

The $p$-value against the one-sided alternative $H_1 : p_1 < p_2$ is hygecdf(a, n, a+c, a+b) and against the alternative $H_1 : p_1 > p_2$ is 1 - hygecdf(a-1, n, a+c, a+b). If the hypothesis is two-sided, then the $p$-value cannot be obtained by doubling one-sided $p$-value due to asymmetry of the hypergeometric distribution. The two-sided $p$-value is obtained as the sum of all probabilities hygepdf(x, n, a+c, a+b), $x = 0, 1, \ldots, \min\{a+b, a+c\}$ which are smaller than or equal to hygepdf(a, n, a+c, a+b).

*Example 14.8.* There are 22 subjects enrolled in a clinical trial and 9 are females. Researchers plan to administer 11 portions of a drug and 11 placebos. Only 2 females are administered the drug. Are the proportions of males and females assigned to the drug significantly different? What are the $p$-values for one- and two-sided alternatives?

```
a = 2; b = 7; c = 9; d =4;
n = a + b + c + d;
T = a;
pval = hygecdf(T,n,a+c,a+b)    %H1: p1<p2    pval=0.0402
%
pa = hygepdf(T,n,a+c,a+b);
for  i = 1:min(a+b, a+c)+1
    p(i) = hygepdf(i-1,n,a+c,a+b) ;
end
pval2 = sum(p(p <= pa))    %H1: p1 ~= p2  pval2=0.1179
```

Since the one-sided $p$-value is less than 5%, we reject the hypothesis of homogeneity of adminstration of a drug versus placebo with respect to gender. For the two-sided alternative, we fail to reject $H_0$.

*Example 14.9.* **The Effect of Passive Smoking on Lung Cancer.** Lawal (2003) considers the following data originally published by Correa et al. (1983)

on the effect of passive smoking on lung cancer. A total of 155 non-smoking ever-married females were tabulated by their lung cancer status and husband's smoking status.

Is the proportion of lung cancer cases homogeneous with respect to the husband's smoking status? Find the $p$-value for both one- and two-sided alternatives.

|  | Smoking status | | |
|---|---|---|---|
|  | Case | Control | Total |
| Spouse smoked | 14 | 61 | 75 |
| Spouse did not smoke | 8 | 72 | 80 |
| Total | 22 | 133 | 155 |

Here $H_0 : p_1 = p_2$ and $H_1 : p_1 > p_2$ or $H_1 : p_1 \neq p_2$.

```
a = 14;  b = 61;  c = 8;  d =72;
n = a + b + c + d;
T = a;
%H1: p1 > p2
pval = 1-hygecdf(T-1,n,a+c,a+b)  %0.0941
%H1: p1 ~= p2
pa = hygepdf(T,n,a+c,a+b);
for  i = 1:min(a+b, a+c)+1
    p(i) = hygepdf(i-1,n,a+c,a+b) ;
end
pval2 = sum(p(p <= pa))              %0.1669
```

Thus, Fisher's exact test fails to reject the null hypothesis at 5% significance level.

Fisher's exact test remains valid for designs with random row totals, random column totals, or tables with random marginals (as in the previous section). In this case the tests are conservative, and more powerful versions exist. A benefit of using Fisher's exact test is that it operates with small cell frequencies, for example, a 0 count in a table cell is a possibility. Since $\chi^2$ or normal approximations assume large $n$ and $np_{ij}$s preferably larger than 5, the reason for the popularity of Fisher's exact test is obvious.

## 14.5 Multiple Tables: Mantel–Haenszel Test

In Chap. 10, p. 380, $2 \times 2$ tables were discussed in the context of comparing two proportions. Here we discuss multiple $2 \times 2$ tables and inference from combined information. The Mantel–Haenszel (Fig. 14.3a,b) methodology can be used in $2 \times 2$ tables to control for a variable that stratifies the data. This leads to multiple tables, one for each level of controlled variable. The Mantel–Haenszel methodology can be used for (i) testing the conditional independence of two factors or (ii) measuring the degree of conditional association (risk ratios). All conditioning is on the variable by which the tables are stratified. There are several other uses of the Mantel–Haenszel methodology such as in survival analysis (longrank tests of Peto and Peto) and in depairing of McNemar's paired designs.

## 14.5.1 Testing Conditional Independence or Homogeneity

Suppose that $k$ independent classifications into a $2 \times 2$ table are observed. We could denote the $i$th such table by

| $a_i$ | $b_i$ | $a_i + b_i$ |
|---|---|---|
| $c_i$ | $d_i$ | $c_i + d_i$ |
| $a_i + c_i$ | $b_i + d_i$ | $n_i$ |

The tables give counts broken down by binary levels of two factors, and the separate tables usually correspond to the levels of a third factor that needs to be controlled. Imagine that we want to test for the independence of two factors, say, political association (Democrat, Republican) and opinion about some social issue (Support, Oppose). The single contingency table may not be significant, but when controlled by gender (two tables, one for males, the other for females) or by age group, the dependence may turn out significant. Thus, multiple tables make inference more precise by controlling for an influential variable.

For each of $k$ tables consider cell at the position $(1, 1)$, so called *pivot*, with $a_i$ counts. If the two tabulated factors are independent, then the counts $a_i$ should be close to "expected" counts $e_i = (a_i + b_i)(a_i + c_i)/n_i$.

The test statistic measuring discrepancies in the pivot cell over all $k$ tables is

$$\chi^2 = \frac{(|A - E| - 1/2)^2}{V}, \text{ where} \tag{14.5}$$

$$A = \sum_{i=1}^{k} a_i, \ E = \sum_{i=1}^{k} e_i, \ V = \sum_{i=1}^{k} \frac{(a_i + b_i)(c_i + d_i)(a_i + c_i)(b_i + d_i)}{n_i^2(n_i - 1)},$$

which has an approximately $\chi^2$-distribution with 1 degree of freedom when the (null) hypothesis of independence/homogeneity is true. Large values of $\chi^2$ are critical for $H_0$.

It is interesting that even for sparse individual tables the $\chi^2$-approximation holds as long as the sum of row totals in all tables is larger than 20, say, $\sum_i (a_i + b_i)$, $\sum_i (c_i + d_i) > 20$.

As in contingency tables, if the marginal sums are fixed in advance, the subsequent inference does not concern the independence; it concerns the homogeneity of one factor within the levels of the other factor. The following example tests for homogeneity of proportions of cancer incidence among smokers and nonsmokers stratified by populations in different cities.

*Example 14.10.* The three $2 \times 2$ tables provide classification of people from three Chinese cities, Zhengzhou, Taiyuan, and Nanchang, with respect to smoking habits and incidence of lung cancer (Liu, 1992).

|                          | Zhengzhou |     |       | Taiyuan |     |       | Nanchang |     |       |
|--------------------------|-----------|-----|-------|---------|-----|-------|----------|-----|-------|
| Cancer Diagnosis:        | yes       | no  | total | yes     | no  | total | yes      | no  | total |
| Smoker                   | 182       | 156 | 338   | 60      | 99  | 159   | 104      | 89  | 193   |
| Nonsmoker                | 72        | 98  | 170   | 11      | 43  | 54    | 21       | 36  | 57    |
| Total                    | 254       | 254 | 508   | 71      | 142 | 213   | 125      | 125 | 250   |

We can apply the Mantel–Haenszel test to decide if the proportions of cancer incidence for smokers and nonsmokers coincide for the three cities, i.e., $H_0 : p_{1i} = p_{2i}$, where $p_{1i}$ is the proportion of incidence of cancer among smokers in city $i$ and $p_{2i}$ is the proportion of incidence of cancer among nonsmokers in city $i$, $i = 1, 2, 3$. We use the two-sided alternative, $H_1 : p_{1i} \neq p_{2i}$, for some $i \in \{1, 2, 3\}$ and fix the type I error rate at $\alpha = 0.10$.

To compute $\chi^2$ in (14.5), we find $A$, $E$, and $V$. From the tables, $A = \sum_i a_i = 182 + 60 + 104 = 346$. Also, $E = \sum_i e_i = 338 \cdot 254/508 + 159 \cdot 71/213 + 193 \cdot 125/250 = 169 + 53 + 96.5 = 318.5$.

$$
\begin{aligned}
V &= \sum_{i=1}^{k} \frac{(a_i + b_i)(c_i + d_i)(a_i + c_i)(b_i + d_i)}{n_i^2 (n_i - 1)} \\
&= \frac{338 \cdot 254 \cdot 170 \cdot 254}{508^2 \cdot 507} + \frac{159 \cdot 71 \cdot 54 \cdot 142}{213^2 \cdot 212} + \frac{193 \cdot 125 \cdot 57 \cdot 125}{250^2 \cdot 249} \\
&= 28.33333 + 9 + 11.04518 = 48.37851.
\end{aligned}
$$

Therefore,

$$
\chi^2 = \frac{(|346 - 318.5| - 0.5)^2}{48.37851} = 15.0687.
$$

Because the statistic $\chi^2$ is distributed approximately as $\chi_1^2$, the $p$-value (via MATLAB m-file ◢ mantelhaenszel.m) is 0.0001.

```
[chi2, pval] = mantelhaenszel([182 156; 72 98; ...
    60 99; 11 43; 104 89; 21 36])
```

```
%chi2 = 15.0687
%pval  = 1.0367e-004
```

In this case, there is clear evidence that the cancer rates are not homogeneous among the three cities.

### 14.5.2 Conditional Odds Ratio

Measures of association in a $2 \times 2$ table have been discussed in "risk" summaries: risk ratio and odds ratio (Chap. 10) are examples. Suppose that $k$ independent classifications into a $2 \times 2$ table are observed.

Let the rows in the $i$th table

|          | Presence | Absence |          |
|----------|----------|---------|----------|
| Group 1  | $a_i$    | $b_i$   | $a_i + b_i$ |
| Group 2  | $c_i$    | $d_i$   | $c_i + d_i$ |
|          | $a_i + c_i$ | $b_i + d_i$ | $n_i$ |

represent Groups 1 and 2 and the columns Presence/Absence of a particular attribute. For example, Groups could be case and control subjects and Presence/Absence can be related to a particular risk factor. The tables can be stratified to control for some other risk factor or demographic feature of the subjects such as gender, age, smoking status, etc.

Then the proportion of subjects with the attribute present for Group 1 is $a_i/(a_i + b_i)$, and the proportion of subjects with the attribute present for Group 2 is $c_i/(c_i + d_i)$.

The observed odds ratio (between the groups) for a single $i$th table is

$$\frac{a_i/b_i}{c_i/d_i} = \frac{a_i d_i}{b_i c_i}.$$

The combined odds ratio for all $k$ strata tables is defined as

$$\widehat{OR}_{MH} = \frac{\sum_i a_i d_i / n_i}{\sum_i b_i c_i / n_i}.$$

The expression $\widehat{OR}_{MH}$ was proposed in Mantel and Haenszel (1959) and represents the weighted average of individual odds ratios $\frac{a_i d_i}{b_i c_i}$ with weights $b_i c_i / n_i$.

An approximation to the variance of the log odds ratio $\widehat{OR}_{MH}$ is given by the RGB formula (Robins et al., 1986):

$$\mathbb{V}\mathrm{ar}(log(\widehat{OR})) = \frac{\sum_i R_i P_i}{2R^2} + \frac{\sum_i P_i S_i + Q_i R_i}{2RS} + \frac{\sum_i S_i Q_i}{2S^2},$$

where $P_i = (a_i + d_i)/n_i$, $Q_i = (b_i + c_i)/n_i$, $R_i = a_i d_i/n_i$, $S_i = b_i c_i/n_i$, $R = \sum_i R_i$, and $S = \sum_i S_i$. If $k$ is equal to 1, the RGB variance formula reduces to the familiar $(\frac{1}{a} + \frac{1}{b} + \frac{1}{c} + \frac{1}{d})$, (p. 383).

The $(1 - \alpha)100\%$ confidence interval for $OR$ is

$$\left[\widehat{OR}_{MH} \cdot \exp\{-z_{1-\alpha/2}\hat{\sigma}\}, \ \widehat{OR}_{MH} \cdot \exp\{z_{1-\alpha/2}\hat{\sigma}\}\right],$$

where $\hat{\sigma} = \sqrt{\mathbb{V}\mathrm{ar}(log(\widehat{OR}))}$.

(a)                          (b)                          (c)

**Fig. 14.3** (a) Nathan Mantel (1919–2002), (b) William Haenszel (1910–1998), and (c) Quinn McNemar (1900–1986).

## 14.6 Paired Tables: McNemar's Test

Another type of table commonly used in dental, opthalmology, and pharmacology trials is matched-pair tables summarizing the designs in which interventions are applied to the same patient. For example, in randomized split-mouth trials comparing the effectiveness of tooth-specific interventions to prevent decay, one tooth in a subject is randomly selected to receive treatment $A$ while the contralateral tooth in the same subject receives treatment $B$. Another example is crossover trials testing the efficacy of drugs. In this design, a patient is randomly administered treatment $A$ or $B$ in the first time period and then administered the remaining treatment in the second time period. The link between the split-mouth design and crossover trials is apparent – the tooth location in the split-mouth design is analogous to time in the crossover design.

The matched-pair design has statistical advantages. Because the control and test groups are subject to the same environment, this design controls for many confounding factors. Thus differences in outcomes between test and control groups are likely attributable to the treatment. Moreover, since control and test groups receive both interventions, matched-pair studies usually require no more than half the number of subjects to produce the same precision as parallel group studies.

McNemar's test (after Quinn McNemar, Fig.14.3c) used for inference in paired tables. Although data for the McNemar test resemble contingency tables, the structure of the tables is quite different and the inference is quite different. For simplicity, assume that measurements at *Before* and *After* on the same subject result in *Positive* and *Negative* responses. From $N$ subjects we obtain $2N$ responses, or $N$ pairs of responses, organized as follows:

|  |  | After | | |
|---|---|---|---|---|
|  |  | Positive | Negative | Total |
| Before | Positive | $A$ | $B$ | $A+B$ |
|  | Negative | $C$ | $D$ | $C+D$ |
|  | Total | $A+C$ | $B+D$ | $N$ |

For example, $A$ is the number of subjects (pairs of responses) where both *Before* and *After* resulted in a positive. More generally, *Before* and *After* could be any two different groups of matched subjects that produce binary responses.

The marginal sum $A+B$ is the number of positives in *Before* and $A+C$ is the number of positives in *After*. The proportion of positives in *Before* is $\hat{p}_1 = \frac{A+B}{N}$, while the proportion of positives in *After* is $\hat{p}_2 = \frac{A+C}{N}$. Let the population counterparts of $\hat{p}_1$ and $\hat{p}_2$ be $p_1$ and $p_2$.

### 14.6.1 Risk Differences

The McNemar test examines the difference between the proportions that derive from the marginal sums and tries to infer if the two population proportions $p_1$ and $p_2$ differ significantly. The difference between this test and the test for two proportions from Chap. 10, p. 381, is that in the paired tables the two proportions are *not independent*. Note that both sample proportions depend on $A$ from the upper left cell of the table. The inference about the difference between $p_1$ and $p_2$ involves only entries $B$ and $C$ from the table, since $A/N$ cancels.

Under the hypothesis $H_0 : p_1 = p_2$, both $B$ and $C$ are distributed as binomial $\mathcal{B}in(B+C,0.5)$. In this case, $\mathbb{E}B = \frac{B+C}{2}$ and $\mathbb{V}ar(B) = \frac{B+C}{4}$. By the CLT,

$$Z = \frac{B - \mathbb{E}B}{\sqrt{\mathbb{V}ar(B)}} = \frac{B - (B+C)/2}{2\sqrt{B+C}}.$$

Thus, after simplifying and squaring, we obtain the statistic

$$\chi^2 = Z^2 = \frac{(B-C)^2}{B+C},$$

which has an approximately $\chi^2$-distribution with 1 degree of freedom. Large values of $\chi^2$ are critical.

When $B+C$ is small, it is recommended to use a continuity correction in $\chi^2$ as $\chi^2 = Z^2 = \frac{(|B-C|-1)^2}{B+C}$.

The confidence interval for the difference in proportions is

$$[\hat{p}_1 - \hat{p}_2 - z_{1-\alpha/2}s, \ \hat{p}_1 - \hat{p}_2 + z_{1-\alpha/2}s],$$

where $s$ is the square root of $s^2$, and

$$s^2 = \frac{\hat{p}_1(1-\hat{p}_1)}{N} + \frac{\hat{p}_2(1-\hat{p}_2)}{N} + \frac{2(\hat{p}_{11}\hat{p}_{22} - \hat{p}_{12}\hat{p}_{21})}{N},$$

where $\hat{p}_{11} = A/N$, $\hat{p}_{12} = B/N$, $\hat{p}_{21} = C/N$, and $\hat{p}_{22} = D/N$. Note that the first two factors in the expression for $s^2$ are for the case of independent proportions, while the third factor accounts for the dependence.

### 14.6.2 Risk Ratios

The risk ratio is defined as

$$RR = \frac{\hat{p}_1}{\hat{p}_2} = \frac{A+B}{A+C},$$

with variance for the log(RR),

$$\mathbb{V}\mathrm{ar}\,(\log(RR)) = \frac{1}{A+B} + \frac{1}{A+C} - \frac{2A}{(A+B)(A+C)}.$$

### 14.6.3 Odds Ratios

To conduct the inference on odds ratios in paired tables, we use the Mantel–Haenszel theory with parallel (unpaired) tables derived from a paired table.

Any table with $N$ paired observations generates $N$ Mantel–Haenszel tables

$$\begin{array}{|c|c|}\hline \mathbf{1}&0\\\hline 0&0\\\hline\end{array} \mapsto \begin{array}{|c|c|}\hline 1&1\\\hline 0&0\\\hline\end{array} \qquad \begin{array}{|c|c|}\hline 0&\mathbf{1}\\\hline 0&0\\\hline\end{array} \mapsto \begin{array}{|c|c|}\hline 1&0\\\hline 0&1\\\hline\end{array} \qquad \begin{array}{|c|c|}\hline 0&0\\\hline \mathbf{1}&0\\\hline\end{array} \mapsto \begin{array}{|c|c|}\hline 0&1\\\hline 1&0\\\hline\end{array} \quad \text{and} \quad \begin{array}{|c|c|}\hline 0&0\\\hline 0&\mathbf{1}\\\hline\end{array} \mapsto \begin{array}{|c|c|}\hline 0&0\\\hline 1&1\\\hline\end{array}$$

where the first table is paired and the second (red) is parallel.

|        |          | After |          |
|--------|----------|----------|----------|
|        |          | Positive | Negative |
| Before | Positive | •        | •        |
|        | Negative | •        | •        |

$\longrightarrow$

|          | Before | After |
|----------|--------|-------|
| Positive | •      | •     |
| Negative | •      | •     |

For example, $\mathbf{1}$ in the paired cell $\begin{array}{|c|c|}\hline 1&0\\\hline 0&0\\\hline\end{array}$ where both Before and After are Positive is translated to 1 for each Before and After in a table $\begin{array}{|c|c|}\hline 1&1\\\hline 0&0\\\hline\end{array}$, contrasting Positives and Negatives for Before and After. From an analysis of Mantel–Haenszel tables one can show that the empirical odds ratio for the probabilities $p_1$ and $p_2$ is

$$OR = \frac{B}{C}.$$

To find the confidence interval on the $\log(OR)$ we use the Miettinen's test-based method (Miettinen, 1976). First note that when $H_0$ is true, McNemar's $\chi^2$-statistic can be expressed as

$$\chi^2 = \frac{(\log(OR)-0)^2}{\mathbb{V}\mathrm{ar}\,[\log(OR)]},$$

in which the only unknown is $\mathbb{V}\mathrm{ar}\,[\log(OR)]$ since $\log(OR) = \log(B/C)$ and $\chi^2 = (|B-C|-1)^2/(B+C)$ are easy to find. By solving it for $\mathbb{V}\mathrm{ar}\,[\log(OR)]$, one can find the confidence interval for $\log(OR)$ as

$$[\log(OR) - z_{1-\alpha/2}s, \quad \log(OR) + z_{1-\alpha/2}s],$$

where $s = \sqrt{\mathbb{V}\mathrm{ar}\,[\log(OR)]}$. Of course, the confidence interval for the $OR$ is derived from the above using antilogs.

Another formula for variance in a matched-pairs table comes from the Robins, Breslow, and Greenland (RGB) estimator for Mantel–Haenszel unmatched cases, $s = \sqrt{\mathbb{V}\mathrm{ar}\,[\log(OR)]} = \sqrt{1/B + 1/C}$. It is easy to see that the estimator of $\log(OR)$ is consistent since when the number of tables goes to infinity, $s \to 0$.

MATLAB function ◢ mcnemart.m computes the estimators and confidence
intervals for McNemar's layout.

```matlab
function [] = mcnemart(matr)
% input matr is matrix [A B; C D]
A=matr(1,1); B= matr(1,2); C= matr(2,1); D=matr(2,2);
if( A*B*C*D==0 )
    matr = matr + 0.5;
end
%
N = A + B + C + D;
%
p1 = (A + B)/N; %prob row
p2 = (A + C)/N; %prob column

p11 = A/N; p12 = B/N; p21 = C/N; p22 = D/N;

% risk difference
diffp1p2 = p1 - p2
    delta = p11*p22 - p12*p21
s = sqrt( p1*(1-p1)/N + p2*(1-p2)/N + 2*delta/N )
ssel = sqrt( ((B+C)*N  - (B-C)^2)/N^3 )
    lbd = p1 - p2 - norminv(0.975)*s;
    ubd = p1 - p2 + norminv(0.975)*s;
cidiff = [lbd, ubd]


% odds ratio
or1 = B/C
% miettinen approx
s2lor1 = (log(B/C))^2 * (B+C)/((abs(B-C)-1)^2 + 0.0000001)
slor1 = sqrt(s2lor1)
    lblor1 = log(or1) - norminv(0.975)*slor1 ;
    ublor1 = log(or1) + norminv(0.975)*slor1 ;
miettlint = [lblor1, ublor1]
miettint = [exp(lblor1), exp(ublor1)]

% RGB approx
slor2 = sqrt(1/B + 1/C)

lblor2 = log(or1) - norminv(0.975)*slor2 ;
ublor2 = log(or1) + norminv(0.975)*slor2 ;
rgblint = [lblor2, ublor2]
rgbint = [exp(lblor2), exp(ublor2)]
% --------- exact method ------------------
 df1l = 2*C +2; df2l = 2*B;
 F1 = finv(0.975, df1l, df2l);
 tl = B/(B+(C+1)*F1 );
 orexactl = tl/(1-tl);

 df1u = 2*B +2; df2u = 2*C;
 F2 = finv(0.975, df1u, df2u);
 tu = (B+1)*F2/(C+(B+1)*F2 );
```

```
orexactu = tu/(1-tu);
exactorci = [orexactl, orexactu]

%--------- approximate prob ---------
phat = B/(B+C);
pl = phat - norminv(0.975)*sqrt(phat * (1-phat)/(B+C)) - 1/(2*B+2*C);
pu = phat + norminv(0.975)*sqrt(phat * (1-phat)/(B+C)) + 1/(2*B+2*C);
intor = [pl/(1-pl), pu/(1-pu)]
```

*Example 14.11.* **Split-Mouth Trials for Dental Sealants.** Randomized split-mouth trials (RSM) are frequently used in dentistry to examine the effectiveness of preventive interventions that impact individual teeth as opposed to the whole mouth. For example, to examine the effectiveness of dental sealants in preventing caries, a permanent first molar is randomly chosen for the intervention while its contralateral tooth serves as the control. Because the control and test teeth are subject to the same oral environment, this design controls for many confounding factors such as diet, tooth morphology, and oral hygiene habits. Thus differences in outcomes between test and control teeth are likely attributable to the treatment. Because of this pairing, adequate power may be achieved with a smaller sample size than if the teeth were independent.

Forss and Halme (1998) report results of a split-mouth study that started in 1988 with 166 children, with the goal of assessing tooth-sealant materials. Participants were children from Finland, aged 5 to 14 years (mean age 11 years). To be included in the study, children had to have a contralateral pair of newly erupted, sound, unsealed permanent first or second molar teeth.

Interventions on the occlusal surfaces of sound first or second permanent molars involved glass ionomer Fuji III sealant as a treatment and third-generation, resin-based, light-cured Delton sealant as control. The results are recorded at the 7-year follow-up involving 97 children (the dropout rate was 42%).

|  |  | Control (Resin) | | |
|---|---|---|---|---|
|  |  | Caries | No Caries | Total |
| Treatment (Fuji III) | Caries | 8 | 15 | 23 |
|  | No Caries | 8 | 66 | 74 |
|  | Total | 16 | 81 | 97 |

Risk differences and risk and odds ratios with confidence intervals are obtained by the MATLAB program ◄ mcnemart.m. Several approaches to confidence intervals (exact, approximate RGB, approximate Miettinen) are presented. The odds ratio is 1.8750, the exact 95% confidence interval for $OR$ is [0.7462, 5.1064], and three approximations are miettint = [0.7003, 5.0198], rgbint = [0.7950, 4.4224], and intor = [0.7724, 6.6080].

The complete output from mcnemart.m is

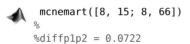

```
  mcnemart([8, 15; 8, 66])
%
%diffp1p2 = 0.0722
```

```
%delta = 0.0434
%s = 0.0646
%ssel = 0.0489
%cidiff = -0.0545   0.1989
%or1 = 1.8750
%s2lor1 = 0.2525
%slor1 = 0.5025
%miettlint = -0.3562   1.6134
%miettint = 0.7003   5.0198
%slor2 = 0.4378
%rgblint = -0.2295   1.4867
%exactorci = 0.7462   5.1064
%intor = 0.7724   6.6080
```

*Example 14.12.* **Schistosoma Mansoni.** *Schistosoma mansoni* is a parasite that is found in Africa, Madagascar, parts of South America (such as Venezuela and Brazil), Puerto Rico, and the West Indies. Among human parasitic diseases, schistosomiasis ranks second behind malaria in terms of socioeconomic and public health importance in tropical and subtropical areas. Some estimate that there are approx. 20,000 deaths related to schistosomiasis yearly. Sleigh et al. (1982) compare results of one Bell and one Kato-Katz examination performed on each of 315 stool specimens from residents in an area in northeastern Brazil endemic for schistosomiasis mansoni. The following table, discussed also in Kirkwood and Sterne (2003), summarizes the findings.

|               | Kato-Katz Positive | Kato-Katz Negative | Total |
|---------------|:------------------:|:------------------:|:-----:|
| Bell Positive | 184                | 54                 | 238   |
| Bell Negative | 14                 | 63                 | 77    |
| Total         | 198                | 117                | 315   |

Is the probability of detecting Schistosomiasis mansoni the same for the two tests? Find the odds ratio and its approximate confidence intervals. From the output of  mcnemart.m we have

```
%or1 = 3.8571
%miettint =   2.2045       6.7488
%exactorci = 2.1130       7.5193
%intor =      2.2327       8.7633
```

*Example 14.13.* **Testing for Salmonella.** Large discrepancies are usually found when different ELISAs for the diagnosis of pig salmonellosis are compared. Mainar-Jaime et al. (2008) explored the diagnostic agreement of two commercial assays: (i) Salmonella Covalent Mix-ELISA (Svanovir), positive OD% =20% denoted as test A and (b) Swine Salmonella Antibody Test Kit (HerdCheck), positive OD% =10% as test B, for the detection of antibodies to Salmonella spp. in slaughter pigs.

Populations of pigs slaughtered in abattoirs from Saskatchewan, Canada.
Population 1: animals from farms marketing <10,000 pigs/year.
Population 2: animals from farms marketing 10,000 pigs/year.

|  | Population 1 | | | Population 2 | | |
|---|---|---|---|---|---|---|
|  | Test B + | Test B − | Total | Test B + | Test B − | Total |
| Test A + | 11 | 16 | 27 | 2 | 5 | 7 |
| Test A − | 6 | 119 | 125 | 1 | 72 | 73 |
| Total | 17 | 135 | 152 | 3 | 77 | 80 |

From McNemar's $\chi^2$-test one concludes that test A (at either OD% = 40% or OD% =20%) significantly differs from test B in the proportion positive at $\alpha = 0.05$.

✏

### 14.6.4 Multicategorical Paired Tables: Stuart–Maxwell Test*

A natural generalization of the McNemar test is a design in which there are $N$ matched pairs where each response can be classified into $k > 2$ different categories. We will be interested in the equality of proportions for categories in the two populations that form the paired responses. For simplicity we discuss only the case of $k = 3$. A general test is given in Maxwell (1970) and implemented as MATLAB function stuartmaxwell.m.

Assume that paired responses come from *Before* and *After* and each response may result in one of the three categories: *Positive*, *Neutral*, and *Negative*.

|  |  | After | | | |
|---|---|---|---|---|---|
|  |  | Positive | Neutral | Negative | Total |
|  | Positive | $n_{11}$ | $n_{12}$ | $n_{13}$ | $n_{1.}$ |
| Before | Neutral | $n_{21}$ | $n_{22}$ | $n_{23}$ | $n_{2.}$ |
|  | Negative | $n_{31}$ | $n_{32}$ | $n_{33}$ | $n_{3.}$ |
|  | Total | $n_{.1}$ | $n_{.2}$ | $n_{.3}$ | $N$ |

Notice that the table counts matched pairs, for example, $n_{23}$ is the number of cases where *Before* resulted in neutral and *After* resulted in negative. Thus, again we have a total of $2N$ responses organized into $N$ matched pairs.

Denote by $p_{ij}$ the population proportion of subjects in population $i$ (one of *Before, After*) that are classified as $j$ (one of *Positive, Neutral*, or *Negative*).

The null hypothesis of interest is

$$H_0: p_{11} = p_{21}, \ p_{12} = p_{22}, \ \text{and} \ p_{13} = p_{23}.$$

This hypothesis states that the population proportions of the three categories are the same for the two groups *Before* and *After*. The alternative is a negation of $H_0$:

$$H_1 : (p_{11}, p_{12}, p_{13}) \neq (p_{21}, p_{22}, p_{23}).$$

The test statistic is

$$\chi^2 = \frac{\bar{n}_{23}(n_{1\cdot} - n_{\cdot 1})^2 + \bar{n}_{13}(n_{2\cdot} - n_{\cdot 2})^2 + \bar{n}_{12}(n_{3\cdot} - n_{\cdot 3})^2}{2(\bar{n}_{12}\bar{n}_{13} + \bar{n}_{12}\bar{n}_{23} + \bar{n}_{13}\bar{n}_{23})}, \tag{14.6}$$

where $\bar{n}_{ij} = (n_{ij} + n_{ji})/2$.

The statistic $\chi^2$ in (14.6) has a $\chi^2$-distribution with 2 degrees of freedom and its large values are critical.

**Rule:** Reject $H_0$ at significance level $\alpha$ if $\chi^2 > \chi^2_{2,1-\alpha}$, where $\chi^2_{2,1-\alpha}$ is a $1 - \alpha$ quantile of a $\chi^2$-distribution with 2 degrees of freedom (chi2inv(1-alpha, 2)). The *p*-value of this test is 1-chi2cdf(chi2, 2) for chi2 as in (14.6).

*Example 14.14.* Speckmann (1965) and Lawal (2003) provide data on the religious affiliation of husbands and wives in 264 marriages in Surinam.

|  |  | Wife | | | |
| --- | --- | --- | --- | --- | --- |
|  |  | Christian | Muslim | Hindu | Total |
|  | Christian | 17 | 1 | 7 | 25 |
| Husband | Muslim | 1 | 66 | 6 | 73 |
|  | Hindu | 9 | 6 | 151 | 166 |
|  | Total | 27 | 73 | 164 | 264 |

Here Hindu counts combine Sanatan Dharm and Arya Samaj affiliations. Are the population proportions of husbands who are Christian, Muslim, or Hindu equal to the proportions of wives who belong to those three religious groups?

Note that the data are paired and the Stuart–Maxwell test is appropriate. For the data in the table, $\chi^2$ in (14.6) is $28/124 = 0.2258$. At 2 degrees of freedom, the *p*-value of the test is 0.8932 (as 1-chi2cdf(0.2258, 2). Thus, $H_0$ is not rejected, the proportions of these three religious groups in the population of Surinam husbands are not different than the proportions among the wives. 🖉

## 14.7 Exercises

14.1. **Amoebas and intestinal disease.** When an epidemic of severe intestinal disease occurred among workers in a plant in South Bend, Indiana, doctors said that the illness resulted from infection by the amoeba *Entamoeba histolytica*. There are actually two races of these amoebas, large and small, and the large ones were believed to be causing the disease. Doctors suspected that the presence of the small ones might help people resist infection by the large ones. To check on this, public health officials chose a random sample of 138 apparently healthy workers and determined if they were infected with either the large or small amoebas. The table below [given by Cohen (1973)] provides the resulting data. Is the presence of the large race independent of the presence of the small one? Test at 5% significance level.

|            | Large race | | |
|------------|---------|--------|-------|
| Small race | Present | Absent | Total |
| Present    | 12      | 23     | 35    |
| Absent     | 35      | 68     | 103   |
| Total      | 47      | 91     | 138   |

14.2. **Drinking and Smoking.** Alcohol and nicotine consumption during pregnancy are believed to be associated with certain characteristics in children. Since drinking and smoking behaviors may be related, it is important to understand the nature of this relationship when fully assessing their influence on children. In a study by Streissguth et al. (1984), 452 mothers were classified according to their alcohol intake prior to pregnancy recognition and their nicotine intake during pregnancy. The data are summarized in the following table.

|                       | Nicotine (mg/day) | | |
|-----------------------|------|------|------------|
| Alcohol (ounces/day)  | None | 1–15 | 16 or more |
| None                  | 105  | 7    | 11         |
| 0.01–0.10             | 58   | 5    | 13         |
| 0.11–0.99             | 84   | 37   | 42         |
| 1.00 or more          | 57   | 16   | 17         |

(a) Calculate the column sums. In what way does the pattern of alcohol consumption vary with nicotine consumption?
(b) Calculate the row sums. In what way does the pattern of nicotine consumption vary with alcohol consumption?
(c) Formulate $H_0$ and $H_1$ for assessing whether or not alcohol consumption and nicotine consumption are independent.
(d) Compute the table of expected counts.

(e) Find the $\chi^2$ statistic. Report the degrees of freedom and the $p$-value.

(f) What do you conclude from the analysis of this table?

14.3. **Alcohol and Marriage.** A national survey was conducted to obtain information on the alcohol consumption patterns of American adults by marital status. A random sample of 1303 residents 18 years old and over yielded the data below. Do the data suggest at a 5% significance level that marital status and alcohol consumption patterns are statistically dependent?

|          | Abstain | 1–60 | Over 60 | Total |
|----------|---------|------|---------|-------|
| Single   | 67      | 213  | 74      | 354   |
| Widowed  | 85      | 633  | 129     | 847   |
| Divorced | 27      | 60   | 15      | 102   |
| Total    | 179     | 906  | 218     | 1303  |

14.4. **Family Size.** A demographer surveys 1000 randomly chosen American families and records their family sizes and family incomes:

|               | Family Size        |
|---------------|--------------------|
| Family Income | 2  3  4  5  6  7   |
| Low           | 145 81 57 22  9  8 |
| Middle        | 151 73 71 33 13 10 |
| High          | 124 60 80 42 13  8 |

Do the data provide sufficient evidence to conclude that family size and family income are statistically dependent?

(a) State the $H_0$ and $H_1$ hypotheses.

(b) Perform the test. Use $\alpha = 0.05$. Comment.

14.5. **Nightmares.** Over the years numerous studies have sought to characterize the nightmare sufferer. From these studies has emerged the stereotype of someone with high anxiety, low ego strength, feelings of inadequacy, and poorer-than-average physical health. A study by Hersen (1971) explored whether gender is independent of having frequent nightmares. Using Hersen's data summarized in the table below, test the hypothesis of independence at level $\alpha = 0.05$.

|                                                    | Men | Women |
|----------------------------------------------------|-----|-------|
| Nightmares often (at least once a month)           | 55  | 60    |
| Nightmares seldom (less than once in a month)      | 105 | 132   |

14.6. **Independence of Segregation.** According to a mathematical model of inheritance, at a given meiosis, the probability of an allele at one locus passing to the gamete is independent of an allele at any locus on another chromosome passing to the gamete. This is usually referred as the law of independent segregation of genes. Also, one allele from the pair at any locus passes to the gamete with probability equal to 1/2.

Roberts et al. (1939) conducted an extensive set of experiments for testing independent segregation of the genes in mice and rats. One of results of a mating of the form $Aa\ Bb\ Dd \times aa\ bb\ dd$ is reported as:

|        | ab  | aB  | Ab  | AB  | Total |
|--------|-----|-----|-----|-----|-------|
| d      | 427 | 440 | 509 | 460 | 1836  |
| D      | 494 | 467 | 462 | 475 | 1898  |
| Total  | 921 | 907 | 971 | 935 | 3734  |

Using $\chi^2$ statistic test for the independence of segregation at $\alpha = 5\%$.
*Hint.* The expected number in each cell is $1/2 \times 1/2 \times 1/2 \times 3734 = 466.75$.

14.7. **Site of Corpus Luteum in Caesarean Births.** Williams (1921) observed that in Caesarean-section births, the corpus luteum was located in the right ovary 23 times and 16 times in the left for male children. For female children the numbers were 13 and 12, respectively. Test for the independence of ovary side and gender of a child.

14.8. **An Easy Grade?** A student wants to take a statistics course with a professor who is an easy grader. There are three professors scheduled to teach the course sections next semester. The student manages to obtain a random sample of grades given by the three professors this past year.

| Observed | Prof A | Prof B | Prof C | Total |
|----------|--------|--------|--------|-------|
| Grades A | 10     | 12     | 28     | 50    |
| Grades B | 35     | 30     | 15     | 80    |
| Grades C | 15     | 30     | 25     | 70    |
| Total    | 60     | 72     | 68     | 200   |

Using a significance level of 1%, test the hypothesis that a student's grade is independent of the professor.

14.9. **Importance of Bystanders.** When a group of people is confronted with an emergency, a diffusion-of-responsibility process can interfere with an individual's responsiveness. Darley and Latané (1968) asked subjects to participate in a discussion carried over an intercom. Aside from the experimenter to whom they were speaking, subjects thought that there were zero, one, or four other people (bystanders) also listening over the intercom. Part way through the discussion, the experimenter feigned serious illness and asked for help. Darley and Latané noted how often the subject sought help for the experimenter as a function of the number of supposed bystanders. The data are summarized in the table:

|                  | Sought assistance | No assistance |
|------------------|-------------------|---------------|
| No bystanders    | 11                | 2             |
| One bystander    | 16                | 10            |
| Four bystanders  | 4                 | 9             |

What could Darley and Latané conclude from the results?
(a) State $H_0$ and $H_1$.
(b) Perform the test at a 5% significance level.

14.10. **Baseball in 2003.** As part of baseball's attempt to make the All-Star Game more meaningful, baseball Commissioner Bud Selig pushed through a proposal to give the winning league home-field advantage in the World Series. The proposal, while controversial to many baseball purists, was just one of many changes instituted by Selig since he became commissioner in 1998. Despite Selig's efforts, which are generally rated positively by baseball fans, sports fans are most likely to say Major League Baseball has the most serious problems of the major professional sports leagues. And nearly 4 in 10 sports fans say baseball is currently in a state of crisis or has major problems. Nevertheless, the majority of sports fans say they are following baseball as much or more than they did 3 years ago. Our interest is in how the opinions of fans who are particular about baseball compare to those of general sports fans regarding the Major League events. The question to 140 declared fans was: *Are you following Major League Baseball more closely than you did three years ago, less closely, or about the same as before?*

|              | More | Less | Same | Total |
|--------------|------|------|------|-------|
| Sports Fans  | 6    | 19   | 32   | 57    |
| Baseball Fans| 13   | 25   | 45   | 83    |
| Total        | 19   | 44   | 77   | 140   |

Test the hypothesis that the type of fan is independent of a change in interest level. Use $\alpha = 5\%$.

14.11. **Psychosis in Adopted Children.** Numerous studies have been done to determine the etiology of schizophrenia. Such factors as biological, psychocultural, and sociocultural influences have been suggested as possible causes of this disorder. To test if schizophrenia has a hereditary component, researchers compared adopted children whose biological mothers are schizophrenic ("exposure") to adopted children whose biological mothers are normal ("nonexposure").

Furthermore, the child-rearing abilities of adoptive families have been assessed to determine if there is a relationship between those children who become psychotic and the type of family into which they are adopted. The families are classified as follows: healthy, moderately disturbed, and severely disturbed.

The following data are from an experiment described in Carson and Butcher (1992).

|           | Type of Adoptive Family | | | | | |
|-----------|---------|--------|---------------------|--------|-------------------|--------|
|           | Healthy | | Moderately Disturbed | | Severely Disturbed | |
| Diagnosis | Exp     | Nonexp | Exp                 | Nonexp | Exp               | Nonexp |
| None      | 41      | 42     | 11                  | 26     | 6                 | 15     |
| Psychotic | 10      | 11     | 18                  | 25     | 38                | 28     |

(a) For moderately disturbed families find the risk difference for a child's psychosis with the mother's schizophrenia as a risk factor. Find a 95% confidence interval for the risk difference.

(b) For severely disturbed families find the risk ratio and odds ratio for psychosis with the mother's schizophrenia as a risk factor. Find a 95% confidence interval for the risk ratio and odds ratio.

(c) Is a mother's schizophrenia a significant factor for a child's psychosis overall? Assess this using the Mantel–Haenszel test. Find the overall odds ratio and corresponding 95% confidence interval.

14.12. **The Midtown Manhattan Study.** The data set below has been analyzed by many authors (Haberman, Goodman, Agresti, etc.) and comes from the study by Srole et al. (1962). The data cross-classifies 1660 young New York residents with respect to two factors: Mental Health and parents' Socioeconomic Status (SES).

The Mental Health factor is classified by four categories: Well, Mild Symptom Formation, Moderate Symptom Formation, and Impaired. The parents' SES has six categories ranging from High (1) to Low (6).

Here is the table:

|          | Well | Mild | Moderate | Impaired |
|----------|------|------|----------|----------|
| 1 (High) | 64   | 94   | 58       | 46       |
| 2        | 57   | 94   | 54       | 40       |
| 3        | 57   | 105  | 65       | 60       |
| 4        | 72   | 141  | 77       | 94       |
| 5        | 36   | 97   | 54       | 78       |
| 6 (Low)  | 21   | 71   | 54       | 71       |

(a) Test the hypothesis that Mental Health and parents' SES are independent factors at a 5% confidence level. You can use ◀ tablerxc.m code.

(b) The table below provides the expected frequencies. Explain what the expected frequencies are. Explain how the number **104.0807** from the table was obtained and show your work.

|          | Well    | Mild       | Moderate | Impaired |
|----------|---------|------------|----------|----------|
| 1 (High) | 48.4542 | 95.0145    | 57.1349  | 61.3964  |
| 2        | 45.3102 | 88.8494    | 53.4277  | 57.4127  |
| 3        | 53.0777 | **104.0807** | 62.5867 | 67.2548  |
| 4        | 71.0169 | 139.2578   | 83.7398  | 89.9855  |
| 5        | 49.0090 | 96.1024    | 57.7892  | 62.0994  |
| 6 (Low)  | 40.1319 | 78.6952    | 47.3217  | 50.8512  |

14.13. **Tonsillectomy and Hodgkin's Disease.** A study by Johnson and Johnson (1972) involved 85 patients with Hodgkin's disease. Each of these had a normal sibling (one who did not have the disease). In 26 of these pairs, both individuals had had tonsillectomies (T); in 37 pairs, both individuals had not had tonsillectomies (N); in 15 pairs, only the normal individual had had a tonsillectomy; in 7 pairs, only the one with Hodgkin's disease had had a tonsillectomy.

|  | Normal/T | Normal/N | Total |
|---|---|---|---|
| Patient/T | 26 | 15 | 41 |
| Patient/N | 7 | 37 | 44 |
| Total | 33 | 52 | 85 |

A goal of the study was to determine whether there was a link between the disease and having had a tonsillectomy: Is the proportion of those who had tonsillectomies the same among those with Hodgkin's disease as among those who do not have it?

14.14. **School Spirit at Duke.** Duke has always been known for its great school spirit and support of its athletic teams, as evidenced by the famous Cameron Crazies. One way that school enthusiasm is shown is by donning Duke paraphernalia including shirts, hats, shorts, and sweatshirts. The students' project in an introductory statistics class was to explore possible links between school spirit, measured by the number of students wearing paraphernalia, and some other attributes. It was hypothesized that men would wear Duke clothes more frequently than women. The data were collected on the Bryan Center walkway starting at a random hour on five different days. Each day 100 men and 100 women were tallied.

|  | Duke Paraphenalia | No Duke Paraphenalia | Total |
|---|---|---|---|
| Male | 131 | 369 | 500 |
| Female | 52 | 448 | 500 |
| Total | 183 | 817 | 1000 |

14.15. **Two Halloween Questions with Easy Answers.** A study was designed to test whether or not aggression is a function of anonymity. The study was conducted as a field experiment on Halloween (Fraser, 1974); 300 children were observed unobtrusively as they made their rounds. Of these 300 children, 173 wore masks that completely covered their faces, while 127 wore no masks. It was found that 101 children in the masked group displayed aggressive or antisocial behavior versus 36 children in the unmasked group.
(a) Are anonymity and aggression independent? Use $\alpha = 0.01$.
(b) If $p_1$ is the (population) proportion for aggressive behavior for subjects wearing a mask and $p_2$ is the proportion for subjects not wearing a mask, find a 95% confidence interval for the odds ratio:

$$\frac{p_1/(1-p_1)}{p_2/(1-p_2)}.$$

14.16. **Runners and Heart Attack.** The influence of running on preventing heart attacks has been studied by a local runners club. The following two-way table classifies 350 people as runners or nonrunners, and whether or not they have had a heart attack. The factors are runner status and history of heart attack.

|  | Heart attack | No heart attack | Total |
|---|---|---|---|
| Runner | 12 | 112 | 124 |
| Nonrunner | 36 | 190 | 226 |
| Total | 48 | 302 | 350 |

(a) Test for the independence of factors. Use $\alpha = 0.05$.
(b) Explain in words (in terms of this problem) what constitutes errors of the first and second kind in the above testing.

14.17. **Perceptions of Dangers of Smoking.** A poll was conducted to determine if perceptions of the hazards of smoking were dependent on whether or not the person smoked. One hundred (100) smokers and 100 nonsmokers were randomly selected and surveyed. The results are given below.

| Smoking is: | Very Dangerous | Dangerous | Somewhat Dangerous | Not Dangerous |
|---|---|---|---|---|
| Smokers | 21 (35.5) | 29 (30) | 29 ( ) | 21 ( ) |
| Nonsmokers | 50 (35.5) | 31 ( ) | 11 ( ) | 8 ( ) |

Test the hypothesis that smoking status does not affect perception of the dangers of smoking at $\alpha = 0.05$ (three theoretical frequencies are given in parentheses).

14.18. **Red Dye No. 2.** Fienberg (1980) discusses an experiment in which the food additive Red Dye No. 2 was fed to two groups of rats at various dosages. Some rats died during the experiment, which lasted 131 weeks, and the remainder were sacrificed at the end of the 131st week. All rats were examined for tumors.

| Age of Death | 0–131 weeks | | Terminal sacrifice | |
|---|---|---|---|---|
| Dosage | Low | High | Low | High |
| Tumor present | 4 | 7 | 0 | 7 |
| Tumor absent | 26 | 16 | 14 | 14 |

Test the dependence among the three factors.

14.19. **Cooper Hawks.** The data in the table below originates from a study designed to assess whether tape-recorded alarm calls enhanced the detectability of Cooper's hawks (*Accipiter cooperii*) (Rosenfield et al., 1988). Observations were made on each sampling transect both with and without recorded calls. Thus, the observations are paired rather than independent.

|  | Detected | Not detected | Total # Transects |
|---|---|---|---|
| Tapes used | 9 | 9 | 18 |
| Tapes not used | 4 | 14 | 18 |
| Total | 13 | 23 | 36 |

According to Engeman and Swanson (2003), Rosenfield et al. (1988) have assumed that the observations under the two conditions from each transect could be considered independent. Their analysis for the nestling stage data uses Fisher's exact test and results in a one-tailed $p$-value of 0.08. They concluded that during the nestling stage, broadcast recordings "can markedly increase the chance of detecting Cooper's hawks near their nests." However, an appropriate analysis would have been McNemar's test for paired data. Engeman and Swanson (2003) continue: "...let us presume that for some reason we could assume that observations with and without recorded calls were independent rather than paired. Then the data set, by all criteria of which we are aware, still is of sufficient size to apply Pearson's $\chi^2$, rather than Fisher's exact test as the authors have done."

We note that rather than size, the design of an experiment is critical for a suggestion to use Fisher's exact test. The marginals should be fixed before the experiment; this is not the case here.

Find the $p$-value for the correct McNemar test.

14.20. **Hepatic Arterial Infusion.** Allen-Mersh et al. (1994), as well as Spiegelhalter et al. (2004), reported the results of a trial in which patients undergoing therapy for liver metastasis were randomized to receive it either systematically (as is standardly done) or via hepatic arterial infusion (HAI). Of 51 randomized to HAI, 44 died, and of 49 randomized to systematic therapy, 46 died. Estimate the log odds ratio and find a 95% confidence interval.

14.21. **Vaccine Efficacy Study.** Consider the data from a vaccine efficacy study (Chan, 1998). In a randomized clinical trial of 30 subjects, 15 were inoculated with a recombinant DNA influenza vaccine and the remaining 15 were inoculated with a placebo. Twelve of the 15 subjects in the placebo group (80%) eventually became infected with influenza, whereas for the vaccine group, only 7 of the 15 subjects (47%) became infected. Suppose $p_1$ is the probability of infection for the vaccine group and $p_2$ is the probability of infection for the placebo group.

What is the one-sided $p$-value for testing the null hypothesis $H_0 : p_1 = p_2$ obtained by Fisher's exact test?

---

| MATLAB FILES USED IN THIS CHAPTER |

http://springer.bme.gatech.edu/Ch14.Tables/

concoef.m, fisherexact.m, hindu.m, kappa1.m, mantelhaenszel.m, mcnemart.m, partialrxcxp.m, PAS.m, ponvexam.m, stuartmaxwell.m, table2x2.m, tablerxc.m, tablerxcxp.m, unmatch.m

---

# CHAPTER REFERENCES

Agresti, A. (2002). *Categorical Data Analysis*, 2nd edn. Wiley, New York.

Allen-Mersh T. G., Earlam, S., Fordy, C., Abrams, K., and Houghton, J. (1994). Quality of life and survival with continuous hepatic-artery floxuridine infusion for colorectal liver metastases. *Lancet*, **344**, 8932, 1255–1260.

Armitage, P. and Berry, G. (1994). *Statistical Methods in Medical Research*, 3rd edn. Blackwell, Oxford.

Chan, I. (1998). Exact tests of equivalence and efficacy with a non-zero lower bound for comparative studies. *Stat. Med.*, **17**, 1403–1413.

Cohen, J. (1960). A coefficient of agreement for nominal scales. *Educ. Psychol. Meas.*, **20**, 37–46.

Cohen, J. E. (1973). Independence of Amoebas. In *Statistics by Example: Weighing Chances*, edited by F. Mosteller, R. S. Pieters, W. H. Kruskal, G. R. Rising, and R. F. Link, with the assistance of R. Carlson and M. Zelinka. Addison-Wesley, Reading, p. 72.

Correa, P., Pickle, L. W., Fontham, E., Lin, Y., and Haenszel, W. (1983). Passive smoking and lung cancer. *Lancet*, **2**, 8350, 595–597.

Darley, J. M. and Latané, B. (1968). Bystander intervention in emergencies: diffusion of responsibility. *J. Personal. Soc. Psychol.*, **8**, 4, 377–383.

Engeman, R. M. and Swanson, G. B. (2003). On analytical methods and inferences for 2 x 2 contingency table data using wildlife examples. *Int. Biodeterior. Biodegrad.*, **52**, 243–246.

Fienberg, S. E. (1970). The analysis of multidimensional contingency tables. *Ecology*, **51**, 419–433.

Fienberg, S. E. (1980). *The analysis of cross-classified categorical data*. MIT Press, Cambridge, MA.

Fisher, R . A. (1935). *The Design of Experiments* (8th edn. 1966). Oliver and Boyd, Edinburgh.

Forss, H. and Halme, E. (1998). Retention of a glass ionomer cement and a resin-based fissure sealant and effect on carious outcome after 7 years. *Community Dent. Oral Epidemiol.*, **26**, 1, 21–25.

Fraser, S. C. (1974). Deindividuation: effects of anonymity on aggression in children. Unpublished technical report. University of Southern California, Los Angeles.

Galton, F. (1892). *Finger Prints*, MacMillan, London.

Johnson, S. and Johnson, R. (1972). Tonsillectomy history in Hodgkin's disease. *New Engl. J. Med.*, **287**, 1122–1125.

Hersen, M. (1971). Personality characteristics of nightmare sufferers. *J. Nerv. Ment. Dis.*, **153**, 29–31.

Kirkwood, B. and Sterne, J. (2003). *Essential Medical Statistics*, 2nd edn. Blackwell, Oxford.

Lawal, H. B. (2003). *Categorical Data Analysis with SAS and SPSS Applications*. Erlbaum, Mahwah.

Mainar-Jaime, R. C., Atashparvar, N., Chirino-Trejo, M., and Blasco, J. M. (2008). Accuracy of two commercial enzyme-linked immunosorbent assays for the detection of antibodies to *Salmonella spp.* in slaughter pigs from Canada. *Prevent. Vet. Med.*, **85**, 41–51.

Mantel, N. and Haenszel, W. (1959). Statistical aspects of the analysis of data from retrospective studies of disease. *J. Natl. Cancer Inst.*, **22**, 719–748.

Maxwell, A. E. (1970). Comparing the classification of subjects by two independent judges. *Br. J. Psychiatry*, **116**, 651–655.

Miettinen, O. S. (1976). Estimability and estimation in case-referent studies. *Am. J. Epidemiol.*, **103**, 226–235.

Roberts, E., Dawson, W. M., and Madden, M. (1939). Observed and theoretical ratios in Mendelian inheritance. *Biometrika*, **31**, 56–66.

Robins, J., Breslow, N., and Greenland, S. (1986). Estimators of the Mantel-Haenszel variance consistent in both sparse data and large-strata limiting models. *Biometrics*, **42**, 311–323.

Rosenfield, R. N., Bielefeldt, J., and Anderson, R. K. (1988). Effectiveness of broadcast calls for detecting Cooper's hawks. *Wildl. Soc. Bull.*, **16**, 210–212.

Schoener, T. W. (1968). The Anolis lizards of Bimini: Resource partitioning in a complex fauna. *Ecology*, **48**, 704–726.

Sleigh, A., Hoff, R., Mott, K., Barreto, M., de Paiva, T., Pedrosa, J., and Sherlock, I. (1982). Comparison of filtration staining (Bell) and thick smear (Kato) for the detection and quantitation of Schistosoma mansoni eggs in faeces. *Trop. Med. Hyg.*, **76**, 3, 403–406.

Speckmann, J. D. (1965). Marriage and kinship among the Indians in Surinam. *Assen: Van-Gorgum*, 32–33.

Spiegelhalter, D. J., Abrams, K. R., and Myles, J. P. (2004). *Bayesian Approaches to Clinical Trials and Health Care Evaluation*. Wiley, New York.

Srole, L., Langner, T., Michael, S., Opler, M., and Rennie, T. (1962). *Mental Health in the Metropolis: The Midtown Manhattan Study*, vol. 1. McGraw-Hill, New York.

Streissguth, A. P., Martin, D. C., Barr, H. M., Sandman, B. M., Kirchner, G. L., and Darby, B. L. (1984). Intrauterine alcohol and nicotine exposure: attention and reaction time in 4-year-old children. *Dev. Psychol.*, **20**, 533–541.

Williams, J. W. (1921). A critical analysis of twenty-one years' experience with cesarean section. *Johns Hopkins Hosp. Bull.*, **32**, 173–184.

Worchester, J. (1971). The relative odds in $2^3$ contingency table. *Am. J. Epidemiol.*, **93**, 145–149.

# Chapter 15
# Correlation

*The invalid assumption that correlation implies cause is probably among the two or three most serious and common errors of human reasoning.*

– Stephen Jay Gould

## WHAT IS COVERED IN THIS CHAPTER

• Calculating the Pearson Coefficient of Correlation. Conditional Correlation
• Testing the Null Hypothesis of No Correlation
• General Test for Correlation and Confidence Intervals. Fisher $z$-Transformation
• Inference for Two Correlation Coefficients
• Nonparametric Correlation Measures: Spearman's and Kendall's Correlation Coefficient

## 15.1 Introduction

Roget's New Thesaurus (3rd Edition, 1995) defines *correlation* as a *logical or natural association between two or more things*, with synonyms *connection, interconnection, interdependence, interrelationship, link, linkage, relation, relationship,* and *tie-in*.

Statistically, correlation is a measure of the particular affinity between two sets of comparable measurements. It is often incorrectly believed that the notions of correlation and statistical dependence and causality coincide. According to an anecdote, it could be true that the number of drownings at a particular large beach in one season is positively and significantly correlated with the number of ice-creams sold at the beach during the same period of time. However, nobody would argue that the relationship is causal. Purchasing an ice-cream at the beach does not increase the risk of drowning – the positive correlation is caused by a latent (lurking) variable, the number of visitors at the beach.

> Correlation is, in informal terms, a constrained dependence. For example, the most common (Pearson) coefficient of correlation measures the strength and direction of the *linear* relationship between two variables.

If measurements are correlated, then they are dependent, but not necessarily vice versa. A simple example involves points on a unit circle. One selects $n$ angles $\varphi_i$, $i = 1,\ldots,n$, uniformly from $[0,2\pi]$. The angles determine $n$ points on the unit circle, $(x_i, y_i) = (\cos\varphi_i, \sin\varphi_i)$, $i = 1,\ldots,n$. Although $x = (x_1,\ldots,x_n)$ and $y = (y_1,\ldots,y_n)$ are functionally dependent (since $x_i^2 + y_i^2 = 1$), their coefficient of correlation is 0 or very close to 0. See also Exercise 15.1. Only for a normal distribution do the notions of correlation and independence coincide: uncorrelated normally distributed measurements are independent.

## 15.2 The Pearson Coefficient of Correlation

In Chap. 2, when discussing the summaries of multidimensional data we discussed correlations between the variables and connected the coefficient of correlation with scatterplots and a descriptive statistical methodology. In this chapter we will take an inferential point of view on correlations. For developing the tests and confidence intervals we will assume that the data come from normal distributions for which the population coefficient of correlation is $\rho$.

Suppose that pairs $(X_1,Y_1),(X_2,Y_2),\ldots,(X_n,Y_n)$ are observed, and assume that $X$s and $Y$s come from normal $\mathcal{N}(0,\sigma_X^2)$ and $\mathcal{N}(0,\sigma_Y^2)$ distributions and that the correlation coefficient between $X$ and $Y$ is $\rho_{XY}$, or simply $\rho$. We are interested in making an inference about $\rho$ when $n$ pairs $(X_i,Y_i)$ are observed.

From the observations the following sums of squares are formed: $S_{xx} = \sum_{i=1}^{n}(X_i - \overline{X})^2$, $S_{yy} = \sum_{i=1}^{n}(Y_i - \overline{Y})^2$, and $S_{xy} = \sum_{i=1}^{n}(X_i - \overline{X})(Y_i - \overline{Y})$. Then, an estimator of $\rho$ is

$$r = \frac{S_{xy}}{\sqrt{S_{xx}\,S_{yy}}}.$$

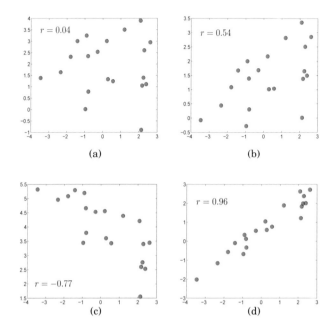

**Fig. 15.1** A simulated example showing scatterplots with correlated components (a) $r =$ 0.04, (b) $r = 0.54$, (c) $r = -0.77$, and (d) $r = 0.96$.

An alternative expression for $r$ is

$$ r = \frac{\sum_i X_i Y_i - n\overline{XY}}{\sqrt{(\sum_i X_i^2 - n(\overline{X})^2)(\sum_i Y_i^2 - n(\overline{Y})^2)}}. $$

This estimator is an MLE for $\rho$, and it is asymptotically unbiased, that is, $\lim_{n\to\infty} \mathbb{E} r_n = \rho$.

Sample correlation coefficient $r$ is always in $[-1,1]$. To show this, consider the Cauchy–Schwartz inequality that states:

*For any two vectors $(u_1, u_2, \ldots, u_n)$ and $(v_1, v_2, \ldots, v_n)$, it holds that $(\sum_i u_i v_i)^2 \leq \sum_i u_i^2 \sum_i v_i^2$.*

Substitutions $u_i = X_i - \overline{X}$ and $v_i = Y_i - \overline{Y}$ prove the inequality $r^2 \leq 1$. Figure 15.1 shows simulated correlated data (green dots) with coefficients (a) $r = 0.04$, (b) $r = 0.54$, (c) $r = -0.77$, and (d) $r = 0.96$.

**Partial Correlation.** Let $r_{xy}, r_{xz}$, and $r_{yz}$ be correlations between the pairs of variables $X$, $Y$, and $Z$. The partial correlation between $X$ and $Y$ if $Z$ is accounted for (or excluding the influence of $Z$) is

$$ r_{xy.z} = \frac{r_{xy} - r_{xz} r_{yz}}{\sqrt{1 - r_{xz}^2}\sqrt{1 - r_{yz}^2}}. $$

When two variables $Z$ and $W$ are excluded from the correlation of $X$ and $Y$, the partial coefficient is

$$r_{xy.zw} = \frac{r_{xy.w} - r_{xz.w}r_{yz.w}}{\sqrt{1-r_{xz.w}^2}\sqrt{1-r_{yz.w}^2}}$$

or, equivalently,

$$r_{xy.zw} = \frac{r_{xy.z} - r_{xw.z}r_{yw.z}}{\sqrt{1-r_{xw.z}^2}\sqrt{1-r_{yw.z}^2}},$$

depending on the order of exclusion of $W$ and $Z$.

### 15.2.1 Inference About $\rho$

To test $H_0 : \rho = 0$ against one of the alternatives $\rho >, \neq, < 0$, the $t$-statistic is used:

$$t = r\sqrt{\frac{n-2}{1-r^2}},$$

which has a $t$-distribution with $df = n - 2$ degrees of freedom.

| Alternative | $\alpha$-level rejection region | $p$-value |
|---|---|---|
| $H_1 : \rho > 0$ | $[t_{df,1-\alpha}, \infty)$ | 1-tcdf(t, df) |
| $H_1 : \rho \neq 0$ | $(-\infty, t_{df,\alpha/2}] \cup [t_{df,1-\alpha/2}, \infty)$ | 2*tcdf(-abs(t), df) |
| $H_1 : \rho < 0$ | $(-\infty, t_{df,\alpha}]$ | tcdf(t, df) |

*Example 15.1.* The table below gives the number of times $X$ rats ran through a maze and the time $Y$ it took them to run through the maze on their last trial.

| Rat | Trials ($X$) | Time ($Y$) |
|---|---|---|
| 1 | 8 | 10.9 |
| 2 | 9 | 8.6 |
| 3 | 6 | 11.4 |
| 4 | 5 | 13.6 |
| 5 | 3 | 10.3 |
| 6 | 6 | 11.7 |
| 7 | 3 | 10.7 |
| 8 | 2 | 14.8 |

(a) Find $r$.

(b) Is the maze learning significant? Test the hypothesis that the population correlation coefficient $\rho$ is 0, versus the alternative that $\rho$ is negative. $\sum_i X_i = 42$, $\sum X_i^2 = 264$, $\sum_i Y_i = 92$, $\sum_i Y_i^2 = 1084.2$, $\sum_i X_i Y_i = 463.8$, $\overline{X} = 5.25$, $\overline{Y} = 11.5$.

$$r = \frac{463.8 - 8 \cdot 5.25 \cdot 11.5}{\sqrt{(264 - 8 \cdot 5.25^2)(1084.2 - 8 \cdot 11.5^2)}} = \frac{-19.2}{\sqrt{43.5 \cdot 26.2}} = -0.5687.$$

For testing $H_0 : \rho = 0$ versus $H_1 : \rho < 0$ we find

$$t = r\sqrt{\frac{n-2}{1-r^2}} = -1.69.$$

For $\alpha = 0.05$ the critical value is $t_{6,0.05} = -1.943$ and the null hypothesis is not rejected.

**Remark:** Sample size is critical for our decision for the matching observed values of $r$. If for the same $r$, the sample size $n$ were 30, then $t = -3.6588$, and $H_0$ would be rejected, since $t < t_{28,0.05} = -1.7$.

In MATLAB:

```
X = [8 9 6 5 3 6 3 2];
Y = [10.9 8.6 11.4 13.6 10.3 11.7 10.7 14.8];
n=8;
cxy=cov(X,Y)
   % cxy =
   %      6.2143      -2.7429
   %     -2.7429       3.7429
rxy=cxy(1,2)/sqrt(cxy(1,1)*cxy(2,2))
   %rxy =    -0.5687
tstat = rxy * sqrt(n-2)/sqrt(1-rxy^2)
   %tstat =   -1.6937
pval = tcdf(tstat, n-2)
   %pval =    0.0706
   %n=8, p-val = 7% > 5% do not reject H_0
   %=============
n=30; rxy=-0.5687;
tstat = rxy * sqrt(n-2)/sqrt(1-rxy^2)
   %tstat =-3.6585
pval = tcdf(tstat, n-2)
   % pval = 5.2070e-004
```

When partial correlations are of interest, the test is based on the statistic $t = r\sqrt{\frac{n-q-2}{1-r^2}}$, which has $df = n - q - 2$ degrees of freedom. Here $q$ is the number of other variables accounted for.

*Example 15.2.* On the basis of $n = 28$ records it was found that the education level of the mother ($X$) was negatively correlated with infant mortality ($Y$)

as $r_{xy} = -0.6$. Socioeconomic status ($Z$) is a variable that was not taken into account. $Z$ is correlated with $X$ and $Y$ as $r_{xz} = 0.65$ and $r_{yz} = -0.7$. Find $r_{xy.z}$ and test for the significance of the population counterpart $\rho_{xy.z}$.

```
rxy=-0.6; rxz=0.65; ryz=-0.7;
rxy_z = (rxy - rxz*ryz)/sqrt( (1-rxz^2)*(1-ryz^2))
%rxy_z = -0.2672

n=28;
2 * tcdf(    rxy * sqrt(n-2)/sqrt(1-rxy^2), n-2)
%p-value for testing H_0: rho_xy = 0, against
%the two sided alternative
%ans = 7.3810e-004

2 * tcdf(    rxy_z * sqrt(n-1-2)/sqrt(1-rxy_z^2), n-1-2)
%p-value for h_0: rho_xy.z = 0, against
%the two sided alternative
%ans = 0.1779
```

Thus, the significant correlation between $X$ and $Y$ ($p$-value of 0.000738) becomes insignificant ($p$-value of 0.1779) when variable $Z$ is taken into account.

### 15.2.1.1 Confidence Intervals for $\rho$.

Confidence intervals for the population coefficient of correlation $\rho$ are obtained not by using direct measurements, but in a transformed domain. This transformation is known as Fisher's $z$-transformation and is introduced next.

Let $(X_{11}, X_{21}), \ldots, (X_{1n}, X_{2n})$ be a sample from a bivariate normal distribution $\mathcal{N}_2(\mu_1, \mu_2, \sigma_1^2, \sigma_2^2, \rho)$, and $\overline{X}_i = \frac{1}{n}\sum_{j=1}^{n} X_{ij}$, $i = 1, 2$.

The Pearson coefficient of linear correlation

$$r = \frac{\sum_{i=1}^{n}(X_{1i} - \overline{X}_1)(X_{2i} - \overline{X}_2)}{\left[\sum_{i=1}^{n}(X_{1i} - \overline{X}_1)^2 \cdot \sum_{i=1}^{n}(X_{2i} - \overline{X}_2)^2\right]^{1/2}}$$

has a complicated distribution involving special functions, e.g., Anderson (1984), p. 113. However, it is well known that the asymptotic distribution for $r$ is normal, $\mathcal{N}(\rho, \frac{(1-\rho^2)^2}{n})$. Since the variance is a function of the mean $[\sigma^2(\rho) = \frac{(1-\rho^2)^2}{n}$, see equation (6.4)], the transformation defined as

$$\vartheta(\rho) = \int \frac{c}{\sigma(\rho)} d\rho$$

$$= \int \frac{c\sqrt{n}}{1-\rho^2} d\rho$$

$$= \frac{c\sqrt{n}}{2} \int \left( \frac{1}{1-\rho} + \frac{1}{1+\rho} \right) d\rho$$

$$= \frac{c\sqrt{n}}{2} \log \left( \frac{1+\rho}{1-\rho} \right) + k$$

stabilizes the variance. It is known as Fisher's $z$-transformation for the correlation coefficient (usually for $c = 1/\sqrt{n}$ and $k = 0$).

Assume that $r$ and $\rho$ are mapped to $w$ and $\zeta$ as

$$w = \frac{1}{2} \log \left( \frac{1+r}{1-r} \right) = \text{arctanh } r, \quad \zeta = \frac{1}{2} \log \left( \frac{1+\rho}{1-\rho} \right) = \text{arctanh } \rho.$$

The distribution of $w$ is approximately normal, $\mathcal{N}(\zeta, \frac{1}{n-3})$, and this approximation is quite accurate when $\rho^2/n^2$ is small and $n$ is as low as 20.

The inverse $z$-transformation is

$$r = \tanh(w) = \frac{\exp\{2w\} - 1}{\exp\{2w\} + 1}.$$

The use of Fisher's $z$-transformation is illustrated on finding the confidence intervals for $\rho$ and testing hypotheses about $\rho$.

To exemplify the above-stated transformations, we generated $n = 30$ pairs of normally distributed random samples with theoretical correlation $\sqrt{2}/2$. This was done by generating two i.i.d. normal samples $a$ and $b$ of length 30 and taking the transformation $x_1 = a + b$, $x_2 = b$. The sample correlation coefficient $r$ was found. This was repeated $M = 10,000$ times. A histogram of 10,000 sample correlation coefficients is shown in Fig. 15.2a. A histogram of the $z$-transformed $r$s is shown in Fig. 15.2b with a superimposed normal approximation $\mathcal{N}(\text{arctanh}(\sqrt{2}/2), 1/(30 - 3))$.

The sampling distribution for $w$ is approximately normal with mean $\zeta = \frac{1}{2} \log \frac{1+\rho}{1-\rho}$ and variance $1/(n - 3)$. This approximation is satisfactory when the number of pairs exceeds 20. Thus, the $(1 - \alpha)100\%$ confidence interval for $\zeta = \frac{1}{2} \log \frac{1+\rho}{1-\rho}$ is

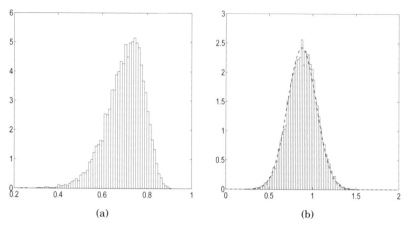

**Fig. 15.2** (a) Simulational run of 10,000 $r$s from a bivariate population having a theoretical $\rho = \sqrt{2}/2$. (b) The same $r$s transformed into $w$s with the normal approximation superimposed.

$$[w_L, w_U] = \left[ w - \frac{z_{1-\alpha/2}}{\sqrt{n-3}}, \quad w + \frac{z_{1-\alpha/2}}{\sqrt{n-3}} \right],$$

where $w = \frac{1}{2} \log \frac{1+r}{1-r}$.

Since $r = \frac{e^{2w}-1}{e^{2w}+1}$, an approximate $(1-\alpha)100\%$ confidence interval for $\rho$ is

$$[r_L, r_U] = \left[ \frac{e^{2w_L}-1}{e^{2w_L}+1}, \quad \frac{e^{2w_U}-1}{e^{2w_U}+1} \right].$$

**Remark.** More accurate approximation for the sampling distribution of $w$ has the mean $\zeta + \rho/(2n-2)$. The correction $\rho/(2n-2)$ is often used in testing and confidence intervals when $n$ is not large. In the above interval the corrected $w$ would be $\frac{1}{2} \log \frac{1+r}{1-r} - \frac{r}{2(n-1)}$.

*Example 15.3.* If $r = -0.5687$ and $n = 8$, then $w = -0.6456$. The bounds for the 95% confidence interval for $\zeta = \frac{1}{2} \log \frac{1+\rho}{1-\rho}$ are $w_L = -0.6456 - 1.96/\sqrt{5} = -1.522$ and $w_U = -0.6456 + 1.96/\sqrt{5} = 0.2309$. The confidence interval for $\rho$ is obtained by back-transforming $w_L$ and $w_U$ using $r = \frac{e^{2w}-1}{e^{2w}+1}$. The result is $[-0.9091, 0.2269]$.

✎

### 15.2.1.2 Test for $\rho = \rho_0$

We saw that when $\rho_0 = 0$, this test is a $t$-test with a statistic given as $t = r\sqrt{\frac{n-2}{1-r^2}}$. When $\rho_0 \neq 0$, then the test of $H_0 : \rho = \rho_0$ versus $H_1 : \rho >, \neq, < \rho_0$ does not have a simple generalization. An approximate test is based on a normal approximation. Under $H_0 : \rho = \rho_0$ the test statistic

$$z = \frac{\sqrt{n-3}}{2}\left[\log\left(\frac{1+r}{1-r}\right) - \log\left(\frac{1+\rho_0}{1-\rho_0}\right)\right],$$

has an approximately standard normal distribution. This implies the following testing rules:

| Alternative | $\alpha$-level rejection region | $p$-value |
|---|---|---|
| $H_1 : \rho > \rho_0$ | $[z_{1-\alpha}, \infty)$ | `1-normcdf(z)` |
| $H_1 : \rho \neq \rho_0$ | $(-\infty, z_{\alpha/2}] \cup [z_{1-\alpha/2}, \infty)$ | `2*normcdf(-abs(z))` |
| $H_1 : \rho < \rho_0$ | $(-\infty, z_\alpha]$ | `normcdf(z)` |

*Example 15.4.* **Interparticular Spacing in Nanoprisms.** There is much interest in knowing and understanding how nanoparticles interact optically. One reason is the use of nanoparticles as plasmon rulers. Plasmon rulers are beneficial to the field of biomedical engineering as they allow researchers to measure changes and differences in DNA or cells at the nano level. This is promising for diagnostics, especially with respect to genetic disorders, which could be potentially identified by data from a plasmon ruler. In order to create a basis for this idea, research must be done to determine the effect of different interparticle spacing on the maximum wavelength of absorbance of the particles. While a linear correlation between measured separation and wavelength is not strong, researchers have found that the correlation between the reciprocal of separation, recsep=1/separation, and the logarithm of a wavelength, logwl=log(wavelength), is strong. The data from the lab of Dr. Mostafa El-Sayed, Georgia Tech, are given in the table below, as well as in nanoprism.dat.

| recsep | logwl | recsep | logwl | recsep | logwl |
|--------|-------|--------|-------|--------|-------|
| 2.9370 | 0.0694 | 2.9284 | 0.0433 | 2.9212 | 0.0331 |
| 2.9196 | 0.0288 | 2.9149 | 0.0221 | 2.9106 | 0.0121 |
| 2.9047 | 0.0080 | 2.9047 | 0.0069 | 2.9031 | 0.0049 |
| 2.9320 | 0.0714 | 2.9154 | 0.0427 | 2.9165 | 0.0336 |
| 2.9085 | 0.0292 | 2.9090 | 0.0240 | 2.9047 | 0.0197 |
| 2.9058 | 0.0052 | 2.9025 | 0.0104 | 2.9025 | 0.0087 |
| 2.8976 | 0.0078 | 2.8971 | 0.0065 | 2.8976 | 0.0062 |

(a) Test the hypothesis that the population coefficient of correlation for the transformed measures is $\rho = 0.96$ against the alternative $\rho < 0.96$. Use $\alpha = 0.05$.

(b) Find a 95% confidence interval for $\rho$.

From the MATLAB code below, we see that $r = 0.9246$ and the $p$-value for the test is 0.0832. Thus, at a 5% significance level $H_0 : \rho = 0.96$ is not rejected. The 95% confidence interval for $\rho$ in this case is $[0.8203, 0.9694]$.

```
%nanoprism.m
load 'nanoprism.dat'
recsep = nanoprism(:,1);
logwl = nanoprism(:,2);
r = corr(recsep, logwl)      %0.9246
n = length(logwl);
fisherz = @(x) atanh(x); invfisherz = @(x) tanh(x);
%fisher z and inverse transformations as pure functions
%Forming z for testing H0: rho = rho0:
z = (fisherz(r) - fisherz(0.96))/(1/sqrt(n-3))   %-1.3840
pval = normcdf(z)  %0.0832
%95% confidence interval

invfisherz([fisherz(r) - norminv(0.975)/sqrt(n-3) ,...
            fisherz(r) + norminv(0.975)/sqrt(n-3)])
   %0.8203    0.9694
```

### 15.2.1.3 Test for the Equality of Two Correlation Coefficients

In some testing scenarios we might be interested to know if the correlation coefficients from two bivariate populations, $\rho_1$ and $\rho_2$, are equal. From the first population the pairs $(X_i^{(1)}, Y_i^{(1)}), i = 1, \ldots, n_1$ are observed. Analogously, from the second population the pairs $(X_i^{(2)}, Y_i^{(2)}), i = 1, \ldots, n_2$ are observed and sample correlations $r_1$ and $r_2$ are calculated. The populations are assumed normal and independent, but components $X$ and $Y$ within each population might be correlated.

The test statistic for testing $H_0 : \rho_1 = \rho_2$ versus $H_1 : \rho_1 >, \neq, < \rho_2$ is expressed in terms of Fisher's $z$-transformations of the sample correlations $r_1$ and $r_2$:

$$z = \frac{w_1 - w_2}{\sqrt{\frac{1}{n_1 - 3} + \frac{1}{n_2 - 3}}},$$

where $w_i = \frac{1}{2} \log \frac{1 + r_i}{1 - r_i}$, $i = 1, 2$ and $n_1$ and $n_2$ are the number of pairs in the first and second sample, respectively.

| Alternative | $\alpha$-level rejection region | $p$-value |
|---|---|---|
| $H_1 : \rho_1 > \rho_2$ | $[z_{1-\alpha}, \infty)$ | `1-normcdf(z)` |
| $H_1 : \rho_1 \neq \rho_2$ | $(-\infty, z_{\alpha/2}] \cup [z_{1-\alpha/2}, \infty)$ | `2*normcdf(-abs(z))` |
| $H_1 : \rho_1 < \rho_2$ | $(-\infty, z_\alpha]$ | `normcdf(z)` |

If the $H_0$ is not rejected, then one may be interested in pooling the two sample estimators $r_1$ and $r_2$. This is done in the domain of $z$-transformed values as

$$w_p = \frac{(n_1 - 3)w_1 + (n_2 - 3)w_2}{n_1 + n_2 - 6}$$

and inverting $w_p$ to $r_p$ via

$$r_p = \frac{1 - \exp\{2w_p\}}{1 + \exp\{2w_p\}}.$$

*Example 15.5.* **Swallowtail Butterflies.** The following data were extracted from a larger study by Brower (1959) on a speciation in a group of swallowtail butterflies (Fig. 15.3). Morphological measurements are in millimeters coded $\times$ 8 ($X$ – length of eighth tergile, $Y$ – length of superuncus).

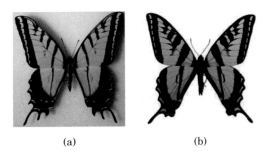

(a)                    (b)

**Fig. 15.3** (a) *Papilio multicaudatus.* (b) *Papilio rutulus.*

| Species | X | Y | X | Y | X | Y | X | Y |
|---|---|---|---|---|---|---|---|---|
| *Papilio* | 24 | 14 | 21 | 15 | 20 | 17.5 | 21.5 | 16.5 |
| *multicaudatus* | 21.5 | 16 | 25.5 | 16 | 25.5 | 17.5 | 28.5 | 16.5 |
| | 23.5 | 15 | 22 | 15.5 | 22.5 | 17.5 | 20.5 | 19 |
| | 21 | 13.5 | 19.5 | 19 | 26 | 18 | 23 | 17 |
| | 21 | 18 | 21 | 17 | 20.5 | 16 | 22.5 | 15.5 |
| *Papilio* | 20 | 11.5 | 21.5 | 11 | 18.5 | 10 | 20 | 11 |
| *rutulus* | 19 | 11 | 20.5 | 11 | 19.5 | 11 | 19 | 10.5 |
| | 21.5 | 11 | 20 | 11.5 | 21.5 | 10 | 20.5 | 12 |
| | 20 | 10.5 | 21.5 | 12.5 | 17.5 | 12 | 21 | 12.5 |
| | 21 | 11.5 | 21 | 12 | 19 | 10.5 | 19 | 11 |
| | 18 | 11.5 | 21.5 | 10.5 | 23 | 11 | 22.5 | 11.5 |
| | 19 | 13 | 22.5 | 14 | 21 | 12.5 | 19.5 | 12.5 |

The observed correlation coefficients are $r_1 = -0.1120$ (for *P. multicaudatus*) and $r_2 = 0.1757$ (for *P. rutulus*). We are interested if the corresponding population correlation coefficients $\rho_1$ and $\rho_2$ are significantly different.

The Fisher $z$-transformations of $r_1$ and $r_2$ are $w_1 = -0.1125$ and $w_2 = 0.1776$. The test statistic is $z = \frac{-0.1125 - 0.1776}{\sqrt{1/17 + 1/25}} = -0.9228$. For this value of $z$ the $p$-value against the two-sided alternative is 0.3561, and the null hypothesis of the equality of population correlations is not rejected. Here is a MATLAB session for the above exercise.

```
PapilioM=[24,    14;    21,    15;    20,    17.5;   21.5, 16.5; ...
          21.5, 16;    25.5, 16;    25.5, 17.5;   28.5, 16.5; ...
          23.5, 15;    22,    15.5; 22.5, 17.5;   20.5, 19;     ...
          21,    13.5; 19.5, 19;    26,    18;      23,    17;    ...
          21,    18;    21,    17;    20.5, 16;      22.5, 15.5];

PapilioR=[20,    11.5; 21.5, 11;    18.5, 10;     20,    11;      ...
          19,    11;    20.5, 11;    19.5, 11;     19,    10.5; ...
          21.5, 11;    20,    11.5; 21.5, 10;     20.5, 12;     ...
          20,    10.5; 21.5, 12.5; 17.5, 12;     21,    12.5; ...
          21,    11.5; 21,    12;    19,    10.5; 19,    11;      ...
          18,    11.5; 21.5, 10.5; 23,    11;     22.5, 11.5; ...
          19,    13;    22.5, 14;    21,    12.5; 19.5, 12.5];
PapilioMX=PapilioM(:,1); % X_m
PapilioMY=PapilioM(:,2); % Y_m

PapilioRX=PapilioR(:,1); % X_r
PapilioRY=PapilioR(:,2); % Y_r

n1=length(PapilioMX);
n2=length(PapilioRX);

r1=corr(PapilioMX, PapilioMY); % -0.1120
r2=corr(PapilioRX, PapilioRY); %  0.1757
%test for rho1 = 0
  pval1 = 2* tcdf(-abs(r1*sqrt(n1-3)/sqrt(1-r1^2)), n1-3);
```

```
% 0.6480
%test for rho2 = 0
pval2 = 2* tcdf(-abs(r2*sqrt(n2-3)/sqrt(1-r2^2)), n2-3);
%0.3806

fisherz = @(x)  1/2*log( (1+x)/(1-x) );
%Fisher z transformation as pure function
w1 = fisherz(r1);   %-0.1125
w2 = fisherz(r2);   %0.1776

%test for rho1 = rho2  vs rho1 ~= rho2
z = (w1 - w2)/sqrt(1/(n1-3) + 1/(n2-3)) %-0.9228
pval = 2 * normcdf(-abs(z)) %0.3561
```

### 15.2.1.4 Testing the Equality of Several Correlation Coefficients

We are interested in testing $H_0 : \rho_1 = \rho_2 = \cdots = \rho_k$.

Let $r_i$ be the correlation coefficient based on $n_i$ pairs, $i = 1, \ldots, k$, and let $w_i = \frac{1}{2} \log \frac{1+r_i}{1-r_i}$ be its Fisher transformation. Define

$$N = (n_1 - 3) + (n_2 - 3) + \cdots + (n_k - 3) = \sum_{i=1}^{k} n_i - 3k, \quad \text{and}$$

$$\overline{w} = \frac{(n_1 - 3)w_1 + (n_2 - 3)w_2 + \cdots + (n_k - 3)w_k}{N}.$$

Then the statistic

$$\chi^2 = (n_1 - 3)w_1^2 + (n_2 - 3)w_2^2 + \cdots + (n_k - 3)w_k^2 - N(\overline{w})^2$$

has a $\chi^2$-distribution with $k - 1$ degrees of freedom.

*Example 15.6.* In testing whether the correlations between sepal and petal lengths differ for the species: setosa, versicolor and virginica, the *p*-value found was close to 0.

```
load fisheriris
%Correlations between sepal and petal lengths
%for species setosa, versicolor and virginica
n1=50; n2=50; n3=50; k=3; N=n1+n2+n3-3*k;
r =[ corr(meas(1:50, 1), meas(1:50, 3)), ...
     corr(meas(51:100, 1), meas(51:100, 3)), ...
     corr(meas(101:150, 1), meas(101:150, 3)) ]
                    %0.2672    0.7540    0.8642
fisherz = @(r)  1/2 * log( (1+r)./(1-r) );
     w=fisherz(r)
                    %0.2738    0.9823    1.3098
wbar =   [n1-3   n2-3   n3-3]/N * w'
                                   %0.8553
```

```
chi2 =    [n1-3,   n2-3, n3-3]*(w.^2)'- N*wbar^2   %26.3581
pval = 1-chi2cdf(chi2, k-1)                        %1.8897e-006
```

### 15.2.1.5  Power and Sample Size in Inference About Correlations

Power and sample size computations for testing correlations use the fact that
Fisher's $z$-transformed sample correlation coefficients have an approximately
normal distribution, as we have seen before.

The statistic $t = r\sqrt{\frac{n-2}{1-r^2}}$, which has a $t$-distribution with $df = n-2$ degrees
of freedom, determines the critical points of the rejection region as

$$r_{\alpha,n-2} = \sqrt{\frac{t^2_{n-2,1-\alpha}}{t^2_{n-2,1-\alpha}+(n-2)}} \, ,$$

where $\alpha$ is replaced by $\alpha/2$ for the two-sided alternative, and where the sign of
the square root is negative if $H_1 : \rho < 0$.

Let $z$-transformations of $r$ and $r_{\alpha,n-2}$ be $w$ and $w_\alpha$, respectively. Then the
power, as a function of $w$, is

$$1-\beta = \Phi\left((w - w_\alpha)\sqrt{n-3}\right).$$

The sample size needed to achieve a power of $1-\beta$, in a test of level $\alpha$
against the one-sided alternative, is

$$n = \left(\frac{z_{1-\alpha}+z_{1-\beta}}{w_1}\right)^2 + 3.$$

Here $w_1$ is $z$-transformation of $\rho_1$ from the specific alternative, $H_1 : \rho = \rho_1$,
and $z_{1-\alpha}$ and $z_{1-\beta}$ are quantiles of a standard normal distribution. When the
alternative is two-sided, $z_{1-\alpha}$ is replaced by $z_{1-\alpha/2}$.

*Example 15.7.* One wishes to determine the size of a sample sufficient to reject
$H_0 : \rho = 0$ with a power of $1-\beta = 0.90$ in a test of level $\alpha = 0.05$ whenever
$\rho \geq 0.4$. Here we take $H_1 : \rho = 0.4$ and find that a sample of size 51 will be
necessary:

```
fisherz = @(x) atanh(x);
w1 = fisherz(0.4) % 0.4236
n =((norminv(0.95)+norminv(0.90))/w1)^2 + 3  %50.7152
```

If $H_0 : \rho = 0$ is to be rejected whenever $|\rho| \geq 0.4$, then a sample of size 62 is
needed:

```
n =((norminv(0.975)+norminv(0.90))/w1)^2 + 3  %61.5442
```

### 15.2.1.6 Multiple Correlation Coefficient

Consider three variables $X_1$, $X_2$, and $Y$. The correlation between $Y$ and the pair $X_1, X_2$ is measured by the multiple correlation coefficient

$$R_{y.x_1 x_2} = \sqrt{\frac{r_{x_1 y}^2 - 2 r_{x_1 x_2} \cdot r_{x_1 y} \cdot r_{x_2 y} + r_{x_2 y}^2}{1 - r_{x_1 x_2}^2}}.$$

This correlation is significant if

$$F = \frac{R_{y.x_1 x_2}^2 / 2}{\left(1 - R_{y.x_1 x_2}^2\right)/(n-3)}$$

is large. Here $n$ is the number of triplets $(X_1, X_2, Y)$, and statistic $F$ has an $F$-distribution with $2, n-3$ degrees of freedom.

The general case $R = R_{y.x_1 x_2 \dots x_k}$ will be covered in the multiple regression section, p. 605; in fact, $R^2$ is called the coefficient of determination and testing its significance is equivalent to testing the significance of the multiple regression of $Y$ on $X_1, \dots, X_k$.

## 15.2.2 Bayesian Inference for Correlation Coefficients

To conduct a Bayesian inference on a correlation coefficient, a bivariate normal distribution for the data is assumed and Wishart's prior is placed on the inverse of the covariance matrix. Recall that the Wishart distribution is a multivariate counterpart of a gamma distribution (more precisely, of a $\chi^2$-distribution) and the model is in fact a multivariate analog of a gamma prior on the normal precision parameter.

To illustrate this Bayesian model, a bivariate normal sample of size $n = 46$ is generated. For this sample Pearson's coefficient of correlation was found to be $r = 0.9908$, with sample variances of $s_x^2 = 8.1088$ and $s_y^2 = 35.0266$.

WinBUGS code ✻corr.odc, as given bellow, is run and Bayes estimators for population correlation $\rho$ and component variances $\sigma_x^2$ and $\sigma_y^2$ are obtained as 0.9901, 8.114, and 34.99. These values are close to the classical estimators since the priors are noninformative. The hyperparameters of Wishart's prior are matrix W and degrees of freedom df, and low degrees of freedom, df=3, make this prior "vague."

As an exercise, compute a classical 95% confidence interval for $\rho$ (p. 577) and compare it with the 95% credible set $[0.9823, 0.9952]$. Are the intervals similar?

```
model{
for( i in 1:nn){
    y[i,1:2] ~ dmnorm( mu[i,], Tau[,] )
    mu[i,1] ~ dnorm(mu.x, tau1)
    mu[i,2] ~ dnorm(mu.y, tau2)
    }
mu.x ~ dnorm(0, 0.0001)
mu.y ~ dnorm(0, 0.0001)
tau1 ~ dgamma(0.001, 0.001)
tau2 ~ dgamma(0.001, 0.001)
Tau[1:2,1:2] ~ dwish( W[,], df )
df <- 3
Sigma[1:2, 1:2]  <- inverse(Tau[,])
rho <- Sigma[1,2]/sqrt(Sigma[1,1]*Sigma[2,2])
}


DATA
list( nn=46, y = structure(.Data = c( 0.5674,    -1.6458,
    -0.4656,    -4.8531,
    1.5253 ,   -2.1200,
    ...
    8.3435,    14.7693,
    11.0151,   18.8988,
    8.9949,    15.3797,
    10.5287,   18.1455), .Dim=c(46,2)) ,
W = structure(.Data = c(1,0,0,1),.Dim=c(2,2) )   )


INIT
list(Tau = structure(.Data = c(1,1,1,1), .Dim = c(2,2)),
 mu.x = 1, mu.y = 1, tau1=1, tau2=1)
```

|            | mean    | sd     | MC error  | val2.5pc | median | val97.5pc | start | sample |
|------------|---------|--------|-----------|----------|--------|-----------|-------|--------|
| Sigma[1,1] | 8.114   | 1.757  | 0.01462   | 5.361    | 7.881  | 12.21     | 1001  | 100000 |
| Sigma[1,2] | 16.68   | 3.625  | 0.03035   | 11.0     | 16.2   | 25.13     | 1001  | 100000 |
| Sigma[2,1] | 16.68   | 3.625  | 0.03035   | 11.0     | 16.2   | 25.13     | 1001  | 100000 |
| Sigma[2,2] | 34.99   | 7.570  | 0.06331   | 23.15    | 33.98  | 52.63     | 1001  | 100000 |
| rho        | 0.9901  | 0.0033 | 3.287E-5  | 0.9823   | 0.9906 | 0.9952    | 1001  | 100000 |

## 15.3 Spearman's Coefficient of Correlation

Charles Edward Spearman (Fig. 15.4) was a late bloomer, academically speaking. He received his Ph.D. at the age of 48, after having served as an officer
in the British army for 15 years. He is most famous in the field of psychology,
where he theorized that "general intelligence" was a function of a comprehensive mental competence rather than a collection of multifaceted mental
abilities. His theories eventually led to the development of factor analysis.

**Fig. 15.4** Charles Edward Spearman (1863–1945).

Spearman (1904) proposed the rank correlation coefficient long before statistics became a scientific discipline. For bivariate data, an observation has two coupled components $(X, Y)$ that may or may not be related to each other. Let $\rho = \text{Corr}(X, Y)$ represent the unknown correlation between two components. In a sample of $n$, let $R_1, \ldots, R_n$ denote the ranks for the first component $X$ and $S_1, \ldots, S_n$ denote the ranks for $Y$. For example, if $x_1 = x_{(n)}$ is the largest value from $x_1, \ldots, x_n$ and $y_1 = y_{(1)}$ is the smallest value from $y_1, \ldots, y_n$, then $(R_1, S_1) = (n, 1)$. Corresponding to Pearson's (parametric) coefficient of correlation, the Spearman coefficient of correlation is defined as

$$\hat{\rho} = \frac{\sum_{i=1}^{n}(R_i - \overline{R})(S_i - \overline{S})}{\sqrt{\sum_{i=1}^{n}(R_i - \overline{R})^2 \cdot \sum_{i=1}^{n}(S_i - \overline{S})^2}}. \tag{15.1}$$

This expression can be simplified. From (15.1), $\overline{R} = \overline{S} = (n+1)/2$ and $\sum(R_i - \overline{R})^2 = \sum(S_i - \overline{S})^2 = nVar(R_i) = n(n^2-1)/12$. Define $D$ as the difference between ranks, i.e., $D_i = R_i - S_i$. With $\overline{R} = \overline{S}$, we can see that

$$D_i = (R_i - \overline{R}) - (S_i - \overline{S})$$

and

$$\sum_{i=1}^{n} D_i^2 = \sum_{i=1}^{n}(R_i - \overline{R})^2 + \sum_{i=1}^{n}(S_i - \overline{S})^2 - 2\sum_{i=1}^{n}(R_i - \overline{R})(S_i - \overline{S}),$$

i.e.,

$$\sum_{i=1}^{n}(R_i - \overline{R})(S_i - \overline{S}) = \frac{n(n^2-1)}{12} - \frac{1}{2}\sum_{i=1}^{n} D_i^2.$$

By dividing both sides of the equation by $\sqrt{\sum_{i=1}^{n}(R_i-\overline{R})^2 \cdot \sum_{i=1}^{n}(S_i-\overline{S})^2}=\sum_{i=1}^{n}(R_i-\overline{R})^2=n(n^2-1)/12$, we obtain

$$\hat{\rho}=1-\frac{6\sum_{i=1}^{n}D_i^2}{n(n^2-1)}. \tag{15.2}$$

Consistent with Pearson's coefficient of correlation (the standard parametric measure of covariance), Spearman's coefficient of correlation ranges between $-1$ and $1$. If there is perfect agreement, i.e., all the differences are 0, then $\hat{\rho}=1$. The scenario that maximizes $\sum D_i^2$ occurs when ranks are perfectly opposite: $R_i=n-S_i+1$.

If the sample is large enough, then Spearman's statistic can be approximated using the normal distribution. It was shown that if $n>10$, then

$$Z=(\hat{\rho}-\rho)\sqrt{n-1}\sim\mathcal{N}(0,1).$$

*Example 15.8.* Stichler et al. (1953) list tread wear for tires, each tire measured by two methods based on (a) weight loss and (b) groove wear.

| Weight | Groove | Weight | Groove |
|--------|--------|--------|--------|
| 45.9 | 35.7 | 41.9 | 39.2 |
| 37.5 | 31.1 | 33.4 | 28.1 |
| 31.0 | 24.0 | 30.5 | 28.7 |
| 30.9 | 25.9 | 31.9 | 23.3 |
| 30.4 | 23.1 | 27.3 | 23.7 |
| 20.4 | 20.9 | 24.5 | 16.1 |
| 20.9 | 19.9 | 18.9 | 15.2 |
| 13.7 | 11.5 | 11.4 | 11.2 |

For this data, $\hat{\rho}=0.9265$. Note that if we opt for the parametric measure of correlation, the Pearson coefficient is 0.948.
✐

**Ties in the Data:** The statistics in (15.1) and (15.2) are not designed for paired data that include tied measurements. If ties exist in the data, a simple adjustment should be made. Define $u'=\sum u(u^2-1)/12$ and $v'=\sum v(v^2-1)/12$ where the $us$ and $vs$ are the ranks for $X$ and $Y$ adjusted (e.g., averaged) for ties. Then

$$\hat{\rho}'=\frac{n(n^2-1)-6\sum_{i=1}^{n}D_i^2-6(u'+v')}{\{[n(n^2-1)-12u'][n(n^2-1)-12v']\}^{1/2}},$$

and it holds that, for large $n$,

$$Z=(\hat{\rho}'-\rho')\sqrt{n-1}\sim\mathcal{N}(0,1).$$

The MATLAB function corr(x,y,'type','Spearman') computes the Spearman correlation coefficient for column vectors $x$ and $y$.

## 15.4 Kendall's Tau

M. G. Kendall (Fig. 15.5) formalized an alternative measure of dependence (originally proposed and used in the nineteenth century) by finding out how many pairs in a bivariate sample are "concordant," which means that the signs between $X$ and $Y$ agree in the pairs. Pairs for which one sign is plus and the other is minus are "discordant." From $(X_i, Y_i)$, $i = 1, \ldots, n$ one can choose $\binom{n}{2}$ different pairs. The pair $(X_i, Y_i), (X_j, Y_j)$ is concordant if either $X_i \leq X_j$ and $Y_i \leq Y_j$ or $X_i \geq X_j$ and $Y_i \geq Y_j$. The pair is called discordant if either $X_i \leq X_j$ and $Y_i \geq Y_j$ or $X_i \geq X_j$ and $Y_i \leq Y_j$. For example, the pairs $(2, 4)$ and $(1, -1)$ are concordant, while the pairs $(-2, 4)$ and $(1, -1)$ are discordant.

**Fig. 15.5** Sir Maurice George Kendall (1907–1983).

Kendall's $\hat{\tau}$-statistic (Kendall, 1938) is defined as

$$\hat{\tau} = \frac{2S_\tau}{n(n-1)}, \quad S_\tau = \sum_{i=1}^{n} \sum_{j=i+1}^{n} \text{sign}\{r_i - r_j\},$$

where $r_i$s are defined via ranks of the second sample corresponding to the ordered ranks of the first sample, $\{1, 2, \ldots, n\}$, i.e.,

$$\begin{pmatrix} 1 & 2 & \ldots & n \\ r_1 & r_2 & \ldots & r_n \end{pmatrix}.$$

In this notation $\sum_{i=1}^{n} D_i^2$ from Spearman's coefficient of correlation becomes $\sum_{i=1}^{n} (r_i - i)^2$. In terms of the number of concordant ($n_C$) and discordant ($n_D = n - n_C$) pairs,

$$\hat{\tau} = \frac{2(n_C - n_D)}{n(n-1)},$$

and in the case of ties, use

$$\hat{\tau} = \frac{n_C - n_D}{n_C + n_D}.$$

*Example 15.9.* **Prevention of Vitreous Loss.** Limbal incisions were made in rabbit eyes to mirror the initial steps of lens extraction. The vitreous body loses water when the eye is open and decreases in weight, as reported by Galin et al. (1971). The results had implications in the context of cataract surgery. The authors measured the vitreous body weight for each eye of 15 New Zealand albino rabbits. One eye has been open for 5 min. $(y)$, while the other served as a control $(x)$. The measurements of vitreous weight (in mg) are provided next:

| Rabbit # | 1 | 2 | 3 | 4 | 5 | 6 | 7 | 8 |
|---|---|---|---|---|---|---|---|---|
| Control eye $(x)$ | 1848 | 1532 | 1460 | 1947 | 1810 | 1718 | 1686 | 1617 |
| Open eye $(y)$ | 1738 | 1440 | 1388 | 1756 | 1692 | 1629 | 1583 | 1499 |
| Rabbit # | 9 | 10 | 11 | 12 | 13 | 14 | 15 | |
| Control eye $(x)$ | 1724 | 1873 | 1928 | 2226 | 1708 | 1605 | 1822 | |
| Open eye $(y)$ | 1596 | 1794 | 1785 | 2044 | 1602 | 1491 | 1702 | |

We will find Kendall's $\hat{\tau}$ and provide an approximate 95% confidence interval.

The sample variance of $\hat{\tau}$ when no ties are present is approximately

$$s^2(\hat{\tau}) = 4 \sum_{i=1}^{n} c_i^2 - 2 \sum_{i=1}^{n} c_i - \frac{(2n-3)\left(\sum_{i=1}^{n} c_i\right)^2}{\binom{n}{2}}.$$

With the presence of ties the expression for sample variance is more complicated, but the above expression can serve as an approximation. Then $(1-\alpha)100\%$ confidence interval is

$$\left[\hat{\tau} - \frac{z_{1-\alpha/2}}{\binom{n}{2}} s(\hat{\tau}), \ \ \hat{\tau} + \frac{z_{1-\alpha/2}}{\binom{n}{2}} s(\hat{\tau})\right] \cap [-1, 1].$$

Using ◢ rabbits.m we found no ties, $n_C = 100$, $n_D = 5$, $\hat{\tau} = 0.9048$, and a 95% confidence interval of $[0.8091, 1.0000]$. Using the difference $T = n_C - n_D$ one can test for independence of the two components. The test has a $p$-value of

$$p = 2\mathbb{P}\left(Z \geq \frac{3(|T|+1)\sqrt{2}}{\sqrt{n(n-1)(2n-5)}}\right)$$

where $Z$ is standard normal. In our example a strong dependence between control and open eye measurements is found

```
T=nc-nd    %95
p = 2 * (1-normcdf(3*(abs(T)+1)*sqrt(2)/sqrt(n*(n-1)*(2*n -5)))))
%1.8965e-008
```

## 15.5 Cum hoc ergo propter hoc

We conclude this chapter with a discussion on the misuses of correlations. The fallacy that correlation implies causation is summarized by Gould's quote at the chapter's beginning (Latin *Cum hoc ergo propter hoc* meaning "With this, therefore because of this"). We already mentioned the "link" between ice-cream sold on the beach and the number of drowning accidents, but the fallacy causes more serious damage to science. Spurious correlations are often misused in medical and health science and attributed to causations. The number of published studies with *voodoo* causations, often conflicted from study to study, is stunning.

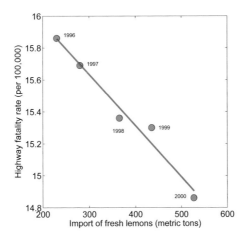

**Fig. 15.6** Fresh lemons imported to USA from Mexico (in metric tons; U.S. Department of Agriculture) and total U.S. highway fatality rate (per 100,000; U.S. NHTSA, DOT HS 810 780).

As an extreme case of spurious correlation we give an example (popular among bloggers on the Web) involving data on imports of fresh lemons from Mexico (1996–2000) and U.S. highway fatality rates (1996–2000), Fig. 15.6. The correlation is $r = -0.986$ and is highly significant ($p < 0.0002$) even with sample size $n = 5$. Some bloggers provided "causal links" citing less expensive car air-fresheners that make drivers happy or slower traffic caused by trucks from Mexico transporting lemons.

There are two possible errors in correlation inference caused by grouping data. The first one is if two separate groups are combined. For each group there may not be correlation, but when the groups are combined, the correlation may be significant and, of course, spurious. Figure 15.7 illustrated this point. Observations represented by red circles (group 1, $r = 0.0643$), as well as the pairs represented by blue circles (group 2, $r = 0.0079$), show no significant correlation. However, when the groups are combined, the correlation increases to $r = 0.7031$, and it is significant with a $p$-value of $1.5 \times 10^{-5}$. Details can be found in ◢ spur.m.

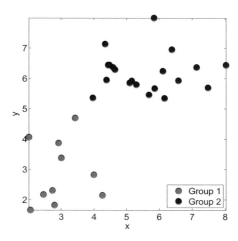

**Fig. 15.7** Spurious correlation when two groups of uncorrelated pairs are combined.

The second error is more subtle. Often, repeated bivariate measurements are considered as independent and an artificial correlation due to a blocking factor is introduced. For example, if for 15 subjects one measures weight ($X$) and skinfold thickness ($Y$) before and after a diet and combines the measurements, then due to the "increased sample size" a significance of correlation between $X$ and $Y$ is more likely.

## 15.6 Exercises

15.1. **Correlation Between Uniforms and Their Squares.** Generate 10,000 uniform random numbers between $-1$ and $1$ in the form of a vector x.

Demonstrate that y=x.2 has a small correlation with x, regardless of their perfect functional relationship.

15.2. **Muscle Strength of "Ethanol Abusers."** It is estimated that 10% of European and North American adults, and up to one-third of acute hospital admissions, are alcoholics. Obviously, the high proportion of alcoholics in the hospitalized population imposes severe financial constraints on health authorities and emphasizes the need for primary caretakers to focus on minimizing alcohol misuse. A staggering two-thirds of chronic ethanol abusers have skeletal muscle myopathy (Martin et al., 1985; Worden, 1976).

Hickish et al. (1989) provide height, quadriceps muscle strength, and age data in 41 male alcoholics, as in the table below. The data are available as alcos.xls or alcos.ascii.

| Height (cm) | Quadriceps muscle strength (N) | Age (years) | Height (cm) | Quadriceps muscle strength (N) | Age (years) |
|---|---|---|---|---|---|
| 155 | 196 | 55 | 172 | 147 | 32 |
| 159 | 196 | 62 | 173 | 441 | 39 |
| 159 | 216 | 53 | 173 | 343 | 28 |
| 160 | 392 | 32 | 173 | 441 | 40 |
| 160 | 98  | 58 | 173 | 294 | 53 |
| 161 | 387 | 39 | 175 | 304 | 27 |
| 162 | 270 | 47 | 175 | 404 | 28 |
| 162 | 216 | 61 | 175 | 402 | 34 |
| 166 | 466 | 24 | 175 | 392 | 53 |
| 167 | 294 | 50 | 175 | 196 | 37 |
| 167 | 491 | 35 | 176 | 368 | 51 |
| 168 | 137 | 65 | 177 | 441 | 49 |
| 168 | 343 | 41 | 177 | 368 | 48 |
| 168 | 74  | 65 | 177 | 412 | 32 |
| 170 | 304 | 55 | 178 | 392 | 49 |
| 171 | 294 | 47 | 178 | 540 | 41 |
| 172 | 294 | 31 | 178 | 417 | 42 |
| 172 | 343 | 38 | 178 | 324 | 55 |
| 172 | 147 | 31 | 179 | 270 | 32 |
| 172 | 319 | 39 | 180 | 368 | 34 |
| 172 | 466 | 53 |     |     |     |

(a) Find the sample correlation between Height and Strength, $r_{HS}$. Test the hypothesis that the population correlation coefficient between the Height and Strength $\rho_{HS}$ is significantly positive at the level $\alpha = 0.01$.

Since an increase in Age is expected to decrease the Strength (negative correlation), find the correlation between Height and Strength when Age is accounted for, that is, find $r_{HS.A}$. Test the hypothesis that $\rho_{HS.A}$ is positive at the level $\alpha = 0.01$.

Find an approximate 95% confidence interval for $\rho_{HS}$.

15.3. **Vending Machine and Pharmacy Errors.** Mr. Joseph Bentley, the owner of a pharmacy store, wants to remove the Coke vending machine standing in front of his store because he believes the vending machine influences the number of errors the store employees make. More precisely, as more Coke is sold outside his store, more errors are made. He provided the following data:

| Errors made | 5 | 3 | 10 | 9 | 5 | 7 | 8 | 4 |
|---|---|---|---|---|---|---|---|---|
| Coke sold | 112 | 100 | 220 | 250 | 100 | 200 | 160 | 100 |

Find the coefficient of correlation. Comment on why this correlation is high. Is there a causation – are Coke sales by themselves influencing the pharmacy employees?

15.4. **Vending Machine and Pharmacy Errors Revisited.** Refer to Exercise 15.3. In addition to Errors and Coke, Mr. Bentley provided the count of people that pass by his store (and the vending machine):

| Errors made | 5 | 3 | 10 | 9 | 5 | 7 | 8 | 4 |
|---|---|---|---|---|---|---|---|---|
| Coke | 112 | 100 | 220 | 250 | 100 | 200 | 160 | 100 |
| People | 10000 | 6000 | 17000 | 20000 | 9000 | 15000 | 14000 | 8000 |

Find the coefficient of correlation between Errors and Coke sales while accounting for the number of people. Comment.

15.5. **Corn Yields and Rainfall.** The following table published by Misner (1928) has been analyzed by Ezekiel and Fox (1959). The variables are years:

rain (X): rainfall measurements in inches, in the six states, from 1890 to 1927. Year 1 in the data below corresponds to 1890.

yield (Y): yearly corn yield in bushels per acre, in six Corn Belt states (Iowa, Illinois, Nebraska, Missouri, Indiana, and Ohio).

| year | X | Y | year | X | Y |
|---|---|---|---|---|---|
| 1 | 9.6 | 24.5 | 20 | 12.0 | 32.3 |
| 2 | 12.9 | 33.7 | 21 | 9.3 | 34.9 |
| 3 | 9.9 | 27.9 | 22 | 7.7 | 30.1 |
| 4 | 8.7 | 27.5 | 23 | 11.0 | 36.9 |
| 5 | 6.8 | 21.7 | 24 | 6.9 | 26.8 |
| 6 | 12.5 | 31.9 | 25 | 9.5 | 30.5 |
| 7 | 13.0 | 36.8 | 26 | 16.5 | 33.3 |
| 8 | 10.1 | 29.9 | 27 | 9.3 | 29.7 |
| 9 | 10.1 | 30.2 | 28 | 9.4 | 35.0 |
| 10 | 10.1 | 32.0 | 29 | 8.7 | 29.9 |
| 11 | 10.8 | 34.0 | 30 | 9.5 | 35.2 |
| 12 | 7.8 | 19.4 | 31 | 11.6 | 38.3 |
| 13 | 16.2 | 36.0 | 32 | 12.1 | 35.2 |
| 14 | 14.1 | 30.2 | 33 | 8.0 | 35.5 |
| 15 | 10.6 | 32.4 | 34 | 10.7 | 36.7 |
| 16 | 10.0 | 36.4 | 35 | 13.9 | 26.8 |
| 17 | 11.5 | 36.9 | 36 | 11.3 | 38.0 |

| 18 | 13.6 | 31.5 | 37 | 11.6 | 31.7 |
|----|------|------|----|------|------|
| 19 | 12.1 | 30.5 | 38 | 10.4 | 32.6 |

Find the sample correlation coefficient $r$ and a 95% confidence interval for the population coefficient $\rho$.

15.6. **Drosophilæ.** Sokoloff (1966) reported the correlation between body weight and wing length in *Drosophila pseudoobscura* as 0.52 in a sample of $n_1 = 39$ at the Grand Canyon, and as 0.67 in a sample of $n_2 = 20$ at Flagstaff, Arizona. Do the correlations in these two populations differ significantly? Use $\alpha = 0.05$.

15.7. **Confidence Interval for the Difference of Two Correlation Coefficients.** Using the results on testing the equality of two correlation coefficients develop a $(1 - \alpha)100\%$ confidence interval for their difference.

15.8. **Oxygen Intake.** The human body takes in more oxygen when exercising than when it is at rest, and to deliver the oxygen to the muscles, the heart must beat faster. Heart rate is easy to measure, but the measurement of oxygen uptake requires elaborate equipment. If oxygen uptake (VO2) is strongly correlated with the heart rate (HR) under a particular set of exercise conditions, then its predicted, rather than measured, values could be used for various research purposes.[1]

| HR | VO2 | HR | VO2 |
|----|-----|----|-----|
| 94 | 0.473 | 108 | 1.403 |
| 96 | 0.753 | 110 | 1.499 |
| 95 | 0.929 | 113 | 1.529 |
| 95 | 0.939 | 113 | 1.599 |
| 94 | 0.832 | 118 | 1.749 |
| 95 | 0.983 | 115 | 1.746 |
| 94 | 1.049 | 121 | 1.897 |
| 104 | 1.178 | 127 | 2.040 |
| 104 | 1.176 | 135 | 2.231 |
| 106 | 1.292 | | |

Find the sample correlation $r$ and calculate a 95% confidence interval for its population counterpart, $\rho$.

15.9. **Obesity and Pain.** Khimich (1997) found that a pain threshold increases in obese subjects and increases with age. Obesity is measured as the percentage over ideal weight ($X$). The response to pain is measured by using the threshold of the nociceptive flexion reflex ($Y$), which is a measure of the pricking pain sensation in an individual. Measurements $X$ and $Y$ are

---

[1] Data provided by Paul Waldsmith from experiments conducted in Don Corrigan's lab at Purdue University, West Lafayette, Indiana.

considered to be normal. We are interested in an inference about the correlation between $X$ and $Y$. The following data were obtained:

| $X$ | 89 | 90 | 75 | 30 | 51 | 75 | 62 | 45 | 90 | 20 |
|---|---|---|---|---|---|---|---|---|---|---|
| $Y$ | 2 | 3 | 4 | 4.5 | 5.5 | 7 | 9 | 13 | 15 | 14 |

(a) Using results $n = 10, \sum_i X_i Y_i = 4461.5, \sum_i X_i = 627, \sum_i X_i^2 = 45141, \sum_i Y_i = 77, \sum_i Y_i^2 = 799.5$ calculate the Pearson coefficient of linear correlation, $r$.
(b) Test the hypothesis that the population coefficient of correlation, $\rho$, is 0, against the alternative $H_1 : \rho < 0$. Use $\alpha = 0.05$.
(c) Let the age $Z$ (in years) of the individuals from the table be as follows (in the corresponding order): 20 18 23 19 44 51 36 47 60 55. Find the partial coefficient of correlation $r_{xy.z}$ if $r_{xz} = -0.2089$ and $r_{yz} = 0.8627$.
(d) Find a 95% confidence interval for $\rho$.

---

**MATLAB AND WINBUGS FILES AND DATA SETS USED IN THIS CHAPTER**

http://springer.bme.gatech.edu/Ch15.Corr/

   corrs.m, errorscoke.m, fisherzsimu.m, histo.m, iriscorr.m, lemon.m, nanoprism.m, ObesityPain.m, rabbits.m, spur.m, variouscorrs.m

corr.odc

nanoprism.dat

---

# CHAPTER REFERENCES

Anderson, T. W., (1984). *An Introduction to Multivariate Statistical Analysis,* 2nd Ed. *Wiley,* New York.

Arvin, D. V. and Spaeth, R. (1998). Trends in Indiana's water use, 1986–1996. *Indiana Department of Natural Resources Special Report*, No. 1.

Brower, L. P. (1959). Speciation in butterflies of the *Papilio glaucus* group. I: Morphological relationships and hybridizations. *Evolution,* **13**, 40–63.

Ezekiel, M. and Fox, K. A. (1959). *Methods of Correlation and Regression Analysis*. Wiley, New York.

Galin, M. A., Robbins, R., and Obstbaum, S. (1971). Prevention of vitreous loss. *Brit. J. Ophthal.*, **55**, 533–537.

Hickish, T., Colston, K., Bland, J. M., and Maxwell J. D. (1989). Vitamin D deficiency and muscle strength in male alcoholics. *Clin. Sci.*, **77**, 171–176.

Kendall, M. (1938). A new measure of rank correlation. *Biometrika*, **30**, 1–2, 81–89.

Khimich, S. (1997). Level of sensitivity of pain in patients with obesity. *Acta Chir. Hung.*, **36**, 166–167.

Martin, R., Ward, K., Slavin, G., Levi, J., and Peters, T. J. (1985). Alcoholic skeletal myopathy, a clinical and pathological study. *Q. J. Med.*, **55**, 233–251.

Misner, E. G. (1928). Studies of the relationship of weather to the production and price of farm products. I: Corn, mimeographed publication, Cornell University, March 1928.

Sokoloff, A. (1966). Morphological variation in natural and experimental populations of *Drosophila pseudoobscura* and *Drosophila persimilis*. *Evolution*, **20**, 49–71.

Spearman, C. (1904). General intelligence: objectively determined and measured. *Am. J. Psychol.*, **15**, 201–293.

Stichler, R. G., Richey, G. G., and Mandel, J. (1953). Measurement of treadware of commercial tires. *Rubber Age*, **73**, 2.

Worden R. E. (1976). Pattern of muscle and nerve pathology in alcoholism. *N.Y. Acad. Sci.*, **273**, 351–359.

# Chapter 16
# Regression

*The experiments showed further that the mean filial regression towards mediocrity was directly proportional to the parental deviation from it.*

– Francis Galton, F.R.S. & c. (1886)

## 16.1 Introduction

The rather curious name *regression* was given to a statistical methodology by British scientist Sir Francis Galton, who analyzed the heights of sons and the average heights of their parents. From his observations, Galton (Fig. 16.1a) concluded that sons of very tall (or short) parents were generally

taller (shorter) than average, but not as tall (short) as their parents. The results were published in 1886 under the title *Regression Towards Mediocrity in
Hereditary Stature*. In the course of time the word *regression* became synony-
mous with the statistical study of the functional relationship between two or
more variables. The data set illustrating Darwin's finding and used by Pearson
is given in ⬚ pearson.dat. The scatterplot and regression fits are analyzed in
◀ galton.m and summarized in Fig. 16.1b. The circles correspond to pairs of
father–son heights, the black line is the line $y = x$, the red line is the regression line, and the green line is the regression line constrained to pass through
the origin. Darwin's findings can be summarized by the observation that the
slope of the regression (red) line was significantly smaller than the slope of the
$45°$-line.

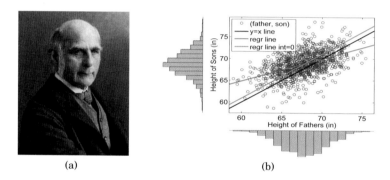

(a)                                                            (b)

**Fig. 16.1** (a) Sir Francis Galton (1822–1911). (b) Galton's father–son height data (used by
Pearson). The *circles* correspond to pairs of father–son heights, the *black line* is the line
$y = x$, the *red line* is the regression line, and the *green line* is the regression line constrained
to pass through the origin.

Usually the response variable $y$ is "regressed" on several predictors or covariates, $x_1, \ldots, x_k$, and this raises many interesting questions involving the
model choice and fit, collinearity among the predictors, and others. When we
have a single predictor $x$ and a linear relation between $y$ and $x$, the regression
is called a simple linear regression.

## 16.2 Simple Linear Regression

Assume that we observed $n$ pairs $(x_1, y_1), \ldots, (x_n, y_n)$, and each observation $y_i$
can be modeled as a linear function of $x_i$, plus an error,

$$y_i = \beta_0 + \beta_1 x_i + \epsilon_i, \ i = 1, \ldots, n.$$

Here $\beta_0$ and $\beta_1$ are the population intercept and slope parameters, respectively, and $\epsilon_i$ is the error. We assume that the errors are not correlated and have mean 0 and variance $\sigma^2$, thus $\mathbb{E}y_i = \beta_0 + \beta_1 x_i$ and $\mathbb{V}\text{ar}\, y_i = \sigma^2$. The goal is to estimate this linear model, that is, estimate $\beta_0$, $\beta_1$, and $\sigma^2$ from the $n$ observed pairs. To put our discussion in context, we consider an example concerning a study of factors affecting patterns of insulin-dependent diabetes mellitus in children.

*Example 16.1.* **Diabetes Mellitus in Children.** Diabetes mellitus is a condition characterized by hyperglycemia resulting from the body's inability to use blood glucose for energy. In type 1 diabetes, the pancreas no longer makes insulin and therefore blood glucose cannot enter the cells to be used for energy.

The objective was to investigate the dependence of the level of serum C-peptide on various other factors in order to understand the patterns of residual insulin secretion. C-peptide is a protein produced by the beta cells of the pancreas whenever insulin is made. Thus, the level of C-peptide in the blood is an index of insulin production.

The part of the data from Sockett et al. (1987), discussed in the context of statistical modeling by Hastie and Tibshirani (1990), is given next. The response measurement is the logarithm of C-peptide concentration (pmol/ml) at the time of diagnosis, and the predictor is the base deficit, a measure of acidity.

| Deficit ($x$) | $-8.1$ | $-16.1$ | $-0.9$ | $-7.8$ | $-29.0$ | $-19.2$ | $-18.9$ | $-10.6$ | $-2.8$ | $-25.0$ | $-3.1$ |
|---|---|---|---|---|---|---|---|---|---|---|---|
| Log C-peptide ($y$) | 4.8 | 4.1 | 5.2 | 5.5 | 5 | 3.4 | 3.4 | 4.9 | 5.6 | 3.7 | 3.9 |

| Deficit ($x$) | $-7.8$ | $-13.9$ | $-4.5$ | $-11.6$ | $-2.1$ | $-2.0$ | $-9.0$ | $-11.2$ | $-0.2$ | $-6.1$ | $-1$ |
|---|---|---|---|---|---|---|---|---|---|---|---|
| Log C-peptide ($y$) | 4.5 | 4.8 | 4.9 | 3.0 | 4.6 | 4.8 | 5.5 | 4.5 | 5.3 | 4.7 | 6.6 |

| Deficit ($x$) | $-3.6$ | $-8.2$ | $-0.5$ | $-2.0$ | $-1.6$ | $-11.9$ | $-0.7$ | $-1.2$ | $-14.3$ | $-0.8$ | $-16.8$ |
|---|---|---|---|---|---|---|---|---|---|---|---|
| Log C-peptide ($y$) | 5.1 | 3.9 | 5.7 | 5.1 | 5.2 | 3.7 | 4.9 | 4.8 | 4.4 | 5.2 | 5.1 |

| Deficit ($x$) | $-5.1$ | $-9.5$ | $-17.0$ | $-3.3$ | $-0.7$ | $-3.3$ | $-13.6$ | $-1.9$ | $-10.0$ | $-13.5$ |
|---|---|---|---|---|---|---|---|---|---|---|
| Log C-peptide ($y$) | 4.6 | 3.9 | 5.1 | 5.1 | 6.0 | 4.9 | 4.1 | 4.6 | 4.9 | 5.1 |

We will follow this example in MATLAB as an annotated step-by-step code/output of ◀ cpeptide.m. For more sophisticated analysis, MATLAB has quite advanced built-in regression tools, regress, regstats, robustfit, stepwise, and many other more or less specialized fitting and diagnostic tools.

After importing the data, we specify p, which is the number of parameters, rename the variables, and find the sample size.

```
Deficit =[-8.1 -16.1 -0.9 -7.8 -29.0 -19.2 -18.9 -10.6 -2.8...
    -25.0 -3.1 -7.8 -13.9 -4.5 -11.6 -2.1 -2.0 -9.0 -11.2 -0.2...
    -6.1 -1 -3.6 -8.2 -0.5 -2.0 -1.6 -11.9 -0.7 -1.2 -14.3 -0.8...
    -16.8 -5.1 -9.5 -17.0 -3.3 -0.7 -3.3 -13.6 -1.9 -10.0 -13.5];

logCpeptide =[ 4.8 4.1 5.2 5.5 5 3.4 3.4 4.9 5.6 3.7 3.9 ...
    4.5 4.8 4.9 3.0 4.6 4.8 5.5 4.5 5.3 4.7 6.6 5.1 3.9 ...
    5.7 5.1 5.2 3.7 4.9 4.8 4.4 5.2 5.1 4.6 3.9 5.1 5.1 ...
    6.0 4.9 4.1 4.6 4.9 5.1];
```

```
%%%%%%%%%%%%%%%%%%%%%%%%%%%%%%%%%%%%%%%%%%%
p = 2; %number of parameters, (beta0, beta1)
       %"Deficit" measurement is "x",  "logCpeptide" is "y".
x = Deficit' ;       %as a column vector
y = logCpeptide' ;  %as a column vector
n = length(x);
```

It is of interest to express the log C-peptide (variable $y$) as a linear function of alkaline deficiency (variable $x$), and the population model $y = \beta_0 + \beta_1 x + \epsilon$ is postulated. Finding estimators for $\beta_0$ and $\beta_1$ is an exercise in calculus – finding the extrema of a function of two variables. The following derivation is known as the least-squares method, which is a broad mathematical methodology for approximate solutions of overdetermined systems, first described by Gauss at the end of the eighteenth century. The best regression line minimizes the sum of squares of errors:

$$L = \sum_{i=1}^{n} \epsilon_i^2 = \sum_{i=1}^{n} (y_i - (\beta_0 + \beta_1 x_i))^2.$$

When pairs $(x_i, y_i)$ are considered fixed, $L$ is a function of $\beta_0$ and $\beta_1$ only.

Minimizing $L$ amounts to solving the so-called *normal* equations

$$\frac{\partial L}{\partial \beta_0} = -2 \sum_{i=1}^{n} [y_i - \beta_0 - \beta_1 x_i] = 0 \quad \text{and}$$

$$\frac{\partial L}{\partial \beta_1} = -2 \sum_{i=1}^{n} [x_i y_i - \beta_0 x_i - \beta_1 x_i^2] = 0,$$

that is,

$$n\beta_0 + \beta_1 \sum_{i=1}^{n} x_i = \sum_{i=1}^{n} y_i \quad \text{and}$$

$$\beta_0 \sum_{i=1}^{n} x_i + \beta_1 \sum_{i=1}^{n} x_i^2 = \sum_{i=1}^{n} x_i y_i. \tag{16.1}$$

Let $\bar{x} = \frac{1}{n} \sum_{i=1}^{n} x_i$ and $\bar{y} = \frac{1}{n} \sum_{i=1}^{n} y_i$ be the sample means of predictor values and the responses. If

$$S_{xy} = \sum_{i=1}^{n} (x_i - \bar{x})(y_i - \bar{y}) = \sum_{i=1}^{n} y_i (x_i - \bar{x}) = \sum_{i=1}^{n} x_i y_i - n \bar{x}\, \bar{y},$$

$$S_{xx} = \sum_{i=1}^{n} (x_i - \bar{x})^2 = \sum_{i=1}^{n} x_i^2 - n\bar{x}^2,$$

and

$$S_{yy} = \sum_{i=1}^{n} (y_i - \bar{y})^2 = \sum_{i=1}^{n} y_i^2 - n\bar{y}^2,$$

then the values for $\beta_0$ and $\beta_1$ minimizing $L$ or, equivalently, solving the normal equations (16.1) are

$$\hat{\beta}_1 = \frac{S_{xy}}{S_{xx}} \text{ and } \hat{\beta}_0 = \bar{y} - \hat{\beta}_1\bar{x}.$$

We will simplify the notation by denoting $\hat{\beta}_0$ by $b_0$ and $\hat{\beta}_1$ by $b_1$. Thus, the fitted regression equation is

$$\hat{y} = b_0 + b_1 x,$$

$$b_1 = \frac{S_{xy}}{S_{xx}} \text{ and } b_0 = \bar{y} - b_1\bar{x}.$$

For values $x = x_i$ the fits $\hat{y}_i$ are obtained as

$$\hat{y}_i = b_0 + b_1 x_i,$$

with the residuals $e_i = y_i - \hat{y}_i$. The residuals are the most important diagnostic modality in regression. They explain how well the predicted data $\hat{y}_i$ fit the observations, and if the fit is not good, residuals indicate what caused the problem.

```
%Sums of Squares
SXX = sum( (x - mean(x)).^2 )   %SXX=2.1310e+003
SYY = sum( (y - mean(y)).^2 )   %SYY=21.807
SXY = sum( (x - mean(x)).* (y - mean(y)) ) %SXY=105.3477
%estimators of coefficients beta1 and beta0
b1 = SXY/SXX                    %0.0494
b0 = mean(y) - b1 * mean(x)     %5.1494
 % predictions
yhat = b0 + b1 * x;
  %residuals
res = y - yhat;
```

We found that yhat=5.1494+0.0494*x. Figure 16.2 shows scatterplot of log C-peptide level ($y$) against alkaline deficiency ($x$) with superimposed regression fit $b_0 + b_1 x$.

Denote by $SSE$ the sum of squared residuals, $SSE = \sum_{i=1}^{n} e_i^2$.

One can show that $SSE = S_{yy} - b_1 S_{xy}$ and that $\mathbb{E}(SSE) = (n-2)\sigma^2$. Thus, the mean square error $MSE = SSE/(n-2)$ is an unbiased estimator of error variance $\sigma^2$. Recall the fundamental ANOVA identity $SST = SSTr + SSE$. In regression terms, the fundamental ANOVA identity has the form

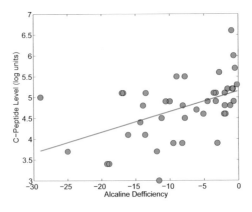

**Fig. 16.2** Scatterplot of log C-peptide level ($y$) against alkaline deficiency ($x$). The regression fit (*red*) is $\hat{y} = 5.1494 + 0.0494x$.

$$SST = SSR + SSE,$$

where $SST = S_{yy}$, $SSR = b_1 S_{xy}$, and $SSE = \sum_{i=1}^{n} e_i^2$. Since

$$\mathbb{E}SSR = \sigma^2 + \beta_1^2 S_{xx},$$

$SSR$ has an associated 1 degree of freedom and the regression mean sum of squares $MSR$ is $SSR/1 = SSR$.

The statistic $MSR$ becomes an unbiased estimator of variance $\sigma^2$ when $\beta_1 = 0$. Thus, to test $H_0 : \beta_1 = 0$ one should have $F = MSR/MSE$ close to 1 since under $H_0$ both $MSR$ and $MSE$ estimate the same quantity, $\sigma^2$. Under $H_0$, the statistic $F = MSR/MSE$ has an $F$-distribution with 1 and $n-2$ degrees of freedom.

Large values of $F$ indicate that there is a contribution of $\beta_1$ in $MSR$ and discrepancy from $H_0$ can be assessed using an $F$-test. The sums of squares, degrees of freedom, mean squares, $F$-statistic and $p$-value associated with observed $F$ are customarily summarized in an ANOVA table:

| Source | DF | SS | MS | F | $p$-value |
|---|---|---|---|---|---|
| Regression | 1 | SSR | $MSR = SSR$ | $F = \frac{MSR}{MSE}$ | $\mathbb{P}(F_{1,n-2} > F)$ |
| Error | $n-2$ | SSE | $MSE = \frac{SSE}{n-2}$ | | |
| Total | $n-1$ | SST | | | |

The $p$-value is associated with testing of $H_0$, which essentially states that covariate $x$ does not influence the response $y$ and the same fit can be obtained by just taking $\bar{y}$ as the model for $y_i$s.

```
%ANOVA Identity
SST = sum((y - mean(y)).^2)     %this is also SYY
SSR = sum((yhat - mean(y)).^2) %5.2079
SSE = sum((y - yhat).^2)        %=sum(res.^2), 16.599
% forming F and testing the adequacy of linear regression
MSR = SSR/(p - 1)   %5.2079
MSE = SSE/(n - p)   %estimator of variance, 0.4049
s = sqrt(MSE)       %0.6363
F = MSR/MSE         %12.8637
pvalue = 1-fcdf(F, p-1, n-p)
%testing H_0: regression has beta1=0,
%that is, there is no need for linear fit, p-val = 0.00088412
```

The above calculations are arranged in the ANOVA table:

| Source | DF | SS | MS | F | $p$-value |
|---|---|---|---|---|---|
| Regression | 1 | 5.2079 | 5.2079 | 12.8637 | 0.0009 |
| Error | 41 | 16.5990 | 0.4049 | | |
| Total | 42 | 21.8070 | | | |

Figure 16.3a shows plot of residuals $y_i - \hat{y}_i$ against $x_i$. A normalized histogram of residuals with superimposed normal distribution $\mathcal{N}(0, 0.6363^2)$, is given in Fig. 16.3b.

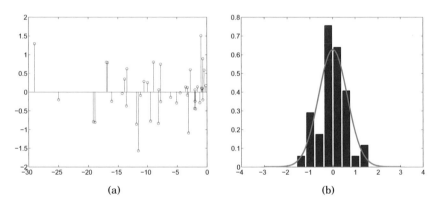

(a)                     (b)

**Fig. 16.3** (a) Plot of residuals $y_1 - \hat{y}_i$ against $x$. (b) Normalized histogram of residuals with superimposed normal distribution $\mathcal{N}(0, s^2)$, with $s$ estimated as 0.6363.

The quantity $R^2$, called the *coefficient of determination*, is defined as

$$R^2 = \frac{SSR}{SST} = 1 - \frac{SSE}{SST}.$$

The $R^2$ in this context coincides with the square of the correlation coefficient between $(x_1, \ldots, x_n)$ and $(y_1, \ldots, y_n)$. However, the representation of $R^2$ via the ratio $SSR/SST$ is more illuminating. In words, $R^2$ explains what proportion of the total variability $(SST)$ encountered in observations is explained or accounted for by the regression $(SSR)$. Thus, a high $R^2$ is desirable in any regression. The adjusted $R^2$ is defined as

$$R^2_{adj} = 1 - \frac{n-1}{n-p} \cdot \frac{SSE}{SST},$$

but it is important only in cases with many predictors $(p > 2)$ since it penalizes inclusion of predictors in the model.

```
%  Other measures of goodness of fit
R2 = SSR/SST        %0.2388
R2adj = 1 - (n-1)/(n-p)* SSE/SST      %0.2203
```

Often, instead of regressing $y_i$ on $x_i$, one regresses $y_i$ on $x_i - \overline{x}$ as

$$y_i = \beta_0^* + \beta_1(x_i - \overline{x}).$$

This is beneficial for several reasons. In practice, we calculate only the estimator $b_1$. Since the fitted line contains the point $(\overline{x}, \overline{y})$, the intercept $\beta_0^*$ is estimated by $\overline{y}$, and our regression fit is

$$\hat{y}_i = \overline{y} + b_1(x_i - \overline{x}).$$

In the Bayesian context estimating $\beta_0^*$ and $\beta_1$ is more stable and efficient than estimating $\beta_0$ and $\beta_1$ directly since $\overline{y}$ and $b_1$ are uncorrelated.

Estimators $b_0$ and $b_1$ are unbiased estimators of population's $\beta_0$ and $\beta_1$. We will show that they are unbiased and that their variance is intimately connected with the variance of responses, $\sigma^2$.

$$\mathbb{E}b_1 = \beta_1 \text{ and } \mathbb{V}\text{ar } b_1 = \frac{\sigma^2}{S_{xx}},$$

$$\mathbb{E}b_0 = \beta_0 \text{ and } \mathbb{V}\text{ar } b_0 = \sigma^2 \left( \frac{1}{n} + \frac{(\overline{x})^2}{S_{xx}} \right).$$

Here is the rationale:

$$\mathbb{E}b_1 = \mathbb{E}\frac{S_{xy}}{S_{xx}} = \frac{1}{S_{xx}}\mathbb{E}\sum_{i=1}^{n} y_i(x_i - \overline{x})$$

$$= \frac{1}{S_{xx}}\mathbb{E}\sum_{i=1}^{n}(\beta_0^* + \beta_1(x_i - \overline{x}) + \epsilon_i)(x_i - \overline{x})$$

$$= \frac{1}{S_{xx}}\left[\sum_{i=1}^{n}\beta_0^*(x_i - \overline{x}) + \sum_{i=1}^{n}\beta_1(x_i - \overline{x})^2 + \mathbb{E}\sum_{i=1}^{n}\epsilon_i(x_i - \overline{x})\right]$$

$$= \frac{1}{S_{xx}}\left[0 + \beta_1 S_{xx} + 0\right] = \beta_1.$$

$$\mathbb{V}\text{ar}\,b_1 = \mathbb{V}\text{ar}\,\frac{S_{xy}}{S_{xx}} = \frac{1}{S_{xx}^2}\sum_{i=1}^{n}\mathbb{V}\text{ar}\,(y_i(x_i - \overline{x}))$$

$$= \frac{1}{S_{xx}^2}\sum_{i=1}^{n}\sigma^2(x_i - \overline{x})^2 = \frac{\sigma^2}{S_{xx}}.$$

Since $b_0 = \overline{y} - b_1\overline{x}$,

$$\mathbb{E}b_0 = \mathbb{E}(\overline{y} - b_1\overline{x}) = \beta_0 + \beta_1\overline{x} - \beta_1\overline{x} = \beta_0$$

and

$$\mathbb{V}\text{ar}\,b_0 = \mathbb{V}\text{ar}\,\overline{y} + \mathbb{V}\text{ar}\,(b_1\overline{x}) - 2\mathbb{C}\text{ov}(\overline{y}, b_1\overline{x}) = \frac{\sigma^2}{n} + (\overline{x})^2\frac{\sigma^2}{S_{xx}} - 2\cdot 0 = \sigma^2\left[\frac{1}{n} + \frac{(\overline{x})^2}{S_{xx}}\right].$$

Sample counterparts of $\mathbb{V}\text{ar}\,b_0$ and $\mathbb{V}\text{ar}\,b_1$ will be needed for the inference in subsequent sections, they are obtained by plugging in the $MSE$ in place of $\sigma^2$.

The covariance between $b_0$ and $b_1$ is

$$\mathbb{C}\text{ov}(b_0, b_1) = \mathbb{C}\text{ov}(\overline{y} - b_1\cdot\overline{x}, b_1) = \mathbb{C}\text{ov}(\overline{y}, b_1) - \overline{x}\cdot\mathbb{V}\text{ar}\,(b_1) = -\overline{x}\cdot\frac{\sigma^2}{S_{xx}},$$

since $\mathbb{C}\text{ov}(\overline{y}, b_1) = 0$.

In MATLAB the estimator of $\sigma$ and sample standard deviations of estimators $b_0$ and $b_1$ from Example 16.1 are as follows:

```
% s, sb0, and sb1
 s = sqrt(MSE)   %s = 0.6363
%Standard errors of parameter estimators
sb1 = s/sqrt(SXX)    %sb1 = 0.0138
sb0 = s * sqrt(1/n + (mean(x))^2/SXX )   %sb0 = 0.1484
```

## 16.2.1 Testing Hypotheses in Linear Regression

To find the estimators of regression parameters and calculate their expectations and variances we do not need distributional properties of errors – except that they are independent, have a mean 0 and a variance that does not vary with $x$. However, to test the hypotheses about the population intercept and slope, and to find confidence intervals, we need to assume that the errors $\epsilon_i$ are i.i.d. normal. In practice, the residual analysis is conducted to verify whether the normality assumption is justified.

### 16.2.1.1 Inference About the Slope Parameter $\beta_1$

For a given constant $\beta_{10}$, the test for

$$H_0 : \beta_1 = \beta_{10}$$

relies on the statistic

$$t = \frac{b_1 - \beta_{10}}{\sqrt{s^2/S_{xx}}},$$

where $s^2 = MSE$. This statistic under $H_0$ has a $t$-distribution with $n-2$ degrees of freedom, and testing is done as follows:

| Alternative | $\alpha$-level rejection region | $p$-value (MATLAB) |
|---|---|---|
| $H_1 : \beta_1 > \beta_{10}$ | $[t_{n-2,1-\alpha}, \infty)$ | `1-tcdf(t,n-2)` |
| $H_1 : \beta_1 \neq \beta_{10}$ | $(-\infty, t_{n-2,\alpha/2}] \cup [t_{n-2,1-\alpha/2}, \infty)$ | `2*tcdf(-abs(t),n-2)` |
| $H_1 : \beta_1 < \beta_{10}$ | $(-\infty, t_{n-2,\alpha}]$ | `tcdf(t,n-2)` |

The distribution of the test statistic is derived from a linear representation of $b_1$ as

$$b_1 = \sum_{i=1}^{n} a_i y_i, \quad a_i = \frac{x_i - \overline{x}}{S_{xx}}.$$

Under $H_0$, $b_1 \sim \mathcal{N}(\beta_{10}, \sigma^2/S_{xx})$. Thus,

$$t = \frac{b_1 - \beta_{10}}{\sqrt{s^2/S_{xx}}} = \frac{b_1 - \beta_{10}}{\sqrt{\sigma^2/S_{xx}}} \times \frac{\sigma}{s} = \frac{\frac{b_1 - \beta_{10}}{\sqrt{\sigma^2/S_{xx}}}}{\sqrt{\frac{SSE}{(n-2)\sigma^2}}},$$

which by definition has a $t_{n-2}$ distribution, as $Z/\sqrt{\frac{\chi^2_{n-2}}{n-2}}$. We also used the fact that $s^2 = MSE = SSE/(n-2)$.

The $(1-\alpha)100\%$ confidence interval for $\beta_1$ is

$$\left[ b_1 - t_{n-2,1-\alpha/2} \frac{s}{\sqrt{S_{xx}}}, \ b_1 + t_{n-2,1-\alpha/2} \frac{s}{\sqrt{S_{xx}}} \right].$$

### 16.2.1.2 Inference About the Intercept Parameter $\beta_0$

For a given constant $\beta_{00}$, the test for

$$H_0 : \beta_0 = \beta_{00}$$

relies on the statistic

$$t = \frac{b_0 - \beta_{00}}{s\sqrt{\frac{1}{n} + \frac{(\overline{x})^2}{S_{xx}}}}.$$

Under $H_0$ this statistic has a $t$-distribution with $n-2$ degrees of freedom and testing is done as follows:

| Alternative | $\alpha$-level rejection region | $p$-value (MATLAB) |
|---|---|---|
| $H_1 : \beta_0 > \beta_{00}$ | $[t_{n-2,1-\alpha}, \infty)$ | 1-tcdf(t,n-2) |
| $H_1 : \beta_0 \neq \beta_{00}$ | $(-\infty, t_{n-2,\alpha/2}] \cup [t_{n-2,1-\alpha/2}, \infty)$ | 2*tcdf(-abs(t),n-2) |
| $H_1 : \beta_0 < \beta_{00}$ | $(-\infty, t_{n-2,\alpha}]$ | tcdf(t,n-2) |

This is based on the representation of $b_0$ as $b_0 = \overline{y} - b_1\overline{x}$ and under $H_0$ $b_0 \sim \mathcal{N}\left(\beta_{00}, \sigma^2\left(\frac{1}{n} + \frac{(\overline{x})^2}{S_{xx}}\right)\right)$. Thus,

$$t = \frac{b_0 - \beta_{00}}{s\sqrt{1/n + (\overline{x})^2/S_{xx}}} = \frac{b_1 - \beta_{00}}{\sigma\sqrt{1/n + (\overline{x})^2/S_{xx}}} \times \frac{\sigma}{s} = \frac{\frac{b_0 - \beta_{00}}{\sigma\sqrt{1/n + (\overline{x})^2/S_{xx}}}}{\sqrt{\frac{SSE}{(n-2)\sigma^2}}},$$

which by definition has a $t_{n-2}$ distribution, as $Z/\sqrt{\frac{\chi^2_{n-2}}{n-2}}$.

The $(1-\alpha)100\%$ confidence interval for $\beta_0$ is

$$\left[ b_0 - t_{n-2,1-\alpha/2} \; s \sqrt{\frac{1}{n} + \frac{(\bar{x})^2}{S_{xx}}} \; , \; b_0 + t_{n-2,1-\alpha/2} \; s \sqrt{\frac{1}{n} + \frac{(\bar{x})^2}{S_{xx}}} \; \right].$$

```
% are the coefficients equal to 0?
t1 = b1/sb1  %3.5866
pb1 = 2 * (1-tcdf(abs(t1),n-p) ) %8.8412e-004
t0 = b0/sb0  %34.6927
pb1 = 2 * (1-tcdf(abs(t0),n-p) ) %0

%test H_0: beta1 = 0.04 vs H_1: beta1 > 0.04
tst1 = (b1 - 0.04)/sb1            %0.6846
ptst1 = 1 - tcdf( tst1, n-p )     %0.2487
% test H_0: beta0 = 5.8 vs  H_1: beta0 < 5.8
tst2 = (b0 - 5.8)/sb0             %-4.3836
ptst2 = tcdf(tst2, n-p )          %3.9668e-005
%%%%%%%%%%%%%%%%%%%%%%%%%%%%%%%%%%%%%%%%%%
% Find 95% CI for beta1
[b1 - tinv(0.975, n-p)*sb1,  b1 + tinv(0.975, n-p)*sb1]
%  0.0216    0.0773
% Find 99% CI for beta0
[b0 - tinv(0.995, n-p)*sb0,  b0 + tinv(0.995, n-p)*sb0]
%  4.7484    5.5503
```

### 16.2.1.3  Inference About the Variance $\sigma^2$

Testing $H_0 : \sigma^2 = \sigma_0^2$ relies on the statistic $\chi^2 = \frac{(n-2)MSE}{\sigma_0^2} = \frac{SSE}{\sigma_0^2}$. This statistic under $H_0$ has a $\chi^2$-distribution with $n-2$ degrees of freedom and testing is done as follows:

| Alternative | $\alpha$-level rejection region | $p$-value (MATLAB) |
|---|---|---|
| $H_1 : \sigma^2 < \sigma_0^2$ | $[0, \chi^2_{n-2,\alpha}]$ | `chi2cdf(chi2,n-2)` |
| $H_1 : \sigma^2 \neq \sigma_0^2$ | $[0, \chi^2_{n-2,\alpha/2}] \cup [\chi^2_{n-2,1-\alpha/2}, \infty)$ | `2*chi2cdf(ch,n-2)` |
| $H_1 : \sigma^2 > \sigma_0^2$ | $[\chi^2_{n-2,1-\alpha}, \infty)$ | `1-chi2cdf(chi2,n-2)` |

where chi2 is the test statistic and ch=min(chi2,1/chi2).
The $(1-\alpha)100\%$ confidence interval for $\sigma^2$ is

$$\left[ \frac{SSE}{\chi^2_{n-2,1-\alpha/2}}, \; \frac{SSE}{\chi^2_{n-2,\alpha/2}} \right].$$

The following part of MATLAB script tests $H_0 : \sigma^2 = 0.5$ versus $H_1 : \sigma^2 < 0.5$ and finds a 95% confidence interval for $\sigma^2$. As is evident, $H_0$ is not rejected ($p$-value 0.1981), and the interval is [0.2741, 0.6583].

```
%test H_0: sigma2 = 0.5 vs H_1: sigma2 < 0.5
ch2 = SSE/0.5    %33.1981
ptst3 = chi2cdf(ch2, n-p)    %0.1981
% Find 95% CI for sigma2
[SSE/chi2inv(0.975, n-p),   SSE/chi2inv(0.025, n-p)]
%    0.2741    0.6583
```

### 16.2.1.4 Inference About the Mean Regression Response for $x = x^*$

Suppose that the regression $\hat{y} = b_0 + b_1 x$ has been found and that we are interested in making an inference about the response $y_m = \mathbb{E}(y|x = x^*) = \beta_0 + \beta_1 x^*$. The statistic for $y_m$ is $\hat{y}_m = b_0 + b_1 x^*$, and it is a random variable since both $b_0$ and $b_1$ are random variables.

The $\hat{y}_m$ is an unbiased estimator of $y_m$, $\mathbb{E}\hat{y}_m = E(b_0 + b_1 x^*) = \beta_0 + \beta_1 x^* = y_m$, as expected. The variance of $\hat{y}_m$ is obtained from representation $\hat{y}_m = b_0 + b_1 x^* = \bar{y} + b_1(x^* - \bar{x})$ and the fact that the correlation between $\bar{y}$ and $b_1$ is zero:

$$\mathbb{V}\text{ar}\,\hat{y}_m = \sigma^2 \left( \frac{1}{n} + \frac{(x^* - \bar{x})^2}{S_{xx}} \right).$$

Thus,

$$\hat{y}_m \sim \mathcal{N}\left( \beta_0 + \beta_1 x^*, \sigma^2 \left( \frac{1}{n} + \frac{(x^* - \bar{x})^2}{S_{xx}} \right) \right),$$

from which we develop the inference.

The test

$$H_0 : y_m = y_0$$

relies on the statistic

$$t = \frac{\hat{y}_m - y_0}{s\sqrt{\frac{1}{n} + \frac{(x^* - \bar{x})^2}{S_{xx}}}}.$$

This statistic under $H_0$ has a $t$-distribution with $n-2$ degrees of freedom and testing is done as in the cases of $\beta_0$ and $\beta_1$.

The $(1-\alpha)100\%$ confidence interval for $y_m = \beta_0 + \beta_1 x^*$ is

$$\left[ \hat{y}_m - t_{n-2,1-\alpha/2}\, s \sqrt{\frac{1}{n} + \frac{(x^*-\bar{x})^2}{S_{xx}}}\; ,\; \hat{y}_m + t_{n-2,1-\alpha/2}\, s \sqrt{\frac{1}{n} + \frac{(x^*-\bar{x})^2}{S_{xx}}} \right].$$

### 16.2.1.5 Inference About a New Response for $x = x^*$

Suppose that the regression $\hat{y} = b_0 + b_1 x$ has been established and that we are interested in predicting the response $\hat{y}_{pred}$ for a new observation, corresponding to a covariate $x = x^*$. Given the value $x = x^*$, the difference between the inference about the mean response $y_m$ discussed in the previous section and the inference about an individual outcome $y_{pred}$ is substantial. As in the previous subsection, $\hat{y}_{pred} = b_0 + b_1 x^*$, and the mean of $\hat{y}_{pred}$ is $\mathbb{E}(\hat{y}_{pred}) = \beta_0 + \beta_1 x^* = y_{pred}$, which is in fact equal to $y_m$.

Where $y_{pred}$ and $y_m$ differ is their variability. The variability of $\hat{y}_{pred}$ has two sources, first, the variance of the distribution of $y$s for $x = x^*$, which is $\sigma^2$, and second, the variance of sampling distribution for $b_0 + b_1 x^*$, which is $\sigma^2 \left( \frac{1}{n} + \frac{(x^*-\bar{x})^2}{S_{xx}} \right)$. Thus, $\mathbb{V}\text{ar}(\hat{y}_{pred}) = MSE + \mathbb{V}\text{ar}(\hat{y}_m)$.

The distribution for $\hat{y}_{pred}$ is normal,

$$\hat{y}_{pred} \sim \mathcal{N}\left( \beta_0 + \beta_1 x^*, \sigma^2 \left( 1 + \frac{1}{n} + \frac{(x^*-\bar{x})^2}{S_{xx}} \right) \right),$$

and the subsequent inference is based on this distribution.

The test

$$H_0 : y_{pred} = y_0$$

relies on the statistic

$$t = \frac{\hat{y}_{pred} - y_0}{s \sqrt{1 + \frac{1}{n} + \frac{(x^*-\bar{x})^2}{S_{xx}}}}.$$

This statistic under $H_0$ has a $t$-distribution with $n-2$ degrees of freedom, which implies the inference.

The $(1-\alpha)100\%$ confidence interval for $y_{pred}$ is

$$\left[\hat{y}_{pred} - t_{n-2,1-\alpha/2}\, s\sqrt{1 + \frac{1}{n} + \frac{(x^* - \overline{x})^2}{S_{xx}}}\ ,\ \hat{y}_{pred} + t_{n-2,1-\alpha/2}\, s\sqrt{1 + \frac{1}{n} + \frac{(x^* - \overline{x})^2}{S_{xx}}}\right].$$

```
% predicting y for the new observation x, CI and PI
newx = 230; %Deficit = 230
y_newx = b0 + b1 * newx  % 16.5195
sym = s * sqrt(1/n + (mean(x) - newx)^2/SXX )
  %st.dev. for mean response, sym = 3.2839
syp = s * sqrt(1 + 1/n + (mean(x) - newx)^2/SXX )
  %st.dev. for the prediction syp = 3.3450
alpha = 0.05;
  %mean response interval
lbym = y_newx - tinv(1-alpha/2, n-p) * sym;
rbym = y_newx + tinv(1-alpha/2, n-p) * sym;
  % prediction interval
lbyp = y_newx - tinv(1-alpha/2, n-p) * syp;
rbyp = y_newx + tinv(1-alpha/2, n-p) * syp;
  % the intervals
[lbym rbym]   % 9.8875    23.1516
[lbyp rbyp]   % 9.7642    23.2749
```

Next, we will find Bayesian estimators of regression parameters in the same example, *Diabetes Mellitus in Children*, by using WinBUGS. On p. 606 we mentioned that taking $x_i - \overline{x}$ as a predictor instead of $x_i$ is beneficial in the Bayesian context. From such a parametrization of regression,

$$y_i = \beta_0^* + \beta_1(x_i - \overline{x}) + \epsilon_i,$$

the traditional intercept $\beta_0$ is easily obtained as $\beta_0^* - \beta_1\overline{x}$.

```
model{
for (i in 1:ntotal){
y[i] ~ dnorm( mui[i], tau )
mui[i] <-  bb.0 + b.1 *(x[i] - mean(x[]))
yres[i] <- y[i] - mui[i]
}
bb.0 ~ dnorm(0, 0.0001)
b.0 <- bb.0 - b.1 * mean(x[])
b.1 ~ dnorm(0, 0.0001)
tau ~ dgamma(0.001, 0.001)
s <- 1/sqrt(tau)
}

DATA
```

```
list(ntotal=43,
y = c(4.8,  4.1,  5.2,  5.5,  5.0,  3.4,  3.4,  4.9,  5.6,  3.7,
      3.9,  4.5,  4.8,  4.9,  3.0,  4.6,  4.8,  5.5,  4.5,  5.3,
      4.7,  6.6,  5.1,  3.9,  5.7,  5.1,  5.2,  3.7,  4.9,  4.8,
      4.4,  5.2,  5.1,  4.6,  3.9,  5.1,  5.1,  6.0,  4.9,  4.1,
      4.6,  4.9,  5.1),
x = c(-8.1,  -16.1,  -0.9,  -7.8,  -29.0,  -19.2,  -18.9,  -10.6,
      -2.8,  -25.0,  -3.1,  -7.8,  -13.9,  -4.5,   -11.6,  -2.1,
      -2.0,  -9.0,  -11.2,  -0.2,  -6.1,  -1.0,  -3.6,  -8.2,
      -0.5,  -2.0,  -1.6,  -11.9,  -0.7,  -1.2,  -14.3,  -0.8,
      -16.8,  -5.1,  -9.5,  -17.0,  -3.3,  -0.7,  -3.3,  -13.6,
      -1.9,  -10.0,  -13.5))
```

INITS
```
list(bb.0 = 0, b.1 = 0, tau=1)
```

The output is given in the table below. It contains Bayesian estimators b.0 for $\beta_0$ and b.1 for $\beta_1$. In the least-squares regression we found $b_1 = S_{xy}/S_{xx} = 0.0494$, $b_0 = \bar{y} - b_1 \cdot \bar{x} = 5.1494$, and $s = \sqrt{MSE} = 0.6363$. Since priors were non-informative, we expect that Bayesian estimators will be close to the classical. Indeed that is the case: b.0 = 5.149, b.1 = 0.0494, and s = 0.6481.

The classical standard errors of estimators for $\beta_0$ and $\beta_1$ are sb0 = 0.1484 and sb1 = 0.0138, while the corresponding Bayesian estimators are 0.1525 and 0.01418.

The classical 95% confidence interval for $\beta_1$ was found to be $[0.0216, 0.0773]$. The Bayesian 95% credible set for $\beta_1$ is $[0.02139, 0.07733]$, as is evident from val2.5pc and val97.5pc in the output below.

| | mean | sd | MC error | val2.5pc | median | val97.5pc | start | sample |
|---|---|---|---|---|---|---|---|---|
| b.0 | 5.149 | 0.1525 | 3.117E-4 | 4.848 | 5.149 | 5.449 | 2001 | 200000 |
| b.1 | 0.0494 | 0.0141 | 3.072E-5 | 0.02139 | 0.04944 | 0.07733 | 2001 | 200000 |
| s | 0.6481 | 0.0734 | 1.771E-4 | 0.5236 | 0.6415 | 0.811 | 2001 | 200000 |
| yres[1] | 0.05111 | 0.09944 | 2.175E-4 | -0.1444 | 0.05125 | 0.2472 | 2001 | 200000 |
| yres[2] | -0.2537 | 0.1502 | 3.459E-4 | -0.5499 | -0.2533 | 0.0418 | 2001 | 200000 |
| yres[3] | 0.09544 | 0.1431 | 2.925E-4 | -0.1861 | 0.09505 | 0.378 | 2001 | 200000 |
| yres[4] | 0.7363 | 0.09957 | 2.167E-4 | 0.5406 | 0.7364 | 0.9325 | 2001 | 200000 |
| ... | | | | | | | | |
| yres[41] | -0.4552 | 0.1333 | 2.727E-4 | -0.7173 | -0.4555 | -0.1919 | 2001 | 200000 |
| yres[42] | 0.245 | 0.1028 | 2.314E-4 | 0.04251 | 0.2451 | 0.4475 | 2001 | 200000 |
| yres[43] | 0.6179 | 0.125 | 2.879E-4 | 0.3718 | 0.6179 | 0.8632 | 2001 | 200000 |

Thus, the Bayesian approach to regression estimation is quite close to the classical when the priors on $\beta_0$ and $\beta_1$ and the precision $\tau = 1/\sigma^2$ are noninformative.

*Example 16.2.* **Hubble Telescope and Hubble Regression.** Hubble's constant (H) is one of the most important numbers in cosmology because it is instrumental in estimating the size and age of the universe. This long-sought number indicates the rate at which the universe is expanding, from the primordial "Big Bang." The Hubble constant can be used to determine the intrinsic brightness and masses of stars in nearby galaxies, examine those same properties in more distant galaxies and galaxy clusters, deduce the amount of dark matter present in the universe, obtain the scale size of faraway galaxy clusters, and serve as a test for theoretical cosmological models.

**Fig. 16.4** Edwin Powell Hubble (1889–1953).

In 1929, Edwin Hubble[1] (Fig. 16.4) investigated the relationship between the distance of a galaxy from the Earth and the velocity with which it appears to be receding. Galaxies appear to be moving away from us no matter which direction we look. This is thought to be the result of the "Big Bang." Hubble hoped to provide some knowledge about how the universe was formed and what might happen in the future. The data collected included distances (megaparsecs[2]) to $n = 24$ galaxies and their recessional velocities (km/sec).

Hubble's law is as follows: Recessional velocity = H × distance,
where H is Hubble's constant (units of H are [km/sec/Mpc]). By working backward in time, the galaxies appear to meet in the same place. Thus 1/H can be used to estimate the time since the Big Bang – a measure of the age of the universe.

| Distance in megaparsecs ([Mpc]) | 0.032 | 0.034 | 0.214 | 0.263 | 0.275 | 0.275 |
|---|---|---|---|---|---|---|
| | 0.45 | 0.5 | 0.5 | 0.63 | 0.8 | 0.9 |
| | 0.9 | 0.9 | 0.9 | 1.0 | 1.1 | 1.1 |
| | 1.4 | 1.7 | 2.0 | 2.0 | 2.0 | 2.0 |
| Recessional velocity ([km/sec]) | 170 | 290 | −130 | −70 | −185 | −220 |
| | 200 | 290 | 270 | 200 | 300 | −30 |
| | 650 | 150 | 500 | 920 | 450 | 500 |
| | 500 | 960 | 500 | 850 | 800 | 1090 |

A regression analysis seems appropriate; however, there is no intercept term in Hubble's law. Can you verify that the constant term of the regression analysis is not significantly different than 0 at any *reasonable*[3] level of $\alpha$. Find

---

[1] Edwin Powell Hubble (b. Nov. 20, 1889, Marshfield, Missouri, U.S., d. Sept. 28, 1953, San Marino, California.), American astronomer who is considered the founder of extragalactic astronomy and who provided the first evidence of the expansion of the universe.

[2] 1 parsec = 3.26 light years

[3] Reasonable here means level $\alpha$ not larger than 0.10

the 95% confidence interval for the slope $\beta_1$, also known as Hubble's constant H, from the given data.

The age of the universe as predicted by Hubble (in years) is about 2.3 billion years.

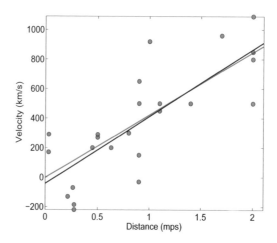

**Fig. 16.5** Hubble's data and regression fits. The *blue line* is an unconstrained regression (with intercept fitted), and the *red line* is a no-intercept fit. The slope for the no-intercept fit is $b_1 = 423.9373$ (=H).

```
%H = 423.9373
secinyear =60*60*24*365   %31536000
kminmps = 3.08568025 * 10^19;
age = 1/H  *  kminmps/secinyear   %2.3080e+009
```

Modern measurements put H at approx. 70, thus predicting the age of the universe to about 14 billion years.

Figure 16.5 showing Hubble's data and regression fits is generated by ◀ hubble.m.

## 16.3 Testing the Equality of Two Slopes*

Let $(x_{1i}, y_{1i})$, $i = 1, \ldots, n_1$ and $(x_{2i}, y_{2i})$, $i = 1, \ldots, n_2$, be the pairs of measurements obtained from two groups, and for each group the regression is estimated as

$$y_{1i} = b_{0(1)} + b_{1(1)} x_{1i} + e_{i(1)}, i = 1, \ldots, n_1, \text{ and}$$
$$y_{2i} = b_{0(2)} + b_{1(2)} x_{2i} + e_{i(2)}, i = 1, \ldots, n_2,$$

where in groups $i = 1, 2$ the statistics $b_{0(i)}$ and $b_{1(i)}$ are estimators of the respective population parameters, intercepts $\beta_{0(i)}$, and slopes $\beta_{1(i)}$. We are interested in testing the equality of the population slopes,

$$H_0 : \beta_{1(1)} = \beta_{1(2)},$$

against the one- or two-sided alternatives.

The test statistic is

$$t = \frac{b_{1(1)} - b_{1(2)}}{s.e.(b_{1(1)} - b_{1(2)})}, \tag{16.2}$$

where the standard error of the difference $b_{1(1)} - b_{1(2)}$ is

$$s.e.(b_{1(1)} - b_{1(2)}) = \sqrt{s^2 \left[ \frac{1}{S_{xx(1)}} + \frac{1}{S_{xx(2)}} \right]},$$

and $s^2$ is the polled estimator of variance,

$$s^2 = \frac{SSE_1 + SSE_2}{n_1 + n_2 - 4}.$$

Statistic $t$ in (16.2) has a $t$-distribution with $n_1 + n_2 - 4$ degrees of freedom and, in addition to testing, could be used for a $(1 - \alpha)\,100\%$ confidence interval for $\beta_{1(1)} - \beta_{1(2)}$,

$$[(b_{1(1)} - b_{1(2)}) \mp t_{n_1+n_2-4, 1-\alpha/2} \times s.e.(b_{1(1)} - b_{1(2)})].$$

*Example 16.3.* **Cadmium Poisoning.** Chronic cadmium poisoning is an insidious disease associated with the development of emphysema and the excretion in the urine of a characteristic protein of low molecular weight. The first signs of chronic cadmium poisoning become apparent following a latent interval after exposure has ended. Respiratory functions deteriorate faster with in age. The data set featured in Armitage and Berry (1994) gives ages (in years) and vital capacity (in liters) for 84 men working in the cadmium industry, ⌨ cadmium.dat|mat|xlsx. The observations with flag exposure equal to 0 denote persons unexposed to cadmium oxide fumes, while flags 1 and 2 correspond to exposed persons. The purpose of the study was to assess the degree of influence of exposure to respiratory functions. Since respiratory functions are influenced by age, regardless of exposure, age as a covariate needs to be taken into account. Thus, the suggested methodology is to test the equality of the slopes in group regressions of vital capacity to age:

$$H_0 : \beta_{1(exposed)} = \beta_{1(unexposed)} \quad \text{versus} \quad H_1 : \beta_{1(exposed)} < \beta_{1(unexposed)}.$$

The research hypothesis is that the regression in the exposed group is "steeper," that is, the vital capacity decays significantly faster with age. This corresponds to a smaller slope parameter for the exposed group since in this case the slopes are negative (Fig. 16.6). The inference is supported by the following MATLAB code.

```
xlsread vitalcapacity.xlsx;
twos = ans;
   x1 = twos( twos(:,3) > 0, 1);   y1 = twos( twos(:,3) > 0, 2);
   x2 = twos( twos(:,3) ==0, 1);   y2 = twos( twos(:,3) ==0, 2);
   n1=length(x1); n2 = length(x2);
SXX1 = sum((x1 - mean(x1)).^2)   %4.3974e+003
SXX2 = sum((x2 - mean(x2)).^2)   %6.1972e+003
SYY1 = sum((y1 - mean(y1)).^2)   %26.5812
SYY2 = sum((y2 - mean(y2)).^2)   %20.6067
SXY1 = sum((x1 - mean(x1)).*(y1 - mean(y1))) %-236.3850
SXY2 = sum((x2 - mean(x2)).*(y2 - mean(y2))) %-189.7116
b1_1 = SXY1/SXX1     %-0.0538
b1_2 = SXY2/SXX2     %-0.0306
SSE1 = SYY1 - (SXY1)^2/SXX1   %13.8741
SSE2 = SYY2 - (SXY2)^2/SXX2   %14.7991
s2 = (SSE1 + SSE2)/(n1 + n2 - 4)   %0.3584
s = sqrt(s2)   %0.5987
seb1b2 = s * sqrt( 1/SXX1 + 1/SXX2 ) %0.0118
t = (b1_1 - b1_2)/seb1b2     %-1.9606
pval = tcdf(t, n1 + n2 - 4)     %0.0267
```

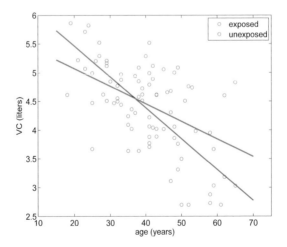

**Fig. 16.6** Samples from exposed (*red*) and unexposed (*green*) groups with fitted regression lines. The slopes of the two regressions are significantly different with a *p*-value smaller than 3%.

Thus, the hypothesis of equality of slopes is rejected with a $p$-value of 2.67%.

**Note.** Since the distribution of $t$-statistic is calculated under $H_0$, which assumes parallel regression lines, the more natural estimator s22, in place of s2, takes into account this fact. The number of degrees of freedom in the $t$-statistic changes to $n_1 + n_2 - 3$. The changes in the inference are minimal, as evidenced from the MATLAB code accounting for s22.

```
%s22 accounts for equality of slopes:
s22 = (SYY1 + SYY2 - ...
 (SXY1 + SXY2)^2/(SXX1 + SXX2))/(n1 + n2 - 3) %0.3710
s = sqrt(s22)   %0.6091
seb1b2 = s * sqrt( 1/SXX1 + 1/SXX2 ) %0.0120
t = (b1_1 - b1_2)/seb1b2        %-1.9270
pval = tcdf(t, n1 + n2 - 3)    %0.0287
```

For the case of the confidence interval, estimator s2 and $n_1 + n_2 - 4$ degrees of freedom should be used.

## 16.4 Multivariable Regression

It is often the case that in an experiment leading to regression analysis more than a single covariate is available. For example, in Chap. 2, p. 31, we discussed an experiment in which two indices of the amount of body fat (Siri and Brozek indices) were calculated from the body density measure. In addition, a variety of body measurements, including weight, height, adiposity, neck, chest, abdomen, hip, thigh, knee, ankle, biceps, forearm, and wrist, were recorded. Recall that the body density measure is complicated and potentially unpleasant, since it is taken by submerging the subject in water. Therefore, it is of interest to ask whether the Brozek index can be well predicted using the nonintrusive measurements.

If $x_1, x_2, \ldots, x_k$ are variables, covariates, or predictors, and we have $n$ joint measurements of covariates and the response, $x_{i1}, x_{i2}, \ldots, x_{ik}, y_i$, $i = 1, 2, \ldots, n$, then multivariable regression expresses the response as a linear combination of covariates, plus an intercept and an additive error:

$$y_i = \beta_0 + \beta_1 x_{i1} + \beta_2 x_{i2} + \cdots + \beta_k x_{ik} + \epsilon_i, i = 1, \ldots, n.$$

The errors $\epsilon_i$ are assumed to be independent and normal with mean 0 and constant variance $\sigma^2$. We will denote by $k$ the number of covariates, but the number of parameters in the model is $p = k + 1$ because the intercept $\beta_0$ should be added. To avoid confusion we will mostly use $p$ – the number of parameters in all expressions that involve dimensions and derived statistics.

As in the univariate case, we will be interested in estimating and testing the coefficients, error variance, and mean and prediction responses. However, multivariable regression brings several new challenges when compared to a simple regression. The two main challenges are (i) the possible presence of *multicollinearity* among the covariates, that is, covariates being correlated among themselves, and (ii) a multitude of possible models and the need to find the *best model* by identifying the "best" subset of predictors. A synonym for multivariable regression is multivariate regression, although the second term is also used when the response $y$ is multivariate, which is not the case here.

### 16.4.1 Matrix Notation

If the regression equations for all $n$ observations $(x_{i1}, x_{i2}, \ldots, x_{ik}, y_i)$, $i = 1, \ldots, n$ are written as

$$y_1 = \beta_0 + \beta_1 x_{11} + \beta_2 x_{12} + \cdots + \beta_k x_{1k} + \epsilon_1,$$
$$y_2 = \beta_0 + \beta_1 x_{21} + \beta_2 x_{22} + \cdots + \beta_k x_{2k} + \epsilon_2,$$
$$\vdots$$
$$y_n = \beta_0 + \beta_1 x_{n1} + \beta_2 x_{n2} + \cdots + \beta_k x_{nk} + \epsilon_n,$$

then one can write the above system in convenient matrix form as

$$y = X\beta + \epsilon,$$

where

$$y = \begin{bmatrix} y_1 \\ y_2 \\ \vdots \\ y_n \end{bmatrix},$$

$$X = \begin{bmatrix} x_1 \\ x_2 \\ \vdots \\ x_n \end{bmatrix} = \begin{bmatrix} 1 & x_{11} & x_{12} & \cdots & x_{1k} \\ 1 & x_{21} & x_{22} & \cdots & x_{2k} \\ \vdots & \vdots & & \vdots \\ 1 & x_{n1} & x_{n2} & \cdots & x_{nk} \end{bmatrix}, \quad \beta = \begin{bmatrix} \beta_0 \\ \beta_1 \\ \vdots \\ \beta_k \end{bmatrix}, \quad \text{and} \quad \epsilon = \begin{bmatrix} \epsilon_1 \\ \epsilon_2 \\ \vdots \\ \epsilon_n \end{bmatrix}.$$

Note that $y$ and $\epsilon$ are $n \times 1$ vectors, $X$ is an $n \times p$ matrix, and $\beta$ is a $p \times 1$ vector. Here $p = k + 1$. To find the least-squares estimator of $\beta$, one minimizes the sum of squares:

$$\sum_{i=1}^{n} (y_i - (\beta_0 + \beta_1 x_{i1} + \cdots + \beta_k x_{ik}))^2.$$

The minimizing solution

$$
b = \begin{bmatrix} b_0 \\ b_1 \\ \vdots \\ b_k \end{bmatrix}
$$

satisfies the system (normal equations)

$$X'Xb = X'y, \tag{16.3}$$

and the least-squares estimator of $\beta$ is

$$b = (X'X)^{-1}X'y.$$

The fitted values are obtained as

$$\hat{y} = Xb = X(X'X)^{-1}X'y,$$

and the residuals are

$$e = y - \hat{y} = y - Xb = y - X(X'X)^{-1}X'y = (I - X(X'X)^{-1}X')y,$$

where $I$ is an $n \times n$ identity matrix.

The matrix $H = X(X'X)^{-1}X'$ that appears in expressions for fitted values and residuals is important in this context; it is called the *hat* matrix. In terms of the hat matrix $H$,

$$\hat{y} = Hy, \quad \text{and} \quad e = (I - H)y.$$

Matrices $H$ and $I - H$ are projection matrices, and the $n$-dimensional vector $y$ is projected to $\hat{y}$ by $H$ and to residual vector $e$ by $I - H$. Any projection matrix $A$ is *idempotent*, which means that $A^2 = A$. Simply put, a projection of a projection will be the same as the original projection. Geometrically, vectors $\hat{y}$ and $e$ are orthogonal since the product of their projection matrices is $\mathbf{0}$. Indeed, because $H$ is idempotent, $H(I - H) = H - H^2 = H - H = \mathbf{0}$.

The errors $\epsilon_i$ are independent and the variance of vector $\epsilon$ is $\sigma^2 I$.

We will illustrate some concepts from multivariable regression using dataset fat.dat; all calculations are part of MATLAB script fatregdiag.m.

```
load 'fat.dat'
      casen = fat(:,1);    %case number
      broz  = fat(:,2);    %dependent variable
      siri  = fat(:,3);    %function of densi
      densi = fat(:,4);    %an intrusive measure
      age = fat(:,5);
%below are the predictors
```

```
weight  = fat(:,6);   height = fat(:,7);
adiposi = fat(:,8);   %adiposity is BMI index=weight/height^2
ffwei   = fat(:,9);   %fat free weight, excluded from predictors
                      % since it involves body fat and brozek
neck  = fat(:,10);    chest  = fat(:,11);   abdomen = fat(:,12);
hip   = fat(:,13);    thigh  = fat(:,14);   knee    = fat(:,15);
ankle = fat(:,16);    biceps = fat(:,17);   forearm = fat(:,18);
wrist = fat(:,19);

vecones = ones(size(broz)); % necessary for the intercept

 disp('=========================================================')
 disp('         p = 15, 14 variables + intercept')
 disp('=========================================================')

Z =[age  weight  height  adiposi   neck     chest     abdomen ...
    hip  thigh   knee    ankle     biceps   forearm   wrist];
X =[vecones Z];
Y = broz
%   X is design matrix, n x p where n is the number of subjects
%   and p is the number of parameters, or number of predictors+1.
%   varnames = ['intercept=0' 'age=1' 'weight=2' 'height=3'
%   'adiposi=4' 'neck=5' 'chest=6' 'abdomen=7' 'hip=8' 'thigh=9'
%   'knee=10' 'ankle=11''biceps=12' 'forearm=13' 'wrist=14'];
[n, p] = size(X)
b = inv(X' * X) * X'* Y;
H = X * inv(X' * X) * X';
max(max(H * H - H));   %0 since H is projection matrix
Yhat = H * Y;          %or Yhat  = X * b;
%---------------------------------------------------------------
```

### 16.4.1.1 Sums of Squares and an ANOVA Table

Sums of squares, $SST$, $SSR$, and $SSE$, for multivariable regression have simple expressions in matrix notation. Here we introduce matrix $\boldsymbol{J}$, which is an $n \times n$ matrix in which each element is 1. The total sum of squares can be calculated as

$$SST = \boldsymbol{y}'\boldsymbol{y} - \frac{1}{n}\boldsymbol{y}'\boldsymbol{J}\boldsymbol{y} = \boldsymbol{y}'\left(\boldsymbol{I} - \frac{1}{n}\boldsymbol{J}\right)\boldsymbol{y}.$$

The error sum of squares is $SSE = \boldsymbol{e}'\boldsymbol{e} = \boldsymbol{y}'(\boldsymbol{I}-\boldsymbol{H})'(\boldsymbol{I}-\boldsymbol{H})\boldsymbol{y} = \boldsymbol{y}'(\boldsymbol{I}-\boldsymbol{H})\boldsymbol{y}$ because $\boldsymbol{I} - \boldsymbol{H}$ is a symmetric projection matrix.

By taking the difference,

$$SSR = SST - SSE = \boldsymbol{y}'\left(\boldsymbol{H} - \frac{1}{n}\boldsymbol{J}\right)\boldsymbol{y}.$$

The number of degrees of freedom for $SST$, $SSR$, and $SSE$ are $n-1$, $p-1$, and $n-p$, respectively. Thus, the multivariable regression ANOVA table is

| Source | DF | SS | MS | F | p-value |
|--------|-----|-----|-----|-----|---------|
| Regression | $p-1$ | $SSR$ | $MSR = \frac{SSR}{p-1}$ | $F = \frac{MSR}{MSE}$ | $\mathbb{P}(F_{-1,n-p} > F)$ |
| Error | $n-p$ | $SSE$ | $MSE = \frac{SSE}{n-p}$ | | |
| Total | $n-1$ | $SST$ | | | |

where large values of $F$ are critical for $H_0$ which states that the covariates $x_1, \ldots, x_k$ do not influence the response $y$. Formally, the null hypothesis is

$$H_0 : \beta_1 = \beta_2 = \cdots = \beta_k = 0,$$

while the alternative is that at least one $\beta_i$, $i = 1, \ldots, k$ is not 0.

As in the simple regression, $R^2$ is called the coefficient of determination,

$$R^2 = \frac{SSR}{SST} = 1 - \frac{SSE}{SST}.$$

Adding more variables to a regression always increases $R^2$, even when the covariates have nothing to do with the experiment. If two models have comparable $R^2$s, then, according to Ockham's razor[4], the simpler model should be preferred and adding new variables to the regression should be penalized. One way to achieve this is via an adjusted coefficient of determination,

$$R^2_{adj} = 1 - \frac{n-1}{n-p} \frac{SSE}{SST},$$

which is one of the criteria for comparing models.

The estimator of the error variance $\sigma^2$ is $MSE$. We can find the confidence intervals of the components of $e$ and $\hat{y}$ using respectively the diagonal elements of covariance matrices $MSE \times (I - H)$ and $MSE \times H$.

```
%Sums of Squares
J=ones(n); I = eye(n);
SSR = Y' * (H - 1/n * J) * Y;
SSE = Y' * (I - H) * Y;
SST = Y' * (I - 1/n * J) * Y;
MSR = SSR/(p-1) %806.7607
MSE = SSE/(n-p) %15.9678
F = MSR/MSE       %50.5243
pval = 1-fcdf(F, p-1, n-p)          %0
Rsq = 1 - SSE/SST                   %0.7490
Rsqadj = 1 - (n-1)/(n-p) * SSE/SST  %0.7342
s = sqrt(MSE)                       %3.9960
%- - - - - - - - - - - - - - - - - - - - - - - - - - - - - - - - - -
```

---

[4] *Pluralitas non est ponenda sine neccesitate*, which translates into English as "Plurality should not be posited without necessity" (William of Ockham, 1287–1347)

## 16.4.1.2 Inference About Regression Parameters and Responses

The covariance matrix for a vector of estimators of regression coefficients $\boldsymbol{b}$ is equal to

$$s_{\boldsymbol{b}}^2 = MSE \times (\boldsymbol{X}'\boldsymbol{X})^{-1}.$$

Its $i$th diagonal element is an estimator of variance for $b_i$, and off-diagonal elements at the position $(i, j)$ are estimators of covariances between $b_i$ and $b_j$.

When finding a confidence interval or testing a hypothesis about a particular $\beta_i$, we use $b_i$ and $s_{b_i}$ in the same way as in the univariate regression, only this time the test statistic $t$ has $n - p$ degrees of freedom instead of $n - 2$. Several subsequent MATLAB scripts are excerpts from the file ◀ fatregdiag.m.

```
sig2 = MSE * inv(X' * X);% covariances among b's
sb=sqrt(diag(sig2));
tstats = b./sb;
pvals = 2 * tcdf(-abs(tstats), n-p);
 disp('-------------------------------------')
 disp('     var#        t          pval      ')
 disp('-------------------------------------')
 [ (0:p-1)'    tstats      pvals    ]
 %------------------------------------
 %       var#        t        pval
 %- - - - - - - - - - - - - - - - - - -
 %      0       -0.9430     0.3467
 %    1.0000     1.8942     0.0594
 %    2.0000    -1.6298     0.1045
 %     ...
 %   12.0000     0.9280     0.3543
 %   13.0000     2.3243     0.0210
 %   14.0000    -2.9784     0.0032
```

Note that in the fat example, the intercept is not significant ($p = 0.3467$), nor is the coefficient for variable #12 (biceps) ($p = 0.3543$), while for #13 (forearm) it is significant ($p = 0.0210$).

The regression response $y_m = \boldsymbol{x}^* \boldsymbol{b}$ evaluated at $\boldsymbol{x}^* = (1 \; x_1^* \; x_2^* \; \ldots \; x_k^*)$ has sample variance

$$s_{y_m}^2 = (\boldsymbol{x}^*)\boldsymbol{s}_{\boldsymbol{b}}(\boldsymbol{x}^*)'.$$

For a prediction, the variance is, as in univariate regression, $s_{y_p}^2 = s_{y_m}^2 + MSE$. For the inference about regression response, the $t$-statistic with $n - p$ degrees of freedom is used. As an example, assume that for a "new" person with co-variates $\boldsymbol{x}_h = (1 \; 38 \; 191 \; 72 \; 26 \; 41 \; 104 \; 95 \; 101.5 \; 66 \; 39 \; 24 \; 31 \; 30 \; 18.5)$ a prediction of the Brozek index is needed. The model gives a prediction of 19.5143, and the variances for mean response and individual response are found as below. This is sufficient to calculate confidence intervals on the mean and individual response.

```
%- - - - - - - - - - - - - - - - - - - - - - - - - - - - - - - - - - - - - - -
% predicting mean responses and individual response
% with 95% confidence/prediction intervals
Xh=[1   38   191   72    26    41   104   ...
     95  101.5   66    39    24    31    30    18.5];
  Yh = Xh * b   %19.5143
  sig2h = MSE *   Xh  * inv(X' * X)   * Xh';
  sig2hpre = MSE * (1 + Xh  * inv(X' * X)   * Xh');
  sigh = sqrt(sig2h);
  sighpre = sqrt(sig2hpre);
  %95% CI's on the mean and individual response
  [Yh-tinv(0.975, n-p)*sigh, Yh+tinv(0.975, n-p)*sigh]
      %[17.4347,     21.5940]
  [Yh-tinv(0.975, n-p)*sighpre, Yh+tinv(0.975, n-p)*sighpre]
      %[11.3721    27.6566]
```

## 16.4.2 Residual Analysis, Influential Observations, Multicollinearity, and Variable Selection*

Three important deficiencies in a multivariable linear model can be diagnosed: (i) the presence of outliers, (ii) the nonconstant error variance, and (iii) a possible suboptimal model selection. Although the exposition level of these diagnostic methods exceeds the level in introductory coverage of regression, multivariable regression modeling is important in practice and provides an important step to understanding more sophisticated nonlinear models such as generalized linear models. For this reason, we provide a basic overview of residual and influence analysis, as well as an assessment of multicollinearity and choice of model. For readers interested in a more comprehensive treatment of multivariable linear models, the book by Rawlings et al. (1998) is a comprehensive resource.

### 16.4.2.1 Residual Analysis and Influence

Residual analysis can be combined with cross-validation in a model assessment. The following table gives several types of residuals used in analysis:

| | |
|---|---|
| 1. Ordinary residuals | $e_i = y_i - \hat{y}_i$ |
| 2. Studentized residuals | $r_i = \dfrac{e_i}{s\sqrt{1-h_{ii}}}$ |
| 3. Externally studentized residuals | $t_i = \dfrac{e_i}{s_{-i}\sqrt{1-h_{ii}}}$ |
| 4. Prediction sum of squares residuals (PRESS) | $e_{i,-i} = \dfrac{e_i}{1-h_{ii}}$ |

The ordinary residuals $e_i = y_i - \hat{y}_i$ are components of $(I - H)y$. The *leverages* $h_{ii}$ are diagonal elements of hat matrix $H$. These are important descriptors of design matrix $X$ and explain how far $x_i$ is from $\bar{x}$. All leverages are bounded $1/n \leq h_{ii} \leq 1$ and their sum is $\sum h_{ii} = p$, the number of regression parameters. The *studentized residual* is the residual divided by its standard deviation and recalls the $t$-statistic. Such residuals are scale-free comparable, and values outside the interval $[-2.5, 2.5]$ are potential outliers. Sometimes these residuals are called *internally studentized* since the standard deviation $s = \sqrt{MSE}$ depends on the $i$th observation.

*Externally studentized residuals* (also called R-Student residuals) are more of a measures of influence of the $i$th observation $(y_i, x_i)$ on the $i$th residual. Instead of $s$, the residuals are studentized by an *external standard deviation*,

$$s_{-i} = \sqrt{\frac{(n-p)s^2 - e_i^2/(1 - h_{ii})}{n - p - 1}}. \tag{16.4}$$

This external estimate of $\sigma$ comes from the model fitted without the $i$th observation; however, the refitting is not necessary due to a simple expression in (16.4). Externally studentized residuals can be tested since they are distributed as a $t$-distribution with $n - p - 1$ degrees of freedom. Of course, if multiple residuals are tested simultaneously, then it should be done in the spirit of multiple hypothesis testing (Sect. 9.7).

*PRESS* residuals $e_{i,-i} = e_i/(1 - h_{ii})$ also remove the impact of the $i$th observation $(y_i, x_i)$ on the fit at $x_i$. This is a cross-validatory residual, and one is interested in how a model built without using the $i$th observation would predict the $i$th response.

For model assessment, the statistic *PRESS* is useful. It is defined as a sum of squares of *PRESS* residuals,

$$PRESS = \sum_i e_{i,-i}^2,$$

and used in defining the *prediction* $R^2$,

$$R_{pred}^2 = 1 - \frac{PRESS}{SST}.$$

Here $SST$ is the total sum of squares $\sum_i(y_i - \bar{y})^2$. The ordinary $R^2$ is defined as $1 - SSE/SST$, and in $R_{pred}^2$ the $SSE$ is replaced by $PRESS$. Since the *average prediction error* is defined as $\sqrt{PRESS/n}$, good models should have small *PRESS*.

*DFBETAS* stands for *difference in betas*. It measures the influence of the $i$th observation on $\beta_j$:

$$DFBETAS_{ij} = \frac{b_j - b_{j(-i)}}{s_{-i}\sqrt{c_{jj}}},$$

where $b_j$ is the estimator of $\beta_j$, $b_{j(-i)}$ is the estimator of $\beta_j$ when the $i$th observation is excluded, and $c_{jj}$ is the $(j+1)$st diagonal element in $(\boldsymbol{X'X})^{-1}$. Large *DFBETAS* may indicate which predictor might be influential. The recommended threshold is $2/\sqrt{n}$. The black boxes in Fig. 16.9 are at combinations $(\beta_0 - \beta_{14}$ vs. indices of observations) for which `abs(DFBetas>2/sqrt(n))`.

Since several *DFBETAS* can be large, it is useful to pay attention to those corresponding to large *DFFITS*. The *DFFITS* measure the influence of the $i$th observation on the prediction $\hat{y}_i$:

$$DFFITS_i = \frac{\hat{y}_i - \hat{y}_{i,-i}}{s_{(-i)}\sqrt{h_{ii}}} = \sqrt{\frac{h_{ii}}{1-h_{ii}}} \frac{e_i}{s_{-i}\sqrt{1-h_{ii}}}.$$

The value $\hat{y}_{i,-i}$ is the prediction of $y_i$ on the basis of a regression fit without the $i$th observation. The observation is considered influential if its *DFFITS* value exceeds $2\sqrt{p/n}$.

A measure related computationally to *DFFITS* is Cook's distance, $D_i$:

$$D_i = (DFFITS_i)^2 \frac{s_{-i}^2}{ps^2}.$$

Cook's distance measures the effect of the $i$th observation on the whole vector $\boldsymbol{b} = \hat{\boldsymbol{\beta}}$. An observation is deemed influential if its Cook's distance exceeds $4/n$.

Figure 16.8 shows ordinary residuals plotted against predicted values $\hat{y}$. The radii of circles are proportional to |Dffits| in panel (a) and to Cook's distance in panel (b).

Influential observations are not necessarily outliers and should not be eliminated from a model only on the basis of their influence.

Influential observations can be identified by their influence on a predicted value. One often finds predictions of the $i$th response $\hat{y}_{i,-i}$ in a regression in which the $i$th case $(y_i, \boldsymbol{x}_i)$ is omitted (Fig. 16.7).

```
%prediction of y_i with ith observation  removed
%hat y_i(-i)
ind = 1:n;
Yhati = [];
for i = 1:n
    indi = find(ind ~= i);
    Yi = Y(indi);
    Xi=X(indi,:);
    bi = inv(Xi' * Xi) * Xi'* Yi;
    Yhatii = X(i,:) * bi;
    Yhati =[Yhati; Yhatii];
end
Yhati  %prediction of y_i without i-th observation
%-------------------------------------------------


%============= residual analysis===================
hii = diag(H);                    %leverages
```

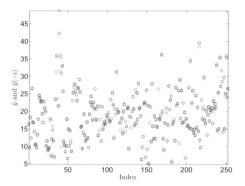

**Fig. 16.7** Predicted responses $\hat{y}_i$ (*blue*) and predicted responses $\hat{y}_{i,-i}$ (*red*). Large changes in prediction signify an influential observation.

```
resid = (I - H)*Y;                %ordinary residuals
sresid = sqrt(MSE .* (1-hii));
stresid = resid./sresid  %studentized residuals
%----------studentized deleted residuals---
di = Y - Yhati; %or  di  = resid./(1-hii)
%di is also called PRESS residual
PRESS =sum(di.^2)
R2pred = 1-PRESS/SST %R^2 predictive

sminusi = sqrt(((n-p)*MSE*ones(n,1) -...
    resid.^2./(1-hii))/(n-p-1));      %stdev(-i)
ti  = resid ./(sminusi .* sqrt(1-hii))
% externally studentized residuals
% outliers based on leverage = hii
outli=hii/mean(hii);
 find(outli > 3)
   % 31 36 39 41 42 86 159 175 206

%influential observations
 Dffits = ti .* sqrt( hii ./(1-hii)) %influ ith to ith
 find(abs(Dffits) > 2 * sqrt(p/n));
 % 31 39 42 82 86 128 140 175 207 216 221 231 250
 %
 CooksD = resid.^2 .* (hii./(1-hii).^2)/(p * MSE)
 % influence if ith to all;
 find(CooksD > 4/n) %find influential
 %31 39 42 82 86 128 175 207 216 221 250


%DFBetas - influence if ith obs on jth coefficient
    cii = diag(inv(X' * X));
    DFBetas =[];
for i = 1:n
    indi = find(ind ~= i);
    Yi = Y(indi);
```

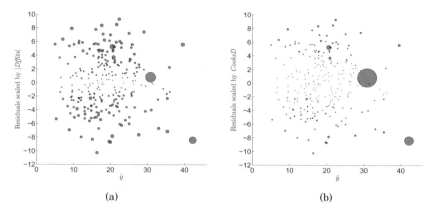

(a)                                             (b)

**Fig. 16.8** Ordinary residuals plotted against predicted values $\hat{y}$. The radii of circles are proportional to |Dffits| in panel (a) and to Cook's distance in panel (b). The observations with the largest *circles* are in both cases the 42nd and the 39th.

```
    Xi=X(indi,:);
    bi   = inv(Xi' * Xi) * Xi'* Yi;
    Hi = Xi * inv(Xi' * Xi)   * Xi';
    SSEi  = Yi' * (eye(n-1) - Hi) * Yi;
    MSEi  = SSEi./(n-p-1);
    DFBetasi = (b - bi)./sqrt(MSEi .* cii) ;
    DFBetas = [DFBetas; DFBetasi'];
end
```

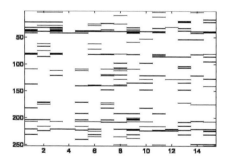

**Fig. 16.9** DFBetas: The $x$-axis enumerates $\beta_0 - \beta_{14}$ while on the $y$-axis are plotted the indices of observations. The *black boxes* are at combinations for which `abs(DFBetas>2/sqrt(n))`.

### 16.4.2.2 Multicollinearity

The multicollinearity problem in regression concerns the correlation among the predictors. Suppose that matrix $X$ contains two collinear columns $x_i$ and $x_j = 2x_i$. Obviously covariates $x_i$ and $x_j$ are linearly dependent and $x_j$ does not bring any new information about the response. This collinearity makes matrix $X$ not of full rank and $X'X$ singular, that is, not invertible, and normal equations (16.3) have no solution. In reality, if multicollinearity is present, then matrix $X'X$ is not singular, but near-singular, in the sense that its determinant is close to 0, making the inversion of $X'X$, and consequently the solution $\hat{\beta}$, very unstable. This happens when either two or more variables are highly correlated or when a variable has a small variance (and, in a sense, is correlated to the intercept).

There are several indices of multicollinearity. We will discuss the *condition index*, which is a global measure, and the *local condition index* and *variance inflation factor*, which are linked to a particular variable.

Let $\lambda_{(1)} \le \lambda_{(2)} \le \cdots \le \lambda_{(p)}$ be ordered eigenvalues of $X'X$. The condition number is defined as the ratio of the largest and smallest eigenvalues:

$$K = \sqrt{\frac{\lambda_{(n)}}{\lambda_{(1)}}}.$$

Concerning values for $K$ starting at around 10, values between 30 and 100 influence the results, and values over 100 indicate a serious collinearity problem.

The (local) condition index for variable $x_i$ is

$$K_i = \sqrt{\frac{\lambda_{(n)}}{\lambda_i}}.$$

Since eigenvalues explain the budget of variances among the variables, a large condition index means the variance in variable $i$ is relatively small, which is a source of multicollinearity. Variables with indices that exceed 30 are problematic.

The variance inflation factor (*VIF*) explains the extent of correlation of a particular variable $x_i$ to the rest of predictors. It is defined as

$$VIF_i = \frac{1}{1 - R_i^2},$$

where $R_i^2$ is the coefficient of determination in regression of $x_i$ to the rest of predictors. *VIF*s exceeding 10 are considered serious. Computationally, one finds the correlation matrix for the predictors. The diagonal elements of this inverse are the *VIF*s. Unfortunately, a *VIF* diagnostic sometimes can miss a problem since the intercept is not included in the analysis.

Multicollinearity can be diminished by excluding problematic variables causing the collinearity in the first place. Alternatively, groups of variables can be combined and merged into a single variable.

Another measure is to keep all variables but "condition" matrix $\boldsymbol{X'X}$ by adding $k\boldsymbol{I}$, for some $k > 0$, to the normal equations. There is a tradeoff: the solutions of $(\boldsymbol{X'X} + k\boldsymbol{I})\hat{\boldsymbol{\beta}} = \boldsymbol{X'y}$ are more stable, but some bias is introduced.

```
lambdas = eig(X' * X);
K = max(lambdas)/min(lambdas)
Ki = max(lambdas)./lambdas
%
 Xprime = [];
    for k=2:p
        Xkbar = mean(X(:,k));
        sk = std( X(:,k));
        Xprimek = 1/sqrt(n-1) .* (X(:,k)- Xkbar * ones(n,1) )/sk;
        Xprime = [Xprime Xprimek];
    end
    RXX = Xprime' * Xprime;
    VIF = diag (inv(RXX));
    VIF'
% 2.2509   33.7869    2.2566   16.1634    4.4307   10.6846   13.3467
%15.1583    7.9615    4.8288    1.9455    3.6745    2.1934    3.3796
```

Alternatively, for most of the above calculations one can use MATLAB's regstats or diagnostics.m,

```
s = regstats(Y,Z,'linear','all');
[index,res,stud_res,lev,DFFITS1,Cooks_D,DFBETAS]=diagnostics(Y,Z);
```

### 16.4.2.3  Variable Selection in Regression

Model selection involves finding a subset of predictors from a large number of potential predictors that is optimal in some sense.

We defined the coefficient of determination $R^2$ as the proportion of model-explained variability, and it seems natural to choose a model that maximizes $R^2$. It turns out that this is not a good idea since the maximum will always be achieved by that model that has the maximal number of parameters. It is a fact that $R^2$ increases when even a random or unrelated predictor is included.

The adjusted $R^2$ penalizes the inclusion of new variables and represents a better criterion for choosing a model. However, with $k$ variables there would be $2^k$ possible candidate models, and even for moderate $k$ this could be a formidable number.

There are two mirroring procedures, forward selection and backward selection, that are routinely employed in cases where checking all possible models is infeasible. Forward selection proceeds in the following way.

**STEP 1.** Choose the predictor that has the largest $R^2$ among the models with a single variable. Call this variable $x_1$.

**STEP 2.** Assume that the model already has $k$ variables, $x_1, \ldots, x_k$, for some $k \geq 1$. Select the variable $x_{k+1}$ that gives the maximal increase to $R^2$ and refit the model.

**STEP 3.** Denote by $SSR(x_1, \ldots, x_k)$ the regression sum of squares for a regression fitted with variables $x_1, \ldots, x_k$. Then $R(x_{k+1}|x_1, \ldots, x_k) = SSR(x_1, \ldots, x_{k+1}) - SSR(x_1, \ldots, x_k)$ is the contribution of the $(k+1)$st variable and it is considered significant if

$$R(x_{k+1}|x_1, \ldots, x_k)/MSE > F_{1, n-k-1, \alpha}. \tag{16.5}$$

If relation (16.5) is satisfied, then variable $x_{k+1}$ is included in the model. Increase $k$ by one and go to **STEP 2**.

If relation (16.5) is not satisfied, then the contribution of $x_{k+1}$ is not significant, in which case go to **STEP 4**.

**STEP 4.** Stop with the model that has $k$ variables. **END**

The $MSE$ in (16.5) was estimated from the full model. Note that the forward selection algorithm is "greedy" and chooses the single best improving variable at each step. This, of course, may not lead to the optimal model since in reality variable $x_1$, which is the best for one-variable models, may not be included in the best two-variable model.

Backward stepwise regression starts with the full model and removes variables with insignificant contributions to $R^2$. Seldom do these two approaches end with the same candidate model.

MATLAB's Statistics Toolbox has two functions for stepwise regression: stepwisefit, a function that proceeds automatically from a specified initial model and entrance/exit tolerances, and stepwise, an interactive tool that allows you to explore individual steps in a process.

An additional criterion for the goodness of a model is the Mallows $C_p$. This criterion evaluates a proposed model with $k$ variables and $p = k + 1$ parameters. The Mallows $C_p$ is calculated as

$$C_p = (n - p)\frac{s^2}{\hat{\sigma}^2} - n + 2p,$$

where $s^2$ is the $MSE$ of the candidate model and $\hat{\sigma}^2$ is an estimator of $\sigma^2$, usually taken to be the best available estimate. The $MSE$ of the full model is typically used as $\hat{\sigma}^2$.

A common misinterpretation is that in $C_p$, $p$ is referred to as the number of predictors instead of parameters. This is correct only for models without the intercept (or when 1 from the vector of ones in the design matrix is declared as a predictor).

Adequate models should have a small $C_p$ that is close to $p$. Typically, a plot of $C_p$ against $p$ for all models is made. The "southwesternmost" points close to the line $C_p = p$ correspond to adequate models. The $C_p$ criterion is also employed in forward and backward variable selection as a stopping rule.

**Bayesian Multiple Regression.** Next we revisit ⬛ fat.dat with some Bayesian analyses. We selected four competing models and compared them using the Laud–Ibrahim predictive criterion, LI. Models with smaller LI are favored.

Laud and Ibrahim (1995) argue that agreement of model-simulated predictions and original data should be used as a criterion for model selection. If for $y_i$ responses $\hat{y}_{i,new}$ are hypothetical replications according to the posterior predictive distribution of competing model parameters, then

$$\mathrm{LI} = \sum_{i=1}^{n} (\mathbb{E}\hat{y}_{i,new} - y_i)^2 + \mathbb{V}\mathrm{ar}\,(\hat{y}_{i,new})$$

measures the discrepancy between the observed and model-predicted data. A smaller LI is better. The file fat.odc performs a Laud–Ibrahim Bayesian model selection and prefers model #2 of the four models analyzed.

```
#fat.odc
model{
for(j in 1:N ){
# four competing models
mu[1, j] <- b1[1] + b1[2] *age[j] + b1[3]*wei[j] + b1[4]*hei[j] +
                b1[5]*adip[j] + b1[6]*neck[j] + b1[7]*chest[j] +
                b1[8]*abd[j] + b1[9]*hip[j] + b1[10]*thigh[j] +
                b1[11]*knee[j]+b1[12]*ankle[j]+b1[13]*biceps[j] +
                b1[14]*forea[j] + b1[15]*wrist[j]
mu[2, j] <- b2[1]+b2[2]*wei[j]+b2[3]*adip[j]+b2[4]*abd[j]
mu[3, j] <- b3[1]+b3[2]*adip[j]
mu[4, j] <- b4[1]*wei[j]+b4[2]*abd[j]+b4[3]*abd[j]+b4[4]*wrist[j]
            }
#LI - Laud-Ibrahim Predictive Criterion. LI-smaller-better
for(i in 1:4 ){
    tau[i] ~ dgamma(2,32)
    LI[i] <- sqrt( sum(D2[i,]) + pow(sd(broz.new[i,]),2))
# data sets 1-4 for different models
            for (j in 1:N) {
            broz2[i,j] <- broz[j]
            broz2[i,j]    ~ dnorm(mu[i,j],tau[i])
            broz.new[i,j] ~ dnorm(mu[i,j],tau[i])
            D2[i,j] <- pow(broz[j]-broz.new[i,j],2)
                }
        }
# Compare predictive criteria between models i and j
# Comp[i,j] is 1 when LI[i]<LI[j], i-th model better.
for (i in 1:3) { for (j in i+1:4)
                {Comp[i,j] <- step(LI[j]-LI[i])}}
# priors
    for (j in 1:15)  { b1[j] ~ dnorm(0,0.001)}
    for(j in 1:4)    { b2[j] ~ dnorm(0,0.001)
                        b4[j] ~ dnorm(0,0.001)}
    for(j in 1:2)    { b3[j] ~ dnorm(0,0.001)}
}

#DATA 1:  Load this first
```

```
list(N = 252)

# DATA2: Then load the variables

broz[] age[] wei[]  hei[]   ...  biceps[] forea[] wrist[]
12.6 23 154.25 67.75 ...     32.0   27.4   17.1
23.4 38.5 93.6 83.00 ...     30.5   28.9   18.2
...248 lines deleted...
25.3 72 190.75 70.50 ...     30.5   29.4   19.8
30.7 74 207.50 70.00 ...     33.7   30.0   20.9
END

#the line behind 'END' has to be empty

# INITS (initialize by loading tau's
#             and generating the rest
list(tau=c(1,1,1,1))
```

The output is given in the table below. Note that even though the posterior mean of LI[4] is smaller that that of LI[2], it is the posterior median that matters. Model #2 is more frequently selected as the best compared to model #4.

| | mean | sd | MC error | val2.5pc | median | val97.5pc | start | sample |
|---|---|---|---|---|---|---|---|---|
| LI[1] | 186.5 | 209.2 | 17.63 | 84.0 | 104.6 | 910.5 | 1001 | 20000 |
| LI[2] | 96.58 | 23.46 | 1.924 | 85.08 | 93.14 | 131.2 | 1001 | 20000 |
| LI[3] | 119.6 | 5.301 | 0.03587 | 109.4 | 119.5 | 130.2 | 1001 | 20000 |
| LI[4] | 94.3 | 4.221 | 0.03596 | 86.33 | 94.2 | 103.0 | 1001 | 20000 |
| Comp[1,2] | 0.2974 | 0.4571 | 0.02785 | 0.0 | 0.0 | 1.0 | 1001 | 20000 |
| Comp[1,3] | 0.5844 | 0.4928 | 0.03939 | 0.0 | 1.0 | 1.0 | 1001 | 20000 |
| Comp[1,4] | 0.3261 | 0.4688 | 0.03008 | 0.0 | 0.0 | 1.0 | 1001 | 20000 |
| Comp[2,3] | 0.9725 | 0.1637 | 0.01344 | 0.0 | 1.0 | 1.0 | 1001 | 20000 |
| Comp[2,4] | 0.5611 | 0.4963 | 0.01003 | 0.0 | 1.0 | 1.0 | 1001 | 20000 |
| Comp[3,4] | 5.0E-5 | 0.007071 | 4.98E-5 | 0.0 | 0.0 | 0.0 | 1001 | 20000 |

More comprehensive treatment of Bayesian approaches in linear regression can be found in Ntzoufras (2009).

## 16.5 Sample Size in Regression

The evaluation of power in a regression with $p-1$ variables and $p$ parameters (when an intercept is present) requires specification of a significance level and precision. Suppose that we want a power such that a total sample size of $n = 61$ would make $R^2 = 0.2$ significant for $\alpha = 0.05$ and the number of predictor variables $p-1 = 3$. Cohen's effect size here is defined as $f^2 = R^2/(1-R^2)$. Recall that values of $f^2 \approx 0.01$ correspond to small, $f^2 \approx 0.0625$ to medium, and $f^2 \approx 0.16$ to large effects. Note that from $f^2 = R^2/(1-R^2)$ one gets $R^2 = f^2/(1+f^2)$, which can be used to check the adequacy of the elicited/required effect size.

The power, similarly as in ANOVA, is found using the noncentral $F$-distribution,

$$1 - \beta = \mathbb{P}(F^{nc}(p-1, n-p, \lambda) > F^{-1}(1-\alpha, p-1, n-p)), \qquad (16.6)$$

where $\lambda = nf^2$ is the noncentrality parameter.

*Example 16.4.* For $p = 4$, $R^2 = 0.2$, that is, $f^2 = 0.25$ (an X-large effect size), and a sample size of $n = 61$, one gets $\lambda = 61 \times 0.25 = 15.25$, and a power of approx. 90%.

```
p=4; n=61; lam=15.25;
1-ncfcdf( finv(1-0.05, p-1, n-p), p-1, n-p, lam)
% ans =   0.9014
```

## 16.6 Linear Regression That Is Nonlinear in Predictors

In linear regression, "linear" concerns the parameters, not the predictors. For instance,

$$\frac{1}{y_i} = \beta_0 + \frac{\beta_1}{x_i} + \epsilon_i, \quad i = 1, \ldots, n$$

and

$$y_i = \epsilon_i \times \exp\{\beta_0 + \beta_1 x_{1i} + \beta_2 x_{2i}\}, \quad i = 1, \ldots, n$$

are examples of a linear regression. There are many functions where $x$ and $y$ can be linearized by an obvious transformation of $x$ or $y$ or both; however, one needs to be mindful that the normality and homoscedasticity of errors is often compromised. In such a case, fitting a regression is simply an optimization task without natural inferential support. Bayesian solutions involving MCMC are generally more manageable as regards inference (Bayes estimators, credible sets, predictions).

An example where errors are not affected by the transformation of variables is a polynomial relationship between $x$ and $y$ postulated as

$$y_i = \beta_0 + \beta_1 x_i + \beta_2 x_i^2 + \cdots + \beta_k x_i^k + \epsilon_i, \quad i = 1, \ldots, n,$$

which is in fact a linear regression. Simply, the $k$ predictors are $x_{1i} = x_i, x_{2i} = x_i^2, \ldots, x_{ki} = x_i^k$, and estimating the polynomial relationship is straightforward. The example below is a research problem in which a quadratic relationship is used.

*Example 16.5.* **Von Willebrand Factor.** Von Willebrand disease is a bleeding disorder caused by a defect or deficiency of a blood clotting protein called von Willebrand factor. This glue-like protein, produced by the cells that line blood vessel walls, interacts with blood cells called platelets to form a plug that prevents bleeding. In order to understand the differential bonding mechanics underlying von Willebrand-type bleeding disorders, researchers in Dr. Larry McIntire's lab at Georgia Tech studied the interactions between the wild-type platelet GPIba molecule (receptor) and wild-type von Willebrand Factor (ligand).

The mean stop time rolling parameter was calculated from frame-by-frame rolling velocity data collected at 250 frames per second. Mean stop time indicates the amount of time a cell spends stopped, so it is analogous to the bond lifetime. This parameter, being an indicator for how long a bond is stopped before the platelet moves again, can be used to assess the bond lifetime and off-rate (Yago et al., 2008).

For the purpose of exploring interactions between the force and the mean stop times, Ficoll 6% is added to increase the viscosity.

Data are courtesy of Dr. Leslie Coburn. The mat file coburn.mat contains the structure coburn with data vectors coburn.fxssy, where x = 0, 6 is a code for Ficoll absence/presence and y = 1, 2, 4, ..., 256 denotes the shear stress (in $dyn/cm^2$). For example, coburn.f0ss16 is a $243 \times 1$ vector of mean stop times obtained with no Ficoll, under a shear stress of 16 $dyn/cm^2$.

| Shear Stress | 2 | 4 | 8 | 16 | 32 | 64 | 128 | 256 |
|---|---|---|---|---|---|---|---|---|
| Shear Number | 1 | 2 | 3 | 4 | 5 | 6 | 7 | 8 |
| Mean Stop Time | f0ss2 | f0ss4 | f0ss8 | f0ss16 | f0ss32 | f0ss64 | f0ss128 | f0ss256 |
| Sample Size | 26 | 57 | 157 | 243 | 256 | 185 | 62 | 14 |

We fit a regression on the logarithm of mean stop time log2mst, with no Ficoll present, as a quadratic function of share stress ($dyn/cm^2$). This regression is linear (in parameters) with two predictors, shear and the squared shear, shear2=shear$^2$.

The regression fit is

$$\text{log2mst} = -6.2423 + 0.8532\ \text{shear} - 0.0978\ \text{shear2}.$$

Regression is significant ($F = 63.9650, p = 0$); however, its predictive power is rather weak, with $R^2 = 0.1137$.

Figure 16.10 is plotted by the script ◀ coburnreg.m.

```
%coburnreg.mat
load 'coburn.mat';
mst=[coburn.f0ss2;  coburn.f0ss4;  coburn.f0ss8;   coburn.f0ss16; ...
     coburn.f0ss32; coburn.f0ss64; coburn.f0ss128; coburn.f0ss256];

shearn = [1 * ones( 26,1); 2 * ones( 57,1); 3 * ones(157,1); ...
          4 * ones(243,1); 5 * ones(256,1); 6 * ones(185,1); ...
          7 * ones( 62,1); 8 * ones( 14,1)];

shearn2 = shearn.^2; %quadratic term
```

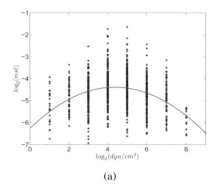 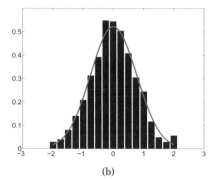

(a)　　　　　　　　　　　　　　　　　　(b)

**Fig. 16.10** (a) Quadratic regression on log mean stop time. (b) Residuals fitted with normal density.

```
%design matrix
X = [ones(length(shearn),1) shearn  shearn2];
[b,bint,res,resint,stats] = regress(log2(mst), X) ;
        %b0=-6.2423,  b1=0.8532,  b2 =-0.0978
stats   %R2,       F,            p,           sigma2
        %0.1137   63.9650    0.0000       0.5856
```

## 16.7 Errors-In-Variables Linear Regression*

Assume that in the context of regression both responses $Y$ and covariates $X$ are measured with error. This is a frequent scenario in research labs in which it would be inappropriate to apply standard linear regression, which assumes that covariates are designed and constant.

This scenario in which covariates are observed with error is called errors-in-variables (EIV) linear regression. There are several formulations for EIV regression (Fuller, 2006). For pairs from a bivariate normal distribution $(x_i, y_i)$, $i = 1,\ldots,n$, the EIV regression model is

$$y_i \sim \mathcal{N}(\beta_0 + \beta_1 \xi_i, \sigma_y^2)$$
$$x_i \sim \mathcal{N}(\xi_i, \sigma_x^2).$$

In an equivalent form, the regression is $\mathbb{E}y_i = \beta_0 + \beta_1 \mathbb{E}x_i = \beta_0 + \beta_1 \xi_i$, $i = 1,\ldots,n$, and the inference on parameters $\beta_0$ and $\beta_1$ is made conditionally on $\xi_i$. To make the model identifiable, a parameter $\eta = \frac{\sigma_x^2}{\sigma_y^2}$ is assumed known. Note that we do not need to know individual variances, just their ratio.

If the observations $(x_i, y_i)$, $i = 1,\ldots,n$ produce sums of squares $S_{xx} = \sum_i (x_i - \bar{x})^2$, $S_{yy} = \sum_i (y_i - \bar{y})^2$, and $S_{xy} = \sum_i (x_i - \bar{x})(y_i - \bar{y})$, then

$$\hat{\beta}_1 = \frac{-(S_{xx} - \eta S_{yy}) + \sqrt{(S_{xx} - \eta S_{yy})^2 + 4\eta S_{xy}^2}}{2\eta S_{xy}}, \tag{16.7}$$

$$\hat{\beta}_0 = \overline{y} - \hat{\beta}_1 \overline{x}.$$

The estimators of the errors are

$$\hat{\sigma}_x^2 = \frac{1}{2n} \times \frac{\eta}{1 + \eta \hat{\beta}_1^2} \sum_{i=1}^{n} \left(y_i - (\hat{\beta}_0 + \hat{\beta}_1 x_i)\right)^2$$

$$\hat{\sigma}_y^2 = \frac{\hat{\sigma}_x^2}{\eta}.$$

When $\eta = 1$ (variances of errors are the same), the solution in (16.7) coincides with numerical least-squares minimization, while for $\eta = 0$ (no errors in covariates, $\sigma_x^2 = 0$), we are back to the standard regression where $\hat{\beta}_1 = \frac{S_{xy}}{S_{xx}}$ (Exercise 16.16).

*Example 16.6.* Griffin et al. (1945) reported plasma volume (in $cc$) and circulating albumin (in $gm$) for $n = 58$ healthy males. Both quantities were measured with error and it was assumed that the variance of plasma measurement exceeded the variance of circulating albumin by 200 times. Using EIV regression, establish an equation that would be used to predict circulating albumin from plasma volume. The data are given in ⬛ circalbumin.dat, where the first column contains plasma volume and the second is albumin. The script ◀ errorinvar.m calculates the following EIV regression equation: $\mathbb{E}y_i = 0.0521 \cdot \mathbb{E}x_i - 13.1619$. This straight line is plotted in black in Fig. 16.11. For comparison, the standard regression is $\mathbb{E}y_i = 0.0494x_i - 5.7871$, and it is plotted in red. Parameter $\eta$ was set to 200, and the variances are estimated as $s_x^2 = 4852.8$ and $s_y^2 = 24.2641$.
✎

## 16.8 Analysis of Covariance

Analysis of covariance (ANCOVA) is a linear model that includes two types of predictors: quantitative (like the regression) and categorical (like the ANOVA). It can be formulated in quite general terms, but we will discuss the case of a single predictor of each kind.

The quantitative variable $x$ is linearly connected with the response, and for a fixed categorical variable the problem is exactly the regression. On the

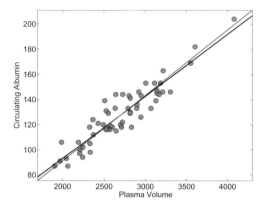

**Fig. 16.11** EIV regression (*black*) and standard regression (*red*) for `circalbumin` data.

other hand, for a fixed variable $x$, the model is ANOVA with treatments/groups defined by the categorical variable.

The rationale behind the merging of the two models is that in a range of experiments, modeling the problem as regression only or as ANOVA only may be inadequate, and that introducing a quantitative covariate to ANOVA, or equivalently groups/treatments to regression, may better account for the variability in data and produce better modeling and prediction results.

We will analyze two illustrative examples. In the first example, which will be solved in MATLAB, the efficacies of two drugs for lowering blood pressure are compared by analyzing the drop in blood pressure after taking the drug. However, the initial blood pressure measured before the drug is taken should be taken into account since a drop of 50, for example, is not the same if the initial pressure was 90 as opposed to 180.

In the second example, which will be solved in WinBUGS, the measured response is the strength of synthetic fiber and the two covariates are the fiber's diameter (quantitative) and the machine on which the fiber was produced (categorical with three levels).

We assume the model

$$y_{ij} = \mu + \alpha_i + \beta(x_{ij} - \overline{x}) + \epsilon_{ij}, \; i = 1, \ldots, a; \; j = 1, \ldots, n, \qquad (16.8)$$

where $a$ is the number of levels/treatments, $n$ is a common sample size within each level, and $\overline{x} = \frac{1}{an} \sum_{i,j} x_{ij}$ is the overall mean of $x$s. The errors $\epsilon_i$ are assumed independent normal with mean 0 and constant variance $\sigma^2$.

For practical reasons the covariates $x_{ij}$ are centered as $x_{ij} - \overline{x}$ since the expressions for the estimators will be more simple. Let $\overline{x}_i = \frac{1}{n} \sum_j x_{ij}$ be the $i$th treatment mean for the $x$s. The means $\overline{y}$ and $\overline{y}_i$ are defined analogously to $\overline{x}$ and $\overline{x}_i$.

In ANCOVA we calculate the sums of squares and mixed-product sums as

$$
\begin{aligned}
S_{xx} &= \sum_{i=1}^{a} \sum_{j=1}^{n} (x_{ij} - \bar{x})^2 \\
S_{xy} &= \sum_{i=1}^{a} \sum_{j=1}^{n} (x_{ij} - \bar{x})(y_{ij} - \bar{y}) \\
S_{yy} &= \sum_{i=1}^{a} \sum_{j=1}^{n} (y_{ij} - \bar{y})^2 \\
T_{xx} &= \sum_{i=1}^{a} (\bar{x}_i - \bar{x})^2 \\
T_{xy} &= \sum_{i=1}^{a} (\bar{x}_i - \bar{x})(\bar{y}_i - \bar{y}) \\
T_{yy} &= \sum_{i=1}^{a} (\bar{y}_i - \bar{y})^2 \\
Q_{xx} &= \sum_{i=1}^{a} \sum_{j=1}^{n} (x_{ij} - \bar{x}_i)^2 = S_{xx} - T_{xx} \\
Q_{xy} &= \sum_{i=1}^{a} \sum_{j=1}^{n} (x_{ij} - \bar{x}_i)(y_{ij} - \bar{y}_i) = S_{xy} - T_{xy} \\
Q_{yy} &= \sum_{i=1}^{a} \sum_{j=1}^{n} (y_{ij} - \bar{y}_i)^2 = S_{yy} - T_{yy}
\end{aligned}
$$

We are interested in finding estimators for the parameters in model (16.8), the common mean $\mu$, treatment effects $\alpha_i$, regression slope $\beta$, and the variance of the error $\sigma^2$.

The estimators are $\hat{\mu} = \bar{y}$, $b = \hat{\beta} = Q_{xy}/Q_{xx}$, and $\hat{\alpha}_i = \bar{y}_i - \bar{y} - b(\bar{x}_i - \bar{x})$.

The estimator of the variance, $\sigma^2$, is $s^2 = MSE = SSE/(a(n-1)-1)$, where $SSE = Q_{yy} - Q_{xy}^2/Q_{xx}$.

If there are no treatment effects, that is, if all $\alpha_i = 0$, then the model is a plain regression and

$$
y_{ij} = \mu + \beta(x_{ij} - \bar{x}) + \epsilon_{ij}, \quad i = 1, \ldots, a; j = 1, \ldots, n.
$$

In this reduced case the error sum of squares is $SSE' = S_{yy} - S_{xy}^2/S_{xx}$, with $an - 2$ degrees of freedom. Thus, the test $H_0 : \alpha_i = 0$ is based on an $F$-statistic,

$$
F = \frac{(SSE' - SSE)/(a-1)}{SSE/(a(n-1)-1)},
$$

that has an $F$-distribution with $a - 1$ and $a(n-1) - 1$ degrees of freedom.

The test for regression $H_0 : \beta = 0$ is based on the statistic

$$
F = \frac{Q_{xy}^2/Q_{xx}}{SSE/(a(n-1)-1)},
$$

which has an $F$-distribution with 1 and $a(n-1) - 1$ degrees of freedom.

Next we provide a MATLAB solution for a simple ANCOVA layout.

*Example 16.7.* Kodlin (1951) reported an experiment that compared two substances for lowering blood pressure, denoted as substances A and B. Two groups of animals are randomized to the two substances and a decrease in pressure is recorded. The initial pressure is recorded. The data for the blood pressure experiment appear in the table below. Compare the two substances by accounting for the possible effect of the initial pressure on the decrease in pressure. Discuss the results of the ANCOVA analysis and compare them with those obtained by ignoring the potential effect of the initial pressure.

| | Substance A | | | Substance B | |
|---|---|---|---|---|---|
| Animal | Decrease | Initial | Animal | Decrease | Initial |
| 1 | 45 | 135 | 11 | 34 | 90 |
| 2 | 45 | 125 | 12 | 55 | 135 |
| 3 | 20 | 125 | 13 | 50 | 130 |
| 4 | 50 | 130 | 14 | 45 | 115 |
| 5 | 25 | 105 | 15 | 30 | 110 |
| 6 | 37 | 130 | 16 | 45 | 140 |
| 7 | 50 | 140 | 17 | 45 | 130 |
| 8 | 20 | 93 | 18 | 23 | 95 |
| 9 | 25 | 110 | 19 | 40 | 90 |
| 10 | 15 | 100 | 20 | 35 | 105 |

We will use $\alpha = 0.05$ for all significance assessments.

```
%codlin.m
%x = Initial; y = Decrease; g=1 for 'A', g=2 for 'B'
x = [135 125 125 130 105 130 140  93 110 100 ...
      90 135 130 115 110 140 130  95  90 105]);
y = [45 45 20 50 25 37 50 20 25 15 ...
     34 55 50 45 30 45 45 23 40 35];
g = [1 1 1 1 1 1 1 1 1 1 2 2 2 2 2 2 2 2 2 2];

a = 2; n = 10;

x1=x(g==1);   x2=x(g==2);
y1=y(g==1);   y2=y(g==2);
x1b=mean(x1); x2b=mean(x2);
y1b=mean(y1); y2b=mean(y2);
xb = mean(x); yb = mean(y);

SXX = sum( (x - xb).^2 );
SXY = sum( (x - xb).* (y - yb) );
SYY = sum( (y - yb).^2 );

QXX = sum( (x1-x1b).^2 + (x2-x2b).^2  );
QXY = sum( (x1-x1b).* (y1 - y1b) + ...
    (x2-x2b).*(y2 - y2b)  );
QYY = sum( (y1-y1b).^2 + (y2-y2b).^2  );
% estimators of model parameters mu, alpha_i, beta
mu = yb          %mu = 36.7
b = QXY/QXX      %b = 0.5175
alpha = [y1b y2b] - [yb yb] - b*([x1b x2b]-[xb xb])
                 %alpha =[-4.8714    4.8714]
TXX = SXX - QXX;
TXY = SXY - QXY;
TYY = SYY - QYY;
```

```
SSE = QYY - QXY^2/QXX;
MSE = SSE/(a * (n-1) - 1 )                    %MSE = 60.6542
SSEp = SYY - SXY^2/SXX;

% F-test for testing $H_0: alpha_i = 0 (all i)
F1 = ((SSEp - SSE)/(a-1)) /MSE           %F1 = 7.6321
pvalF1 = 1- fcdf(F1,a-1, a*(n-1) - 1) %pvalF1=0.0133

% F-test for testing $H_0: beta = 0
F2 = (QXY^2/QXX)/MSE     %F2 = 24.5668
pvalF2 = 1 - fcdf(F2, 1, a * (n-1)-1 )
                        %pvalF2 = 0.00012
```

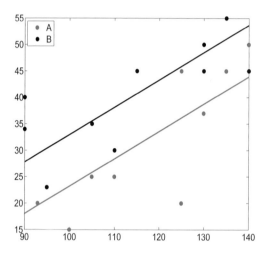

**Fig. 16.12** Scatterplot of decrease ($y$) against the initial value ($x$) for substances $A$ and $B$.

Note that both null hypotheses are found to be significant, but the testing of the equality of treatments seems to be the more important hypothesis in this example. The conclusion is that substances $A$ and $B$ are significantly different at level $\alpha = 0.0133$. Figure 16.12 plots the decrease ($y$) against the initial value ($x$) for substances $A$ and $B$. The ANCOVA fits $\hat{y} = 36.7 \pm 4.8714 + 0.5175 \cdot (x - 116.65)$ are superimposed.

If we conducted the ANOVA test by ignoring the covariates, the test would find no differences among the drugs ($p$-value = 0.2025). Likewise, if both treatments were lumped together and the response regressed on $x$, the regression would be significant but with a $p$-value more than eight times larger than in ANCOVA.

```
oX = [ones(size(x'))  x'];
[b, bint, r, rint, stats] = regress(y', oX);
stats    % 0.4599   15.3268    0.0010   83.0023
[p, table, statan] = anova1(y, g);
p          %0.2025
```

This example shows how important accounting for sensible predictors is and how selecting the right statistical procedure is critical in decision making.

MATLAB has a built-in function aoctool that, when invoked as aoctool(x,y,g), opens a front-end suite with various extended modeling and graphing capabilities. The output stats from aoctool can be imported into multcompare for subsequent pairwise comparison analysis.

The following example is solved in a Bayesian manner.

*Example 16.8.* **ANCOVA Fibers.** Three machines produce monofilament synthetic fiber for medical use (surgery, implants, devices, etc.). The measured response is the strength $y$ (in pounds, lb.) and a covariate is the diameter $x$ (in inches/1000).

```
#ancovafibers.odc
model{
for (i in 1:ntotal){
y[i] ~ dnorm( mui[i], tau )
mui[i] <-  mu + alpha[g[i]] + beta1 *(x[i] - mean(x[]))
}
#alpha[1] <- 0.0;      #CR constraints
alpha[1] <- -sum( alpha[2:a] );#STZ constraints

mu ~ dnorm(0, 0.0001)
beta1 ~ dnorm(0, 0.0001)
beta0 <- mu - beta1 * mean(x[])
alpha[2] ~ dnorm(0, 0.0001)
alpha[3] ~ dnorm(0, 0.0001)
tau ~ dgamma(0.001, 0.001)
var <- 1/tau
}

DATA
list(ntotal=15, a=3,
y = c(36, 41, 39, 42, 49,
      40, 48, 39, 45, 44,
      35, 37, 42, 34, 32),
x = c(20, 25, 24, 25, 32,
      22, 28, 22, 30, 28,
      21, 23, 26, 21, 15),
g = c( 1,  1,  1,  1,  1,
       2,  2,  2,  2,  2,
       3,  3,  3,  3,  3) )

INITS
list(mu=1, alpha=c(NA, 0, 0), beta1=0, tau=1)
```

The output from WinBUGS is

| | mean | sd | MC error | val2.5pc | median | val97.5pc | start | sample |
|---|---|---|---|---|---|---|---|---|
| alpha[1] | 0.182 | 0.6618 | 0.001848 | −1.129 | 0.1833 | 1.494 | 10001 | 100000 |
| alpha[2] | 1.218 | 0.6888 | 0.003279 | −0.1576 | 1.217 | 2.596 | 10001 | 100000 |
| alpha[3] | −1.4 | 0.743 | 0.003757 | −2.876 | −1.399 | 0.07178 | 10001 | 100000 |
| beta0 | 17.18 | 3.086 | 0.01394 | 11.01 | 17.18 | 23.31 | 10001 | 100000 |
| beta1 | 0.9547 | 0.1266 | 5.634E-4 | 0.7019 | 0.9543 | 1.207 | 10001 | 100000 |
| mu | 40.2 | 0.4564 | 0.001372 | 39.29 | 40.2 | 41.11 | 10001 | 100000 |
| var | 3.118 | 1.671 | 0.007652 | 1.276 | 2.716 | 7.348 | 10001 | 100000 |

The regression equations corresponding to the three treatments (machines) are

$$\hat{y}_{1i} = 17.18 + 0.182 + 0.9547 \cdot x_i,$$
$$\hat{y}_{2i} = 17.18 + 1.218 + 0.9547 \cdot x_i, \text{ and}$$
$$\hat{y}_{3i} = 17.18 - 1.4 + 0.9547 \cdot x_i.$$

Note that all three 95% credible sets for alpha contain 0, while the credible set for the slope beta1 does not contain 0. This analysis of credible sets is not a formal Bayesian testing, but agrees with the output from

```
y = [36 41 39 42 49    40 48 39 45 44    35 37 42 34 32];
x = [20 25 24 25 32    22 28 22 30 28    21 23 26 21 15];
g = [1  1   1  1  1     2  2  2  2  2     3  3  3  3  3];
aoctool(x, y, g) %parallel lines option
```

where the $p$-value corresponding to ANOVA part is 0.11808. The test for the regression slope is significant with $p$-value of 4.2645e-06.

✐

## 16.9 Exercises

16.1. **Regression with Three Points.** Points $(x_1, y_1) = (1, 1)$, $(x_2, y_2) = (2, 2)$, and $(x_3, y_3) = (3, 2)$ are given.
(a) Find (by hand calculation) the regression line that best fits the points, and sketch a scatterplot of points with the superimposed fit.
(b) What are the "predictions" at 1, 2, and 3, i.e, $\hat{y}_1$, $\hat{y}_2$, and $\hat{y}_3$?
(c) What are $SST$, $SSR$, and $SSE$?
(d) Find an estimator of variance $\hat{\sigma}^2$.

16.2. **Age and IVF Success Rate.** The highly publicized (recent TV reports) in vitro fertilization (IVF) success cases for women in their late fifties all involve donors' eggs. If the egg is the woman's own, the story is quite different.
IVF, an assisted reproductive technology (ART) procedure, involves extracting a woman's eggs, fertilizing the eggs in the laboratory, and then transferring the resulting embryos to the woman's uterus through the cervix.

Fertilization involves a specialized technique known as intracytoplasmic sperm injection (ICSI).

The table below shows the live-birth success rate per transfer rate from a woman's own eggs, by age of recipient. The data are for the year 1999, published by the CDC at http://www.cdc.gov/art/ARTReports.htm.

| Age ($x$) | 24 | 25 | 26 | 27 | 28 | 29 | 30 | 31 | 32 | 33 | 34 | 35 |
|---|---|---|---|---|---|---|---|---|---|---|---|---|
| Percentage ($y$) | 38.7 | 38.6 | 38.9 | 41.4 | 39.7 | 41.1 | 38.7 | 37.6 | 36.3 | 36.9 | 35.7 | 33.8 |

| Age ($x$) | 36 | 37 | 38 | 39 | 40 | 41 | 42 | 43 | 44 | 45 | 46 |
|---|---|---|---|---|---|---|---|---|---|---|---|
| Percentage ($y$) | 33.2 | 30.1 | 27.8 | 22.7 | 21.3 | 15.4 | 11.2 | 9.2 | 5.4 | 3.0 | 1.6 |

Select the age in the range 33–46 since it shows an almost linear decay of the success rate (Fig. 16.13). Fit the linear model $\hat{y} = b_0 + b_1 x$. Would the quadratic relationship $\hat{y} = b_0 + b_1 x + b_2 x^2$ be more appropriate?

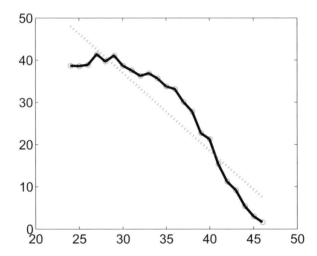

**Fig. 16.13** Success rate versus age.

16.3. **Sharp Dissection and Severity of Postoperative Adhesions.** Postoperative adhesions are formed after surgical cardiac and great vessel procedures as part of the healing process. Scar tissue makes the reentry complex and increases the rate of iatrogenic lesions. Currently, as reoperations are needed in 10 to 20% of heart surgeries, various methods have been investigated to prevent or decrease the severity of postoperative adhesions.

The surgical time spent in the adhesiolysis procedure and amounts of sharp dissection are informative summaries to to predict the severity of pericardial adhesions, as reported by Lopes et al. (2009). For example, the authors reported a linear relationship between the logarithm of the amount of sharp dissection `lasd` and severity score `sesco` assessed by a standard categorization. Use the data in MATLAB format to answer (a)–(e).

```
shdiss = [14   108   39   311    24   112   104 ...
           382    42   74    67   145    21    93 ...
            75   381   36    36    73   239    35   69];
lasd = log(shdiss);

sesco = [6    12    9   18    7   12   12 ...
          17     8   11   14   15    7   12 ...
          13    18    9    9   10   16    7   10];
```

(a) Write down the linear relationship between `lasd` and `sesco`.
(b) What is $R^2$ here and what does it represent?
(c) Test the hypothesis $H_0 : \beta_0 = 0$ versus the alternative $H_1 : \beta_0 < 0$. Use $\alpha = 0.05$. The critical cut points are provided at the back of the problem, or, alternatively, you can report the $p$-value.
(d) Find a 95% confidence interval for the population intercept $\beta_1$.
(e) For `lasd=4` predict the severity score. Find a 99% confidence interval for the mean response.

16.4. **Kanamycin Levels in Premature Babies.** Miller (1980) describes a project and provides data on assessing the precision of noninvasive measuring of kanamycin concentration in neonates.

Premature babies are susceptible to infections, and kanamycin (an aminoglycoside) is used for the treatment of sepsis. Because kanamycin is ineffective at low doses and potentially harmful at high doses, it is necessary to constantly monitor its levels in a premature baby's body during treatment. The standard procedure for measuring serum kanamycin levels is to take blood samples from a heel. Unfortunately, due to frequent blood sampling, neonates are left with badly bruised heels.

Kanamycin is routinely administered through an umbilical catheter. An alternative procedure for measuring serum kanamycin would be to reverse the flow in the catheter and draw a blood sample from it. The concern about this noninvasive method is that the blood drawn from the point close to an infusion may have an elevated level of kanamycin compared to blood samples from more distant points in the body.

In a carefully designed experimental setup, blood samples from 20 babies were obtained simultaneously from an umbilical catheter and a heel venapuncture (using heelstick). If the agreement is satisfactory, physicians would be willing to use the catheter values instead of heelstick values. Here are the data:

| Baby | Heelstick | Catheter | Baby | Heelstick | Catheter |
|------|-----------|----------|------|-----------|----------|
| 1    | 23.0      | 25.2     | 11   | 26.4      | 24.8     |
| 2    | 33.2      | 26.0     | 12   | 21.8      | 26.8     |
| 3    | 16.6      | 16.3     | 13   | 14.9      | 15.4     |
| 4    | 26.3      | 27.2     | 14   | 17.4      | 14.9     |
| 5    | 20.0      | 23.2     | 15   | 20.0      | 18.1     |
| 6    | 20.0      | 18.1     | 16   | 13.2      | 16.3     |
| 7    | 20.6      | 22.2     | 17   | 28.4      | 31.3     |
| 8    | 18.9      | 17.2     | 18   | 25.9      | 31.2     |
| 9    | 17.8      | 18.8     | 19   | 18.9      | 18.0     |
| 10   | 20.0      | 16.4     | 20   | 13.8      | 15.6     |

(a) Model the Heelstick responses with Catheter as predictor in a linear regression.

(b) Are there any unusual observations? Does regression improve when unusual observations are removed from the analysis?

(c) Test $H_0 : \beta_1 = 1$ versus $H_1 : \beta_1 < 1$ at the level $\alpha = 0.05$.

(d) Find a 97% confidence interval for the population intercept.

(e) Find 95% confidence and prediction intervals for the regression response when cath = 20.

(f) Using WinBUGS, estimate the parameters in a Bayesian regression with noninformative priors. Compare Bayesian and the least-squares solutions.

16.5. **Degradation of Scaffolds.** In an experiment conducted at Georgia Tech/ Emory Center for the Engineering of Living Tissues the goal was to find a suitable biomechanical replacement for cartilage, better known as tissue engineered cartilage. There are many factors (dimensional or mechanical) at which the cartilage scaffold is tested to assess whether it is a viable replacement. One of the problems is the degradation of scaffolds as the tissue grows, which affects all of the experimental metrics. The experimental data collected comprise a tissue growth experiment in which no cells were added, thus approximating the degradation of the scaffold over a sequence of 8 days. The dynamic shear summaries capture two physical phenomena, the modulus or the construct's ability to resist deformation under load and the frequency at which the modulus was evaluated. This modulus provides a measure of the extent of interconnectivity within the fibrous scaffold. The table below contains moduli, in 1000s, for the frequency $f = 1$ over 8 days, each day represented by three independent measurements.

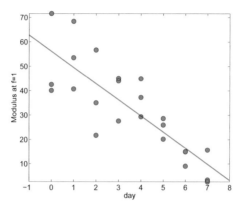

**Fig. 16.14** The modulus against time for frequency $f = 1$.

| Day | Mod at $f = 1$ | Day | Mod at $f = 1$ |
|-----|----------------|-----|----------------|
| 0 | 42.520 | 4 | 44.929 |
| 0 | 71.590 | 4 | 29.348 |
| 0 | 40.063 | 4 | 37.259 |
| 1 | 68.397 | 5 | 28.625 |
| 1 | 53.527 | 5 | 25.956 |
| 1 | 40.676 | 5 | 20.179 |
| 2 | 21.724 | 6 | 14.994 |
| 2 | 35.032 | 6 | 9.051 |
| 2 | 56.687 | 6 | 14.923 |
| 3 | 44.029 | 7 | 2.692 |
| 3 | 45.058 | 7 | 15.688 |
| 3 | 27.579 | 7 | 3.420 |

The data and the code are available in the file ◀ degradation.m.

(a) Use the moduli for frequency $f = 1$. Write down the linear regression model: $mod1 = b_0 + b_1 \cdot day$, where $b_0$ and $b_1$ are estimators of the population intercept and slope (Fig. 16.14). What is $R^2$ for your regression?

(b) Test the hypothesis that the population intercept is equal to 100 versus the alternative that it is smaller than 100.

(c) Find a 96% confidence interval for the population slope.

(d) For day = 5.5 find the prediction of the modulus. What is the standard deviation for this predicted value?

16.6. **Glucosis in Lactococcus Lactis.** The data set ▣ Lactis.dat is courtesy of Dr. Eberhard Voit at the Georgia Institute of Technology and are an excerpt from a larger collection of data dealing with glycolysis in the bacterium *Lactococcus lactis* MG1363 (which is involved in essentially all yogurts, cheeses, etc.). The experiment was conducted in aerobic conditions with cell suspension in a 50-mM KPi buffer with a pH of 6.5, and 20 mM [6-13C] glucose. There are four columns in the file ▣ Lactis.dat: time, glycose, lactate, and acetate.

The levels of extracelular metabolites lactate and acetate are monitored in time. After time $t = 5$ the level of lactate stabilizes around 30 while the level of acetate shows linearly increasing trend in time.

(a) Plot lactate level against time. Take a subset of lactate levels for times $> 5$ (86 observations) and find basic descriptive statistics.

(b) Check the normality of the subset data. Test the hypothesis that the mean lactate level for $t > 5$ is 30 against the two-sided alternative.

(c) Select acetate levels for $t > 5$. Fit linear relationship of acetate level against time and show that the linear model is justified by providing and discussing ANOVA table.

16.7. **Weight and Latency in Rats.** Data consisting of rat body weight (grams) and latency to seizure (minutes) are given for 15 rats (adapted from Kleinbaum et al., 1987)

| Rat # | Weight | Latency | Rat # | Weight | Latency | Rat # | Weight | Latency |
|-------|--------|---------|-------|--------|---------|-------|--------|---------|
| 1 | 348 | 1.80 | 2 | 372 | 1.95 | 3 | 378 | 2.90 |
| 4 | 390 | 2.30 | 5 | 392 | 1.10 | 6 | 395 | 2.50 |
| 7 | 400 | 1.30 | 8 | 409 | 2.00 | 9 | 413 | 1.70 |
| 10 | 415 | 2.00 | 11 | 423 | 2.95 | 12 | 428 | 2.25 |
| 13 | 464 | 3.05 | 14 | 468 | 3.70 | 15 | 470 | 3.62 |

(a) Test the hypothesis $H_0 : \beta_0 = 0$ against the two-sided alternative. Use $\alpha = 0.05$.

(b) Test the hypothesis $H_0 : \beta_1 = 0.02$ against the alternative $H_0 : \beta_1 < 0.02$. Use $\alpha = 0.05$.

(c) Find a 95% confidence interval for the slope $\beta_1$.

(d) For weight $wei = 410$ find the mean latency response, $\hat{y}_m$. Test the hypothesis $H_0 : \hat{y}_m = 3$ versus the alternative $H_1 : \hat{y}_m < 3$. Test the same hypothesis for the predicted response $\hat{y}_{pred}$. In both tests use $\alpha = 0.05$.

16.8. **Rinderpest Virus in Rabbits.** Temperatures (temp) were recorded in a rabbit at various times (time) after the rabbit was inoculated with rinderpest virus (Carter and Mitchell, 1958). Rinderpest (RP) is an infectious viral disease of cattle, domestic buffalo, and some species of wildlife; it is commonly referred to as cattle plague. It is characterized by fever, oral erosions, diarrhea, lymphoid necrosis, and high mortality. In German, *Rinderpest* means cattle plague.

| Time after injection | Temperature |
|:---:|:---:|
| (time in hrs) | (temp in $^\circ$ F) |
| 24 | 102.8 |
| 32 | 104.5 |
| 48 | 106.5 |
| 56 | 107.0 |
| 70 | 105.1 |
| 72 | 103.9 |
| 80 | 103.2 |
| 96 | 102.1 |

(a) Demonstrate that a linear regression with one predictor (time) gives an insignificant $F$-statistic and a low $R^2$.

(b) Include time2 (squared time) as the second predictor, making the regression quadratic in variables (but still linear in coefficients). Show that this regression is significant and has a large $R^2$.

(c) Find the 90% confidence interval for the coefficient of the quadratic term and test the hypothesis that the intercept is equal to 100 versus the one-sided alternative that it is smaller than 100. Use $\alpha = 0.05$.

16.9. **Hemodilution.** Clark et al. (1975) examined the fat filtration characteristics of a packed polyester-and-wool filter used in arterial lines during clinical hemodilution. They collected data on the filter's recovery of solids for ten patients who underwent surgery. The table below shows removal rates of lipids and cholesterol. Fit a regression line to the data, with cholesterol as the response variable and lipids as the covariate.

| | Removal rates, mg/kg/L $\cdot 10^{-2}$ | |
|:---:|:---:|:---:|
| Patient | Lipids $(x)$ | Cholesterol $(y)$ |
| 1 | 3.81 | 1.90 |
| 2 | 2.10 | 1.03 |
| 3 | 0.79 | 0.44 |
| 4 | 1.99 | 1.18 |
| 5 | 1.03 | 0.62 |
| 6 | 2.07 | 1.29 |
| 7 | 0.74 | 0.39 |
| 8 | 3.88 | 2.30 |
| 9 | 1.43 | 0.93 |
| 10 | 0.41 | 0.29 |

Give answers to the following questions using standard regression output.

(a) What test is performed by a given $p$-value? State $H_0$ and $H_1$.

(b) Find the 95% confidence interval for the intercept of the population regression line $(\beta_0)$.

(c) Test the hypothesis that the slope of the population regression line is equal to 1 versus the one-sided alternative that it is less than 1.

(d) Discuss the adequacy of the proposed linear fit. Predict the cholesterol rate for lipids at the level 1.65 mg/kg/L$\times 10^{-2}$. Find the 95% confidence and prediction intervals.

16.10. **Anscombe's Data Sets.** A celebrated classic example of the role of residual analysis and statistical graphics in statistical modeling was created by Anscombe (1973). He constructed four quite different data sets $(x_i, y_i)$, $i = 1, \ldots, 11$ that share the descriptive statistics necessary to establish a linear regression fit $\hat{y} = \hat{\beta}_0 + \hat{\beta}_1 x$.

A linear model is appropriate for Data Set 1; the scatterplots and residual analysis suggest that Data Sets 2–4 seem not to be amenable to linear modeling.

| | | | | | | Set 1 | | | | | |
|---|---|---|---|---|---|---|---|---|---|---|---|
| $x$ | 10 | 8 | 13 | 9 | 11 | 14 | 6 | 4 | 12 | 7 | 5 |
| $y$ | 8.04 | 6.95 | 7.58 | 8.81 | 8.33 | 9.96 | 7.24 | 4.26 | 10.84 | 4.82 | 5.68 |
| | | | | | | Set 2 | | | | | |
| $x$ | 10 | 8 | 13 | 9 | 11 | 14 | 6 | 4 | 12 | 7 | 5 |
| $y$ | 9.14 | 8.14 | 8.74 | 8.77 | 9.26 | 8.10 | 6.13 | 3.10 | 9.13 | 7.26 | 4.74 |
| | | | | | | Set 3 | | | | | |
| $x$ | 10 | 8 | 13 | 9 | 11 | 14 | 6 | 4 | 12 | 7 | 5 |
| $y$ | 7.46 | 6.77 | 12.74 | 7.11 | 7.81 | 8.84 | 6.08 | 5.39 | 8.15 | 6.42 | 5.73 |
| | | | | | | Set 4 | | | | | |
| $x$ | 8 | 8 | 8 | 8 | 8 | 8 | 8 | 19 | 8 | 8 | 8 |
| $y$ | 6.58 | 5.76 | 7.71 | 8.84 | 8.47 | 7.04 | 5.25 | 12.50 | 5.56 | 7.91 | 6.89 |

(a) Using MATLAB fit the regression line for the four sets and provide four ANOVA tables. What statistics are the same? What statistics differ?

(b) Plot the residuals against the fitted values. Discuss the appropriateness of the regressions.

16.11. **Potato Leafhopper.** The length of the developmental period (in days) of the potato leafhopper, *Empoasca fabae*, from egg to adult seems to be dependent on temperature (Kouskolekas and Decker, 1966). The original data were weighted means, but for the purpose of this analysis we shall consider them as though they were single observed values.

| Temp $^\circ$ F | 59.8 | 67.6 | 70.0 | 70.4 | 74.0 | 75.3 | 78.0 | 80.4 | 81.4 | 83.2 | 88.4 | 91.4 | 92.5 |
|---|---|---|---|---|---|---|---|---|---|---|---|---|---|
| Development (days) | 58.1 | 27.3 | 26.8 | 26.3 | 19.1 | 19.0 | 16.5 | 15.9 | 14.8 | 14.2 | 14.4 | 14.6 | 15.3 |

(a) Find a 98% confidence interval for unknown slope $\beta_1$.

(b) Test the hypothesis that the intercept is equal to 60 against the alternative that it is larger than 60. Take $\alpha = 0.01$.

(c) What is the 96% confidence interval for the mean response (mean number of days) if the temperature is 85°F?

16.12. **Cross-validating a Bayesian Regression.** In this simulated exercise covariates $x_1$ and $x_2$ are generated as

```
x1 = rand(1, 40);   x2 = floor(10 * rand(1,40)) + 1;
```

and the response variable $y$ is obtained as

```
y = 2 + 6 * x1 - 0.5 * x2 + 0.8*randn(size(x1))
```

Write a WinBUGS program that selects 20 triples $(x_1, x_2, y)$ to train the linear regression model $\hat{y} = b_0 + b_1 x_1 + b_2 x_2$ and then uses the remaining 20 triples to evaluate the model by comparing the original responses $y_i, i = 21, \ldots, 40$ with regression-predicted values $\hat{y}_i, i = 21, \ldots, 40$. The comparison involves calculating the MSE, the mean of $(y_i - \hat{y}_i)^2$, $i = 21, \ldots, 40$.
This is an example of how a cross-validation methodology is often employed to assess statistical models.
How do the Bayesian estimators of $\beta_0$, $\beta_1$, $\beta_2$, and $\sigma$ compare to the "true" values 2, 6, −0.5, and 0.8?

16.13. **Taste of Cheese.** As cheddar cheese matures, a variety of chemical processes take place. The taste of mature cheese is related to the concentration of several chemicals in the final product. In a study of cheddar cheese from LaTrobe Valley of Victoria, Australia, samples of cheese were analyzed for their chemical composition and were subjected to taste tests. The table below presents data [from the experiments of G. T. Lloyd and E. H. Ramshaw, CISRO Food Research, Victoria, Australia, analyzed in Moore and McCabe (2006)] for one type of cheese manufacturing process. *Taste* is the response variable of interest. The taste scores were obtained by combining scores from several tasters. Three of the chemicals whose concentrations were measured were *acetic acid, hydrogen sulfide,* and *lactic acid*. For acetic acid and hydrogen sulfide log transformations were taken.

| Taste | Acetic | H2S | Lactic | Taste | Acetic | H2S | Lactic |
|-------|--------|------|--------|-------|--------|-------|--------|
| 12.3 | 4.54 | 3.13 | 0.86 | 20.9 | 5.16 | 5.04 | 1.53 |
| 39.0 | 5.37 | 5.44 | 1.57 | 47.9 | 5.76 | 7.59 | 1.81 |
| 5.6 | 4.66 | 3.81 | 0.99 | 25.9 | 5.70 | 7.60 | 1.09 |
| 37.3 | 5.89 | 8.73 | 1.29 | 21.9 | 6.08 | 7.97 | 1.78 |
| 18.1 | 4.90 | 3.85 | 1.29 | 21.0 | 5.24 | 4.17 | 1.58 |
| 34.9 | 5.74 | 6.14 | 1.68 | 57.2 | 6.45 | 7.91 | 1.90 |
| 0.7 | 4.48 | 3.00 | 1.06 | 25.9 | 5.24 | 4.94 | 1.30 |
| 54.9 | 6.15 | 6.75 | 1.52 | 40.9 | 6.37 | 9.59 | 1.74 |
| 15.9 | 4.79 | 3.91 | 1.16 | 6.4 | 5.41 | 4.70 | 1.49 |
| 18.0 | 5.25 | 6.17 | 1.63 | 38.9 | 5.44 | 9.06 | 1.99 |
| 14.0 | 4.56 | 4.95 | 1.15 | 15.2 | 5.30 | 5.22 | 1.33 |
| 32.0 | 5.46 | 9.24 | 1.44 | 56.7 | 5.86 | 10.20 | 2.01 |
| 16.8 | 5.37 | 3.66 | 1.31 | 11.6 | 6.04 | 3.22 | 1.46 |
| 26.5 | 6.46 | 6.92 | 1.72 | 0.7 | 5.33 | 3.91 | 1.25 |
| 13.4 | 5.80 | 6.69 | 1.08 | 5.5 | 6.18 | 4.79 | 1.25 |

(a) Find the equation in multivariable linear regression that predicts Taste using Acetic, H2S, and Lactic as covariates.

(b) For Acetic = 5, H2S = 8, and Lactic = 2 estimate the regression response $\hat{Y}_h$ and find standard deviations for the mean and individual responses. [Ans. 43.37, 6.243, 11.886]

(c) Find the 98% confidence interval for the intercept $\beta_0$.

(d) Construct an ANOVA table.

(e) Find ordinary, studentized, and studentized deleted residuals for the observation $Y_8$.

(f) Find $DFFITS_8$ and $COOKSD_8$.

(g) Find $DFBETAS_8$ on the Lactic coefficient.

(h) Find and discuss the $VIF$.

16.14. **Slowing the Progression of Arthritis.** Arthritis is caused by the breakdown of collagen in joint cartilage by the enzyme MMP-13. The antibiotic doxycycline is a general inhibitor of MMPs and, by inhibiting the activity of MMP-13, is an effective method of slowing the progression of arthritis. At present, doxycycline is used to treat both rheumatiod arthritis and osteoarthritis. The rabbit's HIG-82 synovial cell line was used to model arthritis. MMP-13 was prepared by adding PMA, which guarantees that its presence, and APMA, which activates it. The enzyme MMP-13 was mixed with a quenched substrate. When the enzyme cleaves the substrate, it fluoresces. This fluorescence was used to measure the amount of MMP-13 activity. Doxycycline, which decreases the amount of enzyme activity, was added in increasing concentrations: 0, 25, 50, 75, 100, and 200 micromols. The decrease in the fluorescence produced by the cleaved substrate when doxycycline was present was used to measure the decrease in activity of the enzyme. The same experiment was performed on three different plates, and the data was normalized.

The data set 🖳 arthritis1.mat can be found on the book's Web page. The data file contains 72 observations (rows); the first column represents doxycycline concentration (0, 25, 50, 75, 100 and 200), the second column is the fluorescence response, and the third column is the plate number.

(a) Fit the linear regression where fluorescence is the response and doxycycline concentration is the predictor. Predict the fluorescence if the doxycycline concentration is equal to 125.

(b) Since three plates are present, run aoctool with doxycycline concentration as $x$, fluorescence as $y$, and plate number as $g$. Is there a significant difference between the plates?

16.15. **Insulin on Opossum Liver.** Corkill (1932) provides data on the influence of insulin on opossum liver. In the experimental setup the 20 animals (common gray Australian opossums – *Trichosurus*) fasted for 24 or 36 hours. Ten animals, four from the 24-hour fasting group and six from the 36-hour fasting group, were injected with insulin, while the remaining ten animals served as controls, that is, they received no insulin. After 3 to 4 hours liver glycogen and blood sugar were measured. The weights of the animals were recorded as well.

The goal of the study was to explore the deposition of liver glycogen after the insulin regimen in opossums. In rabbits and cats, for example, it was previously found that insulin induced significant glycogen storage. This study found a slight depletion of liver glycogen after the insulin treatment.

Our goal is to model the liver glycogen based on weight, level of blood sugar, insulin indicator, and fasting regime.

Is the insulin indicator (0 no, 1 yes) an important covariate in the model?

| Animal | Weight | Liver glycogen | Blood sugar | Fasting period | Insulin |
|---|---|---|---|---|---|
| 1 | 1502 | 1.80 | 0.124 | 24 | 0 |
| 2 | 1345 | 0.95 | 0.115 | 24 | 0 |
| 3 | 1425 | 1.12 | 0.128 | 24 | 0 |
| 4 | 1650 | 1.05 | 0.110 | 24 | 0 |
| 5 | 1520 | 0.45 | 0.052 | 24 | 1 |
| 6 | 1300 | 0.48 | 0.050 | 24 | 1 |
| 7 | 1250 | 0.75 | 0.045 | 24 | 1 |
| 8 | 1620 | 0.60 | 0.040 | 24 | 1 |
| 9 | 1725 | 0.76 | 0.130 | 36 | 0 |
| 10 | 1450 | 0.51 | 0.112 | 36 | 0 |
| 11 | 1800 | 0.48 | 0.105 | 36 | 0 |
| 12 | 1685 | 0.34 | 0.121 | 36 | 0 |
| 13 | 1560 | 0.38 | 0.116 | 36 | 0 |
| 14 | 1650 | 0.45 | 0.108 | 36 | 0 |
| 15 | 1650 | 0.65 | 0.032 | 36 | 1 |
| 16 | 1575 | 0.28 | 0.025 | 36 | 1 |
| 17 | 1260 | 0.10 | 0.045 | 36 | 1 |
| 18 | 1485 | 0.26 | 0.050 | 36 | 1 |
| 19 | 1520 | 0.18 | 0.030 | 36 | 1 |
| 20 | 1616 | 0.30 | 0.028 | 36 | 1 |

16.16. **Slope in EIV Regression.** Show that the EIV regression slope in (16.7) tends to $S_{xy}/S_{xx}$ when $\eta \to 0$.

16.17. **Interparticular Spacing and Wavelength in Nanoprisms 2.** In the context of Example 15.4, let $x = (\text{separation})^{-1}$ and $y = \log(\text{wavelength})$. The part of MATLAB regression output for ▣ nonoprism.dat data set is given below.

```
[b, bint, r, rint, stats] = regress(y,[ones(size(x)) x])

%b =
%    -4.7182
%     1.6289
%
%bint =
%    -5.6578    -3.7787
%     1.3061     1.9516
```

```
%
%r =
%      0.0037
%     -0.0084
%      ...
%      0.0058
%      0.0046
%
%rint =
%    -0.0101      0.0176
%    -0.0232      0.0064
%      ...
%    -0.0096      0.0213
%    -0.0110      0.0203
%
%stats =
%     0.8545   111.5735     0.0000     0.0001
```

Using information contained in this output, answer the following questions:
(a) What is the regression equation linking $x$ and $y$?
(b) Predict $y = \log(\text{wavelength})$ for $x = 2.9$. What is the *wavelength* for such $x$?
(c) What is $R^2$ here and how is it interpreted? What is the $F$-statistic here? Is it significant?
(d) The 95% confidence interval for the population slope $\beta_1$ is [1.3061, 1.9516]. Using information in this output construct a 99% confidence interval for $\beta_1$.

---

| MATLAB AND WINBUGS FILES AND DATA SETS USED IN THIS CHAPTER |
| --- |

http://springer.bme.gatech.edu/Ch16.Reg/

adhesions.m, ancova2.m, coburnreg.m, cpeptide.m, degradation.m, diabetes.m, diagnostics.m, dissection.m, errorinvar.m, fatreg.m, fatreg1.m, fatregdiag.m, galton.m, hemo.m, hubble.m, invitro.m, kanamycin.m, kodlin.m, myeb.m, oldfaithful.m, pedometer1.m, ratwei.m, tastecheese.m, vitalcapacity.m

BUGS

ancovafibers.odc, fat.odc, mellitus.odc, regressionpred.odc, vortex.odc

adhesion.xls, alcos.dat|xls, arthritis1.dat|mat, circalbumin.dat, coburn.mat, Cpeptideext.dat|mat, fat.dat, fatreg1.m, galton.dat, galtoncompact.dat, kanamycin.dat, Lactis.dat, nanoprism.dat, pearson.dat, pmr1.mat, vitalcapacity.xlsx

---

# CHAPTER REFERENCES

Anscombe, F. (1973). Graphs in statistical analysis. *Am. Stat.*, **27**, 17–21.

Armitage. P. and Berry, G. (1994). *Statistical Methods in Medical Research.* Blackwell Science, London.

Carter, G. and Mitchell, C. (1958). Methods for adapting the virus of Rinderpest to rabbits. *Science*, **128**, 252–253.

Clark, R., Margraf, H., and Beauchamp, R. (1975). Fat and solid filtration in clinical perfusions. *Surgery*, **77**, 216–224.

Corkill, B. (1932). The influence of insulin on the liver glycogen of the common grey australian "opossum" (Trichosurus). *J. Physiol.*, **75**, 1, 29–32. PMCID: PMC1394507.

Galton, F. (1886). Regression towards mediocrity in hereditary stature. *J. Anthropol. Inst. Great Br. Ireland*, **15**, 246–263.

Griffin, G. E., Abbott, W. E., Pride, M. P., Runtwyler, E., Mautz, F. R., and Griffith, L. (1945). Available (thiocyanate) volume total circulating plasma proteins in normal adults, *Ann. Surg.*, **121**, 3, 352–360.

Hastie, T. and Tibshirani, R. (1990). *Generalized Additive Models.* Chapman & Hall, London.

Kleinbaum, D. G., Kupper, L. L., and Muller, K. E. (1987). *Applied Regression Analysis and Other Multivariable Methods.* PWS-Kent, Boston.

Kodlin, D. (1951). An application of the analysis of covariance in pharmacology. *Arch. Int. Pharmacodyn. Ther.*, **87**, 1–2, 207–211.

Kouskolekas, C. and Decker, G. (1966). The effect of temperature on the rate of development of the potato leafhopper, *Empoasca fabae* (Homoptera: Cicadelidae). *Ann. Entomol. Soc. Am.*, **59**, 292–298.

Laud, P. and Ibrahim, J. (1995). Predictive model selection. *J. R. Stat. Soc. Ser. B*, **57**, 1, 247–262.

Lopes, J. B., Dallan, L. A. O., Moreira, L. P. F., Carreiro, M. C., Rodrigues, F. L. B., Mendes, P. C., and Stol, N. A. G. (2009). New quantitative variables to measure postoperative pericardial adhesions. Useful tools in experimental research. *Acta Cir. Bras.*, **24**, São Paulo.

Moore, D. and McCabe, G. (2006). *Introduction to the Practice of Statistics*, 5th edn. Freeman, San Francisco.

Ntzoufras, I. (2009). *Bayesian Modeling Using WinBUGS*. Wiley, Hoboken.

Rawlings, J. O., Pantula, S., and Dickey, D. (1998). *Applied Regression Analysis: A Research Tool*. Springer Texts in Statistics. Springer, Berlin Heidelberg New York.

Sockett, E. B., Daneman, D., Clarson, C., and Ehrich, R. M. (1987). Factors affecting and patterns of residual insulin secretion during the first year of type I (insulin dependent) diabetes mellitus in children. *Diabetes*, **30**, 453–459.

Yago, T., Lou, J., Wu, T., Yang, J., Miner, J. J., Coburn, L., Lopez, L. A., Cruz, M. A., Dong, J.-F., McIntire, L. V., McEver, R. P., and Zhu, C. (2008). Platelet glycoprotein Iba forms catch bonds with human WT vWF but not with type 2B von Willebrand disease vWF. *J. Clin. Invest.*, **118**, 9, 3195–3207.

# Chapter 17
# Regression for Binary and Count Data

*There are 10 types of people in the world, those who can read binary, and those who can't.*

– Anonymous

## 17.1 Introduction

Traditional uni- or multivariable linear regression assumes a normally distributed response centered at a linear function of the predictors. For example, in regression with a single predictor, the response $y_i$ is modeled as normal $\mathcal{N}(\beta_0 + \beta_1 x_i, \sigma^2)$, where the expectation (conditional on covariate $x_i$) is a linear function of $x_i$.

For some regression scenarios this model is inadequate because the response is not normally distributed. The response could be categorical, for example, with two or more categories ("disease present – disease absent," "sur-

vived – died," "low – medium – high," etc.) or be integer valued ("number with
the disease," "number of failures," etc.), and yet it may still depend on a covariate or a vector of covariates, $x$. In this chapter we discuss logistic and Poisson
regressions that are appropriate for binary and counting responses.

In logistic regression the responses are binary, coded, without loss of gen-
erality, as 0 and 1 (or as 1 and 2 in WinBUGS). In Poisson regression the
responses are nonnegative integers well modeled by a Poisson distribution in
which the rate $\lambda$ depends on one or more covariates. The covariates enter the
model in a linear fashion; however, their connection with the (expected) response is nonlinear.

Both logistic and Poisson regressions are examples of a wide class of models
called generalized linear models (GLMs). The term generalized linear model
refers to models introduced by Nelder and Wedderburn (1972) and popularized
by the monograph of McCullagh and Nelder (1982, second edition 1989). In a
canonical GLM model, the response variable $y_i$ is assumed to follow a distribution from the class of distributions called the exponential family, with mean
$\mu$, which is assumed to depend on covariates via their linear combination. The
exponential family is a rich family of distributions and includes almost all important distributions (normal, Bernoulli, binomial, Poisson, gamma, etc.). This
link between the mean $\mu$ and covariates can be nonlinear, but the distribution
of $y_i$ depends on covariates only via their linear combination. Linear regression is a special case of GLMs for normally distributed responses, in which the
mean is directly modeled by a linear combination of covariates.

## 17.2 Logistic Regression

Assume that a response $y_i$, depending on a covariate $x_i$, is categorical and can
take two possible values. Examples of such responses include male–female,
sick–healthy, alive–dead, pass–fail, success–failure, win–loss, etc. One usually
assigns provisional numerical values to the responses, say 0 and 1, mainly to
simplify notation. Our interest is in modeling the probability of response $y = 1$
given the observed covariates.

Classical least-squares regression $y_i = \beta_0 + \beta_1 x_i + \epsilon_i$ is clearly inadequate
since for unbounded $x_i$ the linear term $\beta_0 + \beta_1 x_i$ is unbounded as well. In
addition, the residuals have only two possible values, and the variance of $y_i$ is
not free of $x_i$.

We assume that $y$ is Bernoulli distributed with $\mathbb{E}y = \mathbb{P}(y = 1) = p$. Since
$\mathbb{E}(y_i | X = x_i) = \mathbb{P}(y_i = 1 | X = x_i) = p_i$ is a number between 0 and 1, it is reasonable to model $p_i$ as $F(\beta_0 + \beta_1 x_i)$ for some probability CDF $F$. Equivalently,

$$F^{-1}(p) = \beta_0 + \beta_1 x.$$

In principle, any monotone cumulative distribution function $F$ can provide a link between the probability $p$ and the covariate(s), but the most used distributions are logistic, normal, and complementary log-log, leading to logistic, probit, and clog-log regressions. The most popular among the three is the logistic regression because its coefficients, measuring the impact of the predictors on the binary response $y$, have convenient interpretations via the log odds of the events $\{y = 1\}$. In logistic regression, $F^{-1}(p) = \log \frac{p}{1-p}$ is called the *logit* and denoted as $\text{logit}(p)$.

### 17.2.1 Fitting Logistic Regression

The basic statistical model for logistic regression is

$$
\begin{aligned}
y_i &\sim \mathcal{B}er(p_i), \\
\text{logit}(p_i) = \log \frac{p_i}{1-p_i} &= \beta_0 + \beta_1 x_i, \ i = 1,\ldots,n
\end{aligned}
\tag{17.1}
$$

when the responses are Bernoulli, 0 or 1.

When multiple measurements correspond to the same covariate, it is convenient to express the responses as binomial counts, that is, $n_i$ responses corresponding to covariate $x_i$ are grouped together and $y_i$ is the number of responses equal to 1:

$$
\begin{aligned}
y_i &\sim \mathcal{B}in(n_i, p_i), \\
\text{logit}(p_i) = \log \frac{p_i}{1-p_i} &= \beta_0 + \beta_1 x_i, \ i = 1,\ldots,k, \ \sum_{i=1}^{k} n_i = n.
\end{aligned}
\tag{17.2}
$$

In principle, it is always possible to express the binomial response model via Bernoulli responses by repeating $y_i$ times the response 1 and $n_i - y_i$ times the response 0 for the same covariate $x_i$. The converse is not possible in general, especially if $y_i$ depends on a continuous covariate. Assessing the goodness of fit for models of type (17.1) is a well-known problem since the asymptotic distributional results do not hold even for arbitrarily large sample sizes.

How does one estimate parameters $\beta_0$ and $\beta_1$ in logistic model (17.1) or (17.2)? The traditional least-squares algorithm that is utilized in linear regression is not applicable. Estimating the model coefficients amounts to solving a nonlinear equation, and this is done by an iterative procedure. The algorithm for a single predictor is illustrated and implemented in the m-file ◢ logisticmle.m. There, the Newton–Raphson method is used to solve nonlinear likelihood equations and calculate coefficients $b_0$ and $b_1$ as estimators of population parameters $\beta_0$ and $\beta_1$. Details regarding the background and convergence of methods for estimating model parameters are beyond the scope of this text and can be found in McCullagh and Nelder (1989).

Once the parameters $\beta_0$ and $\beta_1$ are estimated, the probability $p_i = \mathbb{P}(y = 1|x = x_i)$ is obtained as

$$\hat{p}_i = \frac{\exp\{b_0 + b_1 x_i\}}{1 + \exp\{b_0 + b_1 x_i\}} = \frac{1}{1 + \exp\{-b_0 - b_1 x_i\}}.$$

In addition to the nature of response $y_i$, there is another key difference between ordinary and logistic regressions. For linear regression the variance does not depend on the mean $\beta_0 + \beta_1 x_i$; it is constant for all $x_i$. This is one of the assumptions for linear regression. In logistic regression the variance is not constant; it is a function of the mean. From (17.1), $\mathbb{E}y_i = p_i$ and $\mathbb{V}\mathrm{ar}\, y_i = \mathbb{E}y_i(1 - \mathbb{E}y_i)$.

If $p - 1$ covariates ($p \geq 2$ parameters) are available, as is often the case, then

$$X_i'b = \ell_i = b_0 + b_1 x_{i1} + b_2 x_{i2} + \cdots + b_{p-1} x_{i,p-1}$$

replaces $b_0 + b_1 x_i$, where $X_i'$ is the $i$th row of a design matrix $X$ of size $n \times p$, and $b = (b_0 \ b_1 \ \dots \ b_{p-1})'$. Now,

$$\hat{p}_i = \frac{\exp\{X_i'b\}}{1 + \exp\{X_i'b\}} = \frac{1}{1 + \exp\{-X_i'b\}},$$

with $b$ maximizing the log-likelihood

$$\log L(\beta) = \ell(\beta) = \sum_{i=1}^{n} y_i \cdot (X_i'\beta) - \sum_{i=1}^{n} \log\left(1 + \exp\{X_i'\beta\}\right).$$

*Example 17.1.* **Caesarean-Section Infections.** A Caesarean-section, or C-section, is major abdominal surgery, so moms who undergo C-sections are more likely to have an infection, excessive bleeding, blood clots, more postpartum pain, a longer hospital stay, and a significantly longer recovery. The data in this example comes from Munich's *Klinikum Großharden* (Fahrmeir and Tutz, 1996) and concerns infections in births by C-section. The response variable of interest is the occurrence or nonoccurrence of infection. Three covariates, each at two levels, were considered as important for the occurrence of infection:

**noplan** – The C-section delivery was planned (0) or not planned (1);

**riskfac** – Risk factors for the mother, such as diabetes, overweight, previous C-section birth, etc., are present (1) or not present (0); and

**antibio** – Antibiotics as a prophylaxis are given (1) or not given (0).

Table 17.1 provides the results.

**Table 17.1** Caesarean-section delivery data.

|  | Planned Infection | | | No Plan Infection | | |
|---|---|---|---|---|---|---|
|  | yes | no | total | yes | no | total |
| Antibiotics | | | | | | |
| Risk factor yes | 1 | 17 | 18 | 11 | 87 | 98 |
| Risk factor no | 0 | 2 | 2 | 0 | 0 | 0 |
| No Antibiotics | | | | | | |
| Risk factor yes | 28 | 30 | 58 | 23 | 3 | 26 |
| Risk factor no | 8 | 32 | 40 | 0 | 9 | 9 |

Here is the MATLAB code that uses built-in functions `glmfit` and `glmval` to fit and present the model.

```
infection = [ 1 11   0   0 28 23  8   0];
total =     [18 98   2   0 58 26 40   9];
proportion = infection./total;
noplan =    [ 0  1   0   1  0  1  0   1];
riskfac  =  [ 1  1   0   0  1  1  0   0];
antibio =   [ 1  1   1   1  0  0  0   0];
[b,dev,stats] = glmfit([noplan' riskfac' antibio'],...
               [infection' total'],'binomial','logit');
logitFit = ...
   glmval(b,[noplan' riskfac' antibio'],'logit');
```

The resulting additive model (no interactions) is

$$\log \frac{\mathbb{P}(\text{infection})}{\mathbb{P}(\text{no infection})} = \beta_0 + \beta_1 \cdot \text{noplan} + \beta_2 \cdot \text{riskfac} + \beta_3 \cdot \text{antibio}$$

with estimators of $\beta$s as

| $b_0$ | $b_1$ | $b_2$ | $b_3$ |
|---|---|---|---|
| −1.8926 | 1.0720 | 2.0299 | −3.2544 |

The interpretation of the estimators for $\beta$ coefficients is illuminating if we look at the odds ratio $\frac{\mathbb{P}(\text{infection})}{\mathbb{P}(\text{no infection})}$:

$$\frac{\mathbb{P}(\text{infection})}{\mathbb{P}(\text{no infection})} = \exp(\beta_0) \cdot \exp(\beta_1 \text{ noplan}) \cdot \exp(\beta_2 \text{ riskfac}) \cdot \exp(\beta_3 \text{ antibio}).$$

For example, when `antibio=1`, i.e., when antibiotics are given, the estimated odds of infection $\mathbb{P}(\text{infection})/\mathbb{P}(\text{no infection})$ increase by the factor $\exp(-3.25) = 0.0388$, that is, the odds decrease 25.79 times. Of course, these statements are valid only if the model is accurate. Other competing models (such as probit or clog-log) may result in different changes in risk ratios.

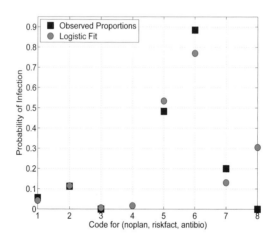

**Fig. 17.1** Caesarean delivery infection predictions. For a triple (noplan, riskfac, antibio), the numbers on the $x$-axis code as follows: **1**=(0,1,1), **2**=(1,1,1), **3**=(0,0,1), **4**=(1,0,1), **5**=(0,1,0), **6**=(1,1,0), **7**=(0,0,0), and **8**=(1,0,0). *Blue squares* are the observed relative frequencies and *green circles* are the model-predicted probabilities of infection. Note that point **4** does not have an observed proportion.

The m-function ◀ `logisticmle.m` also gives standard errors for estimators of $\beta$s. Table provides $t$-values, that is, ratios of coefficients and their standard deviations, for testing if the coefficients are significantly different from 0. These are known as Wald's $Z$ statistics, since they are approximately normal.

**Table 17.2** $t$-ratios (Wald's $Z$ statistic) for the estimators $b = \hat{\beta}$.

|           | $b$      | $s_b$   | $t$      |
|-----------|----------|---------|----------|
| Intercept | −1.8926  | 0.4124  | −4.5893  |
| noplan    | 1.0720   | 0.4253  | 2.5203   |
| riskfac   | 2.0299   | 0.4553  | 4.4588   |
| antibio   | −3.2544  | 0.4812  | −6.7624  |

The *deviance* (p. 665) of this model as a measure of goodness of fit is distributed as $\chi^2$ with 3 degrees of freedom. The number of degrees of freedom is calculated as 7 (number of groups with observations) minus 4 (four estimated parameters $\beta_0 - \beta_3$). Since it is found to be significant,

```
dev = 10.9967;
pval = 1 - chi2cdf(dev, 7-4)    % 0.0117
```

the fit of this model is not good. To improve the fit, one may include the interactions.

One may ask why the regression model was needed in the first place. The probabilities of interest could be predicted by relative frequencies. For example, in the case (noplan = 0, riskfac = 1, antibio = 1), the relative frequency of infection was $1/18 = 0.0556$, just slightly larger than the model-predicted $\hat{p} = 0.0424$. There are two benefits to this approach. First, the model is able to predict probabilities in the cases where no patients are present, such as for (noplan = 1, riskfac = 0, antibio = 1). Second, the predictions for the cases where $y = 1$ is not observed are "borrowing strength" from the model and are not modeled individually. For example, zero as an estimator in the case (noplan = 1, riskfac = 0, antibio = 0) is not reasonable; the model-based estimator $\hat{p} = 0.3056$ is more realistic. Figure 17.1 compares observed and model-predicted infection rates. For a triple of covariates (noplan, riskfac, antibio), the numbers on the $x$-axis code as follows: **1**=(0,1,1), **2**=(1,1,1), **3**=(0,0,1), **4**=(1,0,1), **5**=(0,1,0), **6**=(1,1,0), **7**=(0,0,0), and **8**=(1,0,0). Note that point **4** does not have an observed proportion, however the model-predicted proportion can be found. For computational aspects refer to file ◀ caesarean.m.

Next, we provide a Bayesian solution to this example and compare the model fit with the classical fit above. The comparisons are summarized in Table 17.1.

C-SECTION INFECTIONS

```
model{
  for(i in 1:N){
    inf[i] ~ dbin(p[i],total[i])
    logit(p[i]) <- beta0 + beta1*noplan[i] +
                   beta2*riskfac[i] + beta3*antibio[i]
  }
  beta0 ~dnorm(0, 0.00001)
  beta1 ~dnorm(0, 0.00001)
  beta2 ~dnorm(0, 0.00001)
  beta3 ~dnorm(0, 0.00001)
}
```

DATA

```
list(inf=c(1, 11, 0, 0, 28, 23, 8, 0),
     total = c(18, 98, 2, 0, 58, 26, 40, 9),
     noplan = c(0,1,0,1,0,1,0,1),
     riskfac = c(1,1, 0, 0, 1,1, 0, 0),
     antibio =c(1,1,1,1,0,0,0,0), N=8)
```

INITS

list(beta0 =0, beta1=0, beta2=0, beta3=0)

| | mean | sd | MC error | val2.5pc | median | val97.5pc | start | sample |
|---|---|---|---|---|---|---|---|---|
| beta0 | −1.964 | 0.4258 | 0.001468 | −2.853 | −1.945 | −1.183 | 1001 | 1000000 |
| beta1 | 1.111 | 0.4339 | 8.857E-4 | 0.2851 | 1.102 | 1.986 | 1001 | 1000000 |
| beta2 | 2.104 | 0.4681 | 0.00159 | 1.226 | 2.09 | 3.066 | 1001 | 1000000 |
| beta3 | −3.335 | 0.4915 | 9.756E-4 | −4.337 | −3.322 | −2.411 | 1001 | 1000000 |
| deviance | 32.24 | 2.873 | 0.00566 | 28.67 | 31.59 | 39.49 | 1001 | 1000000 |

**Table 17.3** Comparison of classical and noninformative Bayes estimators $b = \hat{\beta}$, with estimators of standard deviations.

| | $b$ | $s_b$ | $\hat{\beta}_B$ | $\hat{\sigma}_B$ |
|---|---|---|---|---|
| Intercept | −1.8926 | 0.4124 | −1.964 | 0.4258 |
| noplan | 1.0720 | 0.4253 | 1.111 | 0.4339 |
| riskfac | 2.0299 | 0.4553 | 2.104 | 0.4681 |
| antibio | −3.2544 | 0.4812 | −3.335 | 0.4915 |

## 17.2.2 Assessing the Logistic Regression Fit

The measures for assessing the goodness of linear regression fit that we covered in Chap. 16, $R^2$, $F$, MSE, etc., are not appropriate for logistic regression. As in the case of linear regression, there is a range of measures for assessing the performance of logistic regression and we will briefly overview a few.

The significance of model parameters $\beta_0, \beta_1, \ldots$ is tested by the so-called Wald's test. One finds the statistic $Z_i = \frac{b_i}{s(b_i)}$ that has an approximate normal distribution if the coefficient $\beta_i$ is 0. Equivalently, $W_i = \frac{b_i^2}{s^2(b_i)}$ with an approximate $\chi^2$-distribution with 1 degree of freedom can be used. Large values of $|Z_i|$ or $W_i$ are critical for $H_0 : \beta_i = 0$.

The sample variances $s^2(b_i)$ are diagonal elements of $(X'VX)^{-1}$, where

$$X = \begin{bmatrix} 1 & x_{11} & x_{12} & \cdots & x_{1,p-1} \\ 1 & x_{21} & x_{22} & \cdots & x_{2,p-1} \\ & & \cdots & & \\ 1 & x_{n1} & x_{n2} & \cdots & x_{n,p-1} \end{bmatrix}$$

is the design matrix and

$$V = \begin{bmatrix} \hat{p}_1(1-\hat{p}_1) & 0 & \cdots & 0 \\ 0 & \hat{p}_2(1-\hat{p}_2) & \cdots & 0 \\ & & \cdots & \\ 0 & 0 & \cdots & \hat{p}_n(1-\hat{p}_n) \end{bmatrix}.$$

The customary measure for goodness of fit is *deviance*, defined as

$$D = -2\log \frac{\text{likelihood of the fitted model}}{\text{likelihood of the saturated model}}.$$

For the logistic regression in (17.2), where $y_i$ is the number of 1s and $n_i - y_i$ is the number of 0s in class $i$, the likelihood is $L = \prod_{i=1}^k p_i^{y_i}(1-p_i)^{n_i-y_i}$ and the deviance is

$$D = -2\sum_{1=1}^k \left\{ y_i \log\left(\frac{\hat{y}_i}{y_i}\right) + (n_i - y_i)\log\left(\frac{n_i - \hat{y}_i}{n_i - y_i}\right) \right\},$$

where $\hat{y}_i = n_i\hat{p}_i$ is the model fit for $y_i$. The saturated model estimates $p_i$ as $\hat{p}_i = y_i/n_i$ and $\hat{y}_i = y_i$ providing the fit that matches the observations.

The deviance statistic in this case has a $\chi^2$-distribution with $k - p$ degrees of freedom, where $k$ is the number of classes/groups and $p$ is the number of parameters in the model. Recall that in the previous example the deviance of the model was distributed as $\chi^2$ with $k - p = 7 - 4 = 3$ degrees of freedom.

For both Bernoulli and binomial observations, the mean and variance depend on a single parameter, $p$. When the mean is well fitted, the variance could be underfitted (overdispersion in data) or overfitted (underdispersion in data). The ratio $D/df$ is often used to indicate over- or underdispersion in the data.

The traditional $\chi^2$-statistic for the goodness of fit in model (17.2) is defined as

$$\chi^2 = \sum_{i=1}^k \left[ \frac{(y_i - \hat{y}_i)^2}{\hat{y}_i} + \frac{(y_i - \hat{y}_i)^2}{n_i - \hat{y}_i} \right],$$

where $n_i$ is the number of observations in class $i$, $i = 1,\ldots,k$. This statistic has an approx. $\chi^2$-distribution with $k - p$ degrees of freedom.

Goodness of fit measure $G$ is defined as the difference of deviance between the null model (intercept-only model) and the model under consideration. $G$ has a $\chi^2$-distribution with $p - 1$ degrees of freedom, and small values of $G$ are critical, suggesting that the deviance did not improve significantly by adding covariates.

The logistic model can always be expressed in terms of Bernoulli outcomes, where $y_i$ is 0 or 1, as in (17.1). Then $k = n$, $n_i = 1$, the likelihood for the saturated model, is $\prod_{i=1}^n y_i^{y_i}(1 - y_i)^{1-y_i} = 1$ (we assume that $0^0 = 1$), and the deviance for the Bernoulli representation becomes

$$D = -2 \sum_{i=1}^{n} [y_i \log \hat{p}_i + (1 - y_i) \log(1 - \hat{p}_i)].$$

Statistic $D$ does not follow any specific distribution, irrespective of sample size. Likewise, the Pearson $\chi^2$ becomes

$$\chi^2 = \sum_{i=1}^{n} \frac{(y_i - \hat{p}_i)^2}{\hat{p}_i(1 - \hat{p}_i)} \tag{17.3}$$

in model (17.1) and does not follow any specific distribution, either.

To further evaluate the model, several types of residuals are available. Deviance residuals are defined as

$$r_i^D = \text{sign}(y_i - \hat{y}_i) \sqrt{2 \left\{ y_i \log\left(\frac{y_i}{\hat{y}_i}\right) + (n_i - y_i) \log\left(\frac{n_i - y_i}{n_i - \hat{y}_i}\right) \right\}}, \ i = 1, \ldots, k,$$

for model (17.2) and

$$r_i^D = \text{sign}(y_i - \hat{p}_i) \sqrt{2 y_i \log \hat{p}_i + (1 - y_i) \log(1 - \hat{p}_i)}, \ i = 1, \ldots, n,$$

for model (17.1). The deviance $D$ is decomposed to the sum of squares of deviance residuals in an ANOVA-like fashion as $D = \sum (r_i^D)^2$. The squared residual $(r_i^D)^2$ measures the contribution of the $i$th case to the deviance.

Deviance residuals can be plotted against the order of sampling to explore for possible trends and outliers. Also useful for checking the model are half-normal plots where the ordered absolute values $r_i^D$ are plotted against the normal quantiles $\Phi^{-1}\left(\frac{i + n - 1/8}{2n + 1/2}\right)$. These kinds of plots are an extension of Atkinson's (1985) half normal plots in regular linear regression models. Deviation from a straight line in a half-normal plot indicates model inadequacy.

For the model in (17.1), the Pearson residual is defined as

$$r_i^{pea} = \frac{y_i - \hat{p}_i}{\sqrt{\hat{p}_i(1 - \hat{p}_i)}},$$

and the sum of squares of $r_i^{pea}$ constitutes Pearson's $\chi^2$ statistic,

$$\sum_{i=1}^{n} (r_i^{pea})^2 = \sum_{i=1}^{n} \frac{(y_i - \hat{p}_i)^2}{\hat{p}_i(1 - \hat{p}_i)},$$

as in (17.3). This statistic represents a discrepancy measure; however, as we mentioned, it does not follow the $\chi^2$-distribution, even asymptotically.

In the case of continuous covariates, large $n$ and small $n_i$, Hosmer and Lemeshow proposed a $\chi^2$-statistic based on the grouping of predicted values $\hat{p}_i$. All $\hat{p}_i$ are ordered and divided into $g$ approximately equal groups, usually ten. For ten groups, sample deciles of ordered $\hat{p}_i$ could be used.

The Hosmer–Lemeshow statistic is

$$\chi^2_{HL} = \sum_{i=1}^{g} \frac{(n_i - n\overline{p}_i)^2}{n\overline{p}_i},$$

where $g$ is the number of groups, $n_i$ is the number of cases in the $i$th group, and $\overline{p}_i$ is the average of model (predicted) probabilities for the cases in the $i$th group. The $\chi^2_{hl}$-statistic is compared to $\chi^2_{g-2}$ quantiles, and small $p$-values indicate that the fit is poor. In the case of ties, that is, when there are blocks of items with the same predicted probability $\hat{p}$, the blocks are not split but assigned to one of the two groups that share the block. The details of the algorithm can be found in Hosmer and Lemeshow (1989).

In the case of linear regression, $R^2$, as a proportion of model-explained variability in observations, had a strong intuitive appeal in assessing the regression fit. In the case of logistic regression, there is no such intuitive $R^2$. However, there are several proposals of $R^2$-like measures, called pseudo-$R^2$. Most of them are defined in terms of model likelihood or log-likelihood. The model likelihood and log-likelihood are calculated using model logit,

$$\ell_i = b_0 + b_1 x_{i1} + \cdots + b_{p-1} x_{i,p-1},$$

$$LL_p = LL(b_0,\ldots,b_{p-1}) = \sum_{i=1}^{n} (y_i \times \ell_i - \log(1 + \exp\{\ell_i\})),$$

and the model likelihood is $L_p = \exp\{LL_p\}$. The null model is fitted without covariates, and

$$\ell_0 = b_0$$

$$LL_{null} = LL(b_0) = \sum_{i=1}^{n} (y_i \times \ell_0 - \log(1 + \exp\{\ell_0\})).$$

The null model likelihood is $L_{null} = \exp\{LL_{null}\}$.

By analogy to linear regression, $R^2 = \frac{SSR}{SST} = \frac{SST - SSE}{SST}$,

$$R^2_{mf} = \frac{LL_{null} - LL_p}{LL_{null}} = 1 - \frac{LL_p}{LL_{null}},$$

defines McFadden's pseudo-$R^2$. Some other counterparts of $R^2$ are

Cox–Snell:  $R^2_{cs} = 1 - \left[ \frac{L_{null}}{L_p} \right]^{2/n}$ ;

Nagelkerke:  $R^2_n = \dfrac{1 - \left[ \frac{L_{null}}{L_p} \right]^{2/n}}{1 - (L_{null})^{2/n}}$ ;

Effron:  $R^2_e = 1 - \dfrac{\sum_{i=1}^{n} (y_i - \hat{p}_i)^2}{\sum_{i=1}^{n} (y_i - \overline{y})^2}$ ,   $\overline{y} = \dfrac{\sum_{i=1}^{n} y_i}{n}$ .

*Example 17.2.* **Arrhythmia.** Patients who undergo coronary artery bypass graft surgery (CABG) have an approximately 19 to 40% chance of developing atrial fibrillation (AF). AF is a quivering, chaotic motion in the upper chambers of the heart, known as the atria. AF can lead to the formation of blood clots, causing greater in-hospital mortality, strokes, and longer hospital stays. While this can be prevented with drugs, it is very expensive and sometimes dangerous if not warranted. Ideally, several risk factors that would indicate an increased risk of developing AF in this population could save lives and money by indicating which patients need pharmacological intervention. Researchers began collecting data from CABG patients during their hospital stay such as demographics like age and sex, as well as heart rate, cholesterol, operation time, etc. Then the researchers recorded which patients developed AF during their hospital stay. The goal was to evaluate the probability of AF given the measured demographic and risk factors.

The data set ⬛arrhythmia.dat, courtesy of Dr. Matthew C. Wiggins, contains the following variables:

| | |
|---|---|
| $Y$ | Fibrillation |
| $X_1$ | Age |
| $X_2$ | Aortic Cross Clamp Time |
| $X_3$ | Cardiopulmonary Bypass Time |
| $X_4$ | Intensive Care Unit (ICU) Time |
| $X_5$ | Average Heart Rate |
| $X_6$ | Left Ventricle Ejection Fraction |
| $X_7$ | Anamnesis of Hypertension |
| $X_8$ | Gender [1 - Female; 0 - Male] |
| $X_9$ | Anamnesis of Diabetes |
| $X_{10}$ | Previous MI |

The MATLAB script ◢arrhythmia.m provided a logistic regression fit. The script calculates deviance and several goodness-of-fit measures.

```
load 'Arrhythmia.mat'
Y = Arrhythmia(:,1);
X = Arrhythmia(:,2:11);    %Design matrix n x (p-1) without
                           %vector 1 (intercept)
Xdes =[ones(size(Y)) X];   %with the intercept: n x p
n = length(Y); %number of subjects
alpha = 0.05;  %alpha for CIs

[b, dev, stats]=glmfit(X,Y, 'binomial','link','logit')

lin = Xdes * b %linear predictor, n x 1 vector
```

Figure 17.2 shows observed arrhythmia responses (0 or 1) with their logistic fit.

With the linear predictor, fitted probabilities for $\{Y_i = 1\}$ are given as $\hat{p}_i$. Also, the estimators of the $\beta$s and their standard deviations and $p$-values for

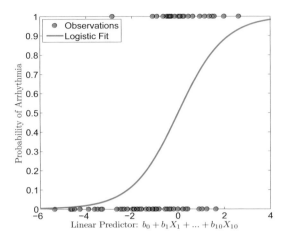

**Fig. 17.2** Arrhythmia responses 0 or 1 with their logistic fit. The abscise axis is the linear predictor lin.

the Wald test are given next. The intercept is significantly nonzero (0.0158), and the variable X1 (*age*) is strongly significant (0.0005). This agrees with the inference based on confidence intervals, only the intervals for $\beta_0$ and $\beta_1$ do not contain 0, or, equivalently, the intervals for the odds ratio, $\exp\{\beta_0\}$ and $\exp\{\beta_1\}$, do not contain 1.

```
phat = exp(lin)./(1 + exp(lin));
V = diag( phat .* (1 - phat) );
sqrtV = diag( sqrt(phat .* (1 - phat) ))
sb = sqrt(   diag( inv( Xdes' * V * Xdes ) ) )
% inv( Xdes' * V * Xdes ) is stats.covb
% Wald tests for parameters beta
z = b./sb  %tests for beta_i = 0, i=0,...,p-1
pvals = 2 * normcdf(-abs(z))   %p-values
%[0.0158;  0.0005;  0.3007;  0.2803;  0.1347;  0.8061
% 0.4217;  0.3810;  0.6762;  0.0842;  0.5942]
%(1-alpha)*100% CI for betas
  CIs = [b - norminv(1 - alpha/2) * sb , b + norminv(1-alpha/2) * sb]
%(1-alpha)*100% CIs for odds ratios
  CIsOR = exp([b-norminv(1-alpha/2)*sb , b+norminv(1-alpha/2)*sb])
%      0.0000    0.1281
%      1.0697    1.2711
%      0.9781    1.0744
%      ...
%      0.8628   10.3273
%      0.4004    4.9453
```

Figure 17.3 shows estimators of $\beta_0 - \beta_{10}$ (as green circles) and 95% confidence bounds. Since the intervals for $\beta_0$ and $\beta_1$ do not contain 0, both the intercept and covariate *age* are important in the model. It is tempting to do variable/model selection based on outcomes of Wald's test – but this is not advisable. Exclusion of a parameter/variable from the model will necessarily change the estimators and confidence intervals for the remaining parameters and previously insignificant parameters may become significant. As in linear regression, best subset, forward, and backward variable selection procedures exist and may be implemented.

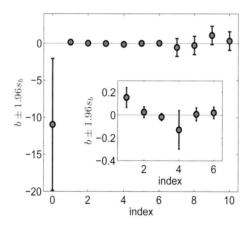

**Fig. 17.3** Estimators of $\beta_0 - \beta_{10}$ are shown as *green circles*, and 95% confidence intervals are given. For comparison, the intervals for $\beta_1 - \beta_6$ are shown separately on a different scale.

Next, we find the log-likelihoods for the model and null model. The model deviance is 78.2515, while the difference of deviances between the models is 26.1949. This would be a basis for a likelihood ratio test if the response were grouped. Since in a Bernoulli setup the distributions of deviance and $G$ are not $\chi^2$, the testing needs to be done by one of the response-grouping methods, such as the Hosmer–Lemeshaw method.

```
%Log-likelihood
loglik = sum( Y  .* lin - log( 1 + exp(lin) ))   %-39.1258
%fitting null model.
[b0, dev0, stats0] = glmfit(zeros(size(Y)),Y,'binomial','link','logit')
  %b0=-0.6381,  dev0=104.4464, stats=... (structure)
loglik0 = sum( Y  .* b0(1) - log(1 + exp(b0(1))) ) %-52.2232
%
G = -2 * (loglik0 -  loglik)  % 26.1949
dev0 - dev %26.1949, the same as G, difference of deviances
%
%model deviance
devi = -2 * sum( Y  .* log( phat + eps) + (1-Y ).*log(1 - phat + eps) )
```

```
                   %78.2515,   directly
dev                %78.2515,   glmfit output
-2 * loglik        %78.2515,   as a link between loglik and deviance
```

Several measures correspond to $R^2$ in the linear regression context: Mac-Fadden's pseudo-$R^2$, Cox–Snell $R^2$, Nagelkerke $R^2$, and Effron's $R^2$. All measures fall between 0.25 and 0.4.

```
%McFadden Pseudo R^2, equivalent expressions
mcfadden  = -2*(loglik0-loglik)/(-2*loglik0) %0.2508
1 - loglik/loglik0 %0.2508
%
coxsnell =1-(exp(loglik0)/exp(loglik))^(2/n) %0.2763
%
nagelkerke=(1-(exp(loglik0)/exp(loglik))^(2/n))/...
    (1-exp(loglik0)^(2/n))      %0.3813
%
effron=1-sum((Y-phat).^2)/sum((Y-sum(Y)/n).^2) %0.2811
```

Next we find several types of residuals: ordinary, Pearson, deviance, and Anscombe.

```
ro = Y  - phat;  %Ordinary residuals

%Deviance Residuals
rdev = sign(Y  - phat) .* sqrt(-2 * Y .* log(phat+eps) - ...
                       2*(1 - Y) .* log(1 - phat+eps));
%Anscombe Residuals
ransc = (betainc(Y,2/3,2/3) - betainc(phat,2/3,2/3) ) .* ...
                   ( phat  .* (1-phat) + eps).^(1/6);
%
% Model deviance is recovered as
%the sum of squared dev. residuals
sum(rdev.^2)  %78.2515
```

Figure 17.4 shows four kinds of residuals (ordinary, Pearson, deviance, and Anscombe), plotted against $\hat{p}$.

If the model is adequate, the smoothed residuals should result in a function close to 0. Figure 17.5 shows Pearson's residuals smoothed by a loess smoothing method, ( loess.m).

Influential and outlying observations can be detected with a plot of absolute values of residuals against half-normal quantiles. Figure 17.6 was produced by the script below and shows a half-normal plot. The upper and lower bounds (in red) show an empirical 95% confidence interval and were obtained by simulation. The sample of size 19 was obtained from Bernoulli $\mathscr{B}er(\hat{p})$, where $\hat{p}$ is the model fit, and then the minimum, mean, and maximum of the absolute residuals of the simulated values were plotted.

```
k = 1:n;
q = norminv((k + n - 1/8)./(2 * n + 1/2));
plot( q, sort(abs(rdev)),  'k-','LineWidth',1.5);

% Simulated Envelope
```

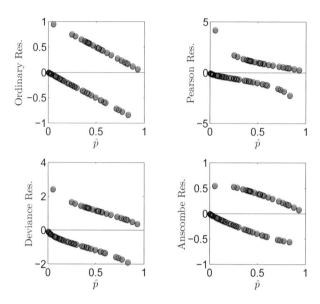

**Fig. 17.4** Ordinary, Pearson, deviance, and Anscombe residuals plotted against $\hat{p}$.

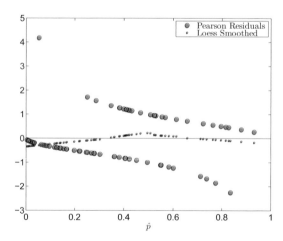

**Fig. 17.5** Pearson's residuals (*green circles*) smoothed. The *red circles* show the result of smoothing.

```
rand('state',1)
env =[];
for i = 1:19
   surrogate = binornd(1, phat);
   rdevsu = sign(surrogate - phat).*sqrt(- 2*surrogate .* ...
   log(phat+eps)- 2*(1 - surrogate) .* log(1 - phat+eps) );
   env = [env   sort(abs(rdevsu))];
end
envel=[min(env'); mean(env' ); max(env' )]';
hold on
plot( q , envel(:,1), 'r-');
plot( q , envel(:,2), 'g-');
plot( q , envel(:,3), 'r-');
xlabel('Half-Normal Quantiles','Interpreter','LaTeX')
ylabel('Abs. Dev. Residuals','Interpreter','LaTeX')
h=legend('Abs. Residuals','Simul. $95%$ CI','Simul. Mean',2)
set(h,'Interpreter','LaTeX')
axis tight
```

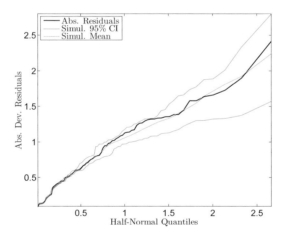

**Fig. 17.6** Half-normal plot for deviance residuals.

To predict the mean response for a new observation, we selected a "new person" with specific covariate values. For this person the estimator for $\mathbb{P}(Y = 1)$ is 0.3179 and 0 for a single future response. A single future response is in fact a classification problem: individuals with a specific set of covariates are classified as either 0 or 1.

```
%Probability of  Y=1 for a new observation
Xh =[1   72   81   130   15   78   43   1   0   0   1]' ;
% responses for a new person
pXh = exp(Xh' * b)/(1 + exp(Xh' * b) ) %0.3179
%(1-alpha) * 100% CI
ppXh = Xh' * b  %-0.7633
```

```
s2pXp = Xh' * inv( Xdes' * V * Xdes ) * Xh  %0.5115
spXh = sqrt(s2pXp)   %0.7152
% confidence interval on the linear part
li = [ppXh-norminv(1-alpha/2)*spXh  ...
 ppXh+norminv(1-alpha/2)*spXh]  %-2.1651    0.6385
% transformation to the CI for the mean response
exp(li)./(1 + exp(li)) %0.1029    0.6544

%Predicting single future observation
cutoff = sum(Y)/n %0.3457
%
pXh > cutoff  %Ynew = 0
```

Next, we provide a Bayesian solution to Arrhythmia logistic model (Arrhythmia.odc) and compare classical and Bayesian model parameters.

```
model{
 eps <- 0.00001
   for(i in 1:N){
   Y[i] ~ dbern(p[i])
   logit(p[i]) <- beta[1] + beta[2] * X1[i]+
             beta[3] * X2[i] + beta[4] * X3[i]+ beta[5] * X4[i] +
             beta[6] * X5[i] + beta[7] * X6[i]  + beta[8] * X7[i] +
             beta[9] * X8[i] + beta[10] * X9[i] + beta[11] * X10[i]
devres[i] <- 2*Y[i]* log(Y[i]/p[i] +eps) +
                2*(1 - Y[i])*log((1-Y[i])/(1-p[i])+eps)
 }
 for(j in 1:11){
 beta[j] ~ dnorm(0, 0.0001)
       }
 dev  <-  sum(devres[])
}
```

DATA + INITS (see Arrhythmia.odc)

The classical and Bayesian model parameters are shown in the table:

| | $\hat{\beta}_0$ | $\hat{\beta}_1$ | $\hat{\beta}_2$ | $\hat{\beta}_3$ | $\hat{\beta}_4$ | $\hat{\beta}_5$ | $\hat{\beta}_6$ | $\hat{\beta}_7$ | $\hat{\beta}_8$ | $\hat{\beta}_9$ | $\hat{\beta}_{10}$ | Dev. |
|---|---|---|---|---|---|---|---|---|---|---|---|---|
| Classical | −10.95 | 0.1536 | 0.0248 | −0.0168 | −0.1295 | 0.0071 | 0.0207 | −0.5377 | −0.2638 | 1.0936 | 0.3416 | 78.25 |
| Bayes | −13.15 | 0.1863 | 0.0335 | −0.0236 | −0.1541 | 0.0081 | 0.0025 | −0.6419 | −0.3157 | 1.313 | 0.4027 | 89.89 |

## 17.2.3 Probit and Complementary Log-Log Links

We have seen that for logistic regression,

$$\hat{p}_i = F(\ell_i) = \frac{\exp\{\ell_i\}}{1 + \exp\{\ell_i\}},$$

where $\ell_i = b_0 + b_1 x_{i1} + \cdots + b_{p-1} x_{i,p-1}$ is the linear part of the model.

A *probit* regression uses a normal distribution instead,

$$\hat{p}_i = \Phi(\ell_i),$$

while for the complementary log-log, the distribution

$$F(x) = 1 - \exp\{-e^x\}$$

is used.

The complementary log-log link interprets the regression coefficients in terms of the hazard ratio rather than the log odds ratio. It is defined as

$$\text{clog-log} = \log(-\log(1-p)).$$

The clog-log regression is typically used when the outcome $\{y = 1\}$ is rare. Probit models are popular in a bioassay context. A disadvantage of probit models is that the link $\Phi^{-1}$ does not have an explicit expression, although approximations and numerical algorithms for its calculation are readily available.

Once the linear part $\ell_i$ in a probit or clog-log model is fitted, the probabilities are estimated as

$$\hat{p}_i = \Phi(\ell_i) \quad \text{or} \quad \hat{p}_i = 1 - \exp(-\exp(\ell_i)),$$

respectively. In MATLAB, the probit and complementary log-log links are optional arguments, $'link', 'probit'$ or $'link', 'comploglog'$.

*Example 17.3.* **Bliss Data.** In his 1935 paper, Bliss provides a table showing a number of flour beetles killed after 5 hours' exposure to gaseous carbon disulfide at various concentrations. This data set has since been used extensively by statisticians to illustrate and compare models for binary and binomial data.

**Table 17.4** Bliss beetle data.

| Dose $(\log_{10} CS_2 \ mgl^{-1})$ | Number of Beetles | Number Killed |
|---|---|---|
| 1.6907 | 59 | 6 |
| 1.7242 | 60 | 13 |
| 1.7552 | 62 | 18 |
| 1.7842 | 56 | 28 |
| 1.8113 | 63 | 52 |
| 1.8369 | 59 | 53 |
| 1.8610 | 62 | 61 |
| 1.8839 | 60 | 60 |

The following Bayesian model is applied on Bliss' data a probit fit is provided (💾 bliss.odc).

```
model{
    for( i in 1 : N ) {
y[i] ~ dbin(p[i],n[i])
        probit(p[i]) <- alpha.star + beta * (x[i] - mean(x[]))
    yhat[i] <- n[i] * p[i]
    }
    alpha <- alpha.star - beta * mean(x[])
    beta ~ dnorm(0.0,0.001)
    alpha.star ~ dnorm(0.0,0.001)
}

DATA
list( x = c(1.6907, 1.7242, 1.7552, 1.7842,
                1.8113, 1.8369, 1.8610, 1.8839),
  n = c(59, 60, 62, 56, 63, 59, 62, 60),
  y = c(6, 13, 18, 28, 52, 53, 61, 60), N = 8)
INITS
list(alpha.star=0, beta=0)
```

| | mean | sd | MC error | val2.5pc | median | val97.5pc | start | sample |
|---|---|---|---|---|---|---|---|---|
| alpha | −35.03 | 2.652 | 0.01837 | −40.35 | −35.01 | −29.98 | 1001 | 100000 |
| alpha.star | 0.4461 | 0.07724 | 5.435E-4 | 0.2938 | 0.4461 | 0.5973 | 1001 | 100000 |
| beta | 19.78 | 1.491 | 0.0104 | 16.94 | 19.77 | 22.78 | 1001 | 100000 |
| yhat[1] | 3.445 | 1.018 | 0.006083 | 1.757 | 3.336 | 5.725 | 1001 | 100000 |
| yhat[2] | 10.76 | 1.69 | 0.009674 | 7.643 | 10.7 | 14.26 | 1001 | 100000 |
| yhat[3] | 23.48 | 1.896 | 0.01095 | 19.77 | 23.47 | 27.2 | 1001 | 100000 |
| yhat[4] | 33.81 | 1.597 | 0.01072 | 30.62 | 33.83 | 36.85 | 1001 | 100000 |
| yhat[5] | 49.59 | 1.623 | 0.01208 | 46.28 | 49.63 | 52.64 | 1001 | 100000 |
| yhat[6] | 53.26 | 1.158 | 0.008777 | 50.8 | 53.33 | 55.33 | 1001 | 100000 |
| yhat[7] | 59.59 | 0.7477 | 0.00561 | 57.91 | 59.68 | 60.82 | 1001 | 100000 |
| yhat[8] | 59.17 | 0.3694 | 0.002721 | 58.28 | 59.23 | 59.71 | 1001 | 100000 |

If instead of probit, the clog-log was used, as cloglog(p[i]) <- alpha.star + beta * (x[i] - mean(x[])), then the coefficients are

| | mean | sd | MC error | val2.5pc | median | val97.5pc | start | sample |
|---|---|---|---|---|---|---|---|---|
| alpha | −39.73 | 3.216 | 0.02195 | −46.24 | −39.66 | −33.61 | 1001 | 100000 |
| beta | 22.13 | 1.786 | 0.01214 | 18.73 | 22.09 | 25.74 | 1001 | 100000 |

For comparisons we take a look at the classical solution ( ◢ beetleBliss2.m). Figure 17.7 shows three binary regressions (logit, probit and clog-log) fitting the Bliss data.

```
disp('Logistic Regression 2: Bliss Beetle Data')
lw = 2.5;
set(0, 'DefaultAxesFontSize', 16);
fs = 15;
msize = 10;
beetle=[...
1.6907  6 59; 1.7242 13 60; 1.7552 18 62; 1.7842 28 56;...
```

```
1.8113 52 63; 1.8369 53 59; 1.8610 61 62; 1.8839 60 60];
%%%%%%%%%%%%%%%%%%%
xi = beetle(:,1);  yi=beetle(:,2); ni=beetle(:,3);
figure(1)
[b, dev, stats] = glmfit(xi,[yi ni],'binomial','link','logit');
[b1, dev1, stats1] = glmfit(xi,[yi ni],'binomial','link','probit');
[b2, dev2, stats2] = glmfit(xi,[yi ni],'binomial','link','cloglog');

xs = 1.5:0.01:2.0;
ys  = glmval(b, xs, 'logit');
y1s = glmval(b1, xs, 'probit');
y2s = glmval(b2, xs, 'cloglog');
% Plot
plot(xs,ys,'r-','LineWidth',lw)
hold on
plot(xs,y1s,'k--','LineWidth',lw)
plot(xs,y2s,'-.','LineWidth',lw)
plot(xi, yi./ni, 'o','MarkerSize',msize,...
       'MarkerEdgeColor','k','MarkerFaceColor','g')
axis([1.5 2.0 0 1])
grid on
xlabel('Log concentration')
ylabel('Proportion killed')
legend('Obs. proportions','Logit','Probit','Clog-log',2)
```

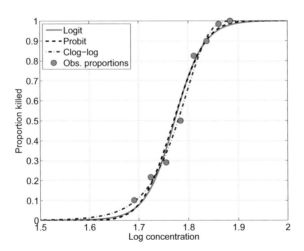

**Fig. 17.7** Bliss data (*green* dots). Regression fit with logit link *red*, probit link *black*, and clog-log link *blue*.

The table below compares the coefficients of the linear part of the three models. Note that classical and Bayesian results are close because the priors in the Bayesian model are noninformative.

| Link | Classical | | | Bayes | | |
|---|---|---|---|---|---|---|
| | Logit | Probit | Clog-log | Logit | Probit | Clog-log |
| Intercept | −60.72 | −34.94 | −39.57 | −60.78 | −35.03 | −39.73 |
| Slope | 34.27 | 19.73 | 22.04 | 34.31 | 19.78 | 22.13 |

## 17.3 Poisson Regression

Poisson regression models the counts $y = \{0, 1, 2, 3, \ldots\}$ of rare events in a large number of trials. Typical examples are unusual adverse events, accidents, incidence of a rare disease, device failures during a particular time interval, etc. Recall that Poisson random variable $Y \sim \mathscr{P}oi(\lambda)$ has the probability mass function

$$f(y) = \mathbb{P}(Y = y) = \frac{\lambda^y}{y!} \exp\{-\lambda\}, \ y = 0, 1, 2, 3, \ldots$$

with both mean and variance equal to the rate parameter $\lambda > 0$.

Suppose that $n$ counts of $y_i$, $i = 1, \ldots, n$ are observed and that each count corresponds to a particular value of a covariate $x_i$, $i = 1, \ldots, n$. A typical Poisson regression can be formulated as follows:

$$y_i \sim \mathscr{P}oi(\lambda_i), \tag{17.4}$$
$$\log(\lambda_i) = \beta_0 + \beta_1 x_i, i = 1, \ldots, n,$$

although other relations between $\lambda_i$ and the linear part $\beta_0 + \beta_1 x_i$ are possible as long as $\lambda_i$ remains positive. More generally, $\lambda_i$ can be linked to a linear expression containing $p - 1$ covariates and $p$ parameters as

$$\log(\lambda_i) = \beta_0 + \beta_1 x_{i1} + \beta_2 x_{i2} + \cdots + \beta_{p-1} x_{i,p-1}, i = 1, \ldots, n.$$

In terms of model (17.4) the Poisson rate $\lambda_i$ is the expectation, and its logarithm can be expressed as $\log \mathbb{E}(y_i | X = x_i) = \beta_0 + \beta_1 x_i$. When the covariate $x_i$ gets a unit increment, $x_i + 1$, then

$$\log \mathbb{E}(y_i | X = x_i + 1) = \beta_0 + \beta_1 x_i + \beta_1 = \log \mathbb{E}(y_i | X = x_i) + \beta_1.$$

Thus, parameter $\beta_1$ can be interpreted as the increment to log rate when the covariate gets an increment of 1. Equivalently, $\exp\{\beta_1\}$ is the ratio of rates,

$$\exp\{\beta_1\} = \frac{\mathbb{E}(y_i | x_i + 1)}{\mathbb{E}(y_i | x_i)}.$$

The model-assessed mean response is $\hat{y}_i = \exp\{b_0 + b_1 x_i\}$, where $b_0$ and $b_1$ are the estimators of $\beta_0$ and $\beta_1$. Strictly speaking, the model predicts the rate $\hat{\lambda}_i$, but the rate is interpreted as the expected response.

The deviance of the model, $D$, is defined as

$$D = 2 \sum_{i=1}^{n} \left( y_i \log \frac{y_i}{\hat{y}_i} - (y_i - \hat{y}_i) \right),$$

where $y_i \log y_i = 0$ if $y_i = 0$. As in logistic regression, the deviance is a measure of goodness of fit of a model and for a Poisson model has a $\chi^2$-distribution with $n - p$ degrees of freedom.

Deviance residuals, defined as

$$r_i^{dev} = \text{sign}(y_i - \hat{y}_i) \times \sqrt{2 y_i \log \frac{y_i}{\hat{y}_i} - 2(y_i - \hat{y}_i)},$$

satisfy $D = \sum_{i=1}^{n} \left( r_i^{dev} \right)^2$.

Pearson's residuals are defined as

$$r_i^{pea} = \frac{y_i - \hat{y}_i}{\sqrt{\hat{y}_i}}.$$

Then the Pearson goodness-of-model-fit statistic $\chi^2 = \sum_{i=1}^{n} (r_i^{pea})^2$ also has a $\chi^2$-distribution with $n - p$ degrees of freedom.

Freedman–Tukey residuals are defined as

$$r_i^{ft} = \sqrt{y_i} + \sqrt{y_i + 1} - \sqrt{4 \hat{y}_i + 1}$$

and Anscombe residuals as

$$r_i^a = \frac{3}{2} \times \frac{y_i^{2/3} - \hat{y}_i^{2/3}}{\hat{y}_i^{1/6}}.$$

Some additional diagnostic tools are exemplified in the following case study ( ◢ ihga.m).

*Example 17.4.* **Case Study: Danish IHGA Data.** In an experiment conducted in the 1980s (Hendriksen et al., 1984), 572 elderly people living in a number of villages in Denmark were randomized, 287 to a control (C) group (who received standard care) and 285 to an experimental group (who received standard care plus IHGA: a kind of preventive assessment in which each person's medical and social needs were assessed and acted upon individually). The important outcome was the number of hospitalizations during the 3-year life of the study.

```
% IHGA
% data
```

**Table 17.5** Distribution of number of hospitalizations in IHGA study.

| Group | # of hospitalizations | | | | | | | | $n$ | Mean | Variance |
|---|---|---|---|---|---|---|---|---|---|---|---|
| | 0 | 1 | 2 | 3 | 4 | 5 | 6 | 7 | | | |
| Control | 138 | 77 | 46 | 12 | 8 | 4 | 0 | 2 | 287 | 0.944 | 1.54 |
| Treatment | 147 | 83 | 37 | 13 | 3 | 1 | 1 | 0 | 285 | 0.768 | 1.02 |

```
x0 = 0 * ones(287,1);   x1 = 1 * ones(285,1);
   %covariate 0-no intervention, 1- intervention
y0 = [0*ones(138,1); 1*ones(77,1); 2*ones(46,1);...
   3*ones(12,1); 4 * ones(8,1);  5*ones(4,1); 7*ones(2,1)];
y1 = [0*ones(147,1); 1*ones(83,1); 2*ones(37,1);...
   3*ones(13,1); 4 * ones(3,1);  5*ones(1,1); 6*ones(1,1)];
   %response # of hospitalizations
x =[x0; x1];   y=[y0; y1];
xdes = [ones(size(y)) x];
[n p] = size(xdes)

[b dev stats] = glmfit(x,y,'poisson','link','log')
yhat = glmval(b, x,'log') %model predicted responses

% Pearson residuals
rpea  = (y - yhat)./sqrt(yhat);
% deviance residuals
rdev = sign(y - yhat) .* sqrt(-2*y.*log(yhat./(y + eps))-2*(y - yhat));
% Friedman-Tukey residuals
rft = sqrt(y) + sqrt(y + 1) - sqrt(4 * yhat + 1)
% Anscombe residuals
ransc = 3/2 * (y.^(2/3) - yhat.^(2/3) )./(yhat.^(1/6))
```

Figure 17.8 shows four our types of residual in Poisson regression fit of IHGA data: Pearson, deviance, Friedman–Tukey, and Anscombe. The residuals are plotted against responses $y$.

```
loglik = sum(y   .* log(yhat+eps) - yhat   - log(factorial(y)));
%
[b0, dev0, stats0] = glmfit(zeros(size(y)),y,'poisson','link','log')
yhat0 = glmval(b0, zeros(size(y)),'log');
loglik0  = sum( y .* log(yhat0 + eps) - yhat0 - log(factorial(y)))

G   = -2 * (loglik0 -  loglik) %LR test, nested model chi2  5.1711
dev0 - dev % the same as G, difference of deviances       5.1711
pval = 1-chi2cdf(G,1)      %0.0230
```

Under $H_0$ stating that the model is null (model with an intercept and no covariates), the statistic $G$ will have $df = p - 1$ degrees of freedom, in our case $df = 1$. Since this test is significant ($p = 0.0230$), the covariate contributes significantly to the model.

Below are several ways to express the deviance of the model. Note that the sum of squares of the deviance residuals simplifies to $-2\sum_{i=1}^{n} y_i \log(\hat{y}_i/y_i)$, since in Poisson regression $\sum_{i=1}^{n}(y_i - \hat{y}_i) = 0$.

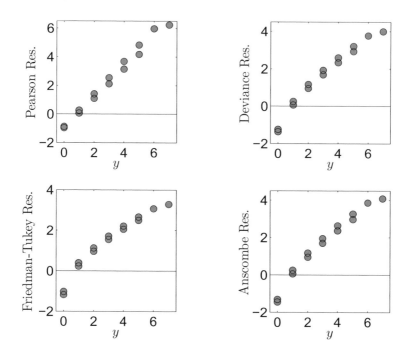

**Fig. 17.8** (a) Four types of residual in a Poisson regression fit of IHGA data: Pearson, deviance, Friedman–Tukey, and Anscombe. The residuals are plotted against responses $y$.

```
%log-likelihood for saturated model
logliksat = sum(y.*log(y+eps)-y-log(factorial(y))) %-338.1663
m2LL = -2 * sum( y  .* log(yhat./(y + eps)) ) %819.8369
deviance = sum(rdev.^2) %819.8369
dev %819.8369    from glmfit
-2*(loglik - logliksat) % 819.8369
```

The following is a Bayesian model fit in WinBUGS (![icon] geriatric.odc).

```
model
{
for (i in 1:n)
    {
        y[i] ~ dpois(lambda[i] )
        log( lambda[i]) <- beta.0 + beta.1 * x[i]
    }
beta.0 ~ dnorm(0, 0.0001)
beta.1 ~ dnorm(0, 0.0001)
lambda.C <- exp(beta.0)
lambda.E <- exp(beta.0 + beta.1 )
diff <- lambda.E - lambda.C
meffect   <- exp( beta.1 )
}
```

```
DATA
list( y = c(0, 0, 0, 0, 0, 0, 0, 0, 0, 0, 0, 0, 0, 0, 0, 0, 0,
0, 0, 0, 0, 0, 0, 0, 0, 0, 0, 0, 0, 0, 0, 0, 0, 0, 0, 0, 0,
                            ...
2, 2, 2, 2, 2, 2, 2, 2, 2, 3, 3, 3, 3, 3, 3, 3, 3, 3, 3, 3, 3,
4, 4, 4, 4, 4, 4, 4, 4, 5, 5, 5, 5, 7, 7, 0, 0, 0, 0, 0, 0, 0,
0, 0, 0, 0, 0, 0, 0, 0, 0, 0, 0, 0, 0, 0, 0, 0, 0, 0, 0, 0, 0,
                            ...
2, 2, 2, 2, 2, 2, 2, 2, 3, 3, 3, 3, 3, 3, 3, 3, 3, 3, 3, 3, 3,
                                            4, 4, 4, 5, 6 ),
x = c(0, 0, 0, 0, 0, 0, 0, 0, 0, 0, 0, 0, 0, 0, 0, 0, 0, 0, 0,
0, 0, 0, 0, 0, 0, 0, 0, 0, 0, 0, 0, 0, 0, 0, 0, 0, 0, 0, 0, 0,
                            ...
0, 0, 0, 0, 0, 0, 0, 0, 0, 0, 0, 0, 0, 0, 0, 0, 0, 0, 0, 0, 0,
0, 0, 0, 0, 0, 0, 0, 0, 0, 0, 0, 0, 0, 0, 0, 0, 0, 0, 0, 0, 0,
0, 0, 0, 0, 0, 0, 0, 0, 0, 0, 0, 0, 0, 0, 1, 1, 1, 1, 1, 1, 1,
1, 1, 1, 1, 1, 1, 1, 1, 1, 1, 1, 1, 1, 1, 1, 1, 1, 1, 1, 1, 1,
                            ...
1, 1, 1, 1, 1, 1, 1, 1, 1, 1, 1, 1, 1, 1, 1, 1, 1, 1, 1, 1, 1,
1, 1, 1, 1, 1 ), n = 572 )
```

```
INITS
list( beta.0 = 0.0, beta.1 = 0.0 )
```

|          | mean     | sd      | MC error | val2.5pc | median   | val97.5pc | start | sample |
|----------|----------|---------|----------|----------|----------|-----------|-------|--------|
| beta.0   | -0.05915 | 0.06093 | 2.279E-4 | -0.1801  | -0.05861 | 0.05881   | 1001  | 100000 |
| beta.1   | -0.2072  | 0.09096 | 3.374E-4 | -0.3861  | -0.2071  | -0.02923  | 1001  | 100000 |
| deviance | 1498.0   | 2.012   | 0.006664 | 1496.0   | 1498.0   | 1504.0    | 1001  | 100000 |
| diff     | -0.1764  | 0.07737 | 2.872E-4 | -0.3285  | -0.1763  | -0.02493  | 1001  | 100000 |
| lambda.C | 0.9443   | 0.05749 | 2.137E-4 | 0.8352   | 0.9431   | 1.061     | 1001  | 100000 |
| lambda.T | 0.7679   | 0.05188 | 1.784E-4 | 0.6693   | 0.7668   | 0.8732    | 1001  | 100000 |
| meffect  | 0.8162   | 0.07437 | 2.76E-4  | 0.6797   | 0.8129   | 0.9712    | 1001  | 100000 |

*Example 17.5.* **Cellular Differentiation Data.** In a biomedical study of the immunoactivating ability of the agents TNF (tumor necrosis factor) and IFN (interferon) to induce cell differentiation, the number of cells that exhibited markers of differentiation after exposure to TNF or IFN was recorded (Piergorsch et al., 1988; Fahrmeir and Tutz, 1994). At each of the 16 dose combinations of TNF/IFN, 200 cells were examined. The number *y* of differentiating cells corresponding to a TNF/IFN combination are given in Table 17.6.

**Table 17.6** Cellular differentiation data.

| Number cells diff | Dose of TNF (U/ml) | Dose of IFN (U/ml) | Number cells diff | Dose of TNF (U/ml) | Dose of IFN (U/ml) |
|---|---|---|---|---|---|
| 11 | 0 | 0 | 31 | 10 | 0 |
| 18 | 0 | 4 | 68 | 10 | 4 |
| 20 | 0 | 20 | 69 | 10 | 20 |
| 39 | 0 | 100 | 128 | 10 | 100 |
| 22 | 1 | 0 | 102 | 100 | 0 |
| 38 | 1 | 4 | 171 | 100 | 4 |
| 52 | 1 | 20 | 180 | 100 | 20 |
| 69 | 1 | 100 | 193 | 100 | 100 |

The suggested model is Poisson with the form

$$\lambda = \mathbb{E}(y|\text{TNF, IFN}) = \exp\{\beta_0 + \beta_1\ \text{TNF} + \beta_2\ \text{IFN} + \beta_3\ \text{TNF}\times\text{IFN}\}.$$

From

```
load 'celular.dat'
number = celular(:,1);
TNF =    celular(:,2);
IFN =    celular(:,3);
[b, dev, stats] = glmfit([TNF IFN   TNF.*IFN],number,...
     'poisson','link','log')
```

the estimators for $\beta_0$–$\beta_3$ are

| $b_0$ | $b_1$ | $b_2$ | $b_3$ |
|---|---|---|---|
| 3.43463 | 0.01553 | 0.00895 | −0.0000567 |

Since $b_3 < 0.0001$, it is tempting to drop the interaction term. However, since the standard error of $b_3$ is s.e.$(b_3) = 0.000013484$, Wald's $Z = b_3/\text{s.e.}(b_3)$ statistic is −4.2050, suggesting that the term TNF×IFN might be significant. However, the overdispersion parameter, which theoretically should be stats.s=1, is estimated as stats.s=3.42566, and the Wald statistic should be adjusted to $Z' = -0.0000567/(3.42566 \times 0.000013484) = 1.2275$. Since the $p$-value is 2 ∗ normcdf(-1.2275) = 0.2196, after all, the interaction term turns out not to be significant and the additive model could be fit:

```
[b, dev, stats] = glmfit([TNF IFN],number,...
     'poisson','link','log')
```

which gives estimates $b_0 = 3.57311$, $b_1 = 0.01314$, and $b_3 = 0.00585$. Details can be found in ◢ celular.m.

The additive model was also fit in a Bayesian manner.

```
model
{
for (i in 1:n)
   {
     numbercells[i] ~ dpois(lambda[i])
     lambda[i] <- exp(beta0 + beta1 * tnf[i]  + beta2 * ifn[i])
   }
beta0 ~ dnorm(0, 0.00001)
beta1 ~ dnorm(0, 0.00001)
beta2 ~ dnorm(0, 0.00001)
}


DATA
list(n=16,
numbercells = c(11,18,20,39,22,38,52,69,31,68,69,128,102,171,180,193),
tnf = c(0,0,0,0,   1,1,1,1,    10,10,10,10,   100,100,100,100),
ifn = c(0,4,20,100,   0,2,20,100,   0,4,20,100,   0,4,20,100 ) )

INITS
list(b0=0, b1=0, b2=0)
```

|    | mean | sd | MC error | val2.5pc | median | val97.5pc | start | sample |
|----|------|------|----------|----------|--------|-----------|-------|--------|
| b0 | 3.573 | 0.05139 | 0.001267 | 3.473 | 3.574 | 3.672 | 1001 | 100000 |
| b1 | 0.01313 | 5.921E-4 | 1.18E-5 | 0.01197 | 0.01314 | 0.01431 | 1001 | 100000 |
| b2 | 0.00585 | 6.399E-4 | 1.142E-5 | 0.004585 | 0.005855 | 0.007086 | 1001 | 100000 |

Note that, because of the noninformative priors on $\beta_0$, $\beta_1$, and $\beta_3$, the Bayesian estimators almost coincide with the MLEs from glmfit.

## 17.4 Log-linear Models

Poisson regression can be employed in the context of contingency tables (Chap. 14). The logarithm of the table cell count is modeled in a linear fashion.

Let in an $r \times c$ contingency table the probability of a cell $(i, j)$ be $p_{ij}$. If $n$ subjects are cross-tabulated, let $n_{ij}$ be the number of subjects classified in the cell $(i, j)$. The count $n_{ij}$ is realization of random variable $N_{ij}$. We assume that the sample size $n$ is random, since in that case the cell frequency $N_{ij}$ is a Poisson random variable with the intensity $\mu_{ij}$. If the sample size $n$ is fixed, then the $N_{ij}$s are realizations of a multinomial random variable. In Fisher's exact test context (p. 546), we saw that if in addition the marginal counts are fixed, the $N_{ij}$s are hypergeometric random variables.

Then the expected table is

| | 1 | 2 | $\cdots$ | $c$ | Total |
|---|---|---|---|---|---|
| 1 | $np_{11}$ | $np_{12}$ | | $np_{1c}$ | $np_{1\cdot}$ |
| 2 | $np_{21}$ | $np_{22}$ | | $np_{2c}$ | $np_{2\cdot}$ |
| | | | | | |
| $r$ | $np_{r1}$ | $np_{r2}$ | | $np_{rc}$ | $np_{r\cdot}$ |
| Total | $np_{\cdot1}$ | $np_{\cdot2}$ | | $np_{\cdot c}$ | $n$ |

Note that both $n_{ij}$ and $e_{ij} = n_{i\cdot} \times n_{\cdot j}/n$ involve observations and empirical marginal probabilities defined by the observed marginal frequencies. If we denote $\mu_{ij} = \mathbb{E}N_{ij} = np_{ij}$, then both $n_{ij}$ and $e_{ij}$ are estimators of $\mu_{ij}$; the first is unconstrained and the second is constrained when the independence of factors is assumed, $p_{ij} = p_{i\cdot}p_{\cdot j}$. This point is important since $e_{ij}$ are "expected" under independence once the table is observed and fixed, while $\mu_{ij}$ are expectations of the random variables $N_{ij}$.

The log-linear model for the expected frequency $\mu_{ij}$ is given as

$$\log\mu_{ij} = \lambda_0 + \lambda_i^R + \lambda_j^C + (\lambda^{RC})_{ij}, \ i = 1,\ldots,r; j = 1,\ldots,c,$$

where $\lambda_i^R$ and $\lambda^C$ are contributions by row and column, respectively, and $\lambda_{ij}^{RC}$ is a row–column interaction term.

This model is called a *saturated log-linear model* and is similar to the two-factor ANOVA model. Note that an unconstrained model has $(1 + r + c + r \times c)$ free parameters but only $r \times c$ observations, $n_{ij}$. To make this overparameterized model identifiable, constraints on parameters are imposed. A standard choice is STZ: $\sum \lambda_i^R = 0, \sum_j \lambda_j^C = 0, \sum_i(\lambda^{RC})_{ij} = \sum_j(\lambda^{RC})_{ij} = 0$. Thus, with the constraints we have $1 - (r-1) + (c+1) + (r-1)(c-1) = r \times c$ free parameters, which is equal to the number of observations, and the model gives a perfect fit for the observed frequencies. This is the reason why this model is called *saturated.*

The hypothesis of independence in a contingency table that was discussed Chap. 14 has simple form:

$$H_0 : \lambda_{ij}^{RC} = 0, \ \ i = 2,\ldots,r; \ j = 2,\ldots,c.$$

Under $H_0$ the log-linear model becomes additive:

$$\log\mu_{ij} = \lambda_0 + \lambda_i^R + \lambda_j^C, \ i = 1,\ldots,r; \ j = 1,\ldots,c.$$

The MLEs of components in the log-linear model [not derived here, but see Agresti (2002)] are

$$\hat{\lambda}_0 = \frac{\sum_{i,j} \log n_{ij}}{rc},$$

$$\hat{\lambda}_i^R = \frac{\sum_j \log n_{ij}}{c} - \hat{\lambda}_0,$$

$$\hat{\lambda}_j^C = \frac{\sum_i \log n_{ij}}{r} - \hat{\lambda}_0,$$

$$\hat{\lambda}_{ij}^{RC} = \log n_{ij} - (\hat{\lambda}_0 + \hat{\lambda}_i^R + \hat{\lambda}_j^C).$$

If any $n_{ij}$ is equal to 0, then *all* entries in the table are replaced by $n_{ij} + 0.5$.

Traditional analysis involves testing that particular $\lambda$ components are equal to 0. One approach, often implemented in statistical software, would be to find the variance of $\hat{\lambda}$ (using a hypergeometric model), $\mathbb{V}\mathrm{ar}(\hat{\lambda})$, and use the statistic $(\hat{\lambda})^2/\mathbb{V}\mathrm{ar}(\hat{\lambda})$ that has a $\chi^2$-distribution with one degree of freedom ($\lambda$ here is any of $\lambda_0$, $\lambda_i^R$, $\lambda_j^C$, or $\lambda_{ij}^{RC}$). Large values of this statistic are critical for $H_0$.

Next, we focus on the Bayesian analysis of a log-linear model.

*Example 17.6.* In Exercise 14.9 concerning a psychological experiment, seeking assistance (help) was dependent on a subject's perception of the number of bystanders. The resulting chi-square statistic $\chi^2 = 7.908$ was significant with a *p*-value of about 2%. We revisit this exercise and provide a Bayesian solution using a loglinear approach. The WinBUGS program bystanders.odc is used to conduct statistical inference.

```
model{
    for (i in 1:r) for (j in 1:c)  n[i,j] ~ dpois(mu[i,j]);
    log(mu[i,j]) <- lambda0+ lambdaR[i]+lambdaC[j]+lambdaRC[i,j]
                    }

    lambda0 ~ dnorm(0,prec)

    lambdaR[1] <- 0
    lambdaC[1] <- 0
    for (i in 2:r) { lambdaR[i] ~ dnorm(0,prec) }
    for (i in 2:c) { lambdaC[i] ~ dnorm(0,prec) }

    for (j in 1 : c) {  lambdaRC[1, j] <- 0 }
    for (i in 2 : r) {  lambdaRC[i, 1] <- 0;
    for (j in 2 : c) {  lambdaRC[i, j] ~ dnorm(0, prec)}
}

DATA
list(r=3,c=2,prec=0.0001,
    n=structure(.Data=c(11,2,16,10,4,9),.Dim=c(3,2)))
```

```
INITS
list(lambda0 = 0,lambdaR = c(NA,0,0), lambdaC=c(NA,0),
lambdaRC = structure(.Data = c(NA, NA, NA,0,NA,0),
                      .Dim = c(3,2)) )
```

|            | mean   | sd     | MC error | val2.5pc | median | val97.5pc | start | sample |
|------------|--------|--------|----------|----------|--------|-----------|-------|--------|
| lambda0    | 2.351  | 0.3075 | 0.003347 | 1.707    | 2.366  | 2.908     | 2001  | 100000 |
| lambdaC[2] | -1.922 | 0.8446 | 0.01376  | -3.796   | -1.85  | -0.481    | 2001  | 100000 |
| lambdaR[2] | 0.3895 | 0.3977 | 0.004085 | -0.372   | 0.3842 | 1.186     | 2001  | 100000 |
| lambdaR[3] | -1.089 | 0.6145 | 0.006468 | -2.375   | -1.061 | 0.03655   | 2001  | 100000 |
| lambdaRC[2,2] | 1.435 | 0.9372 | 0.01451 | -0.2453  | 1.38   | 3.465     | 2001  | 100000 |
| lambdaRC[3,2] | 2.801 | 1.051  | 0.01653 | 0.9045   | 2.744  | 5.041     | 2001  | 100000 |

The hypothesis of independence is assessed by testing that all $\lambda_{RC}$ are equal to 0. In this case lambdaRC[1,1], lambdaRC[1,2], and lambdaRC[2,1] are set to 0 because of identifiability constraints. The 95% credible set for interaction lambdaRC[2,2] contains 0 – therefore, this interaction is not significant. On the other hand, lambdaRC[3,2] is significantly positive since the credible set [0.9045, 5.041] does not contain 0.

*Example 17.7.* **Upton–Fingleton Square.** An example from Upton and Fingleton (1985) concerns finding directional trends in spatial count data. The simple count data set is given as follows:

| 0 | 0 | 0 | 1 | 0 |
|---|---|---|---|---|
| 3 | 3 | 5 | 2 | 0 |
| 2 | 3 | 6 | 2 | 5 |
| 1 | 5 | 4 | 6 | 7 |
| 6 | 2 | 4 | 3 | 4 |

We are interested in testing if there are any north–south or east–west trends present in the spatial pattern. The idea of Upton and Fingleton was to establish a Poisson regression where the response was the intensity $n_{ij}$ in the location $(i, j)$ and the covariates are $x$-coordinate, related to the east–west trend, and $y$-coordinate, related to the north–south trend.

Thus, the observed frequencies $n_{ij}$ will be modeled as

$$\mathbb{E}(n_{ij}) = \exp\{\beta_0 + \beta_1 x_i + \beta_2 y_j\}, \; i, j = 1, \ldots, 5,$$

where $x_i = i$ and $y_j = j$. Significant $\beta_1$ or $\beta_2$ in a well-fitting Poisson regression will indicate the presence of corresponding trends (see ◀ UptonFingleton.m for details).

## 17.5 Exercises

17.1. **Blood Pressure and Heart Disease.** This example is based on data
given by Cornfield (1962). A sample of male residents of Framingham, Massachusetts, aged 40–59, were chosen. During a 6-year follow-up period they
were classified according to several factors, including blood pressure and
whether they had developed coronary heart disease. The results for these
two variables are given in the table below. The covariate for blood pressure
represents the interval, for example, 122 stands for blood pressure in the
interval 117–126.

| Blood pressure | No disease | Disease | Total |
|---|---|---|---|
| 112 (<117) | 156 | 3 | 159 |
| 122 (117–126) | 252 | 17 | 269 |
| 132 (127–136) | 284 | 12 | 296 |
| 142 (137–146) | 271 | 16 | 287 |
| 152 (147–156) | 139 | 12 | 151 |
| 162 (157–166) | 85 | 8 | 93 |
| 177 (167–186) | 99 | 16 | 115 |
| 197 (> 186) | 43 | 8 | 51 |

Using logistic regression, estimate the probability of disease for a person
with an average blood pressure equal to 158.

17.2. **Blood Pressure and Heart Disease in WinBUGS.** Use data from the
previous exercise. Parameter $p$ represents the probability of development
of coronary disease and can be estimated as

$$\hat{p} = \frac{e^{b_0 + b_1 BP}}{1 + e^{b_0 + b_1 BP}},$$

where $b_0$ and $b_1$ are Bayes estimators of $\beta_0$ and $\beta_1$ obtained by WinBUGS.
Use noninfromative priors on $\beta_0$ and $\beta_1$.
(a) What are values $b_0$ and $b_1$? Provide plots of posterior densities for $b_0$
and $b_1$.
(b) What are the 95% credible intervals for $\beta_0$ and $\beta_1$?
(c) Estimate the probability of disease for a person with an average blood
pressure of 158.

17.3. **Sex of Turtles and Incubation Temperature.** Below are data on the
relationship between the proportion of male turtles and incubation temperature for turtle eggs from New Mexico.

| Temp in ° C | Male | Female |
|---|---|---|
| 27.2 | 0 | 10 |
| 28.3 | 8 | 4 |
| 29.9 | 8 | 2 |

Predict the probability of female turtle for a temperature of 28° C.

17.4. **Health Promotion.** Students at the University of the Best in England (UBE) investigated the use of a health promotion video in a doctor's surgery. Covariates **Age** and **Amount of Weekly Exercise** for a sample of 30 men were obtained, and each man was asked a series of questions on the video. On the basis of the responses to these questions the psychologist simply recorded whether the promotion video was **Effective** or **Not Effective**. The data collected are as follows:

| Number | Age | Exercise | Code of response | Response |
|--------|-----|----------|------------------|----------|
| 1 | 27 | 3 | 0 | Not Effective |
| 2 | 26 | 5 | 0 | Not Effective |
| 3 | 28 | 10 | 1 | Effective |
| 4 | 40 | 4 | 1 | Effective |
| 5 | 19 | 3 | 0 | Not Effective |
| 6 | 36 | 14 | 0 | Not Effective |
| 7 | 41 | 5 | 1 | Effective |
| 8 | 27 | 5 | 0 | Not Effective |
| 9 | 33 | 1 | 0 | Not Effective |
| 10 | 34 | 2 | 1 | Effective |
| 11 | 51 | 8 | 1 | Effective |
| 12 | 33 | 21 | 1 | Effective |
| 13 | 23 | 12 | 0 | Not Effective |
| 14 | 41 | 19 | 1 | Effective |
| 15 | 38 | 2 | 0 | Not Effective |
| 16 | 25 | 9 | 0 | Not Effective |
| 17 | 40 | 6 | 0 | Not Effective |
| 18 | 36 | 25 | 1 | Effective |
| 19 | 40 | 13 | 0 | Not Effective |
| 20 | 39 | 3 | 0 | Not Effective |
| 21 | 45 | 10 | 1 | Effective |
| 22 | 39 | 5 | 0 | Not Effective |
| 23 | 40 | 2 | 1 | Effective |
| 24 | 20 | 1 | 0 | Not Effective |
| 25 | 47 | 9 | 1 | Effective |
| 26 | 31 | 17 | 1 | Effective |
| 27 | 37 | 7 | 1 | Effective |
| 28 | 30 | 0 | 0 | Not Effective |
| 29 | 32 | 13 | 1 | Effective |
| 30 | 25 | 10 | 0 | Not Effective |

(a) Use the fitted model to estimate the probability that a male of age 40 who exercises 10 hours a week would find the video "effective."
**Comment:** The linear predictor is

$$-7.498 + 0.17465 \text{ Age} + 0.16324 \text{ Exercise}.$$

The positive coefficient with covariate Age means that older subjects tend to respond "video is effective" with higher probabilities. Similarly, the positive coefficient with predictor Exercise indicates that increasing the values of Exercise also increases the probabilities for the response "video is effective."
(b) Comment on the fit of the model based on deviance and residuals.

17.5. **PONV.** Despite advances over the past decade, including the advent of 5-$HT_3$ receptor antagonists, combination therapy, and multimodal strategies, postoperative nausea and vomiting (PONV) remains a serious and frequent adverse event associated with surgery and anesthesia. PONV can be very distressing for patients, can lead to medical complications, and impose economic burdens. A meta-analysis of several studies gives rates of 37% for nausea and 20% for vomiting in patients undergoing general anesthesia. However, indiscriminate prophylaxis is not recommended (the "prevent-or-cure" dilemma).

There are considerable variations in the reported incidence of PONV, which can be attributed to a number of factors. Risk factors for PONV can be divided into patient risk factors, procedural risk factors, anesthetic risk factors, and postoperative risk factors. The main and well-understood risk factors are gender, history of motion sickness/PONV, smoking status, and use of postoperative opioids.

A data set 🖳 PONV.xls or PONV.mat (courtesy of Jelena Velickovic, MD anesthesiologist from Belgrade) contains records of 916 patients consisting of some demographic, anamnetic, clinical, and procedural variables. Several variables are of interest to be modeled and controlled: manifestation of PONV from 0 to 2 hours after surgery, PONV from 2 to 24 hours, and PONV from 0 to 24 hours (PONV0to2, PONV2to24, and PONV0to24).

Three score variables (SinclairScore, ApfelScore, and LelaScore) summarize the relevant demographic and clinical information prior to surgery with the goal of predicting PONV.

The starter file ponv.m gives a basic explanation of variables and helps to read all the variables into MATLAB.

Fit a logistic model for predicting PONV0to24 based on a modified Sinclair Score defined as MSS = SinclairScore + 1/20 * LelaScore. What is the probability of PONV0to24 for a person with MSS = 1.3?

17.6. **Mannose-6-phosphate Isomerase.** McDonald (1985) counted allele frequencies at the mannose-6-phosphate isomerase (Mpi) locus in the amphipod crustacean *Megalorchestia californiana*, which lives on sandy beaches of the Pacific coast of North America. There were two common alleles, Mpi90 and Mpi100. The latitude of each collection location, the number of each allele, and the proportion of the Mpi100 allele are shown here:

| location | latitude | Mpi90 | Mpi100 | prop Mpi100 |
|---|---|---|---|---|
| Port Townsend, WA | 48.1 | 47 | 139 | 0.748 |
| Neskowin, OR | 45.2 | 177 | 241 | 0.577 |
| Siuslaw River, OR | 44.0 | 1087 | 1183 | 0.521 |
| Umpqua River, OR | 43.7 | 187 | 175 | 0.483 |
| Coos Bay, OR | 43.5 | 397 | 671 | 0.628 |
| San Francisco, CA | 37.8 | 40 | 14 | 0.259 |
| Carmel, CA | 36.6 | 39 | 17 | 0.304 |
| Santa Barbara, CA | 34.3 | 30 | 0 | 0.000 |

Estimate the probability of allele Mpi100, taking into account latitude as a covariate, using logistic regression.

17.7. **Arthritis Treatment Data.** The data were obtained from Koch and Edwards (1988) for a double-blind clinical trial investigating a new treatment for rheumatoid arthritis. In this data set, there were 84 subjects of different ages who received an active or placebo treatment for their arthritis pain, and the subsequent extent of improvement was recorded as marked, some, or none. The dependent variable **improve** was an ordinal categorical observation with three categories (0 for none, 1 for some, and 2 for marked). The three explanatory variables were **treatment** (1 for active or 0 for placebo), **gender** (1 for male, 2 for female), and **age** (recorded as a continuous variable). The data in ▨ arthritis2.dat is organized as a matrix, with 84 rows corresponding to subjects and 5 columns containing ID number, treatment, gender, age, and improvement status:

$$
\begin{array}{ccccc}
57 & 1 & 1 & 27 & 1 \\
9 & 0 & 1 & 37 & 0 \\
46 & 1 & 1 & 29 & 0 \\
\cdots & & & & \\
15 & 0 & 2 & 66 & 1 \\
1 & 0 & 2 & 74 & 2 \\
71 & 0 & 2 & 68 & 1
\end{array}
$$

Dichotomize the variable **improve** as improve01=improve>0; and fit the binary regression with **improve01** as a response and **treatment**, **gender**, and **age** as covariates. Use the three links logit, probit, and comploglog and compare the models by comparing the deviances.

17.8. **Third-degree Burns.** The data for this exercise, discussed in Fan et al. (1995), refer to $n = 435$ adults who were treated for third-degree burns by the University of Southern California General Hospital Burn Center. The patients were grouped according to the area of third-degree burns on the body. For each midpoint of the groupings "log(area +1)," the number of patients in the corresponding group who survived, and the number who died from the burns.

| log(area+1) | Survived | Died |
|---|---|---|
| 1.35 | 13 | 0 |
| 1.60 | 19 | 0 |
| 1.75 | 67 | 2 |
| 1.85 | 45 | 5 |
| 1.95 | 71 | 8 |
| 2.05 | 50 | 20 |
| 2.15 | 35 | 31 |
| 2.25 | 7 | 49 |
| 2.35 | 1 | 12 |

• Fit the logistic regression on the probability of death due to third-degree burns with the covariate log(area+1).

• Using your model, estimate the probability of survival for a person for which log(area + 1) equals 2.

17.9. **Diabetes Data.** The data repository of Andrews and Herzberg (1985) features a data set containing measures of blood glucose, insulin levels, relative weights, and clinical classifications of 145 subjects, 🖵 diabetes.dat:

| 1 0.81 | 80 | 356 | 124 | 55 1 |
|---|---|---|---|---|
| 2 0.95 | 97 | 289 | 117 | 76 1 |
| 3 0.94 | 105 | 319 | 143 | 105 1 |
| ... | | | | |
| 143 0.90 | 213 | 1025 | 29 | 209 3 |
| 144 1.11 | 328 | 1246 | 124 | 442 3 |
| 145 0.74 | 346 | 1568 | 15 | 253 3 |

The columns represent the following variables:

| Variable | Meaning |
|---|---|
| relwt | Relative weight |
| glufast | Fasting plasma glucose |
| glutest | Test plasma glucose |
| instest | Plasma insulin during test |
| sspg | Steady-state plasma glucose |
| group | Clinical group: (3) overt diabetic; (2) chem. diabetic; (1) normal |

From the variable group form the variable dia and set it to 1 if diabetes is present (group=1,2) and 0 if the subject has no diabetes (group=1). Find the regression of the five other variables on dia.

17.10. **Remission Ratios over Time.** A clinical trial on a new anticancer agent produced the following remission ratios for 40 patients on trial at each of the six stages of the trial:

| 9/40 14/40 22/40 29/40 33/40 35/40 |
|---|

Fit the logistic model for the probability of remission if the stages are measured in equal time units. Give the probability that a new patient on this regiment will be in remission at stage 4 and discuss how this probability compares to 29/40.

17.11. **Death of Sprayed Flour Beetles.** Hewlett (1974) and Morgan (2000) provide data that have been considered by many researchers in bioassay theory. The data consist of a quantal bioassay for pyrethrum in which the mortality of adult flour beetles (*Tribolium castaneum*) was measured over time under four dose levels. The columns are cumulative numbers of dead adult flour beetles exposed initially to pyrethrum, a well-known plant-based insecticide. Mixed with oil, the pyrethrum was sprayed at the given dosages over small experimental areas in which the groups of beetles were confined but allowed to move freely. The beetles were fed during the experiment in order to eliminate the effect of natural mortality.

| Dose(mg/cm2) | | 0.20 | | 0.32 | | 0.50 | | 0.80 | |
|---|---|---|---|---|---|---|---|---|---|
| | Sex | M | F | M | F | M | F | M | F |
| Day | 1 | 3 | 0 | 7 | 1 | 5 | 0 | 4 | 2 |
| | 2 | 14 | 2 | 17 | 6 | 13 | 4 | 14 | 9 |
| | 3 | 24 | 6 | 28 | 17 | 24 | 10 | 22 | 24 |
| | 4 | 31 | 14 | 44 | 27 | 39 | 16 | 36 | 33 |
| | 5 | 35 | 23 | 47 | 32 | 43 | 19 | 44 | 36 |
| | 6 | 38 | 26 | 49 | 33 | 45 | 20 | 46 | 40 |
| | 7 | 40 | 26 | 50 | 33 | 46 | 21 | 47 | 41 |
| | 8 | 41 | 26 | 50 | 34 | 47 | 25 | 47 | 42 |
| | 9 | 41 | 26 | 50 | 34 | 47 | 25 | 47 | 42 |
| | 10 | 41 | 26 | 50 | 34 | 47 | 25 | 48 | 43 |
| | 11 | 41 | 26 | 50 | 34 | 47 | 25 | 48 | 43 |
| | 12 | 42 | 26 | 50 | 34 | 47 | 26 | 48 | 43 |
| | 13 | 43 | 26 | 50 | 34 | 47 | 27 | 48 | 43 |
| Group Size | | 144 | 152 | 69 | 81 | 54 | 44 | 50 | 47 |

Using WinBUGS, model the proportion of dead beetles using sex, dosage, and day as covariates. The data set in WinBUGS format can be found in 🔆tribolium.odc.

17.12. **Mortality in Swiss White Mice.** An experiment concerning the influence of diet on the rate of *Salmonella enteritidis* infection (mouse typhoid) among W-Swiss mice was conducted by Schneider and Webster (1945). Two diets, one of whole wheat and whole dried milk [coded 100] and the other synthetic [coded 191], are compared against two doses of bacilli, 50K and 500K. The experimental results are summarized in the following table.

| | Dose 50K | | Dose 500K | |
|---|---|---|---|---|
| | Number | Number of | Number | Number of |
| Diet | of mice | surviving | of mice | surviving |
| **100** | 293 | 194 | 144 | 54 |
| **191** | 296 | 120 | 141 | 25 |

The authors conclude that diet is able to condition natural resistance, but one of the factors was the genetic constitution of the mice employed.

Model the probability of survival using an additive logistic regression with the covariates Diet and Dose.

17.13. **Kyphosis Data.** The measurements in 🖳 kyphosis.dat are from Hastie and Tibshirani (1990, p. 301) and were collected on 83 patients undergoing corrective spinal surgery (Bell et al., 1994). The objective was to determine the important risk factors for kyphosis, or the forward flexion of the spine at least 40 degrees from vertical following surgery. The covariates are age in month, the starting vertebrae level of the surgery, and the number of vertebrae involved.

The data set 🖳 kyphosis.dat has four columns:

| | | | |
|---|---|---|---|
| 71 | 5 | 3 | 0 |
| 158 | 14 | 3 | 0 |
| 128 | 5 | 4 | 1 |
| ... | | | |
| 120 | 13 | 2 | 0 |
| 42 | 6 | 7 | 1 |
| 36 | 13 | 4 | 0 |

where the first column is the **age** (in month), the second is the starting vertebrae **start** of the surgery, the third is the **number** of vertebrae involved, and the fourth column is binary indicator for the presence of kyphosis, 0 if absent and 1 if present.

Using logistic regression, model the probability of kyphosis given the risk factors age, start, and number.

17.14. **Prostate Cancer.** The Prostate Cancer clinical trial data of Byar and Green (1980) is given in the file 🖳 prostatecanc.dat with description of variables in 🖳 prostatecanc.txt or ◢ prostatecanc.m. There are 475 observations with 12 measured covariates. The response is the stage (3 or 4) of a patient assessed by a physician.

Propose a model for predicting the stage by a subset of predictors.

17.15. **Pediculosis Capitis.** An outbreak of *Pediculosis capitis* is being investigated in a girls' school containing 291 pupils. Of 130 children who live in a nearby housing estate, 18 were infested, and of 161 who live elsewhere, 37 were infested. Thus, the school girls are stratified by the housing attribute into two groups: (A) the nearby housing estate and (B) elsewhere.

(a) Test the hypothesis that the population proportions of infested girls for groups A and B are the same.

(b) Run a logistic regression that predicts the probability of a girl being infested including the predictor housing that takes value 1 if the girl is from group A and 0 if she is from group B. All you need are the sample sizes from groups A and B and the corresponding incidences of of infestation. You

might need to recode the data and represent the summarized data as 291
individual cases containing the incidence of infestation and housing status.
(c) A sample of 26 girls from group A and 34 from group B are randomly
selected for more detailed modeling analysis. The instances of infestation
(0-no, 1-yes), housing (A=1, B=0), family income (in thousands), family size,
and girl's age are recorded. The data are available in the file 📊 lice.xls.
Propose the logistic model that predicts the probability that a girl who is
infested will possess some or all of the predictors (housing, income, size,
age). This is an open-ended question, and you are expected to defend your
proposed model.
(c.1) According to your model, what is the probability of a girl being infested
if housing is A, family income is 74, family size is 4, and age is 12? Of course,
you will use only the values of the predictors included in your model.
(c.2) If the family size is increased by 1 and all other covariates remain the
same, how much do the odds of infestation change?
(d) The 55 affected girls were divided randomly into two groups of 29 and
26. The first group received a standard local application and the second
group a new local application. The efficacy of each was measured by clearance of the infestation after one application. By this measure the standard
application failed in ten cases and the new application in five. Is the new
treatment more effective? This part may mimic the methodology in (a).

17.16. **Finney Data.** In a controlled experiment to study the effect of the rate
and volume of air inspired on a transient reflex vasoconstriction in the skin
of the digits, 39 tests under various combinations of rate and volume of
air inspired were conducted (Finney, 1947). The end point of each test was
whether or not vasoconstriction occurred.

| Volume | Rate | Response | Volume | Rate | Response | Volume | Rate | Response |
|---|---|---|---|---|---|---|---|---|
| 3.70 | 0.825 | Constrict | 3.50 | 1.09 | Constrict | 1.80 | 1.50 | Constrict |
| 1.25 | 2.50 | Constrict | 0.75 | 1.50 | Constrict | 0.95 | 1.90 | No constrict |
| 0.80 | 3.20 | Constrict | 0.70 | 3.50 | Constrict | 1.90 | 0.95 | Constrict |
| 0.60 | 0.75 | No constrict | 1.10 | 1.70 | No constrict | 1.60 | 0.40 | No constrict |
| 0.90 | 0.75 | No constrict | 0.90 | 0.45 | No constrict | 2.70 | 0.75 | Constrict |
| 0.80 | 0.57 | No constrict | 0.55 | 2.75 | No constrict | 2.35 | 0.03 | No constrict |
| 0.60 | 3.00 | No constrict | 1.40 | 2.33 | Constrict | 1.10 | 1.83 | No constrict |
| 0.75 | 3.75 | Constrict | 2.30 | 1.64 | Constrict | 1.10 | 2.20 | Constrict |
| 3.20 | 1.60 | Constrict | 0.85 | 1.415 | Constrict | 1.20 | 2.00 | Constrict |
| 1.70 | 1.06 | No constrict | 1.80 | 1.80 | Constrict | 0.80 | 3.33 | Constrict |
| 0.40 | 2.00 | No constrict | 0.95 | 1.36 | No constrict | 0.95 | 1.90 | No constrict |
| 1.35 | 1.35 | No constrict | 1.50 | 1.36 | No constrict | 0.75 | 1.90 | No constrict |
| 1.60 | 1.78 | Constrict | 0.60 | 1.50 | No constrict | 1.30 | 1.625 | Constrict |

Model the probability of vasoconstriction as a function of two covariates,
Volume and Rates. Use MATLAB and compare the results with Win-
BUGS output. The data in MATLAB and WinBUGS formats is provided
in 📊 finney.dat.

17.17. **Shocks.** An experiment was conducted (Dalziel et al., 1941) to assess the
effect of small electrical currents on farm animals, with the eventual goal

of understanding the effects of high-voltage power lines on livestock. The experiment was carried out with seven cows using six shock intensities, 0, 1, 2, 3, 4, and 5 milliamps (shocks on the order of 15 milliamps are painful for many humans). Each cow was given 30 shocks, 5 at each intensity, in random order. The entire experiment was then repeated, so each cow received a total of 60 shocks. For each shock the response, mouth movement, was either present or absent. The data as quoted give the total number of responses, out of 70 trials, at each shock level. We ignore cow differences and differences between blocks (experiments).

| Current (milliamps) $x$ | Number of responses $y$ | Number of trials $n$ | Proportion of responses $p$ |
|---|---|---|---|
| 0 | 0 | 70 | 0.000 |
| 1 | 9 | 70 | 0.129 |
| 2 | 21 | 70 | 0.300 |
| 3 | 47 | 70 | 0.671 |
| 4 | 60 | 70 | 0.857 |
| 5 | 63 | 70 | 0.900 |

Using logistic regression and noninformative priors on its parameters, estimate the proportion of responses after a shock of 2.5 milliamps. Find 95% credible set for the population proportion.

17.18. **Ants.** The data set ⬛ ants.dat, discussed in Gotelli and Ellison (2002), provides the ant species richness (number of ant species) found in 64-square-meter sampling grids in 22 forests (coded as 1) and 22 bogs (coded as 2) surrounding the forrests in Connecticut, Massachusetts, and Vermont. The sites span 3° of latitude in New England. There are 44 observations on 4 variables (columns in data set): Ants – number of species, Habitat – forests (1) and bogs (2), and Elevation – in meters above sea level.
Using Poisson regression, model the number of ant species (Ants) with covariates Habitat and Elevation.

17.19. **Sharp Dissection and Postoperative Adhesions Revisited.** In Exercise 16.3 we fitted a linear relationship between the logarithm of the amount of sharp dissection lasd (predictor) and severity score sesco (response). Criticize this linear model. Model this relationship using Poisson regression and graphically compare the linear and Poisson fits.

17.20. **Airfreight breakage.** A substance used in biological and medical research is shipped by air freight to users in cartons of 1000 ampules. The data below, involving ten shipments, were collected on the number of times a carton was transferred from one aircraft to another over the shipment route ($X$) and the number of ampules found to be broken upon arrival ($Y$).

| $X$ | 1 | 0 | 2 | 0 | 3 | 1 | 0 | 1 | 2 | 0 |
|---|---|---|---|---|---|---|---|---|---|---|
| $Y$ | 16 | 9 | 17 | 12 | 22 | 13 | 8 | 15 | 19 | 11 |

Using WinBUGS, fit $Y$ by Poisson regression, with $X$ as a covariate. According to your model, how many packages will be broken if the number of shipment routes is $X = 4$?

17.21. **Body Fat Affecting Accuracy of Heart Rate Monitors.** In the course *Problems in Biomedical Engineering I* at Georgia Tech, a team of students investigated whether the readings of heart rate from a chest strap monitor (Polar T31) were influenced by the subject's percent body fat.

Hand counts facilitated by a stethoscope served as the gold standard. The absolute differences between device and hand counts (AD) were regressed on body fat (BF) measurements. The measurements for 28 subjects are provided below.

| Subj. | BF | AD | Subj. | BF | AD | Subj. | BF | AD | Subj. | BF | AD |
|---|---|---|---|---|---|---|---|---|---|---|---|
| 1 | 17.8 | 4 | 8 | 18.8 | 1 | 15 | 25.1 | 3 | 22 | 24.1 | 6 |
| 2 | 13.2 | 3 | 9 | 13.4 | 0 | 16 | 18.3 | 2 | 23 | 12.9 | 2 |
| 3 | 7.7 | 3 | 10 | 39.4 | 7 | 17 | 16.9 | 3 | 24 | 30.1 | 6 |
| 4 | 11.8 | 1 | 11 | 6.8 | 1 | 18 | 27.8 | 6 | 25 | 17.1 | 4 |
| 5 | 23.9 | 0 | 12 | 25.0 | 6 | 19 | 36.0 | 5 | 26 | 18.4 | 4 |
| 6 | 27.2 | 0 | 13 | 19.9 | 0 | 20 | 31.9 | 1 | 27 | 14.6 | 4 |
| 7 | 27.6 | 0 | 14 | 23.0 | 9 | 21 | 17.4 | 2 | 28 | 26.8 | 3 |

A significant non-constant representation of AD as a function of BF can be translated as a significant influence of percent of body fat on the accuracy of the device.

(a) Why the linear regression is not adequate here?

(b) Fit the Poisson regression AD $\sim \mathcal{P}oi(\exp\{b_0 + b_1 \text{BF}\})$. Is the slope $b_1$ from the linear part significantly positive?

(c) Using WinBUGS find 95% credible set for the slope $b_1$ in the linear part of a Bayesian Poisson regression model. Use the non-informative priors.

17.22. **Miller Lumber Company Customer Survey.** Kutner et al. (2005) analyze a data set from a survey of customers of the Miller Lumber Company. The response is the total number of customers (in a representative 2-week period) coming from a tract of a metropolitan area within 10 miles from the store. The covariates include five variables concerning the tracts: number of housing units, average income in dollars, average housing unit age in years, distance to nearest competitor in miles, and distance to store in miles. Fit and assess a Poisson regression model for the number of customers as predicted by the covariates. The data are in ◀ lumber.m.

17.23. $SO_2$, $NO_2$, **and Hospital Admissions.** Fan and Chen (1999) discuss a public health data set consisting of daily measurements of pollutants and other environmental factors in Hong Kong between January 1, 1994 and December 31, 1995. The association between levels of pollutants and the number of daily hospital admissions for circulation and respiratory problems is of particular interest.

The data file ⌨ hospitaladmissions.dat consists of six columns: (1) year, (2) month, (3) day in month, (4) concentration of sulfur dioxide $SO_2$, (5) concentration of pollutant nitrogen $NO_2$, and (6) daily number of hospital admissions.

(a) Using logistic regression, determine how the probability of a high level of sulfur dioxide (with values $> 20\,\mu g/m^3$) is associated with the level of pollutant nitrogen $NO_2$.

(b) Using a Poisson regression model, explore how the number of hospital admissions varies with the level of $NO_2$.

17.24. **Kidney Stones.** Charig et al. (1986) provide data on the success rates of two methods of treating kidney stones: open surgery methods and percutaneous nephrolithotomy. There are two predictors: Size of stone at two levels: $< 2\,cm$ in diameter, coded as Small, and $> 2\,cm$ in diameter, coded as Large. The two methods are coded as A (open surgery) and B (percutaneous nephrolithotomy). The outcome of interest is the outcome of the treatment (Success, Failure).

| Count | Size | Method | Outcome |
|-------|------|--------|---------|
| 81 | Small | A | Success |
| 6 | Small | A | Failure |
| 234 | Small | B | Success |
| 36 | Small | B | Failure |
| 192 | Large | A | Success |
| 71 | Large | A | Failure |
| 55 | Large | B | Success |
| 25 | Large | B | Failure |

There are four combinations of the covariates: Small A, Small B, Large A, and Large B. Find the relative frequencies of the outcome "Success" and compare them with the model-predicted probabilities using logistic regression.

Show that these data hide a Simpson paradox.

---

| **MATLAB AND WINBUGS FILES AND DATA SETS USED IN THIS CHAPTER** |
http://springer.bme.gatech.edu/Ch17.Logistic/

arrhythmia.m, arthritis2.m, beetleBliss1.m, beetleBliss2.m, caesarean.m, celular.m, counterr.m, dmdreg.m, ihga.m, kyphosis.m, logisticmle.m, logisticSeed.m, lumber.m, outbreak.m, ponv.m, prostatecanc.m, UptonFingleton.m

accidentssimple.odc, arrhythmia.odc, beetles.odc, bliss.odc, bystanders.odc, caesarean.odc, celldifferentiation.odc, errors1.odc, geriatric.odc, loglinear1.odc, microdamage.odc, raynaud.odc, remission.odc, tribolium.odc

ants.dat, Arrhythmia.mat, arrhythmiadata.txt, arthritis2.dat, birthweight.dat, cardiac.mat|txt, celular.dat, diabetes.dat, dmd.dat|mat, finney.dat, hospitaladmissions.dat, ihgadat.m, kyphosis.dat|txt, lowbwt.dat, microdamage.mat|txt, outbreak.mat, pima.dat, PONV.mat|xls, programm.mat, prostatecanc.dat, tribolium.mat

---

# CHAPTER REFERENCES

Andrews, D. F. and Herzberg, A. M. (1985). *Data: A Collection of Problems from Many Fields for the Student and Research Worker*. Springer, Berlin Heidelberg New York.

Atkinson, A. C. (1985). *Plots, Transformations and Regression: An Introduction to Graphical Methods of Diagnostic Regression Analysis*. Oxford University Press, New York.

Bell, D. F., Walker, J. L., O'Connor, G., and Tibshirani, R. (1994). Spinal deformity after multiple-level cervical laminectomy in children. *Spine*, **19**, 4, 406–411.

Bliss, C. I. (1935). The calculation of the dose-mortality curve. *Ann. Appl. Biol.*, **22**, 134–167.

Byar, D. P. and Green, S. B. (1980). The choice of treatment for cancer patients based on covariate information: Application to prostate cancer. *Bull. Cancer (Paris)*, **67**, 477–488.

Charig, C. R., Webb, D. R., Payne, S. R., and Wickham, J. A. E. (1986). Comparison of treatment of renal calculi by open surgery, percutaneous nephrolithotomy, and extracorpeal shockwave lithotripsy. *Br. Med. J.*, **292**, 879–882.

Cornfield, J. (1962). Joint dependence of risk of coronary heart disease on serum cholesterol and systolic blood pressure: a discriminant function analysis. *Federat. Proc.*, **21**, 58–61.

Dalziel, C. F., Lagen, J. B., and J. L. Thurston, J. L. (1941). Electric shocks. *Trans. IEEE*, **60**, 1073–1079.

Fahrmeir, L. and Tutz, G. (1994). *Multivariate Statistical Modelling Based on Generalized Linear Models*. Springer, Berlin Heidelberg New York.

Fan, J. and Chen, J. (1999). One-step local quasi-likelihood estimation. *J. R. Stat. Soc. Ser. B*, **61**, 4, 927–943.

Fan, J., Heckman, N. E., and Wand, M. P. (1995). Local polynomial kernel regression for generalized linear models and quasi-likelihood functions. *J. Am. Stat. Assoc.*, **90**, 429, 141–150.

Finney, D. J. (1947). The estimation from individual records of the relationship between dose and quantal response. *Biometrika*, **34**, 320–334.

Gotelli, N. J. and Ellison, A. M. (2002). Biogeography at a regional scale: determinants of ant species density in bogs and forests of New England. *Ecology*, **83**, 1604–1609.

Hastie, T. and Tibshirani, R. (1990). *Generalized Additive Models*. Chapman & Hall, London.

Hendriksen, C., Lund, E., and Stromgard, E. (1984). Consequences of assessment and intervention among elderly people: a 3 year randomised, controlled trial. *Br. Med. J.*, **289**, 1522–1544.

Hewlett, P. S. (1974). Time from dosage to death in beetles Tribolium castaneum, treated with pyrethrins or DDT, and its bearing on dose-mortality relations. *J. Stored Prod. Res.*, **10**, 27–41.

Hosmer, D. and Lemeshow, S. (1989). *Applied Logistic Regression* (2nd edn. 2000). Wiley, New York.

Koch, G. and Edwards, S. (1988). Clinical efficiency trials with categorical data. In: Peace, K. E. (ed) *Biopharmaceutical Statistics for Drug Development*. Dekker, New York, pp. 403–451.

Kutner, M. H., Nachtsheim, C. J., Neter, J., and Li, W. (2005). *Applied Linear Statistical Models*. McGraw-Hill/Irwin, New York.

McCullagh, P. and Nelder, J. A. (1989). *Generalized Linear Models*, 2nd edn. Monographs on Statistics and Applied Probability. Chapman & Hall/CRC, Boca Raton.

Morgan, B. J. T. (2000). *Applied Stochastic Modelling*. Arnold Texts in Statistics. Oxford University Press, New York.

Piergorsch, W. W., Weinberg, C. R., and Margolin, B. H. (1988). Exploring simple independent action in multifactor table of proportions. *Biometrics*, **44**, 595–603.

Schneider, H. A. and Webster, L. T. (1945). Nutrition of the host and natural resistance to infection. I: The effect of diet on the response of several genotypes of musculus to salmonella enteritidis infection. *J. Exp. Med.*, **81**, 4, 359–384.

Upton, G. J. G. and Fingleton B. (1985). *Spatial Data Analysis by Example*, vol. 1. Wiley, Chichester.

# Chapter 18
# Inference for Censored Data and Survival Analysis

*The first condition of progress is the removal of censorship.*

– George Bernard Shaw

## 18.1 Introduction

Survival analysis models the survival times of a group of subjects (usually with some kind of medical condition) and generates a survival curve, which shows how many of the subjects are "alive" or survive over time.

What makes survival analysis different from the standard regression methodology is the presence of censored observations; in addition, some subjects may have left the study and may be lost to follow-up. Such subjects were known to have survived for some amount of time (up until the time we last saw them), but we do not know how much longer they might ultimately have survived. Several methods have been developed for using this "at least this long" infor-

701

mation to finding unbiased survival curve estimates, the most popular being the nonparametric method of Kaplan and Meier.

An observation is said to be *censored* if we know only that it is less than (or greater than) a certain known value. For instance, in clinical trials, one could be interested in patients' survival times. Survival time is defined as the length of time between diagnosis and death, although other "start" events (such as surgery), and other "end" events (such as relapse of disease, increase in tumor size beyond a particular threshold, rejection of a transplant, etc.) are commonly used. Because of many constraints, trials cannot be run until the endpoints are observed for all patients. Just because for a particular subject the time to endpoint is not fully observed, partial information is still available: the patient survived up to the end of the observational period, and this should be incorporated into the analysis. Such observations are called *right censored*. Observations can also be *left censored*, for example, an assay may have a detection threshold. In order to utilize information contained in censored observations, special methods of analysis are required.

Biomedical engineers are often interested in the reliability of medical devices, and most of the methodology from survival analysis is applicable in device reliability analyses. There are of course important differences. In the reliability of multicomponent systems an important aspect is optimization of the number and position of components. Analogous considerations with organs or parts of organs as components in living systems (animals or humans) are impossible. Methods such as "accelerated life testing" commonly used in engineering reliability are inappropriate when dealing with human subjects.

However, in comparing the lifetimes of subjects in clinical trials involving different treatments (humans, animals) or different engineering interventions (systems, devices), the methodology that deals with censored observations is shared.

## 18.2 Definitions

Let $T$ be a random variable with CDF $F(t)$ representing a lifetime. The survival (or survivor) function is the tail probability for $T$, i.e., $S(t) = 1 - F(t)$, $t > 0$. The function $S(t)$ gives the probability of surviving up to time $t$, that is,

$$S(t) = \mathbb{P}(T > t).$$

The hazard function or hazard rate is defined as

$$h(t) = \frac{f(t)}{S(t)} = \frac{f(t)}{1 - F(t)},$$

when $T$ has a density $f(t)$. Note that $S'(t) = -f(t)$. It is insightful to represent $h(t)$ in limit terms,

$$\frac{S(t) - S(t + \Delta t)}{\Delta t} \times \frac{1}{S(t)} = \frac{F(t + \Delta t) - F(t)}{\Delta t} \times \frac{1}{S(t)}$$
$$= \frac{\mathbb{P}(t < T \leq t + \Delta t)}{\Delta t \, \mathbb{P}(T > t)}$$
$$= \frac{\mathbb{P}(T \leq t + \Delta t | T > t)}{\Delta t},$$

when $\Delta t \to 0$. It represents "an instantaneous" probability that an event that was not observed up to time $t$ will be observed before $t + \Delta t$, when $\Delta t \to 0$.

Cumulative hazard is defined as $H(t) = \int_0^t h(s) ds$. Both hazard and cumulative hazard uniquely determine the distribution of lifetime $F$, $F(t) = 1 - \exp\{-\int_0^t h(s) ds\}$, i.e., $F(t) = 1 - \exp\{-H(t)\}$. Cumulative hazard can also be connected to the survival function as $H(t) = -\log S(t)$, or

$$S(t) = \exp\{-H(t)\}. \tag{18.1}$$

*Example 18.1.* The hazard function for an exponential distribution with density $f(t) = \lambda e^{-\lambda t}$, $t \geq 0, \lambda > 0$ is constant in time, $h(t) = \lambda$.

*Example 18.2.* The hazard rate for a one-parameter Weibull distribution with CDF $F(t) = 1 - \exp\{-t^\gamma\}$ and density $f(t) = \gamma t^{(\gamma-1)} \exp\{-t^\gamma\}$, $t \geq 0, \gamma > 0$ is $h(t) = \gamma t^{\gamma-1}$. The parameter $\gamma$ is called a *shape* parameter. Depending on the "shape parameter," $\gamma$, the hazard function $h(t)$ could model various types of survival analyses.

The two-parameter version of Weibull is used more frequently. It is defined as $F(t) = 1 - \exp\{-\lambda t^\gamma\}$, with density $f(t) = \lambda \gamma t^{(\gamma-1)} \exp\{-\lambda t^\gamma\}$, $t \geq 0, \gamma > 0, \lambda > 0$. The parameter $\lambda$ is called a "rate" parameter. In this case $h(t) = \lambda \gamma t^\gamma$.

Two important summaries in the parametric case (where the survival distribution is specified up to a parameter) are mean residual life (mrl) and median life, defined respectively as

$$\mathrm{mrl}(t) = \frac{\int_t^\infty S(x)dx}{S(t)},$$

$$t_{0.5}: S(t_{0.5}) = 0.5.$$

*Example 18.3.* For an exponential lifetime, the expected lifetime and mrl coincide. Indeed, $\mathbb{E}T = 1/\lambda$, while

$$\mathrm{mrl}(t) = \frac{\int_t^\infty e^{-\lambda x}dx}{e^{-\lambda t}} = \frac{1}{\lambda}.$$

At first glance this looks like a paradox, the expected total lifetime $\mathbb{E}T$ is equal to the residual lifetime $\mathrm{mrt}(t)$ irrespective of $t$. This is an example of the "inspection paradox" that follows from the memoryless property of the exponential distribution. The median lifetime of an exponential distribution is $t_{0.5} = (\log 2)/\lambda$.

In MATLAB, dfittool, normfit, wblfit, and other commands for fitting parametric distributions can be applied to censored data by specifying the censoring vector at input.

## 18.3 Inference with Censored Observations

We will consider the case of right-censored data (the most common type of censoring) and two approaches: a parametric approach, in which the survival function $S(t)$ would have a specific functional form, and a nonparametric approach, in which no such functional form is assumed.

### 18.3.1 Parametric Approach

In the parametric approach the models depend on the parameters, and the parameters are estimated by taking into account both uncensored and censored observations. We will show how to find an MLE in the general case and illustrate it on an exponential lifetime distribution.

Let $(t_i, \delta_i)$, $i = 1,\ldots,n$ be observations of a lifetime $T$ for $n$ individuals, with $\delta \in \{0,1\}$ indicating censored and fully observed lifetimes, and let $k$ observations be fully observed while $n-k$ are censored. Suppose that the underlying lifetime $T$ has a density $f(t|\theta)$ with survival function $S(t|\theta)$. Then the likelihood is

$$L(\theta|t_1,\ldots,t_n) = \prod_{i=1}^{n}(f(t_i|\theta)^{\delta_i} \times (S(t_i|\theta))^{1-\delta_i} = \prod_{i=1}^{k}(f(t_i|\theta) \times \prod_{i=k+1}^{n} S(t_i|\theta).$$

Since $h(t_i|\theta) \times (S(t_i|\theta)) = f(t_i|\theta)$, then

$$L(\theta|t_1,\ldots,t_n) = \prod_{i=1}^{n}(h(t_i|\theta)^{\delta_i} \times S(t_i|\theta). \tag{18.2}$$

*Example 18.4.* We will show that for an exponential lifetime in the presence of right-censoring, the MLE for $\lambda$ is

$$\hat{\lambda} = \frac{k}{\sum_{i=1}^{n} t_i}, \tag{18.3}$$

where $k$ is the number of noncensored data and $\sum_{i=1}^{n} t_i$ is the sum of *all* observed and censored times. From (18.2), the likelihood is $L = \lambda^k \exp\{-\lambda\sum_{i=1}^{n} t_i\}$. By taking the log and differentiating one gets the MLE as the solution to $\frac{k}{\lambda} - \sum_{i=1}^{n} t_i = 0$.

The variance of the MLE $\hat{\lambda}$ is $k/\left(\sum_{i=1}^{n} t_i\right)^2$ and can be used to find the confidence interval for $\lambda$ (Exercise 18.2).

*Example 18.5.* **Immunoperoxidase and BC.** Data analyzed in Sedmak et al. (1989) and also in Klein and Moeschberger (2003) represent times to death (in months) for breast cancer patients with different immunohistochemical responses. Out of 45 patients in the study, 9 were immunoperoxidase positive while the remaining 36 were negative (+ denotes censored time).

| |
|---|
| Immunoperoxidase Negative |
| 19, 25, 30, 34, 37, 46, 47, 51, 56, 57, 61, 66, 67, 74, 78, 86, 122+, 123+, 130+, 130+, 133+, 134+, 136+, 141+, 143+, 148+, 151+, 152+, 153+, 154+, 156+, 162+, 164+, 165+, 182+, 189+ |
| Immunoperoxidase Positive |
| 22, 23, 38, 42, 73, 77, 89, 115, 144+ |

Assume that lifetimes are exponentially distributed and that rates $\lambda_1$ (for Immunoperoxidase Negative) and $\lambda_2$ (for Immunoperoxidase Positive) are to be estimated. The following MATLAB code finds MLEs of $\lambda_1$ and $\lambda_2$, first directly by using (18.3) and then by using MATLAB's built-in function mle with option 'censoring'.

```
ImmPeroxNeg=[...
19, 25, 30, 34, 37, 46, 47, 51, 56, 57, 61, 66, 67, 74, 78, 86,...
122, 123, 130, 130, 133, 134, 136, 141, 143, 148, 151, 152,...
153, 154, 156, 162, 164, 165, 182, 189];
CensorIPN=[0,0,0,0,0,0,0,0,0,0,0,0,0,0,0,0,...
  1,1,1,1,1,1,1,1,1,1,1,1,1,1,1,1,1,1,1,1];

ImmPeroxPos=[...
22, 23, 38, 42, 73, 77, 89, 115, 144];
CensorIPP=[0,0,0,0,0,0,0,0,1];

%number of observed (non-censored)
k1 = sum(1-CensorIPN)    %16
k2 = sum(1-CensorIPP)    %8

% MLEs of rate lambda for 2 samples.
hatlam1 = k1/sum(ImmPeroxNeg)  %0.0042
hatlam2 = k2/sum(ImmPeroxPos)  %0.0128

[reclambdahat1 lamci1] = mle(ImmPeroxNeg, ...
   'distribution','exponential','censoring',CensorIPN)
   % 237.6250
   % 153.6769   415.7289
[reclambdahat2 lamci2] = mle(ImmPeroxPos, ...
   'distribution','exponential','censoring',CensorIPP)
   % 77.8750
   % 43.1959   180.3793
%(MATLAB parametrization) scale to rate
lambdahat1 = 1/reclambdahat1  %0.0042
lambdahat2 = 1/reclambdahat2  %0.0128
```

In conclusion, the patients who are immunoperoxidase positive are at increased risk since the rate $\hat{\lambda}_2 = 0.0128$ exceeds $\hat{\lambda}_1 = 0.0042$.

### 18.3.2 Nonparametric Approach: Kaplan–Meier Estimator

Assume that individuals in the study are assessed at discrete time instances $t_1, t_2, \ldots, t_k$, which may not be equally spaced. Sometimes, the times are selected when failures occur. If we want to calculate the probability of survival up to time $t_i$, then by the chain rule of conditional probabilities and their Markovian property,

$$\hat{S}(t_i) = \mathbb{P}(\text{ surviving to time } t_i\ ) = \mathbb{P}(\text{ survived up to time } t_1\ )$$
$$\times\ \mathbb{P}(\text{ surviving to time } t_2\ |\ \text{ survived up to time } t_1\ )$$
$$\times\ \mathbb{P}(\text{ surviving to time } t_3\ |\ \text{ survived up to time } t_2\ )$$
$$\cdots$$
$$\times\ \mathbb{P}(\text{ surviving to time } t_i\ |\ \text{ survived up to time } t_{i-1}\ ).$$

Suppose $r_i$ subjects are at risk at time $t_{i-1}$ and are not censored at time $t_{i-1}$. In the $i$th interval $(t_{i-1}, t_i)$ among these $r_i$ subjects $d_i$ have an event, $\ell_i$ are censored, and $r_{i+1}$ survive. The $r_{i+1}$ subjects will be at risk at the beginning of the $(i+1)$th time interval $(t_i, t_{i+1})$, that is, at time $t_i$. Thus, $r_i = d_i + \ell_i + r_{i+1}$. We can estimate the probability of survival up to time $t_i$, given that one survived up to time $t_{i-1}$, as $1 - d_i/(r_{i+1} + d_i + \ell_i) = 1 - d_i/r_i$.

The $\ell_i$ subjects censored at time $t_i$ do not contribute to the survival function for times $t > t_i$.

$$\hat{S}(t) = \left(1 - \frac{d_1}{r_1}\right) \times \left(1 - \frac{d_2}{r_2}\right) \times \cdots \times \left(1 - \frac{d_i}{r_i}\right)$$

$$= \prod_{t_i \le t} \left(1 - \frac{d_i}{r_i}\right), \text{ for } t > t_1;$$

$$\hat{S}(t) = 1, \text{ for } t < t_1.$$

This is the celebrated Kaplan–Meier or product-limit estimator (Kaplan and Meier, 1958). This result has been one of the most influential developments in the past century in statistics; the paper by Kaplan and Meier (Fig. 18.1) is the most cited paper in the field of statistics (Stigler, 1994).

(a)          (b)

**Fig. 18.1** (a) Edward Kaplan (1920–2006) and (b) Paul Meier (b. 1924). Kaplan and Meier never actually met during the time their article was published. They both submitted their idea for the "product-limit estimator" to the *Journal of the American Statistical Association* at approximately the same time, so their results have been merged through mail correspondence.

For uncensored observations, the Kaplan–Meier estimator is identical to the regular MLE. The difference occurs when there is a censored observation – then the Kaplan–Meier estimator takes the "weight" normally assigned to that observation and distributes it evenly among all observed values to the

right of the observation. This is intuitive because we know that the true value
of the censored observation must be somewhere to the right of the censored
value, but information about what the exact value should be is lacking.

The variance of Kaplan–Meier estimator is estimated by Greenwood's formula:

$$\text{Var}\left(\hat{S}(t)\right) = \left(\hat{S}(t)\right)^2 \times \sum_{t_i \leq t} \frac{d_i}{r_i(r_i - d_i)}.$$

The pointwise confidence intervals (for a fixed time $t_*$) for the survival
function $S(t_*)$ can be found in several ways. The two most popular confidence
intervals are

*linear*

$$\left[\hat{S}(t_*) - z_{1-\alpha/2}\sqrt{\text{Var}\left(\hat{S}(t_*)\right)}, \hat{S}(t_*) + z_{1-\alpha/2}\sqrt{\text{Var}\left(\hat{S}(t_*)\right)}\right]$$

and *log-transformed*

$$\left[(\hat{S}(t_*))^{1/v}, (\hat{S}(t_*))^v\right], \quad v = \exp\left\{\frac{z_{1-\alpha/2}\sqrt{\text{Var}\left(\hat{S}(t_*)\right)}}{\log \hat{S}(t_*)}\right\}.$$

Although not centered at $\hat{S}(t_*)$, the log-transformed interval is considered superior to the linear.

The above pointwise confidence intervals differ from simultaneous confidence bounds on $S(t)$ for which the confidence of $1 - \alpha$ means that the probability that *any* part of the curve $S(t)$ will fall outside the bounds does not
exceed $\alpha$. Such general bounds are naturally wider than those generated by
pointwise confidence intervals since the overall confidence is controlled. Two
important types of confidence bands are Nair's equal precision bands and
the Hall–Wellner bands. These bounds fall beyond the scope of this text; see
Klein and Moeschberger (2003), p. 109, for further discussion and implementation. The bounds computed in MATLAB's [f,t,flo,fup]=ecdf(...) also return
lower and upper confidence bounds for the CDF. These bounds are calculated
by using Greenwood's formula and are not simultaneous confidence bounds.

The Kaplan–Meier estimator also provides an estimator for the cumulative
hazard $H(t)$ as

$$\hat{H}(t) = -\log\big(\hat{S}(t)\big).$$

Better small-sample performance in estimating the cumulative hazard could be achieved by the Nelson–Aalen estimator,

$$\tilde{H}(t) = \sum_{t_i < t} d_i/r_i, \text{ for } t > t_1, \ \tilde{H}(t) = 0, \text{ for } t \le t_1,$$

with an estimated variance $\sigma_{\tilde{H}}^2(t) = \sum_{t_i < t} d_i/r_i^2$. Using $\tilde{H}(t)$ and $\sigma_{\tilde{H}}^2(t)$, pointwise confidence intervals on $H(t)$ can be obtained.

*Example 18.6.* **Catheter Complications in Peritoneal Dialysis.** The following example is from Chadha et al. (2000). The authors studied a sample of 36 pediatric patients undergoing acute peritoneal dialysis through Cook Catheters. They wished to examine how long these catheters performed properly. They noted the date of complication (either occlusion, leakage, exit-site infection, or peritonitis).

Half of the subjects had no complications before the catheter was removed. Reasons for removal of the catheter in this group of patients were that the patient recovered ($n = 4$), the patient died ($n = 9$), or the catheter was changed to a different type electively ($n = 5$). If the catheter was removed prior to complications, that represented a censored observation, because they knew that the catheter remained complication free at least until the time of removal.

| Day | Censored, $\ell$ | Fail, $d$ | At Risk, $r$ | 1−(Failures/At Risk) | KM |
|---|---|---|---|---|---|
| 1 | 8 | 2 | 36 | $1 - 2/36 = 0.944$ | 0.9444 |
| 2 | 2 | 2 | $36 - 8 - 2 = 26$ | $1 - 2/26 = 0.92$ | $0.92 \cdot 0.944 = 0.8718$ |
| 3 | 1 | 2 | $26 - 2 - 2 = 22$ | $1 - 2/22 = 0.91$ | $0.91 \cdot 0.872 = 0.7925$ |
| 4 | 1 | 1 | $22 - 1 - 2 = 19$ | $1 - 1/19 = 0.95$ | $0.95 \cdot 0.793 = 0.7508$ |
| 5 | 6 | 3 | $19 - 1 - 1 = 17$ | $1 - 3/17 = 0.82$ | 0.6183 |
| 6 | 0 | 2 | $17 - 6 - 3 = 8$ | $1 - 2/8 = 0.75$ | 0.4637 |
| 7 | 0 | 1 | $8 - 0 - 2 = 6$ | $1 - 1/6 = 0.83$ | 0.3865 |
| 10 | 0 | 2 | $6 - 0 - 1 = 5$ | $1 - 2/5 = 0.60$ | 0.2319 |
| 12 | 0 | 2 | $5 - 0 - 2 = 3$ | $1 - 2/3 = 0.33$ | 0.0773 |
| 13 | 0 | 1 | $3 - 0 - 2 = 1$ | $1 - 1/1 = 0.00$ | 0.0000 |

MATLAB script ◢ chada.m finds the Kaplan–Meier estimator and generates Fig. 18.2. plots

```
%chada.m
times=[1,1,1,1,1,1,1,1,1,1,2,2,2,2,...
       3,3,3,4,4,5,5,5,5,5,5,5,5,5,...
       6,6,7,10,10,12,12,13];
censored =[1,1,1,1,1,1,1,1,0,0,1,1,...
           0,0,1,0,0,1,0,1,1,1,1,1,...
           1,0,0,0,0,0,0,0,0,0,0,0];

% Calculate and plot empirical
% cdf and confidence bounds
[f,x,flo,fup] = ecdf(times,'censoring',censored);
(1-f)'
```

```
%1.0000  0.9444  0.8718  0.7925  0.7508
%0.6183  0.4637  0.3865  0.2319  0.0773  0
 x'
%1  1  2  3  4  5  6  10  12  13
stairs(x,1-f,'LineWidth',2)
hold on
stairs(x,1-flo,'r:','LineWidth',2)
stairs(x,1-fup,'r:','LineWidth',2)
legend('Empirical','LCB','UCB',...
       'Location','NE')
hold off
```

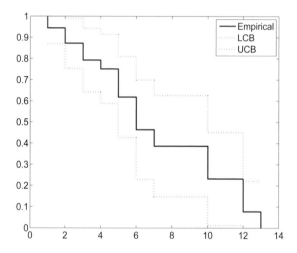

**Fig. 18.2** Kaplan–Meier estimator for Catheter Complications data.

*Example 18.7.* **Strength of Weathered Cord.** Data from Crowder et al. (1991) lists strength measurements (in coded units) for 48 pieces of weathered cord. Seven of the pieces of cord were damaged and yielded strength measurements that are considered right-censored. That is, because the damaged cord was taken off test, we know only the lower limit of its strength. In the MAT-LAB code below, the vector data represents the strength measurements, and the vector censor indicates (with a zero) if the corresponding observation in data is censored.

```
data=[36.3,41.7,43.9,49.9,50.1,50.8,51.9,52.1,52.3,52.3,...
      52.4,52.6,52.7,53.1,53.6,53.6,53.9,53.9,54.1,54.6,...
      54.8,54.8,55.1,55.4,55.9,56.0,56.1,56.5,56.9,57.1,...
      57.1,57.3,57.7,57.8,58.1,58.9,59.0,59.1,59.6,60.4,...
      60.7,26.8,29.6,33.4,35.0,40.0,41.9,42.5];
```

```
    censor=[ones(1,41),zeros(1,7)];
[kmest,sortdat,sortcen]= KMcdfSM(data',censor',0);
plot(sortdat,1-kmest,'k');
```

The table below shows how the Kaplan–Meier estimator is calculated for the first 16 measurements, which includes 7 censored observations. Figure 18.3 shows the estimated survival function for the cord strength data.

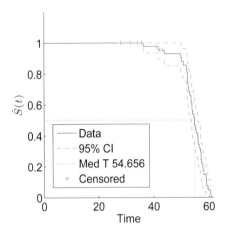

**Fig. 18.3** Kaplan–Meier estimator cord strength (in coded units).

| Uncensored | $x_j$ | $m_j$ | $d_j$ | $\frac{m_j-d_j}{m_j}$ | $1-F_{KM}(x_j)$ |
|---|---|---|---|---|---|
| | 26.8 | 48 | 0 | 1.000 | 1.000 |
| | 29.6 | 47 | 0 | 1.000 | 1.000 |
| | 33.4 | 46 | 0 | 1.000 | 1.000 |
| | 35.0 | 45 | 0 | 1.000 | 1.000 |
| 1 | 36.3 | 44 | 1 | 0.977 | 0.977 |
| | 40.0 | 43 | 0 | 1.000 | 0.977 |
| 2 | 41.7 | 42 | 1 | 0.976 | 0.954 |
| | 41.9 | 41 | 0 | 1.000 | 0.954 |
| | 42.5 | 40 | 0 | 1.000 | 0.954 |
| 3 | 43.9 | 39 | 1 | 0.974 | 0.930 |
| 4 | 49.9 | 38 | 1 | 0.974 | 0.905 |
| 5 | 50.1 | 37 | 1 | 0.973 | 0.881 |
| 6 | 50.8 | 36 | 1 | 0.972 | 0.856 |
| 7 | 51.9 | 35 | 1 | 0.971 | 0.832 |
| 8 | 52.1 | 34 | 1 | 0.971 | 0.807 |
| 9 | 52.3 | 33 | 2 | 0.939 | 0.758 |
| ⋮ | ⋮ | ⋮ | ⋮ | ⋮ | ⋮ |

### 18.3.3 Comparing Survival Curves

In clinical trials it is often important to compare survival curves calculated for cohorts undergoing different treatments. Most often one is interested in comparing the new treatment to the existing one or to a placebo. In comparing two survival curves we are testing whether the corresponding hazard functions $h_1(t)$ and $h_2(t)$ coincide:

$$H_0 : h_1(t) = h_2(t) \quad vs \quad H_1 : h_1(t) >, \neq, < h_2(t).$$

The simplest comparison involves exponential lifetime distributions where the comparison between survival/hazard functions is simply a comparison of constant rate parameters. The statistic is calculated using logarithms of hazard rates as

$$Z = \frac{\log \lambda_1 - \log \lambda_2}{\sqrt{1/k_1 + 1/k_2}},$$

where $k_1$ and $k_2$ are numbers of observed (uncensored) survival times in the two comparison groups. Now, the inference relies on the fact that statistic $Z$ is approximately standard normal.

*Example 18.8.* In Example 18.5 the rate for immunoperoxidase-positive patients was larger than that of immunoperoxidase-negative patients? Was this difference significant?

```
z = (log(lambdahat1)-log(lambdahat2))/sqrt(1/k1+1/k2)   %-2.5763
p = normcdf(z)   %0.0050
```

As is evident from the code, the hazard rate $\lambda_2$ (and, in the case of exponential distribution, hazard function) for the immunoperoxidase-positive patients is significantly larger than the rate for negative patients, $\lambda_1$, with a *p*-value of 1/2 percent.
✎

**LogRank Test.** Tests for comparing survival functions in nonparametric fashion use Hanszel–Mantel theory on survival data represented as $2 \times 2$ tables. Another popular term is logrank test.

Let $(r_{11}, d_{11})$, $(r_{12}, d_{12}), \dots$, $(r_{1k}, d_{1k})$ be the number of people at risk and the number of people who died at times $t_{11}, t_{12}, \dots, t_{1k}$ in the first cohort, and $(r_{21}, d_{21})$, $(r_{22}, d_{22}), \dots$, $(r_{2m}, d_{2m})$ be the number of people at risk and the number of people who died at times $t_{21}, t_{22}, \dots, t_{1m}$ in the second cohort. We merge the two data sets together with the corresponding times. Thus there will be $D = k + m$ time points if there are no ties, and each time point corresponds to a death from either the first or second cohort. For example, if times of events in the first sample are 1, 4, and 10 and in the second 2, 3, 7, and 8, then in the merged data sets the times will be 1, 2, 3, 4, 7, 8, and 10.

For a time $t_i$ from the merged data set, let $r_{1i}$ and $r_{2i}$ correspond to the number of subjects at risk in cohorts 1 and 2, respectively, and let $r_i = r_{1i} + r_{2i}$ be the number of subjects at risk in the combined sample. Analogously, let $d_{1i}$, $d_{2i}$, and $d_i = d_{1i} + d_{2i}$ be the number of events at time $t_i$.

Then, if $H_0 : h_1(t) = h_2(t)$ is true, then $d_{1i}$ has a hypergeometric distribution with parameters $(r_i, d_i, r_{1i})$.

| | Event | No event | At risk |
|---|---|---|---|
| Treatment 1 | $d_{1i}$ | $r_{1i} - d_{1i}$ | $r_{1i}$ |
| Treatment 2 | $d_{2i}$ | $r_{2i} - d_{2i}$ | $r_{2i}$ |
| Merged | $d_i$ | $r_i - d_i$ | $r_i$ |

Since $d_{1i} \sim \mathcal{HG}(r_i, d_i, r_{1i})$, the expectation and variance of $d_{1i}$ are

$$\mathbb{E}d_{1i} = r_{1i} \times \frac{d_i}{r_i},$$

$$\mathbb{V}\mathrm{ar}(d_{1i}) = \frac{r_{i1}}{r_i}\left(1 - \frac{r_{i1}}{r_i}\right)\left(\frac{r_i - d_i}{r_i - 1}\right)d_i.$$

Note that in the terminology of the Kaplan–Meier estimator, the number of subjects with no event at time $t_i$ is equal to $r_i - d_i = r_{i+1} + \ell_i$, where $\ell_i$ is the number of subjects censored in the time interval $(t_{i-1}, t_i)$ and $r_{i+1}$ is the number of subjects at risk at the beginning of the subsequent interval $(t_i, t_{i+1})$.

The test statistic for testing $H_0 : h_1(t) = h_2(t)$ against the two-sided alternative $H_1 : h_1(t) \neq h_2(t)$ is

$$\chi^2 = \frac{\left(\sum_{i=1}^{D}(d_{i1} - \mathbb{E}(d_{1i}))\right)^2}{\sum_{i=1}^{D}\mathbb{V}\mathrm{ar}(d_{1i})},$$

which has a $\chi^2$-distribution with 1 degree of freedom. The continuity correction 0.5 can be added to the numerator of the $\chi^2$-statistic as $(|\sum_{i=1}^{D}(d_{i1} - \mathbb{E}(d_{1i}))| - 0.5)^2$ when the sample size is small. If the statistic is calculated as chi2, then its large values are critical and the $p$-value of the test is equal to 1-chi2cdf(chi2,1).

If the alternative is one-sided, $H_1 : h_1(t) < h_2(t)$ or $H_1 : h_1(t) > h_2(t)$, then the preferable statistic is

$$Z = \frac{\sum_{i=1}^{D}(d_{i1} - \mathbb{E}(d_{1i}))}{\sqrt{\sum_{i=1}^{D}\mathbb{V}\mathrm{ar}(d_{1i})}}$$

and the $p$-values are `normcdf(Z)` and `1-normcdf(Z)`, respectively. A more general statistic is of the form

$$Z = \frac{\sum_{i=1}^{D} W(t_i)(d_{i1} - \mathbb{E}(d_{1i}))}{\sqrt{\sum_{i=1}^{D} W^2(t_i)\mathbb{V}\mathrm{ar}(d_{1i})}},$$

where $W(t_i) = 1$ (as above), $W(t_i) = r_i$ (Gehan's statistic), and $W(t_i) = \sqrt{r_i}$ (Tarone–Ware statistic).

*Example 18.9.* **Histiocytic Lymphoma.** The data (from McKelvey et al., 1976; Armitage and Berry, 1994) given below are survival times (in days) since entry to a trial by patients with diffuse histiocytic lymphoma. Two cohorts of patients are considered: (1) with stage III and (2) with stage IV of the disease. The observations with + are censored.

| Stage III | 6 | 19 | 32 | 42 | 42 | 43+ | 94 | 126+ | 169+ | 207 | 211+ |
|---|---|---|---|---|---|---|---|---|---|---|---|
| | 227+ | 253 | 255+ | 270+ | 310+ | 316+ | 335+ | 346+ | | | |
| Stage IV | 4 | 6 | 10 | 11 | 11 | 11 | 13 | 17 | 20 | 20 | 21 |
| | 22 | 24 | 24 | 29 | 30 | 30 | 31 | 33 | 34 | 35 | 39 |
| | 40 | 41+ | 43+ | 45 | 46 | 50 | 56 | 61+ | 61+ | 63 | 68 |
| | 82 | 85 | 88 | 89 | 90 | 93 | 104 | 110 | 134 | 137 | 160+ |
| | 169 | 171 | 173 | 175 | 184 | 201 | 222 | 235+ | 247+ | 260+ | 284+ |
| | 290+ | 291+ | 302+ | 304+ | 341+ | 345+ | | | | | |

Using a logrank test we will assess the equality of the two survival curves. Figure 18.4 and table

```
%------------------------------------------------------------
%   L          S.E.        z          p-value      alpha
%------------------------------------------------------------
% 8.65786    3.35325    2.43283      0.01498      0.050
%------------------------------------------------------------
% The survival functions are statistically different
```

are outputs from ◀ `logrank.m` (Cardillo, 2008).

# 18.4 The Cox Proportional Hazards Model

We often need to take into account that survival is influenced by one or more covariates, which may be categorical (such as the kind of treatment a patient received) or continuous (such as the patient's age, weight, or the dosage of a drug). For simple situations involving a single factor with just two values (such as drug versus placebo) we discussed a method for comparing the survival curves for the two groups of subjects. But for more complicated situations a regression-type model that incorporates the effect of each predictor on the shape of the survival curve is needed.

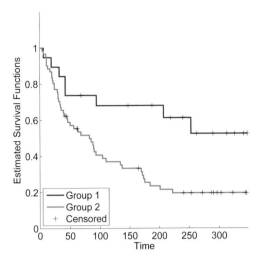

**Fig. 18.4** Kaplan–Meier estimators of the two survival functions. Here group 1 corresponds to Stage III and group 2 to Stage IV cohorts.

Assume that the log hazard for subject $i$ can be modeled via a linear relationship:

$$\log h(t, x_i) = \beta_0 + \beta_1 x_{1,i} + \cdots + \beta_p x_{p,i},$$

where $x_i = x_{1,i}, \ldots, x_{p,i}$ is $p$-dimensional vector of covariates associated with subject $i$. The Cox model assumes that $\beta_0$ is a log baseline hazard, $\log h_0(t) = \beta_0$, i.e., log hazard for a "person" for whom all covariates are 0 (Cox, 1972; Cox and Oakes, 1984). Alternatively, one can set the baseline hazard to correspond to a person for whom all covariates are averages of covariates from all subjects in the study. For the Cox model,

$$\log h(t, x_i) = \log h_0(t) + \beta_1 x_{1,i} + \cdots + \beta_p x_{p,i},$$

or, equivalently,

$$h(t, x_i) = h_0(t) \times \exp\{\beta_1 x_{1,i} + \cdots + \beta_p x_{p,i}\} = h_0(t) \times \exp\{x_i' \beta\}.$$

Inclusion of an intercept would lead to nonidentifiability because

$$h_0(t) \times \exp\{x_i' \beta\} = (h_0(t) e^{-\alpha}) \times \exp\{\alpha + x_i' \beta\}.$$

This allows for some freedom in choosing the baseline hazard.

For two subjects, $i$ and $j$, the ratio of hazard functions

$$h(t,x_i)/h(t,x_j) = \exp\{(x_i' - x_j')\beta\}$$

is free of $t$, motivating the name *proportional*. Also, for a subject $i$,

$$S(t,x_i) = (S_0(t))^{\exp\{x_i'\beta\}},$$

where $S_0(t)$ is the survival function corresponding to the baseline hazard $h_0(t)$. This follows directly from (18.1) and $H(t,x_i) = H_0(t)\exp\{x_i'\beta\}$.

In MATLAB, coxphfit fits the Cox proportional hazards regression model, which relates survival times to predictor variables. The following example uses coxphfit to fit Cox's proportional hazards model.

*Example 18.10.* **Mayo Clinic Trial in PBC.** Primary biliary cirrhosis (PBC) is a rare but fatal chronic liver disease of unknown cause, with a prevalence of about 1/20,000. The primary pathologic event appears to be the destruction of interlobular bile ducts, which may be mediated by immunologic mechanisms.

The PBC data set available at StatLib is an excerpt from the Mayo Clinic trial in PBC of the liver conducted between 1974 and 1984. From a total of 424 patients that met eligibility criteria, 312 PBC patients participated in the double-blind, randomized, placebo-controlled trial of the drug D-penicillamine. Details of the trial can be found in Markus et al. (1989).

Survival statuses were recorded for as many patients as possible until July 1986. By that date, 125 of the 312 patients had died and 187 were censored.

The variables contained in the data set 📂 pbc.xls|dat are described in the following table:

```
casen = pbc(:,1);          %case number 1-312
lived = pbc(:,2);          %days lived (from registration to study date)
indicatord = pbc(:,3);     %0 censored, 1 death
treatment = pbc(:,4);      %1 - D-Penicillamine, 2 - Placebo
age = pbc(:,5);            %age in years
gender = pbc(:,6);         %0 male, 1 female
ascites= pbc(:,7);         %0 no, 1 yes
hepatomegaly=pbc(:,8);     %0 no, 1 yes
spiders = pbc(:,9);        %0 no, 1 yes
edema = pbc(:,10);         %0 no, 0.5 yes/no therapy, 1 yes/therapy
bilirubin = pbc(:,11);     %bilirubin [mg/dl]
cholesterol = pbc(:,12);   %cholesterol [mg/dl]
albumin = pbc(:,13);       %albumin [gm/dl]
ucopper =pbc(:,14);        %urine copper [mg/day]
aphosp =pbc(:,15);         %alcaline phosphatase [U/liter]
sgot = pbc(:,16);          %SGOT [U/ml]
trig =pbc(:,17);           %triglycerides [mg/dl]
platelet = pbc(:,18);      %# platelet count [#/mm^3]/1000
prothro = pbc(:,19);       %prothrombin time [sec]
histage = pbc(:,20);       %hystologic stage [1,2,3,4]
```

To illustrate the CPH model, in this example we selected four predictors and formed a design matrix $X$ as

```
X = [treatment age gender edema];
```

The treatment has two values, 1 for treatment by D-penicillamine and 2 for placebo. The variable edema takes three values: 0 if no edema is present, 0.5 when edema is present but no diuretic therapy was given or edema resolved with diuretic therapy, and 1 if edema is present despite administration of diuretic therapy.

The variable lived is the lifetime observed or censored, a censoring vector is 1-indicatord, and a baseline hazard is taken to be a hazard for which all covariates are set to 0.

```
[b,logL,H,stats] = coxphfit(X,lived,...
   'censoring',1-indicatord,'baseline',0);
```

The output H is a two-column matrix as a discretized cumulative hazard estimate. The first column of H contains values from the vector lived, while the second column contains the estimated baseline cumulative hazard evaluated at lived.

To illustrate the model, we selected two subjects from the study to find survival curves corresponding to their covariates. Subject #100 is a 51-year-old male with no edema who received placebo while subject #275 is a 38-year-old female with no edema who received D-penicillamine treatment.

```
X(100,:) %2.0000      51.4689    0           0
%          placebo;   51 y.o.;   male;    no edema;
X(275,:) %1.0000      38.3162    1.0000      0
% D-Penicillamine;    38 y.o.;   female;  no edema;
```

First we find cumulative hazards at the mean values of predictors, as well as for subjects #100 and #275, as

$$H(t,\overline{x}) = H_0(t) \times \exp\{\beta_1\overline{x}_1 + \cdots + \beta_4\overline{x}_4\},$$
$$H(t,x_i) = H_0(t) \times \exp\{\beta_1 x_{1,i} + \cdots + \beta_4 x_{4,i}\}, \ i = 100, \ 275$$

```
Hmean(:,2)      =   H(:,2) .* exp(mean(X)*b);   %c.haz. average
Hsubj100(:,2) =   H(:,2) .* exp(X(100,:)*b); %subject #100
Hsubj275(:,2) =   H(:,2) .* exp(X(275,:)*b); %subject #275
```

Here, the estimators of coefficients $\beta_1,\ldots,\beta_4$ are

```
b'
%      0.0831       0.0324    -0.3940      2.2424
```

Note that the treatment coefficient $0.0831 > 0$ indicates that, given all other covariates fixed, the placebo increases the risk over the treatment. Also note that age and edema statuses also increase the risk, while the risk for female subjects is smaller.

Next, from cumulative hazards we find survival functions

```
Smean      = exp(-Hmean(:,2));
Ssubj100 = exp(-Hsubj100(:,2));
Ssubj275 = exp(-Hsubj275(:,2));
```

The subsequent commands plot the survival curves for an "average" subject (blue), as well as for subjects #100 (black) and #275 (red), see Fig. 18.5.

```
stairs(H(:,1),Smean,'b-','linewidth',2)
hold on
stairs(H(:,1),Ssubj100,'k-')
stairs(H(:,1),Ssubj275,'r-')
xlabel('$t$ (days)','Interpreter','LaTeX')
ylabel('$\hat S(t)$','Interpreter','LaTeX')
legend('average subject','subject #100', 'subject #275', 3)
axis tight
```

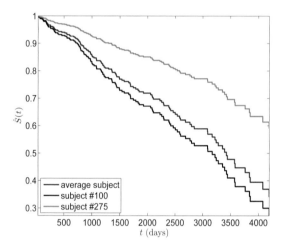

**Fig. 18.5** Cox model for survival curves for a "subject" with average covariates, subject #100 (51-year-old male on placebo, no edema), and subject #275 (38-year-old female on treatment, no edema).

## 18.5 Bayesian Approach

We will focus on parametric models in which the lifetime distributions are specified up to unknown parameters. The unknown parameters will be assigned prior distributions and the inference will proceed in a Bayesian fashion. Nonparametric Bayesian modeling of survival data is possible; however, the methodology is advanced and beyond the scope of this text. For a comprehensive coverage see Ibrahim et al. (2001).

Let survival time $T$ have distribution $f(t|\theta)$, where $\theta$ is unknown parameters. For $t_1,\ldots,t_k$ observed and $t_{k+1},\ldots,t_n$ censored times, the likelihood is

$$L(\theta|t_1,\ldots,t_n) = \prod_{i=1}^{k} f(t_i|\theta) \times \prod_{i=k+1}^{n} S(t_i|\theta).$$

If the prior on $\theta$ is $\pi(\theta)$, then the posterior is

$$\pi(\theta|t_1,\ldots,t_n) \propto L(\theta|t_1,\ldots,t_n) \times \pi(\theta).$$

The Bayesian estimator of hazard is

$$\hat{h}_B(t) = \int h(t|\theta)\pi(\theta|t_1,\ldots,t_n)d\theta$$

and the survival function is

$$\hat{S}_B(t) = \int S(t|\theta)\pi(\theta|t_1,\ldots,t_n)d\theta.$$

*Example 18.11.* In Example 18.4 we showed that for the exponential lifetime in the presence of censoring, the MLE for $\lambda$ is

$$\hat{\lambda} = k / \sum_{i=1}^{n} t_i,$$

where $k$ is the number of uncensored data and $\sum_{i=1}^{n} t_i$ is the sum of *all* observed and censored times and that the likelihood was $L(\lambda) = \lambda^k \exp\{-\lambda \sum_{i=1}^{n} t_i\}$. If a gamma $\mathcal{G}a(\alpha,\beta)$ prior on $\lambda$ is adopted, $\pi(\lambda) \propto \lambda^{\alpha-1} \exp\{-\beta\lambda\}$, then the conjugacy leads to the posterior

$$\pi(\lambda|t_1,\ldots,t_n) \propto \lambda^{k+\alpha-1} \exp\{-(\beta + \sum_{i=1}^{n} t_i)\, \lambda\},$$

from which the Bayes estimator of $\lambda$ is the posterior mean,

$$\hat{\lambda}_B = \frac{k+\alpha}{\beta + \sum_{i=1}^{n} t_i}.$$

One can show that the posterior predictive distribution of future failure time $t_{n+1}$ is

$$f(t_{n+1}|t_1,\ldots,t_n) = \int_0^\infty \lambda e^{-\lambda t_{n+1}} \times \pi(\lambda|t_1,\ldots,t_n)d\lambda$$

$$= \frac{(k+\alpha)(\beta + \sum_{i=1}^{n} t_i)^{\alpha+k}}{(\beta + \sum_{i=1}^{n} t_i + t_{n+1})^{\alpha+k+1}}, \quad y_{n+1} > 0.$$

This distribution is known as an inverse beta distribution.

The Bayes estimator of hazard function coincides with the Bayes estimator of $\lambda$,

$$\hat{h}_B(t) = \frac{k + \alpha}{\beta + \sum_{i=1}^{n} t_i},$$

while the Bayes estimator of the survival function is

$$\hat{S}_B(t) = \left(1 + \frac{t}{\beta + \sum_{i=1}^{n} t_i}\right)^{-(k+\alpha)}.$$

The expression for $\hat{S}_B(t)$ can be derived from the moment-generating function of a gamma distribution.

When the posterior distribution is intractable, one can use WinBUGS.

### 18.5.1 Survival Analysis in WinBUGS

WinBUGS uses two arrays to define censored observations: observed (uncensored) times and censored times. For example, an input such as

```
list(times = c(0.5, NA, 1, 2, 6, NA, NA),
t.censored = c(0,  0.9, 0, 0, 0,  9, 12))
```

corresponds to times $\{0.5, 0.9+, 1, 2, 6, 9+, 12+\}$.

In WinBUGS, direct time-to-event modeling is possible with exponential, Weibull, gamma, and log-normal densities.

There is a multiplier $I$ that is used to implement censoring. For example, if Weibull dweib(r, mu) observations are on the input, the multiplier exceeded time t.censored[i]:

```
t[i] ~ dweib(r, mu) I(t.censored[i],)
```

For uncensored data, t.censored[i] = 0. The above describes right-censoring. Left-censored observations are modeled using the multiplier I(,t.censored[i]).

*Example 18.12.* **Bayesian Immunoperoxidase and BC.**  In Example 18.5 MLEs and confidence intervals on $\lambda_1$ and $\lambda_2$ were found. In this example we find Bayes estimators and credible sets.

Before discussing the results, note that when observations are censored, their values are unknown parameters in the Bayesian model and predictions can be found. On the other hand, all censored observations need to be initialized, and the initial values should exceed the censoring times. Here is the WinBUGS program that estimates $\lambda_1$ and $\lambda_2$.

```
model{
for(i in 1:n1)  {
   ImmPeroxNeg[i] ~ dexp(lam1) I(CensorIPN[i], )
```

```
    }
for(i in 1:n2)  {
    ImmPeroxPos[i] ~ dexp(lam2) I(CensorIPP[i], )
    }
lam1 ~ dgamma(0.001, 0.001)
lam2 ~ dgamma(0.001, 0.001)
}
```

DATA

```
list( n1 = 36, n2 = 9,
ImmPeroxNeg=c(19, 25, 30, 34, 37, 46, 47, 51,
56, 57, 61, 66, 67, 74, 78, 86,
NA, NA, NA, NA, NA,  NA, NA, NA, NA, NA,
NA, NA, NA, NA, NA,  NA, NA, NA, NA, NA),
CensorIPN = c(0,0,0,0,0,0,0,0,0,0,0,0,0,0,0,0,0,
122, 123, 130, 130, 133, 134, 136, 141, 143, 148,
151, 152, 153, 154, 156, 162, 164, 165, 182, 189),
ImmPeroxPos = c(22, 23, 38, 42, 73, 77, 89, 115, NA),
CensorIPP= c(0,0,0,0,0,0,0,0,144))
```

INITS

```
list(lam1=1, lam2 = 1,
ImmPeroxNeg=c(NA, NA, NA, NA, NA, NA, NA, NA,
                    NA, NA, NA, NA, NA, NA, NA, NA,
 200, 200, 200, 200, 200, 200, 200, 200, 200, 200,
 200, 200, 200, 200, 200, 200, 200, 200, 200, 200),
ImmPeroxPos = c(NA, NA, NA, NA, NA, NA, NA, NA, 200)  )
```

|  | mean | sd | MCrror | val2.5pc | median | val97.5pc | start | sample |
|---|---|---|---|---|---|---|---|---|
| ImmPeroxNeg[17] | 374.8 | 270.8 | 0.4261 | 128.0 | 289.7 | 1107.0 | 1001 | 500000 |
| ImmPeroxNeg[18] | 376.2 | 270.6 | 0.4568 | 129.0 | 291.2 | 1107.0 | 1001 | 500000 |
| ImmPeroxNeg[19] | 383.2 | 270.6 | 0.4342 | 136.0 | 298.0 | 1115.0 | 1001 | 500000 |
|  | ... |  |  |  |  |  |  |  |
| ImmPeroxNeg[35] | 435.1 | 270.2 | 0.4319 | 188.1 | 349.9 | 1166.0 | 1001 | 500000 |
| ImmPeroxNeg[36] | 442.3 | 270.6 | 0.4272 | 195.0 | 357.1 | 1173.0 | 1001 | 500000 |
| ImmPeroxPos[9] | 233.1 | 102.8 | 0.1730 | 146.0 | 200.5 | 508.2 | 1001 | 500000 |
| lam1 | 0.0042 | 0.0010 | 2.849E-6 | 0.0024 | 0.0041 | 0.0065 | 1001 | 500000 |
| lam2 | 0.0128 | 0.0045 | 7.496E-6 | 0.0055 | 0.0123 | 0.0231 | 1001 | 500000 |

The Bayes estimator of $\lambda_1$ is $\hat{\lambda}_{1,B} = 0.004211$, and a 95% credible set is [0.002404, 0.006505]. The Bayes estimator and interval are close to classical (Exercise 18.2).

*Example 18.13.* **Smoking Cessation Experiment.** The data set for this example comes from a clinical trial discussed in Banerjee and Carlin (2004). A number of smokers entered into a smoking cessation study, and 263 of them quit. These 263 quitters were monitored and checked to see if and when they relapsed. RelapseT is the time to relapse; it is either observed or censored, with the censoring indicator contained in the vector censored.time. The independent covariates are Age (age of the individual), AgeStart (age when he/she started smoking), SexF (Female=1, Male=0), SIUC (whether the individual received an intervention or not), and F10Cigs (the average number of cigarettes smoked per day).

A logistic distribution is constrained to a nonnegative domain to model RelapseT. The parameters of dlogis(mu,tau) are the mean mu, which depends on the linear combination of covariates, and tau, which is a rate parameter. The standard deviation is $\pi/(\sqrt{3}\tau) \approx 1.8138/\tau$.

```
model {
for (i in 1:N)
     {
     RelapseT[i] ~ dlogis(mu[i],tau) I(censored.time[i],)
     mu[i]  <-  beta[1] + beta[2] * Age[i] + beta[3] * AgeStart[i] +
     beta[4] * SexF[i] + beta[5] * SIUC[i] + beta[6] * F10Cigs[i]
     }
for( j in 1:6){
beta[j] ~ dnorm(0, 0.01)
               }
tau ~ dgamma(1,0.01)
meanT <- mean(mu[])
sigma <- 1.8138/tau       #1.8138 ~ pi/sqrt(3)


# Evaluate Survival Curve for a Subject with covariates:
Ag <- 50; AgSt <- 18; SxF <- 0; S <- 1; Cigs <- 20;
fmu    <-  beta[1] + beta[2] * Ag  + beta[3] * AgSt +
               beta[4] * SxF  + beta[5] * S    + beta[6] *  Cigs
for(i in 1:100) {
     time[i]  <-  i/10
     Surv[i] <- 1/(1 + exp(tau*(time[i] - fmu)))
     }
}
# Data and Inits omitted (see Smoking.odc)
```

| | mean | sd | MCrror | val2.5pc | median | val97.5pc | start | sample |
|---|---|---|---|---|---|---|---|---|
| beta[1] | 2.686 | 2.289 | 0.1254 | −2.026 | 2.701 | 7.284 | 1001 | 100000 |
| beta[2] | 0.07817 | 0.03844 | 0.002078 | 0.009544 | 0.07815 | 0.1509 | 1001 | 100000 |
| beta[3] | −0.07785 | 0.07222 | 0.003745 | −0.2181 | −0.07965 | 0.06802 | 1001 | 100000 |
| beta[4] | −0.9555 | 0.5382 | 0.009515 | −2.02 | −0.9494 | 0.08981 | 1001 | 100000 |
| beta[5] | 1.666 | 0.5993 | 0.01482 | 0.4886 | 1.672 | 2.817 | 1001 | 100000 |
| beta[6] | −0.02347 | 0.02459 | 9.21E-4 | −0.07072 | −0.02394 | 0.0281 | 1001 | 100000 |
| meanT | 5.314 | 0.3382 | 0.00604 | 4.712 | 5.292 | 6.034 | 1001 | 100000 |
| sigma | 3.445 | 0.345 | 0.005367 | 2.841 | 3.421 | 4.187 | 1001 | 100000 |
| tau | 0.5317 | 0.05256 | 8.15E-4 | 0.4332 | 0.5302 | 0.6384 | 1001 | 100000 |
| Surv[1] | 0.9628 | 0.01129 | 5.199E-4 | 0.9366 | 0.9642 | 0.9808 | 1001 | 100000 |
| Surv[2] | 0.9609 | 0.01173 | 5.459E-4 | 0.9336 | 0.9623 | 0.9797 | 1001 | 100000 |
| ... | | | | | | | | |
| Surv[99] | 0.1402 | 0.05249 | 0.003402 | 0.06123 | 0.1334 | 0.2645 | 1001 | 100000 |
| Surv[100] | 0.1342 | 0.05118 | 0.003311 | 0.05765 | 0.1274 | 0.2557 | 1001 | 100000 |

The Surv values (ordinate, mean, std, median, and quantiles) are exported to MATLAB as data file  smokingoutbugs.mat. The file smokingbugs.m reads in the data and plots the posterior estimator of the survival curve (Fig. 18.6). Note that the survival curve is $S(t|\mu,\tau) = 1/(1 + \exp\{\tau(t - \mu)\})$. The posterior distribution of $S(t|\mu,\tau)$ is understood as a distribution of a function of $\mu$ and $\tau$ for $t$ fixed.

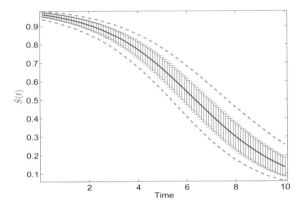

**Fig. 18.6** Bayesian estimator of survival curve. The *green* bands are the 0.025 and 0.975 percentiles of the posterior distribution for $S(t)$, while the *blue* errorbars have a size of posterior standard deviations.

*Example 18.14.* **Duration of Remissions in Acute Leukemia.** A data set analyzed by Freireich et al. (1963), and subsequently by many authors, comes from a trial of 42 leukemia patients (under age 20) treated in 11 US hospitals.

The effect of 6-mercaptopurine (6-MP) therapy on the duration of remissions induced by adrenal corticosteroids has been studied as a model for testing new agents. Some patients were treated with 6-MP and the rest were controls. The trial was designed as matched pairs. The matching was done with respect to remission status (partial = 1, complete = 2). Randomization to 6-MP or control arms was done within a pair. Patients were followed until leukemia relapsed or until the end of the study. Overall survival was not significantly different for the two treatment programs since patients maintained on placebo were treated with 6-MP when relapse occurred.

| # | Status | Contr | 6-MP | # | Status | Contr | 6-MP | # | Status | Contr | 6-MP |
|---|--------|-------|------|---|--------|-------|------|---|--------|-------|------|
| 1 | 1 | 1 | 10 | 8 | 2 | 11 | 34+ | 15 | 2 | 8 | 17+ |
| 2 | 2 | 22 | 7 | 9 | 2 | 8 | 32+ | 16 | 1 | 23 | 35+ |
| 3 | 2 | 3 | 32+ | 10 | 2 | 12 | 25+ | 17 | 1 | 5 | 6 |
| 4 | 2 | 12 | 23 | 11 | 2 | 2 | 11+ | 18 | 2 | 11 | 13 |
| 5 | 2 | 8 | 22 | 12 | 1 | 5 | 20+ | 19 | 2 | 4 | 9+ |
| 6 | 1 | 17 | 6 | 13 | 2 | 4 | 19+ | 20 | 2 | 1 | 6+ |
| 7 | 2 | 2 | 16 | 14 | 2 | 15 | 6 | 21 | 2 | 8 | 10+ |

```
#Duration of Steroid-induced Remissions in Acute Leukemia
model {
for (i in 1:n) {
            log(mu[i]) <- b0+ b1[treat[i]] + b2[status[i]]
            t[i] ~ dweib(r,mu[i])  I(t.cen[i],)
            S[i]   <- exp(-mu[i]*pow(t[i],r));
            f[i]   <- mu[i]*r*pow(t[i],r-1)*S[i]
            Lik[i] <- pow(f[i],1-delta[i])*pow(S[i],delta[i]);
            logLik[i] <- log(Lik[i])
            }

b0 ~ dnorm(0,0.00001)
b1[1] <- 0
b1[2] ~ dnorm(0,0.001)
b2[1] <- 0
b2[2] ~ dnorm(0,0.001)
r ~ dgamma(0.01,0.01)

    Dev <-   -2*sum(logLik[]) #deviance

}

DATA

list(n=42,
t = c(1, 22, 3, 12, 8, 17, 2, 11, 8, 12,
      2, 5, 4, 15, 8, 23, 5, 11, 4, 1, 8,
      10, 7, NA, 23, 22, 6, 16, NA, NA, NA,
      NA, NA, NA, 6, NA, NA, 6, 13, NA, NA, NA),
t.cen = c(0,0,0,0,0,0,0,0,0,0,0,0,0,0,0,
          0,0,0,0,0,0,0,0,0,32,0,0,0,0,
          34,32,25,11,20,19,0,17,35,0,0,9,6,10),
treat = c(1,1,1,1,1, 1,1,1,1,1, 1,1,1,1,1, 1,1,1,1,1,1,
          2,2,2,2,2, 2,2,2,2,2, 2,2,2,2,2, 2,2,2,2,2,2),
status = c(1,2,2,2,2,1,2,2,2,2,2,1,2,2,2,1,1,2,2,2,2,
           1,2,2,2,2,1,2,2,2,2,2,1,2,2,2,1,1,2,2,2,2),
delta = c(0,0,0,0,0, 0,0,0,0,0, 0,0,0,0,0, 0,0,0,0,0,0,
          0,0,1,0,0, 0,0,1,1,1, 1,1,1,0,1, 1,0,0,1,1,1) )

INITS

list(r=1,b0 = 0, b1=c(NA,0),b2=c(NA,0),
t = c(NA,NA,NA,NA,NA,NA,NA,NA,NA,NA,NA,NA,NA,NA,NA,
      NA,NA,NA,NA,NA,NA,NA,NA,32,NA,NA,NA,NA,34,32,
      25,11,20,19,NA,17,35,NA,NA,9,6,10))
```

|       | mean   | sd     | MCrror   | val2.5pc | median | val97.5pc | start | sample |
|-------|--------|--------|----------|----------|--------|-----------|-------|--------|
| Dev   | 241.2  | 7.527  | 0.02825  | 228.5    | 240.6  | 257.8     | 1001  | 200000 |
| b0    | −3.209 | 0.6867 | 0.009883 | −4.636   | −3.181 | −1.949    | 1001  | 200000 |
| b1[2] | −1.772 | 0.4241 | 0.003255 | −2.629   | −1.76  | −0.9696   | 1001  | 200000 |
| b2[2] | 0.1084 | 0.4306 | 0.002543 | −0.6967  | 0.09381| 0.9995    | 1001  | 200000 |
| r     | 1.37   | 0.2069 | 0.003338 | 0.9877   | 1.361  | 1.798     | 1001  | 200000 |
| t[24] | 57.34  | 29.86  | 0.1341   | 32.62    | 48.55  | 132.9     | 1001  | 200000 |
| t[29] | 58.96  | 29.15  | 0.1345   | 34.6     | 50.17  | 134.7     | 1001  | 200000 |
|       | ...    |        |          |          |        |           |       |        |
| t[41] | 37.21  | 30.96  | 0.1128   | 7.161    | 29.06  | 115.5     | 1001  | 200000 |
| t[42] | 39.68  | 31.09  | 0.1177   | 10.95    | 31.45  | 117.3     | 1001  | 200000 |

*Example 18.15.* **Photocarcinogenicity.** Grieve (1987) and Dellaportas and Smith (1993) explored photocarcinogenicity in four treatment groups with 20 rats in each treatment group. Treatment 3 is the test drug and the others are some type of control. Response is time to death or censoring time.

We will find Bayes estimators for median survival times for the four treatments. The survival times are modeled as Weibull $\mathcal{W}ei(r, \mu_i)$, $i = 1, \ldots, 4$. The WinBUGS code specifies the censoring and priors on the Weibull model. The data part provides observed and censored times. Posterior densities for median survival times are given in Fig. 18.7.

```
model{
for(i in 1 : M) {
        for(j in 1 : N) {
t[i, j] ~ dweib(r, mu[i])I(t.cen[i, j],)
                            }
        mu[i] <- exp(beta[i])
        beta[i] ~  dnorm(0.0, 0.001)
        median[i] <-  pow(log(2) * exp(-beta[i]), 1/r)
                    }
r ~ dexp(0.001)
}

DATA

list( t = structure(.Data = c(12, 1, 21, 25, 11, 26, 27, 30, 13,
12, 21, 20, 23, 25, 23, 29, 35, NA, 31, 36,       32, 27, 23, 12,
18, NA, NA, 38, 29, 30, NA, 32, NA, NA, NA, NA, 25, 30,
37, 27,      22, 26, NA, 28, 19, 15, 12, 35, 35, 10, 22, 18, NA,
12, NA, NA, 31, 24, 37, 29,      27, 18, 22, 13, 18, 29, 28, NA,
16, 22, 26, 19, NA, NA, 17, 28, 26, 12, 17, 26), .Dim = c(4,20)),
t.cen = structure(.Data = c(0, 0, 0, 0, 0, 0, 0, 0, 0, 0, 0, 0,
0, 0, 0, 0, 0, 40, 0, 0, 0, 0, 0, 0, 0, 40, 40, 0, 0, 0, 40, 0, 40,
40, 40, 40, 0, 0, 0, 0, 0, 0, 10, 0, 0, 0, 0, 0, 0, 0, 0, 0, 24, 0,
40, 40, 0, 0, 0, 0, 0, 0, 0, 0, 0, 0, 0, 20, 0, 0, 0, 0, 29, 10, 0,
0, 0, 0, 0, 0), .Dim = c(4, 20)), M = 4, N = 20)

INITS
```

```
list( r=1, beta=c(0,0,0,0) )   #generate the rest
```

|          | mean  | sd MCrror | val2.5pc | median | val97.5pc | start | sample |
|----------|-------|-----------|----------|--------|-----------|-------|--------|
| median[1] | 23.82 1.977 0.02301 | | 20.20 | 23.74 | 27.96 | 1001 | 100000 |
| median[2] | 35.11 3.459 0.01650 | | 29.23 | 34.79 | 42.82 | 1001 | 100000 |
| median[3] | 26.80 2.401 0.02169 | | 22.46 | 26.66 | 31.94 | 1001 | 100000 |
| median[4] | 21.38 1.845 0.01446 | | 18.09 | 21.26 | 25.34 | 1001 | 100000 |

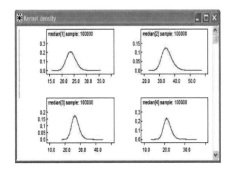

**Fig. 18.7** Posterior densities of median survival times median[1]–median[4].

## 18.6 Exercises

18.1. **Simulation of Censoring.**

```
y = exprnd(10,50,1); % Random failure times exponential(10)
d = exprnd(20,50,1); % Drop-out times exponential(20)
t = min(y,d);        % Observe the minimum of these times
censored = (y>d);    % Observe whether the subject failed
```

Using MATLAB's ecdf calculate and plot empirical CDF and confidence bounds for arguments t and censored.

18.2. **Immunoperoxidase.** In the context of Example 18.5 find a confidence interval for $\lambda_1$, the rate parameter for immunoperoxidase-negative patients. Use the fact that for MLE $\hat{\lambda}_1$,

$$\frac{\hat{\lambda}_1 - \lambda_1}{\hat{\lambda}/\sqrt{k_1}} \text{ or}$$

$$\sqrt{k_1}\,(\log\hat{\lambda}_1 - \log\lambda_1),$$

both have an approximately standard normal distribution.

18.3. **Massachusetts Data.** This exercise will illustrate the survival analysis of outcome predictions from DNA microarrays. Such survival analyses are often found in DNA microarray papers that concern the diagnosis or prognosis of human cancers. A description is given in Bhattacharjee et al. (2001) which is available at: http://www.pnas.org/content/98/24/13790.full.
A publicly available data set of 125 adenocarcinomas can be downloaded from: http://www.genome.wi.mit.edu/MPR/lung.

18.4. **Expected Lifetime.** Let $T$ be a lifetime with survival function $S(t)$. Using integration by parts in the definition of $\mathbb{E}T$ show

$$\mathbb{E}T = \int_0^\infty S(t)dt.$$

18.5. **Censored Rayleigh.** The lifetime (in hours) of a certain sensor has Rayleigh distribution, with survival function

$$S(t) = \exp\left\{-\frac{1}{2}\lambda t^2\right\}, \ \lambda > 0.$$

Twelve sensors are placed under test for 100 hours, and the following failure times are recorded 23, 40, 41, 67, 69, 72, 84, 84, 88, 100+, 100+. Here + denotes a censored time.
(a) If failure times $t_1, \ldots, t_r$ are observed, and $t_{r+1}^+, \ldots, t_n^+$ are censored, show that the MLE of $\lambda$ is

$$\hat{\lambda} = \frac{2r}{\sum_{i=1}^r t_i + \sum_{i=r+1}^n t_i^{+2}}.$$

Evaluate the MLE for the given data. Consult Example 18.4.
(b) Calculate and plot Kaplan-Meier estimator and superimpose $S(t)$ evaluated at $\hat{\lambda}$.

18.6. **MLE for Equally Censored Data.** A cohort of $n$ subjects are monitored in the time interval $[0, T]$, where $T$ is fixed in advance. Suppose that $r$ failures are observed ($r$ can be any number from 0 to $n$) at times $t_1, t_2, \ldots, t_r \leq T$. There are $(n - r)$ subjects that survived the entire period $[0, T]$, and their failure times are not observed.
Suppose that $f(t)$ is the density of a lifetime. The likelihood is

$$L = C \prod_{i=1}^r f(t_i) \, (1 - F(T))^{n-r}$$

for some normalizing constant $C$.

(a) Express the likelihood $L$ for the exponential lifetime distribution, that is, $f(t) = \lambda e^{-\lambda t}$, $t \geq 0$, and $F(t) = 1 - e^{-\lambda t}$, $t \geq 0$.
(b) Take the log of the likelihood, $\ell = \log(L)$.
(c) Find the derivative of $\ell$ with respect to $\lambda$. Set the derivative to 0 and solve for $\lambda$. If the solution $\hat{\lambda}$ maximizes $\ell$, $(\ell''(\hat{\lambda}) < 0)$, then $\hat{\lambda}$ is the MLE of $\lambda$.
(d) Show that the MLE is

$$\hat{\lambda}_{mle} = \frac{r}{\sum_{i=1}^{r} t_i + (n-r)T}.$$

(e) If in the interval $[0,8]$ four subjects failed at times $t_1 = 2$, $t_2 = 5/2$, $t_3 = 4$, and $t_4 = 5$ and two subjects survived without failure, find the estimator $\lambda$, assuming that the lifetime has an exponential distribution. Check your calculations with

```
data = [2  2.5  4  5  8  8];
cens = [0  0    0  0  1  1];
lamrec = mle(data, 'distribution','exponential','censoring',cens)
lammle = 1/lamrec    %MATLAB uses reciprocal parametrization
```

(f) What is the MLE of $\lambda$ if the unobserved failure times are ignored, that is, only four observed failure times, $t_1$, $t_2$, $t_3$, and $t_4$, are used.

18.7. **Malignant Melanoma.** Survival times of 256 males with malignant melanoma who had metastases on admission to the MD Anderson Clinic are reported (McDonald, 1963). The period of admission was between 1944 and 1960. The table shows survival times (no censoring) of the patients on a year scale.
Find the mean survival time.

| Survival time (in years) | Number of patients admitted in time interval | Number of patients dying in time interval |
|---|---|---|
| 0–1 | 256 | 167 |
| 1–2 | 89 | 48 |
| 2–3 | 41 | 23 |
| 3–4 | 18 | 6 |
| 4–5 | 12 | 3 |
| 5–6 | 9 | 6 |
| 6–7 | 3 | 1 |
| 7–8 | 2 | 1 |
| 8–9 | 1 | 1 |
| 9+ | 0 | 0 |

18.8. **Rayleigh Survival Times.** It was observed that in clinical studies dealing with cancer survival times follow Rayleigh distribution with pdf

$$f(x) = 2\lambda t e^{-\lambda t^2}, \quad t \geq 0, \ \lambda > 0.$$

(a) Show that the hazard function is linear.

(b) Find the mean survival time as a function of $\lambda$.

(c) For $t_1, \ldots, t_k$ observed and $t_{k+1}, \ldots, t_n$ censored times, show that the likelihood is proportional to

$$L \propto \lambda^k \exp\{-\lambda \sum_{i=1}^{n} t_i^2\}.$$

If the prior on $\lambda$ is gamma $\mathcal{G}(\alpha, \beta)$, show that the posterior is gamma $\mathcal{G}(\alpha + k, \beta + \sum_{i=1}^{n} t_i^2)$.

(d) Show that the Bayes estimators of the hazard and survival functions are

$$\hat{h}_B(t) = \frac{2(k+\alpha)t}{\beta + \sum_{i=1}^{n} t_i^2} \quad \text{and} \quad \hat{S}_B(t) = \left(1 + \frac{t^2}{\beta + \sum_{i=1}^{n} t_i^2}\right)^{-(k+\alpha)}.$$

18.9. **Western White Clematis.** Muenchow (1986) tested whether male or female flowers (of *Clematis ligusticifolia*) were equally attractive to insects. The data in table represent waiting times (in minutes), which includes censored data (observations with +).

| Male flowers | | | Female flowers | | |
|---|---|---|---|---|---|
| 1 | 9 | 27 | 1 | 19 | 57 |
| 1 | 9 | 27 | 2 | 23 | 59 |
| 2 | 9 | 30 | 4 | 23 | 67 |
| 2 | 11 | 31 | 4 | 26 | 71 |
| 4 | 11 | 35 | 5 | 28 | 75 |
| 4 | 14 | 36 | 6 | 29 | 75+ |
| 5 | 14 | 40 | 7 | 29 | 78+ |
| 5 | 14 | 43 | 7 | 29 | 81 |
| 6 | 16 | 54 | 8 | 30 | 90+ |
| 6 | 16 | 61 | 8 | 32 | 94+ |
| 6 | 17 | 68 | 8 | 35 | 96 |
| 7 | 17 | 69 | 9 | 35 | 96+ |
| 7 | 18 | 70 | 14 | 37 | 100+ |
| 8 | 19 | 83 | 15 | 39 | 102+ |
| 8 | 19 | 95 | 18 | 43 | 105+ |
| 8 | 19 | 102+ | 18 | 56 | |
| | | 104+ | | | |

Compare survival functions for Male and Female flowers.

---

**MATLAB AND WINBUGS FILES AND DATA SETS USED IN THIS CHAPTER**
http://springer.bme.gatech.edu/Ch18.Survival/

chada.m, cordband.m, ImmunoPerox.m, kmcdfsm.m, KMmuenchow.m, kmplot.m, limphomaLR.m, logrank.m, logrankdata.mat, logrankmod.m, Muenchow.m, PBC.m, simulation1.m, simulation2.m, smokingbugs.m, weibullsim.m

ibrahim1.odc, ibrahim2.odc, Immunoperoxidase.odc, Leukemia.odc, photocar.odc, Smoking.odc

gehan.dat KMmuenchow.txt limphoma.mat pbc.xls pbcdata.dat prostatecanc.dat smokingoutbugs.mat

---

# CHAPTER REFERENCES

Armitage, P. and Berry, G. (1994). *Statistical Methods in Medical Research*, 3rd edn. Blackwell, London.

Banerjee, S. and Carlin, B. P. (2004). Parametric spatial cure rate models for interval-censored time-to-relapse data. *Biometrics*, **60**, 268–275.

Bhattacharjee, A., Richards, W. G., Staunton, J., Li, C., Monti, S., Vasa, P., Ladd, C., Beheshti, J., Bueno, R., Gillette, M., Loda, M., Weber, G., Mark, E. J., Lander, E. S., Wong, W., Johnson, B. E., Golub, T. R., Sugarbaker, D. J., and Meyerson, M. (2001). Classification of human lung carcinomas by mRNA expression profiling reveals distinct adenocarcinoma subclasses. *Proc. Natl. Acad. Sci. U.S.A.*, **98**, 24, 13790–13795.

Cardillo, G. (2008). LogRank: comparing survival curves of two groups using the log rank test, http://www.mathworks.com/matlabcentral/fileexchange/22317.

Chadha, V., Warady, B. A., Blowey, D. L., Simckes, A. M., and Alon, U. S. (2000). Tenckhoff catheters prove superior to Cook catheters in pediatric acute peritoneal dialysis. *Am. J. Kidney Dis.*, **35**, 6, 1111–1116.

Cox, D. R. (1972). Regression models and life tables (with discussion). *J. R. Stat. Soc. Ser. B*, **34**, 187–220.

Cox, D. R. and Oakes, D. (1984). *Analysis of Survival Data*. Chapman & Hall, New York.

Crowder, M. J., Kimber, A. C., Smith, R. L. and Sweeting, T. J. (1991). *Analysis of Reliability Data*. Chapman and Hall, London

Dellaportas, P. and Smith, A. F. M. (1993). Bayesian inference for generalized linear and proportional hazards models via Gibbs sampling. *Appl. Stat.*, **42**, 443–460.

Freireich, E., Gehan, E., Frei, E. III, Schroeder, L., Wolman, I., Anbari, R., Burgert, E., Mills, S., Pinkel, D., Selawry, O., Moon, J., Gendel, B., Spurr, C., Storrs, R., Haurani, F., Hoogstraten, B., and Lee, S. (1963). The effect of 6 mercaptopurine on the duration of steroid-induced remissions in acute leukemia. *Blood*, **21**, 6, 699–716.

Glasser, M. (1967). Exponential survival with covariance. *J. Am. Stat. Assoc.*, **62**, 561–568.

Grieve, A. P. (1987). Applications of Bayesian software: two examples. *Statistician*, 36, 283–288.

Ibrahim, J. G., Chen, M. H., and Sinha, D. (2001). *Bayesian Survival Analysis*. Springer, Berlin Heidelberg New York.

Kaplan, E. L. and Meier, P. (1958). Nonparametric estimation from incomplete observations. *J. Am. Stat. Assoc.*, **53**, 457–481.

Klein, J. P. and Moeschberger, M. L. (2003). *Survival Analysis: Techniques for Censored and Truncated Data*, 2nd edn. Springer, Berlin Heidelberg New York.

MacDonald, E. J. (1963). The epidemiology of melanoma. *Ann. N.Y. Acad. Sci.*, **100**, 4–17.

Markus, B. H., Dickson, E. R., Grambsch, P. M., Fleming, T. R., Mazzaferro, V., Klintmalm, G. B., Wiesner, R. H., Van Thiel, D. H., and Starzl, T. E. (1989). Efficiency of liver transplantation in patients with primary biliary cirrhosis. *New Engl. J. Med.*, **320**, 26, 1709–1713.

McKelvey, E. M., Gottlieb, J. A., Wilson, H. E., Haut, A., Talley, R. W., Stephens, R., Lane, M., Gamble, J. F., Jones, S. E., Grozea, P. N., Gutterman, J., Coltman, C., and Moon, T. E. (1976). Hydroxyldaunomycin (Adriamycin) combination chemotherapy in malignant lymphoma. *Cancer*, **38**, 4, 1484–1493.

Muenchow, G. (1986). Ecological use of failure time analysis. *Ecology*, **67**, 246–250.

Sedmak, D. D., Meineke, T. A., Knechtges, D. S., and Anderson, J. (1989). Prognostic significance of cytokeratin-positive breast cancer metastases. *Mod. Pathol.*, **2**, 516–520.

# Chapter 19
# Bayesian Inference Using Gibbs Sampling
# – BUGS Project

*Beware: MCMC sampling can be dangerous!*

– Disclaimer in WinBUGS User Manual

## WHAT IS COVERED IN THIS CHAPTER

- Where to find WinBUGS, How to Install, Resources
- Step-by-step Example
- Built-in Functions and Common Distributions in BUGS
- MATBUGS: A MATLAB Interface to BUGS

## 19.1 Introduction

BUGS is a freely available software for constructing and evaluating Bayesian statistical models using simulation approaches based on the Markov Chain Monte Carlo methodology.

BUGS and WINBUGS are distributed freely and are the result of many years of development by a team of statisticians and programmers at the Medical Research Council Biostatistics Research Unit in Cambridge (BUGS and WinBUGS), and from recently by a team at the University of Helsinki (OpenBUGS); see the project pages http://www.mrc-bsu.cam.ac.uk/bugs/ and http://www.openbugs.info/w/.

Models are represented by a flexible language, and there is also a graphical feature, DOODLEBUGS, that allows users to specify their models as directed graphs. For complex models DOODLEBUGS can be very useful (Lunn et al., 2000). As of April 2011, the latest version are WinBUGS 1.4.3 and OpenBUGS 3.1.2. A comprehensive overview of WinBUGS programming and applications can be found in Ntzoufras (2009) and Congdon (2001, 2003, 2005).

## 19.2 Step-by-Step Session

We start this tutorial chapter on WinBUGS with a simple regression example. Consider the model

$$y_i | \mu_i, \tau \sim \mathcal{N}(\mu_i, \tau), \ i = 1, \dots, n,$$
$$\mu_i = \alpha + \beta(x_i - \bar{x}),$$
$$\alpha \sim \mathcal{N}(0, 10^{-4}),$$
$$\beta \sim \mathcal{N}(0, 10^{-4}),$$
$$\tau \sim \mathcal{G}a(0.001, 0.001).$$

The normal distribution is parameterized by a *precision* parameter $\tau$ that is the reciprocal of the variance, $\tau = 1/\sigma^2$. Natural priors for precision parameters are gamma, and small values of the precision reflect the flatness (noninformativeness) of the priors. Assume that $(x, y)$ pairs $(1, 1)$, $(2, 3)$, $(3, 3)$, $(4, 3)$, and $(5, 5)$ are observed.

Estimators in classical, least-squares regression of $y$ on $x - \bar{x}$ are given in the following MATLAB output:

```
y  = [1 3 3 3 5]'; %response
xx = [1 2 3 4 5]';
X  = [ones(size(xx))  xx-mean(xx)];
[b.b,b.int,res.res,res.int,stats] = regress(y,X);

b.b'
%    3.0000    0.8000
stats
%    0.8000   12.0000    0.0405    0.5333
```

Thus, the estimators are $\hat{\alpha} = \bar{y} = 3$, $\hat{\beta} = 0.8$, and $\hat{\tau} = 1/\hat{\sigma}^2 = 1/0.5333 = 1.875$.

What about Bayesian estimators? We will find the estimators by MCMC simulation, as empirical means of the simulated posterior distributions. Assume that the initial parameter values are $\alpha_0 = 0.1$, $\beta_0 = 0.6$, and $\tau = 1$. Start WinBUGS and input the following code in [**File > New**]:

```
# A simple regression
model{
    for (i in 1:N) {
```

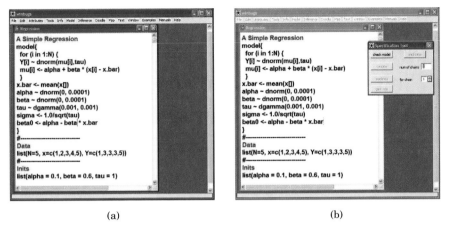

**Fig. 19.1** (a) Opening WinBUGS front end with a simple regression task. The simple regression program is opened or typed in. (b) The front end after selecting **Specification** from the **Model** menu.

```
Y[i] ~ dnorm(mu[i],tau)
mu[i] <- alpha + beta * (x[i] - x.bar)
}
x.bar <- mean(x[])
alpha ~ dnorm(0, 0.0001)
beta ~ dnorm(0, 0.0001)
tau ~ dgamma(0.001, 0.001)
sigma <- 1.0/sqrt(tau)
}
#- - - - - - - - - - - - - - - - - - - - - - - - - - -
DATA
list(N=5, x=c(1,2,3,4,5), Y=c(1,3,3,3,5))
#- - - - - - - - - - - - - - - - - - - - - - - - - - -
INITS
list(alpha = 0.1, beta = 0.6, tau = 1)
```

Next, make sure that the cursor is somewhere within the scope of "model," that is, somewhere between the first open and the last closed curly bracket. Go to the **Model** menu and open **Specification**. The **Specification Tool** window will pop out (Fig. 19.1b). Next, press **check model** in the Specification Tool window. If the model is correct, the response on the lower left border of the window should be: **model is syntactically correct** (Fig. 19.2a). Next, data are read in. Highlight the "list" statement in the data part of your code (Fig. 19.2b). In the Specification Tool window, select **load data**. If the data are in the correct format, you should receive a response in the lower left corner of the WinBUGS window: **data loaded** (Fig. 19.3a). You will need to compile your model in order to activate the **inits** buttons.

Select **compile** in the Specification Tool window. The response should be: **model compiled** (Fig. 19.3b), and the **load inits** and **gen inits** buttons become active. Finally, highlight the "list" statement in the initials part of your

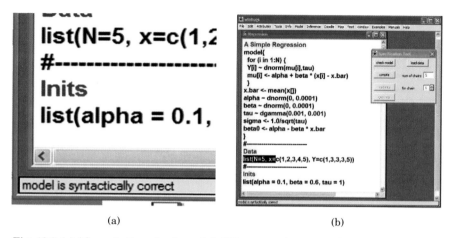

**Fig. 19.2** (a) After selecting **check model**, if the syntax is correct, the response is **model is syntactically correct.** (b) Highlighting the list in the data prior to reading data in.

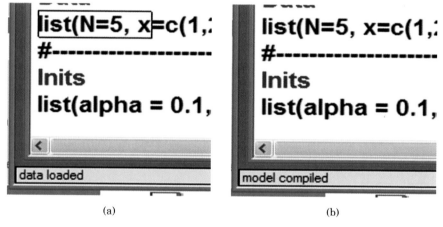

**Fig. 19.3** WinBUGS' responses to (a) **load data** and (b) **compile** in the model specification tool.

code, and in the Specification Tool window, select **load inits** (Fig. 19.4a). The response should be: **model is initialized** (Fig. 19.4b), and this completes the reading in of the model. If the response is **initial values loaded but this or another chain contains uninitialized variables,** click on the **gen inits** button. The response should be: **initial values generated, model initialized**.

Now you are ready to burn in some simulations and at the same time check if the program works. Recall that burning in the Markov chain model is necessary for the chain to "forget" the initialized parameter values. In the **Model** menu, choose **Update...** and open **Update Tool** to check if your model updates (Fig. 19.5a).

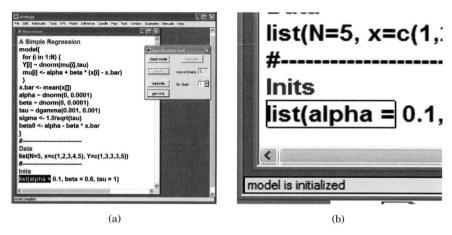

(a)            (b)

**Fig. 19.4** (a) Highlighting the list to initialize the model. (b) WinBUGS confirms that the model (in fact a Markov chain) is initialized.

From the **Inference** menu, open **Samples...**. A window titled **Sample Monitor Tool** will pop out (Fig. 19.5b). In the **node** subwindow, input the names of the variables you want to monitor. In this case, the variables are alpha, beta, and tau. If you correctly input the variable name, the **set** button becomes active and you should set the variable. Do this for all three variables of interest. In fact, sigma as a transformation of tau is available to be set as well.

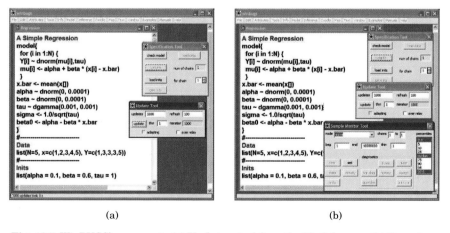

(a)            (b)

**Fig. 19.5** WinBUGS' response to (a) **Update...** tool from the **Model** menu, (b) **Samples...** from the **Inference** menu.

Now choose alpha from the subwindow in **Sample Monitor Tool**. All of the buttons (**clear, set, trace, history, density, stats, coda, quantiles, bgr**

**diag, auto cor**) are now active. Return to **Update Tool** and select the desired number of simulations, say 100,000, in the **updates** subwindow. Press the **update** button (Fig. 19.6a).

Return to **Sample Monitor Tool** and check **trace** for the part of the MC trace for $\alpha$, **history** for the complete trace, **density** for a density estimator of $\alpha$, etc. For example, pressing the **stats** button will produce something like the following table:

|       | mean  | sd     | MC error | val2.5pc | median | val97.5pc | start | sample |
|-------|-------|--------|----------|----------|--------|-----------|-------|--------|
| alpha | 2.996 | 0.5583 | 0.001742 | 1.941    | 2.998  | 4.041     | 1001  | 100000 |

The mean 2.996 is the Bayes estimator (as the mean from the sample from the posterior for $\alpha$). There are two precision outputs, sd and MC error. The former is an estimator of the standard deviation of the posterior and can be improved by increasing the sample size but not the number of simulations. The latter is the simulation error and can be improved by additional simulations. The 95% credible set (1.941, 4.041) is determined by val2.5pc and val97.5pc, which are the 0.025 and 0.975 (empirical) quantiles from the posterior. The empirical median of the posterior is given by median. The outputs start and sample show the starting index for the simulations (after burn-in) and the available number of simulations.

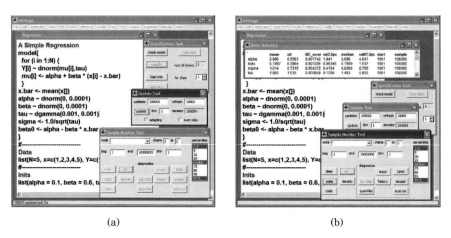

(a)                                                    (b)

**Fig. 19.6** (a) Select the simulation size and update. (b) After the simulation is done, check the stats node.

For all parameters a comparative table (Fig. 19.6b) is as follows:

|       | mean   | sd     | MC error | val2.5pc | median | val97.5pc | start | sample |
|-------|--------|--------|----------|----------|--------|-----------|-------|--------|
| alpha | 2.996  | 0.5583 | 0.001742 | 1.941    | 2.998  | 4.041     | 1001  | 100000 |
| beta  | 0.7987 | 0.3884 | 0.001205 | 0.06345  | 0.7999 | 1.537     | 1001  | 100000 |
| sigma | 1.014  | 0.7215 | 0.004372 | 0.4134   | 0.8266 | 2.765     | 1001  | 100000 |
| tau   | 1.865  | 1.533  | 0.006969 | 0.1308   | 1.463  | 5.852     | 1001  | 100000 |

We recall the least squares estimators from the beginning of this session: $\hat{\alpha} = 3$, $\hat{\beta} = 0.8$, and $\hat{\tau} = 1.875$, and note that their Bayesian counterparts are very close.

Densities (smoothed histograms) and traces for all parameters are given in Fig. 19.7.

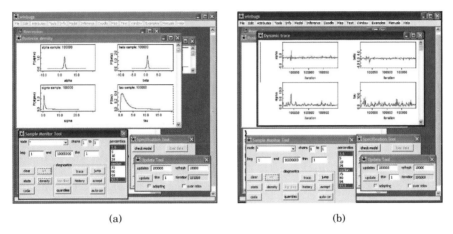

(a)                                          (b)

**Fig. 19.7** Checking (a) `density` and (b) `trace` in the **Sample Monitor Tool**.

If you want to save the trace for $\alpha$ in a file and process it in MATLAB, select **coda** and the data window will open with an information window as well. Keep the data window active and select **Save As** from the **File** menu. Save the $\alpha$s in `alphas.txt`, where it will be ready to be imported into MATLAB. Later in this chapter we will discuss the direct interface between WinBUGS and MATLAB called MATBUGS.

## 19.3 Built-in Functions and Common Distributions in WinBUGS

This section contains two tables: one with the list of built-in functions and another with the list of available distributions.

A first-time WinBUGS user may be disappointed by the selection of built-in functions – the set is minimal but sufficient. The full list of distributions in WinBUGS can be found in **Manuals>OpenBUGS User Manual**. WinBUGS also allows for the inclusion of distributions for which functions are not built in. Table 19.2 provides a list of important continuous and discrete distributions, with their syntax and parametrizations. WinBUGS has the capability to define custom distributions, both as a likelihood and as a prior, via the so-called *zero-tricks* (p. 296).

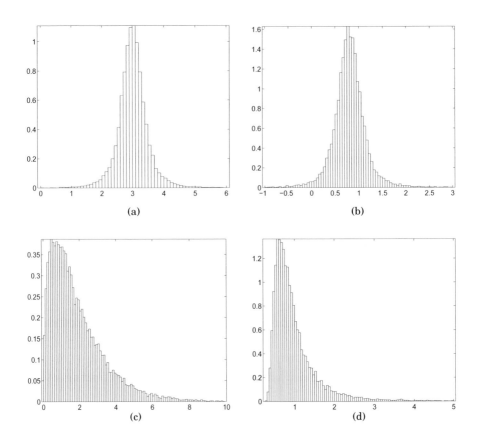

**Fig. 19.8** Traces of the four parameters from a simple example: (a) $\alpha$, (b) $\beta$, (c) $\tau$, and (d) $\sigma$ from WinBUGS. Data are plotted in MATLAB after being exported from WinBUGS.

## 19.4 MATBUGS: A MATLAB Interface to WinBUGS

There is strong motivation to interface WinBUGS with MATLAB. Cutting and pasting results from WinBUGS is cumbersome if the simulation size is in millions or if the number of simulated parameters is large. Also, the data manipulation and graphical capabilities in WinBUGS are quite rudimentary compared to MATLAB.

**Table 19.1** Built-in functions in WinBUGS

| WinBUGS code | Function |
|---|---|
| abs(y) | $|y|$ |
| cloglog(y) | $\ln(-\ln(1-y))$ |
| cos(y) | $\cos(y)$ |
| equals(y, z) | 1 if $y = z$; 0 otherwise |
| exp(y) | $\exp(y)$ |
| inprod(y, z) | $\sum_i y_i z_i$ |
| inverse(y) | $y^{-1}$ for symmetric positive–definite matrix $y$ |
| log(y) | $\ln(y)$ |
| logfact(y) | $\ln(y!)$ |
| loggam(y) | $\ln(\Gamma(y))$ |
| logit(y) | $\ln(y/(1-y))$ |
| max(y, z) | $y$ if $y > z$; $y$ otherwise |
| mean(y) | $n^{-1} \sum_i y_i$, $n = dim(y)$ |
| min(y, z) | $y$ if $y < z$; $z$ otherwise |
| phi(y) | standard normal CDF $\Phi(y)$ |
| pow(y, z) | $y^z$ |
| sin(y) | $\sin(y)$ |
| sqrt(y) | $\sqrt{y}$ |
| rank(v, s) | number of components of $v$ less than or equal to $v_s$ |
| ranked(v, s) | $s$th smallest component of $v$ |
| round(y) | nearest integer to $y$ |
| sd(v) | standard deviation of components of $y$ ($n - 1$ in denom.) |
| step(y) | 1 if $y \geq 0$; 0 otherwise |
| sum(y) | $\sum_i y_i$ |
| trunc(y) | greatest integer less than or equal to $y$ |

MATBUGS is a MATLAB program that communicates with WinBUGS. The program matbugs.m was written by Kevin Murphy and his team and can be found at: http://code.google.com/p/matbugs.

We now demonstrate how to solve Jeremy's IQ problem in MATLAB by calling WinBUGS. First we need to create a simple text file, say, jeremy.txt:

```
model{
for(i in 1 : N)
    {
        scores[i] ~ dnorm(theta, tau)
    }
  theta ~ dnorm(mu, xi)
```

and then run the MATLAB file:

```
dataStruct = struct( ...
    'N', 5, ...
    'tau',1/80,...
    'xi',1/120,...
    'mu',110,...
    'scores',[97 110 117 102 98]);
```

**Table 19.2** Built-in distributions with WinBUGS names and their parameterizations.

| Distribution | WinBUGS code | Density |
|---|---|---|
| Bernoulli | x ~ dbern(p) | $p^x(1-p)^{1-x}, x=0,1; 0\le p\le1$ |
| Binomial | x ~ dbin(p, n) | $\binom{n}{x}p^x(1-p)^{n-x}, x=0,\ldots,n; 0\le p\le1$ |
| Categorical | x ~ dcat(p[]) | $p[x], x=1,2,\ldots,\dim(p)$ |
| Poisson | x ~ dpois(lambda) | $\frac{\lambda^x}{x!}\exp(-\lambda), x=0,1,2,\ldots, \lambda>0$ |
| Beta | x ~ dbeta(a,b) | $\frac{1}{B(a,b)}x^{a-1}(1-x)^{b-1}, 0=x\le1, a,b>-1$ |
| Chi-square | x ~ dchisqr(k) | $\frac{x^{k/1-1}\exp(-x/2)}{2^{k/2}\Gamma(k/2)}, x\ge0, k>0$ |
| Double exponential | x ~ ddexp(mu, tau) | $\frac{\tau}{2}\exp(-\tau|x-\mu|), x\in R, \tau>0, \mu\in R$ |
| Exponential | x ~ dexp(lambda) | $\lambda\exp(-\lambda x), x\ge0, \lambda\ge0$ |
| Flat | x ~ dflat() | constant; not a proper density |
| Gamma | x ~ dgamma(a, b) | $\frac{b^a x^{a-1}}{\Gamma(a)}\exp(-bx), x,a,b>0$ |
| Normal | x ~ dnorm(mu, tau) | $\sqrt{\tau/(2\pi)}\exp\{-\frac{\tau}{2}(x-\mu)^2\}, x,\mu\in R, \tau>0$ |
| Pareto | x ~ dpar)alpha,c) | $ac^a x^{-(a+1)}, x>c$ |
| Student-t | x ~ dt(mu, tau, k) | $\frac{\Gamma((k+1)/2)}{\Gamma(k/2)}\sqrt{\frac{\tau}{k\pi}}[1+\frac{\tau}{k}(x-\mu)^2]^{-(k+1)/2}, x\in\mathbb{R}, k\ge2$ |
| Uniform | x[] ~ dunif(a, b) | $\frac{1}{b-a}, a\le x\le b$ |
| Weibull | x[] ~ dweib(v, lambda) | $v\lambda x^{v-1}\exp(-\lambda x^v), x,v,\lambda>0$, |
| Multinomial | x[] ~ dmulti(p[], N) | $\frac{(\sum_i x_i)!}{\prod_i x_i!}\prod_i p_i^{x_i}, \sum_i x_i=N, 0<p_i<1, \sum_i p_i=1$ |
| Dirichlet | p[] ~ ddirch(alpha[]) | $\frac{\Gamma(\sum_i a_i)}{\prod_i\Gamma(a_i)}\prod_i p_i^{a_i-1}, 0<p_i<1, \sum_i p_i=1$ |
| Multivariate normal | x[] ~ dmnorm(mu[], T[,]) | $(2\pi)^{-d/2}|T|^{1/2}\exp\{-1/2(x-\mu)'T(x-\mu)\}, x\in R^d$ |
| Multivariate Student-t | x[] ~ dmt(mu[], T[,], k) | $\frac{\Gamma((k+d)/2)}{\Gamma(k/2)}\frac{|T|^{1/2}}{k^{d/2}\pi^{d/2}}[1+\frac{1}{k}(x-\mu)'T(x-\mu)]^{-(k+d)/2}, x\in\mathbb{R}^d, k\ge2$ |
| Wishart | x[,] ~ dwish(R[,], k) | $|R|^{k/2}|x|^{(k-p-1)/2}\exp\{-1/2Tr(Rx)\}$ |

```
 initStruct = struct( ...
     'theta', 100 );

cd('C:\MyBugs\matbugs\')
[samples, stats] = matbugs(dataStruct, ...
fullfile(pwd, 'jeremy.txt'), ...
'init', initStruct, ...
        'nChains', 1, ...
'view', 0, ...
        'nburnin', 2000, ...
        'nsamples', 50000, ...
'thin', 1, ...
'monitorParams', {'theta'}, ...
        'Bugdir', 'C:/Program Files/BUGS');

baymean = mean(samples.theta)
frmean=mean(dataStruct.scores)

  figure(1)
  [p, x] = ksdensity(samples.theta);
  plot(x, p);
```

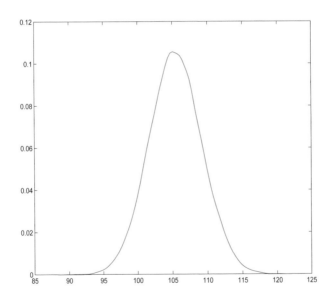

**Fig. 19.9** Posterior for Jeremy's data set. Data are plotted in MATLAB after being exported from WinBUGS by MATBUGS.

## 19.5 Exercises

19.1. **A Coin and a Die.** The following WinBUGS code simulates flips of a coin. The outcome **H** is coded by 1 and **T** by 0. Mimic this code to simulate rolls of a fair die.

```
#coin
model{
flip ~ dcat(p.coin[])
coin <- flip - 1
}
DATA
list(p.coin=c(0.5, 0.5))
#just generate initials
```

19.2. **De Mere Paradox in WinBUGS.** In 1654 the Chevalier de Mere asked Blaise Pascal (1623–1662) the following question: *In playing a game with three dice, why is the sum 11 advantageous to the sum 12 when both are the result of six possible outcomes?* Indeed, there are six favorable triplets for each of the sums **11** and **12**:

| | | | | | | |
|---|---|---|---|---|---|---|
| **11**: | (1, 4, 6), | (1, 5, 5), | (2, 3, 6), | (2, 4, 5), | (3, 3, 5), | (3, 4, 4) |
| **12**: | (1, 5, 6), | (2, 4, 6), | (2, 5, 5), | (3, 3, 6), | (3, 4, 5), | (4, 4, 4) |

19.3. **Simulating the Probability of an Interval.** Consider an exponentially distributed random variable $X$, $X \sim \mathscr{E}\left(\frac{1}{10}\right)$, with density $f(x) = \frac{1}{10}\exp\{-x/10\}$, $x > 0$. Compute $\mathbb{P}(10 < X < 16)$ using (a) exact integration, (b) MATLAB's expcdf, and (c) WinBUGS.

19.4. **WinBUGS as a Calculator.** WinBUGS can approximate definite integrals, solve nonlinear equations, and even find values of definite integrals over random intervals. The following WinBUGS program finds an approximation to $\int_0^{\pi} \sin(x)dx$, solves the equation $y^5 - 2y = 0$, and finds the integral $\int_0^R z^3(1-z^4)dz$, where $R$ is a beta $\mathscr{B}e(2,2)$ random variable. The solution is given by the following code:

```
model{
F(x) <- sin(x)
int <- integral(F(x), 0, pi, 1.0E-6)
pi<- 3.141592659

y0 <- solution(F(y), 1,2, 1.0E-6)
F(y)  <- pow(y,5) - 2*y
zero <-  pow(y0, 5)-2*y0

randint <- integral(F(z), 0, randbound, 1.0E-6)
F(z) <- pow(z,3)*(1-pow(z,4))
randbound ~ dbeta(2,2)
}

NO DATA
```

```
INITS
    list(x=1, y=0, z=NA, randbound=0.5)
```

After model checking, one should go directly to compiling (no data to load in) and initializing the model. There is NO need to update the model, to go to the Inference tool, to set the variables for monitoring or to sample. One simply goes to the **Info** menu and checks **Node Info**. In the Node Info tool one specifies int for the approximation of an integral, y0 for the solution of an equation, zero for checking that y0 satisfies the equation (approximately), and randint for the value of a random interval.

---

**MATLAB AND WINBUGS FILES AND DATA SETS USED IN THIS CHAPTER**

http://springer.bme.gatech.edu/Ch19.WinBUGS/

 simple.m

 DeMere.odc, jeremy.odc, Regression1.odc, Regression2.odc, simulationd.odc

 alpha.txt, beta.txt, sigma.txt, tau.txt

---

# CHAPTER REFERENCES

Congdon, P. (2001). Bayesian Statistical Modelling. Wiley, Hoboken.

Congdon, P. (2003). Applied Bayesian Models. Wiley, Hoboken.

Congdon, P. (2005). Bayesian Models for Categorical Data. Wiley, Hoboken.

Lunn, D. J., Thomas, A., Best, N., and Spiegelhalter, D. (2000). WinBUGS – a Bayesian modelling framework: concepts, structure, and extensibility. *Stat. Comput.*, **10**, 325–337.

Ntzoufras, I. (2009). *Bayesian Modeling Using WinBUGS*. Wiley, Hoboken.

# Index

Printed in the United States of America